$ 96 (8A)
AF771239

K.R. Kerry · G. Hempel (Eds.)

ANTARCTIC ECOSYSTEMS

Ecological Change and Conservation

With 197 Figures

Springer-Verlag Berlin Heidelberg New York
London Paris Tokyo Hong Kong Barcelona

Dr. K. R. Kerry
Australian Antarctic Division, Department of the Arts, Sport, the Environment, Tourism and Territories
Channel Highway, Kingston
Tasmania, Australia 7050

Professor
Dr. G. Hempel
Alfred-Wegener-Institut für
Polar- und Meeresforschung
Columbusstraße
2850 Bremerhaven, FRG

ISBN 3-540-52101-1 Springer-Verlag Berlin Heidelberg New York
ISBN 0-387-52101-1 Springer-Verlag New York Berlin Heidelberg

Library of Congress Cataloging-in-Publication Data
Antarctic ecosystems : ecological change and conservation / K. R. Kerry, G. Hempel (eds.).
"Compilation of papers which were presented at a Symposium on Antarctic Biology ... which was the fifth in a series held under the auspices of the Scientific Committee on Antarctic Research (SCAR) ... at the University Centre, University of Tasmania, Hobart, Australia, 29 August–3 September 1988"
ISBN 3-540-52101-1 (Berlin)
ISBN 0-387-52101-1 (New York)
1. Ecology–Antarctic regions–Congresses. 2. Natural history–Antarctic regions–Congresses. 3. Man–Influence on nature–Antarctic regions–Congresses. 4. Animal populations–Antarctic regions–Congresses. I. Kerry, K. R. (Knowles R.). II. Hempel, Gotthilf. III. Symposium on Antarctic Biology (SCAR) (5th : 1988 : University of Tasmania) IV. International Council of Scientific Unions. Scientific Committee on Antarctic Research. QH84.2.A583 1990 574.5′2621–dc20

This work is subject to copyright. All rights are reserved, whether the whole or part of the material is concerned, specifically the rights of translation, reprinting, re-use of illustrations, recitation, broadcasting, reproduction on microfilms or in other ways, and storage in data banks. Duplication of this publication or parts thereof is only permitted under the provisions of the German Copyright Law of September 9, 1965, in its current version. And a copyright fee must always be paid. Violations fall under the prosecution act of the German Copyright Law.

© Springer-Verlag Berlin Heidelberg 1990
Printed in Germany

The use of registered names, trademarks, etc. in this publication does not imply, even in the absence of a specific statement, that such names are exempt from the relevant protective laws and regulations and therefore free for general use.

Typesetting, Printing and Binding: Triltsch, Würzburg
2131/3145-543210 Printed on acid-free paper

Preface

The Fifth Symposium on Antarctic Biology

This book is a compilation of papers which were presented at a Symposium on Antarctic Biology and which addressed specifically the Symposium theme of *Ecological Change and the Conservation of Antarctic Ecosystems*. This Symposium which was the fifth in a series held under the auspices of the Scientific Committee on Antarctic Research (SCAR) was held at the University Centre, University of Tasmania, Hobart, Australia, 29 August – 3 September 1988. Earlier symposia were held in Paris, France 1962, Cambridge, England 1968, Washington DC, USA 1974 and Wilderness, South Africa 1983. The proceedings of each of these meetings were published by Carrick and Prevost (1964), Holdgate (1970), LLano (1977) and Siegfried, Condy and Laws (1985), respectively.

The theme of the symposium was concerned with short- and long-term changes in ecosystem and community structure caused by natural and human factors. Knowledge of these changes contributes to the understanding of the ecological processes which take place in a changing environment and the variability of ecological factors which are important for the development of realistic monitoring strategies and sound conservation practices.

Although the symposium focused on the special theme, papers were accepted in all fields of biological research on the Antarctic Continent, the Antarctic and Subantarctic islands and in the Southern Ocean.

The meeting was attended by 212 registrants representing 30 nations. A total of 80 papers including 9 keynote papers were presented orally. A further 93 poster papers were displayed and described in brief presentations to the plenary. This number of papers offered for presentation was considerably more than expected and indicates the interest in Antarctic biology and conservation. Originally, it was intended to publish all papers presented (subject to review), however, because of the very large number of papers offered it was not possible and so the Steering Committee selected those oral contributions and posters which were considered directly relevant to the theme. Unfortunately, this selection meant that many excellent and scientifically stimulating papers could not be included. These are now being published in scientific journals, particularly POLAR BIOLOGY and ANTARCTIC SCIENCE.

The 45 papers which comprise this book have all been subjected to an extensive peer review process and scientific editing.

Organization of the symposium was guided by an International Steering Committee, comprising Professor G. Hempel (FRG) Chairman, Professor J. Valencia (Chile), Professor R. Siegfried (South Africa), W. N. Bonner (UK) and Dr. K. R. Kerry (Australia), and a local organizing committee including Dr. K. R. Kerry, Dr. R. Seppelt, Dr. H. Tyndale-Biscoe. Dr. L. Hammond, Dr. R. G. Chittleborough and Dr. P. Quilty. Special assistance was provided by L. Denham (overall organization), G. Hosie (poster sessions) and P. Waddington (secretarial assistance). The artwork for the meeting and the cover design for this book were provided by Jane James and the index for this volume by Clodagh Jones.

The meeting was sponsored by SCAR, the Scientific Committee on Oceanic Research (SCOR), the Australian Academy of Sciences and the Australian Antarctic Division. Financial support was provided towards travel costs by SCAR, SCOR and the International Council of Scientific Unions (ICSU). Additional financial support was provided by the Australian Antarctic Division who underwrote the meeting. Funds for the subvention for publication and the editing of the proceedings were provided by the Australian Antarctic Division and the Alfred-Wegener-Institut für Polar- und Meeresforschung in the Federal Republic of Germany.

The style and conventions adopted in this volume follow the requirements of the journal *POLAR BIOLOGY*. Where such conventions were not specified, for instance English versus American spelling, or the use of a particular common name for species, it was aimed to maintain consistency within chapters. Where possible the euphausiids (Euphausiacea) are referred to by their specific rather than their common name. The word krill is used as the collective name for members of the Euphausiacea and Krill or Antarctic Krill for *Euphausia superba* specifically.

Finally, we thank all who contributed to the symposium, particularly those who presented papers and those who worked behind the scenes. We thank also all the authors for their cooperation, E. Kerry, S. Marschall and C. Willeweit for their assistance in the preparation of the volume and those listed on page XII who reviewed the manuscripts.

June 1990
Kingston, Bremerhaven

K. R. Kerry, G. Hempel

References

Carrick R, Holdgate M, Prevost J (eds) (1964) Biologie Antarctique. Hermann, Paris
Holdgate MW (ed) (1970) Antarctic ecology. 2 Vols. Academic Press, New York
Llano GA (ed) (1977) Adaptations within Antarctic ecosystems. Gulf Publishing Company, Houston 1252 pp
Siegfried WR, Condy PR, Laws R (eds) (1985) Antarctic nutrient cycles and food webs. Springer, Berlin Heidelberg New York, 700 pp

Contents

Contributors

You will find the addresses at the beginning of the respective contribution

Referees

The following scientists provided reviews of one or more of the papers considered for publication.

Adams NJ
Adelung D
Ashton D
Arntz W
Bannash R
Bayly I
Bengston JL
Benninghof WS
Best P
Block W
Bonner WN
Bryden MM
Boveng PL
Burton H
Cooper J
Costa D
Croxall JP
Davis LS
Dieckmann G
Drew E
Ellis-Evans JC
Everson I
El-Sayed S
Fryxall GA
Fütterer D
Gallardo V
Hirsch P
Hopins TL
Hosie G
Hubold G
Hureau J-C
Jacobson L
Jouventin P
Kappen L
Kerry KR
Klages NTW
Koch K-H
Lewis-Smith RI
Lindner G
Line MA
Marchant HJ
Mehlum F
Miller DGM
Nemoto T
Nicol S
North AW
Piatowski U
Piepenberg D
Quilty PG
Ryan P
Schneider R
Scott J
Seppelt R
Shaughnessy P
Siegfried WR
Simmons GM
Siniff DB
Smetacek V
Spindler M
Schiel S
Stoddart M
Thomas D
Tilzer MM
Tingey RJ
Trehan P
Trivelpiece WL
Tyler P
Walton DW
Whitehead MD
Williams RJ
Wilson K-J
Woehler E
Wynn-Williams DD

Long- and Medium-Term Changes in Antarctic Environments

Significance of Evidence for Changes in the Antarctic Marine Environment Over the Last 5 Million Years

P. G. QUILTY [1]

Summary. Recent discoveries in the Vestfold Hills, in the Larsemann Hills and near the South Pole, of fossil vertebrates, microfossils and wood in Pliocene sediments all indicate that Antarctica was considerably warmer at various times in the Pliocene than at present. However, data are sparse, the results very tentative, and there is conflict in places between onshore and offshore evidence. Results suggest that changes towards the present glacial regime were later and more rapid than had been envisaged earlier.

While the above examples provide some firm data on Pliocene palaeoenvironments, other studies, for example of diatoms from sediments near Casey Station, provide tantalising glimpses of support for the concept of warm environments.

1 Introduction

For the purpose of discussion *biology* is divided into two parts: *neontology*, covering some 200–1000 years (±), and *palaeontology* to encompass the previous 3.8 billion years (Fig. 1) that period for which life has been on earth. Palaeontology teaches many lessons not readily gained from neontology (or what present-day biologists often regard as biology). Similarly, palaeontological studies would be very much the poorer if neontology had not advanced beyond where it was 200 years ago and provided a sound basis for interpreting the fossil record. The two facets are complementary and closely allied, not separate as much modern practice would indicate.

Palaeontology shows that life on earth has never been static and that there was probably never a 'static' *balance of nature*. Organisms and the balance between them have been continually changing, even without humanity's intervention. To maintain the world as it is in a single, short time frame is unrealistic. From palaeontology it is also apparent that some changes are very fast, perhaps almost instantaneous, e.g. the Cretaceous/Tertiary extinction event may have been due to a catastrophic, external in influence (Kauffman 1984).

The aim of this chapter is to describe examples of natural change in Antarctic ecology over the last 4.5 million years (Ma), and to use them as lessons to be borne in mind when examining modern ecological changes, some of which may have been caused by human activities.

Some changes in the global environment are most noteworthy in the polar regions and can be very well documented there. In addition, they can be shown to have caused changes in the biota. Perhaps we have reached a time when one element of the biota (our own species) causes environmental change, and may even be able to control the environment.

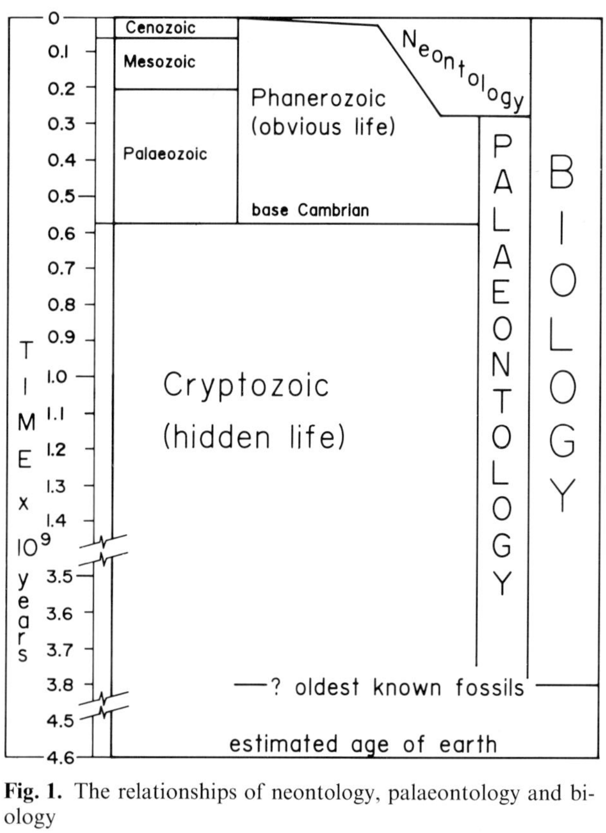

Fig. 1. The relationships of neontology, palaeontology and biology

[1] Australian Antarctic Division, Channel Highway, Kingston, Tasmania 7050, Australia.

Antarctic Ecosystems. Ecological Change and Conservation
(ed. by K. R. Kerry and G. Hempel)
© Springer-Verlag Berlin Heidelberg 1990

Remarkable discoveries of some very significant fossils in and near Antarctica in the last 5 years have led to inferences that rates of natural change were very rapid, and much faster than previously envisaged. The resultant new data and information from many sources lead to a finer understanding of the 120 000–150 000 year (Ka) cycles of glaciation, now commonly explained by the Milankovitch hypothesis (Imbrie et al. 1984). This review will in particular discuss events in the Pliocene, the time interval between 5.3 Ma ago, and the beginning of the Pleistocene, as defined by the first appearance datum (FAD) of the foraminiferid *Globorotalia* (*Truncorotalia*) *truncatulinoides* (d'Orbigny), and variously dated as 1.6 Ma (Berggren et al. 1985) to 1.85 or 1.95 Ma B.P.

Our understanding of the evolutionary processes that lead to modern Antarctica over the last few million years is based on very incomplete data, many from offshore Antarctica or even farther away. That understanding has depended largely on drilling programs in the Ross Sea area such as the Deep Sea Drilling Project (DSDP) and Dry Valleys Drilling Project (DVDP). It will be enhanced by the results from the Ocean Drilling Program (ODP) which has recently drilled in the Weddell Sea, Prydz Bay and on Kerguelen Plateau. In addition to marine and onshore rock drilling, there has been deep ice drilling to identify events over the more recent past (150 000 years) (Lorius et al. 1985).

A key component of these initiatives has been the study of oxygen isotopes, because the isotopic composition of sea water (and its HCO_3^- content) and of ice reflects very clearly the advance and retreat of polar ice caps, and variations in marine temperatures.

The classic work as its relates to Antarctic and sub-Antarctic regions is by Shackleton and Kennett (1975) based on DSDP cores from the South Pacific Ocean. These authors suggested a steady intensification of Antarctic glaciation since the mid-Miocene (i.e. about 15 Ma ago) without any major reversals. This hypothesis has recently been revised (Kennett 1985).

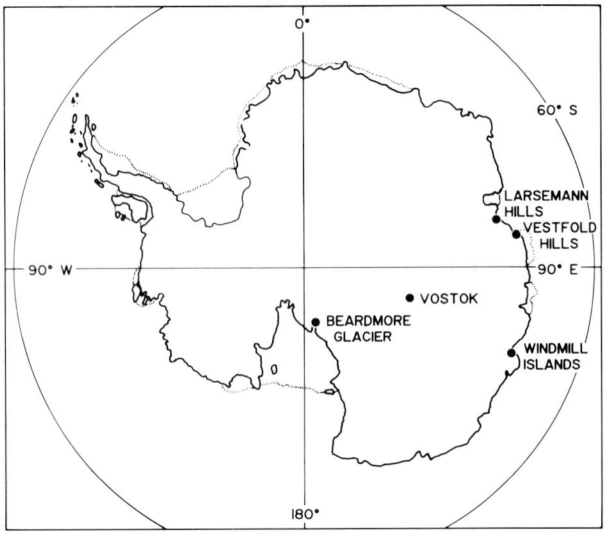

Fig. 2. Localities mentioned in the text

2 Pliocene Vertebrates and Invertebrates

In the Vestfold Hills, Adamson and Pickard (1986) have shown that certain unconsolidated sediments in the Marine Plain region some 10 km southwest of Davis Station (Fig. 2) are early to mid-Pliocene in age. The Pliocene age is confirmed (Adamson and Pickard 1986; Pickard et al. 1986, 1988) by studies of molluscs, diatoms and the racemisation of protein from the molluscs. It now appears to be in the range of 3.5–4.5 Ma (Pickard et al. 1988). Younger, unconsolidated glacial deposits have been shown to be Holocene (10 000 years or less) in age by ^{14}C techniques (Adamson and Pickard 1986; Zhang and Peterson 1985); most dates are in the 5000–8000 year range.

When visiting Marine Plain briefly in early 1985 in order to sample the section for foraminiferids, I recovered the skull and upper jaw of a highly evolved dolphin (Fordyce and Quilty in prep.) characterised by a long, delicate, edentate upper jaw probably adapted for a diet of squid, and belonging to a new genus. The dolphin had the facility for high frequency echo location, consistent with the squid eating hypothesis.

Another visit in early 1986 confirmed that such fossils are common in the 10 km^2 area of outcrop and that the locality (Quilty in press) will become a major source of vertebrate fossils. While only a dolphin has been indentified so far, bones of other forms have been found.

Preliminary oxygen isotope data derived from molluscs in the same sediments suggest that water temperatures were significantly above those of today, perhaps of the order of 5 °C (Quilty in press). Pickard et al. (1986, 1988), even without the oxygen isotope data, indicated that glaciation was less extensive than now (we are now in an interglacial interval) and that the ice front was considerably inland of its present location. However, palynological investigation of Marine Plain sediments (Truswell pers. commun.; Quilty in press) has not yet yielded evidence of Pliocene terrestrial vegetation.

Marine Plain is the only area of Antarctica to yield post-Eocene (36–54 Ma) vertebrate fossils. Eocene vertebrate fossils are known from the Antarctic Peninsula region, but at that time the Antarctic ecosystem was not isolated from that of the rest of the southern hemisphere. By the Pliocene the Circumpolar Current had developed, Antarctica had become glaciated, and an isolated, distinctly Antarctic ecosystem had evolved.

Stainforth et al. (1975) compiled data worldwide and postulated that the Pliocene epoch had a generally warm and equable climate following a colder, late Miocene. Vail et al. (1977) and Haq et al. (1987) have generated sea level variation curves based on the analysis of coastal onlap and demonstrated that sea level at 4.5 Ma was the highest since the Miocene and higher than that of today. Their hypothesis is consistent with a lower level of glaciation than now and thus consistent with the evidence of the Marine Plain fauna. The Pliocene is seen as a time of markedly fluctuating sea level, both highs and lows becoming lower with time after the 4.5 Ma high peak, which may have been 75–100 m higher than at present (Haq et al. 1987).

3 Fossil Wood

An even more significant discovery was made in the Oliver Bluff region of the Beardmore Glacier (Fig. 2), Transantarctic Mountains, in 1986 when fossil wood was recovered (Webb and Harwood 1987). Study has shown that it is *Nothofagus*, similar to the Chilean and Argentinian *N. betuloides* (Mirb.) Blume and *N. gunnii* (Hook.) Oerst. of Tasmania (Carlquist 1987).

The rock unit that yielded the essentially in situ wood fragments is the Sirius Formation (Webb and Harwood 1987). It is presently 1800–1900 m above sea level, although at the time of the growth, an altitude nearer sea level has been postulated. Its age is not well defined but is believed to be younger than 3.1 Ma, and probably younger than 2.5 Ma, i.e. towards the youngest part of the Pliocene, even earliest Pleistocene. The Sirius Formation contains reworked microfossils of several ages, some of which lived as recently as 2.5 or 3.1 Ma. The rock unit may be even younger. Diatoms are the most diagnostic fossils in this instance.

The wood appears to have come from shrublike plants and is in fragments up to 10 cm long. Carlquist (1987) discussed the material and suggested floristic poverty, perhaps the presence of only a single species in the flora. The wood is common and seems to have grown in glacial sediments during intervals of ice withdrawal. As with the Marine Plain fossils, the wood provides evidence of an interval, more recent than that of the dolphin, during which conditions were much warmer than at present.

Since the discovery of the dolphin and wood, Pliocene sediments have been found in the Larsemann Hills (Quilty 1990) and it is possible that scattered sediments in the Windmill Islands region are of Pliocene age (Harwood pers. commun.). The Larsemann Hills material is approximately coeval with that from Marine Plain but palaeoclimatic information has not yet been extracted. The Windmill Islands material seems to represent warm water conditions but its age is very poorly defined. In both the Larsemann Hills and Windmill Islands the sediments are poorly indurated shale and mudstone containing foraminiferids (Larsemann Hills only), diatoms and sponge spicules. Other poorly documented discoveries of sediments of apparently similar age have also been reported (e.g. Anonymous 1988).

Kennett (1985) presented oxygen and carbon isotope data from better sections than were available to Shackleton and Kennett (1975) for the Miocene and Pliocene. These data indicate that the interval variously reported in that paper as 5 – 4 and 4.5 – 3.4 Ma was a relatively warm one, at the end of which, there was a marked increase in $\delta\,^{18}O$ of the oceans, suggesting an increase in the global ice volume. These indications from marine ^{18}O are consistent with the data from the dolphin locality.

However, Kennett's (1985) data for the younger Pliocene and early Pleistocene are not consistent with the climatic indications from Antarctic wood. The wood suggests warm conditions but the oxygen isotope data indicate an intensified glaciation at 2.6 to 2.4 Ma at approximately the time of growth of the wood. The key question revolves around a more precise dating of the wood.

In summary, there is now ample evidence that during the Pliocene, Antarctica had significant intervals when it was much less glaciated than now, that vegetation grew in places and even that much of East Antarctica may have been covered by extensive epicontinental seas (Webb et al. 1984). This evidence contrasts markedly with some earlier proposals (e.g. Blank and Margolis 1975) that the interval 4.5 – 3.5 Ma was as cold as the glacial intervals of the Pleistocene.

The evidence presented here is perhaps highly biased towards those intervals which were warm, of high sea level and low ice volume and thus underrepresent periods of lowered sea level, more extensive ice and, in turn, of less widespread sedimentary sections to retain a climatic record. Deep-sea sedimentary sections may eventually fill in the gaps.

4 Quaternary

Over the last 750 000 years, the so-called Milankovitch cycles can be recognised through high resolution studies of oxygen isotope records (Emiliani 1978) from deep-sea deposits. Each cycle consists of a 120 000–150 000 year record reflecting a slow increase in $\delta^{18}O$ (a result of a slow expansion of the northern and southern hemisphere ice caps) followed by a 10 000–20 000 year reversal related to a "quick" collapse of the ice caps. This rapid decay is generally followed by an interval of stability (we are in one now) before the next cycle. The total amount of ice in Arctic and Antarctic ice caps when both existed (e.g. 25 000 years ago) is enough to vary sea level by approximately 200 m.

The last 150 000 year interval has been studied in detail by the analysis of a 2300-m ice core from the Soviet drilling program near Vostok Station (Lorius et al. 1985). Furthermore, there are Australian plans to drill through a thick ice column 700 km south of Casey in the early to mid-1990s and to conduct a very high resolution analysis of an ice column representing perhaps 0.5–1.0 Ma of climatic history. Four to eight Milankovitch cycles may be represented and detailed correlation of the ice core record with marine core sections will provide a more effective calibration of the marine sections. This, in turn, may allow the Milankovitch hypothesis to be examined in earlier sediments, hopefully linking the 750 000 year record now known with that in earlier sections, including of course that of the Pliocene.

Data so far available indicate that the Milankovitch cycles coincide with variations in atmospheric compo-

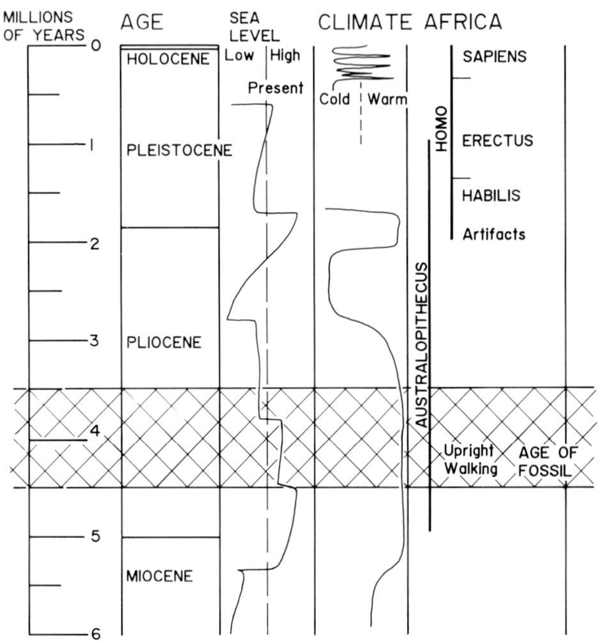

Fig. 3. Variation in sea level over the last 5 million years and oxygen isotope anomalies for the last 750 000 years

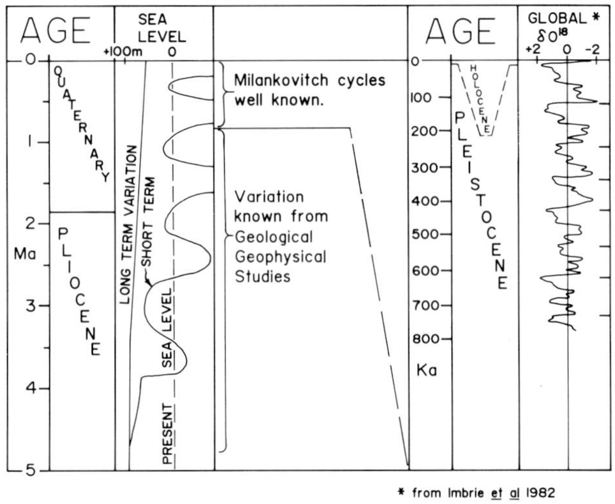

Fig. 4. Correlation of events at 3.5–4.5 Ma

sition, although this has been adequately documented only for the last cycle. This, in turn, suggests variation in sea ice cover, sea level, water temperature, atmospheric constituents, marine productivity and, in summary, of the ecosystem. It is not known what the effects on the fauna and flora may be but they must have been considerable and noteworthy of study by biologists (in the broad sense of that concept discussed earlier).

The last glacial maximum was reached 18 000–25 000 years ago (Lorius et al. 1985). Since then, the northern hemisphere ice cap has largely melted, contributing as much as 40×10^6 km^3 of water to the oceans and causing a 120–140 m rise in sea level. This occurred between 18 000 and 8000 years ago; since that time sea level has been approximately stable (Figs. 3, 4).

5 The Future

The history of the Milankovitch cycles would suggest that, under natural conditions, sea level and climate would remain stable for another few thousand (perhaps 3000–5000) years after which another ice buildup would begin; probably including the development of a northern hemisphere ice cap larger than the Antarctic, causing Europe, Canada and parts of the USA to be overrun by thick ice. The consequences for civilisation in those areas would be considerable.

Humanity has intervened in the process, causing the "Greenhouse effect" due to burning of enormous quantities of oil, gas and coal. This, in turn, may thwart nature and cause, instead of an ice buildup, a rise in seawater temperature and potentially an ice meltdown with a consequent sea level rise, with equally drastic consequences for cities on low level coasts.

The results outlined above have related directly to water temperature and sea level (and variation in ice volume). While these are still poorly known, other environmental factors will also change but there are no independent data concerning them. Thus, a major sea level fall will expose vast areas of continental shelf globally, which will then be inundated by a sea level rise. Such events would influence the evolution of benthic organisms which must be able to adapt quickly to regular change. Fluctuation in ice volume causes a fluctuation of several percent in oceanic salinity and all dependent water parameters.

Increased pole-equator atmospheric gradients during glacial intervals cause increased wind velocities with impact on sea roughness and absorption of O_2 and CO_2 from the atmosphere. This, in turn, will affect the position of the carbonate compensation depth (CCD). Variation in winter ice cover causes variation in the rate of production of Antarctic Bottom Water (AABW) causing a related variation in the ability to transport surface chemistry to ocean depths.

None of these factors is properly understood as yet, but it is likely that the impact of many of them will be greater in polar regions than elsewhere. They are the subject of a substantial research effort, which, in the near future, should focus especially on the ways that organisms are influenced during periods of change.

6 Conclusion

The Antarctic environment has changed quite dramatically over the last few million years and even over the last 750 000 years in a cyclic way. These changes are reflected in the fossil record and show that even the Antarctic ecosystem is subject to dramatic short-term changes, and that long-term stability is not to be expected. When large-scale change occurs, the ecosystem seldom, if ever, returns to the original state (Chittleborough 1984).

In the next few thousand years, rather than entering a new ice age as the Milankovitch cycle hypothesis would suggest, the world probably will warm because of the "Greenhouse effect", causing a major perturbation to the natural cycle. Perhaps a controlled atmosphere can be generated by humanity to ensure an approximately constant environment.

Research is needed on the effects on organisms of rapid changes in various elements, separately and together, of the Antarctic ecosystem.

References

Adamson DA, Pickard J (1986) Cainozoic history of the Vestfold Hills. In: Pickard J (ed) Antarctic oasis. Academic Press, Sydney, pp 63–97

Anonymous (1988) Newly discovered fossils suggest ancient marine environment in West Antarctica. Antarct J US June: 23(2):21

Berggren WA, Kent DV, van Couvering JA (1985) Neogene geochronology and chronostratigraphy. In: Snelling NJ (ed) The chronology of the geological record. Mem Geol Soc 10:211–250

Blank RG, Margolis SV (1975) Pliocene climatic and glacial history of Antarctica as revealed by southeast Indian Ocean deep-sea cores. Bull Geol Soc Am 86:1058–1066

Carlquist S (1987) Pliocene *Nothofagus* wood from the Transantarctic Mountains. Aliso 11:571–583

Chittleborough G (1984) Nature, extent and management of Antarctic living resources. In: Harris S (ed) Australia's Antarctic policy options. Centre for resource and environmental studies. Australian National University Monograph, No 11, pp 135–161

Emiliani C (1978) The cause of the ice ages. Earth Planetary Sci Lett 37:349–352

Fordyce RE, Quilty PG A new family of fossil whales from the Pliocene of East Antarctica (in prep.)

Haq BU, Hardenbol J, Vail PR (1987) Chronology of fluctuating sea levels since the Triassic. Science 235:1156–1167

Imbrie J, Hays JD, Martinson A, McIntyre AC, Mix JJ, Morley NG, Pisias NG, Prell WL, Shackleton NJ (1984) The orbital theory of Pleistocene climate: support from a revised chronology of the marine δO^{18} record. In: Imbrie J (ed) Milankovitch and climate. NATO ASI Series, 2 vols, pp 269–305

Kauffman EG (1984) The fabric of Cretaceous marine extinctions. In: Berggren WA, van Couvering JA (eds) Catastrophes and earth history. Princeton University Press, Princeton, pp 151–248

Kennett JP (1985) Miocene to early Pliocene oxygen and carbon isotope stratigraphy in the southwest Pacific, Deep Sea Drilling Project Leg 90. Initial Reports Deep Sea Drilling Project, US Govt Printing Office, Washington 90:1383–1411

Lorius C, Jouzel J, Ritz C, Merlivat L, Barkov NI, Korotkevich YS, Kotlyakov VM (1985) A 150 000-year climatic record from Antarctic ice. Nature 316:591–596

Pickard J (1986) Antarctic oasis. Terrestrial environments and history of the Vestfold Hills. Academic Press, Sydney, 367 pp

Pickard J, Adamson DA, Harwood DM, Miller GH, Quilty PG, Dell RK (1986) Early Pliocene marine sediments in the Vestfold Hills, East Antarctica. S Afr J Sci 82:520–521

Pickard J, Adamson DA, Harwood DM, Miller GH, Quilty PG, Dell RK (1988). Early Pliocene sediments: coastline, and climate of East Antarctica. Geology 16:158–161

Quilty PG The geology of Marine Plain, Vestfold Hills, East Antarctica. In: Thomson MRA, Crame JA, Thomson JW (eds) Geological evolution of Antarctica. Cambridge University Press (in press)

Quilty PG 1990 Ammoelphidiella from the Pliocene of Larsemann Hills, East Antarctica. J Foramin Res 20:1–7

Shackleton NJ, Kennett JP (1975) Paleotemperature history of the Cenozoic and the initiation of Antarctic glaciation: oxygen and carbon isotope analyses in DSDP sites 277, 279 and 281. Initial Reports Deep Sea Drilling Project, US Govt Printing Office, Washington 29:743–755

Stainforth RM, Lamb JL, Luterbacher H, Beard JH, Jeffords RM (1975) Cenozoic planktonic foraminiferal zonation and characteristics of index forms. Paleont Contrib Univ Kansas Article 62:1–425
Vail PR, Mitchum RM, Thompson S (1977) Seismic stratigraphy and global changes of sea level. Mem Am Assoc Petroleum Geol 26:83–97
Webb P-N, Harwood DM (1987) The terrestrial flora of the Sirius Formation: its significance in interpreting Late Cenozoic glacial history. Antarct J US (4) 22:7–11
Webb P-N, Harwood DM, McKelvey BC, Mercer JH, Stott LD (1984) Cenozoic marine sedimentation and ice-volume variation on the East Antarctic craton. Geology 12:287–291
Zhang Q, Petersen JA (1985) A geomorphology and late Quaternary geology of the Vestfold Hills, Antarctica. ANARE Sci Rep 133:1–84

Temperature and Evolution: Southern Ocean Cooling and the Antarctic Marine Fauna

A. CLARKE[1]

Summary. Temperature can influence the physiology of marine organisms on a variety of time scales ranging from short-term fluctuations (tidal, vertical migration) to long-term climatic change. During the past 60 Ma shallow water marine organisms living at high southern latitudes around the margins of the continental fragments of Gondwana have experienced a decrease in mean seawater temperature from about 15 °C in the early Tertiary to the present range of roughly + 2° to − 1.8 °C. The early Cretaceous fauna around Gondwana was relatively rich and diverse. Despite the influence of glaciation the present fauna is rich in biomass and can show a very high within-site diversity. Some groups, however, notably fish and groups with calcareous skeletons such as bivalves and gastropods, are low in species richness. Evidence from physiology suggests that adaptation to low temperature is not a particularly severe evolutionary problem. The concept that the tropics are more equable than the polar regions is purely anthropocentric and entirely inappropriate for marine organisms. Polar organisms generally have a low cost of maintenance allowing higher growth efficiencies and thus affording a distinct energetic advantage over warmer water forms. Relating periods of extinction to a lowering of seawater temperature leads to a paradox in that the rates of cooling are so much slower (by several orders of magnitude) than those with which living marine organisms can cope, that it is difficult to see why previous marine communities could not adapt to track the change in temperature. One explanation is that altough a long-term change in mean temperature is often slow, this may be accompanied by severe short-term changes with which the fauna cannot cope. Also, it is unlikely that temperature change alone causes widespread extinction, but temperature varying with other ecological factors. If temperature change is indeed a problem then the direction is immaterial, climatic 'amelioration' is just as much a problem as climatic 'deterioration'. Clearly physiology, ecology and palaeontology have much to teach each other.

1 Introduction

Environmental parameters influencing the biology of marine organisms can vary over a range of time scales. Because they are easily measured and visualised, the most obvious of these are tidal, daily or seasonal fluctuations. These variations are generally of shorter duration than the life span of many animals and plants, obvious exceptions being microbial organisms and small invertebrates with several generations within a season.

Some aspects of the environment, however, also vary over much longer time scales (tens of millions of years). Examples include temperature, salinity, the chemical and isotopic composition of seawater, the degree of seasonality and possibly even the intensity of incident solar radiation. In this chapter the history of the Southern Ocean shallow-water marine fauna is examined in relation to long-term changes in seawater temperature, particularly over the past 60 million years (Ma).

2 Southern Ocean Cooling

In late Cretaceous times, roughly 65 Ma B.P., the high latitude waters around Gondwana were temperate with no evidence of sea level ice. At the same time the tropics were cooler. Ocean temperatures thus appear to have been much more even across the globe and there was only a small latitudinal temperature gradient (Valentine 1984).

During the Cenozoic, the high latitude seas cooled and came under the influence of ice. The least equivocal data for Southern Ocean palaeotemperatures comes from analyses of the stable oxygen isotope composition of foraminifera in sediment cores. Although the measurement of oxygen isotope composition is relatively straightforward, interpretation of the data is

[1] British Antarctic Survey, High Cross, Madingley Road, Cambridge CB3 0ET, UK.

Antarctic Ecosystems. Ecological Change and Conservation
(ed. by K. R. Kerry and G. Hempel)
© Springer-Verlag Berlin Heidelberg 1990

complicated by factors other than temperature which influence the fractionation observed. Particularly important is the amount of water frozen into ice sheets, for this influences the isotopic composition of the seawater and hence of the foraminiferal shells deposited at that time. Corrections for this ice can be made, and a generally agreed picture for the history of Southern Ocean temperature is shown in Fig. 1. This picture has been updated in detail, particularly with more precise data for the last 5 Ma (Quilty, this Vol.); nevertheless the overall picture is believed to be broadly correct.

Both surface and bottom waters in southern latitudes have cooled from mild Cretaceous temperatures of about 15 °C, to present levels of between + 2° and − 1.8 °C. Superimposed on this general trend, however, were periods of more rapid change, notably at the Eocene/Oligocene boundary (35–40 Ma B.P.) and in the late Pliocene (4–5 Ma B.P.). There was also a less dramatic period of cooling in the middle Miocene (15–20 Ma B.P.).

Although in geological terms these were rapid changes in temperature (perhaps 8 °C in 3 Ma, equivalent to only 0.003 °C every 1000 years), they were very slow in relation to typical oceanographic variation (Table 1.). Normal geological rates of temperature change are usually considerably slower than this, sometimes by several orders of magnitude (Schopf 1984). On the face of it these slow changes in mean temperature would not seem to pose a particularly severe evolutionary challenge to marine faunas. Before examining this question, however, it is necessary to look at the geological context and the palaeontological record. These are discussed in more detail in a companion paper elsewhere (Clarke and Crame 1989) and so will only be summarised here.

3 The Geological Context

The tectonic history of the southern hemisphere is not fully understood, but we do know enough to draw a broad picture. In late Jurassic times southern latitudes were dominated by the huge landmass of Gondwana, formed from what are now South America, Africa, India, Australasia and Antarctica. This was surrounded by the Tethys Sea to the north, and the proto-Pacific Ocean to the south. Gondwana had started to fragment by the late Cretaceous, with South America, Africa and India all separating from the Antarctica/Australia block at about this time. It is not clear when South America finally separated from western Antarctica but the best current estimate for the opening of the Drake Passage is 25–30 Ma B.P. (Oligocene). The separation of Australia from eastern Antarctica occurred at about the same time, and together these

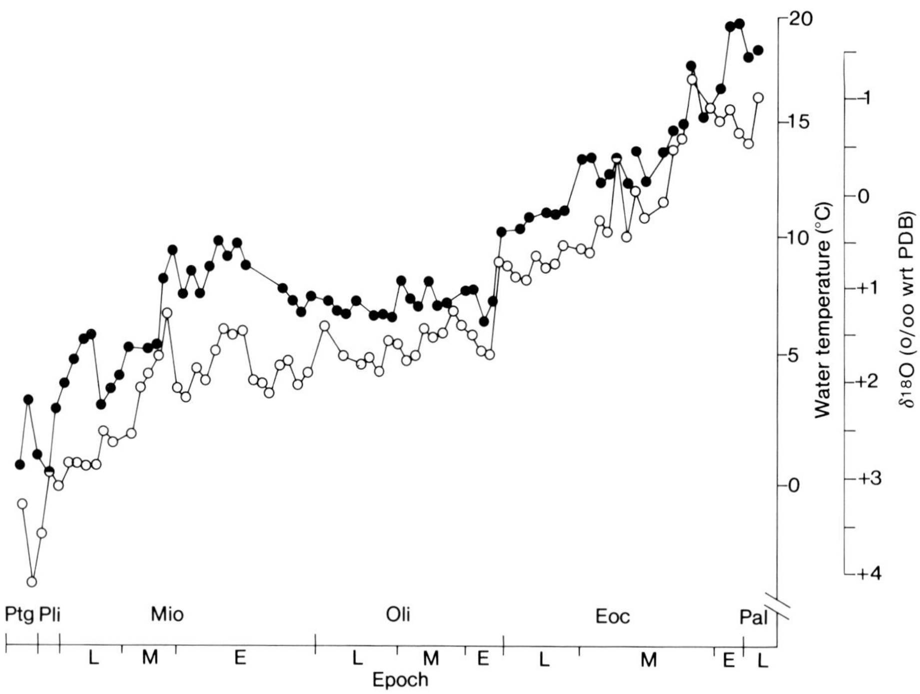

Fig. 1. Temperature of the Southern Ocean during the Tertiary as indicated by the oxygen isotope composition of foraminifera in sediment cores from DSDP sites 277, 279 and 281. ●, planktonic species; ○, benthic species. Data are expressed as oxygen isotope fractionation with respect to PDB (belemnite) standard (Redrawn from Kennett 1977)

Table 1. Typical rates of change in temperature encountered by marine organisms in different circumstances

Event	Rate of change in temperature	Examples
RAPID		
Tidal influx	15 °C in 60 min	Killifish *Fundulus heteroclitus* in salt marshes (Sidell et al. 1983)
Submergence of exposed intertidal organisms	Rapid (and in both directions)	Barnacles, mussles, gastropods, algae
Vertical migration through thermocline in tropics	Perhaps 15 °C in 2–3 h	Many vertical migrators, including myctophid fish and decapods
MODERATE (SEASONAL)		
Seasonal change in temperate waters	Typically 4 °C in winter to 15 °C in summer (but range may be as much as 20 °C)	Many temperate water species
SLOW/VERY SLOW		
Long-term oceanographic change	About 0.1 °C per year	Current global warming in low latitudes (Strong 1989)
Geologically 'fast'	5 °C in 100 000 years	Rapid changes in the Southern Ocean, for example at the Eocene/Oligocene boundary
Geologically 'slow'	Perhaps 1 °C every 3 million years	Gradual changes in mean ocean temperature (Schopf 1984)

resulted in the establishment of the circum-Antarctic current which isolates the Southern Ocean marine fauna to this day.

The large-scale distribution of landmass outlined above gives only limited information. To understand the evolution of the Southern Ocean fauna we also need to know the distribution of shallow-water habitats through time (for it is in these areas that most of the biomass and many of the species are found), and how these have been influenced by glaciation. The presence of shallow-water habitats has to be inferred from the relative distribution of the major landmasses and the presence of shallow-water deposits in the fossil record. Present knowledge is summarised in Fig. 2. It can be seen that at the present time the area of shallow water available to marine organisms is very small, due to mainly the extensive ice cap. The weight of this ice has also depressed the mean depth of the continental shelves.

Prior to the present glaciation there appear to have been extensive areas of shallow water. In particular there were a number of major embayments lying between eastern and western Antarctica (Fig. 2 B). Even further back there were again extensive areas of shallow water (Fig. 2 A).

4 Glaciological History

The distribution of potential shallow-water habitats is only part of the story. We also need to know whether the ice sheets have at any time extended so far as to scrape the continental shelves free of biota. Although evidence from sediment structure, ice-rafted debris and continental shelf topography, all indicate that ice has previously extended over many areas of continental shelf and even eroded it in some areas, it is not clear whether these extensions were simultaneous and whether they would have been sufficient to eradicate the organisms there (Clarke and Crame in press).

5 Palaeontological Evidence

The fossil record of high southern latitude marine organisms is not good and this greatly limits our understanding of the history of the marine fauna. Of particular importance are the early Tertiary deposits found on Seymour Island and James Ross Island. Fossils from these deposits have established clearly that in late Cretaceous and Eocene times there existed around Gondwana a rich shallow-water marine invertebrate fauna which included solitary corals, polychaetes, bivalves, gastropods, scaphopods, crabs, echinoids, asteroids and ophiuroids. There was also a diverse fish fauna (Eastman and Grande 1989). It is striking that several groups which are either lacking or represented only poorly in the modern fauna appear to have been quite diverse in the early Tertiary around Gondwana (notably decapods and teleost fish).

It has long been the accepted view that the tropics have been the major centre of evolution of shallow-water marine organisms, with new forms spreading later into higher latitudes (see review by Crame 1986). However, the rich late Cretaceous and early Tertiary faunas of Antarctica include several groups which predate their appearance in lower latitude fossil communities by as much as 15 million years. This earlier appearance at high latitudes (termed *high latitude heterochroneity*) indicates that the shallow waters around Gondwana may have been active in speciation, producing many forms which later moved into warmer or

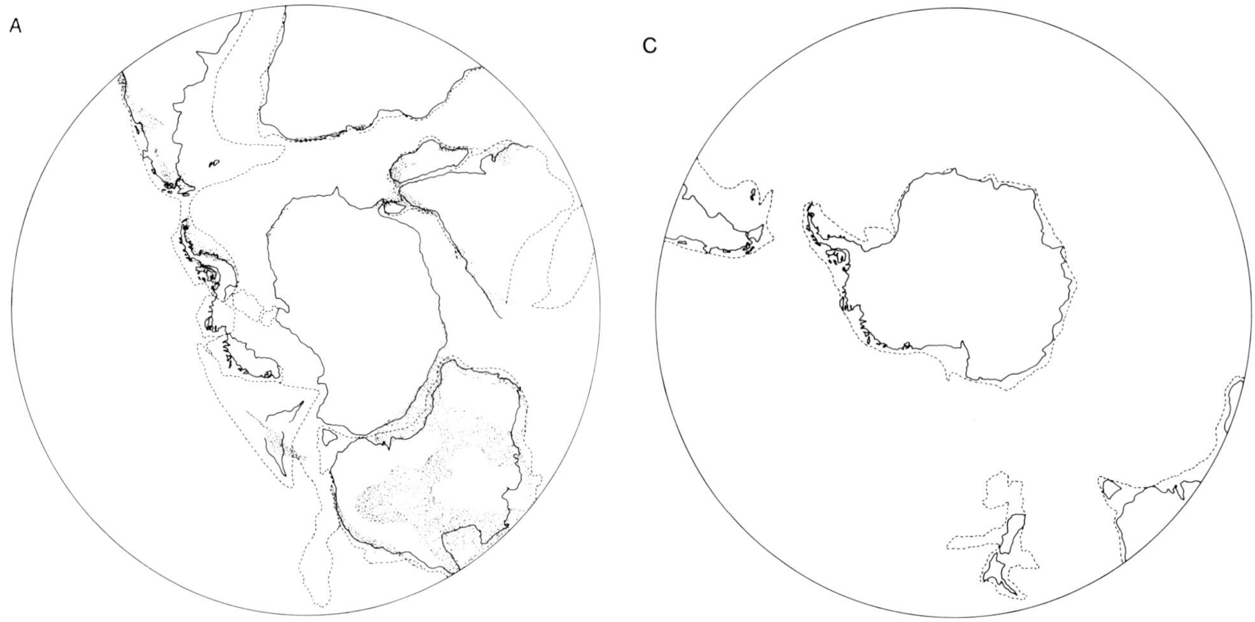

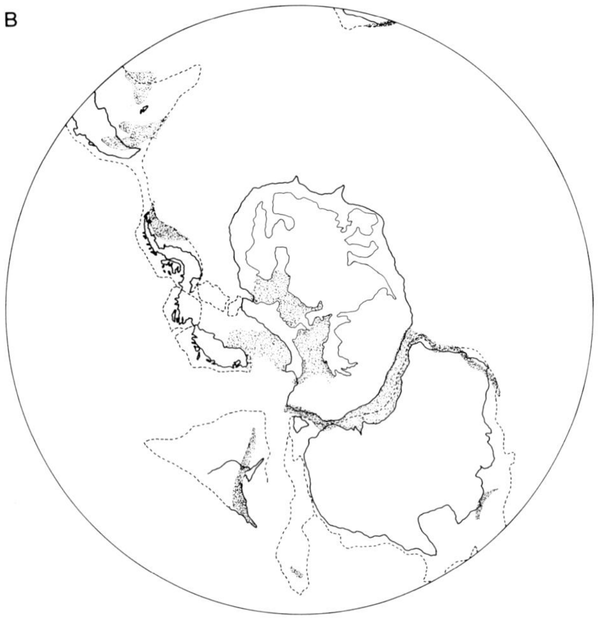

Fig. 2. Reconstruction of Gondwana (**A, B**) and the present disposition of continental blocks (**C**), with shallow-water areas indicated by *stippling*. **A** 120 Ma B.P. (early Cretaceous); **B** 60 Ma BP (Eocene); **C** present day. (Clarke and Crame 1989; based on Gondwana reconstruction of Lawver et al. 1985)

deeper water (Zinsmeister and Feldmann 1984). Particularly good evidence for a long history of shallow-water marine fauna in high southern latitudes is provided by gastropods. Indeed there is some eivdence that this area may have been a centre of dispersal in predatory gastropods. Overall the evidence suggests that the present Southern Ocean shallow-water marine fauna is the result of a long period of evolution in situ (Lipps and Hickman 1982; Clarke and Crame 1989).

6 The Present Thermal Regime

At the present day there is a particularly marked temperature gradient from the tropics to the poles. In the early Cenozoic this gradient was far less steep (Valentine 1984). However, not only does mean temperature vary along a latitudinal gradient, but so does the degree of seasonal variation in temperature. Taking these features together it is possible to divide the world ocean into a number of different thermal environ-

Table 2. A classification of benthic marine habitats based on temperature characteristics

Habitat/zone	Major features
Tropical	High temperature year-round, with little seasonal variation. Temperatures generally in range 25° to 30 °C, increasing towards equator. Production levels low and generally seasonal (except in upwelling areas)
Temperate	Temperatures vary seasonally, sometimes by as much as 20 °C. Seasonal cycles of temperature and production usually tightly linked
Polar	Temperatures low or very low (always below 4 °C and can be as low as −1.9 °C) with only little seasonal variation (in high southern latitudes variation less than ± 0.2 °C). Intensely seasonal production
Bathyal	Temperatues low (being determined in many areas by high latitude bottom water), with little seasonal variation. No primary production in situ (except around hydrothermal vents). Food availability may be seasonal or aseasonal depending on trophic niche and pattern of vertical flux

(The temperature characteristics of pelagic habitats depend on depth. Above the thermocline temperature patterns are similar to shallow-water benthic habitats at the same latitude, though generally less extreme; below the thermocline temperatures are generally stable and less than 4 °C, very similar to bathyal habitats.)

ments (Table 2). This classification is crude, but it does express some features of relevance to present-day faunas and their evolutionary history.

It is important to recognise that these various thermal environments have differing physiological consequences. An organism living in a thermally stable environment has the opportunity to evolve a physiology which is finely tuned to operate at a given temperature, for it is never called upon to do so at a greatly different temperature. In polar organisms this fine tuning is associated with a limited thermal tolerance (stenothermy), for many species will die at temperatures above about 6 °C (see discussion in Clarke 1987a). However, an organism living in an environment where the temperature changes either rapidly or seasonally cannot afford such fine tuning. Instead its physiology must be able to cope with a range of temperatures. The exact mechanisms by which this is achieved are not fully understood; they may involve eurythermal metabolism, or in some cases switching from 'cold water' to 'warm water' enzyme variants (for a recent review see Clarke 1987a).

Clearly there will be a significant difference in the physiology of a tropical organism adapted to cope with a warm, stable temperature and a temperate organism adapted to cope with a temperature which may be only a few degrees lower on average, but which varies seasonally by perhaps 15 °C. This difference in the seasonality of temperature may explain in part why the distribution of many shallow-water benthic organisms stop at the tropical/temperate divide (Valentine 1973). [These remarks refer, of course, only to ectothermic (poikilothermic) organisms. Endotherms (homoiotherms) maintain a more or less constant internal body temperature by a mixture of behavioural, physiological and anatomical means; as a result their metabolic rate and other aspects of their physiology are to a large extent independent of external temperature.]

Before discussing the relevance of these differences in physiology to the evolution of the Southern Ocean shallow-water marine fauna, it is necessary to discuss first some broader aspects of the modern fauna.

7 The Modern Fauna

It is traditional to view high latitude marine communities as poor in species and composed of slow-growing, infrequently reproducing organisms which are somehow scraping a living in the face of a harsh environment. It is now clear that this picture is quite incorrect. In many areas diversity is high, biomass can be very high and physiological studies have revealed that cold water organisms have some distinct energetic advantages over relatives living in warmer water.

7.1 Biomass

The Southern Ocean is now known to contain areas as rich in biomass as anywhere in the world (see, for example: Hardy 1972; White and Robins 1972; Lowry 1975; Dayton and Oliver 1977; Richardson and Hedgpeth 1977; Arnaud et al. 1986; Jazdzewski et al. 1986; Voß 1988). The most comprehensive data for Antarctica are from Arthur Harbour, Anvers Island (Fig. 3). The data available indicate that both hard and soft substrates in Antarctica support a high biomass and a large number of individuals.

7.2 Diversity

Ecologists have long sought to explain why some areas contain more species than others. A long-standing practical difficulty has been deciding on a suitable measure for expressing this. Clearly a simple count of the number of species in a sample will not do, since this

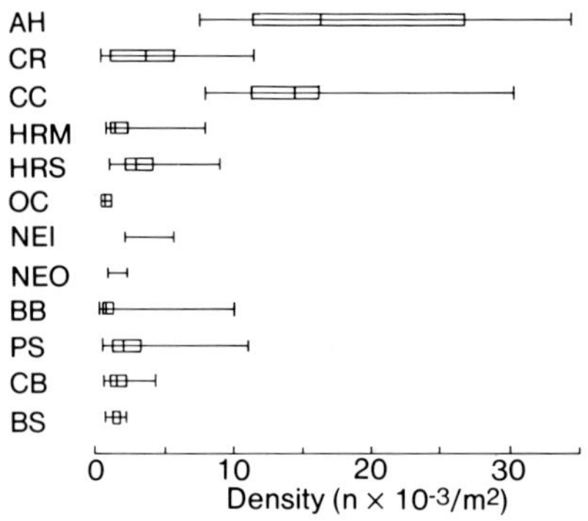

Fig. 3. Density (thousands of individuals per m^2) of organisms retained by a 0.5-mm screen from shallow-water soft substrate in Arthur Harbour, Anvers Island, Antarctica. Data are plotted as median (*vertical line*), quartiles (*box*) and range (*outer vertical lines*) for 14 replicate samples. Data also shown, for comparison, from other (mainly temperate) sites. Abbreviations: *AH* Arthur Harbour; *CR* Columbia River; *CC* Cape Cod Bay; *HRM* Hampton Roads (mud); *HRS* Hampton Roads (sand); *OC* Oregan Coast; *NEI* New England inner shelf; *NEO* New England outer shelf; *BB* Bramble Bay; *PS* Puget Sound; *CB* Chesapeake Bay; *BS* Western Beaufort Sea; (Redrawn from Richardson 1976)

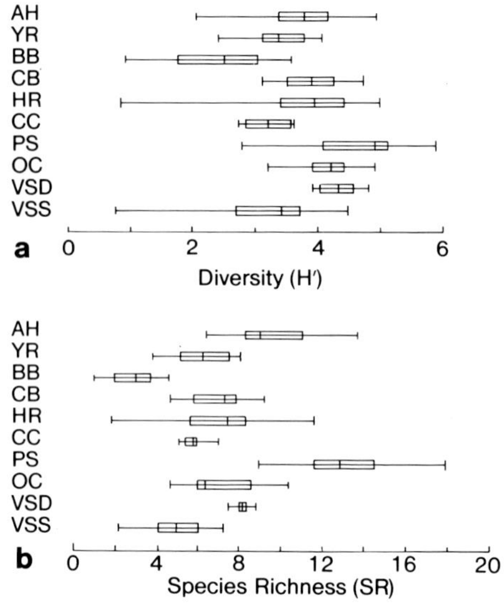

Fig. 4a, b. *Above*: Diversity (Shannon-Wiener *H'*); *below* species richness (Margalef *SR*) for soft-bottom fauna from Arthur Harbour, Anvers Island, Antarctica, compared with data for similar habitats elsewhere in the world. *YR* York River; *HR* Hampton Roads; *VSD* Virginia deep shelf; *VSS* Virginia shallow shelf; further abbreviations as in Fig. 3. Redrawn from Richardson (1976). Note that Margalef's formula for SR is quoted incorrectly in this reference (and in Richardson and Hedgpeth 1977), although the computation is correct

is somewhat dependent on the size of the sample taken (the more individuals examined, the greater is the chance of detecting rare species). Instead a number of diversity indices have been used, many of them based on ideas from information theory. These have the disadvantage, however, that the value of the index depends both on the number of species present and the distribution of individuals amongst those species (evenness or equitability).

In recent years the measurement of diversity indices has somewhat gone out of fashion amongst marine ecologists. And yet there remains an intuitive feeling that the diversity of a fauna, in the sense of the numbers of species present in a given area, must be telling us something. The calculation of a diversity index for a marine community involves a great deal of tedious and time-consuming work, and in the Southern Ocean data are available only for a soft-bottomed community in Arthur Harbour, Anvers Island (Richardson 1976; Richardson and Hedgpeth 1977). At this site diversity (as measured by the Shannon-Wiener index, *H'*) and species richness were greater than at many similar soft-bottom habitats elsewhere in the world (Fig. 4) and close to the theoretical maximum (Lowry 1975).

Although this is the only site in Antarctica for which there are data, the high within-site diversity is matched by taxonomic studies which reveal high species richness in many groups of Southern Ocean invertebrates. Examples include sponges, bryozoans, polychaetes and amphipods (summarised by Dell 1972).

Despite this overall tendency towards high species richness in the Southern Ocean there are several groups with a strikingly low diversity. These include decapods (Yaldwyn 1965) and teleost fish (Eastman and Grande 1989). Other groups appear to have an intermediate level of species richness, not as poor as decapods but well below levels found in the tropics. Examples are gastropod and bivalve molluscs, although (at least in gastropods) present estimates of species richness may well be underestimates because so many species in the Southern Ocean are very small with only a poorly documented taxonomy.

7.3 Latitudinal Diversity Clines

Gastropod and bivalve molluscs in particular have been instrumental in the development of the concept of a latitudinal cline in the diversity of marine organisms. In several groups of organisms (notably bivalves, gastropods, corals, decapods and foraminifera) tropical communities contain many more species than temperate or polar communities (Stehli et al. 1967, 1969; Stehli 1968).

Data for bivalves are shown in Fig. 5. At all three taxonomic levels, richness peaks in the tropics and

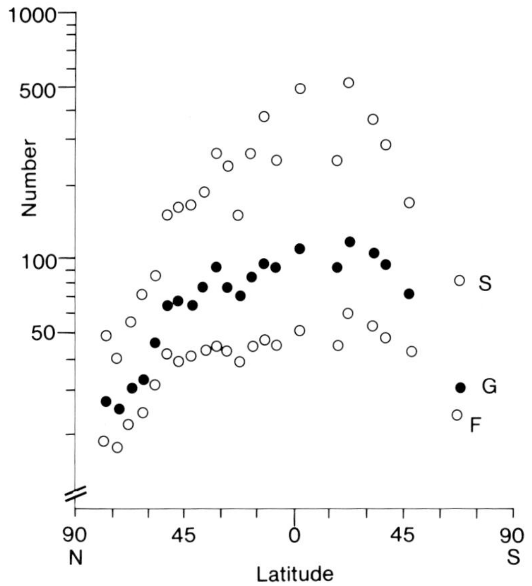

Fig. 5. Latitudinal clines in bivalve diversity. Each *point* represents the mean number of species (*S*), genera (*G*) or families (*F*) of bivalve in classes of 5 °C latitude (Data replotted from Stehli et al. 1967). Note that the vertical scale is logarithmic

decreases towards both northern and southern high latitudes. The data for the northern hemisphere are clearly more complete than for the southern hemisphere. In fact, the Antarctic is represented by only a single point, based on Powell (1960). Powell's taxonomic work, however, was thorough and utilised samples from a number of Antarctic locations. Apart from the difficulties with the smaller forms alluded to above, it is therefore reasonably complete. Clearly, in constrast to groups such as sponges, bryozoa and polychaetes, bivalves are relatively poor in species in the Southern Ocean. Before discussing possible reasons for the differences, however, it is necessary to discuss first some aspects of the physiology of Southern Ocean organisms.

8 Temperature and Physiology

A change in temperature affects many (indeed most) of the physical parameters of seawater including density, viscosity, pH and gas solubility (Clarke 1983). Since ectothermic marine organisms perforce must operate at the temperature of the seawater in which they live, temperature also has profound effects on all aspects of their physiology. Thus temperature influences all energetic processes which occur within the organism, including enzyme activity, the activity of contractile proteins and microtubules, ATP generation and intermediary metabolism, the maintenance of osmotic balance, neural activity, circulatory processes and the synthesis of gonads and new tissue. The effects of temperature are complex and by no means easy to predict. It is clear, however, that evolution has been able to answer (at least partially) the physiological challenges set by living at low temperature, for the Southern Ocean marine fauna contains examples of all but a few minor marine phyla, with many taxa represented by a large number of species and a high biomass. What is not clear, however, are the mechanisms by which this adaptation to temperature has been achieved. Neither do we know the full extent of the compensation for temperature. A complete review of the effects of low temperature on the physiology of Southern Ocean ectotherms is outside the scope of this work (for a recent general review see Clarke 1987a; for details of particularly well-studied systems see Johnston 1985 and Dietrich et al. 1987). However, two aspects are of particular relevance to understanding the evolution of the Southern Ocean fauna. These are the relationships of both basal metabolism and calcification to temperature.

8.1 Basal Metabolism

Basal metabolism is essentially the cost of staying alive. It is the total cost to the organism of all necessary maintenance processes such as protein turnover, membrane lipid turnover, osmotic work, DNA repair, basal neural and muscular activity (to maintain muscle tonus, vigilance and, in the case of a pelagic organism, position in the water column) and basal respiratory and circulatory processes (to maintain a flow of oxygen to, and waste products from, the tissues). The relative importance of these various process is not known for any marine organism, and indeed may vary from time to time. However, it is generally considered that the two most expensive processes are likely to be protein turnover and osmotic work (for a more complete discussion see Clarke 1983, 1987a).

It is almost impossible to make a precise measurement of basal metabolism. However, it is generally assumed that a good approximation is the metabolic rate of an inactive, non-feeding, unstressed individual that is not actively synthesising gametes. This is usually termed standard (or resting) metabolic rate.

Measurements of standard metabolic rate have been made on a vast number of marine invertebrates from tropical, temperatue and polar waters. Several authors have proposed relationships between standard metabolic rate and temperature, and all of these show a roughly exponential increase in metabolism with temperature, equivalent to Q_{10} values of between 1.8 and 3.0 (see summary in Clarke 1983, in press).

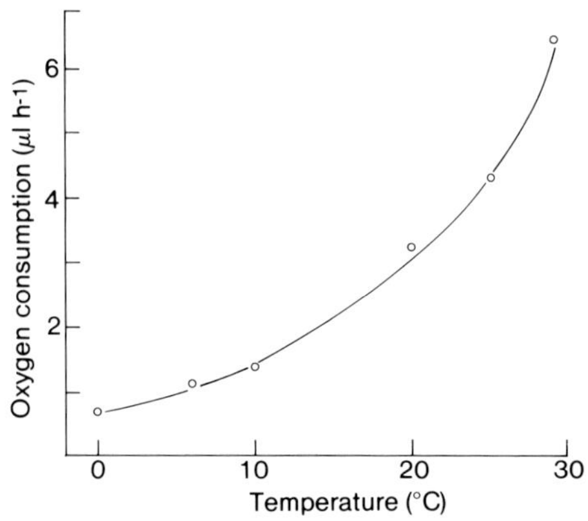

Fig. 6. Relationship between the oxygen consumption of a representative marine crustacean of 1 mg dry mass ($\mu l\, h^{-1}$) and estimated mean environmental temperature. These data were obtained from the parameters a and k in the relationship $QO_2 = aW^k$ (where QO_2 in oxygen uptake in $\mu l\, h^{-1}$ and W is the dry mass in mg). Values of a and k were calculated by Ivleva (1980) from all respiration data for marine crustaceans (both benthic and pelagic) available from the literature, pooled into six (unequal) temperature blocks between 0 °C and 30 °C. These six values of a and k were then used to calculate the estimated oxygen consumptions plotted here, taking W as 1 mg dry mass. Because of the wide variety of techniques employed by the different investigators, these data only approximate standard metabolism and the *fitted line* represents an upper boundary on the relationship between basal metabolism and temperature. The total number of determinations pooled was > 5000. For further details see Ivleva (1980) and Clarke (1983)

The most comprehensive study is that of Ivleva (1980), summarised in Fig. 6. This relationship shows that there is a distinct energetic advantage in living in cold water; the cost of basal metabolism is roughly six times greater at 30 °C than at 0 °C. Since basal metabolism represents energy that is ecologically wasted in the sense that it cannot be used for growth, activity or reproduction, it is clear that for a given intake of energy from food relatively less is wasted at low temperatures. This means that, all other things being equal, ecological growth efficiencies will be higher in polar waters (Clarke 1987b).

There is, as yet, no clear explanation of the shape of the relationship between basal metabolism and temperature (Fig. 6). It is possible that this is determined in part by the costs of protein turnover, but this hypothesis has yet to be tested. This relationship, of course, tells us nothing about the costs of other aspects of metabolism (such as locomotor activity, growth or reproduction), only about the cost of maintenance. The evidence available suggests that most of these other costs are roughly comparable in polar, temperate and tropical marine invertebrates and fish.

8.2 Calcification

One metabolic cost, however, that does vary with temperature is that of calcification. The free energy of precipitiation of calcium carbonate from seawater (which must be met by the organism, whatever the metabolic route employed) increases approximately linearly from 30° to 0 °C (Clarke 1983). What is not clear is the energetic significance of this to the organism. If the cost of producing a shell in a typical marine mollusc is only, say, 1% of the total energy budget, then a temperature dependence of this magnitude in the cost of carbonate precipitation is probably irrelevant. If the cost is, say, 15% of the total energy budget, then the temperature dependence is very significant indeed.

A number of lines of evidence point to the metabolic cost of calcification being energetically significant in marine molluscs. These include the tendency for cold water species to be small, with thin or poorly calcified shells bearing relatively little ornamentation (Nicol 1967). A striking example of this has been provided for marine gastropods by Graus (1974). In species from four sites in the western Atlantic there was a strong positive correlation between the mean calcification index and estimated mean water temperature (Fig. 7). Gastropods from the warmest waters (Puerto Rico) were carrying shells on average almost half as heavy again as similarly sized species from the cooler waters of Woods Hole. As well as mean calcification

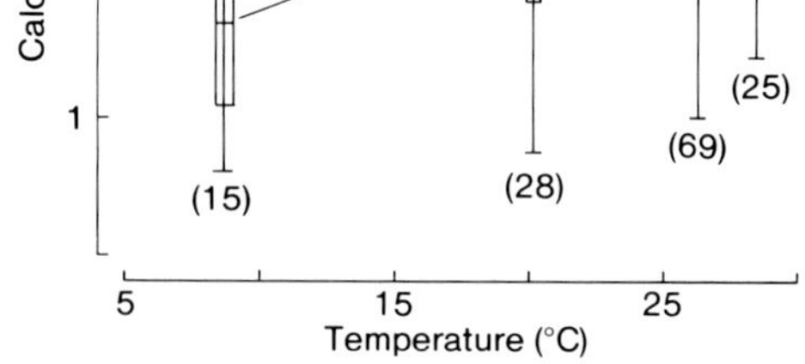

Fig. 7. Relationship between calcification index and estimated mean temperature in shallow-water marine gastropods from the western Atlantic. Calcification index was calculated as shell dry mass/shell internal volume. Data are plotted as mean with one standard deviation (*box*) and 95% confidence intervals, with number of species examined in *parentheses*. Sampling locations, in order of increasing temperature, were Woods Hole (42 °N), Beaufort (35 °N), South Florida (26 °N), and Puerto Rico (18 °N). Redrawn from Graus (1974)

index, both the range of values and the maximum increased towards the tropics.

8.3 Growth Rate

It has long been known that most species of polar marine organism grow slowly in comparison with related species of similar size and ecology from warmer waters (see summary in Clarke 1983). This slow growth has long been felt to be a direct consequence of the low temperature. Such an explanation, however, leads to a paradox for organisms would then appear to have evolved a compensation for the rate-limiting effects of temperature in some aspects of their physiology (for example locomotion) but not others (such as growth).

In many species with slow annual growth rates, growth is limited to the summer months with virtually no growth in winter. When growth actually occurs, it often proceeds quite rapidly. This suggests that growth is limited not by temperature, but by food availability (Clarke 1988). Only when the limitation of growth rate by resource seasonality has been allowed for can we judge the extent of temperature compensation in the growth process (Clarke in press).

A particularly striking example of this is shown by the venerid bivalve genus *Protothaca* (Harrington 1986). Molluscs preserve a record of growth in their shells. With care annual growth rings can be detected and, between these, daily growth increments. It is therefore possible to estimate the number of days in the year on which a given individual bivalve has been able to feed. In Fig. 8 data are shown for *Protothaca* from Panama to the Gulf of Alaska. It can be seen that in subtropical waters *Protothaca* is able to feed almost all year-round, whereas in the Gulf of Alaska feeding was possible for only half the year. As a result, bivalves in Alaska grew much more slowly than those in Panama. This had nothing (directly) to do with temperature; it was food availability that governed growth rate.

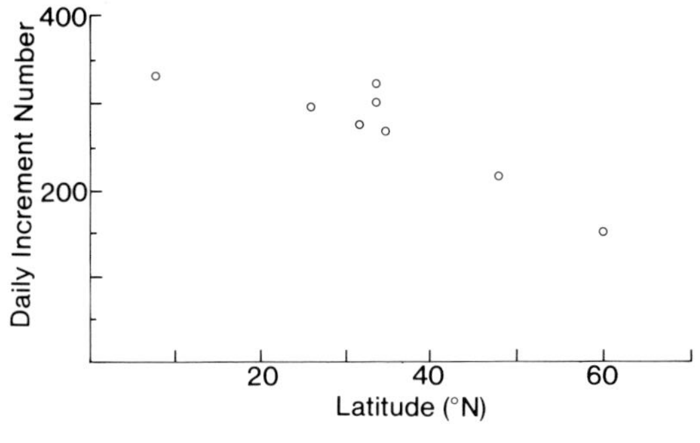

Fig. 8. Daily growth increments during the first year of life in the venerid bivalve genus *Protothaca*. Samples were obtained at points along a latitudinal gradient in the eastern Pacific from Panama to the Gulf of Alaska. Each *data point* represents the mean number of daily increments indicating feeding in a sample of 5 individuals from each site (Redrawn from Harrington 1986)

8.4 A Summary: Adaptation to Temperature

Clearly, an understanding of how organisms adapt to temperature at the molecular and physiological levels is a prerequisite for attempting to speculate on how long-term changes in temperature might influence patterns of evolution. It was emphasised earlier that the temperature experienced by an organism can vary over a wide range of time scales (Table 1). Since the pattern of seasonal temperature variation itself varies in different parts of the ocean (Table 2), it is necessary to distinguish different types of response (Table 3). Physiological studies of polar marine invertebrates and fish suggest that evolutionary adaptation to the low seawater temperature has presented no insuperable problems. A wide variety of organisms are present, often at high biomass. Detailed studies of locomotor ability and growth rate have indicated that the low temperature does not impair performance unduly, although very fast swimming may be precluded at very low temperatures. Slow growth rates appear to be the

Table 3. Different types of response to temperature in organisms (Clarke 1987a)

Response	Definition
Acute	The immediate short-term physiological response to a rapid change in temperature (mostly in an experiment). Can include heat shock and/or torpor
Acclimatisation	The adjustment of an organism's physiology to changes in temperature. These may be tidal, diurnal or seasonal and hence are generally on a time scale shorter than the organism's life span
Acclimation	The adjustment of an organism's physiology to a new temperature in the laboratory. (Note that in laboratory acclimation it is usual to modify only one variable, for example temperature, keeping all others constant. An organism undergoing acclimatisation in the field is subject to coincident variation in a whole range of environmental variables)
Adaptation	The evolutionary adjustment of an organism's physiology to temperature. This can, of course, involve adjustment to a seasonal or daily variation in temperature requiring acclimatisation

result of a severely seasonal shortage of food rather than low temperature per se.

9 The Influence of Seasonality

The influence of seasonality on growth rate has been outlined above. However, the intense seasonality of high latitude marine systems also greatly influences reproduction since, like growth, the production of eggs and sperm requires intake of energy. In general, the intensely seasonal pattern of primary production (and hence food availability) limits gametogenesis to the summer months. Production of an ovary may therefore take two successive seasons, with little storage of yolk during the intervening winter (although gametes may continue to develop).

It has long been known that marine invertebrates living at high latitudes tend to produce large eggs that hatch out either advanced larvae or miniature adults (thus avoiding free-living larval stages altogether). Although this tendency is *correlated* with low water temperature, it is more likely related to the seasonality of the environment. The period of food availability for larval stages at high latitudes is so short that many larval forms would be unable to mature before the winter, when they would presumably starve. It is possible that the low temperature may accentuate this effect by slowing the rate of larval development. This would imply, of course, that little (or no) compensation for temperature had evolved in marine invertebrates, in contrast to the physiology of adults (see above). Although there is some evidence for compensation in the larval development of crustaceans (Clarke 1982), careful studies of echinoderms have shown that compensation is lacking in the embryonic development (Bosch et al. 1987).

10 Discussion: Temperature and the Evolution of the Southern Ocean Marine Fauna

Fossil evidence indicates that the late Cretaceous and early Tertiary the shallow waters of southern high latitudes were considerably warmer than today, with no sea-level ice. The marine fauna appears to have been relatively rich and diverse, and to have included some groups that are only poorly represented in the present fauna (notably decapods and teleost fish). Since then seawater temperature has decreased, and the area available for shallow water fauna has fluctuated widely as a result of tectonic movements and periods of glaciation (see summary in Clarke and Crame 1989).

These patterns have some relevance to current discussions concerning the dynamics of extinction and speciation in marine communities. Indeed the tectonic and thermal history of the high southern latitudes make this area an historical laboratory of enormous value for testing fundamental ideas concerning the evolution of faunas.

The reduction in the diversity of calcareous forms as the temperature decreased is of particular interest. Without a better fossil record we cannot say whether this was a gradual decrease or whether a specific event (for example the sharp cooling in the Pliocene) was responsible. The microfossil record suggests a relatively sudden shift from calcareous to siliceous forms about 35 Ma B.P. (Kennett 1977) but this may not have applied to macrofossils. Perhaps the extra energetic costs in secreting a skeleton at lower temperatures is the major reason for the relatively poor representation of molluscs in the Southern Ocean. Certainly there is a relative lack of medium-sized gastropods and bivalves in the Southern Ocean (although they are not lacking entirely, and some representatives can be extremely abundant locally). Nevertheless many species are very small; of the 51 species of gastropods in the depth range 0–100 m at the South Orkney Islands 38 (75%) are less than 10 mm maximum size (Picken 1985). Although this suggests the influence of the cost of calcification, it is possible that other factors are also at work. One such factor might be predation, for a thin and poorly calcified shell offers little protection against crushing or drilling predators. It is possible that small size is in part a response to predation (although predators can themselves be quite small); unfortunately there are few data with which to test this hypothesis.

Certainly gastropods and bivalves have received considerable taxonomic attention, and these have been among the taxa primarily responsible for the establishment of the concept of a latitudinal cline in diversity. Since several other groups seem to be particularly diverse in the Southern Ocean (for example polychaetes, bryozoans, amphipods and sponges) it is perhaps premature to extend the concept to the fauna as a whole. More detailed taxonomic work is required.

The loss of the decapods and essentially all but a single suborder of fish is difficult to explain. Certainly the classical pattern of adaptive radiation shown by the nototheniid fishes in the Southern Ocean has all the features of the occupation of an essentially empty habitat by an immigrant group. The demise of the earlier teleost fauna cannot be ascribed to cold alone; the development of antifreeze and the associated physiology appears to be a relatively simple evolutionary problem, for there is evidence that this has occurred on a number of separate occasions (Scott et al. 1986). Perhaps in the decapods and fish we are looking simply at biogeographic and historical factors; a glacial advance causing extinction of the earlier fauna and recolonisation from outside Antarctica made difficult by the isolating barrier of the Polar Frontal Zone.

10.1 Temperature and Extinction

The history of the earths marine fauna has been punctuated by periods of greater than normal extinction. These so-called mass extinctions have attracted a great deal of scientific interest, but are still essentially without convincing explanation. Two of the more widely accepted explanations are:

1. A decrease in the area of shallow water.
2. A decrease in temperature.

It is therefore interesting that the present Southern Ocean marine fauna is living at a period when the available area of shallow water is greatly reduced compared with earlier in the Tertiary (Fig. 2) and following a period of particularly sharp temperature drop in the Pliocene (Fig. 1).

Unfortunately a detailed sequence of fossils around the period of Pliocene cooling is lacking, thus it is not known in detail what happened as the temperature decreased. However, the high biomass and diversity of the present fauna indicate that low temperature per se presents no insuperable physiological problems. Indeed in purely energetic terms it is more economical for marine ectotherms to live at the poles than at the tropics (Fig. 6). The ecological problems in living at high latitudes centre not on temperature but on the intensely seasonal availability of resources, leading to slow (but efficient) growth and delayed maturity.

Since the rates of changes in temperature with which many marine organisms can cope in the short term, and to which they can evolve adaptation, are so rapid compared with typical geological rates even at their fastest (by several orders of magnitude), it is pertinent to ask whether temperature change is ever a problem on the geological time scale. The answer seems to be that it is, for there are well-documented cases of the extinction of many species associated with decreases in temperature. What is not clear is whether the decrease in temperature is always the direct *cause* of that extinction.

A particularly severe bout of extinction affected the Pliocene molluscan fauna of the western north Atlantic and affords an excellent example of temperature-related extinction. In the early Pliocene, Florida was even warmer than today and a rich molluscan fauna flourished in the shallow tropical seas. This fauna, like modern tropical faunas, contained a mix of species with differing thermal ranges. Some stenothermal species were found only in tropical waters; other more eurythermal species had ranges that extended into more temperate waters. When the water temperature around Florida decreased in the late Pliocene many species died out. Interestingly it was generally the exclusively tropical species that died out; the more eurythermal species survived. (Strictly the difference in survival was between those of wide and narrow *geographical range*; the associated thermal tolerances are only inferred, but the inference is a reasonable one.) This extinction event has been described in detail by Stanley (1984), who has termed this effect of differential survival a *thermal filter*. The reason for the differential survival of different organisms may thus lie in their differing thermal tolerances. As discussed earlier both tropical and polar marine ectotherms have to cope with a relatively narrow range of temperatures, and hence can evolve a physiology finely tuned to a particular temperature. Temperate water organisms, however, need a more eurythermal physiology to cope with seasonal changes in temperature.

Although varying thermal tolerances in marine invertebrates with differing geographical ranges is an intuitively appealing explanation for the differential survival of taxa during periods of seawater cooling, it does lead to something of a paradox. Which is that even in periods of temperature change that are rapid in palaeoclimatic terms, the rate of change of mean temperature is so slow in biological terms as to appear to be nothing of a problem. For example an estimate of the rate of fall of seawater temperature in the Pliocene of 8 °C in 3 Ma is equivalent to only 0.003 °C every 1000 years. A typical stenothermal polar organism can survive within a range of $-2°$ to $+4$ °C, a range of 6 °C, and a typical polar marine invertebrate might have a life span of 20 years. This geologically rapid rate of temperature change thus amounts to only 0.00006 °C per generation, or 0.001 % of the range of tolerance per generation. This is a low rate of evolution compared with typical rates reported for morphological characters (see discussion in Levinton 1988), and emphasises the paradox: *why does temperature appear to cause extinctions when the rates of temperature change are so slow compared with those that living marine invertebrates tolerate without stress?*

This intriguing problem has received little attention, and yet it is central to any attempt to explain patterns of extinction in relation to climatic change. Perhaps the answer lies in the precise pattern of temperature change. Isotopic data such as that illustrated in Fig. 1

can only indicate the *mean* temperature. However, it is possible that what is important biologically are the extremes. In other words, as mean temperature drops slowly (and at a rate with which evolution should theoretically be able to cope), perhaps there are episodic low temperatures which are below the lower lethal temperature of some organisms. A sudden cold snap may thus result in widespread mortality (as is well known for living benthic organisms) even though the change in *mean* temperature recorded in foraminiferan shells is only slight. What is not clear is whether this mechanism can explain total extinction of species. Perhaps a complete explanation must involve other ecological factors acting in concert with temperature, with the key feature varying with taxon, time and place. A final point is that although research may indicate a high degree of adaptiveness in the physiology of living polar organisms, these are the survivors of the thermal filter. We are unable to examine those species that could not adapt, and we are unlikely to be able to estimate what fraction of the previous fauna they comprised.

10.2 Climatic Amelioration and Climatic Deterioration

Almost without exception the invocation of temperature stress as an agent of extinction occurs when there is a period of cooling (climatic 'deterioration'). However, if it is indeed a change in temperature that is causing extinction, then the direction is immaterial. A rise in temperature (so-called amelioration) is just as severe a stress as a fall; the direction does not matter, for it is the move away from the temperature to which the organism is adapted that forms the supposed stress. The idea that warm waters are equable and somehow less stressful than cold waters is entirely anthropocentric. The terms climatic amelioration and deterioration should be abandoned, for they have no meaning to marine invertebrates. Indeed, purely in energetic terms it is the tropical environment that may be regraded as the more stressful, since so much energy must be wasted solely in maintenance (Fig. 6). (This ignores, of course, the effect of increased seasonality at high latitudes. It also takes no account of the fact that the *nature* of the physiological stress induced by temperature increase and decrease differs; see Clarke in press.)

10.3 Extinction or Migration?

Although physiological studies have revealed a significant capacity for adaptative response in marine invertebrates, the response of many marine organisms to climatic change is often migration rather than evolution. This is not to say that individual organisms migrate (any more than individual organisms evolve), for the time scale of significant temperature change is frequently longer than the life span of the individual. Rather it is a case of the boundaries of distribution shifting by processes including mortality of adults and differential survival of young stages. Thus over a long period of time (although often very rapidly on a geological time scale) the population shifts. When the fossil record is poor, it is then difficult to distinguish migration from extinction.

Migration is only possible, of course, when there are suitable areas to go to. This may explain why the western Atlantic Pliocene extinction was so extensive, particularly amongst purely tropical forms: there were simply no tropical refugia to which one could retreat (Stanley 1984). For shallow-water marine invertebrates around the continental fragments of Gondwana there was equally nowhere to escape the cooling of the water. Migration into deeper water was no answer for the temperature of the deep water is dictated by that generated at high southern latitudes. (Equally organisms from the deep sea were not 'pre-adapted' to cold temperatures, as has frequently been suggested, for they were subject to precisely the same temperature regime as the high latitude shallow-water fauna.)

The shallow-water fauna around Antarctica may have had to migrate into deeper water as a response to a widespread extension of the continental ice sheet (Clarke and Crame in press), but this is a separate issue from migration as a response to temperature.

10.4 Temperature or Seasonality?

The Cooling of the Southern Ocean appears to have been accompanied by an increase in the degree of seasonality. It is not clear whether this was a smooth increase, or whether seasonality increased suddenly. One piece of evidence that the increase in seasonality may have been relatively sudden, and possibly associated with the onset of the present oceanographic regime, comes from echinoid fossils in Australia. At about the time of the onset of the circum-Antarctic current there appears to have been a switch from species that free-spawned, to species that brooded (Foster 1974). This has been interpreted as a response to a sudden decrease in seawater temperature but it is more likely a response to an increase in seasonality. It would be extremely valuable to obtain data on the history of seasonality of the Southern Ocean from daily growth increments presence in molluscan shells at different ages.

10.5 Evolution at High Latitudes

A corollary of the view that tropical climates are somehow more equable than colder waters has been the idea that the tropics were the centres of speciation. Species evolved in the tropics, migrating to lower, cooler latitudes with time.

As well as the slight inconsistency in that tropical forms were apparently able to spread to cooler waters, whereas a lowering of global water temperatures has led to widespread extinction, there is no obvious physiological reason why any one particular thermal regime should provide a focus of speciation. Indeed, the recent demonstration of the first appearance of many forms at high latitudes (heterochroneity: Zinsmeister and Feldmann 1984) suggests that speciation proceeds just as effectively in cooler waters.

There is good evidence that some groups may have originated in the cooler, high latitudes (for example predatory gastropods: Taylor and Taylor 1977), and several groups appear to be actively speciating at the present time (for example buccinid and littorinid gastropods: Arnaud and Bandel 1976). One factor that may be influencing speciation at high latitudes is the prevalence of non-pelagic development. The type of development (pelagic versus non-pelagic) can often be determined in fossil gastropods from the structure of the protoconch. Detailed study of fossil gastropods has shown that species which reproduce without a pelagic larval phase tend to speciate more rapidly than those with pelagic larvae (reviewed by Jablonski and Lutz 1983). Indeed, it has recently been suggested that the explosive radiation of the cheilostome Bryozoa in the Cenomanian and after may have been triggered by the evolution of brooded non-pelagic larvae (Taylor 1988). The reason for the increased speciation is believed to be the reduced gene flow associated with non-dispersing larvae. A reduced gene flow across the Polar Frontal Zone may also explain the high degree of endemism in the Southern Ocean fauna (Dell 1972).

Many marine invertebrates in the Southern Ocean have non-pelagic larvae, although pelagic larvae are found in a few species. Interestingly these species are frequently among the most common and widespread of species (for example the limpet *Nacella concinna*). It would clearly be valuable to compare the genetic identity of widely separated populations of marine invertebrates with pelagic and non-pelagic development, using both traditional electrophoretic methods as well as more sensitive techniques such as mitochondrial DNA.

Acknowledgements. I am grateful to Alistair Crame for many useful and stimulating discussions during the preparation of this paper.

References

Arnaud PM, Bandel K (1976) Comments on six species of marine Antarctic Littorinacea (Mollusca, Gastropoda). Tethys 8:213–230

Arnaud PM, Jazdzewski , Presler P, Sicinski J (1986) Preliminary survey of benthic invertebrates collected by Polish Antarctic expeditions in Admiralty Bay (King George Island, South Shetland Islands, Antarctica). Pol Polar Res 7:7–24

Bosch I, Beauchamp KB, Steele ME, Pearce JS (1987) Development, metamorphos, and seasonal abundance of embryos and larvae of the Antarctic sea urchin *Sterechinus neumayeri*. Biol Bull 173:126–135

Clarke A (1982) Temperature and embryonic development in polar marine invertebrates. Int J Invert Repr Develop 5:71–82

Clarke A (1983) Life in cold water: the physiological ecology of polar marine ectotherms. Oceanogr Mar Biol Ann Rev 21:341–453

Clarke A (1987a) The adaptation of aquatic animals to low temperatures. In: Grout BWW, Morris GJ (eds) The effects of low temperatures on biological systems. Edward Arnold, London, pp 315–348

Clarke A (1987b) Temperature, latitude and reproductive effort. Mar Ecol Progr Ser 38:89–99

Clarke A (1988) Seasonality in the Antarctic marine environment. Comp Biochem Physiol 90B:461–473

Clarke A (in press) What is cold adaptation and how should we measure it? Am Zool

Clarke A, Crame JA (1989) The origin of the Southern Ocean marine fauna. In: Crame JA (ed) Origins and evolution of the Antarctic biota. Geological Society, Spec Publ 47:253–268

Crame JA (1986) Polar origins of marine invertebrate faunas. Palaios 1:616–617

Dayton PK, Oliver JS (1977) Antarctic soft-bottom benthos in oligotrophic and eutrophic environments. Science 197:55–58

Dell RK (1972) Antarctic benthos. In: Russell FS, Yonge M (eds) Advances in marine biology, vol 10. Academic Press, London, pp 1–216

Dietrich HW, Prasad V, Luduena RF (1987) Cold-stable microtubules from Antarctic fishes contain unique alpha tubulins. J Biol Chem 262:8360–8366

Eastman JT, Grande L (1989) Evolution of the Antarctic fish fauna with emphasis on the Recent nototheniods. In: Crame JA (ed) Origins and evolution of the Antarctic biota. Geological Society, Spec Publ 47:241–252

Foster RJ (1974) Eocene echinoids and the Drake Passage. Nature 249:751

Graus RR (1974) Latitudinal trends in the shell characteristics of marine gastropods. Lethaia 7:303–314

Hardy P (1972) Biomass estimates for some shallow-water infaunal communities at Signy Island, South Orkney Islands. Bull Br Antarct Surv 31:93–106

Harrington RJ (1986) Growth patterns within the genus *Protothaca* (Bivalvia: Veneridae) from the Gulf of Alaska to Panama: palaeotemperatures, palaeobiogeography and palaeolatitudes. PhD Thesis, University of California, Santa Barbara, 235 pp

Ivleva IV (1980) The dependence of crustacean respiration on body mass and habitat temperature. Int Rev Ges Hydrobiol 65:1–47

Jablonski D, Lutz RA (1983) Larval ecology of marine benthic invertebrates: palaeobiological implications. Biol Rev 58:21–89

Jazdezwski K, Jurasz W, Kittel W, Presler E, Presler P, Sicinski J (1986) Abundance and biomass estimates for benthic fauna of the Admiralty Bay, King George Island, South Shetland Islands. Polar Biol 6:5–16

Johnston IA (1985) Temperature adaptation of enzyme function in fish muscle. In: Laverack MS (ed) Physiological adaptations of marine animals. Society for Experimental Biology (The Company of Biologists), Cambridge. Symp Soc Exp Biol 34:95–122

Kennett JP (1977) Cenozoic evolution of Antarctic glaciation, the circum-Antarctic Ocean, and their impact on global paleoceanography. J Geophys Res 82:3843–3860

Lawver LA, Sclater JG, Meinke L (1985) Mesozoic and Cenozoic reconstructions of the South Atlantic. Tectonophysics 114:233–254

Levinton JS (1988) Genetics, palaeontology and macroevolution. Cambridge University Press, Cambridge

Lipps JH, Hickman CS (1982) Origin, age and evolution of Antarctic and deep-sea faunas. In: Ernst WG, Morin JG (eds) The environment of the deep-sea (Rubey Vol II). Prentice-Hall, Englewood Cliffs, New Jersey, pp 325–356

Lowry JK (1975) Soft bottom macrobenthic community of Arthur Harbour, Antarctica. Antarct Res Ser (American Geophysical Union) 23:1–19

Nicol D (1967) Some characteristics of cold-water marine pelecypods. J Palaeontology 41:1330–1340

Picken GB (1985) Marine habitats – benthos. In: Bonner WN, Walton DWH (eds) Antarctica. Pergamon Press, Oxford, pp 154–172

Powell AWB (1960) Antarctic and Subantarctic Mollusca. Auckland Inst Mus Recs 5:117–193

Richardson MD (1976) The classification and structure of marine macrobenthic assemblagies at Arthur Harbour, Anvers Island, Antarctica. PhD Thesis, Oregon State University, 142 pp

Richardson MD, Hedgpeth JW (1977) Antarctic soft-bottom macrobenthic community adaptations to a cold, stable, highly productive, glacially affected environment. In: Llano GA (ed) Adaptations within Antarctic ecosystems. The Smithsonian Institution, Washington DC, pp 181–196

Schopf TJM (1984) Climate is only half the story in the evolution of organisms through time. In: Brenchley PJ (ed) Fossils and climate. John Wiley, Chichester, pp 278–289

Scott GK, Fletcher GL, Davies PL (1986) Fish antifreeze proteins: recent gene evolution. Can J Fish Aquat Sci 43:1028–1034

Sidell BD, Johnston IA, Moerland TS, Goldspink G (1983) The eurythermal myofibrillar protein complex of the mummichog (*Fundulus heteroclitus*): adaptation to a fluctuating thermal environment. J Comp Physiol 153:167–173

Stanley SM (1984) Marine mass extinctions: a dominant role for temperature. In: Nitecki MH (ed) Extinctions. University of Chicago Press, Chicago, pp 69–117

Stehli FG (1968) Taxonomic diverstiy gradients in pole location: the recent model. In: Drake ET (ed) Evolution and environment. Yale University Press, New Haven, pp 163–227

Stehli FG, Douglas RG, Newell ND (1969) Generation and maintenance of gradients in taxonomic diversity. Science 164:947–949

Stehli FG, McAlester AL, Helsley CE (1967) Taxonomic diversity of recent bivalves and some implications for geology. Bull Geol Soc Am 78:455–466

Strong AE (1989) Greater global warming revealed by satellite-derived sea-surface-temperature trends. Nature 338:642–645

Taylor JD, Taylor CN (1977) Latitudinal distribution of predatory gastropods on the eastern Atlantic shelf. J Biogeogr 4:73–81

Taylor PD (1988) Major radiation of cheilostome bryozoans: triggered by the evolution of a new larval type? Hist Biol 1:45–64

Valentine JW (1973) Evolutionary paleoecology of the marine biosphere. Prentice-Hall, Englewood Cliffs, New Jersey, 511 pp

Valentine JW (1984) Climate and evolution in the shallow sea. In: Brenchley PJ (ed) Fossils and climate. John Wiley, Chichester, pp 265–277

Voß J (1988) Zoogeographic and Gemeinschaftsanalyse des Makrozoobenthos des Weddellmeeres (Antarktis). Berichte zur Polarforschung 45:145 pp

White MG, Robins MW (1972) Biomass estimates from Borge Bay, Signy Island, South Orkney Islands. Bull Br Antarct Surv 31:45–50

Yaldwyn JC (1965) Antarctic and Subantarctic decapod Crustacea. In: Van Meighem J, Van Oye P (eds) Biogeography and ecology in Antarctica. Monographiae Biologicae 15, Junk, The Hague, pp 323–332

Zinsmeister WJ, Feldmann RM (1984) Cenozozic high latitude heterochroneity of southern hemisphere marine faunas. Science 224:281–283

Environmental and Biological Variability in the McMurdo Ice Shelf Ecosystem

C. Howard-Williams[1], R. D. Pridmore[2], P. A. Broady[3] and W. F. Vincent[1]

Summary. The McMurdo Ice Shelf is an ablation region on the northwestern side of the Ross Ice Shelf. The surface forms the largest non-marine aquatic ecosystem in the McMurdo Sound region with an interlinking system of lakes, pools and streams occuring across more than 1500 km^2. Two major types of ice shelf morphology with different physical and biological characteristics were distinguished: "Pinnacle Ice" with many small interconnected pools and streams, and "Undulating Ice" with continuous moraine cover and discrete pools and lakes up to 10^4 m^2. The flora of these is dominated by benthic rather than planktonic communities. Cyanobacteria which coat the base of the pools with mats and films of varying thickness are generally the most commonly occurring and abundant organisms. Benthic diatoms and coccoid chlorophytes are also found throughout the system but tend to dominate in the waters of the Pinnacle Ice rather than the Undulating Ice. There are large variations in the conductivity and nutrient content of these waters, with a marine tidal influence in some parts. The water bodies are subject to continual change as the Ice Shelf moves, with marked temporal variability in environmental conditions on a diel, seasonal and long-term (years to decades) basis. This area contains the most extensive microbial growths in southern Victoria Land and is a potential inoculum source of micro-organisms for the entire region.

1 Introduction

The large ice shelf regions of Antarctica have not been explored in terms of their ecosystem properties. The surfaces of some of these melt to varying degrees each summer to form reticulated networks of free water. Some of these networks are very extensive and the water may persist for several weeks, e.g. McMurdo Ice Shelf, George VI Ice Shelf and the Amery Ice Shelf (Debenham 1920; Reynolds 1981; Mellor and McKinnon 1960). The McMurdo Ice Shelf has been the most intensively studied of these, particularly with respect to the diatom flora (Brady 1980; Kellogg and Kellogg 1987) but its general ecological and environmental properties are little known.

The McMurdo Ice Shelf (Fig. 1, Plate 1) is an ablation region in the north western corner of the Ross Ice Shelf. It is a region of some 1500–2000 km^2 of ice with a highly varied topography (Plates 2, 3). During mid- to late summer, melt pools and streams appear across the Ice Shelf. Widespread deposits of moraine and marine sediment thought to be transported up from the seabed as trapped material in anchor ice, as well as some wind-blown sediment from ice-free areas

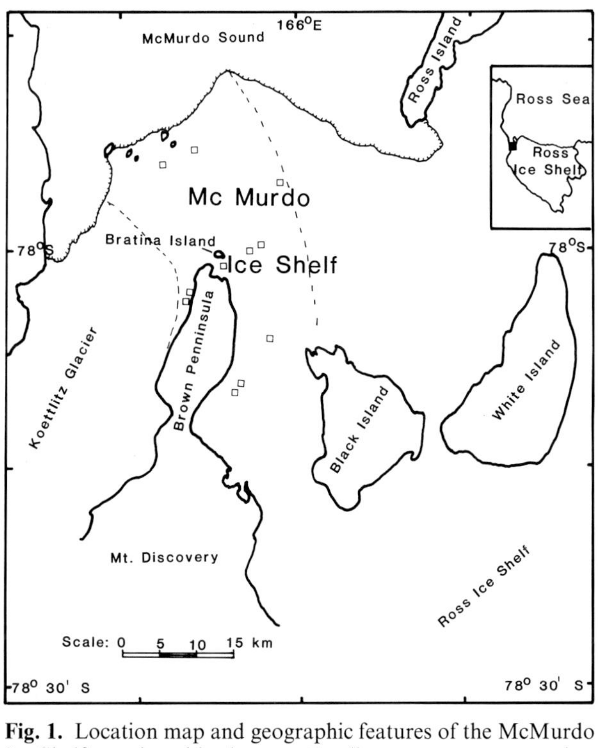

Fig. 1. Location map and geographic features of the McMurdo Ice Shelf mentioned in the text. *Small squares* are transect sites for the large-scale variability study. *Inset* Location of McMurdo Ice Shelf in the Ross Sea region

[1] Taupo Research Laboratory, DSIR, P.O. Box 415, Taupo, New Zealand.
[2] Water Quality Centre, DSIR, Private Bag, Hamilton, New Zealand.
[3] Department of Plant and Microbial Sciences, University of Canterbury, Christchurch, New Zealand.

Antarctic Ecosystems. Ecological Change and Conservation
(ed. by K. R. Kerry and G. Hempel)
© Springer-Verlag Berlin Heidelberg 1990

Plate 1. Oblique aerial photograph of the McMurdo Ice Shelf looking north from near Minna Bluff. The open water of McMurdo Sound appears in the distance. The dark sediment-covered regions with ponds and lakes constitute the Undulating Ice (Reproduced by permission of the US Navy)

Plate 2. Typical small pond in the Undulating Ice. A fairly large ice-covered pond can be seen behind as can the undulating relief of the sediment-covered ice shelf

Plate 3. Typical broken terrain of the Pinnacle Ice with small interconnected ponds between ice pinnacles

occur across the shelf (Debenham 1920, Swithinbank 1970). Lenses of mirabilite (predominantly $Na_2SO_4 \cdot IOH_2O$) occur in the ice resulting in surface saline deposits at several locations (Brady and Batts 1981). The McMurdo Ice Shelf is 10 to 50 m thick over much of its area (Swithinbank 1970) with a surface relief of up to 20 m in places; some of the ponds are at, or possibly below, sea level.

Ice movement throughout the McMurdo Ice Shelf is slow, 2 $m\,y^{-1}$ being a typical rate for the area between Black Island and Brown Peninsula (Swithinbank 1970). This is an order of magnitude lower than for the adjacent Ross Ice Shelf. However, there is widespread evidence of continual movement and change of ponds and watercourses. Although a small amount of algal material collected from ponds on the Ice Shelf by members of Scott's "Discovery" expedition was reported on by Fritsch (1912) it was not until 1980 that the abundance of algae in these ponds was recognised. Brady (1980) noted that the largest non-marine biota in southern Victoria Land exists on the McMurdo Ice Shelf. The work of Brady (1980) and more recently by Kellogg and Kellogg (1987) has focussed on the diatom flora of the ponds of the Ice Shelf which span a wide range of salinities. These authors, however, only made passing mention of the presence of other algae.

The objective of this study, carried out in 1987–88 summer, was to characterise the broad ecosystem properties of the Ice Shelf and then, as part of a multidisciplinary environmental programme on the Ice Shelf, to examine differences in the distribution and abundance of nutrients (nitrogen and phosphorus) and the microbial flora of the meltwaters.

2 Methods

Sampling was based on a series of long (up to 1 km) transects widely distributed across the Ice Shelf (Fig. 1) and in a series of ponds in the vicinity of Bratina Island. Water samples were collected in acid-washed polythene bottles. All samples were either filtered on site or within a few hours of collection using Whatman GF/C papers pre-rinsed with 100 ml of the sample water. Samples were packed in ice, and stored in dark boxes in the ice for periods of up to 1 week and then deep frozen for transport to New Zealand.

With the exception of conductivity which was measured on a Radiometer Model Com 2E meter corrected to 25 °C, all water analyses were carried out on a Technicon II auto-analyser system. Methods are given in Downes (1988) and Howard-Williams et al. (1986). For saline waters (conductivity $> 3000\ \mu S\,cm^{-1}$) a modification of Solorzano's (1979) NH_4-N method was used and nitrate was reduced to nitrite by cadmium reduction. Detection limits with these analytical procedures were as follows: 0.2 mg DRP m^{-3}, 0.4 mg $NO_3-N\,m^{-3}$, 0.5 mg $NH_4-N\,m^{-3}$, 1 mg TDP m^{-3}, 1 mg TDN m^{-3}. Dissolved organic nitrogen (DON) and dissolved organic phosphorus (DOP) were estimated by difference from total dissolved N and P and the inorganic forms (Downes et al. 1986).

Benthic algal samples were collected with a 2.5 cm^2 corer. Samples for biomass (chlorophyll *a*) were analysed at our field hut, usually within a few hours of collection. Extraction of chlorophyll *a* was carried out in boiling 90% ethanol for 10 min. Absorbance was read on the GF/F filtered pigment extract at 750 and

665 nm using a Shimadzu UV/120/02 spectrophotometer. Corrections for phaeophytin were made following acidification of the sample to 7.5×10^{-3} mol l^{-1} HCl. The detection limit in the field was 0.1 μg Chl.*a* cm^{-2}. Chlorophyll *a* extracted in this way was checked in New Zealand using a frozen duplicate subset of samples and extracting with cold dimethyl sulfoxide (Shoaf and Lium 1976). There was no significant difference ($P > 0.05$) in chlorophyll *a* concentrations determined from the field and later laboratory extraction techniques.

Samples for floral community analysis were analysed in the field by light microscopy at 400 × and 1000 × magnification. Presence or absence of species was noted in one preparation from each sample site and an assessment of abundance was made on the following scale: 0 = not observed, 1 = rare, 2 = occasional, 3 = frequent, 4 = abundant, dominant or co-dominant, 5 = very abundant, totally dominating the sample. For six ponds of varying salinity in the vicinity of Bratina Island an estimate was made of the relative abundance of different species in benthic mats. One hundred random individuals were counted in a preparation made from each of five random samples from each pond. The relative abundance of each species was then calculated.

3 Results

3.1 General Topography

Extensive transect surveys of the McMurdo Ice Shelf revealed two visually and morphologically distinct types of system. These can be described as "Undulating Ice" and "Pinnacle Ice" (Plates 2, 3). Undulating Ice is generally covered by a layer of moraine and marine sediment 10 to 20 cm thick. The general relief of the undulations was up to 20 m and the hollows were frequently occupied by water bodies of varying sizes from ponds of 1 m^2 to lakelets up to 30 000 m^2 (Plate 2). Most of the ponds were occupied by benthic algal mats ranging in surface colour from yellow-green to orange and almost black. Exposed ice was rare (Plate 2). Streams in the Undulating Ice environment were also rare, and flowing waters were usually confined to short connections between ponds. Evidence of old ponds and of continual change in the system was seen in thick layers of old dried algal mats at all heights in the ice undulations above present ponds. Many of these were at steep angles, even vertical. The thickness of some of the layers, up to 50 mm, indicates that the ponds they inhabited had a long life span. Visual estimates from oblique aerial photographs (e.g. Plate 1) indicate that Undulating Ice terrain accounts for approximately 30% of the McMurdo Ice Shelf.

Pinnacle Ice, in contrast, had a flatter relief but was highly variable in small-scale topography (Plate 3). Sediment cover was patchy and a substantial proportion of the area was exposed ice. Ponds were usually small, flowing waters were common and included some large, long (km length scale) fast flowing (< 0.5 m s^{-1} velocity) streams and many small trickles. Algae were neither as abundant, nor did they commonly form cohesive felts as occurred in undulating ice ponds. Rather, they formed a loosely bound community of cyanobacterial filaments and diatoms amongst gravel and sediment similar to the cryoconite community of the alpine glaciers of the region (Vincent 1988). The Pinnacle Ice was apparently a less stable regime than the Undulating Ice. Ice columns collapsed continually during our traverses, flowing waters were rapidly undercutting and downcutting through the ice, small ponds were draining out and new occurrences of moraine slumping were forming new ponds. Although dried benthic algal mats were sometimes visible above present water levels, they were thin and not as common as on Undulating Ice.

A further feature of the Pinnacle Ice was the presence, in some areas, of long north-south orientated ponds which had no visible outflow but received inflowing waters from all directions. These were presumably connected through to the sea below. We observed, for instance, a seal which emerged in one of these ponds and then disappeared again not to reappear. The pond was 25 km south of the Ice Shelf edge.

A single diurnal tide rises and falls around the edges of the land bordering the Ice Shelf. This had a tidal range of 1 m in the lagoon on the edge of Bratina Island (Fig. 1). On the Ice Shelf itself tidal variations in pond water levels were not observed. In those ponds which were connected through to the sea below, the pond level was presumably determined by the piezometric sea level.

3.2 Conductivity and Nutrients

The distributions of conductivity and nutrients were in all cases strongly positively skewed (Fig. 2). Statistical treatments of these data sets were therefore based on non-parametric methods (Campbell 1967).

Conductivity of the waters of the two types of terrain varied from 57 to 4000 μS cm^{-1} in Pinnacle Ice and from 130–56 000 μS cm^{-1} in the Undulating Ice (Fig. 2). Although the median conductivity of the Pinnacle Ice was lower than that in the Undulating Ice (cf. 600 vs 1200 μS cm^{-1}) the high variability in both data

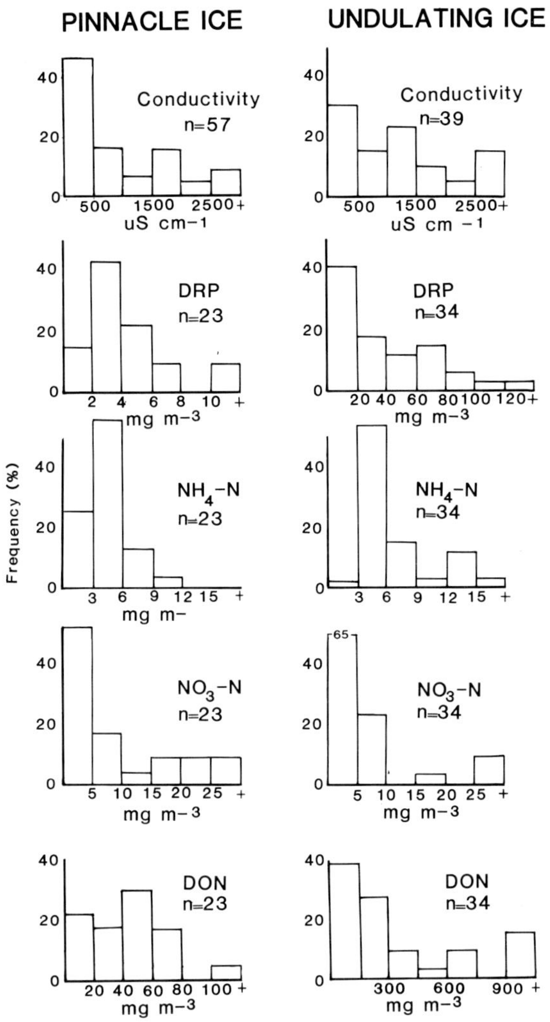

Fig. 2. Frequency diagrams of conductivity ($\mu S\ cm^{-1}$) and nitrogen and phosphorus compounds ($mg\ m^{-3}$) in the waters of the Pinnacle and Undulating Ice. Note the order of magnitude differences in the DRP and DON intervals between the two ice types

sets is such that the sets are not statistically different ($P < 0.05$, Mann-Whitney U-test). However, it is noticeable that high conductivities ($> 4000\ \mu S\ cm^{-1}$) were found only in waters on Undulating Ice.

There were some clear contrasts in the nutrient chemistry of the two types of water. Dissolved reactive phosphorus (DRP) values of the Pinnacle Ice ponds were very much lower ($P < 0.001$) than in those of the Undulating Ice (Fig. 2). The distributions scarcely overlap and 90% of the Pinnacle Ice DRP values were below 8 $mg\ m^{-3}$ while 62% of the Undulating Ice values were above 8 $mg\ m^{-3}$. While no significant differences in NO_3-N were apparent between the two types of water, ($P > 0.05$) the NH_4-N values in the Undulating Ice were significantly higher ($P < 0.001$). Concentrations of dissolved inorganic nitrogen (DIN) species were, in general, low compared with ice melt from the alpine glaciers of the adjacent dry valleys (Howard-Williams et al. 1986) where DIN values in excess of 100 $mg\ m^{-3}$ were common. The inorganic N/P ratios of the McMurdo Ice Shelf were very low with a mean of 3.1 in the Pinnacle Ice and 1.3 in Undulating Ice.

Dissolved organic nitrogen (DON) and phosphorus (DOP) values were very different in the two types of water bodies ($P < 0.001$). In the Pinnacle Ice 77% of the DON values were below 80 $mg\ m^{-3}$ (Fig. 2) whilst in the Undulating Ice 73% were above 80 $mg\ m^{-3}$. Ninety-five percent of DOP values in the Pinnacle Ice waters were below 3 $mg\ m^{-3}$ and 77% in the Undulating Ice were above 3 $mg\ m^{-3}$. In general DRP, NH_4-N, DOP and DON were significantly higher in water of the Undulating Ice although there was considerable variability in the data sets.

3.3 Taxonomic Composition

All waters were relatively clear with the pond flora dominated by the benthic mats rather than planktonic populations. In total, 58 algal taxa were identified from the McMurdo Ice Shelf of which at least 17 occurred in more than 25% of the water bodies sampled. The frequency distribution of the number of species in the two types of water body followed an approximately normal distribution (Fig. 3). The mean number of species per sample was significantly lower ($P < 0.001$) in the Pinnacle than in the Undulating Ice.

The taxonomic composition of the waters is summarised in Table 1, which includes those species found in more than 25% of the ponds sampled. The cyanobacteria *Phormidium autumnale, P. laminosum, Oscillatoria deflexa, O. limosa* and *Nodularia* sp. were frequently present in the Undulating Ice ponds with large ($> 10\ \mu m$) coccoid chlorophytes and the diatoms *Navicula muticopsis* and *Pinnularia cymatopleura* also common. The waters of the Pinnacle Ice, in contrast, had large and small coccoid chlorophytes, the diatoms *P. cymatopleura, N. muticopsis* and *Nitzschia antarctica* and the cyanobacterium *Phormidium laminosum* as the most frequently occurring species. Other cyanobacteria found in more than 50% of the Pinnacle Ice waters were *P. fragile, P. autumnale* and *Nodularia* sp.

Certain species were significantly more common in Undulating than Pinnacle Ice ponds. These were the two *Oscillatoria* species, *Phormidium autumnale* and the heterocystous N-fixing species *Anabaena* sp. and *Nodularia* sp. Conversely, the small coccoid chlorophytes, the diatoms *Nitzschia antarctica* and *Pinnularia cymatopleura* and the cyanobacteria *Phormidium*

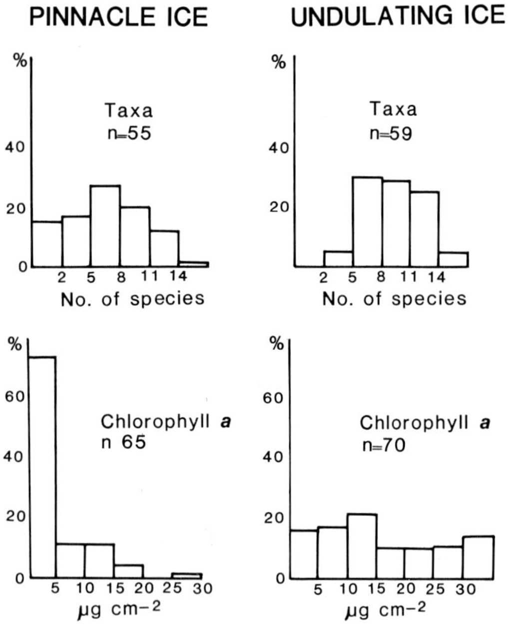

Fig. 3. Frequency diagrams of the total number of species per sample and biomass (spectrophotometrically determined chlorophyll *a*) in the waters of the Pinnacle and Undulating Ice

frigidum and *P. fragile* occurred more frequently in the Pinnacle Ice.

In terms of abundance (measured as dominance, Table 1) within ponds, the Undulating Ice was mostly dominated by the cyanobacteria *Phormidium laminosum*, *Oscillatoria deflexa*, *O. limosa* and *Nodularia* sp., whilst in the Pinnacle Ice frequent dominants were coccoid chlorophytes and the diatom *Pinnularia cymatopleura* as well as *Phormidium frigidum* and *P. fragile*. While over 70% of both Pinnacle and Undulating Ice ponds had cyanobacteria as dominant or co-dominant species, diatoms dominated in 34% of Pinnacle Ice ponds and only 10% of Undulating Ice ponds. Likewise, the coccoid chlorophytes dominated in 21% of Pinnacle Ice ponds and only 9% of Undulating Ice ponds. The taxonomic composition of the ponds on the Undulating Ice varied with the salinity (Table 2) where a wide range of saline conditions occurred. *Nostoc* sp. and *Anabaena* sp. were not found under saline conditions. Oscillatoriacae in fresher waters mostly comprised those with narrow (< 3 μm) trichomes while in saline waters those with larger trichomes dominated. *Oscillatoria* cf. *priestleyi* was the dominant taxon in the saline waters of Salt Pond. The diatoms can also be ordered along a salinity gradient (Table 2) with *Navicula* sp. *B* and *Tropidoneis laevissima* characteristic of saline conditions. *Pinnularia cymatopleura* and *Nitzschia antarctica* and *Navicula* sp. *A* were the most abundant of the freshwater species.

Table 1. Percent occurrence and dominance of algal taxa in 55 benthic samples from Undulating Ice and 49 from Pinnacle Ice collected across the McMurdo Ice Shelf

Species[a]	Occurence (%)		Dominance (%)[b]	
	Undulating	Pinnacle	Undulating	Pinnacle
Cyanobacteria				
Phormidium autumnale	86*	53	14	8
P. fragile	31	53*	7	16*
P. frigidum	27	43*	0	5*
P. laminosum	73	67	21	29
Oscillatoria deflexa	79*	33	38*	13
O. limosa	58*	35	16*	5
Nodularia sp.	76*	51	14*	3
Nostoc sp.	29	25	4	0
Anabaena sp.	42*	28	2	0
Chlorophyta				
Coccoid chlorophytes				
" > 10 μm	79	71	5	13*
" < 10 μm	31	69*	2	9*
Chrysophyta				
Achnanthes brevipes	18	25	2	3
Hantzschia amphioxys	45*	14	4	2
Navicula muticopsis	62	59	0	6*
N. shackletoni	35*	26	0	0
Nitzschia antarctica	42	53*	2	5
Pinnularia cymatopleura	51	65*	9	17*

[a] Species which were found in less than 25% of samples are not shown.

[b] Dominance is the percent of ponds in which the species was considered dominant or co-dominant with abundance ratings of 4 or 5 (see Sect 2).

* Species occurring in a significantly greater number of samples than in the other ice type.

3.4 Benthic Biomass

Significant differences ($P < 0.001$, Mann-Whitney U-test) occurred in benthic biomass between the two types of waters and the distributions of the data were positively skewed (Fig. 3). Seventy-three percent of the waters on the Pinnacle Ice had a biomass value of less than 5 μg Chl *a* cm^{-2} (Fig. 3) while 84% of the samples from Undulating Ice ponds had a biomass greater than this.

No significant associations (Kendalls rank correlation) could be found for biomass versus conductivity or biomass versus nutrients in Pinnacle and Undulating Ice when treated separately. However, biomass

Table 2. Relative abundance of benthic algal and cyanobacteria taxa from 6 ponds of contrasting salinity on the Undulating Ice (salinity measured as conductivity, K25°)

	Pond[a]					
	Fresh	Skua	IRP	P70	Brack	Salt
Conductivity ($\mu S\ cm^{-1}$)	163	818	1415	2800	10 560	56 200
Cyanobacteria						
Nostoc sp.	0.2	–	0.9	5.2	–	–
Anabaena sp.	3.4	4.6	6.5	1.0	–	–
Nodularia sp.	4.8	0.6	5.6	5.8	3.2	4.0
Oscillatoriaceae						
″ < 3 µm	75.6	70.8	40.8	73.8	61.2	19.4
″ 3–5 µm	3.8	2.0	4.8	7.2	19.0	12.8
″ > 5 µm	3.2	1.8	6.6	4.8	9.8	38.4*
Chrysophyta						
Stauroneis aniceps	0.6	2.0	–	–	–	–
Pinnularia cymatopleura	2.2	8.4	5.2	–	–	–
Navicula sp. *A*	0.2	1.4	2.0	–	–	–
Hantzschia amphioxys	0.6	2.0	–	0.2	–	–
Achnanthes brevipes	–	–	–	0.2	–	–
Navicula muticopsis	0.2	0.8	1.8	0.4	0.6	
N. shackeltoni	0.2	–	1.8	–	0.2	–
N. sp. *C*	–	1.6	–	–	0.6	–
N. sp. *B*					2.8	16.2
Tropidoneis laevissima	–	–	–	–	0.6	2.4
Cryptophyta	0.2	0.6	0.6	0.6	1.0	5.0
Chlorophyta						
coccoid forms						
″ < 10 µm	0.2	0.2	2.8	0.2	–	–
″ > 10 µm	0.4	0.2	–	0.4	0.2	–
Chlamydomonas sp.	–	–	2.4	–	–	–

[a] The ponds were within 200 m of each other and were all between 100 m^2 (P70) and 800 m^2 (Fresh) in area.
* Mostly *Oscillatoria priestleyi*. Oscillatoriaceae grouped according to trichome width.
– Taxa not recorded in the samples.

was positively associated with DRP ($P < 0.001$) and NH_4-N ($P < 0.05$, Kendalls rank correlation) in the combined data set. This is probably due to the Undulating Ice having significantly higher biomass, DRP and NH_4-N than Pinnacle Ice rather than any cause and effect relationship. The levels of DON and DOP were strongly related to biomass ($P < 0.001$), due probably to the increased production of dissolved organics as the biomass increases.

3.5 Scales of Temporal Variability

Short-term (hours, days) temporal variability is governed by tidal cycles and diel changes in solar radiation. Tidal changes in water level are restricted to lagoons along Brown Peninsula, around Bratina Island and at Black Island. In some of these, extensive areas of mud are regularly inundated and exposed with the single daily tide (amplitude ca. 1 m). Diel changes across the Ice Shelf are reflected in pond water temperatures which can vary between 1° and 6 °C over 24 h. The greatest diel variations we recorded were in the saline ponds. Short-term temporal variability in the waters here is confined to summer when the ponds have melted.

Seasonal variability is expressed by the annual freeze-thaw cycle. While there was little surface free water in mid-November, most ponds on the Undulating Ice were completely thawed by mid-December and in the region of Bratina Island the thaw was still evident at the end of January. However, in the southern section of the McMurdo Ice Shelf many ponds remained with a substantial ice cover through January. The major melt period for the McMurdo Ice Shelf in 1987/88 was from early December through into February (probably about 80 days), but there are likely to be large interannual variations in the volume and duration of melt each summer.

Long-term variability is seen in the continuously changing pond and stream systems over time scales of years. Ice movement causes cracks, and streams melt down through pond walls causing many ponds to drain out each summer. The water from these either fills new ponds or begins new watercourses. Signs of old ponds were visible throughout the Undulating Ice, and exposed dried cyanobacterial mats were a common occurrence. Some of these were up to 50 mm thick. Even the old mats (> 1-year-old) contained chlorophyll and some recognisable trichomes of Oscillatoriaceae and unicellular chlorophytes. This dried mat material readily fragmented and began to blow around in the wind as soon as it became desiccated.

4 Discussion

The McMurdo Ice Shelf ecosystem can be divided into two contrasting subsystems with very different morphological, chemical and biological properties (Plates 2, 3). These were the Undulating Ice (ca. 30% of the total area) and the Pinnacle Ice. The different properties of the two ice types are dependent on the factors which control the unusual surface topography of this Ice Shelf. Unlike most other ice shelves, the ice of the McMurdo Ice Shelf is of marine origin (Debenham 1920; Swithinbank 1970). Gow (1967) suggested that while surfaces formed by ablation of glacier ice are smooth, those formed by ablation of sea ice are characteristically block-like or tabular. When these blocks are variably coated with sediment or debris, the block-like structures are transformed to "Pinnacle Ice" by differential melting. Overriding this pinnacle effect is the influence of the predominant southerly wind which imparts a north-south orientation to lines of sediment (see e.g. Swithinbank 1970). A similar orientation of lines of high and low ablation would follow resulting in distinct parallel ridges. In areas which are covered by very thick layers (e.g. > 100 mm) of sediment such as the undulating ice, ablation will be considerably retarded (Ostrem 1959). Under such conditions, the factor controlling surface topography may be pressure waves formed when part of the slow moving ice shelf becomes grounded, particularly in winter. Confirmation of this would require a greater knowledge of the seafloor bathymetry beneath the Ice Shelf.

While the waters of the Undulating Ice were almost all ponds and small lakes, usually supporting dense benthic cyanobacterial communities, the more dynamic Pinnacle Ice had many flowing waters and most of the ponds were small. The benthic communities there had a significantly lower biomass with unicellular chlorophytes and diatoms more likely to dominate.

Kellogg and Kellogg (1987) put forward suggestions to account for local variability in diatom communities on the Ice Shelf. These include the presence of salt lenses in the ice and different patterns of marine aerosol deposition. Our nutrient analyses showed that dissolved N and P compounds were in greater concentrations in waters on the Undulating than Pinnacle Ice (Fig. 2). However, there were no correlations between spectrophotometrically determined biomass as chlorophyll *a* and inorganic nutrient levels in either system when examined separately. The significantly greater occurrence and abundance of N_2-fixing cyanobacteria (*Anabaena* sp., *Nodularia* sp. and *Nostoc* sp.) on Undulating Ice may well be a consequence of higher DRP levels in these ponds. With inorganic N/P ratios of less than 1 in many of the waters of the Undulating Ice, an N_2-fixing ability would be a potential advantage.

Dissolved nutrient levels in these waters are dictated by a variety of mechanisms including basal ice melt, weathering of the moraine and marine sediment cover on the ice, and metabolic activities of the benthic algal mats. Nutrient accumulation in the undulating ice ponds may also result from the predominantly endorheic nature of these ponds (e.g. Plate 2), and the thicker moraine sediment cover over the basal ice. Freeze concentration of nutrients and ablation of surface ice in winter over a period of years, as well as a greater sediment cover for weathering products would increase nutrients in the more stable undulating ice ponds. In contrast, continual water movement, small unstable ponds and relatively thin sediment in Pinnacle Ice would not allow for nutrient accumulation there. We suggest that it is the stability of the ponds in the Undulating Ice relative to the Pinnacle Ice which allows for a long-term accumulation of biomass, and, coincidentally, may allow for an accumulation of nutrients to higher levels as well.

The development of thick algal mats will be constrained by continuous disturbance (i.e. freezing, collapse, draining and washouts), rather than by grazing. Heterotrophic consumers are restricted to bacteria protozoa and a few species of nematodes, tardigrades, and rotifers (Vincent 1988; Suren 1990). We might therefore expect the communities of the Pinnacle Ice to reflect an early successional stage in mat development. This was borne out by floristic determinations. Diatoms and coccoid chlorophytes rapidly colonised areas of sandy sediment which were exposed by lift off of cyanobacterial mats and where sediment was distributed by wave action and small collapses of the bands around the periphery of ponds (P. Broady unpubl data). These same organisms were more frequent dominants in the ponds on the Pinnacle Ice (Table 1), so the relative instability of the ponds there would appear to at least partly account for the lower biomass and different floral community.

The large area (1500–2000 km^2) occupied by the McMurdo Ice Shelf ecosystem and its location immediately adjacent to the ice-free areas of southern Victoria Land imply that this Ice Shelf could be an important inoculum source of biota for the inland freshwater and saline habitats of the region. The continual draining and/or uplifting of the ponds by ice movement results in a mosaic of dried algal material which is blown around in the wind. Winds over the Ice Shelf can blow from any direction with the strongest (ca. 90 $km\,h^{-1}$) from the south (NZ Met. Service data). Such wind-blown material may also contribute "biogenic nuclei" of the type described by Saxena (1982) in the formation of ice clouds over the Ross Ice Shelf. The wind transport of minute organisms and electrostatically charged particles in the McMurdo Sound region has been examined by Benninghoff and Benninghoff (1985). Their airborne particle trap near McMurdo Station contained, in addition to filamentous micro-flora, fragments of animal remains which they speculate were derived from the McMurdo Ice Shelf some 30–50 km away. These authors suggest several mechanisms which could aid further long-distance dispersal of wind-blown material. Parker et al. (1982) described the wind removal of lacustrine benthic mats brought to the surface of the McMurdo dry valley lakes. Previous work on the ephemeral streams of southern Victoria Land has demonstrated that similar cyanobacterial mats remain viable from year to year, and were able to rapidly re-initiate metabolic activity after overwintering in an almost freeze-dried condition (Vincent and Howard-Williams 1986a, b). Brady (1980) suggested that the Ice Shelf could maintain a biotic pool for the Victoria Land lake systems during glacial periods when "terrestrial" lakes (i.e. in dry valleys) have evaporated, but our present observations suggest the region may also be an important source of biota during interglacial periods such as the present.

The McMurdo Ice Shelf has a variety of habitats and probably contains the largest concentration of non-marine biota in southern Victoria Land. Variability occurs on a broad scale associated with the speed of physical changes in the system and on a small scale with both physical and chemical factors. As yet we do not know whether the broad ecosystem features we have identified on the McMurdo Ice Shelf can be attributed to other Antarctic ice shelves, although Vincent (1988) mentions that several others show lakes, pools and streams. The George VI Ice Shelf between Alexander Island and the Antarctic Peninsula has a series of melt pools and lakes (Reynolds 1982; Heywood 1977), some of which have been attributed to dust contamination of the ice surface (Wager 1972). Lakes and pools exist on the Amery Ice Shelf including an epishelf lake that consists of freshwater dammed by ice overlying tidal seawater (Mellor and McKinnon 1960). Parts of the Ice Shelf associated with the glacier ice tongues in Terra Nova Bay appear to have an origin like that of the McMurdo Ice Shelf (Debenham 1920). There are almost certainly others around the continent of which we have no knowledge, but our work on the McMurdo Ice Shelf and that of Kellogg and Kellogg (1987) and Brady (1980) suggests that these areas may be important regional foci of biological activity.

Acknowledgements. We thank DSIR Antarctic Division, particularly Malcolm Mcfarlane, Jim Bakrer and Eric Saxby for logistic support. Squadron VXE-6, US Navy provided the valuable helicopter operations, and gave us permission to reproduce Plate 1. Our field colleagues Steve de Mora, Rob Whitehead, Alastair Suren and John Roberts are thanked for assistance and discussions. Max Gibbs, Stuart Pickmere, Lynell May, Malcolm Downes and Virginia Reid provided analytical and laboratory services in New Zealand. The manuscript was typed by Janet Simmiss and we thank Dr E. White for review.

References

Benninghoff WS, Benninghoff AS (1985) Wind transport of electrostatically charged particles and minute organisms in Antarctica. In: Siegfried WR, Condy PR, Laws RH (eds) Antarctic nutrient cycles and food webs. Springer, Berlin Heidelberg New York, pp 592–596

Brady HT (1980) Palaeoenvironmental and biostratigraphic studies in the McMurdo and Ross Sea Regions, Antarctica. PhD Thesis, Macquarie University, Sydney, Australia, 248 pp

Brady HT, Batts B (1981) Large salt beds on the surface of the Ross Ice Shelf near Black Island, Antarctica. J Glaciol 27:11–18

Campbell RD (1967) Statistics for biologists. Cambridge University Press, Cambridge, 242 pp

Debenham F (1920) A new mode of transportation by ice: the raised marine muds of South Victoria Land (Antarctica). Quart J Geol Soc Lond 75:51–76

Downes MT (1988) Taupo research laboratory chemical methods manual. Taupo Research Laboratory Report 102:70 pp

Downes MT, Howard-Williams C, Vincent WF (1986) Sources of organic nitrogen, phosphorus and carbon in Antarctic streams. Hydrobiologia 134:215–225

Fritsch FE (1912) Freshwater algae. National Antarctic Expedition, Natural History, vol 6. British Museum of Natural History, pp 1–6

Gow AJ (1967) Antarctic glaciological studies. Antarct J U S 2:121–122

Heywood RB (1977) Limnological survey of the ablation point area, Alexander Island, Antarctica. Philos Trans R Soc Lond, Ser B 279:39–54

Howard-Williams C, Vincent CL, Broady PA, Vincent WF (1986) Antarctic stream ecosystems: variability in environmental properties and algal community structure. Int Rev Gesamten Hydrobiol 71:511–544

Kellogg DE, Kellogg TB (1987) Diatoms of the McMurdo Ice Shelf, Antarctica: implications for sediment and biotic reworking. Palaeogeog Palaeoclimatol Palaeoecol 60:77–96

Mellor M, McKinnon G (1960) The Amery Ice Shelf and its hinterland. Pol Rec 10:30–34

Ostrem G (1959) Ice melting under a thin layer of moraine and the existence of ice-cores in moraine ridges. Geografisker Annaler 41:228–230

Parker BC, Simmons Jr GM, Wharton Jr RA, Seaburg KG, Love FG (1982) Removal of organic and inorganic matter from Antarctic lakes by aerial escape of blue green algal mats. J Phycol 18:72–78

Reynolds J (1981) Lakes on George IV Ice Shelf, Antarctica. Pol Rec 20:425–432

Saxena VK (1982) Biogenic nuclei involvement in clouds over the Ross Ice Shelf. Antarct J U S 17:212–214

Shoaf WT, Lium BW (1976) Improved extraction of chlorophyll *a* and *b* from algae using dimethylsulphoxide. Limnol Oceanogr 21:926–928

Solorzano L (1979) Determination of ammonia in natural waters by the phenolhypochlorite method. Limnol Oceanogr 24:799–801

Suren A (1990) Microfauna associated with algae mats in melt ponds of the Ross Ice Shelf. Polar Biology 10:329–336

Swithinbank C (1970) Ice movement in the McMurdo Sound area of Antarctica. Proc Int Symp Antarctic Glaciological Exploration, Hannover, New Hampshire, USA 1970, pp 472–482

Vincent WF (1988) Microbial ecosystems of Antarctica. Cambridge Univ Press, Cambridge

Vincent WF, Howard-Williams C (1986a) Antarctic stream ecosystems: physiological ecology of a blue-green algal eplithon. Freshwater Biol 16:219–233

Vincent WF, Howard-Williams C (1986b) Microbial ecology of Antarctic streams. Proc 4th Symp Int Soc Microbial Ecology, pp 201–206

Wager AC (1972) Flooding of the ice shelf in George VI Sound. Br Antarct Surv Bull 28:71–74

Signy Island as a Paradigm of Biological and Environmental Change in Antarctic Terrestrial Ecosystems

R. I. Lewis SMITH [1]

Summary. Terrestrial ecosystems throughout much of Antarctica are exhibiting local changes in their environment which are, either directly or indirectly, influencing biological processes. Central to these changes is the present climatic warming being experienced, especially in the more northerly regions, which is causing substantial recession of glaciers and ice fields and creating new surfaces on which biota may become established. Other temporal changes occur as a function of the natural process of colonisation, growth and ecosystem development but which may be strongly influenced by minor variations in climate or other components of the environment. Examples of such environmental, and consequently of biological, changes are presented for one climatically and ecologically sensitive locality, Signy Island in the South Orkney Islands, to illustrate the diversity of changes which may affect the structure and dynamics of Antarctic ecosystems in general. These are discussed in terms of ecological change resulting from long-term climatic trends, short-term climatic (especially summer temperature) fluctuations, plant colonisation and growth, community development, and environmental perturbation. A plea is made for implementing long-term monitoring studies to determine the direction and rate of environmental and ecological changes, with particular regard to assessing the resiliance of ecosystems to and their recovery from these phenomena. The Antarctic environment offers probably the most significant baseline to which global atmospheric changes may be related. The predicted trend in global warming implies disturbing consequences for the future integrity of Antarctica's, or indeed the world's, environment and biota. However, it offers ideal opportunities to study the cause-and-effect relationship of ecological change and, from this, to develop a strong management policy for the active use and conservation of the Antarctic biome.

[1] British Antarctic Survey, High Cross, Madingley Road, Cambridge CB3 0ET, UK.

Antarctic Ecosystems. Ecological Change and Conservation
(ed. by K. R. Kerry and G. Hempel)
© Springer-Verlag Berlin Heidelberg 1990

1. Introduction

All biological systems are dynamic to a greater or lesser extent, the degree and rate of change being dependent on characters inherent in individual organisms, interspecific interactions and their response to particular environmental variables.

By comparison with most other terrestrial biomes worldwide, ecological processes in Antarctic ecosystems are generally considered to proceed slowly. Thus, despite the strong prevailing westerly wind circulation, immigration of organisms from more northerly continental landmasses is inhibited because of the physical barriers to dispersal, mainly because of the great distances to be traversed. When aerobiological input is successful and viable organisms or propagules are deposited on ice-free land, notably in coastal areas, colonisation proceeds slowly because most environmental conditions combine to create a situation which is unfavourable for rapid establishment. Subsequent development of distinct homogeneous plant assemblages can be very slow and is dependent on relatively stable conditions throughout the short almost aseral succession to the climax community types. Because of Antarctica's isolation from potential sources of immigrant propagules and the severe conditions imposed by the environment on any viable propagules which are successful in reaching a suitable substratum, the terrestrial macroscopic flora and fauna are restricted to taxa which are preadapted to metabolising at low temperatures and tolerant of frequent and rapid freeze-thaw and hydration-dehydration cycles. These are the prime constraints imposed on the composition and survival of the terrestrial biota, which are restricted almost exlcusively to the remarkably successful stress-tolerant non-vascular cryptogams, microarthropods and lower phyla. As yet there is no evidence to suggest that there has been any major change in the composition of these biota since the retreat of the great ice sheets at the end of the last glacial maximum when the first habitats became available for Holocene colonisation.

Under Antarctic conditions autogenic modification of the environment by primary colonists and the succession which follows is minimal, at least in terms of other biomes. However, in the context of the Antarctic biome, the almost invisible process of primary colonisation by microbial, algal and microlichen populations and their role in stabilising soils and secreting organic and inorganic nutrients are vitally important in creating conditions favourable for secondary macroscopic colonists. It is these secondary colonists, in association with the prevailing environmental conditions, which quickly establish the direction of succession and determine the composition and structure of the climax community which ultimately develops. In the Antarctic such successions are generally simple and involve very few seral stages because of the absence of higher organisms and the consequent lack of complex interactions. In most instances the conditions of a particular site favour a select group of colonists and it is these which develop a specific community or climax type (Fig. 1). In a stable situation the only major change which may occur is a shift in abundance of certain species leading to a possible change in dominance; otherwise, the species composition remains remarkably static. Significant change in an otherwise stable climax community results mainly from some event which deflects the climax from its equilibrium state. Under such circumstances the vegetation may recover and either resume its former structure in the short or long term or, depending on the degree of perturbation, undergo a major change in direction and develop towards a new climax.

Because of the low diversity of species and life-forms and the relatively small range of environmental variables on which they depend, the resultant simple Antarctic terrestrial ecosystems are particularly sensitive to even small fluctuations in the prime driving force, namely climate, either directly or indirectly. Often, these systems have little resiliance when the ecological balance is disturbed and response to sudden changes may be immediate and irreversible, while more gradual changes are reflected in the ecosystem by a change in process rate or in the direction of community development.

To illustrate some of the major causes and effects of environmental and biological change which are experienced throughout the Antarctic biome, examples have been selected from one locality, namely Signy Island in the South Orkney Islands, maritime Antarctic (60° 43′ S lat., 45° 38′ W long.). Signy Island is mountainous with an ice cap, temperate glaciers, rugged topography, and complex geology and pedology, all of which provide a wide range of terrestrial habitats and associated biota which have developed a diversity of ecosystems (Smith 1972). However, because of its small size (ca. 20 km^2) and isolated oceanic setting at a comparatively low latitude, Signy Island probably experiences greater fluctuations in climate than most other comparable localities in Antarctica. Consequently, ecological change proceeds relatively rapidly by comparison with continental regions. Nevertheless, the examples discussed here, many of which

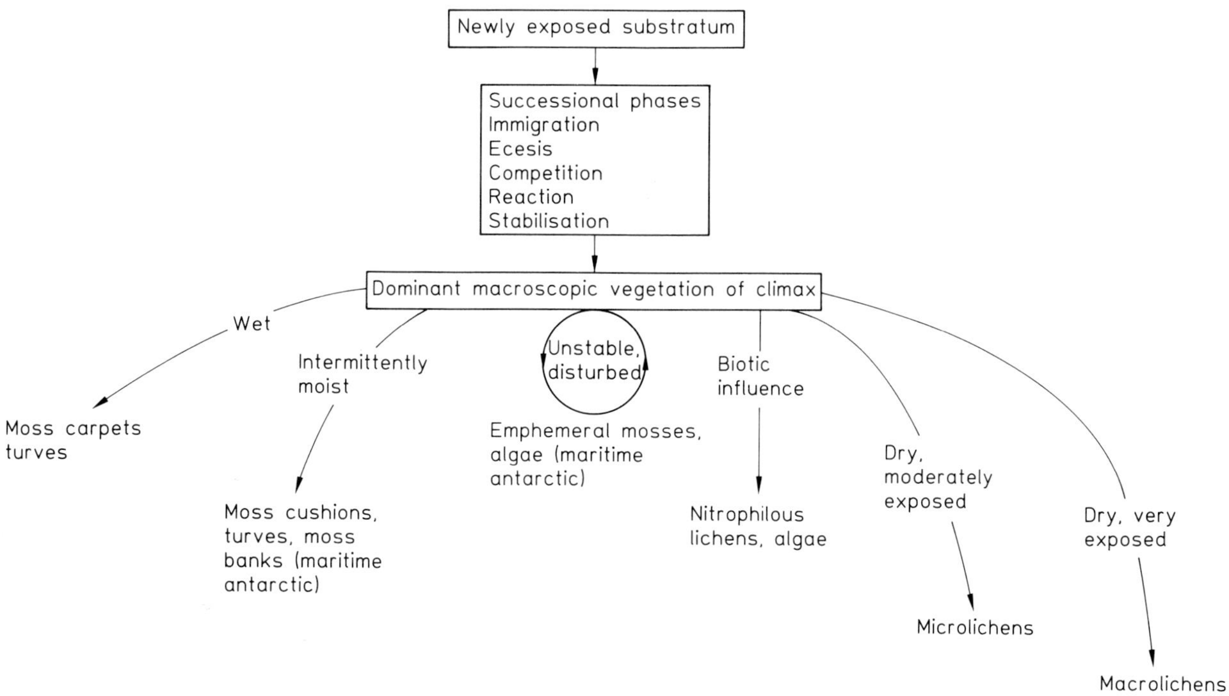

Fig. 1. Summary of idealised development of principal terrestrial climax communities on newly created substrata in Antarctica

relate to the author's as yet unpublished data, are relevant, to a greater or lesser degree, to the biome as a whole. Thus, Signy Island serves as a useful case study to demonstrate various aspects of ecological change in terms of both the environment and its biota. These changes are considered in five broad categories, all of which are temporal (rather than spatial) in nature and result from long-term climatic trends, short-term climatic fluctuations, colonisation and growth, community development, and environmental perturbation.

2 Ecological Change at Signy Island

2.1 Ecological Change Resulting from Long-Term Climatic Trends

In order to examine the early history of Signy Island's terrestrial environment and ecosystem development a sequence of tentative reconstructions is proposed which follow the glacial events since the end of the Pleistocene about 18 000 years ago.

Figure 2 is a reconstruction of the western part of the South Orkney Islands, including Signy Island, at the end of the Late-Wisconsin glacial maximum ca. 17 000–18 000 years B.P. The boundary of the former coastline is based on the 130–150 m marine contour which glaciologists and geomorphologists generally accept as having been the estimated mean global sea level at the end of the Pleistocene (Sugden and Clapperton 1977; Clapperton et al. 1978; Stuiver et al. 1981; Hughes 1983). The postulated South Orkney Islands ice cap was proposed by Sugden and Clapperton (1977). The deep indentations into the coastline represent deep troughs incised by major glacier extensions. Figure 3 is a north-south profile through Signy Island. Although a very small area of Signy and a much more extensive part of Coronation Island would have been above the main ice cap, most of these higher

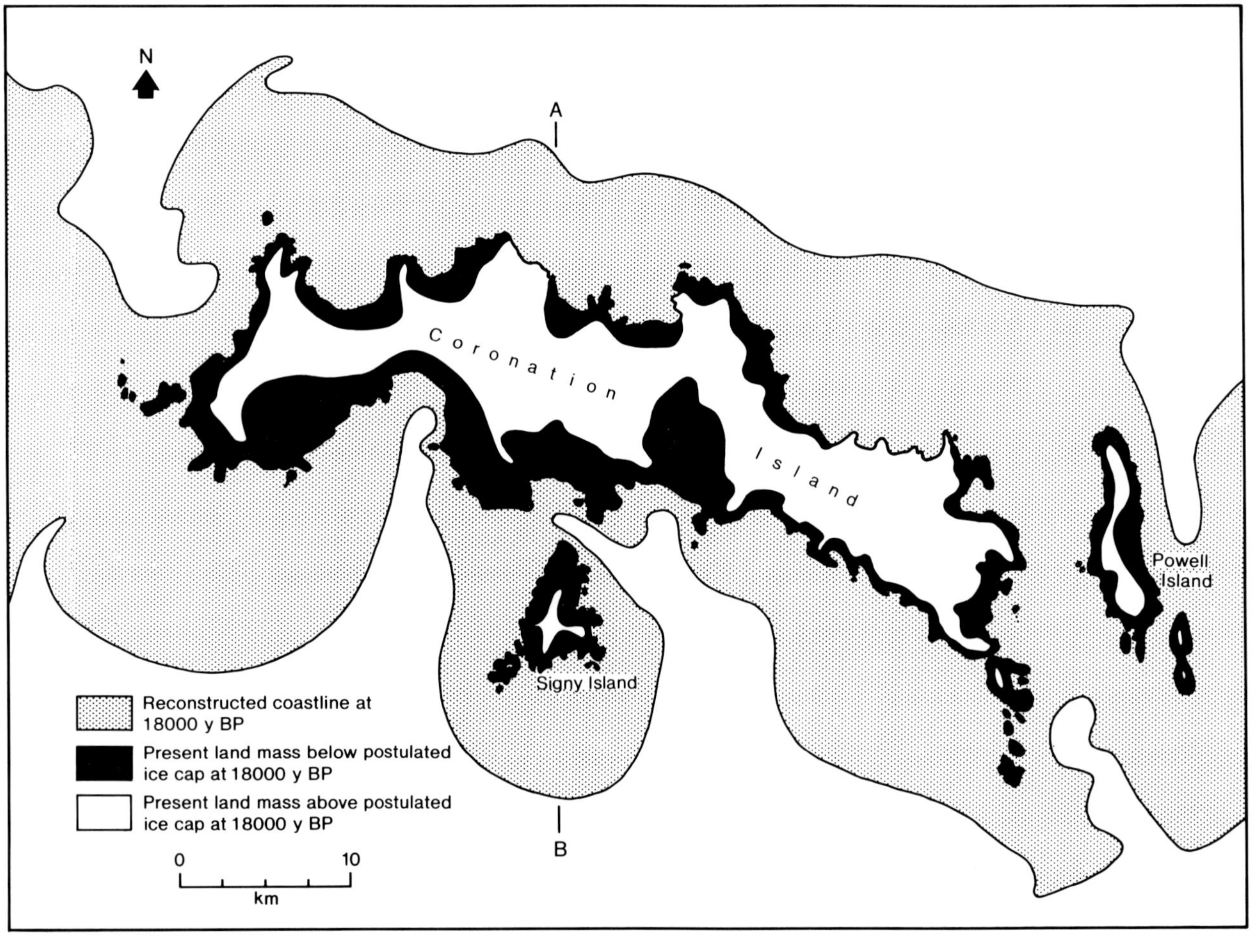

Fig. 2. Reconstructed coastline of the western end of the South Orkney Islands during the Late-Wisconsin glacial maximum (ca. 18 000 years B.P.), corresponding approximately to the 150 m depth contour, at the end of the Pleistocene (derived from data in Harrington 1968 and Sugden and Clapperton 1977). Profile through AB is shown in Fig. 3

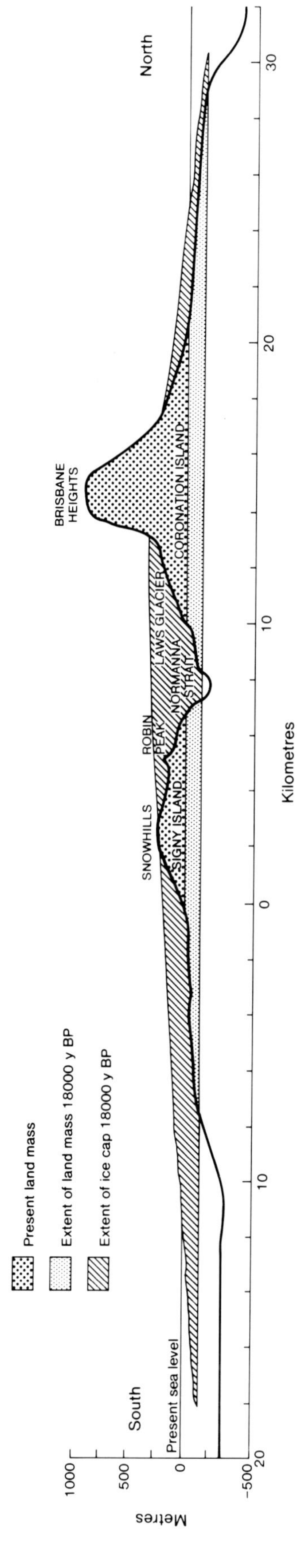

Fig. 3. North-south profile through Signy and Coronation Islands (AB in Fig. 2) at the end of the Pleistocene, showing the postulated ice cap (derived from data in Harrington 1968 and Sugden and Clapperton 1977)

areas would probably have been covered by minor ice fields. It is very unlikely that any ice-free areas would have served as refugia for pre-Pleistocene biota.

During the next 10 000 years there was a gradual global warming and the Late-Wisconsin ice sheets receded. However, according to some estimates (e.g. Nakada and Lambeck 1988) the major period of melting of the Antarctic ice sheet was between 9000 and 6000 years B.P. Figure 4a–f illustrates the possible sequence of events once Signy Island began to emerge as a terrestrial environment. Evidence of the earliest biota colonising ice-free terrain on the island exists in the form of deep banks of peat, formed almost exclusively by the moss *Chorisodontium aciphyllum*, on many north-facing coastal slopes (Fenton 1982a; Fenton and Smith 1982). The oldest radiocarbon date so far obtained for the basal peat in one of these banks is about 5500 years B.P. (corrected from Fenton 1980). Therefore, from about 4000–2000 B.C. until present these moss communities have accumulated peat to a depth of up to 2 m without apparently being destroyed. Their distribution around the island provides a baseline against which the possible maximum advance of the Signy Island ice cap during the past five millenia can be measured (Fig. 4b). All the island's present penguin colonies, and known subfossil colonies, also occur just beyond the postulated boundary of maximum ice advance.

How the island's ice cover fluctuated between the time of establishment of these moss banks and the present is highly speculative and based on a series of radiocarbon dates obtained for surface samples of *in situ* moss peat exposed by the recent recession of ice margins. The existence of these deposits around the island (see Fenton 1982b) reveals a former vegetation which existed at different times in different places, depending on the amount and duration of ice cover, but at all times less extensive than at present. During climatic events of increasing cold, the ice fields increased in extent and depth covering adjacent vegetation *in situ*, but seldom removing it; moss bank surfaces have remained intact, as have individual moss cushions and lichen thalli. The radiocarbon age of the upper 1 cm of such peat deposits gives an indication of when that vegetation was covered by advancing ice during a deterioration of the climate. The 32 dates plotted in Fig. 5 suggest three or four periods of Neoglacial advance over the past ca. 1500 years, although there may have been several minor glacial fluctuations between about 1250 and 1600 A.D., which coincide with early events of the Northern Hemisphere Little Ice Age, and possibly two during the Southern Hemisphere Little Ice Age of ca. 1700–1850 A.D. These periods of increased warming and cooling are fairly consistent with those reported elsewhere

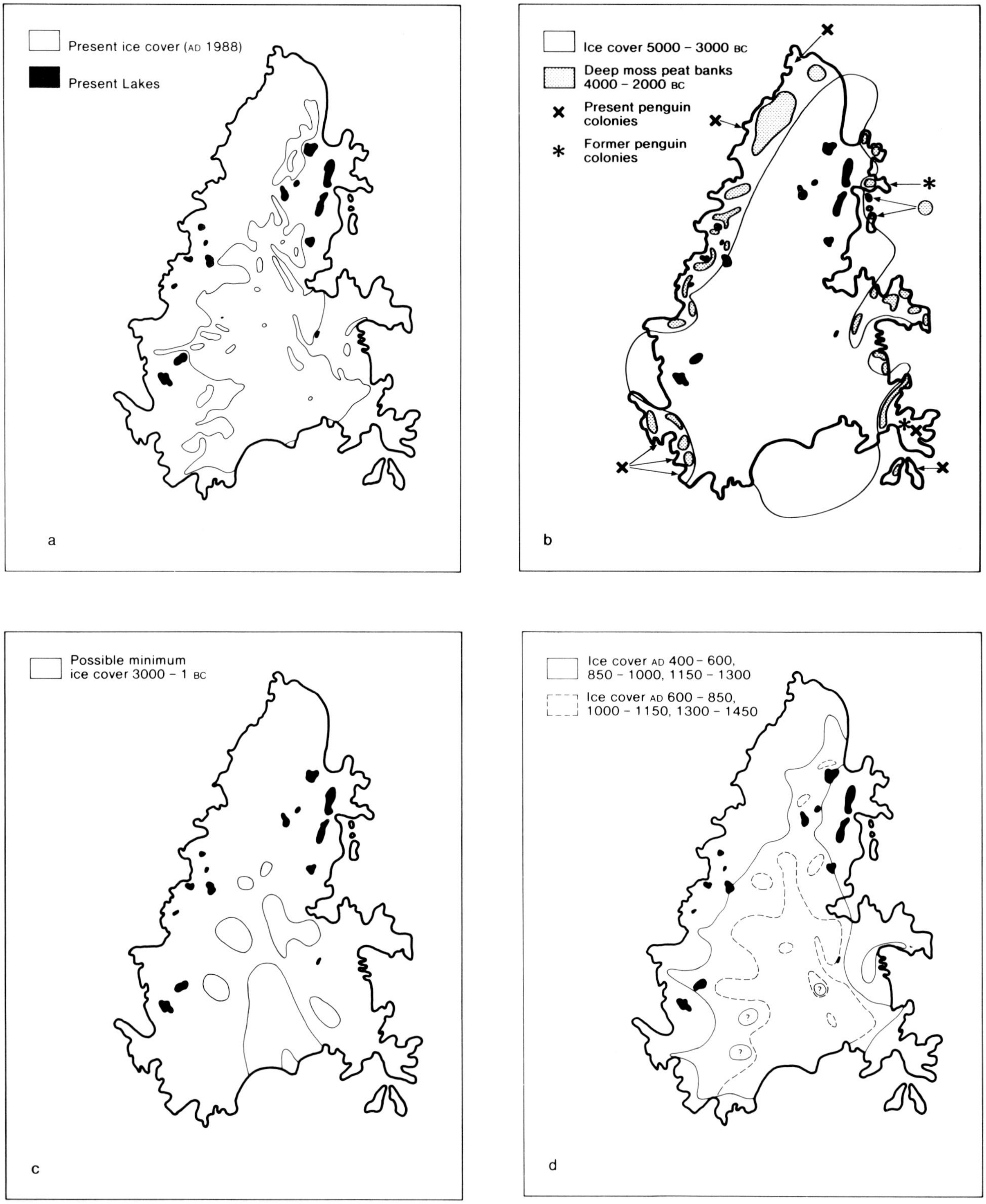

Fig. 4. Present **(a)** and postulated maximum and minimum **(b–f)** ice cover on Signy Island over the past 7000 years during various apparent glacial events, reconstructed from corrected radiocarbon ages of re-exposed moss peat

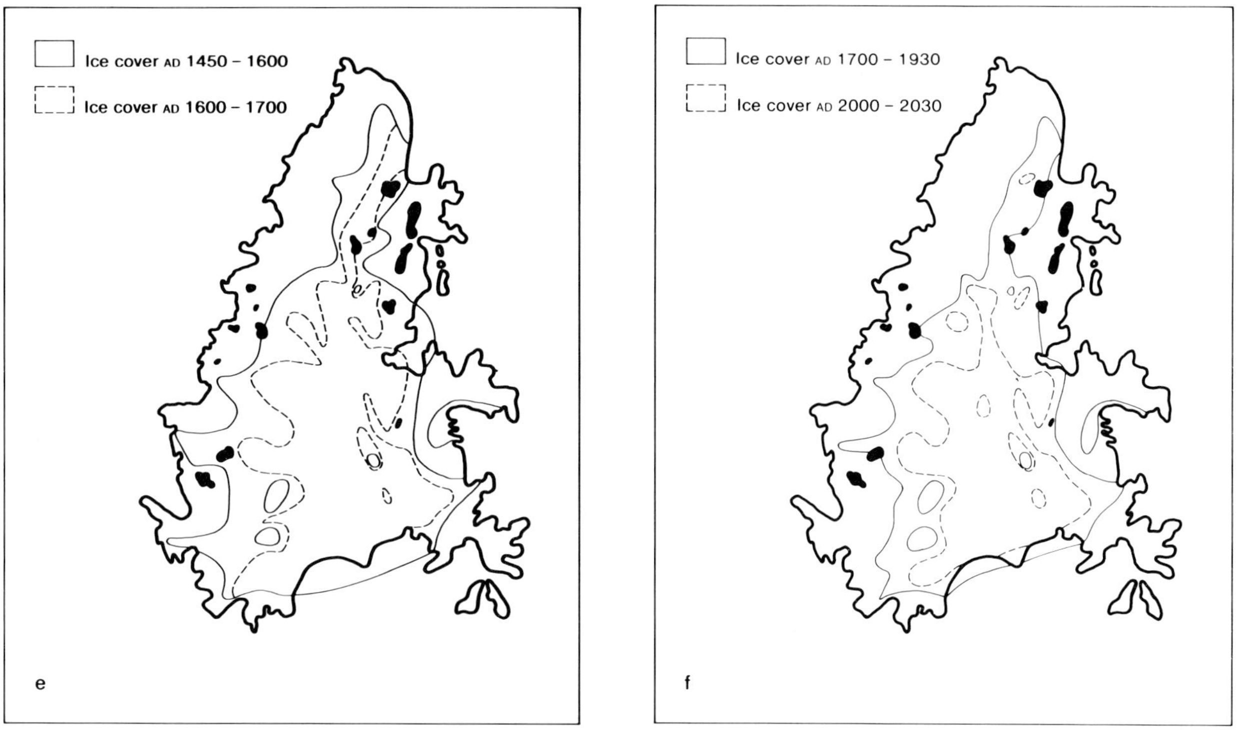

Fig. 4 (continued)

(e.g. Gribbin and Lamb 1978; Mercer 1978; Clapperton and Sugden 1988). From these data (and taking into account contours, present ice depths and the position of deep moss bank) a tentative reconstruction is proposed for the Signy Island ice cover between around 3000 B.C. and present (Fig. 4a–f), with an indication of the putative ice cap about 50 years from now, if the current rate of climatic warming continues (Fig. 4f).

2.2 Ecological Change Resulting from Short-Term Climatic Fluctuations

Evidence for the long-term climatic changes described above was derived from former vegetation re-exposed by the recent retreat of Signy Island's ice fields, resulting from a slight warming of the climate in the region over the past 35 years (see also Collins 1976).

The status of the ice cap is dependent to a very large extent on the air temperature attained during summer (particularly December to February), although the amount of cloud cover and precipitation as rain can also be critical. The relatively thin mantle of low density ice on a small mid-oceanic island such as Signy is very sensitive to minor fluctuations in summer temperature. An increase in mean summer air temperature of $< 1\,^{\circ}C$ may significantly alter the balance between accumulation of ice and loss through increased melt. Consecutive mean annual summer air temperatures for Signy Island and Laurie Island (50 km to the east) do not convey a clear picture of recent climatic trends (Fig. 6a), but when converted to 5-year running means (Fig. 6b) distinct warm and cold events are clearly revealed. The deposition of recent moraines (dated lichenometrically) coincides with the cold spells of the early 1910s, early 1920s to mid-1930s, and late 1940s. However, since 1951 there has been a dramatic increase in mean summer air temperatures and the trend is continuing upwards. Comparison of the island's ice cover, based on ground surveys, in 1949 and in 1989 has revealed a 35% reduction over the 40 years.

Between the mid-1960s and mid-1980s ice margins have receded by over 100 m in many places and the thickness of the island's glaciers reduced by 7–8 m. In recent years several rock exposures, possessing perfectly preserved colonies or individuals of various mosses and lichens, have appeared in the island's ice fields. One rocky knoll in McLeod Glacier was first exposed about 1963 and by 1988 it protruded over 8 m above the ice and covered an area of 0.5 ha. Many rock faces adjacent to glaciers and ice fields now reveal trimlines above which lichens and mosses are abundant, but below which they are absent, having been killed by increasing ice thickness during the Little Ice Age or earlier glacial advances. These fresh surfaces are now being slowly recolonised, mainly by lichens.

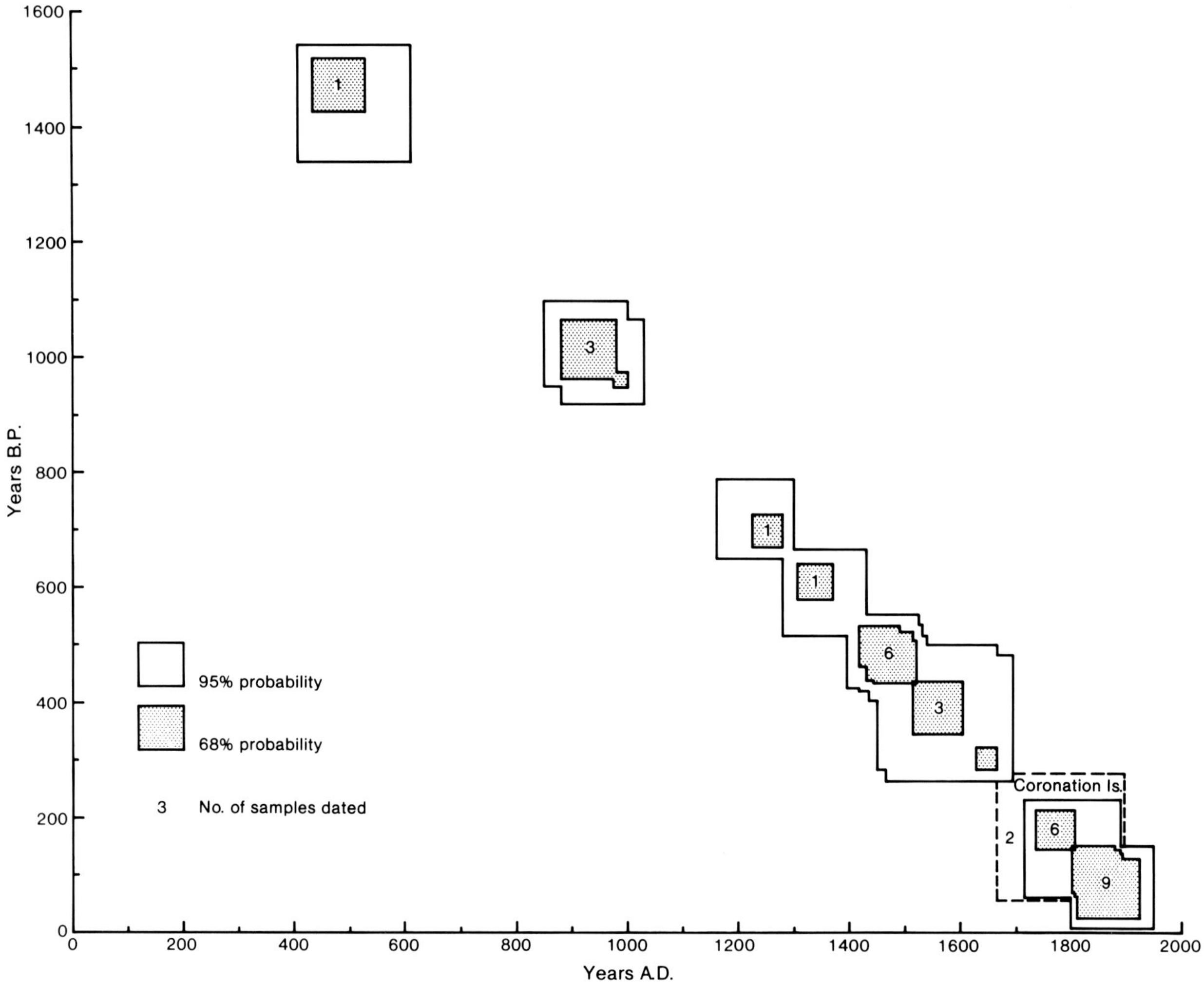

Fig. 5. Corrected radiocarbon ages of re-exposed moss peat surfaces on Signy Island. The calibrated data are expressed in terms of the standard deviation of the C^{14} age (as years B.P. and A.D.) at the 68 and 95% probability level, after reducing the derived age by 30 years to correct for the discrepancy in Southern Hemisphere atmospheric CO_2 (see Stuiver and Becker 1986)

Areas of re-exposed moribund vegetation, killed by ice cover, and bare soil are quickly being recolonised by a visible succession of cyanobacteria, algae, bryophytes and lichens. Some moss species show evidence of regenerating from viable tissue despite having been buried by ice for many decades or possibly even for centuries. The significance of this as a survival strategy is clearly of great importance, but has yet to be confirmed by detailed investigation and tested experimentally. Generally, these colonising species are the same as those comprising the re-exposed remains, suggesting that the original communities are being recreated. Occasionally, the deposition of calcium-rich glacial flour has increased the pH and base status of the soil, and more calcicolous species (e.g. of *Bartramia*, *Bryum*, *Encalypta*, *Pottia*, *Tortula*) are colonising these substrata than existed there previously.

The retreat of ice during recent times has also affected the island's hydrology. Naturally, in any environment with changing ice cover, drainage systems will fluctuate. However, on Signy Island there are examples of major changes to several lakes. One of the most dramatic has probably been the ice-damming of Paternoster Valley causing Changing Lake to rise about 10–12 m above its present level and directing the outflow stream through a col into Three Lakes Valley. Former shorelines and the outflow stream bed are still clearly visible. A similar deep, ice-dammed lake below the glacier tongue north-west of Tioga Hill had its outflow through a valley running west; today there is no lake and the stream flows north through another valley. Both these lakes have probably drained during the past one to two centuries. A small shallow ice-dammed lake north of Jane Peak drained in the early

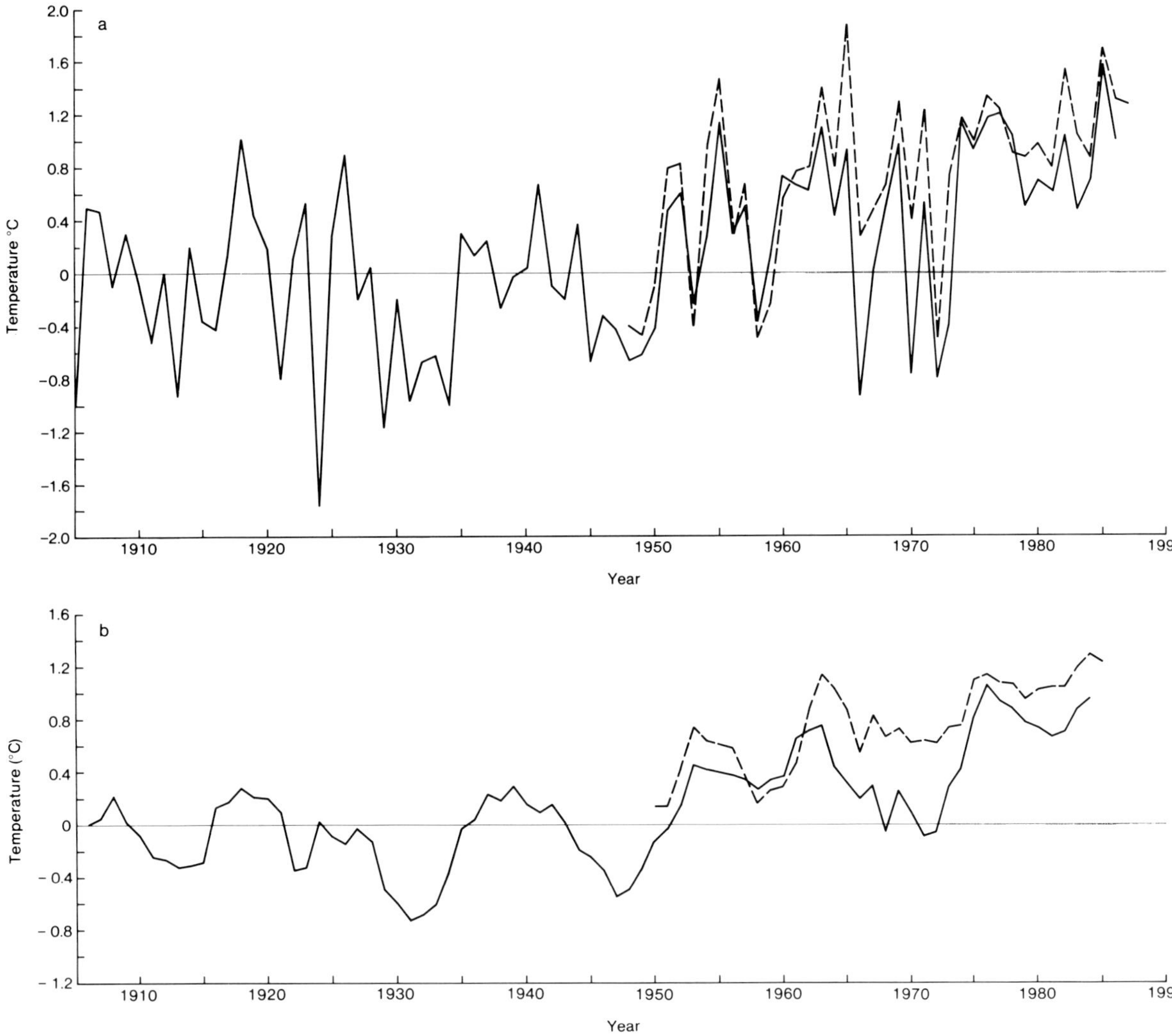

Fig. 6. Summer (December–February) air temperatures for Laurie Island (*continuous lines*) and Signy Island (*dashed lines*), South Orkney Islands, expressed as annual means (**a**) and as 5-year running means (**b**)

1960s when the ice receded, while substantial retreat of ice at the north-east margin of McLeod Glacier has created two rock basins which are now occupied by large pools and which have rapidly become colonised by biota during the past decade.

2.3 Ecological Change Resulting from Colonisation and Growth

Once a newly available substratum has been rendered suitable by weathering and by the primary microbial and algal colonists, the secondary colonists generally comprise an assemblage of species, the composition of which varies little with time. However, as the most aggressive species assume dominance the abundance of the less competitive species continually changes, but few, if any, are lost. There is little or no distinct succession in many communities and the only development towards a climax system is the often slow increase in abundance and biomass of the component species. This "founder effect" implies that, in general, Antarctic plant populations are in the "*r* phase" in the development of communities (see Harper 1977). Qualitative and quantitative monitoring studies made by the author over a period of 25 years on Signy Island have indicated that, under the prevailing climatic regime, colonisation, growth and community development may, in fact, be proceeding faster than has been generally accepted. For example, the deep moss banks represent a distinctive community type but, although several thousand years old, they may reach their climax state within a century, perhaps considerably less (i.e. including their associated epiphytic macrolichens). The only major change would have been the very slow accumulation of peat and upward movement of the permafrost which remains at ca. 25–30 cm below the surface (Fenton 1980). Similarly, the extensive mature fellfield communities dominated by a dense cover of large thalli of *Himantormia lugubris*, *Umbilicaria antarctica*, *Usnea antarctica*, *U. aurantiaco-atra*, various crustose lichens, and associated cushions of turves of moss appear to establish the basis of their assemblage within a decade or two of the substratum becoming suitably weathered. The development of that community towards a climax is largely by the increase in size and number of individuals and their gradual turnover, in what quickly becomes a self-perpetuating system.

In situations where newly exposed ground becomes available for colonisation it may take a long time before the primary community of microbial and algal colonists creates conditions suitable for immigrant propagules to develop. By the time the substrate texture, nutrient status and hydrologic regime are adequate for germination and growth, the pool of cryptogam propagules in the soil may be large and diverse and provide almost immediately conditions that are favourable to a relatively complex secondary community (Smith and Coupar 1986) which may, in fact, change little as it develops towards a climax community. This is clearly demonstrated by an area of fine black detritus which has accumulated by downwashing from the island's ice cap along the ice margin on Jane Col. This mineral soil (organic content $< 2\%$ dry wt) is totally devoid of macroscopic vegetation. However, soil covered by ventilated clear plastic cloches (thereby affording protection from wind and raising soil and air temperature and moisture content) yielded a prominent growth of mosses after 1 year, and almost complete cover by about eight species after 3 years (Smith in press). Adjacent, unprotected control plots remained barren. The regular addition of nutrients did not significantly enhance development of these microcosms, probably because of the relatively large input of organic and inorganic nutrients from the breakdown of both snow and cryokonite algae which have been washed onto the site during each spring melt. However, one species (*Bryum algens*), which is rarely fertile on Signy Island, produced large numbers of very tall sporophytes in the nutrient-treated cloche. There was also a significantly greater development of cyanophytes (species of *Nostoc* and *Phormidium*) within the cloches (Wynn-Williams in press). Such cloche experiments have revealed widely differing floras, notably of algae and cyanophytes, over distances of a few metres within a site. The community which eventually develops comprises those species best adapted to the prevailing conditions at that site. The potential for plant colonisation is therefore often present within a substratum, but is not realised until suitable conditions become available (in this instance as increaed temperature and moisture).

Lichen colonisation may also proceed quite rapidly once the substrate conditions are favourable. Freshly exposed rock surfaces at 17 sites (each ca. 1 m^2) on Signy Island have been photographically monitored since 1973. After only a few years those close to the shore or to penguin colonies had become colonised by numerous tiny lichen thalli, and after 15 years almost all exhibited extensive colonisation (up to ca. 60% cover), especially by crustose and foliose taxa. Lithophilic mosses appear to be relatively late colonists. Those sites in more montane habitats are being colonised much more slowly by nitrophobous and halophobous species of mainly crustose and fruticose lichens and occasional mosses.

To assess the growth of individual crustose lichen thalli and their communal development on rock surfaces, over 250 plants representing 12 species at 21 sites have been monitored photographically every 2 years since 1972–74. Thalli were accurately silhouetted on the prints and their area determined by TV image analysis as described by Wynn-Williams (1988) but using a macro-lens in place of a microscope to obtain the image for quantification. This study has revealed relatively rapid annual areal increases in several typical coastal species, e.g. 41% per annum in colonising thalli of *Umbilicaria antarctica* (over 15 years; Fig 7a), 32% in mature circular thalli of *Buellia latemarginata* (over 14 years; Fig. 7b), and 13–17% in mature irregular thalli of *Caloplaca* spp. (over 11 years), *Acarospora macrocyclos* (over 4 years) and *Xanthoria elegans* (over 4 years; Fig. 7c). Within only a few years adjacent thalli frequently coalesce and can quickly cover an area of rock. On the other hand, nitrophobous and halophobous species typical of inland or montane habitats, such as *Rhizocarpon geographicum*, exhibit an annual areal increase of only 0.5–0.8% (over 13 years; Fig. 7d). This is equivalent to a radial increase of barely 5 mm per century, confirming results obtained by Hooker (1980a), but considerably less than growth rates calculated from known-age moraines by Lindsay (1973) and Curl (1980). This compares with ca. 0.6 mm per annum for *B. latemarginata*. Hooker (1980a) found similar rates for this species and *X. elegans* (0.5 mm $year^{-1}$), but 1.2 mm $year^{-1}$ for *A. macrocyclos* (all over 4 or 5 years), and *C.* cf. *cinericola* on old concrete structures.

The change in size and abundance of lichen thalli varies very much according to species, thallus age and habitat characteristics. All previous reports of lichen growth rates (excluding lichenometric studies) in Antarctica have been based on relatively short-term studies over a few years (e.g. Hooker 1980a–d).

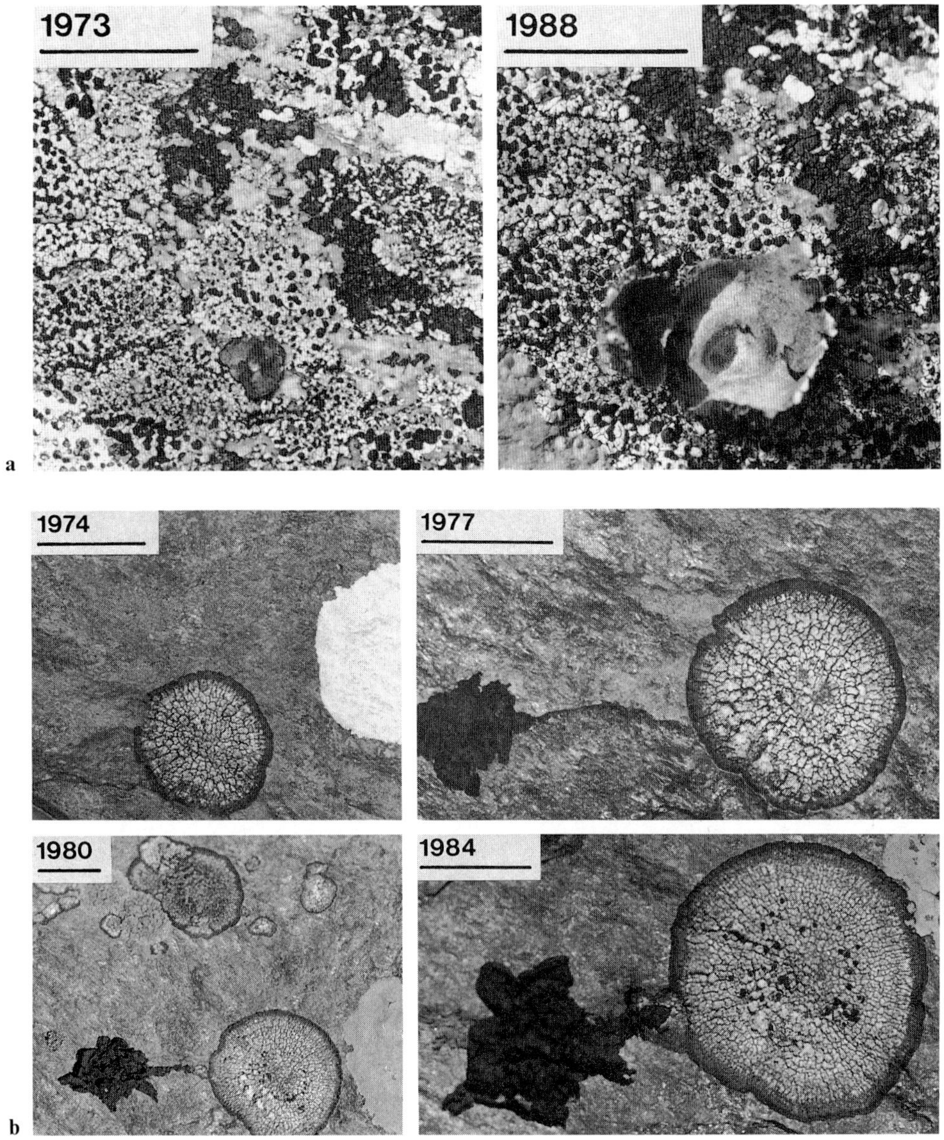

Fig. 7. Change in area of lichen thalli in: **a** *Umbilicaria antarctica* over 15 years (areal increase 615%, discounting three-dimensional growth), **b** *Buellia latemarginata* over 14 years (areal increase 452%); note development of black apothecia and encroachment of foliose thallus of *Mastodia tesselata*

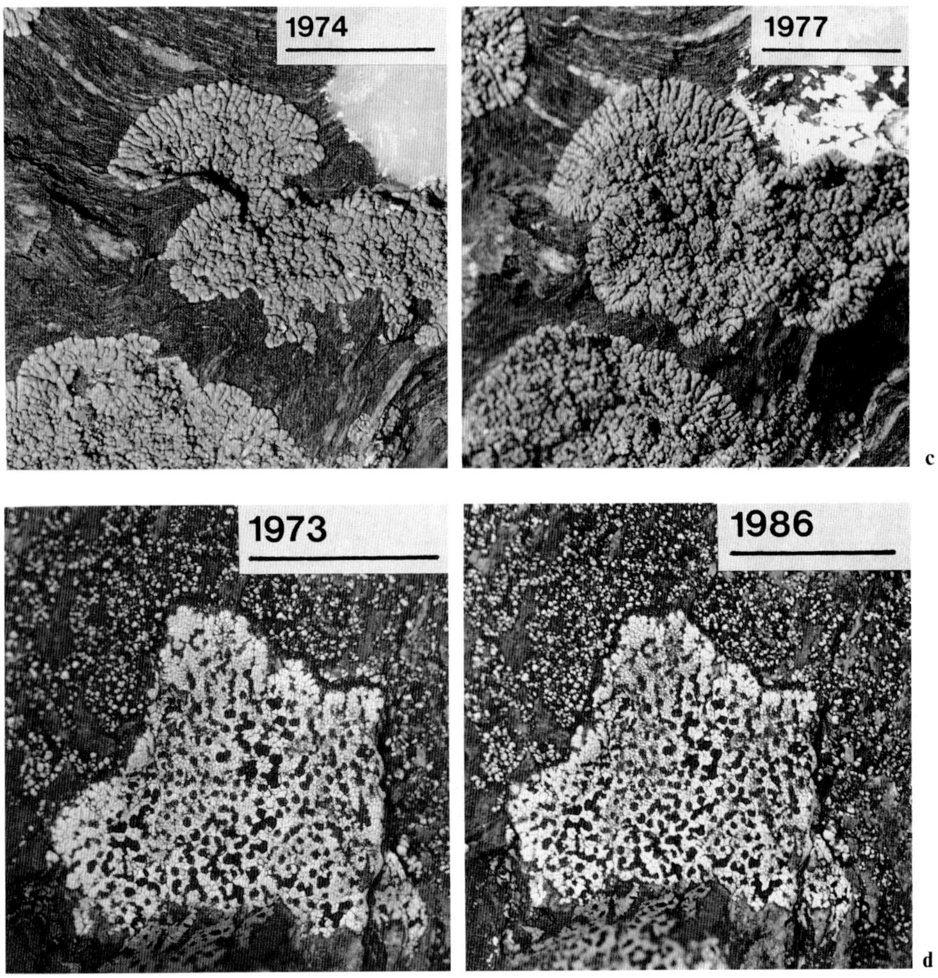

Fig. 7. c, d. *Xanthoria elegans* over 3 years (52%); **d** *Rhizocarpon geographicum* over 13 years (10%). In each photograph the *bar* represents 10 mm

Hooker's (1980d) analysis of growth in the fruticose species *Usnea antarctica* revealed a low production rate in mature thalli, but a much higher rate in juvenile plants. This has been confirmed by monitoring the change in cover abundance over a 20-year period along an environmental gradient at Knife Point (Fig. 8). During this time there has been a substantial increase in *U. antarctica* biomass on the stony ground formerly occupied by a small colony of giant petrels (*Macronectes giganteus*) but abandoned in the early 1950s (see below). The area below the nest site representing the take-off arena for the birds has also undergone a change in the moss cover, particularly in the increase of *Polytrichum alpestre*, at the expense of lichens growing epiphytically on the dominant mosses as the site has recovered from the biotic impact.

2.4 Ecological Change Resulting from Community Development

There are many instances on Signy Island of small-scale floristic change within established communities over short periods of time (see also Fig. 8). To illustrate this wide-scale phenomenon Fig. 9 demonstrates the natural expansion and contraction of colonies of mosses and lichens over a 20-year period within a 1 m^2 area of typical fellfield on predominantly quartz-mica-schist glacial debris. Areas of bare ground become colonised while new gaps appear as plants die and/or are removed. Aggressive species increase in area while less competitive ones are reduced in abundance or disappear. A prominant assemblage of calcicolous mosses and lichens associated with a marble erratic has

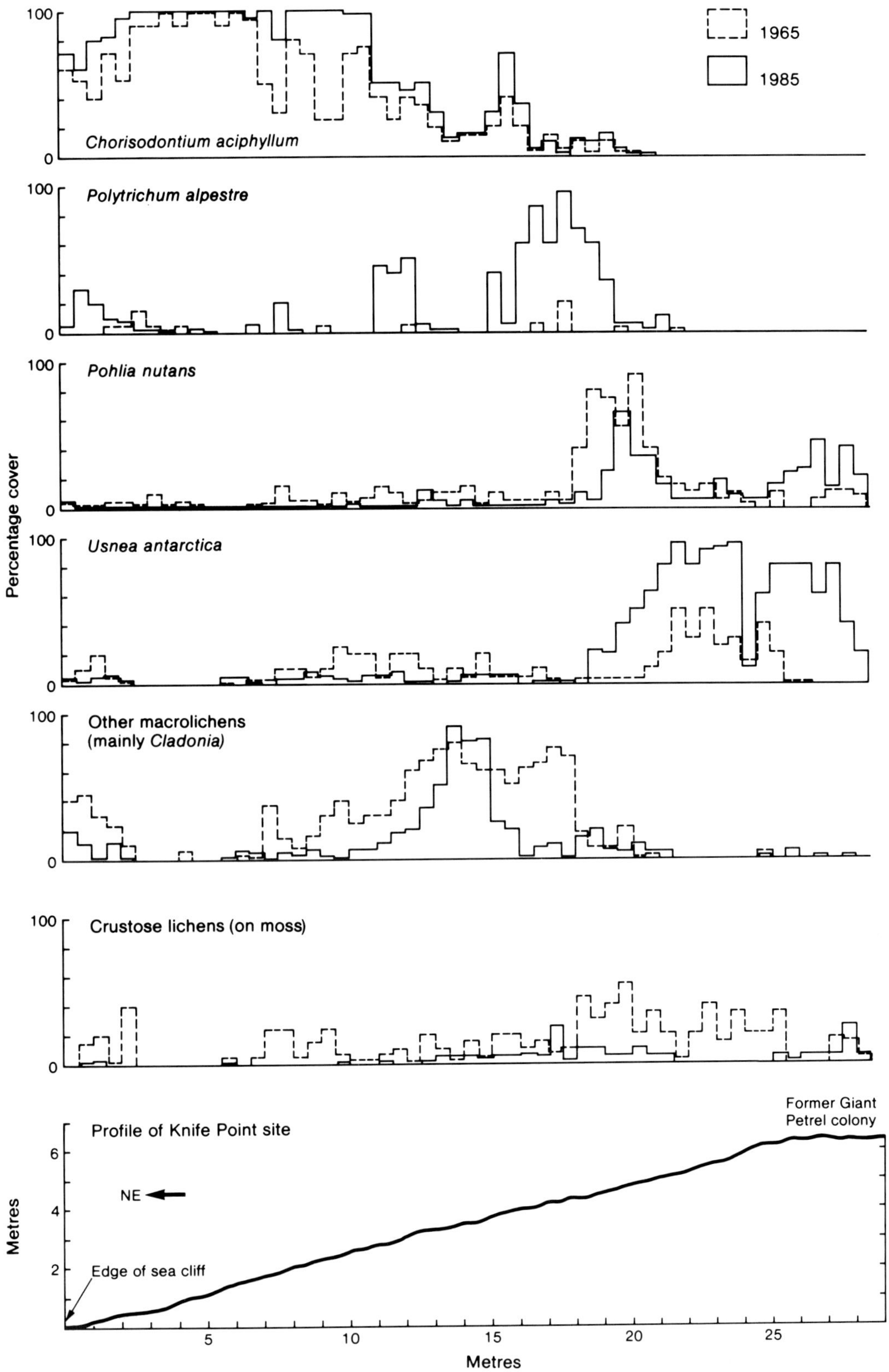

Fig. 8. Change in plant species abundance over 20 years along a transect at Knife Point, Factory Cove, Signy Island. The highest area was occupied by a small giant petrel colony until the early 1950s

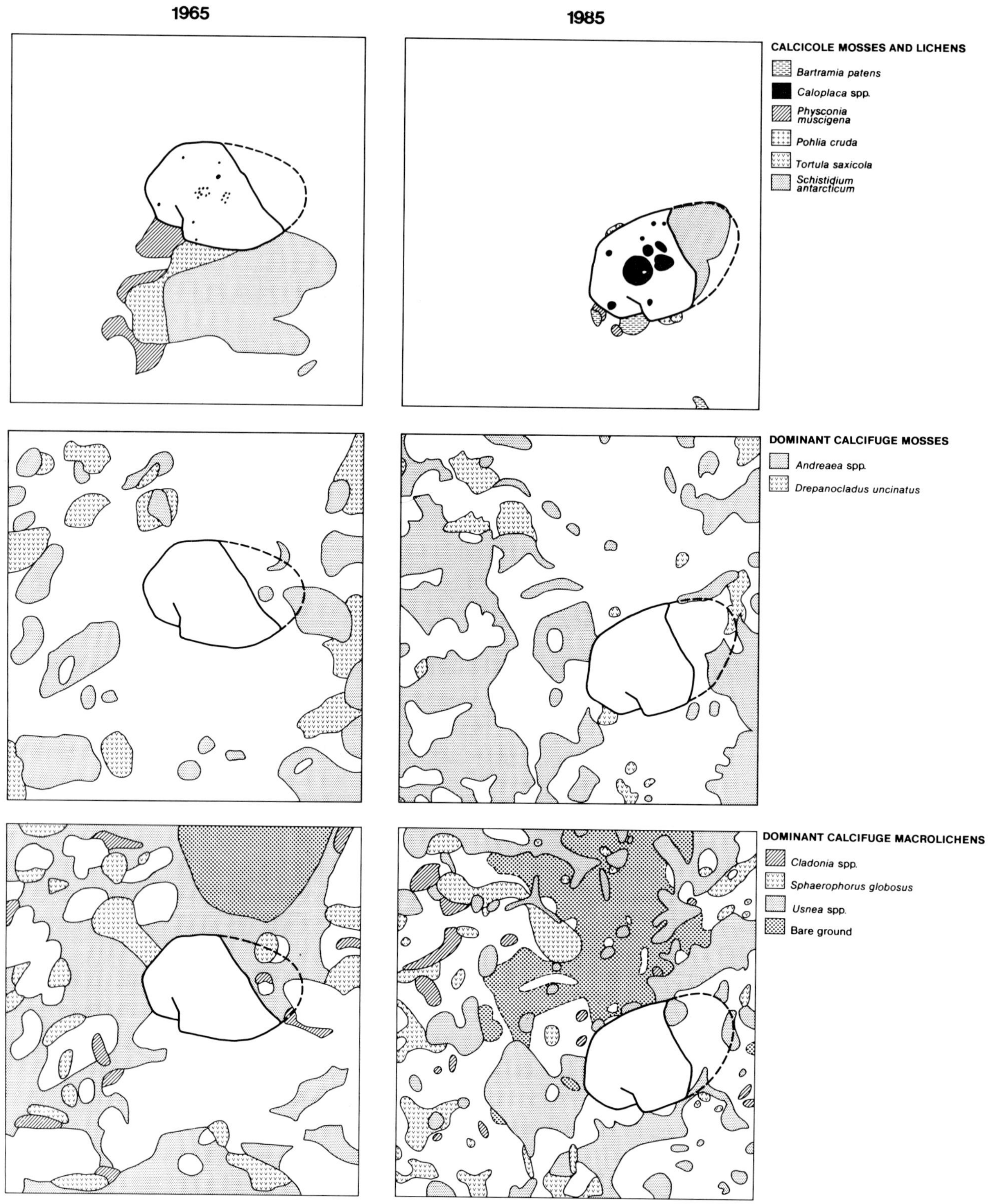

Fig. 9. Change in the distribution of the principal plant species over 20 years within a 1 m^2 quadrat centred on a marble erratic embedded in predominantly quartz-mica-schist debris on a hillside south of Mooring Point, Signy Island. By 1985 the boulder had moved 25 cm downhill, thereby destroying a large proportion of the calcicole plants which were associated with it in 1965 (1965 data adapted from Smith 1972, p. 83)

been almost obliterated as a result of the boulder slipping 25 cm downhill over much of this vegetation. The crustose lichen thalli on the boulder have grown considerably, some coalescing and others dying in their older centres to develop an open ring structure.

As a probable consequence of the ameliorating summer climate at Signy Island during the past 25 years many of the small stands of Antarctic hair grass, *Deschampsia antarctica*, have increased in both area and plant density, and several new colonies have become established. However, there is no evidence of improved seed viability or of an increase in the status of the other native vascular plant, *Colobanthus quitensis*.

2.5 Ecological Change Resulting from Environmental Perturbation

With the rapid increase in Man's activity in the Antarctic, human impact is giving cause for concern for safeguarding the fragile environment and its biota and for preserving the integrity of the sensitive ecosystems. As the number and size of research stations and associated installations, and volume of air, land and sea transport increases, the extent and nature of environmental perturbation and pollution will inevitably grow. There is now an urgent need to monitor and analyse the effects these activities are having on representative samples of ecosystems under such threat. An accurate assessment of the ecological changes caused by human impact is crucial in the preparation of management policies required for the most efficient operation of individual stations and for proposing adequate conservation measures for the protection of the local environment. It is vital that research on ecosystem and species resiliance to perturbation and pollution, the causes of ecosystem degradation and their ability to recover is undertaken now so that answers may be available when advice is sought in connection with the development of a potentially viable minerals regime in Antarctica.

Few data are yet available for ecosystem restoration in areas affected or destroyed by the construction and operation of Antarctic stations. However, the following qualitative examples may serve to provide an indication of the ecological recovery of such areas. Excavated ground close to Signy research station has, over a period of 20 years, become patchily recolonised by prominent native moss species (especially *Ceratodon* cf. *purpureus* on the drier soil, and *Drepanocladus uncinatus* on the wetter areas, the former being renowned as a primary colonist with a cosmopolitan distribution). In one area close to the shore an area destroyed during the construction of the fuel tank in 1964–65 had, by 1980, become densely colonised by a typical primary colonist, *Pottia austro-georgica*. This native ephemeral moss is abundantly fertile, but previously the nearest known population was ca. 250 m across Factory Cove. Having become established, pre-

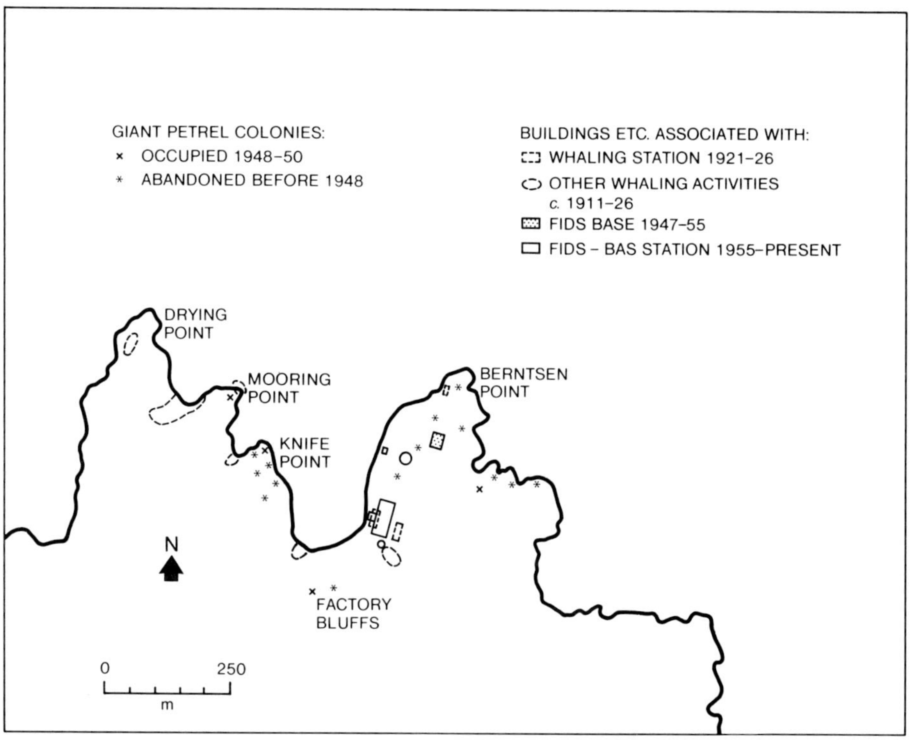

Fig. 10. Distribution of giant petrel nest sites around Factory Cove before they were abandoned as a result of human impact in the vicinity

sumably by spores, it has stabilised the soil surface and the stand is being invaded by other local bryophytes and some lichens. Also within a few metres of some buildings, and near the exhaust vent from the diesel generators, scattered plants of *Deschampsia antarctica* have become established; *Prasiola crispa* is also frequent. Elsewhere, weathered timber (generally > 20 years old) on the older buildings is slowly being colonised by crustose (especially *Rhizoplaca aspidophora* and species of *Buellia*, *Lecidea* and *Lecanora*) and fruticose (notably *Usnea antarctica*) lichens. Whale bones (55–75 years old) and concrete structures (35–65 years old) are abundantly covered by *Xanthoria elegans*, *Caloplaca* spp. and other crustose lichens (Lindsay 1973), while white gloss paint on boulders used to mark bird nest sites in the early 1950s is also colonised extensively by crustose lichens (notably an unidentified black species). As yet there is no evidence, either at Signy Island or elsewhere in the Antarctic, of alien plants becoming established from accidentally introduced propagules. However, the potential for adequately preadapted vascular plants and soil invertebrates to become established has been demonstreated by both controlled experimental introductions (Edwards 1980) and, more seriously, by unintentional invertebrate introductions (Block et al. 1984).

The influence of human presence around Factory Cove has had a varying impact on the local seabird populations. The cliff-breeding snow petrels (*Pagodroma nivea*) and cape petrels (*Daption capense*) appear to have been unaffected, as have the ground nesting dove prions (*Pachyptila desolata*) and brown skuas (*Catharacta lonnbergii*). Wilson's storm petrels (*Oceanites oceanicus*), displaced during the construction of the station, have returned to their former nest sites to within a few metres of the buildings. However, the much more sensitive giant petrels have permanently abandoned all their small colonies in the vicinity of the station, at least 14 of which resulted from disturbance from whaling activities in the 1910s and 1920s. Another four colonies (possibly re-established since the 1920s) were abandoned within a few years of the research station being established in 1947 (Fig. 10). After only 10 years the pebbles of the former nests were densely colonised by juvenile thalli of several lichens, especially *Usnea antarctica*, and occasional mosses. After a further 20 years the sites had reverted to typical fellfield (see Fig. 8). Meanwhile, no attempt to reoccupy any of the nest sites has been made by the birds.

Of far greater immediate concern on Signy Island than anthropogenic impact is the unprecedented threat to the lowland terrestrial and freshwater ecosystems by a biological agent, namely a rapidly increasing

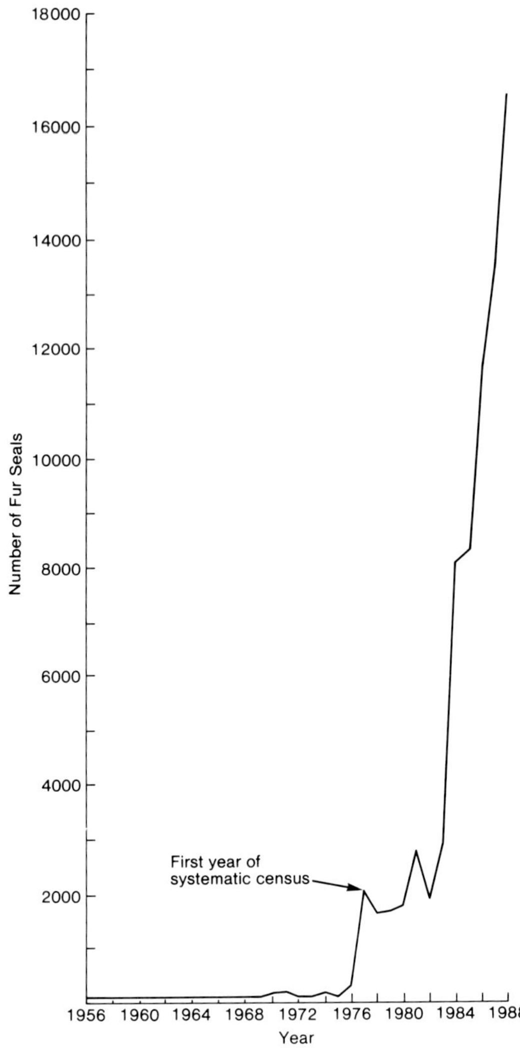

Fig. 11. Increase in the number of fur seals on Signy Island between 1956 and 1988. All censuses since 1977 were made around 24 February

summer population of fur seals (*Arctocephalus gazella*) (Fig. 11). This massive annual immigration represents an overspill of young (2–5 years) males from sub-Antarctic South Georgia, 900 km to the northeast. There is no evidence that fur seals were ever abundant in the South Orkney Islands in the past (Smith 1988). This population explosion began in earnest in 1977 and by 1985 a large proportion of the vegetation on the eastern side of the island and several less extensive areas on the west coast had been destroyed (see Smith 1988). Several giant petrel colonies have been abandoned because of fur seal disturbance. By 1988 the extent and severity of damage had increased and in that year some extensive swards of grass (*Deschampsia antarctica*) on a Specially Protected Area (Lynch Island) 1.5 km north of Signy Island were largely destroyed. As a result of fur seal impact

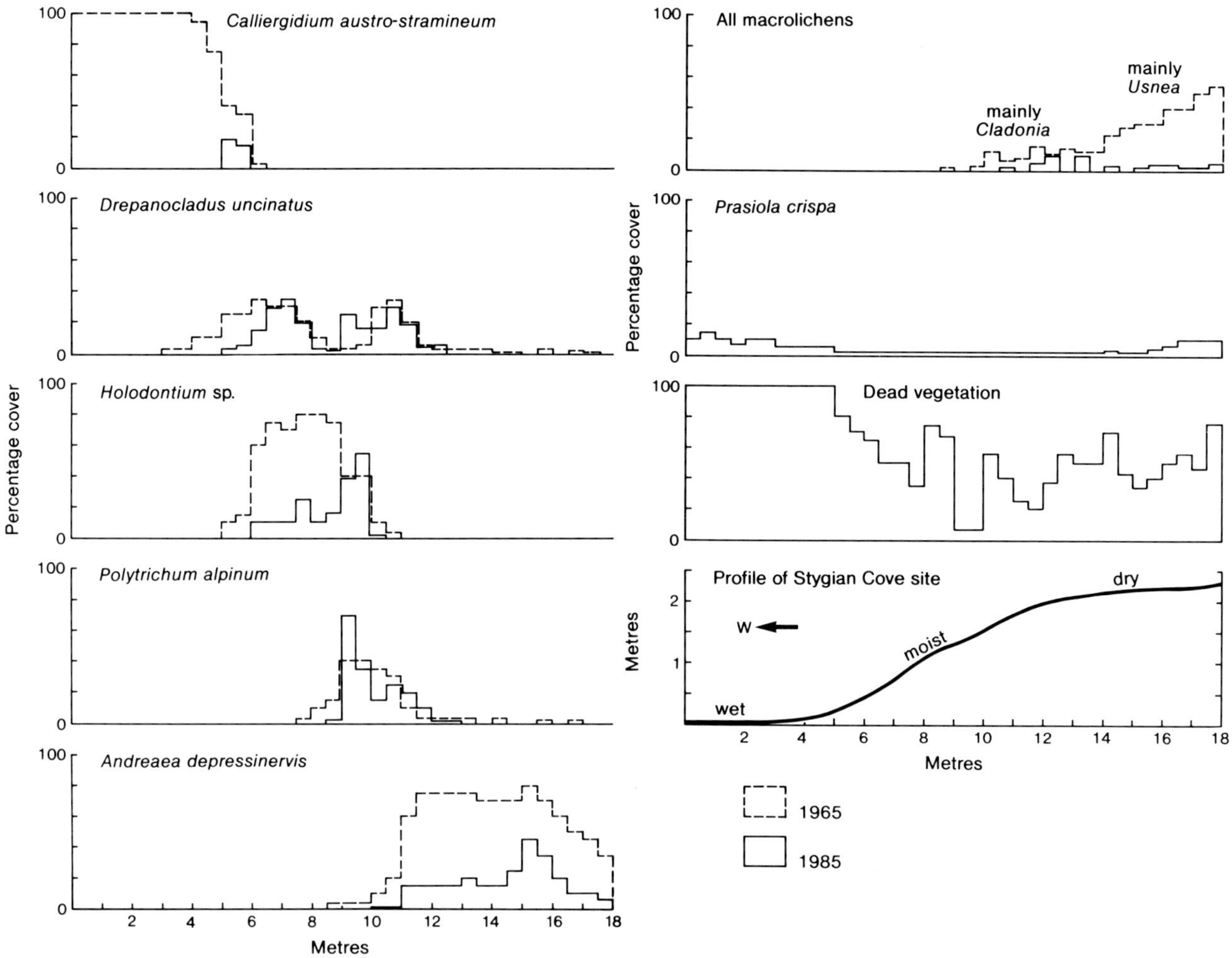

Fig. 12. Change in the distribution of the principal plant species along a transect from a wet moss-dominated mire to a dry moss and lichen dominated fellfield community, south-east of Stygian Cove, Signy Island, over 20 years. The considerable reduction in plant cover was caused by the action of fur seals between 1981 and 1984. By 1988 the moss and lichen cover had been reduced even further

large areas of vegetation have been destroyed and often physically removed, and the composition of the communities drastically altered. A major positive result has been the colonisation of the most actively perturbed areas by the nitrophilous foliose green alga *Prasiola crispa*. Several examples of the vegetational change along transects have been given by Smith (1988), but Fig. 12 also serves to illustrate this impact. Maximum damage has been caused to the originally closed moss carpet in the wet zone of this site. The zone of *Holodontium* sp. was the largest known stand of this rare moss in Antarctica but, unfortunately, most of this population has now been destroyed. In terms of ecosystem stability this exceptional increase in fur seal numbers is causing a sudden and catastrophic ecological transformation. If the mean annual increase of 725% (for the years 1982–88) is maintained, much of the terrestrial environment of Signy Island is undergoing possibly the greatest and most rapid change since it emerged from the Pleistocene ice sheet (unless similar population explosions have occurred in the past, and there is no evidence of this).

Also around Signy Island there has been a substantial increase in the number of breeding penguins and in the extent of their colonies during the past four decades (Croxall et al. 1981). Since 1980 several small new colonies have been established. This has resulted in the destruction of several hectares of vegetation, especially around some of the older colonies.

3 Conclusions

The aim of this account has been to draw attention to the temporal nature of ecological change in the Antarctic terrestrial biome by examining a range of examples from Signy Island, South Orkney Islands.

Because of the intensive biological research programme that has been undertaken there for the past 40 years, an extensive data base exists which has provided the information on which the island's environmental and biological history has been deduced. The examples discussed here serve to illustrate what changes have been or are being experienced throughout the Antarctic biome, albeit on a different scale to that at Signy Island. Although the processes of environmental and biological change have been relatively rapid on this maritime Antarctic island, they can be expected to have proceeded much more slowly as continentality increases.

Climate is undoubtedly the prime force responsible for determining ecological change and fluctuations, particularly in mean summer air temperatures, have probably played the greatest role, either directly or indirectly, in defining the direction and rate of change. Signy Island is currently experiencing a major change resulting, firstly, from the recent rise in summer (but not as yet winter) air temperatures and the consequent reduction in the area of its permanent ice fields and, secondly, from the recent dramatic increase during summer of fur seals. So far as is known, no other areas of Antarctica are being ecologically transformed so quickly, but many are undergoing some degree of change in their environment. How this is affecting their biota is as yet unknown, but it offers important opportunities for research.

Sites such as Signy Island, where there is a wealth of ecological information, much of it dating from when the present climatic change commenced, are especially important for the assessment of both the natural and human-induced trends currently being experienced not only here but also throughout the Antarctic biome. Because of its geographical situation in relation to continental Antarctica, to the Antarctic Convergence, to vagaries of the Weddell Sea current, and to temperate continental South America, the glacial status and, consequently, biological stability of Signy Island (or, indeed, comparable sites in the South Shetland Islands) are particularly sensitive to even minor fluctuations in climate. Its relatively impoverished biota and simple ecosystems make it an outstanding site at which to monitor the consequences resulting from such changes and to allow valid and reasoned predictions based on testable hypotheses to be made of future trends and potential impacts, and of the resiliance of systems to and their recovery from these phenomena. However, this can only be achieved by the implementation of carefully designed long-term experimental monitoring studies to detect and measure change. This crucial and urgent development in Antarctic research is admirably propounded and discussed by Abbott and Benninghoff (this Vol).

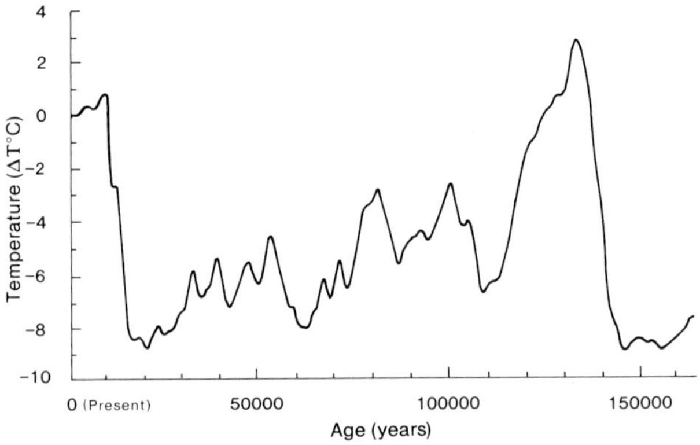

Fig. 13. Smoothed isotope temperature record, expressed as the deviation (°C) from the current mean annual surface temperature value (− 55.5 °C), derived from the 2200-m deep ice core from Vostok station (Lorius et al. 1988)

What does the future hold in store? At present, we are at or close to the peak of the Holocene Interglacial when mean global air temperature is the highest since the Last Interglacial 120 000 years ago (Fig. 13). However, according to climatologists, we are now experiencing the early stages of the CO_2-induced "Super Interglacial" (Mitchell 1977). Palaeoclimatic evidence from the Vostok deep ice core clearly demonstrates that there has been a close correlation between near-surface air temperatures and the CO_2 concentration of the atmosphere during the past 160 000 years (Lorius et al. 1988). Also, it is now firmly established that the present steadily increasing CO_2 levels are caused, directly or indirectly, by human activities, especially over the past century, and that this has been responsible for raising global air temperatures. However, much of this effect is absorbed by the oceans which inhibit any immediate large-scale melting of the polar ice caps. Based on this correlation the latest models, which take into account the time lags in the climatic response as a result of the oceans' heat capactiy, predict a global change of between 0.1° and 0.8 °C per decade from 1985. According to present climatic trends, the most probable scenario will be an increase of ca. 0.3 °C/decade, thereby raising the global mean temperature by almost 2 °C by 2050 A.D. (Fig. 14). However, by then the polar regions may be expected to experience increases of 1–1.4 °C in summer, and 4–4.8 °C, with enhanced precipitation, in winter (World Meteorological Organization 1988). The region most vulnerable to the impact of the increasing "greenhouse effects" of CO_2-induced warming is the West (Lesser) Antarctic ice sheet (Bentley 1983; Budd et al. 1987). While this influence will be greatest in the Antarctic islands and more northerly continental Antarctic coasts where ice is relatively "warm" and thin, the interior environment of Antarctica should be

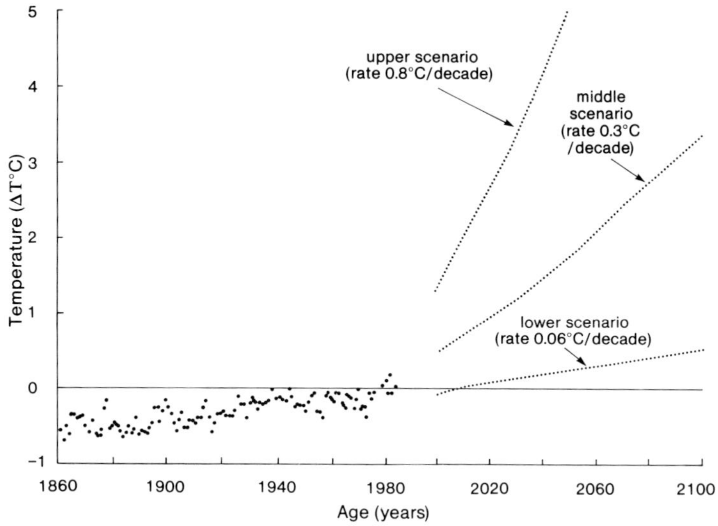

Fig. 14. Mean global surface air temperature from 1860 to 1985 with three possible scenarios for predicted change (plotted as differences from the 1985 value) throughout the 21st century (World Meteorological Organization 1988)

largely buffered against these effects, at least from an ecological point of view.

Field experiments using cloches at Signy Island have shown that a mean increase of ca. 3 °C in summer can have a dramatic effect on the rate of development of *in situ* propagules and may be indicative of what may happen as the "greenhouse effect" gathers momentum (Smith in press). Even allowing for a substantial uncertainty factor in the predictions, we can expect some major ecological changes throughout much of coastal Antarctica over the next few decades, not least at glaciologically sensitive sites such as Signy Island. This gives cause for serious concern for the future equilibrium of Antarctica's environment and biota. In order to assess and predict the implications of such changes we must invest in carefully planned and executed monitoring programmes. These will also provide ideal opportunities for studying ecological processes and ecosystem dynamics, from which we may learn how to better manage and preserve the physical and biological Antarctic environment.

Acknowledgements. I am indebted to many colleagues in B.A.S. for their help in maintaining long-term experiments at Signy Island. I am also grateful to Dr. C. Lorius, Laboratorie de Glaciologie et de Geophysique de l'Environment, St-Martin-d'Hères, France, and two unknown referees, all of whose comments on the manuscript were greatly appreciated. My thanks are also due to Dr. D. D. Harkness, N.E.R.C. Radiocarbon Laboratory, Glasgow, for providing radiocarbon dates for the peat samples.

References

Abbott SB, Benninghoff WS (This Vol.) New directions for biology in the study and conservation of Antarctic ecosystems.

Bentley CR (1983) The West Antarctic ice sheet: diagnosis and prognosis. In: Carbon dioxide, science and consensus. Proc Carbon Dioxide Res Conf, US Department of Energy, pp V.3–IV.50

Block W, Burn AJ, Richard KJ (1984) An insect introduction to the maritime Antarctic. Biol J Linn Soc 23:33–39

Budd WF, McInnes BJ, Jenssen D, Smith N (1987) Modelling the response of the West Antarctic ice sheet to a climatic warming. In: van der Veen CJ, Oerlemans J (eds) Dynamics of the West Antarctic ice sheet. D Reidel, Dordrecht, pp 321–358

Clapperton CM, Sugden DE (1988) Holocene glacier fluctuations in South America and Antarctica. Quat Sci Revs 7:185–198

Clapperton CM, Sugden DE, Birnie RV, Hanson JD, Thom G (1978) Glacier fluctuations in South Georgia and comparison with other island groups in the Scotia Sea. In: van Zinderen Bakker EM (ed) Antarctic glacial history and world palaeoenvironments. AA Balkema, Rotterdam, pp 95–104

Collins NJ (1976) The development of moss-peat banks in relation to changing climate and ice cover on Signy Island in the maritime Antarctic. Bull Br Antarct Surv 43:85–102

Croxall JP, Rootes DM, Price RA (1981) Increases in penguin populations at Signy Island, South Orkney Islands. Bull Br Antarct Surv 54:47–56

Curl JE (1980) A glacial history of the South Shetland Islands, Antarctica. Rep Ohio Inst Polar Stud 63:1–129

Edwards JA (1980) An experimental introduction of vascular plants from South Georgia to the maritime Antarctic. Bull Br Antarct Surv 49:73–80

Fenton JHC (1980) The rate of peat accumulation in Antarctic moss banks. J. Ecol 68:211–228

Fenton JHC (1982a) The formation of vertical edges on Antarctic moss-peat banks. Arct Alp Res 14:21–26

Fenton JHC (1982b) Vegetation re-exposed after burial by ice and its relationship to changing climate in the South Orkney Islands. Bull Br Antarct Surv 51:247–255
Fenton JHC, Smith RI Lewis (1982) Distribution, composition and general characteristics of the moss banks of the maritime Antarctic. Bull Br Antarct Surv. 51:215–236
Gribbin J, Lamb HH (1978) Climatic change in historical times. In: Gribbin J (ed) Climatic change. Cambridge University Press, Cambridge, pp 68–82
Harper JL (1977) Population biology of plants. Academic Press, London
Harrington PK (1968) A geophysical survey near the South Orkney Islands. PhD Thesis, University of Birmingham (unpubl)
Hooker TN (1980a) Factors affecting the growth of Antarctic crustose lichens. Bull Br Antarct Surv 50:1–19
Hooker TN (1980b) Lobe growth and marginal zonation in crustose lichens. Lichenologist 12:313–323
Hooker TN (1980c) Growth and production of *Cladonia rangiferina* and *Sphaerophorus globosus* on Signy Island, South Orkney Islands. Bull Br Antarct Surv 50:27–34
Hooker TN (1980d) Growth and production of *Usnea antarctica* and *U. fasciata* on Signy Island, South Orkney Islands. Bull Br Antarct Surv 50:35–49
Hughes TJ (1983) The stability of the West Antarctic ice sheet: what has happened and what will happen. In: Carbon dioxide, science and consensus. Proc Carbon Dioxide Res Conf, US Department of Energy, pp IV.51–IV.73
Lindsay DC (1973) Estimates of lichen growth rates in the maritime Antarctic. Arct Alp Res 5:341–346
Lorius C, Barkov NI, Jouzel J, Korotkevich YS, Kotlyakov VM, Raynaud D (1988). Antarctic ice core: CO_2 and climate change over the last climatic cycle. Eos 69 (26):681–684
Mercer JH (1978) Glacial development and temperature trends in the Antarctic and in South America. In: van Zinderen Bakker (ed) Antarctic glacial history and world palaeoenvironments. AA Balkema, Rotterdam, pp 73–93
Mitchell JM (1977) Carbon dioxide and future climate. Environ Data Serv, pp 3–9
Nakada M, Lambeck K (1988) The melting history of the late Pleistocene Antarctic ice sheet. Nature 333 (6168):36–40
Smith RI Lewis (1972) Vegetation of the South Orkeny Islands with particular reference to Signy Island. Sci Rep Br Antarct Surv 68:1–124
Smith RI Lewis (1988) Destruction of Antarctic terrestrial ecosystems by a rapidly increasing fur seal population. Biol Conserv 45:55–72
Smith RI Lewis Bryophyte propagule banks: Case study of an Antarctic fellfield soil. In: Walton DWH, Miles J (eds) Primary succession on land. Blackwell, Oxford (in press)
Smith RI Lewis, Coupar AM (1986) The colonization potential of bryophyte propagules in Antarctic fellfield soils. Comm Nat Fr Res Ant 58:189–204
Stuiver M, Becker B (1986) High-precision decadal calibration of the radiocarbon time scale AD 1950–2500 B.C. Radiocarbon 28 (2B):863–910
Stuiver M, Denton GH, Hughes TJ, Fastook JL (1981) History of the marine ice sheet in West Antarctica during the last glaciation: a working hypothesis. In: Denton GH, Hughes TJ (eds) The last great ice sheets. John Wiley & Sons, New York, pp 319–282
Sudgen DE, Clapperton CM (1977) The maximum ice extent on island groups in the Scotia Sea, Antarctica. Quat Res 7:268–282
World Meteorological Organization (1988) WMO/TD No. 225. World climate programme. Developing policies for responding to climatic change. A summary of the discussions and recommendations of the workshops held in Villach 28 September – 2 October 1987 and Bellagio 9–13 November 1987 under the auspices of the Beijer Institut, Stockholm (WMO/WCIP1), Geneva, World Meteorological Organization
Wynn-Williams DD (1988) Television image analysis of microbial communities in Antarctic fellfields. Polarforschung 58 (2/3):239–249
Wynn-Williams DD Microbial colonisation processes in Antarctic fellfield soils – an experimental overview. Proc 11th Symp Polar Biol, Tokyo (in press)

Colonization of Terrestrial Habitats – Organisms, Opportunities and Occurrence

D. W. H. WALTON [1]

Summary. The polar regions offer unrivalled opportunities to study the processes of primary colonization and succession in land habitats, an important yet poorly documented area of ecology. Present glacial recession is extending the limited land area available for colonization whilst human activity is increasing the potential for the introduction of alien species.

This chapter considers briefly the range of organisms already established and the evidence for both long-distance and local dispersal. Propagule characteristics are examined in relation to dispersal potential. The process of colonization is examined for lichens from data gathered from outside Antarctica to illustrate the potential for application to polar sites.

A generalized model is proposed based on habitat favourability for the colonizing event, and its heterogeneity in time and space. The inadequacies of data for this model and primary colonization in general are noted and three major fields identified for future research.

1 Introduction

For over 25 million years the Antarctic continent has had no terrestrial connection with any of the other southern continental landmasses. During much of this period it has been almost completely covered by permanent snow and ice and even now, in what is commonly described as a warm interglacial period (Phillpot 1985) less than 2 % of the continental area is ice-free in summer. The largest single area of snow-free ground is in south Victoria Land at ca. 77° S, whilst the Transantarctic Mountains provide significant areas of rocky substrate much further towards the South Pole. Elsewhere, and especially in the Peninsula area, the proportion of summer snow-free ground increases with decreasing latitude so that some of the subantarctic islands are completely deglaciated. Even these islands, however, are remote (Walton 1985), and small with limited species diversity (Lewis Smith 1984a). Despite the fossil records of a rich and diverse Antarctic flora, glaciers during the glacial periods have virtually swept the continent and many of the islands clean of flora and fauna. Species now present are mostly all immigrant, having arrived in these remote localities presumably by wind and birds. Indeed, Lindsay (1974) stressed the importance of both bipolar and cosmopolitan elements in the Antarctic lichen flora, with few endemic species. The pattern for subantarctic mosses does not appear to be dissimilar, but with a stronger Southern Hemisphere component. Walton (1985) concluded that although the subantarctic island floras showed clear evidence for long-distance dispersal, as first suggested by Hooker (1879), biogeographical explanations for different groups could only be regarded as tentative until taxonomic revisions had been completed. Only then will it be possible to assess realistically the possibilities that some species survived the last glaciation in Antarctic refugia, a possibility raised by several authors over the past 150 years and most recently by Kappen and Straka (1988).

Thus the present Antarctic and subantarctic flora and fauna should offer some valuable information on those groups of organisms which have successfully overcome the long-distance dispersal barrier and are adapted to colonization and establishment in a difficult environment. In addition, there is also considerable potential for the experimental study of the process of colonization within these disjunct snow-free sites. Rudolph and Benninghoff (1977) reviewed the earlier information on long-distance dispersal and colonization in some detail. This chapter aims to provide an assessment of progress since that review, a theoretical framework for Antarctic colonization and some suggestions for specific research projects.

2 Organisms

Recent reviews by Lewis Smith (1984a) and Block (1984) have provided lists of all the groups of terrestrial organisms known from the Antarctic, whilst Hey-

[1] British Antarctic Survey, High Cross, Madingley Road, Cambridge CB3 0ET, UK.

Antarctic Ecosystems. Ecological Change and Conservation
(ed. by K. R. Kerry and G. Hempel)
© Springer-Verlag Berlin Heidelberg 1990

wood (1984) has similarly treated the freshwater habitats. The range of organisms is very restricted, especially for continental areas. Except for *Deschampsia antarctica* and *Colobanthus quitensis,* which extend south to 68° 43′ S at favourable sites on the west coast of the Antarctic Peninsula (Lewis Smith and Poncet 1987), higher plants are almost completely restricted to the subantarctic islands. The truly Antarctic flora is limited to mosses, liverworts, lichens, fungi, algae and bacteria. The fauna shows a similar pattern with the most advanced invertebrates being found in the subantarctic, and the continental animals being mainly Protozoa, Rotifera, Tardigrada and Nematoda together with species of Acarina, Collembola and Mallophaga. Increasing human activity must have a direct effect on alien introductions as documented for microbial populations by Cameron et al. (1977).

With the exception of the two phanerogams all these Antarctic organisms can be regarded as relatively simple in phylogenetic terms but perhaps that is not their most significant feature. More importantly they, or their propagules, must be both small and light enough to be easily carried across the Southern Ocean by wind or by birds.

What features of an organism make it especially suitable for long-distance dispersal?

3 Propagule Characteristics

For some groups of organisms there are data on the principal features of their propagules describing the shape, size, density and frequency of production of airborne propagules of viruses, bacteria, algae, protozoa, fungi, bryophytes, insects and other microfauna. Some of these features are gathered in Table 1.

As long ago as 1946, Wolfenbarger found that for most small propagules the dispersal ability was closely related to the logarithm of the distance from the source population. This general result is the integration of many variables. The periodicity of propagule production and release is controlled by the interaction between organism development and climate. The relationship of this with both local and mesoscale meteorological phenomena, and the possibilities of damage or death of the propagule during transport each add further uncertainty to the dispersal process. For all small organisms, including Antarctic invertebrates, the whole dispersal process can be regarded as a passive one over which the organism itself exerts no control at all.

In general terms there are three different routes for the production of dispersal units. For many cryptogams fragmentation of the adult plant or thallus

Table 1. Some general features of dispersal units for particular groups of organisms

	Shape	Size (μm)	Maximum viability	Independent dispersal	Form	Comment
Viruses	Spherical, polyhedral, rod-like	0.02–0.45	?	Sometimes	Particles	Generally co-dispersed with host cells?
Bacteria	Spherical, rod-like, spiral	0.3–15	?	Rarely, usually attached to other particles	Entire cells, spores	Often dispersed attached to other articles
Algae	Spherical, ellipsoidal, irregular	1–150	?	Yes	Vegetative cells, spores and cysts	Generally aggregated
Protozoa	Ovoid to irregular	2–50	?	Probably	Cysts	
Fungi	Spherical, coiled, branched, rod-like	< 5–50 +	5 h to months	Yes	Spores	
Bryophytes	Globose, tetrahedral	6–20	2 months to 5+ years	Yes	Spores, Gemmae	Vegetative fragments very common for polar sp.
Lichens	Spherical, irregular	30–60	?	Yes	Ascospores and algae independently, Soredia, Isidia	Thallus fragments very common for polar sp.

produces viable vegetative units which allow localized dissemination. The equivalent for some of the simpler animals is the dispersal of a whole animal. The other two routes are distinguished by the production of a special propagule (e.g. seed, cyst, egg, spore) and also by the manner in which it is produced (sexual or asexual). Sexually produced seeds are likely to be the principal propagules for the two flowering plants in the maritime Antarctic (Edwards 1974) but vegetative proliferation has been found in the subantarctic *Festuca contracta* (Tallowin 1977). Asexual propagules are common in lower plants whilst most animals produce sexual propagules. It is important to bear in mind that for some species all three types of propagules may be produced.

All the propagules have to satisfy one primary requirement – establishment at a new site. They thus have to be:

1. Capable of being moved by some agency (winds, animals, water);
2. Viable long enough to reach the new site (surviving possible drought, low temperatures, ionic stress, increased UV);
3. Capable of carrying enough reserves to allow growth at the new sites.

These attributes are essential for both long- and short-distance dispersal. The definition of long and short distance must be arbitrary. However, for Antarctic terrestrial organisms short could be defined as within the local snow-free area whilst long would require the propagule to cross significant distances of sea or snow. Propagule characteristics will determine which species are capable of surviving local dispersal and the more demanding long-distance dispersal. This part of the process can be distinguished from colonization *per se* which begins when any viable propagule arrives at a new site. It seems likely that studies of the colonization process at newly available sites can be more easily studied with locally derived propagules and the results applied to propagules derived from distant areas.

All of these studies, however, are still hampered by the inadequate taxonomic background for most Antarctic organisms. If we cannot accurately name the species involved in Antarctic communities, it will clearly be difficult to determine the specific relationships with populations outside the Antarctic which may be the primary source of propagules. Indeed, it will be even more difficult in many instances to identify the propagules if cultivation is not possible, since at present there are no published keys to Antarctic diaspores. The production of illustrated guides based on spore ornamentation etc. would be a valuable step forward.

4 Colonization – A Poorly Understood Process?

Having established some of the principal features common to organisms which have apparently overcome the problems of long-distance dispersal we now need to consider their fate after deposition in the new environment. The limited diversity of organisms present suggest that many of the species, which may be dispersed in the subantarctic and especially the Antarctic, fail to colonize. Both transplant experiments with flowering plants (Holdgate 1964; Edwards and Green 1973; Edwards 1980) and accidental introduction of invertebrates (Block et al. 1984), lichens (Lindsay 1973) and flowering plants (Longton 1966; Walton 1975) have indicated that the range of species which can survive in both the subantarctic and the Antarctic environment is greater than those currently considered as native.

There are many possible explanations for the failure of a species to become established but we must start with a definition of colonization as a dynamic process. Colonization can be considered as the establishment of a species population in a geographical or ecological space not occupied by that species. Colonization in these terms is a continuing and active process all over the globe; the Antarctic is simply one of the more difficult areas for establishment. Furthermore, if it is assumed that all species are potentially capable of colonization at some time, we can dispense with the class of "colonizing species" and consider the process as one of colonizing episodes for any species.

The arrival of a colonizing propagule at a new site must be regarded as a random event. There are no data to indicate what proportion of this random propagule 'rain' for any species will eventually establish, nor are there any studies which have characterized an acceptable niche for the establishment of any Antarctic species. Indeed, it is evident that there is little direct data on the processes of colonization and establishment for Antarctica.

At the Third SCAR Symposium Rudolph and Benninghoff (1977) stated "A basic biological problem concerning Antarctic terrestrial organisms has been, and continues to be, the understanding of their mode of invasion and colonization."

Fourteen years later this continues to be true for bryophytes and lichens (Longton 1988) and for microbes (Vincent 1988). A combination of small advances from elsewhere, together with some recent initiatives in the Antarctic, suggests considerable progress may now be possible.

The first assumption that there has been a continuing particle rain onto all uncolonized areas of Antarctica is documented in many Antarctic studies of deep snow and ice cores (e.g. Thompson 1977; Ram and Gayley 1988). At the microbial level there have been frequent suggestions that bacteria, fungi and microalgae must be dispersed to the Antarctic by wind but there is only limited, direct evidence to show that this happens. Some of the particles from very old ice have been shown to be viable bacteria whilst others are pollen and spores (Abyzov et al. 1982; Cameron and Morelli 1974). Part of this rain of exotic pollen and spores also falls onto subantarctic islands (Barrow and Smith 1983; Salas 1983; Scott and van Zinderen Bakker 1985). Churchill (1973) recorded a range of exotic pollens from peat on Signy Island whilst more recently Kappen and Straka (1988) reported a similar range from King George Island peat. They also examined moss cushions from Wilkes Land and compared these with the data from the maritime Antarctic islands. From this they concluded that whilst there was good circumstantial evidence that the present cryptogamic vegetation in the maritime Antarctic is due to immigration from South America and the subantarctic, for the continental flora recruitment may have been from refugia on the continent itself. This is in direct contrast to the conclusions of Broady et al. (1987). Their work on the fumarole biota of Mt. Melbourne and earlier work on Mt. Erebus (Broady 1984) both suggest that it is possible for mosses, liverworts and algae to reach remote continental sites from temperate areas. The appearance of *Funaria hygrometrica* around new fumaroles on Deception Islands after the eruption (Collins 1969) is an indication that propagules of new species are available, either in the soil or regularly deposited by wind, and waiting for a chance to establish. Heal (1967) noted that nutrient agar plates exposed on Signy Island produced cultures of fungi and bacteria not isolated from the local soil microflora, but Broady (1979) found no evidence for non-native algal colonists in a study there. The study by Rudolph (1970) at Cape Hallatt was concerned with local distribution of propagules and was also unable to show a non-Antarctic origin for the materials examined. Recent studies by Saxena (1982), however, have shown the presence of biogenic particles in clouds over the Ross Ice Shelf, but their identification and origin awaits further investigation. These studies highlight the inadequacy of present data.

The very limited offshore collections of insects and other arthropods from ships (Clagg 1966) have not produced any specimens in areas within the Antarctic Convergence so, at present, there is no direct evidence for long-distance transport of animals by wind.

There has been much speculation on the possible role of birds in the dispersal of propagules. Lewis Smith (1982) suggested that South Polar skuas and Dominican gulls were probably important in dispersing *Deschampsia antarctica*. Kok (1975) recorded many instances of *Acaena magellanica* and *Uncinia compacta* seeds on birds at Marion Island. Schlitchting et al. (1978) sampled the feet and feathers of South Polar skuas, southern black-backed gulls, Arctic terns, cape pigeons and giant fulmars, mainly collected at sea, and found a wide variety of algae and protozoa. Much more work is needed to establish frequency, viability and possible origins of bird-transported material but these, and other studies in temperate regions, do suggest this may be an important route for the introduction of some species.

The second assumption is that a proportion of the propagules are viable when they reach the new habitats and manage to establish. The evidence from Deception Island (Lewis Smith 1984b) and Mt. Melbourne suggests that this is true but the proportion of the available propagules to which this corresponds is unknown. An important feature of this phase of the process may be that the most successful dispersal units contain either more than one propagule or more than one species, for instance protozoan cysts stuck to algal cells. Broady (1979) noted that algal growth appearing around windborne soil particles often consisted of more than one species.

The third assumption is that after establishment the populations survive long enough to establish themselves at other sites by the local dissemination of propagules. Direct evidence is lacking but can be inferred from observations on the presence of apparently locally derived propagules in simple aerobiological sampling (Broady 1979) and from the extent of the propagule bank in bare, fell-field soils (Lewis Smith and Coupar 1986).

5 Colonization by Specific Groups of Organisms

To illustrate how primary colonization might occur for all groups is beyond the scope of this chapter, but details for one group, i.e. lichens, is illustrative of some of the problems associated with research in this field.

Lichens have long been renowned as pioneer colonizers of fresh substrates in polar and alpine regions (Brodo 1973). There is little direct information about the colonization process for lichens in Antarctica, although some limited indication of substrate type for particular species can be gained from floras, e.g. Lindsay (1974). The only lichen community that has been subjected to detailed investigation is the Beacon Sandstone cryptoendolithic communities in Victoria Land (Friedmann 1982) and these are atypical of most Antarctic communities. Chasmoendolithic communities, however, are widespread and certainly merit detailed investigation.

In a study on the growth of Antarctic crustose lichens Hooker (1980) examined colonization of dated concrete surfaces. He concluded that for *Calopaca* cf. *cinericola* the weathering of the concrete surfaces on Signy Island took from 14 to 40 years before colonization occurred and attributed this to the differing textures and chemical compositions of the concrete used. Lindsay (1973) recorded six lichen species on introduced timbers in South Georgia. He concluded that all had probably been introduced by human agency and since they were obligately lignicolous and corticolous they were limited to introduced timber as the only available substrate on the island.

It is generally assumed that reproduction in lichens will always favour vegetative propagules, e.g. soredia, isidia, which carry both the mycobiont and phycobiont and allow the immediate production of a new thallus. Sexual reproduction by ascospores of the mycobiont requires not only the establishment of the mycelium but relichenization with the appropriate alga. Lindsay (1977) stated that although apothecia are abundant in many Antarctic lichens, mature ascospores are uncommon.

Recent work on temperate species by Ott (1987a) e.g. *Xanthoria parietina*, has shown that in the early stages of development the mycobiont can associate with foreign algae and only later form a permanent symbiosis with, in this case, *Pseudo-trebouxia*. If this developmental sequence proves to be common, a general reassessment will be required of the likelihood of sexual reproduction in lichens, and therefore the potential for Antarctic colonization by this means.

Even in the case of asexual reproduction by soredia, probably the most common route for Antarctic lichens (Lewis Smith 1984a), there is very little detail known on the developmental steps from soredia to mature thallus for any lichens. Observations on the development of soredia from *Hypogymnia physodes*, *Physcia tenella* and *Parmelia sulcata* at temperate sites (Ott 1987c) revealed several general points:

1. Small depressions and fissures were preferred but soredia could anchor themselves on a relatively smooth surface;
2. Attachment is by excretion of a mucilaginous material and the rapid growth of hyphae;
3. The extending hyphal mat incorporated any foreign algae lying on the surface;
4. The soredia secretes further mucilage, develops into a clump of tissue and then begins to differentiate into thallus lobes.

The production of the mucilaginous sheath appears to be all-important in providing initial attachment, later uniting hyphae into a compact layer and possibly serving to protect the hyphae from water stress. Further studies by Ott (1987b) have shown that the timing of propagule dispersal is important in determining the rate of development but unfavourable conditions only delayed and did not prevent the colonization of new habitats. Observations by Schuster et al. (1985) on the earliest stages of growth of a range of lichen species show that it is common for several soredia to fuse during thallus development, suggesting that many thalli are likely to contain various genetic strains. Moreover, young plants of different species were also observed to merge together, giving rise to genetically and morphologically heterogeneous individuals. Could this be responsible for the considerable variability in Antarctic species?

We have no direct data on the colonization process for Antarctic lichens. The same is true for virtually all other groups of organisms.

6 The Antarctic Habitat Template

To help in defining the attributes of colonization in Antarctic terms, it is useful to consider the problem in a theoretical framework. Southwood (1977) examined habitat as a key to the ecological strategies of organisms. Although the focus of his paper was reproductive success and not colonization, several aspects of his approach can be used for examining this particular process in Antarctica and in so doing can indicate potentially productive lines of research for the future.

Southwood used three matrices: the 'favourableness' matrix, the expectancy of 'being there' matrix, and 'uncertainty' matrix. From these he derived the 'reproductive success' matrix. Figure 1 shows the general format for all these matrices. Especially significant in the application of these matrices is the recognition that habitat and resources vary both in time and space. Let us examine how this model approach can be applied to the present problem.

First, consider the 'being there' matrix. The likelihood of the viable propagule being at a given place at a given time is constrained by distance, weather patterns (for those dispersed both by wind or attached to birds), timing, frequency and volume of production of a particular propagule type.

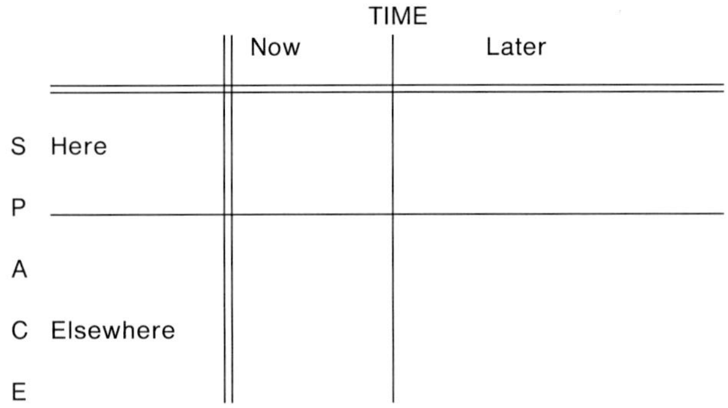

Fig. 1. Basic matrix for colonization success

Second, consider the 'favourableness' matrix. This is a measure of the value of the available microhabitat and its environment to an incoming propagule. The favourableness will be determined by surface roughness, patterns of water availability, light and temperature, availability of nutrients, etc.

Thirdly, consider the 'uncertainty' matrix. This is the synthesis of propagule presence and habitat favourability. How often will the right propagule land in the right microhabitat and actually establish? The more specific the microhabitat requirements, the lower is the probability of occurrence of these features at any given moment in time; the broader the requirements, the more frequently a suitable microhabitat will be found. This is effectively the probability of success in any colonization event for any particular species.

The resolution of these matrices can only be in probability terms but what can be done to provide estimates? A framework is needed within which to organize these variables. Both habitats and species must be considered in terms of time and space components. Evidence of temporal availability in resources and its consequence for organisms can be seen in many Antarctic studies.

Heterogeneity of habitats in time is a crucial feature of a strongly seasonal environment. Figure 2 illustrates diagramatically how "favourableness" (r) can vary with time. F (length of the favourable period) could be equated for many organisms with periods when the microhabitat is free of snow. This period and its converse (L) can vary dramatically. If L is assumed to be a period permitting the continued existence of organisms already established, but not permitting new establishment then $F + L = H$, the period during which an unfilled microhabitat remains to be filled. Success in this instance depends on the match between length of propagule viability (τ) and F.

What of the time characteristics of F and L? Figure 3 gives three types of time classes: predictable, unpredictable and ephemeral. Predictable could be equated with daily meltdown of snow cover in summer, unpredictable could be considered as a daily duration of meltwater at any particular site in any specific season, whilst ephemeral could constitute a random bird dropping on a fell-field area. In Fig. 3 these are shown against a single value of τ as an illustration of the interaction between habitat availability and propagule viability. Variability in H can remain high even when the variability in L is small.

Heterogeneity in space is equally important. Figure 4 relates dispersal to patch distribution. Consider the ability of propagules to be dispersed. Some move only locally (R_t) within an area of favourable microsites whilst others move long distances (R_m) migrating over unfavourable areas to new favourable patches. The interaction between the propagule potential for dispersal (R_t and R_m) and the size and spacing of favour-

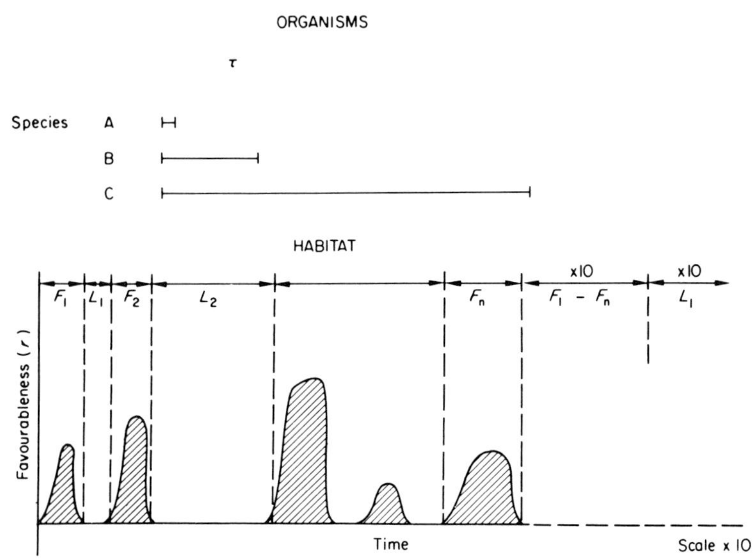

Fig. 2. Heterogeneity of habitats with time. τ length of viability period for propagule (examples are given for three species with differing periods of propagule viability, *A*, *B*, and *C*, illustrating the interactions between viability period and habitat favourability); F length of favourable period for colonization; L length of unfavourable period for colonization (After Southwood 1977)

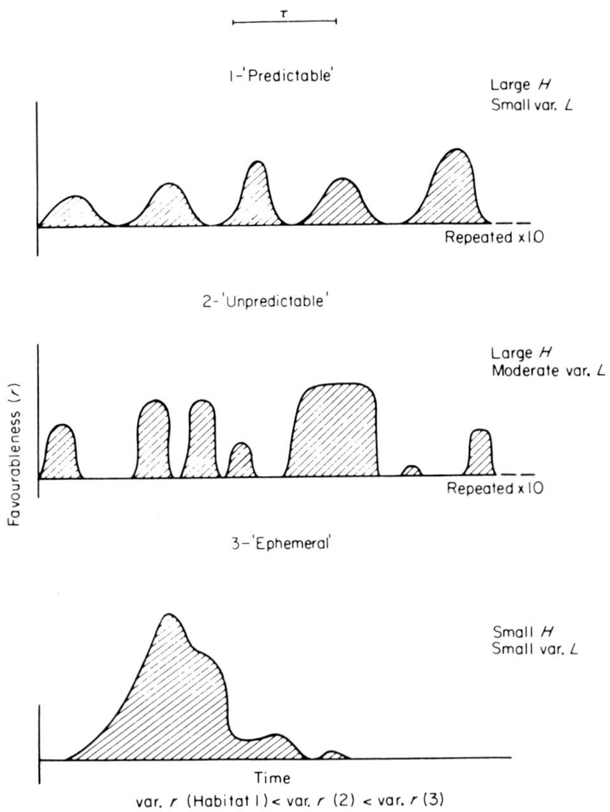

Fig. 3. Timing characteristics for habitat favourability compared to the viability period (r) for one propagule type. H length of time a location is suitable for colonization; L length of time a location is unsuitable colonization, but allows continued existence of those species already established (After Southwood 1977)

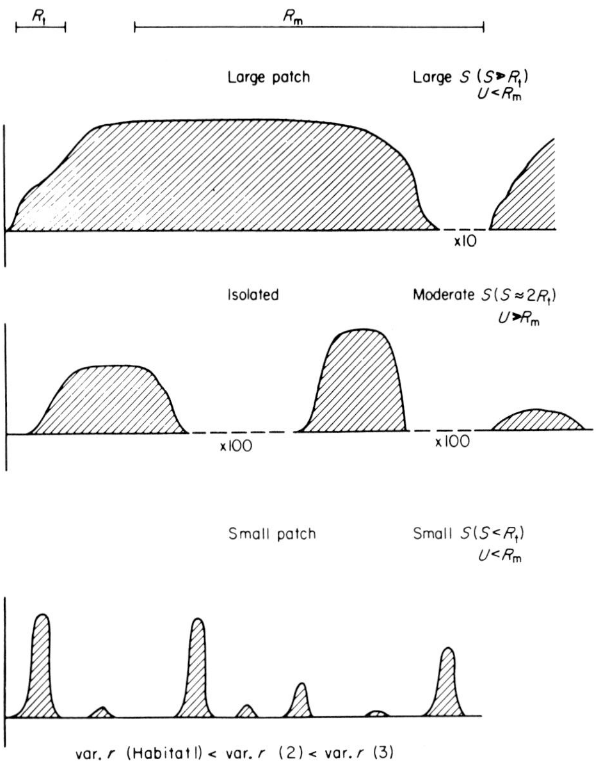

Fig. 4. Heterogeneity of habitats in space. R_t Local dispersal; R_m long-range dispersal; S favourable patch size; U unfavourable interpatch area; variance r = uncertainty matrix (After Southwood 1977)

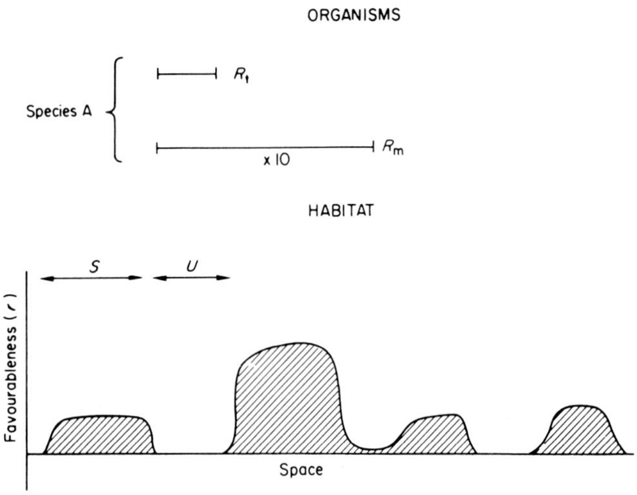

Fig. 5. Space characteristics of favourable habitats. Abbreviations as in Fig. 4 (After Southwood 1977)

able habits (S and U) provides one important determinant of success or failure in colonization.

Finally, what of the qualities of these patches? They can be large, small, isolated etc. (Fig. 5). Thus a large favourable patch (high value of S) may have a size value greater than the R_t distance making it easy for the species to spread locally since there are many favourable habitats within immediate reach. The interpatch distance (U) may also be less than the long-distance viability (R_m) again making spread possible to new habitats outside the immediate patch. Where patches are isolated and/or very small, S can be close to or less than R_t making local spread difficult but, as long as U is less than R_m long-distance spread can still occur. Of course, all these classes represent only conveniently partitioned units in a multi-dimensional continuum, but they do constitute a framework within which comparisons can be made and unifying descriptors sought.

This model framework can be seen as a formalization of the obvious yet it is capable of being more than that. The interactions between propagule biological variables and the time/space continuum of favourable and unfavourable habitats is only likely to be amenable to general investigation and characterization if quantitative measurement can be made. Defining "favourability" is going to prove difficult in the first instance but beyond that estimation of probabilities for events does seem an attainable goal.

Table 2. A checklist of Antarctic habitat availability for colonizer

1.	How far is it to the habitat area from the propagule source area?
2.	How frequent are the habitat patches in the area?
3.	How big are the habitat patches in the area?
4.	How long is each patch available?
5.	How frequently is each patch available?
6.	How frequently is an available patch suitable?
7.	How predictable is the occurrence, duration, availability and suitability of the habitat patches?

Thus, habits are heterogeneous both in time and space – the favourableness matrix. In the Antarctic seasonality is the dominant feature of terrestrial availability. Only in the snow-free summer period is more bare ground available. Water availability is also a key constraint on colonization and establishment, as are the number of hours that the temperature is above the specified baseline level essential for 'food' acquisition by any particular organism. Most of the remaining requirements are species-specific, for example, the roughness of the surface, the extent and availability of key nutrients, etc. Table 2 lists some of the features that could be investigated to provide basic data for the habitat favourability matrix. So, for a hypothetical colonizing moss we would need to sample the air flora to determine the frequency of viable propagules with time in a specific area, examine the area for particular substrates and determine the possibility of a propagule not only landing but remaining on the substrate, and finally determine the proportion of those suitable substrates that combine the required temperature, water and nutrient patterns. This assessment is not, however, presently possible for any Antarctic species.

7 Conclusions

To develop the presently limited understanding of both dispersal and colonization, and to utilize the attributes of Antarctica effectively, it is essential that several approaches are undertaken concurrently. Progress will only be achieved if data from a wide variety of disciplines can be brought together. At least as important is the need for biologists to ask the right questions.

One route, yet to be fully explored, is represented by mathematical analyses to establish correlations. Abbot (1974) applied multiple regression techniques to establish the relationships between geographical parameters and subantarctic biogeography. Although the data base used had serious defects, this approach showed promise. Data of the type in Table 1 provide the essential initial material. Smith (1985) has recently applied similarity coefficients to establish long-term trends in the colonization of volcanic tephra by protozoa. Further developments here might provide some insight into co-dispersal phenomena.

Transport of propagules by wind, over all except the shortest distance, relies on a synchrony of biological and physical factors. Long-distance transport is governed by weather, not climate or prevailing winds which only indicate the most frequent conditions. Thus, it is not sufficient to ask – is the wind generally from the west? Biologists will need to pose more specific questions to meteorologists on the origin of particular air masses, their probable trajectories, the likelihood of active convection within them etc. Direct sampling from aircraft could provide valuable, direct evidence (e.g. Saxena 1982).

Transport of propagules by animals must not only be established by extensive sampling of bird feet and feathers (cf. Schlichting et al. 1978), but must also be related to feeding areas, migration routes and even behaviour patterns, for instance frequency of preening, length of time at sea etc.

For all organisms involved in transport and colonization there is the problem of source area identification as well as the not inconsiderable problem of species identification for many of the propagules. As stated earlier the taxonomy of several important groups is seriously inadequate. Besides making improvements in this field, it will be necessary to prepare identification guides for propagules from most groups. The new tools for genetic analyses may prove to be the best way to establish the relationships between Antarctic immigrants and possible source populations.

Future studies on dispersal and colonization in Antarctic terrestrial and freshwater habitats should focus on three principal themes: (1) dispersal; (2) propagule production and availability; (3) microhabitat description –

1. Aerobiological sampling to establish propagule type, frequency and viability. This needs to be done at a continent-wide level – and here the recent proposals under the BIOTAS programme are an important step forward – as well as at a local level on newly deglaciated areas. Critical assessment under Antarctic conditions of the range of available techniques is an essential first step. Standardization of sampling, synthesis of data and reliability of identification are all crucial.

2. Characterization of the factors controlling propagule availability and viability for at least some species in each of the major groups.

3. Experimental investigation of microhabitats to provide general descriptors applicable to field obser-

vations. Establishing these environmental features will require innovative instrumentation.

Cameron et al. (1977) demonstrated that human activity has produced a significant range of microbiological introductions to Victoria Land. Since then human presence has greatly increased in many areas of Antarctica. It is still possible to make comparisons between "pristine" areas and those under various degrees of human use to assess the scale of human impact in terms of alien introductions and the likelihood of establishment of these new species. Here, studies of colonization become directly relevant to the wider concerns of environmental impact and conservation.

Finally, the general relationships between sources and sinks, and success or failure to colonize can best be established in a model. The calculation of the probability of successful colonization for a given microhabitat must be the long-term aim since this would be based on a secure understanding of the processes involved, and would make a significant contribution to general ecological theory.

Acknowledgements. I am grateful to Dr. Paul Broady for his detailed comments on this work and to my colleagues of the British Antarctic Survey for much useful discussion.

References

Abbot I (1974) Numbers of plant, insects and land bird species on nineteen islands in the southern hemisphere. Biol J Linn Soc 6:143–152

Abyzov SS, Bobin NE, Kudryashov BB (1982) Quantitative analysis of microorganisms during the microbiological investigation of Antarctic glaciers. Biol Bull Acad Sci USSR 9:558–564

Barrow CJ, Smith RIL (1983) Palynological studies in South Georgia. II. Two profiles in Sphagnum Valley, Cumberland West Bay. Br Antarct Surv Bull 58:15–42

Block WB (1984) Terrestrial microbiology, invertebrates and ecosystems. In: Laws RM (ed) Antarctic ecology, vol 1, Academic Press, London, pp 163–236

Block W, Burn AJ, Richard KJ (1984) Insect introduction to the maritime Antarctic. Biol J Linn Soc 23 (1):33–39

Broady PA (1979) Wind dispersal of terrestrial algae at Signy Island, South Orkney Islands. Br Antarct Surv Bull 48:99–102

Broady PA (1984) Taxonomic and ecological investigations of algae on steam-warmed soil on Mt Erebus, Ross Island, Antarctica. Phycologia 23:257–271

Broady PA, Given D, Greenfield L, Thompson K (1987) The biota and environment of fumaroles on Mt Melbourne, northern Victoria Land. Pol Biol 7:97–113

Brodo IM (1973) Substrate ecology. In: Ahmadjian V, Hale ME (eds) The lichens, Academic Press, London, pp 401–441

Cameron RE, Honour RC, Morelli FA (1977) Environmental impact studies of Antarctic sites. In: Llano GA (ed) Adaptations within antarctic ecosystems. Proc. Third SCAR Symp Antarct Biol Smithsonian Institution, Washington DC, pp 1157–1176

Cameron RE, Morelli FA (1974) Viable micro-organisms from ancient Ross Island and Taylor Valley drill cores. Antarct J US 9:113–115

Churchill DM (1973) The ecological significance of tropical mangroves in the early Tertiary floral of southern Australia. Spec Publ Geol Soc Aust 4:79–86

Clagg HB (1966) Trapping of air-borne insects in the Atlantic-Antarctic area. Pac Insects 8:455–466

Collins NJ (1969) The effects of volcanic activity on the vegetation of Deception Island. Br Antarct Surv Bull 21:79–94

Edwards JA (1974) Studies in *Colobanthus quitensis* (Kunth) Bartl. and *Deschampsia antarctica* Desv.: VI Reproductive performance on Signy Island. Br Antarct Surv Bull 39:67–86

Edwards JA (1980) Experimental introduction of vascular plants from South Georgia to the maritime Antarctic. Br Antarct Surv Bull 49:73–80

Edwards JA, Greene DM (1973) The survival of Falkland Islands transplants at South Georgia and Signy Island, South Orkney Islands. Br Antarct Surv Bull 33, 34:33–34

Friedmann EI (1982) Endolithic micro-organisms in the Antarctic cold desert. Science 215:1045–1053

Heal OW (1967) Discussion on the preceding papers. In: Smith JE (organiser) A discussion on the terrestrial Antarctic ecosystem. Phil Trans R Soc Ser B 252:212

Heywood RB (1984) Antarctic inland waters. In: Laws RM (ed) Antarctic ecology, vol 1. Academic Press, London, pp 279–344

Holdgate MW (1964) An experimental introduction of plants to the Antarctic. Br Antarct Surv Bull 3:13–16

Hooker JD (1879) Observations on the botany of Kerguelen Island. Phil Trans R Soc 168:9–16

Hooker TN (1980) Factors affecting the growth of Antarctic crustose lichens. Br Antarct Surv Bull 50:1–19

Kappen L, Straka H (1988) Pollen and spores transport into the Antarctic. Pol Biol 8:173–180

Kok OB (1975) Verspreiding van veervasgehegte sade duer voels. Ostrich 46:261–263

Lewis Smith RI (1982) Farthest south and highest occurrences of vascular plants in the Antarctic. Pol Rec 21:170–173

Lewis Smith RI (1984a) Terrestrial plant biology of the sub-Antarctic and Antarctic. In: Laws RM (ed) Antarctic ecology, vol 2. Academic Press, London, pp 61–162

Lewis Smith RI (1984b) Colonization and recovery by cryptogams following recent volcanic activity on Deception Island, South Shetland Islands. Br Antarct Surv Bull 62:25–51

Lewis Smith RI, Coupar AM (1986) The colonization potential of bryophyte propagules in Antarctic fellfield soils. Com Natl Fr Rech Antarct 58:189–204

Lewis Smith RI, Poncet (1987) *Deschampsia antarctica* and *Colobanthus quitensis* in the Terra Firma Islands. Br Antarct Surv Bull 74:31–35

Lindsay DC (1973) Probable introductions of lichens to South Georgia. Br Antarct Surv Bull 33834:169–172

Lindsay DC (1974) The macrolichens of South Georgia. Br Antarct Surv Sci Rep 89:1–91

Lindsay DC (1974) Lichens of cold deserts. In: Seaward MRD (ed) Lichen ecology. Academic Press, London, pp 183–209

Longton RE (1966) Alien vascular plants on Deception Island, South Shetland Islands. Br Antarct Surv Bull 9:55–60

Longton RE (1988) Biology of polar bryophytes and lichens. Cambridge University Press, Cambridge, 391 pp

Ott S (1987a) Sexual reproduction and developmental adaptations in *Xanthoria parietina*. Nord J Bot 7:219–228

Ott S (1987b) Differences in the developmental rate of lichens. Ann Bot Fenn 24:385–393

Ott S (1987c) The juvenile development of lichen thalli from vegetative diaspores. Symbiosis 3:57–74

Phillpot HR (1985) Physical geography – climate. In: Bonner WN, Walton DWH (eds) Key environments – Antarctica. Pergamon Press, London, pp 23–38
Ram M, Gayley RI (1988) Insoluble particles in Antarctic ice: background aerosol size distribution and diatom concentration. J Geophys Res 93 (D7):8378–8392
Rudolph ED (1970) Local dissemination of plant propagules in Antarctica. In: Holdgate MW (ed) Antarctic ecology, vol 2. Academic Press, London, pp 812–817
Rudolph E, Benninghoff WS (1977) Competitive and adaptative responses of invading versus indigenous biotas in Antarctica – a plea for organized monitoring. In: Llano GA (ed) Adaptations within antarctic ecosystems. Proc Third SCAR Symp Antarct Biol, Smithsonian Institution, Washington DC, pp 1211–1225
Salas MR (1983) Long-distance pollen transport over the southern Tasman Sea: evidence from Macquarie Island. N Z J Bot 21:285–292
Saxena VK (1982) Biogenic nuclei involvement in clouds over the Ross Ice Shelf. Antarct J U S 17:212–214
Schlichting JR, Speziale BJ, Zink RM (1978) Dispersal of algae and protozoa by Antarctic flying birds. Antarct J US 13 (4):147–149
Schuster G, Ott S, Jahns HM (1985) Artificial cultures of lichens in the natural environment. Lichenologist 17 (3):247–253
Scott L, van Zinderen Bakker EM (1985) Exotic pollen and long distance dispersal at a subantarctic island. Grana 24:45–54
Smith HG (1985) The colonization of volcanic tephra on Deception Island by Protozoa: long-term trends. Br Antarct Surv Bull 66:19–33
Southwood TRE (1977) Habitat, the templet for ecological strategies? J Anim Ecol 46:337–365
Tallowin JRB (1977) Vegetative proliferation in *Festuca contracta* T Kirk on South Georgia. Br Antarct Surv Bull 45:13–17
Thompson LG (1977) Variations in microparticle concentration, size distribution and elemental composition found in Camp Century, Greenland and Byrd Station, Antarctica: deep ice cores. In: Isotopes and impurities in snow and ice: Proc Grenoble Symp August–September 1975. IAHS 118:351–364
Vincent WF (1988) Microbial ecosystems of Antarctica. Cambridge University Press, Cambridge, 304 pp
Walton DWH (1975) European weeds and other alien species in the subantarctic. Weed Res 15:271–282
Walton DWH (1985) The sub-Antarctic islands. In: Bonner WN, Walton DWH (eds) Key environments – Antarctica. Pergamon Press, Oxford, pp 293–317
Wolfenbarger DO (1946) Dispersion of small organisms. Distance dispersion rates of bacteria, spores, seed, pollen, and insects: incidence rates of diseases and injuries. Am Midl Nat 35:1–12
Wynn-Williams DD (1990) Microbial colonization processes in Antarctic fellfield soils. Eleventh Symp Polar Biology, Dec 1988, Tokyo (in press)

Changes in Vegetation on Heard Island 1947–1987

J. J. Scott [1]

Summary. The Heard and McDonald Islands are the only subantarctic island group which appears to be free of human-introduced animals and plants. Vegetation changes in its species-poor flora are therefore likely to be due to natural factors. Significant glacial recession has exposed new areas for colonisation over the past 40 years. Analysis of vegetation transect data from seven glacier retreat zones and adjacent areas indicates four main patterns of primary colonisation, with moisture availability and effects of animal disturbance being major differentiating environmental factors. Vegetation colonisation can be rapid under the most favourable environmental conditions, for example abundant surface drainage from springs or snow melt with or without effects of nutrient enrichment by animals. It can be expected that with continuing climatic amelioration and glacial recession, the size of vegetated areas will expand. Changes in distribution of some vascular plant species around the island have been noted and tentatively linked with climatic warming, and additional changes are predicted. Future effects of changing trends in population numbers of animals utilising and interacting with terrestrial vegetation communities are uncertain. Further changes can now be monitored from recently established reference points.

1 Introduction

The subantarctic Heard and MacDonald Islands group is situated at 53° 05′ S, 73° 30′ E, and is an Australian External Territory. The group is well south of the Antarctic Convergence, and the closest land is Isles Kerguelen 520 km to the northwest (Veenstra and Manning 1982). Heard Island is composed of volcanic rocks with some marine limestones, and is 70–80% ice-covered. It is dominated topographically by the active volcano Big Ben, altitude 2745 m (Fig. 1).

[1] Dept of Geography & Environmental Studies, University of Tasmania, Sandy Bay 7005, Australia.

The ice-free coastal areas are separated by glaciers descending from Big Ben. The glaciers are believed to be shallow and fast-moving and hence sensitive indicators of climatic change (Allison and Keage 1986). Widespread glacier retreat has been documented on Heard Island since 1947 and is discussed by Allison and Keage (1986).

Vegetation is best developed on the low-lying ice-free areas along the coast (Fig. 1) which have not been subject to recent glacial activity, for example because of topographic barriers inland. These small, scattered areas are subject to intensive use by seals and seabirds during the summer breeding season, and animal disturbance appears to play an integral role in vegetation community dynamics (J. Scott unpubl. data). There are only 11 species of vascular plants reported for the island (Table 1), as a result of isolation and limited land area available for colonisation. Three of these were recorded for the first time in 1986 (Scott 1989). Current work on the moss flora reports 42 taxa (Bergstrom and Seppelt in press). Studies on the hepatic and lichen floras are not yet completed, although between 20–30 hepatic species, and 40–50 lichen species, are likely to be reported (R. Seppelt pers. commun.). Terrestrial and freshwater algae have not been studied to date.

Six major vascular plant communities have been described in general terms: maritime or tussock grassland, herbfield, meadow, pool complex, cushion carpet and fellfield (Hughes 1987; Seppelt and Hughes

Table 1. Heard Island vascular plants

Acaena magellanica (Lam.) Vahl.[a]
Azorella selago Hook f.
Callitriche antarctica Engelm ex Hegelm
Colobanthus kerguelensis Hook f.
Deschampsia antarctica Desv.[a]
Montia fontana L.
Poa annua L.[a]
Poa cookii Hook f.
Poa kerguelensis Steud.
Pringlea antiscorbutica R. Br. ex Hook f.
Ranunculus biternatus Sm.[a]

[a] Recorded from southeast end of island only.

Antarctic Ecosystems. Ecological Change and Conservation
(ed. by K. R. Kerry and G. Hempel)
© Springer-Verlag Berlin Heidelberg 1990

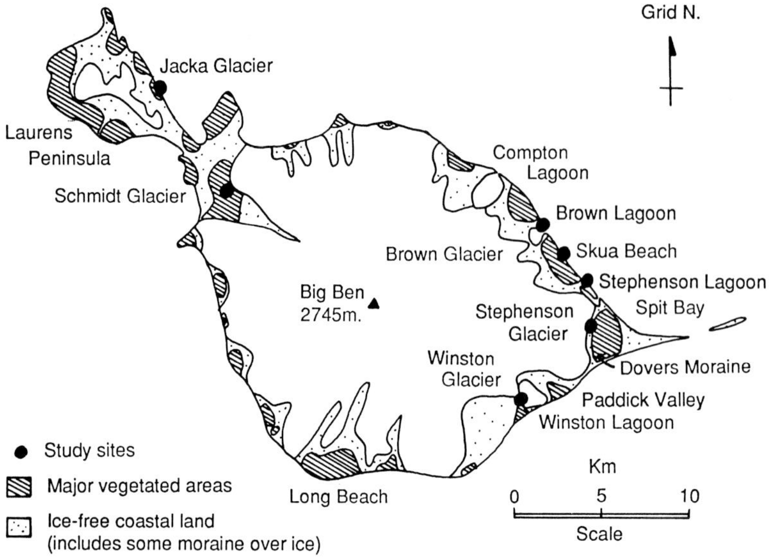

Fig. 1. Heard Island: ice-free coastal ares, vegetated areas and location of study sites

1988). A number of additional communities are in the process of being described (J. Scott unpubl. data). Notable features of the vegetation include the large range of habitats occupied by a small number of widely distributed vascular species, and the considerable extent of variation in morphology and habit of these species under different conditions. As a result, the major recognised vegetation community types are not always clear-cut either in the field or during analysis. Another consideration, especially when interpreting early succession data, is the fact that the majority of the vascular species which have managed to arrive and spread around the island are efficient colonisers almost by definition.

Although Heard Island was occupied by sealing gangs for some decades after 1855, and by the Australian National Antarctic Research Expeditions (ANARE) between 1947 and 1955, the island group appears to be free from human-introduced flora and fauna, which makes it unique in the subantarctic region. In recognition of its biological value, a detailed management plan is being prepared by the Australian Antarctic Division as part of an environmental protection and management ordinance under the Heard and MacDonald Islands Act 1953. It is also listed on the Register of the National Estate in Australia.

The near-pristine status of Heard Island makes it an unusually valuable location for studying subantarctic vegetation dynamics, as any changes in the vegetation can be attributed to natural factors. This is in contrast to the majority of other subantarctic islands, where human-introduced biota have often had marked effects on the vegetation (e.g. see Brothers and Copson 1988 and Scott 1988, Macquarie Island; Leader-Williams et al. 1987, South Georgia; Lesel and Derenne 1975, Iles Kerguelen). Such effects may mask or obscure responses of the vegetation to regional environmental changes.

Over the past 35 years Heard Island has only been visited irregularly, although ANARE scientific expeditions worked on the island for the three summers between 1985 and 1987–88.

1.1 Increase in Area of Ice-Free Land and Implications for Vegetation

The changes in Heard Island glaciers since 1947 are linked with a general increase in surface temperatures for Southern Ocean islands over the same period, reported in Jacka et al. (1984), and are consistent with the general warming trend documented for the Southern Hemisphere since 1950 (Karoly 1988). The mean annual temperature for Isles Kerguelen has increased by 1.5 °C over the past 20 years. Incomplete data for Heard Island indicate a similar increase, and the greatest retreat in its glaciers appears to have occurred since 1970 in response to this (Allison and Keage 1986). The most extensive retreat has been on the Laurens Peninsula in the northwest, and on the southeast half of the island in the lee of Big Ben. Differences in retreat patterns are related primarily to topography and the resulting local variations in climate (Allison and Keage 1986).

The increase in area of ice-free land provides new substrates for primary colonisation by vegetation. The associated climatic amelioration may have contributed to the establishment of at least one additional vascular species and increased the range of several species

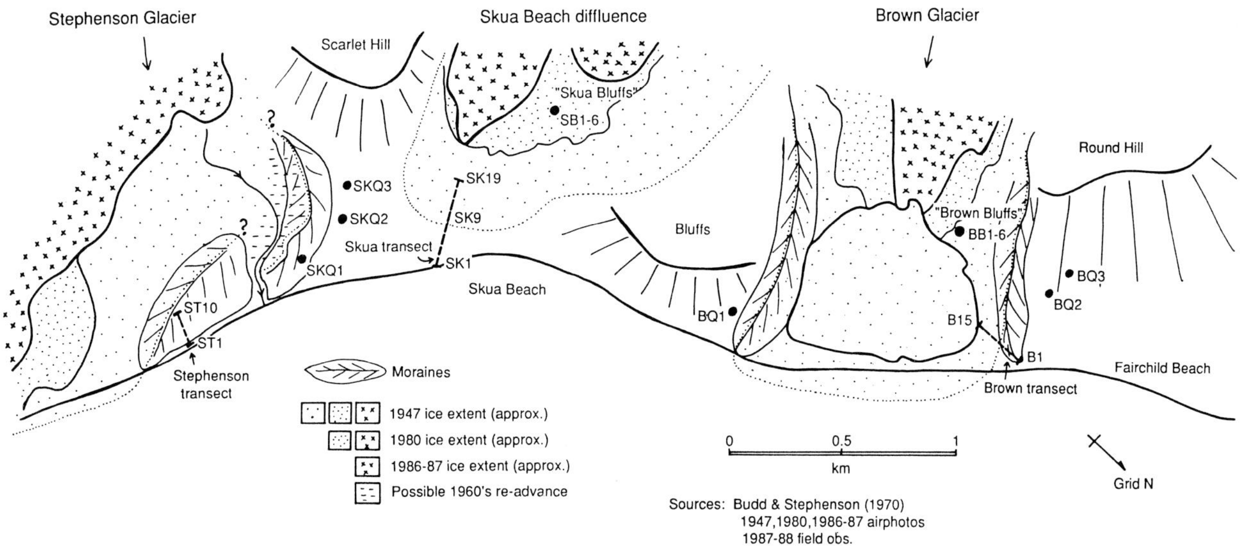

Fig. 2. Sketch map showing ice recession at the Stephenson Glacier, Skua Beach and Brown Glacier, and location of transects and quadrats

already present. Aspects of these changes were investigated during the 1986–87 and 1987–88 austral summers, and are discussed below.

2 Methods of Study of Glacier Retreat Areas

Black and white oblique aerial photographs obtained by an ANARE expedition in December 1947 provide the first reliable documentation of glacier extent on Heard Island. Additional aerial coverage exists for February 1980 (Australian Department of National Mapping, now AUSLIG) and for December–January 1986–87 (Australian Antarctic Division). For the 40-year period 1947–1987 the approximate increase in the area of ice-free land can therefore be determined where reasonable photographic coverage exists. Five such locations were investigated at the southeast end of the island at Skua Beach, Brown Glacier, Stephenson Glacier (Dovers Moraine and Stephenson Moraine) and Winston Glacier, and two sites at the northwest end of the island at the Jacka and Schmidt Glaciers (Fig. 1). Despite inadequate aerial coverage of the latter two glacier fronts, past documentation including anecdotal observations allowed an approximation of the recent ice-free extent to be estimated.

A transect line was positioned in the vicinity of each receding glacier front so as to include ice-free land exposed since 1947. Locations of transects in relation to deduced ice retreat lines are shown in Figs. 2–6, and their characteristics are summarised in Table 2. Actual positioning of the transects was done after

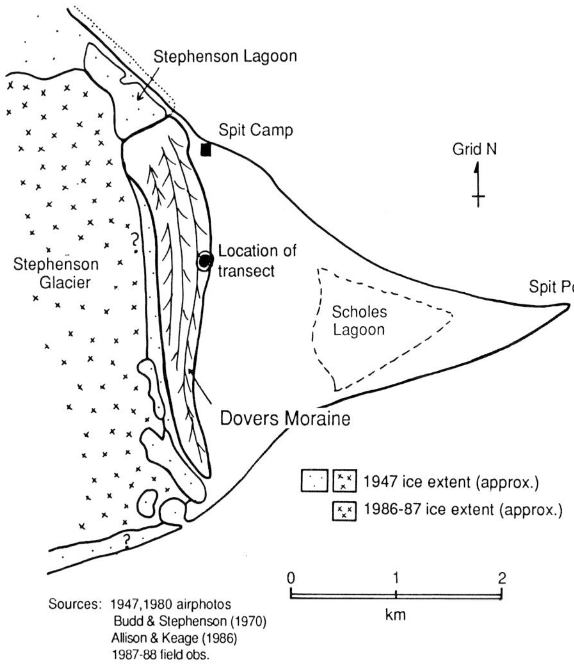

Fig. 3. Sketch map showing ice recession at the Stephenson Glacier (Dovers Moraine), and location of Dovers transect

careful study of the available aerial photos and documentation, and field reconnaissance. The Brown, Jacka and Schmidt transects were the hardest to position, as photographic documentation of the Jacka and Schmidt glacier retreats is sparse, and on the 1947 aerial photos of the Brown Glacier it was difficult to separate the lateral edges of the glacier from what appeared to be late-lying snowdrifts. However, in each

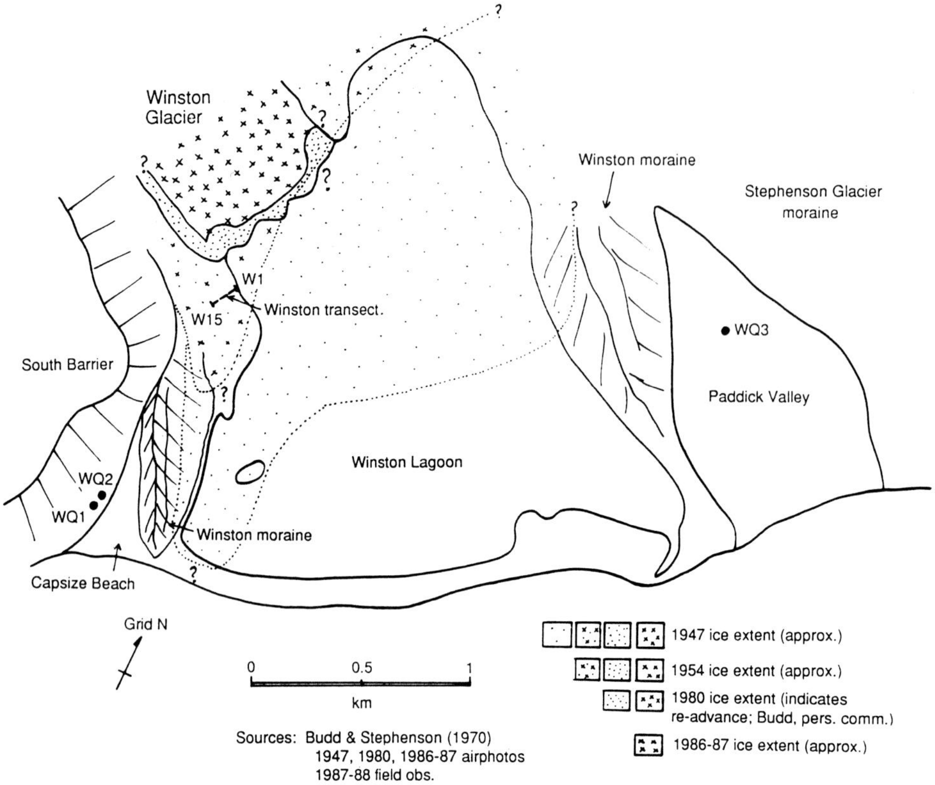

Fig. 4. Sketch map showing ice recession at Winston Glacier, and location of transect and quadrats

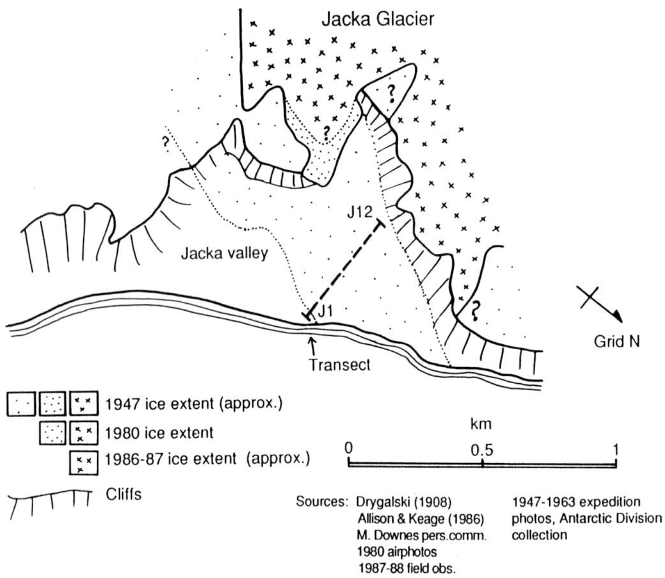

Fig. 5. Sketch map showing ice recession at the Jacka Glacier, and location of transect

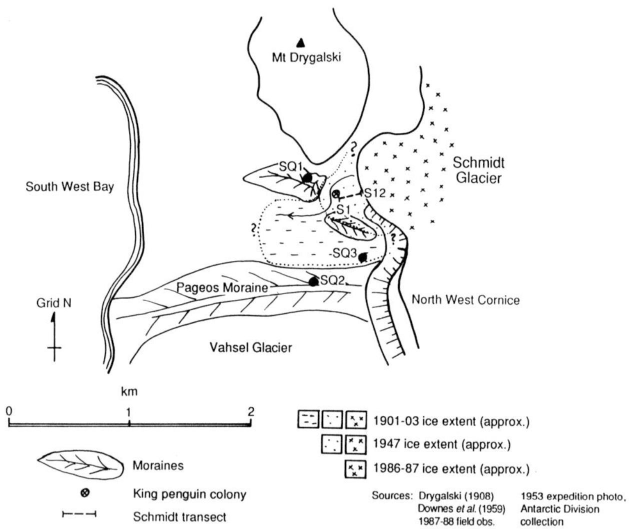

Fig. 6. Sketch map showing ice recession at the Schmidt Glacier, and location of transect and quadrats

case the inland section of the transect (closest to the current ice edge) would definitely have been ice-covered in 1947. Where possible, transect lines also included vegetation on land known to be ice-free prior to 1947 (Skua, Brown and Jacka transects). The Schmidt and Winston transects are wholly located on recently exposed land ("post-1947"), while the Dovers and Stephenson transects are wholly located on land exposed "pre-1947".

Average transect length was 238 m, ranging from 160 to 360 m, apart from the Dovers transect which was 16 m in length (Table 2). Vegetation along each transect was described using random stratified sampling (three quadrats randomly placed along the tran-

Table 2. Transect characteristics

Location	Length (m)	No. quadrats		Altitude (approx.) (m.a.s.l.)	Aspect true N
		"pre-1947"	"post-1947"		
Skua Beach	310	9	10?	2–20	NE-SE
Brown Glacier	235	8	7?	2–20	SE-SW
Stephenson Moraine	160	10	–	2–30	NE
Dovers Moraine	16	7	–	10–20	SE
Winston Glacier	160	–	15	2–10	NE
Jacka Glacier	360	2	20?	2–20	SE
Schmidt Glacier	200	–	12	20–30	NW

sect every 50 m, with seven randomly placed quadrats along the 16-m Dovers transect).

For comparison with the transect data, a number of additional sites were sampled both at ice-free locations known to have been glaciated within the last 20 years, and on "old" areas known to have been ice-free for a considerable period of time. These sites were located in the vicinity of the Skua, Brown, Stephenson, Winston and Schmidt Glaciers, and are shown on Figs. 2, 4 and 6. The recent sites comprise six quadrats in an area of flush vegetation on "Skua Bluffs", which aerial photos indicate was covered by ice in 1980; and six quadrats in similar vegetation on "Brown Bluffs", estimated to have been ice-covered within the last 20 years (Fig. 2). Twelve "old" sites are described, by 12 quadrats. These quadrats are denoted by the letter "Q" in their identifier code (Figs. 2, 4 and 6). The quadrats were located on a random basis within a range of recognised vegetation community types.

One square meter quadrats were used and the percentage cover abundance of each plant species was visually estimated using a standard five-point scale (e.g. see Kershaw 1975). Identification of a number of the mosses was carried out to generic level only. Description of the Heard Island bryophyte flora is currently in progress (Bergstrom and Seppelt in press; Seppelt pers. commun.). In the present study *Ditrichum* spp. includes *D. conicum* and *D. australe*, although not *D. immersum*. *Dicranoweisia* spp. includes *D. brevipes, D. breviseta* and possibly *D. antarctica*. Species of *Grimmia* and *Schistidium* are grouped together, and are referred to as *Grimmia* spp. in the discussion. Both the lichen and hepatic floras are currently under study (R. Seppelt pers. commun.). Since lichens formed a very minor component of the quadrat data they were grouped in the analysis. Hepatics were uncommon in the quadrat data, and with the exception of *Marchantia berteroana* and *Fossombronia australis* they were grouped in the "other bryophytes" category if they occurred.

Nine additional environmental variables were measured for each quadrat, including intensity of use by seals and penguins. This was subjectively estimated using a four-point scale which relied on visual estimation of relative disturbance such as extent of vegetation trampling, manuring and deposition of skin, hair and feathers. The scales of measurement for the different variables are presented with Tables.

The vegetation data, totalling 121 quadrats, were classified using the polythetic divisive computer program TWINSPAN (Hill 1979). The analysis produced an ordered two-way table based on 22 divisions of the data. The nine environmental variables were added after the analysis, plus the age category for each quadrat (i.e. "pre-1947" or "post-1947"). The resulting table was examined, and after a minimal amount of hand-sorting (seven quadrats) a number of groups at the third, fourth and fifth levels of division were chosen for interpretation (Tables 3 and 4, Fig. 7). The levels of significance of environmental differences between groups at different levels of division were calculated using chi-square contingency tables, and are shown in Fig. 7.

3 Transect Site Descriptions

Study site locations are shown on Fig. 1. The transect sites are situated at low altitudes (approximately 0–30 m) close to the coast, apart from the Schmidt and Winston Glacier sites which are around 1.5 km inland. Sampling of all transects began at the coastal end.

The Skua and Brown transects (Fig. 2) initially traverse areas which were ice-free before 1947. The vegetation of these areas has the characteristics of recognised community types. The Brown transect (235 m length) passes through *Poa cookii*-dominated maritime grassland (quadrats B3–B7), while the Skua transect (310 m length) passes through herbfield (SK1) then pool complex vegetation (SK2–SK9). Both transects are subject to disturbance by moulting elephant seals and fur seals in these sections. The second half of both transects covers land which appears

Table 3. Vegetation on "pre-1947" sites

Quadrat group[a]	1	2	3[c]	4[c]	5[c]	6
Quadrat identifier[b]	S S S S S S S S S S B S W K K K K K T T T T K Q C Q 1 1 1 1 1 2 3 5 1 Q 3 Q 3 1 2 3 4 6 0 3 1	J S S S B B B D D D D S S 3 K K K 1 1 1 3 4 5 6 T T 1 1 1 0 1 3 4 9 0 5 7	* * B S B S S S 1 K Q C T T Q 2 Q 7 8 2 2	* S W B K Q Q 1 1 1	* * B B B B B S S J J 3 4 5 6 7 K C 2 1 Q 1 1	S S S S S S S S S B C K K K K K K K K 2 Q 2 3 4 5 6 7 8 9 3
Species						
Poa kerguelensis	1 1 1 1 – 2 1 2 2 1 2 2 2	1 2 1 1 2 2 3 1 1 2 4 1 1	– – – – 2 2	– – –	– – – – – – 1 – –	– – – – – – – – – –
Polytrichum alpinum	4 3 3 2 2 – – – 1 3 1 1 2	1 2 2 2 2 2 2 2 – 2 1 1 1	– 1 – – 1 1	– – –	– – – – – – – – –	– – – – – – – – – –
Ditrichum spp.	– – – – – – – – – – – – –	– 1 1 2 2 2 2 3 – 2 3 – 1	– – – – – –	– – –	– – – – – – – – –	– – – – – – – – – –
Grimmia spp.	1 2 2 1 1 – – – – – 1 – –	1 1 1 1 – 1 1 2 1 1 1 – –	– 1 – 1 – –	– – –	– – – – – – – – –	– – – – – – – – – 1
Marchantia berteroana	– – – – – – – – – 2 3 – 1	– – – – – – – – – – – – –	– – – – – –	– – –	– – – – – – – – –	– – – – – – – – – –
Azorella selago	3 3 2 2 2 2 2 2 – 3 4 3 3	2 4 2 2 3 3 3 1 3 1 3 1 2	5 5 6 5 5 5	1 3 –	1 3 1 – 2 – 1 – –	– 2 2 3 1 4 5 – – 3
Fossombronia australis	– – – – – – – – – – 1 3 2	– – – – – – 1 – – – – – –	– – 1 1 2 1	– – –	– – – – – – – – –	– – – – – 1 1 – – –
Lichens	– – – – – – – – – – – 2 –	2 – 1 – – – – – – – – – –	– – – 1 – –	– – –	– – – – – – – – –	– – 1 – – – – – – –
Colobanthus kerguelensis	1 1 – 1 – 1 1 1 1 2 1 1 1	1 1 – – – 1 1 1 – 1 1 1 1	– – 1 – 2 2	– – –	– – – – – 1 1 – –	– 1 – 2 – 1 1 – – 1
Ceratodon purpureus	– – – – – – – – – – – – –	– – – – – – – – – – – – –	– – – – – –	– – –	– – – – – – – – –	– – – – – 1 1 – – 1
Dicranoweisia spp.	– – – – – – – – – – – – –	1 – – – 1 1 1 – – – – – –	– – – – – –	– – –	– – – – – – – – –	– – – – – – – – – –
Other bryophytes	– – – – – – – – – – 1 – –	1 – – – – – 1 1 – 1 – – 2	– – 1 1 – –	– – 1	– – – – – – – – –	– – – – – – – – – –
Brachythecium austro-salebrosum	– – – – – – – – – – – – –	– – – – – – – – – – – – –	– – – 1 – –	– – –	– – – – – – – – –	– – – – – – – – – –
Bryum pseudotriquetrum	– – – – – – – – – – – – –	– – – – – – – – – – – – –	– – – – – –	– – –	– – – – – – – – –	– – – – – – – – – –
Rhacomitrium crispulum	– – – – – – – – – – – – –	– – – – – – – – – – – – –	– – – – – –	– – –	– – – – – – – – –	– – – – – – – – – –
Pringlea antiscorbutica	– – – – – – – – – – – – –	1 – – – 1 1 1 1 1 1 1 1 –	1 – 1 – 3 2	5 4 5	1 – – – – 1 – – 2	3 3 – – 3 – – – – –
Bartramia patens	– – – – – – – – – – – – –	– – – – – – – 1 – 1 – – –	– – – – 1 –	– – –	– – – – – – – – –	4 – – – – – – – – –
Brachythecium paradoxum	– – – – – – – – – – – – –	– – – – – – – – – – – – –	– – – – – –	– – 1	– – – – – – – – –	4 – – – – – – – – –
Pohlia wahlenbergii	– – – – – – – – 1 – – – –	– – – – – – – – – – – – –	– – – – 2 –	– – 2	– – – – – – – – –	– – – – – – – – – –
Acaena magellanica	– – – – – – – – – – – – –	– – – – – – – – – – – – –	– 4 – – 1 –	– 4 –	– – – – – 3 – – –	– – – – – – – – – –
Callitriche antarctica	– – – – – – – – – – – – –	– – – – – – – – – – – – –	– – – – – –	– – 1	– – – – – 1 – – –	2 1 3 3 3 4 2 4 4 1
Poa cookii	– – – – – – – – – – – – –	1 – – – – – – – – – – – –	2 – – – 1 –	– 3 4	6 5 5 5 5 5 3 4 6	4 4 5 5 5 5 4 5 5 5
Prasiola crispa	– – – – – – – – – – – – –	– – – – – – – – – – – – –	– – – – – –	– – –	– – – – – – 4 1 –	– – – – – – – – – –
Deschampsia antarctica	– – – – – – – – – – – – –	– – – – – – – – – – – – –	– – – – – –	– – –	– – – – – – – – –	– – – – – – – – – –
Montia fontana	– – – – – – – – – – – – –	– – – – – – – – – – – – –	– – 2 – 2 –	– – 1	– – – – – 2 – – –	– 3 3 3 4 3 2 3 3 –
Other algae	– – – – – – – – – – – – –	– – – – – – – – – – – – –	– – – – – –	– – –	– – – – – – – 3 –	3 – 2 – 1 2 1 1 – –
Environmental variables						
Aspect[d]	2 2 4 2 4 1 1 1 1 1 2 3 4	4 4 4 2 2 2 2 2 2 2 2 1 1	2 1 2 2 1 1	2 2 2	3 2 2 3 2 1 1 1 2	1 2 1 1 2 1 2 2 2 3
Altitude[e]	4 4 4 4 4 2 3 3 5 4 5 5 5	2 4 4 4 2 2 2 4 4 4 4 3 4	4 4 5 5 4 4	2 5 3	3 2 2 2 2 5 5 2 1	4 4 4 4 4 4 4 4 4 4
Slope[f]	1 2 1 1 1 3 4 2 3 1 6 4 3	2 2 1 1 1 1 1 5 6 6 5 2 3	1 3 6 3 2 6	6 6 6	4 2 2 2 2 5 2 2 3	1 3 2 1 2 1 1 1 1 3
Vegetation ht.[g]	1 1 1 1 1 1 1 1 1 1 1 1 1	1 1 1 1 1 2 1 1 1 1 1 1 1	2 1 3 1 4 2	6 6 6	4 4 4 3 3 5 3 3 4	3 6 4 3 6 3 2 3 3 6
Soil depth[h]	2 2 3 3 3 2 1 2 1 1 2 1 3	2 2 1 3 1 2 1 2 2 2 3 2 1	3 6 6 6 2 2	4 4 3	1 2 2 1 1 6 3 1 2	6 5 4 5 6 5 6 6 4 4
Fauna use[i]	0 0 0 0 0 0 0 0 0 0 0 0 0	1 0 0 0 0 1 1 0 0 0 0 0 0	1 0 0 0 1 0	0 1 1	1 3 3 3 3 2 3 3 3	0 1 1 1 1 1 1 1 1 1
Bare ground	2 1 3 1 2 3 3 3 3 2 3 3 3	2 3 1 3 2 2 1 2 4 3 3 3 4	2 2 2 4 3 0	4 0 0	3 3 4 4 3 3 4 4 3	0 4 2 3 2 2 3 1 0 1
Rocks	5 6 6 6 6 6 6 6 6 6 5 4 5	6 5 6 6 6 5 5 6 5 6 4 6 5	4 0 1 0 4 0	0 0 0	1 3 3 3 3 0 3 3 1	0 3 0 0 0 0 0 0 0 4
Open water	0 0 0 0 0 0 0 0 0 0 0 0 0	0 0 0 0 0 0 0 0 0 0 0 0 0	0 0 0 0 0 0	0 0 0	0 0 0 0 0 0 0 0 0	0 0 0 0 1 2 3 3 3 0
Age of site[j]	1 1 1 1 1 0 0 0 0 0 0 0 0	0 1 1 1 1 1 1 0 0 0 0 0 0	0 0 0 0 0 0	0 0 0	0 0 0 0 0 0 1 0 0	0 0 0 0 0 0 0 0 0 0
No. of species	10	13	15	9	10	13

[a] Plant cover-abundance: **1** isolated, rare; **2** < 5%; **3** 5–20%; **4** 20–50%; **5** 50–75%; **6** > 75%; [b] Quadrats: **SK** Skua; **B** Brown; **ST** Stephenson; **D** Dovers; **W** Winston; **J** Jacka; **S** Schmidt; **SB** Skua Bluffs; **BB** Brown Bluffs; Quadrats with "Q" – see Figs. 2–6 for locations. [c] *Hand-sorted quadrats [d] Aspect (deg.): **1** 0–90; **2** 91–180; **3** 181–270; **4** 271–360; [e] Altitude (m): **1** 0–2; **2** 3–5; **3** 6–10; **4** 11–20; **5** > 20; [f] Slope (deg.): **1** 0–5; **2** 6–10; **3** 11–15; **4** 16–20; **5** 21–25; **6** > 25; [g] Vegetation height (cm): **1** 0–5; **2** 6–10; **3** 11–15; **4** 16–20; **5** 21–25; **6** > 25; [h] Soil depth (cm): **1** 0–10; **2** 11–20; **3** 21–30; **4** 31–40; **5** 41–50; **6** > 50; [i] Intensity of fauna use: **0** no signs; **1** slight; **2** moderate; **3** heavy; [j] Age: **O** Pre-1947; **I** Post-1947.

Table 4. Vegetation colonisation patterns on "post-1947" sites (for explanations to footnotes, see Table 3)

Quadrat group[a]	7	8	9	10[c]	11[c]	12[c]
				* (last column)		* (last column)
Quadrat identifier[b]	W S S B B B B D D S S W B	J J D	J J J J S J J J J J J J J J J	S S S S S S S S S B S S B B B S W S	W W W W W W W	S W W W B S S W W B W
	1 K K 9 1 1 1 2 7 T T 1 B	1 8 1	9 7 6 4 C 1 1 1 1 5 2 2 2 1 1	C C C C C B C C C B C C 8 B B B Q B	1 2 5 6 3 4 7	B 1 1 1 B B B 8 1 B 9
	5 1 1 2 4 5 1 6 0 3	7	1 8 6 5 4 2 1 0 9 3	1 1 9 8 5 6 7 6 3 2 4 2 4 5 2 2 1		5 1 3 4 6 3 4 2 1
	8 9		2	1 0		
Species						
Poa kerguelensis	1 1 1 2 3 2 1 1 3 1 2 2 2	– – –	– – 1 – – – – – – – – – – – –	1 – – – – – 2 1 1 1 1 1 – 1 1 1 – 1	– – – – – – –	– – – – 1 – – – – – –
Polytrichum alpinum	2 1 – – – – – – 2 – 1 – 1	– – –	– – – – 1 – – – – – – – – – –	– – – – – – – 1 1 – – 1 – 1 – – – 1	– – – – – – –	– – – – – – – – – – 1
Ditrichum spp.	2 1 1 – 1 – – – 3 – – 1 2	– – –	– 1 – 1 – – – – – – – – – – –	1 1 1 1 1 1 1 1 – 1 – – – – – – – –	– – – – – – –	– 1 – – – – – 1 1 1 3
Grimmia spp.	– – 1 – – – – – – – – – –	– – –	– – – – – 1 – 1 – – – – – – –	– – – – – – – 1 – 1 – – – – – – 2 –	– – – – – – –	– – – – – – – – – 1 –
Marchantia berteroana	– – – – – – – – – – – 2 4	– – –	– – – – – – – – – – – – – – –	– – – – – – – – – – – – – 2 2 – – –	– – – – – – –	– – – – – – – – – – –
Azorella selago	2 2 2 2 2 – – 1 2 – 2 2 2	– – –	1 – 2 2 1 – – – 1 1 – – 1 – –	1 1 1 1 2 – 2 2 2 2 2 2 4 3 1 1 5 1	– – – – – – –	– – – – 1 – – 3 – 2 3
Fossombronia australis	1 – – – – – – – – – – – 1	– – –	– – 1 – – – – – – – – – – – –	– – – – – – – – – – – – – 1 – – – –	– – – – – – –	– – – – – – – 1 – – 2
Lichens	– – – – – – – – – – – – –	– – –	– – – 2 – 1 – – – – – – – – –	– – – – – – – – – – – – – – – – – –	– – – – – – –	– – – – – – – – – – –
Colobanthus kerguelensis	1 – – 1 1 – – – 1 1 1 2 2	– – –	1 2 1 – – – 1 2 1 1 – – – – –	– 1 – 1 1 – 1 1 1 1 1 2 – 1 1 1 – –	– – – – – – –	– – – – 1 – 1 1 2 1 1
Ceratodon purpureus	– 1 1 – – – – – – – 1 – 4	– – –	– – 2 2 1 – – – – – – – – – –	– – – – – – – – – – 1 – – 1 – 1 1 1	– – – – – – –	– – – – – – – – – – –
Dicranoweisia spp.	– – – – – – – – – – – – –	– – –	2 1 2 2 1 1 1 2 2 2 1 1 1 1 1	– 1 1 1 1 – – 1 4 – 2 4 – – – – – –	– – – – – – –	– – – – – – – – – – –
Other bryophytes	2 – 1 1 – – – – 1 – 2 – 1	– – –	1 1 1 1 – 1 – 1 1 – – – 1 – –	– 1 – – – – 1 1 – – – 2 – 1 1 1 1 –	– – – – – – –	– – – – 2 – – – – – 1
Brachythecium austro-salebrosum	– – – – – – – – – – – – –	– – –	– – – – – 1 1 1 – – 1 – 1 – –	1 1 1 1 – – 1 1 – – – – – – – – 1 –	– – – – – – –	– – – – – – – 1 1 – –
Bryum pseudo-triquetrum	– – – – – – – – – – – – 1	– – –	– – – – – – – – – – – – – – –	– – 1 1 – 1 1 – 1 – 1 2 – 1 – 1 2 –	– – – – – – –	– – – – – – – 1 – 2 –
Rhacomitrium crispulum	– – – – – – – – – – – – –	– – –	– – 1 – – – – – – – – – – – –	– – – – – – – 1 – 1 – – – – – – – –	– – – – – – –	– – – – – – – – – – 1
Pringlea antiscorbutica	1 1 – 1 1 – 1 1 1 1 1 1 2	1 1 1	3 2 1 3 1 – 1 2 2 1 – – – 1 1	2 1 1 2 – 1 2 – 1 3 1 2 2 3 2 1 2 1	– – 1 – – – –	– 2 3 3 2 1 2 1 3 3 1
Bartramia patens	– – – – – – – – 1 – – – 1	– – 1	– 1 1 1 – – – – 1 – – – – – –	– – – – – – 1 – – 1 – – – – – 1 – 1	– – – – – – –	– – – – – – – – 1 1 1
Brachythecium para-doxum	– – – – – – – – – – – – –	– – –	– 1 1 – 1 – – 1 1 – – – – – –	– 1 1 – – – – – – – 1 – – – – – – –	– – – – – – –	– – – – – – – – – 1 –
Pohlia wahlenbergii	– 1 – – – – – – – – – – 1	– – –	– 1 – – – – – – – – 1 – 1 1 1	2 2 3 3 – 3 4 2 2 4 4 3 – 4 6 2 3 1	1 – 3 1 3 2 4	4 6 4 3 6 4 4 4 4 5 3
Acaena magellanica	– – – – – – – – – – – 1 –	– – –	– – – – – – – – – – – – – – –	– – – – – 1 – – – – – – – – – – – 1	– – – – – – –	– – – – – – – 1 – – 1
Callitriche antarctica	– – – – – – – – – – – – –	– – –	– – – – – – – – – – – – – – –	– – – – – – – – – – – – – – – – – –	3 3 2 – 4 2 1	– – – – – – – 2 – – –
Poa cookii	– – – – – – – – – – – – –	– – –	3 2 2 2 1 – 1 1 – – – – – – –	1 1 – 1 1 – 1 – – – – 1 2 1 2 – 2 1	4 1 3 2 – – –	– – – – 2 1 – – – 2 –
Prasiola crispa	– – – – – – – – – – – – –	– – –	– – – – – – – – – – – – – – –	– – – – – – – – – – – – – – – – – –	– – – – – – –	– – – – – – – – – – –
Deschampsia antarctica	1 – – – – – – – – – – 2 –	– – –	– – – – – – – – – – – – – – –	– – – – – – – – – – – – – – – 1 – –	4 4 4 4 4 5 4	– – 3 3 2 – – 4 3 – 3
Montia fontana	– – – – – – – – – – – 1 –	– – –	– – – – – – – – – – – – – – –	– – – – – – – – – 1 – – – – 1 – – –	2 2 2 2 2 – 2	– 3 2 3 1 – – 2 2 2 2
Other algae	– – – 1 – – – – – – – – –	– – –	– – – – – – – – – – – – – – –	– – 2 3 – – – – – – – – – – – – – –	– – – – – – –	4 – – 2 1 – – 1 2 2 1
Environmental variables						
Aspect[d]	1 4 2 2 3 3 3 2 2 1 1 1 2	1 2 2	1 4 4 1 4 2 2 2 2 1 2 2 2 2 2	4 4 4 4 1 2 2 4 4 2 4 4 2 2 2 2 3 2	1 1 1 1 1 1 1	2 1 1 1 2 2 2 1 1 2 1
Altitude[e]	3 4 4 2 2 1 1 4 4 2 4 3 4	3 2 4	2 2 2 2 5 3 3 3 2 2 4 4 4 3 2	5 5 5 5 5 5 5 5 5 5 5 5 2 4 4 5 5 5	1 1 1 1 1 1 1	5 3 3 3 4 5 5 1 3 5 3
Slope[f]	1 4 1 3 3 6 1 5 5 6 2 2 1	2 1 6	2 1 1 1 6 1 1 1 1 1 3 4 3 3 1	4 3 1 1 5 3 2 3 1 2 1 1 1 2 2 2 6 1	1 1 1 1 1 1 3	2 1 1 1 3 2 2 3 1 2 2
Vegetation ht.[g]	1 1 1 1 1 1 1 1 1 1 1 1 2	1 1 1	6 2 1 2 1 1 1 2 3 1 1 1 1 1 1	1 1 1 1 1 1 3 1 1 4 1 1 5 4 2 1 3 1	2 1 1 3 1 1 1	1 2 2 4 3 2 5 2 3 6 1
Soil depth[h]	2 2 3 1 2 1 1 3 3 1 1 2 1	2 1 3	2 1 3 1 2 1 1 1 2 1 1 1 1 3 1	1 1 1 2 2 1 2 1 1 1 1 1 2 1 1 1 3 1	1 1 1 1 1 1 2	1 2 2 2 1 1 1 2 3 1 1
Fauna use[i]	0 0 0 1 0 1 1 0 0 0 0 1 0	0 0 0	2 0 1 1 0 0 0 0 0 0 0 0 0 0 0	0 0 0 0 0 0 0 0 1 0 1 1 2 0 0 0 0 0	3 3 3 3 3 3 1	0 0 0 0 0 0 0 1 0 0 1
Bare ground	3 2 2 4 2 3 3 5 3 5 4 3 0	1 3 5	2 3 3 1 4 3 2 2 2 1 2 2 2 3 3	3 4 3 3 4 3 2 4 1 1 4 1 4 2 1 1 0 2	3 4 3 4 3 3 4	1 3 3 3 1 1 1 2 2 1 4
Rocks	4 6 6 5 6 6 6 4 5 4 5 6 4	6 6 4	6 6 6 6 5 6 6 6 6 6 6 6 6 6 6	6 5 5 5 5 6 5 5 4 5 4 3 3 4 3 6 0 6	3 3 3 4 3 3 1	4 4 4 4 4 5 6 2 4 4 4
Open water	0 0 0 0 0 0 0 0 0 0 0 0 0	0 0 0	0 0 0 0 0 0 0 0 0 0 0 0 0 0 0	0 0 0 0 0 1 0 0 0 0 0 0 3 0 0 0 0 0	2 2 3 0 2 1 2	3 1 1 3 0 2 1 2 3 3 1
Age of site[j]	1 1 1 1 1 1 1 0 0 0 0 1 1	1 1 0	1 1 1 1 1 1 1 1 1 1 1 1 1 1 1	1 1 1 1 1 1 1 1 1 1 1 1 0 1 1 1 0 1	1 1 1 1 1 1 1	1 1 1 1 1 1 1 1 1 1 1
No. of species	18	2	18	23	6	19

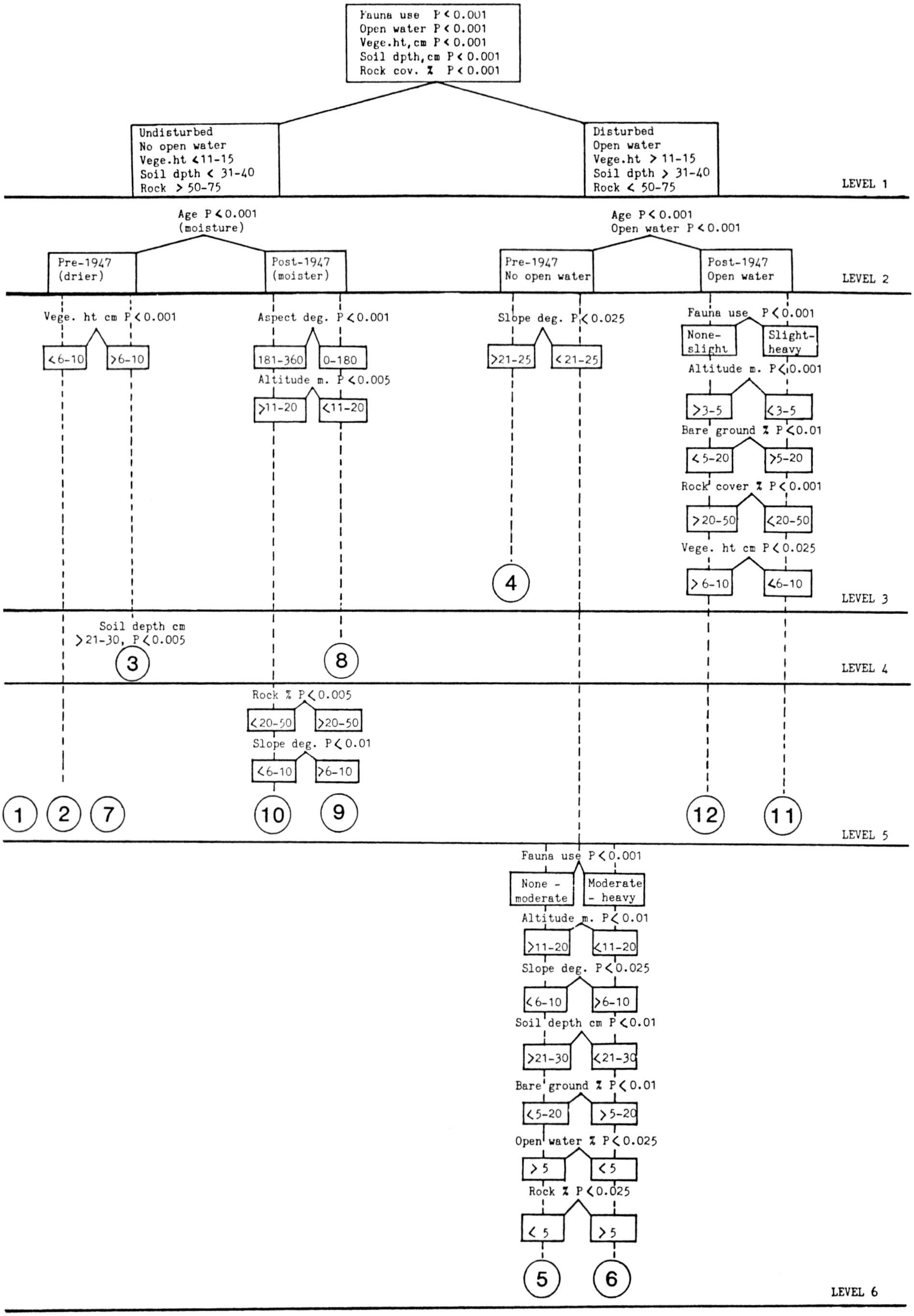

Fig. 7. Summary of TWINSPAN levels of division showing significant differentiating environmental factors and positions of quadrat groups

to have become ice-free since 1947. Mean retreat of the Brown Glacier snout between 1947–1980 is calculated at 600 m (Allison and Keage 1986). It has left massive lateral moraines and a pro-glacial lagoon, and the snout in December 1987 was resting on smoothed bedrock several metres behind the lagoon back beach (S. Goldsworthy pers. commun.). The ice patch behind Skua Beach is a diffluence area or spillover from the two main glaciers above, the Brown and the Stephenson. Although probably never a glacier itself, a series of low terminal moraines near the limit of the 1947 ice extent show that it was moving material forward at some stage.

The 16-m-long Dovers Moraine transect (Fig. 3) runs north-south along the outer slope of Dovers Moraine, a large series of terminal moraines of the Stephenson Glacier. The inner series of moraine ridges is ice-cored, and the moraine complex contains some of the oldest glacial deposits on Heard Island (Budd and Stephenson 1970). According to Allison and Keage (1986), the ice boundary retreated about 100 m uniformly along the inner length of Dovers Moraine between 1947 and 1980. The ice retreat lines shown in Fig. 3 are very approximate due to difficulties in interpreting the patchy aerial photo coverage of this area; for example the 1986 ice edge shown does not differentiate the extensive areas of stagnant moraine-covered ice which occur inland of the meltwater lakes at the southern end. Despite this, the retreat lines serve to indicate that the transect site has not been subject to glaciation for a considerable period.

The Stephenson transect (Fig. 2) is 160 m long and traverses upslope on the seaward side of the Stephenson northwestern lateral moraine complex. Evidence for the ice retreat line is from the 1947, 1980 and 1986–87 aerial photos, Budd and Stephenson (1970) and 1987–88 field observations. The entire Winston transect (160 m length) lies on land which has been recently exposed, probably since 1954 (Budd and Stephenson 1970). Mean retreat of the Winston Glacier snout between 1947 and 1980 is calculated at 1400 m (Allison and Keage 1986). Part of the transect nearest the lagoon (quadrats W1–W10) lies in an elephant seal moult area. The majority of the Jacka transect (360 m length) is located on land presumed to be exposed since 1947 (Fig. 5), apart from the first 20–50 m in from the coast which is part of an older terminal moraine vegetated with *Poa cookii*-dominated maritime grassland. This area is used by moulting elephant seals and gentoo penguins. Allison and Keage (1986) calculate a mean retreat of 310 m for the Jacka Glacier snout between 1947 and 1980.

Lack of aerial coverage of the Schmidt Glacier makes estimates of recent ice-free areas uncertain. The only documented evidence for its former extent appears to be a sketch map of the island drawn by the German South Polar Expedition of 1901–1903, which shows the glacier projecting approximately 1 km further seawards than today (von Drygalski 1908), and an approximation of its extent in 1951 shown on the island map included in Downes et al. (1959). Photographic evidence from 1951 (M. Downes pers. commun.) also suggests significant retreat since the 1950s although no quantitative information exists. It is likely that the majority of the Schmidt transect (200 m length) lies on recently exposed land. The transect commences at a small king penguin colony located approximately 300 m from the cliff and overhanging glacier at the head of the valley.

4 Results and Discussion

4.1 Introduction

The TWINSPAN analysis initially divided the quadrats into two groups, one characterised by *Poa cookii, Callitriche antarctica* and *Montia fontana*, and the other by *Poa kerguelensis, Polytrichum alpinum, Azorella selago* and *Dicranoweisia* spp. The former group is subject to animal disturbance and has significantly greater soil depth, vegetation height and occurrence of open water, either as creeks or standing water ($P < 0.001$ for all four variables) than the latter group, which has a significantly greater ($P < 0.001$) proportion of bare rock substrate (Fig. 7).

The second level of division differentiates quadrats on sites which were ice-free prior to 1947 (old or established sites) from those which have been exposed since 1947 (new or colonising sites). The quadrats at this level also appear to be differentiated on the basis of moisture availability, although the only relevant environmental variable measured was the presence of open water, representing an "extreme" of soil moisture occurrence. The presence or absence of open water was significant for differentiation between Groups 4, 5 and 6, and Groups 11 and 12 ($P < 0.001$; Fig. 7). The remaining groups had already been differentiated on the absence of open water at the first level of division; however, the indicator species for Groups 1, 2, 7 and 3 at this level are characteristic of dry sites (*Poa kerguelensis* and *Polytrichum alpinum*), and for Groups 8, 9 and 10, moist sites (*Dicranoweisia* spp. and *Pohlia wahlenbergii*).

At the third level the quadrats are divided into eight groups characterised by a variety of environmental factors. Several of the latter may not reflect significant differences in reality, if a group becomes "biased" by containing a large number of quadrats from one transect with uniform environmental conditions. For example, although aspect and altitude differentiate between Groups 8 and 9, and 10 (Fig. 7), Groups 8 and 9 consist almost entirely of quadrats from the Jacka transect. Similarly, a large number of environmental factors are significant in differentiating Group 11 from Group 12, but may mainly reflect the fact that Group 12 is composed of the first seven quadrats along the Winston transect. At the fourth and fifth levels environmental relationships become less clear, apart from several instances where slope and percentage cover of bare rock are significant (Fig. 7).

The divisions at Level 2 were used as a basis for arranging the classified data into two tables to illustrate patterns of recent vegetation colonisation more clearly. The TWINSPAN two-way table was divided by hand into 12 quadrat groups with a minimal amount of additional hand-sorting of quadrats. The groups are equivalent to divisions at the third, fourth and fifth levels, and their positions in the classification hierarchy are shown in Fig. 7. Table 4 illustrates patterns of colonisation on areas which have been deglaciated in the past 40 years, while Table 3 shows vegetation communities on surrounding areas which have been ice-free for a considerable period of time. The order of species and the order of quadrats within each group (apart from the seven hand-sorted quadrats, which are marked) remain unchanged from the original TWINSPAN classification. The quadrat groups within each table are arranged in the approximate order of increasing soil moisture availability and degree of animal disturbance.

4.2 Vegetation on "Pre-1947" Ice-Free Areas

Table 3 contains six quadrat groups representing vegetation on areas which have been ice-free for longer than 40 years. The quadrats were located on several of the transects, and also in the surrounding areas on sites which have not been glaciated for a considerable period due to topographic barriers inland (e.g. the seaward side of Scarlet Hill, Round Hill and the bluffs at the north end of Skua Beach; Fig. 2). The locations of the quadrats are shown in Figs. 2–6. On a general level, the groups in Table 3 correspond to the basic vascular plant communities described for Heard Island by Hughes (1987). They do not cover the full range of vegetation community types present on the island, although they are representative of the main vegetation communities in the areas where the transects are located.

Groups 1 and 2 are fellfield communites (Hughes 1987) on well-drained exposed sites characterised by a high proportion of bare substrate (mainly rocks) and sparse low-lying vegetation with *Azorella selago, Poa kerguelensis, Polytrichum alpinum, Colobanthus kerguelensis* and *Grimmia* spp. being common. Little or no animal disturbance is evident. *Ditrichum* spp. and *Pringlea antiscorbutica* are common in Group 2, which has a higher species complement (13 species) than Group 1 (10 species) and is likely to represent slightly more sheltered and moister conditions.

Group 1 represents the "oldest" fellfield type, containing all four quadrats from "non-glaciated" areas as well as four quadrats from the pre-1947 Stephenson moraine transect (62% of the group's quadrats). The post-1947 data comprise five quadrats from the Skua transect. These quadrats are located farthest away from the ice edge, nearest to the edge of a series of low moraine ridges which about well-developed pool complex vegetation of the beach terrace. It is possible that the 1947 ice-edge boundary shown in Fig. 2, which was located from a 1947 oblique aerial photo taken a considerable distance offshore, may be inaccurate to the extent that the first 50–100 m of moraine inland from the boundary could have been ice-free in 1947. For similar reasons the Skua, Brown and Jacka quadrats in Group 2 may also have been ice-free in 1947, and although the group contains approximately equal numbers of "pre-1947" quadrats (54%) and "post-1947" quadrats (46%), it was included as part of Table 3 on this basis.

Group 3 represents cushion carpet vegetation (Hughes 1987) dominated by *Azorella selago* and located in slightly more sheltered sites than the fellfield groups, from which it is differentiated by vegetation height (> 6–10 cm, $P < 0.001$; Fig. 7). Group 4 consists of herbfield (Hughes 1987) dominated by *Pringlea antiscorbutica, Poa cookii* and *Acaena magellanica* and located on steep sheltered slopes. It is differentiated from Groups 5 and 6 on the basis of slope (> 21–25°, $P < 0.025$). Group 5 consists of tussock grassland or maritime grassland (Hughes 1987; Seppelt and Hughes 1988) dominated by *Poa cookii* and subject to moderate to high levels of disturbance by moulting elephant seals and fur seals. Group 6 represents pool complex vegetation (Hughes 1987) subject to low intensity use by moulting seals, with *Poa cookii, Azorella selago, Montia fontana* and *Callitriche antarctica*, and aquatic algae (unidentified) common in pools. Although these two groups were not separated out to Level 5, they are characterised by a number of significant environmental differences at Level 6 (Fig. 7).

4.3 Patterns of Vegetation Colonisation on "Post-1947" Ice-Free Areas

Table 4 illustrates four main patterns of primary colonisation on the recently exposed sites under study.

The first colonisation pattern (Groups 7 and 8) illustrates colonisation on well-drained moraines with little available surface moisture. Group 7 fits the Hughes (1987) description of Heard Island fellfield vegetation on the basis of its most common four vascular species, *Poa kerguelensis, Azorella selago, Colobanthus kerguelensis* and *Pringlea antiscorbutica.* It was included in Table 4 since the majority of its quadrats (69%) are from post-1947 ice-free areas. They are located at the inland end of the Skua and Brown transects (i.e. closest to the present ice edge), and on the basis of the aerial photo evidence are highly likely to have been ice-covered in 1947. Several other quadrats in this group, from the Winston transect and Brown Bluffs, are likely to have been ice-covered until at least 1954 (Winston) and the 1960s (Brown Bluffs).

The vegetation of the majority of quadrats in Groups 7 and 8 is 5 cm or less in height. *Azorella selago, Poa kerguelensis* and *Colobanthus kerguelensis* rarely grow higher than 5 cm on Heard, and although *Pringlea antiscorbutica* is common in Group 7, it only occurs as scattered seedlings and small plants. The vascular species mostly establish in crevices between rocks or in the lee of boulders, presumably because of greater moisture availability, or protection from wind, or both. Of the bryophytes, *Polytrichum alpinum, Bartramia patens* and *Marchantia berteroana* grow in crevices between stones, while *Ditrichum* spp. and *Grimmia* spp. colonise stone surfaces. *Fossombronia australis* and *Ceratodon purpureus* grow either in the edges of small *Azorella* and *Poa kerguelensis* clumps or in sheltered rock crevices where some soil and organic matter has accumulated. The overall percentage vegetation cover in Groups 1, 2 and 7 does not differ significantly, although vegetation in Group 7 differs from the more established fellfield vegetation of Groups 1 and 2 by its lesser abundance of *Polytrichum alpinum, Grimmia* spp. and *Azorella selago*, and its higher species diversity (18 species).

Group 8 contains three quadrats whose only vegetation is occasional tiny two-lobed *Pringlea antiscorbutica* seedlings and a fragment of *Bartramia patens* in Quadrat D1. This quadrat, from the pre-1947 Dovers moraine transect, is located on a steep unstable scree slope subject to constant movement of rock and gravel, and in this context could be classified as "newly exposed". The other two quadrats, from the Jacka transect on recently exposed, locally well-drained sites covered in tumbled boulders, support occasional *Pringlea* seedlings in sheltered crevices between rocks.

The second pattern of colonisation is illustrated by Group 9, comprising sites which, although well drained, are subject to continual moisture from the spray of nearby meltwater waterfalls and by surface runoff from a network of small creeks. The group contains 14 quadrats from the Jacka transect and one quadrat from the Schmidt transect. The Jacka and Schmidt glaciers have retreated upslope, leaving cliffs at the heads of their valleys. Both transects end near the cliffs, where the substrate consists of boulders and fallen rock fragments and the ground is continually moist from the spray of waterfalls from the glaciers on the cliffs above. A detailed drawing from a photograph taken during the German South Polar Expedition in 1901–1903 (von Drygalski 1908) shows that the majority of the vertical cliffs in the valley were ice-free at this time, with the glacier capping the cliff tops and cascading down at least one section of the cliffs as an icefall and spreading out in the valley below. Today, these and adjacent cliffs are colonised by a variety of bryophytes, notably dense growths of *Dicranoweisia* spp. with *Brachythecium* spp. and *Pohlia wahlenbergii* on wet ledges. Based on the 1901–1903 drawing, parts of the cliffs may have been colonised by bryophytes for a considerable period of time, i.e. prior to the present glacier retreat phase which has left the valley floor ice-free.

The main bryophyte species which colonise the cliffs have a scattered distribution over the recently exposed ground on the valley floor below. They occur here either as wind- and water-borne fragments which establish in damp crevices (e.g. *Brachythecium* spp. and *Pohlia wahlenbergii*), or as moss clumps already attached to fallen rocks (e.g. *Dicranoweisia* spp. and *Grimmia* spp.), and apart from occasional *Pringlea* and *Azorella* seedlings in sheltered crevices they are the only plants which occur at the inland end of the transect (e.g. quadrats J20–22). Group 9 has the highest percentage cover of rock of all groups in Table 4 combined (> 75% cover, $P < 0.001$), and is differentiated from Group 10 on this basis (> 20–50% cover, $P < 0.01$).

Small *Poa cookii* and *Pringlea* plants occur scattered over the valley floor, becoming more frequent as the transect moves away from the glacier (e.g. quadrats J9, J7, J6 and J4). Quadrat J9 is located on a small knoll which is used by skuas as a roosting and killing area, and the local abundance and height of *Poa cookii* probably reflects nutrient enrichment from this source.

The final two patterns describe colonisation on recently exposed areas with abundant clear-water surface drainage (creeks or impeded drainage) either from snow melt or springs emerging from the moraine. Groups 10 and 12 illustrate this, while Group 11 shows the influence of animal disturbance on the same pattern of colonisation. The majority of quadrats are from the Schmidt transect and the Skua Bluffs and Brown Bluffs sites. The vegetation is undisturbed by animals, and is dominated by flushes of *Pohlia wahlenbergii*. *Bryum pseudotriquetrum*, another moss characteristic of semi-aquatic conditions, occurs commonly in Group 10 and only rarely in the other groups. *Pringlea* and *Azorella* are common, with mature *Pringlea* plants reaching up to 30 cm in height. The species diversity in Group 10 is higher than in any other group, with 23 of the possible total of 26 species present. The missing species are lichens, *Callitriche antarctica* and *Prasiola crispa*. Lichens are generally absent from the vegetation sampled for this study, while the latter two species are characteristic of nutrient-enriched environments resulting from animal disturbance (Smith 1978; Lewis Smith 1984).

The pattern shown in Groups 10 and 12 demonstrates the rapidity with which vegetation colonisation can occur under favourable conditions. Of the four sites whose quadrats are in Group 10, two sites (representing 8 of the 16 quadrats in the group) have become ice-free only very recently; the Skua Bluffs site (three quadrats) and the Brown Bluffs site (five quadrats). There is a marked contrast between sites supporting flushes of this species-rich vegetation and surrounding moraine areas without surface drainage, which support vegetation characteristic of Groups 7 (e.g. Skua and Brown transects) and Group 9 (e.g. Schmidt and Jacka transects).

Group 12 represents wetter conditions than Group 10, and is characterised by the dominance of *Pohlia wahlenbergii* and the presence of *Deschampsia antarctica, Montia fontana* and also algae (species undetermined) in running water. Vegetation height is significantly higher than for the other five groups in Table 4 combined (height > 6–10 cm, $P < 0.001$), and represents luxuriant growth of *Pringlea* plants. Quadrats are from the Skua Bluffs and Brown Bluffs sites and the Winston transect, all sites which have become ice-free between approximately 1954 and 1986.

The second colonisation pattern of this type is illustrated by Group 11, which also demonstrates the rapid nature of colonisation under favourable conditions, as the group comprises eight quadrats from the Winston transect (ice-free probably since 1954). The vegetation is subject to seasonal disturbance by moulting elephant seals. It is dominated by *Deschampsia antarctica*, a species characteristic of "meadow" vegetation (Hughes 1987) in the south-east of the island, with *Pohlia wahlenbergii, Callitriche antarctica* and *Montia fontana* common. The vegetated area is fed by springs emerging from the moraine which appear to originate in bands of lava in the lower cliffs of South Barrier (Fig. 4). It is located within 200 m of the current snout of Winston Glacier. The low species diversity (six species) can be attributed partly to physical disturbance by elephant seals, and partly to isolation from nearby sources of colonising plant propagules. Bare ground comprising a mixture of seal-compacted peat and gravel is common, although in terms of percentage cover it does not differ significantly from the other Table 4 groups, where it occurs as inorganic substrate (fine gravel and sand).

The dominance of *Deschampsia antarctica* in Group 11 is unusual on Heard Island in seal-disturbed shore-fringing environments, which are almost invariably dominated by *Poa cookii* (Group 5, Table 3). *Poa cookii* is present at the Winston site as a small number of scattered clumps near the shoreline. It could be postulated that *Poa cookii* is a more recent arrival on the site than either *Deschampsia* or moulting seals, and that *Deschampsia* has been favoured by disturbance at the expense of species such as *Pohlia, Pringlea, Azorella* and *Colobanthus*, as well as a number of other bryophytes which could be expected on the basis of species present in Group 12 (6 of the 11 quadrats in Group 12 are also located on the Winston transect, immediately inland and slightly upslope of the seal-affected quadrats in Group 11). In this context, *Deschampsia* may represent a transitional stage between undamaged flush vegetation and future dominance by *Poa cookii* in a seasonally disturbed environment.

Callitriche antarctica is only abundant in Group 11 (Table 4), which is likely to be a consequence of nutrient enrichment by animal manuring in a moist environment. Group 6 (Table 3), which represents pool complex vegetation, is the only other group in the study where *Callitriche* is abundant, and is also characterised by animal disturbance and the present of open water.

4.4 Comparison of "Pre-1947" and "Post-1947" Vegetation patterns

When Tables 3 and 4 are considered together, a number of similarities and differences between "established" and "colonising" vegetation become apparent. In terms of floristics, nine of the vascular species which have managed to arrive and survive on Heard Island are present in both the established and colonising patterns of vegetation. The remaining two species have a restricted distribution (see below) and do not occur at the sites studied. The most common vascular species in both Tables are *Azorella selago, Colobanthus kerguelensis, Pringlea antiscorbutica* and *Poa kerguelensis*. Of these, *Pringlea* is the most common in colonising vegetation, being consistently present in all six groups apart from Group 11, where it has been almost eliminated by mechanical disturbance from seals. *Pringlea* occurs mainly in Groups 7, 8 and 9 as small plants and seedlings, and in Groups 10 and 12 in a range of size classes up to mature plants 30 cm in height. *Pringlea* seeds prolifically and during the summer tiny two-lobed seedlings, mostly short-lived, are common in a wide variety of environments. Mature *Pringlea* plants are uncommon in established fellfield, and although small plants and seedlings are common in Group 2 they are absent in Group 1.

Small *Azorella, Colobanthus* and *Poa kerguelensis* seedlings are common amongst soil and organic debris in damp crevices between stones on well-drained moraines (e.g. Skua and Brown transects), and also on wetter sites (e.g. Jacka and Schmidt transects). *Poa cookii* sheds large quantities of seed, and small scattered seedlings are frequent in moist environments even when little soil is available (e.g. Jacka and Schmidt sites), although the success of further establishment probably depends on soil nutrient availability.

Access to plant propagules, comprising seeds, bryophyte and lichen fragments and vegetatively reproducing portions of species such as *Azorella selago* and *Poa cookii*, is likely to play an important role in colonisation patterns. Most of the vascular plants appear to set seed, and frequent strong, gusty winds probably play an active role in dispersal. Not all of the mosses appear to fruit, but distribution of fragments appears to be widespread. The Winston and Schmidt sites both support *Poa cookii* on animal-disturbed areas, despite being a considerable distance inland from coastal sources of this species. Animals are likely to play some role in dispersal, for example at the Schmidt site *Poa cookii* may have been introduced on the feet of king penguins regularly walking in from the coast. The Jacka and Schmidt sites receive bryophyte fragments from the cliffs above, via a combination of rockfall, wind and water.

The presence and status of soil micro-organisms, including algae, were not investigated in this study, although it is acknowledged that their role as initial colonisers of recently deglaciated habitats is likely to be an important one (Lewis Smith 1984).

In established vegetation communities, bryophytes are uncommon in maritime grassland and pool complex vegetation fringing the coast (Groups 5 and 6), probably due to factors such as wind-blown salt spray and animal disturbance. They are most frequent in the fellfield and cushion carpet communities (Groups 1, 2 and 3), especially *Polytrichum alpinum, Ditrichum* spp. and *Grimmia* spp. A number of different bryophyte species, especially *Dicranoweisia* spp. and *Pohlia wahlenbergii*, are common in the colonising groups in Table 4. Group 11 is a notable exception, with an absence of bryophytes (apart from *Pohlia*) probably being a direct result of animal disturbance.

The establishment of *Poa cookii* is almost invariably linked with animal disturbance (e.g. Groups 3, 4, 5 and 6 in Table 3 and Groups 9, 10 and 11 in Table 4), while *Callitriche antarctica* almost always occurs where animal disturbance and open water occur (Group 6, Table 3 and Group 11, Table 4). In this way animal disturbance, in the form of nutrient enrichment, can result in a blurring of the age boundaries defined by Tables 3 and 4. An example is quadrat SC1 in Group 5 (Table 3) which is classified with "established" *Poa cookii* grassland. In fact it is located at the edge of the king penguin colony at the start of the Schmidt transect (Fig. 6) where clumps of dark green *Poa cookii* and patches of the coprophilous alga *Prasiola crispa* occur, although they are absent in the surrounding area. Work on the effects of manuring on soils and vegetation on subantarctic Marion Island (Smith 1976, 1978) has indicated marked effects on vegetation due to nutrient enrichment by seals and seabirds, especially from enhanced levels of N and P. These effects consist of increased plant vitality and colour and changes in the abundance of species present, including an increase in coprophilous species (Smith 1978).

In addition to nutrient enrichment, animal disturbance leads to physical removal of vegetation leaving bare ground, mostly compacted peat in established vegetation (Groups 5 and 6) and compacted peat and gravel in less established vegetation with shallower soils (Group 11). Percentage cover of bare ground is not a significant factor differentiating "established" from colonising vegetation, mainly due to the effects of fauna use of established communities. Percentage rock cover is significant, however ($> 20-50\%$ rock cover for colonising groups, $P > 0.001$). The colonising groups are also characterised by shallower soils ($< 21-30$ cm, $P < 0.001$) and lower vegetation height ($< 11-15$ cm, $P < 0.005$) than the "old" groups of Table 3.

Along with moisture regime and drainage, soil nutrient status and nature of the substrate, wind exposure can be expected to play a significant role in differentiating colonising patterns of vegetation (e.g. see Lewis Smith 1984). Several environmental variables were measured which could indicate wind exposure (aspect and altitude), but with the exception of differentiating Groups 8, 9 and 10 these varables were not significant in the analysis. Although the prevailing winds reaching Heard are from the SSW-NW sector, local topography is expected to affect local climate significantly (Allison and Keage 1986). Qualitative field observations certainly confirm this expectation, and frequent, sometimes unpredictable changes in local wind exposure are experienced. The east-facing Skua, Brown and Jacka glacier fronts are particularly exposed to strong winds being channelled down from the mountains above, but no significant differences can be discerned between quadrat data from these and other sites on this basis. On the basis of species preferences, however, wind exposure and/or moisture availability are possibly significant factors in several divisions, for example in differentiating the three fellfield Groups 1, 2 and 7 from the cushion carpet Group 3 (Fig. 7), where vegetation height is the significant environmental variable. The "exposed" (shorter) vegetation (Groups 1, 2 and 7) is characterised by *Poa kerguelensis* and *Ditrichum* spp., while the taller vegetation (Group 3) is characterised by *Fossombronia australis, Acaena magellanica, Montia fontana* and *Poa cookii*.

The transect data illustrate a number of vegetation colonisation patterns on recently deglaciated land. Availability of surface water drainage, influence of animal disturbance, and degree of access to plant propagules disseminated by wind, water and animals appear to be major factors producing different patterns of colonisation. Establishment and growth of plants along stable drainage lines and on moist animal-affected areas can be particularly rapid.

5 Changes in Distribution of Vascular Plants

It is interesting to speculate on the extent to which a gradual warming of the regional climate is influencing vegetation distribution on Heard Island generally. Based on a number of recent observations the southeast end of the island has a milder local climate than the north and west sectors (e.g. Ledingham 1987; Allison and Keage 1986; unpubl. 1987–88 expedition records), although these observations are unsubstantiated by quantitative data. The southeast end appears to be the most favourable for establishment of vascular plant species, with 4 of the 11 species only found there (Table 1). Two of these, *Poa annua* and *Ranunculus biternatus*, have very restricted known distributions on Heard Island (Scott 1989). Observations since 1986 indicate that both species appear to be spreading outwards from their greatest areas of concentration. *Montia fontana*, which is widely distributed at the eastern end, was first found in two small patches at the northwest end in 1986 and 1987. Both patches are in sheltered sites adjacent to penguin and seal colonies, and it is possible that *Montia* was transported to the sites by skuas or giant petrels (Scott 1989). Although these findings partially reflect a lack of previous systematic botanical survey, the small size of the populations might suggest recent establishment.

Anecdotal evidence from past ANARE expeditions indicates changes in the distribution of *Acaena magellanica* since the 1950s (Budd 1970; Hughes 1987; M. Downes pers. commun.). The species appears to be becoming more widespread since observations in the 1950s that it grew only in the Spit Bay region. It extends from Long Beach to Spit Bay to Compton Lagoon (Budd 1970), with its most prolific growth at Paddick Valley east of Winston Lagoon, Spit Bay, and Skua Beach. The species has tenacious hooked fruits, and it seems surprising that it has not been recorded at the northwest end of the island. The potential ease of transport by birds is demonstrated by an observation of a fruiting *Acaena* head attached to the breast of a kelp gull 10 km off the Kerguelen coast (Smith 1986a).

The occurrence of *Poa annua* on Heard Island is particularly interesting. It was first recorded in 1986 and 1987 growing in two small isolated populations on moraine areas at Winston Lagoon and Stephenson Lagoon (Fig. 1). Both areas were ice-covered or at the ice-moraine interface in 1947. The species appears well-adapted to its present environment and is evidently spreading. Based on observations of animal usage of the two sites, it seems that skuas are the most likely carriers of *Poa annua* seed over long distances (probably from Iles Kerguelen, where it is comon), while elephant seals and skuas both spread seed locally once the plants are established. This seems a more likely explanation for the species presence than introduction by humans. *Poa annua* does not occur at any of the known sites of former human occupation, despite similar suitable habitat conditions at all such sites (Scott 1989). It can be expected that further spread will occur, probably via a combination of wind, seals and skuas. It is also possible that human presence on the island during the past three summer seasons will have inadvertantly spread seeds.

There is cause for speculation about why this opportunistic species, which is one of the commonest naturalised alien plants in the subantarctic region, should have become established only at these two sites and only in the past 40 years. Possible reasons include chance; the lack of introduced herbivores, which greatly contribute to its spread on other subantarctic islands; and the competition-free seedbed provided by the newly deglaciated land surface (as discussed by Smith 1986b), in an ameliorating climatic environment.

6 Predictions for Further Changes

With continuing moderation of the climate and larger areas of ice-free land becoming available for colonisation, the size of vegetated areas will expand. Introduction and establishment of additional vascular plant species is a distinct possibility, and vascular species until now confined to the southeast of the island may become more widely established. The extent of human influence in spreading these species is uncertain.

Another uncertain factor, not necessarily linked with moderating climate, is the future effect on vegetation of changing trends in animal populations which utilise and interact with terrestrial vegetation communities. Both southern giant petrel and elephant seal populations on the island have shown substantial declines over the past 30–40 years (M. Downes unpubl. data; H. Burton unpubl. data, respectively). Rapid increases in the populations of king penguins (Gales and Pemberton 1988) and Antarctic fur seals (S. Goldsworthy unpubl. data) have been noted.

A dramatic increase in Antarctic fur seal numbers over the past two decades compares with other subantarctic and Antarctic islands where the species is increasing. During February and March 1988 numerous areas of *Pringlea antiscorbutica,* a species sensitive to trampling, were observed being damaged by moulting fur seals in the Spit Bay and Skua Beach region. Damage involved breakage of the stems and leaves of mature plants. The main fur seal breeding areas show evidence of disturbance to *Poa cookii* tussocks, as noted by Bonner (1985) on Bird Island, South Georgia where their impact has been intense. Lewis Smith (1988) documents significant damage to vegetation communities by an expanding Antarctic fur seal population on Signy Island.

A number of fixed photo-points have been set up in the Spit Bay area for future monitoring of fur seal impact on selected areas. Other baseline data for use in monitoring future vegetation changes include fixed vegetation transects at the two *Poa annua* sites, fixed vegetation quadrats and photo-points at a number of recently deglaciated areas, and detailed aerial photo mapping of vegetation communities over the majority of the island (J. Scott unpubl. data). It is hoped that re-measurement of these sites can be undertaken at regular intervals, although uncertainty and irregularity of visits to Heard Island means that timing of re-measurements will probably be opportunistic.

The present low level of human impact on the Heard Island environment seems likely to continue, and comprehensive management controls are being formulated for the Territory. These factors make it a unique location for the study of subantarctic vegetation change and the establishment of chance introductions to an island environment.

Acknowledgements. I thank the Antarctic Division of the Department of Arts, Sport, Environment, Tourism and Territories and the Antarctic Science Advisory Committee (ASAC) for making my fieldwork on Heard Island possible, and to Monash University, Botany Department for logistic support during 1988. Thanks also do Dr. R. Seppelt, Antarctic Division and D. Bergstrom, Macquarie University for helping with moss identification; and to Prof. J. B. Kirkpatrick and R. Fensham, University of Tasmania, for invaluable discussions and help with data analysis.

References

Allison IF, Keage PL (1986) Recent changes in the glaciers of Heard Island. Polar Rec 23(144):255–271

Bergstrom DM, Seppelt RD, The moss flora of Heard Island: revised checklist, annotations and phytogeographical considerations. Lindbergia (in press)

Bonner WN (1985) Impact of fur seals on the terrestrial environment at South Georgia. In: Siegfried WR, Condy PR, Laws RM (eds) Antarctic nutrient cycles and food webs. Springer, Berlin Heidelberg New York, pp 641–646

Brothers NP, Copson GR (1988) Macquarie Island flora and fauna management – interpreting progress and predictions for the future. Pap Proc Roy Soc Tas 122(1):129–135

Budd GM (1970) Heard Island reconnaissance, 1969. Polar Rec 15(9):335–336

Budd GM, Stephenson PJ (1970) Recent glacier retreat on Heard Island. In: Gow AJ et al. (eds) Proc Int Symp Antarctic Glaciological Exploration (ISAGE) IAHS Publication 86. Hanover NH, 1968, pp 449–458

Downes MC, Ealey EHM, Gwynn AM, Young PS (1959) The birds of Heard Island. ANARE Reports no 51, Antarctic Division, Commonwealth of Australia

Gales R, Pemberton D (1988) Recovery after exploitation of the king penguin, *Aptenodytes patagonicus* population on Heard Island. Aust Wild Res 15:579–585

Hill MO (1979) TWINSPAN – a FORTRAN program for arranging multivariate data in an ordered two-way table by classification of the individuals and attributes. Section of Ecology and Systematics, Cornell University, Ithaca, New York

Hughes JMR (1987) The distribution and composition of vascular plant communities on Heard Island. Polar Biol 7:153–162

Jacka TH, Christou L, Cook BJ (1984) A data bank of mean monthly and annual surface temperatures for Antarctica, the Southern Ocean and South Pacific Ocean. ANARE Research Notes 22, Antarctic Division, Commonwealth of Australia

Karoly DJ (1988) Evidence of recent temperature trends in the Southern Hemisphere. In: Pearman GI (ed) Greenhouse: planning for climate change. CSIRO, Australia, pp 52–59

Kershaw K (1975) Quantitative and dynamic plant ecology. Second edn, Edward Arnold, London

Leader-Williams N, Lewis Smith RI, Rothery P (1987) Influence of introduced reindeer upon the vegetation of South Georgia: results from a long-term exclusion experiment. J Appl Ecol 24:801–822

Ledingham R (ed) (1987) ANARE expedition to Heard Island 1986–87. Unpublished expedition field report. Antarctic Division, Commonwealth of Australia

Lesel R, Derenne PH (1975) Introducing animals to Iles Kerguelen. Polar Rec 17(110):485–494

Lewis Smith RI (1984) Terrestrial plant biology of the sub-Antarctic and Antarctic. In: Laws RM (ed) Antarctic ecology. Academic Press, London, vol 1, pp 61–162

Lewis Smith RI (1988) Destruction of Antarctic terrestrial ecosystems by a rapidly increasing fur seal population. Biol Conserv 45:55–72

Scott JJ (1988) Rabbit distribution history and related land disturbance, Macquarie Island. Pap Proc Roy Soc Tas 122(1):255–263

Scott JJ (1989) New records of vascular plants at Heard Island. Polar Rec 25(152):37–42

Seppelt RD, Hughes JMR (1988) Contrasts in vegetation patterns: Heard Island and Macquarie Island. 2nd Colloque sur les ecosystemes subantarctiques, Paimpont, France, 8–11 September 1986

Smith JMB (1986a) Specks in the Southern Ocean. Dept Geography and Planning, University of New England, NSW, Australia

Smith JMB (1986b) Origins of Australasian tropicalpine and alpine floras. In: Barlow B (ed) Flora and fauna of alpine Australasia. Ages and origins. CSIRO, Melbourne, pp 109–128

Smith VR (1976) The effect of burrowing species of Procellaridae on the nutrient status of inland tussock grasslands on Marion Island. J S Afr Bot 42(2):265–272

Smith VR (1978) Animal-plant-soil nutrient relationships on Marion Island (subantarctic). Oecologia (Berl) 32:239–253

Veenstra C, Manning J et al. (eds) (1982) Expedition to the Australian Territory of Heard Island and McDonald Islands 1980. Division of National Mapping, Canberra, Tech Rep 31

von Drygalski E (1908) Geographie von Heard-Eiland. In: Deutsche Sudpolar Exped, 1901–1903. Band 11, Geographie und Geologie, vol 3, pp 225–239

Evidence for Change in the Chemistry of Maritime Antarctic Heywood Lake

J. C. ELLIS-EVANS [1]

Summary. Data collected at monthly intervals over the past 16 years from Heywood Lake (Signy Island, South Orkney Islands) reveal that, within a given year, physical, chemical and biological variables generally interacted predictably. The elephant seal population of the catchment has remained static over the entire period of study but fur seal numbers have been increasing steadily within the catchment over the past 10 years. A pronounced increase in peak winter concentrations of both dissolved reactive phosphate (DRP) and ammonium released from anoxic sediments is evident over the same period. A clear trend in recent years towards increasing summer, open-water concentrations of ammonium-N indicates an influx of nutrients via surface runoff in addition to winter sediment release. This would also seem to be a consequence of the increasing fur seal presence in the catchment area. No evidence of increasing summer chlorophyll-a concentration was found, which conforms with the current hypothesis that light, rather than nutrient status, controls phytoplankton population dynamics in Heywood Lake during the open-water period.

1 Introduction

High latitude regions offer opportunities to study fundamental limnological principles in what are essentially simple ecosystems. The range of aquatic environments is considerable, encompassing virtually abiotic hypersaline environments through to nutrient-enriched freshwater lakes with dense phytoplankton (Heywood 1977). Pronounced seasonality (notably of light) is a feature of high latitudes and is a considerable asset in studying biological processes in particular. Catchments are essentially simple and readily defined so potentially facilitating more precise definition of geochemical and hydrological processes.

As interest in the Antarctic increases, and the influence of human activities extends beyond the immediate vicinity of the research stations, so the likelihood of perturbation of the, as yet, largely pristine Antarctic environment must also increase. A critical element in evaluating and predicting these new environmental pressures is our appreciation of natural change, particularly within Antarctic terrestrial and limnetic systems, which are amongst those environments most at risk.

Selected results are presented from the long-term lake monitoring programme at Signy Island which reveal evidence of significant changes in the chemistry of Heywood Lake in recent times.

2 The Study Site

Signy Island (60° 43′ S, 45° 38′ W) is part of the South Orkney Islands and is the main site for biological research by the British Antarctic Survey in the maritime Antarctic. There are 16 recognized lakes on Signy Island, all of which are fresh and ice-covered to a depth of 1 m for 8–12 months each year. Lake temperatures are consistently low, ranging from 0–6 °C on an annual basis. Most are oligotrophic (Heywood et al. 1980) but several are enriched with nutrients from the catchment and of these Heywood Lake is the best known. It has been the subject of a number of investigations (see Ellis-Evans 1985: Hawes 1985 a; Heywood 1987) and is, with oligotrophic Sombre Lake, the main BAS study site for long-term change. The lake comprises two basins separated by a shallow neck region. The larger North Basin has a maximum depth of 6.4 m and a mean depth of 2 m. The lake lies only 4 m above sea level and is separated from the sea by 200 m of gently sloping ground. This permits ready access to the catchment by elephant seals (*Mirounga leonina*) and fur seals (*Arctocephalus gazella*). Nutrient enrichment has resulted in substantial phytoplankton populations developing under spring ice cover and during summer open water (Light et al. 1981). The resulting turbidity reduces light penetration and restricts phytobenthos to the lake shallows.

[1] British Antarctic Survey, High Cross, Madingley Road, Cambridge CB3 0ET, UK.

Antarctic Ecosystems. Ecological Change and Conservation
(ed. by K. R. Kerry and G. Hempel)
© Springer-Verlag Berlin Heidelberg 1990

3 Methods

All water samples were obtained at the North Basin deep spot with a hand-operated diaphragm pump and rigid ABS plastic tubing as described in Hawes (1985a). All samples were returned to the laboratory within 2–3 h and analyzed for a wide range of cations and anions, including dissolved reactive phosphate (DRP), ammonium and chloride, using the methods outlined in Hawes (1983a). Chlorophyll-a (GF/C filtered) was analyzed using hot 95% methanol extraction (pre-1980) or cold 95% acetone. The equations outlined in Vollenweider (1969) were used to calculate chlorophyll-a concentrations.

Although the lake was sampled over the entire water column on each occasion, only the mid-depth data are considered here, unless otherwise indicated, to simplify presentation.

For most years since 1977, there has been a census of all seals present on Signy Island. This census has always been undertaken in the third week of February when seal numbers are highest (Lewis Smith 1988). The island is, for convenience, divided into a number of geographical areas and a detailed count of every fur, elephant, Weddell, crabeater and leopard seal in each of these areas made, involving virtually the entire Signy Base complement. The total data set has been discussed in Lewis Smith (1988) but for the purpose of this chapter only the census counts for the area Stygian Cove – Cemetery Flats, of which Heywood Lake catchment is a major component, are considered here. Additional count data presented here were obtained for the Stygian Cove – Cemetery Flats area by the author or, in two instances, derived from unpublished BAS marine and freshwater scientific reports.

4 Results and Discussion

4.1 Seasonal Variation

Results presented in Fig. 1 represent mean monthly values for mid-depth water samples calculated from

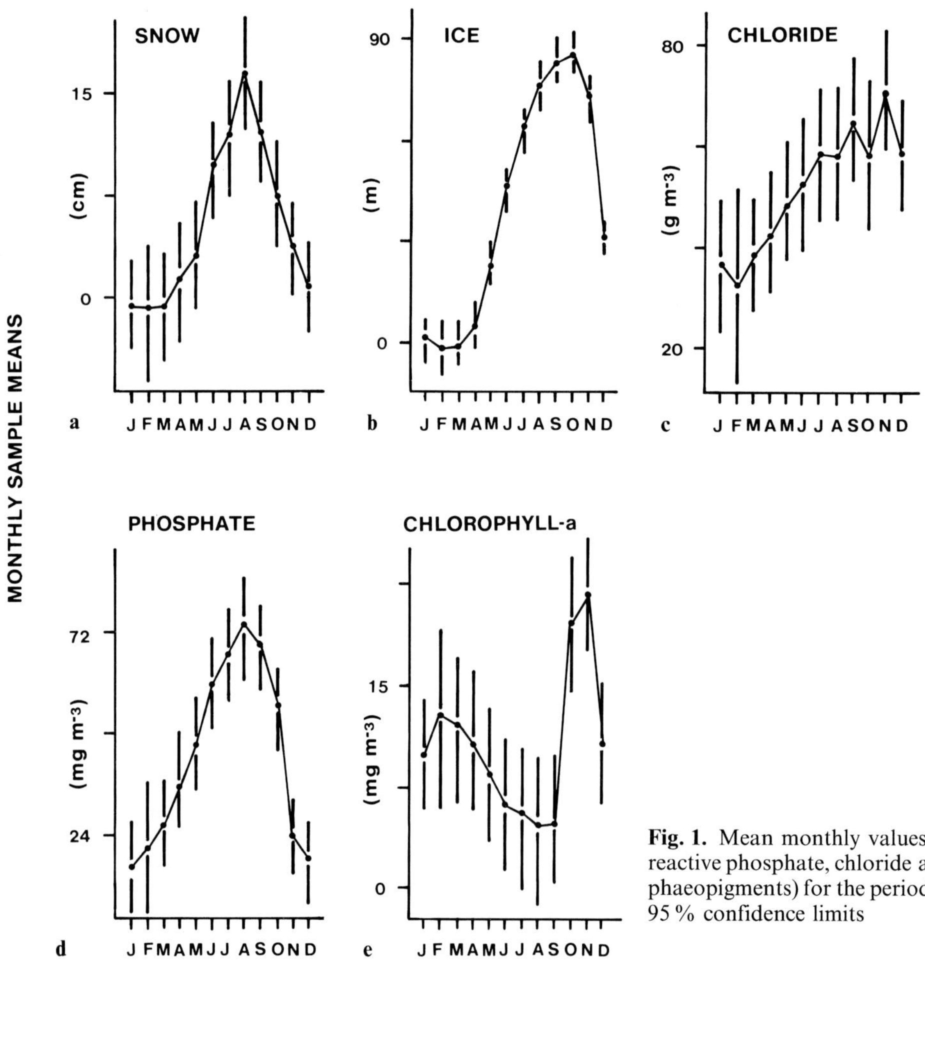

Fig. 1. Mean monthly values of ice and snow depth, dissolved reactive phosphate, chloride and chlorophyll-a (uncorrected for phaeopigments) for the period 1970–1987. *Vertical lines* indicate 95% confidence limits

16 years of data. Figure 1a indicates that snow cover reaches maximum depth in August, and thereafter declines with increasing air temperatures, whereas the ice cover (Fig. 1b) continues to increase in thickness until October. Using chloride as a conservative element (Fig. 1c) the significant concentrating effect of ice freeze-out in this shallow lake is clearly illustrated. The dilution effect of ice melt and influx of substantial volumes of snow meltwater from September onwards is also indicated by fluctuations in chloride data late in the year. Comparison of chloride and phosphorus data (Fig. 1e) indicate that the winter phosphorus increase can be attributed only in part to freeze-out processes and that upward transport of the nutrient from anoxic bottom sediments also contributed significantly.

That biological activity is largely a function of snow cover is clearly illustrated by the phosphorus and chlorophyll-a graphs (Fig. 1d, e); highest phosphorus and lowest chlorophyll-a values being associated with highest snow depth values (August). As snow cover disappears, allowing the percentage light penetration to increase, so algal activity resumes and phosphorus, being utilized by the algae, decreases. Chlorophyll-a concentrations peak around October–November when the ice cover is still intact but decline rapidly as nutrients become scarce in November–December. A second smaller peak of chlorophyll-a is discernible in these data, in February, due to the development of a second population of algae adapted to the higher light levels, but lower nutrient levels, associated with open-water conditions, as described in Hawes (1985a, b).

The data presented in Fig. 1 indicate that over the 16-year period considered, the seasonal changes in various physical, chemical and biological variables followed a predictable pattern. However, if one compares individual years, the scale of seasonal change in specific parameters can vary considerably.

4.2 Interannual Variation

Comparison of the 95% confidence limits for snow and ice cover on Heywood Lake (Fig. 1) clearly demonstrates the greater interannual variability of the former and Arctic lake data agree with these observations (Welch et al. 1986). Arguably, snow cover is the major environmental factor separating maritime and continental Antarctic lakes, with the latter receiving little or no snow fall year-round (Heywood 1977). In lakes of the South Victoria Land dry valleys, for example, factors such as variation in summer inflow rates are far more important (Chinn 1981). The data presented in Fig. 1 indicate that whilst a pronounced seasonality is imposed on maritime Antarctic lakes by snow cover it is also a highly variable factor at Signy Island, on an interannual basis. Snow may disappear in a melt, be blown on or off the snow cover by wind action or modified by physical processes such as wind action and freeze-thaw cycles. Such events can occur at any time during the winter, exerting a profound effect on microbiological processes (Ellis-Evans 1982) and light climate (Hawes 1985b). Work by Roulet and Adams (1986), in subarctic Labrador, indicates that marked changes in the spectral characteristics of under-ice light also occur as a result of variations in snow cover.

Snow melt-fed inflow streams are also observed to flow intermittently into the lake in some winters, dependent on fluctuations in air temperatures. These inflows have been shown to destabilize the water column of Heywood Lake delaying deoxygenation in early winter and thus limiting sediment release of nutrients (Hawes 1983b). The same work showed that in late winter the same processes serve to destabilize the water column making deep water nutrients available to algae within the photic zone of the upper waters.

4.3 Longer Term Variation

Analysis of metereological records show no obvious trend towards warmer mean annual air temperatures at Signy Island since the start of meterological records in the 1950's and certainly not in the past two decades (Fig. 2b). The ice cover (Fig. 2a) also shows no obvious variation in either thickness or duration which could be attributed to changes in the lake heat budget.

If, however, one considers the graphs of DRP and ammonium (Fig. 2c, d) a much clearer pattern is discernible. With the exception of 1973, DRP peak winter values were always around 50 mg m^{-3} until 1981 when levels doubled and continued at this higher level up to the present time. The very high 1973 peak was an exceptional event caused by the development of a marked temperature gradient in the water column (Hawes 1983b). As a result bottom waters remained warmer than usual throughout the 1973 winter which promoted increased microbial activity and thus enhanced the anoxic release of nutrients. Analysis of vertical temperature and conductivity profiles over the period 1972–1989 shows that such a stable profile has never occurred since.

Ammonium-N analyses were only initiated in 1980–81 but the data (Fig. 2c) show obvious signs of steady increase over the intervening period. The one exception to this trend is 1986 when, due to other commitments in that year, regular sampling was

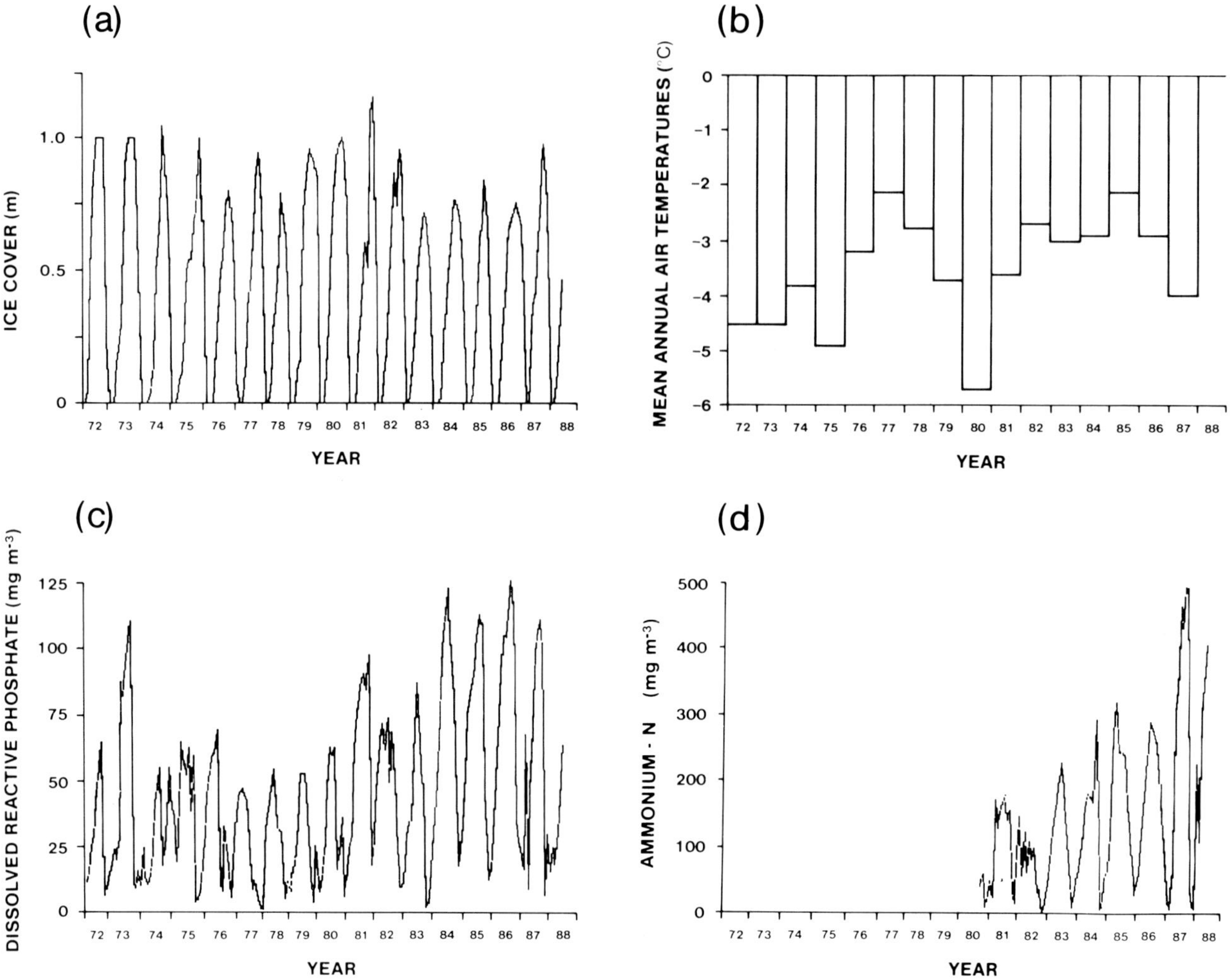

Fig. 2. Interannual fluctuations of **a** ice depth; **b** air temperatures; **c** ammonium-N; **d** dissolved reactive phosphorus, over the period 1972–88. **a, c, d** monthly data; **b** annual means; **c, d** mid-depth (3.0 m) samples

severely curtailed and almost certainly missed the peak of ammonium-N.

Figure 2 indicates that the increased ammonium-N and DRP winter values recorded since 1981 cannot be explained by physical factors such as temperature or ice/snow cover. The possibility of errors in analyses has also been discounted by rigorous methodologial checks. This suggests that biological and/or chemical factors are the source of the observed changes.

Recent work has shown that the consistently low temperatures of Signy Island lakes impose an overall restriction on microbial activity but that the amount of nutrient released from anoxic sediments is broadly related to the amount of readily degradable organic matter sedimenting to the lake bottom (Ellis-Evans 1982; Ellis-Evans and Sanders 1988). It is clear from Fig. 3 that chlorophyll-a values for phytoplankton, during summer open water, have barely changed over the past 18 years despite an order of magnitude change in ammonium-N concentrations in the past 8 years. Summer DRP has not shown a comparable increase. Work by Hawes (1983a, 1985b) has indicated that whilst the plankton of this lake was N rather than P-limited before this observed increase in ammonium-N, the single most important factor governing summer phytoplankton development is self-shading. The results presented in Fig. 3 and the fact that zooplankton numbers have not increased (unpubl. data) would seem to support this contention, so increased organic input to drive increased anoxic nutrient release must be allochthonous in origin. The most obvious source is fur seals whose numbers in the Heywood Lake catchment area have increased markedly since 1974, and particularly since 1984 (Fig. 4). Over the same

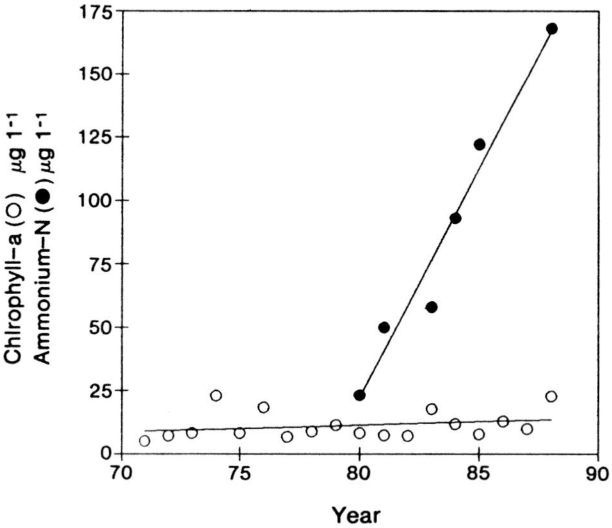

Fig. 3. Changes in summer, open-water concentrations of chlorophyll-a and ammonium-N at mid-depth. Linear regression yielded a highly significant line for ammonium vs time ($P < 0.001$) whilst no significant relationship was found for chlorophyll-a

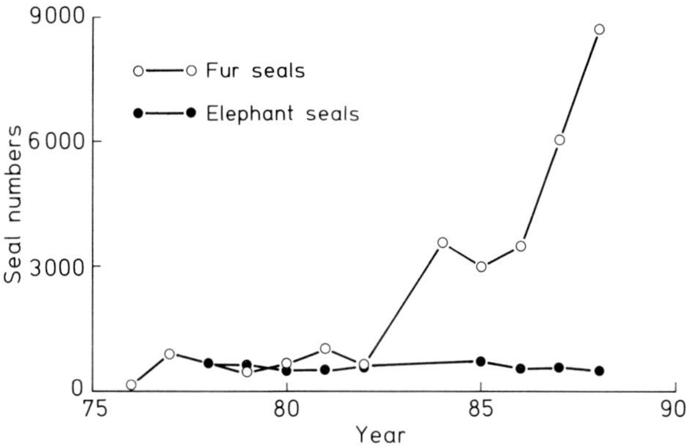

Fig. 4. Changes in fur seal and elephant seal numbers in the vicinity of Heywood Lake over the period 1976–1988

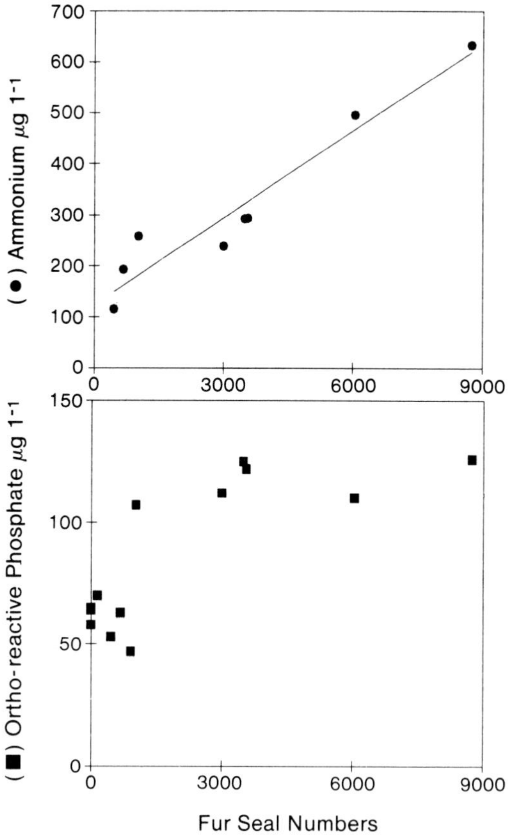

Fig. 5. Scatter plots of the relationship between fur seal numbers and peak winter concentrations of ammonium-N and DRP. The regression line for ammonium-N vs seal numbers was highly significant ($P < 0.001$)

period elephant seal numbers, probably the original source of catchment enrichment, have remained more or less constant.

Elephant seals have been observed to restrict their activities to certain defined low-lying areas of the catchment, but fur seals have steadily expanded throughout the entire catchment and in far greater numbers, so that their impact has been far more profound. The once widespread moss carpet has been compacted and is now largely moribund and has been recolonized by a community dominated by the nitrophilous alga *Prasiola crispa* which is characteristic of sites frequented by seals and birds. The hydric mosses have also been eradicated and even bare ground is now heavily colonized by *Prasiola* and other resistant algae which previously barely featured in the areas' plant community structure. A recent paper by Lewis Smith (1988) documents the ongoing destruction of Signy Island's terrestrial vegetation in some detail.

Work on the hydrology of Heywood Lake catchment has suggested that seepage through the moss carpet areas typically represents well over 50 % of the total nutrient input to the lake (Hawes 1983a). The acidic conditions of the moss carpets which dominated the catchment have been largely replaced by almost neutral pH and this reduces the pH-dependent charge on humus colloids and thus the ability to bind exchangeable ammonium ions (Brady 1974). In addition, compaction of the vegetation means that much of the water flow is probably over the surface instead of percolating through the vegetation. Enrichment of such areas will therefore inevitably markedly increase the input of certain organic and inorganic nutrients to the lake.

Whilst summer nutrient increases can be attributed directly to allochthonous sources, winter increases are only indirectly attributable to allochthonous carbon input. However, if peak winter nutrient concentrations are regressed against fur seal numbers, the influence of the increasing seal numbers is obvious (Fig. 5). Once again there is a significant linear relationship for ammonium-N ($r = 0.965$, $n = 8$, $p < 0.001$) and a reasonable correlation ($r = 0.692$, $n = 12$, $p < 0.01$) for a straight line relationship between seal numbers and DRP is also possible. However, it is clear from the scatter plot that a straight line fit would be misleading in the latter case. Recent work by Gallagher (1985) and Ellis-Evans and Lemon (1989) has shown that DRP release from sediments is tightly coupled to redox-related events involving iron colloids, whereas ammonium-N is unaffected. Thus a simple relationship between ammonium-N and seal numbers was not entirely unexpected. The relationship of winter DRP with seal numbers appears to be two-phase and may describe an initial enrichment response followed by plateauing as some factor such as higher iron concentrations or competition for organic substrates come to bear as catchment characteristics changed to the current status. This relationship merits further study as it has considerable significance for oligotrophic lakes where DRP is normally limiting.

Acknowledgements. I wish to thank the various freshwater field technicians who collected the data constituting the long-term data set, often under extremely difficult conditions. Dr. I Hawes (BAS) kindly commented on the manuscript, Dr. J Shanklin (BAS) provided the air temperature summaries for Signy Island and Dr. J Croxall (BAS) provided much of the seal census data. I am also grateful to the two referees, Dr. Ian Bayly (Monash Univ.) and Harry Burton (ANARE) for most useful criticisms.

References

Brady NC (1974) The nature and properties of soils. Macmillan, New York

Chinn TJH (1981) Hydrology and climate in the Ross Sea area. J R Soc NZ 11:373–386

Ellis-Evans JC (1982) Seasonal microbial activity in Antarctic freshwater lake sediments. Polar Biol 1:129–140

Ellis-Evans JC (1985) Decomposition processes in maritime Antarctic lakes. In: Seigfried WR, Laws RM, Condy PR (eds) Antarctic nutrient cyles and food webs. Proc 4th Symp Antarct Biol, Wilderness, September 1983. Springer, Berlin Heidelberg New York, pp 253–260

Ellis-Evans JC, Sanders MW (1988) Observations on microbial activity in a seaonally anoxic, nutrient-enriched maritime Antarctic lake. Polar Biol 8:311–318

Ellis-Evans JC, Lemon ECG (1989) Some aspects of iron cycling in maritime Antarctic lakes. Hydrobiologia 172:149–164

Gallagher JB (1985) The influence of iron and manganese on nutrient cycling in shallow freshwater Antarctic lakes. In: Seigfried WR, Laws RM, Condy PR (eds) Antarctic nutrient cycles and food webs. Springer, Berlin Heidelberg New York, pp 234–237

Hawes I (1983a) Nutrients and their effects on phytoplankton populations in ice-covered lakes on Signy Island, Antarctica. Polar Biol 2:115–126

Hawes I (1983b) Turbulence and its consequences for phytoplankton development in two ice-covered Antarctic lakes. Br Antarct Surv Bull 60:69–81

Hawes I (1985a) Factors controlling phytoplankton populations in maritime Antarctic lakes. In: Seigfried WR, Laws RM, Condy PR (eds) Antarctic nutrient cycles and food webs. Proc 4th SCAR Symp Antarct Biol, Wilderness, September 1983. Springer, Berlin Heidelberg New York, pp 245–252

Hawes I (1985b) Light climate and phytoplankton photosynthesis in maritime Antarctic lakes. Hydrobiologia 123:69–79

Heywood RB (1977) Antarctic limnology – review and synthesis. In: Llano GA (ed) Adaptations within Antarctic ecosystems. Proc 3rd SCAR Symp Antarct Biol, Washington 1974. Smithsonian Institution, Washington DC, pp 801–828

Heywood RB (1987) Limnological studies in the Antarctic Peninsula region. In: El Sayed SZ (ed) Antarctic aquatic biology. Biomass, vol. 7. SCAR, Cambridge, pp 157–173

Heywood RB, Dartnall HJD, Priddle J (1980) Characteristics and classification of the lakes of Signy Island, South Orkney Islands, Antarctica. Freshwater Biol 10:47–59

Lewis Smith RI (1988) Destruction of Antarctic terrestrial ecosystems by a rapidly increasing fur seal population. Biol Conservation 45:55–72

Light JJ, Ellis-Evans JC, Priddle J (1981) Phytoplankton ecology in an Antarctic lake. Freshwater Biol 11:11–16

Roulet NT, Adams WP (1986) Spectral distribution of light under a subarctic lake cover. Hydrobiologia 134:89–95

Vollenweider RA (1969) A manual on methods for measuring primary production in aquatic environments. IBP Handbook Number 12. Blackwell, Oxford

Welch HE, Legault JA, Bergmann MA (1986) Effects of snow and ice on the annual cycles of heat and light in Saqvaqjuac Lakes. Can J Fish Aquat Sci 44:1451–1461

Eutrophication and Vegetation Development in Maritime Antarctic Lakes

I. Hawes [1]

Summary. Differences in water chemistry and in the composition, abundance and activity of benthic and planktonic plant communities were observed in freshwater lakes of different trophic status at Signy Island, South Orkney Islands, maritime Antarctica. Phytoplankton density increased with increasing nutrient concentration. Picoplankton (< 2 μm) contributed most chlorophyll-a and numbers to the phytoplankton in all lakes, but the nanoplankton fraction (2–20 μm) was disproportionally active in photosynthesis. Some algae only occurred in the eutrophic lakes, notably a large *Chlamydomonas* sp., while smaller flagellated chlorophytes, chrysophytes and cryptophytes predominated in the nanoplankton of oligotrophic lakes. Partitioning of photosynthate in protein, polysaccharide, lipid and metabolite fractions showed little difference between the lakes.

1 Introduction

The 16 small, shallow lakes of Signy Island (60° 43′ S, 45° 38′ W), South Orkney Islands, maritime Antarctica, have formed following progressive deglaciation of the island. The catchments of these lakes are changing. Continuing recession of the ice sheets exposes more and more rock and scree surfaces which are then available for colonisation by terrestrial plants and animals. In some areas, a near complete "turf" of moss and lichen has developed. In coastal areas, exposed ground is often frequented by birds and seals which can import considerable quantities of nutrients from the productive inshore marine ecosystem (Allen et al. 1967; Hawes 1983). In recent years, numbers of fur seals (*Archtocephalus gazella*) have increased dramatically at Signy Island and these have affected several lake catchments (Smith 1988; Ellis-Evans this Vol.).

The characteristics of a lake are profoundly affected by the catchment which drains into it. Consequently, one might expect the differences between the catchments of Signy Island lakes to be reflected in the lakes themselves. Priddle and Heywood (1980) produced a largely hypothetical account of the evolutionary development of lakes currently found on the island. In their scheme, lakes begin as highly oligotrophic pools on the melting ice sheets or nestling between the ice edge and moraine dams. As catchments become more ice-free, these proglacial lakes can undergo shifts in their nutrient status.

At one extreme, the lake may remain oligotrophic with a rocky, poorly vegetated catchment providing few nutrients. Phytoplankton populations are scant, limited by nutrient availability, and the vegetation mostly comprises slow growing, long-lived benthos, notably aquatic mosses. At the other extreme, the catchment undergoes nutrient enrichment as a result of bird and seal intrusions and the subsequent enrichment of the lake water enhances the growth of phytoplankton at the expense of the phytobenthos. This chapter examines the effect of eutrophication on the vegetation of Signy Island lakes by examining five lakes of contrasting character.

2 Study Areas

Most of the extant lakes on Signy Island have been described by Heywood et al. (1980). Five of these have been selected for this study. The selected lakes all lie on the east coast of the island (Fig. 1) and morphometric data are given in Table 1. If altitude is taken as an indicator of lake age, then they can be arranged in order of increasing age as Khyber, Moss, Sombre, Heywood and Wallows. This order coincidently reflects the degree to which each is accessible to seals during summer.

Khyber Lake (unofficial name) is the youngest lake, having formed in the 1984/85 summer as a pool of meltwater adjacent to the Mcleod Glacier. Its depth is currently 3 m but this is likely to increase as approximately 70 % of the lake is still ice-floored. Its catchment comprises snow and ice with very little exposed rock and no bird or seal influence. Meltwater drains directly into the lake at its northern end.

[1] British Antarctic Survey, High Cross, Madingley Road, Cambridge CB3 0ET, UK.

Antarctic Ecosystems. Ecological Change and Conservation
(ed. by K. R. Kerry and G. Hempel)
© Springer-Verlag Berlin Heidelberg 1990

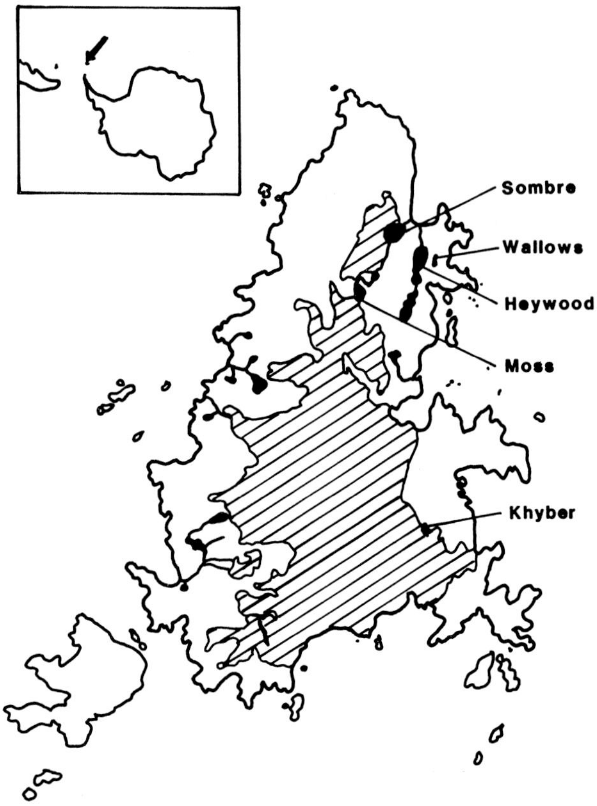

Fig. 1. Map of Signy Island to show location of study lakes and areas of permanent snow cover (*hatched*). Location of Signy Island is shown *inset*

Moss Lake is the uppermost of a chain of three lakes draining Paternoster Valley. It is large, relative to other Signy Island lakes, but lies in a small catchment dominated by bare rock and scree; substantial amounts of snow and ice are still present. The lake is extremely oligotrophic, phytoplankton development minimal and there is an extensive cover of moss, cyanophytes and algae on the lake floor (Priddle 1980). Priddle and Heywood (1980) considered this lake to represent the final stage of development of a maritime Antarctic lake which does not receive nutrient enrichment.

Sombre Lake lies at the seaward end of the chain of lakes which begins with Moss Lake. It is only 10 m above sea level, but the rocky nature of the catchment on the seaward side has, until recently, precluded the incursion of seals. The increasing number of fur seals on the island, however, has resulted in their movements into previously unaffected areas and during 1987/88 over 200 seals were frequently found in the catchment of Sombre Lake or swimming in the lake itself. The area around the lake on the eastern shore is well vegetated with moss and lichen, while areas away from the lake comprise mostly bare rock, scree and ice.

Heywood Lake has a catchment containing almost no permanent snow but has an extensive cover of terrestrial mosses. Elephant seals (*Mirounga leonina*) have occupied the catchment since records began (Laws 1956) and fur seals have been common since 1976. The lake supports a high density of phytoplankton and the water is very turbid (Hawes 1983).

The Wallows (unofficial name) is a shallow (0.6 m deep) pool which drains a small catchment almost entirely made up of a series of elephant seal wallows. It is extremely turbid during the summer months.

3 Methods

Water samples were collected from approximately 0.25 m depth, close to the centre of each lake. Samples were pumped through rigid plastic tubing using a diaphragm pump into opaque polythene bottles. Subsamples were taken for total nitrogen (TN) and total phosphorus (TP) before the samples were filtered (Whatman GF/C). The filtrate was analysed for nitrate (Mackereth et al. 1978), ammonia (Chaney and Marbach 1962), filterable reactive and total filterable phosphorus (FRP and TFP, Eisenreich et al. 1975). TN and TP were determined on the unfiltered samples

Table 1. Features of the five study lakes and their catchments[a]

Lake	MLD (m)	Z_{max} (m)	Area (ha)	Catch (ha)	CA/A	Dist (m)	Alt (m)	Snow	Moss	Ans
Khyber	60	3.0	0.13	4	31	700	150	5	0	0
Moss	225	10.5	1.54	9	6	700	48	3	1	0
Sombre	210	11.2	2.66	81	30	150	10	3	2	1
Heywood	427	6.4	4.50	41	11	200	4	0	4	4
Wallows	80	0.6	0.21	3	14	150	2	0	3	5

[a] *MLD* maximum linear dimension; Z_{max} maximum depth; *Area* lake surface area; *Catch* catchment area; *CA/A* ratio of catchment area:lake area; *Dist* distance from sea; *Alt* altitude. *Snow, Moss* and *Ans* indicate importance of permanent snowfields, terrestrial moss cover and marine animals in the catchments on a scale of 0–5.

using methods described by Mackereth et al. (1978) and Eisenreich et al. (1975) respectively. pH and alkalinity were determined on separate, unfiltered samples, using a PTI 55 pH meter and the Gran titration method described by Talling (1973).

Phytoplankton samples were collected in the same manner as water samples. Three replicate subsamples were concentrated by sedimentation after the addition of Lugol's solution for enumeration of common taxa on 21/1/88 and 14/2/88 and an overall mean calculated. Identification was to genus level only, using keys provided by Bourrelly (1972).

For chlorophyll-a estimation, the phytoplankton was separated into three size fractions by sequential filtration. The filters used had pore sizes of 20 (polyester mesh), 2 (Nuclepore) and 0.2 μm (Nuclepore), corresponding to net, nano and pico size fractions. Filters were extracted overnight into 10 ml of 95% methanol and chlorophyll-a estimated spectrophotometrically using equations given in Marker et al. (1980). Extracts were corrected for phaeopigments by acidification to 0.003 N with HCl and subsequent neutralisation with phenethylamine (Marker and Jinks 1982).

For estimation of photosynthetic uptake of carbon dioxide the ^{14}C bicarbonate method, modified after Vollenweider (1969) was used. Lake water was enclosed in 75 ml polycarbonate flasks, enriched with 5 μCi of ^{14}C bicarbonate (Amersham International, specific activity 56 mCi $mmol^{-1}$) and incubated either in situ or in the laboratory under controlled light and temperature conditions. For logistic reasons, in situ incubations for all samples, regardless of origin, were carried out in Heywood Lake at depths of 0, 1, 2, 4 and 6 m. Illumination for laboratory incubations was provided by a bank of cool white fluorescent tubes and temperature maintained at 2 ± 1 °C in water baths. Incubations lasted 1–6 h, depending on the source of the sample, and were terminated by rapid filtration through a series of 20, 2 and 0.2 μm filters.

^{14}C incorporated into material retained on the filters was fractionated into four intracellular pools (low molecular weight metabolites, lipid, protein, polysaccharide) using the solvent extraction system of Li et al. (1980). Radioactivity of the fractions was measured by liquid scintillation counting using an LKB Rackbeta model 1209 scintillation counter and the counting efficiency determined with an external standard.

Examination of lake benthos was carried out by scuba divers. Divers swam a series of transects across the lakes, and noted the type of vegetation at each depth and visually estimated the importance of the various assemblages. Samples were taken for identification purposes, but these were not quantitative. No experiments were undertaken on benthic material.

4 Results

Concentrations of most determinants, including all species of nitrogen and phosphorus increase from Moss, to Sombre, to Heywood lakes, and again to the Wallows (Table 2). This order agrees with the estimates of degree of catchment enrichment estimated from abundance of seals within catchments (Table 1). Khyber Lake provides the major exceptions to this pattern. The inorganic nitrogen components and concentrations of TFP, chloride and chlorophyll-a of this lake are all unexpectedly high.

The increased nutrient loading which has resulted from biotic enrichment has tended to raise concentrations of TP rather more than TN. Consequently the atomic ratio of TN:TP decreases along the enrichment gradient from 69.5 in Khyber to 20.6 in Moss and 3.8 in Heywood Lake. This increase in TP is parallelled by an increase in total chlorophyll-a concentration. Again Khyber Lake proves an exception to this simple trend, with higher chlorophyll-a concentrations than might be expected (Table 2). Size fractionation of chlorophyll-a shows picoplankton (0.2–2 μm) to comprise the major component in all water bodies (Fig. 2). Net phytoplankton (> 20 μm) has the highest relative abundance in The Wallows and Khyber Lake, in contrast to the nanoplankton (2–20 μm) which is slightly more abundant in the more oligotrophic Sombre and Moss Lakes.

Table 2. Water chemistry of the five study lakes[a]

Lake	TN	TP	NO_3^-	NH_4^+	TFP	FRP	TN:TP	CHL-a
Khyber	521	7.5	297	158	5.5	0.6	69.5	3.3
Moss	181	8.8	23	1	1.1	0.6	21.4	0.4
Sombre	246	13.6	71	2	4.8	0.8	5.2	5.4
Heywood	532	140.2	95	18	74.6	7.3	3.8	48.2
Wallows	13 705	830.0	5260	837	623	590	16.5	247

[a] All data are mg m^{-3}. *TN* total nitrogen; *TP* total phosphorus; NO_3^- nitrate; NH_4^+ ammonia; *TFP* total filtrable phosphorus; *FRP* filtrable reactive phosphorus; *CHL-a* chlorophyll-a

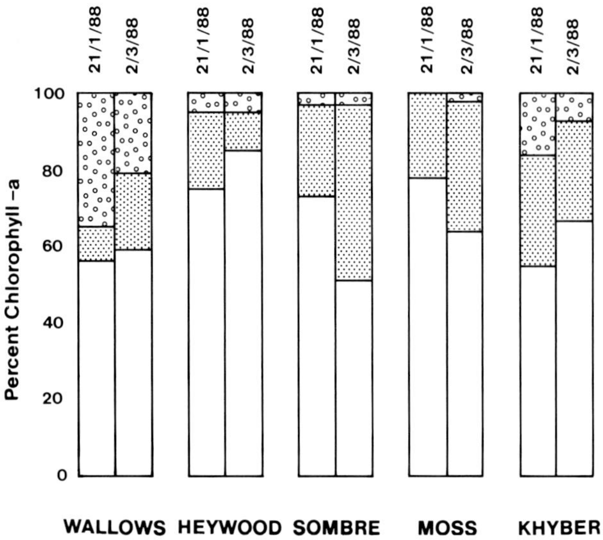

Fig. 2. Size fractionation of chlorophyll-a from the five study lakes on two sampling occasions. Percent total chlorophyll-a in three size categories is shown; net > 20 μm (*large dots*); nano 2–20 μm (*small dots*); pico < 12 μm (*open dots*)

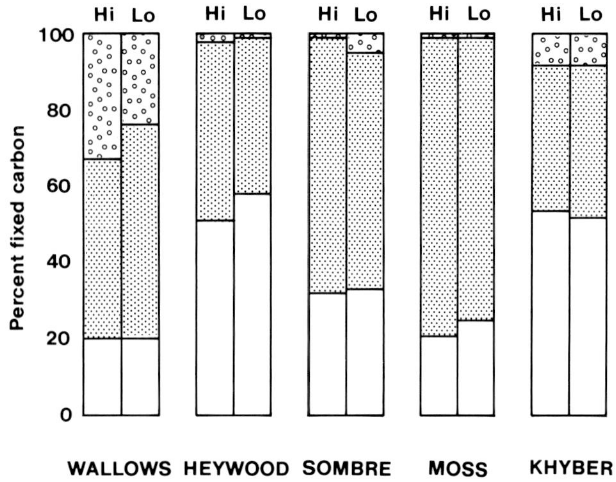

Fig. 3. Size fractionation of ^{14}C uptake under laboratory conditions. Size fractions shown as in Fig. 2. *Hi* and *Lo* refer to incubations at saturating and limiting radiation fluxes (100 and 18 μmol photons $m^{-2}\,s^{-1}$ respectively). Incubation temperature was 2 °C

The species composition of phytoplankton varied between the five water bodies (Table 3). In most cases, small (< 2 μm diameter) cyanophytes, tentatively identified as a *Synechococcus* sp., were numerically most abundant, but their dominance was most complete in Heywood Lake and the Wallows. The exception was Sombre Lake, where a small *Chlorella* was as abundant. The high proportion of net phytoplankton chlorophyll-a in the Wallows (Fig. 2) is attributable in Table 3 to larger, flagellated chlorophytes, notably *Chlamydomonas*.

The photosynthetic activity of the phytoplankton also increased with increasing nutrient concentrations. Maximum rates of photosynthesis for all size fractions were recorded in the Wallows and the lowest in Moss Lake (Table 4). Although picoplankton contributes most of the biomass (as chlorophyll-a and cell numbers; Fig. 2 and Table 3), laboratory incubations of phytoplankton samples on 2/3/88 show a disproportionately high uptake of carbon into the nanoplankton at both saturating and limiting radiation fluxes (Fig. 3). Varying proportions of fixed carbon entered

Table 3. Composition of phytoplankton samples from the five study sites[a]

Taxon	Size	Lake				
		Wallows	Heywood	Sombre	Moss	Khyber
Synechococcus	Pi	1.4×10^8	1.5×10^6	1.5×10^5	4.2×10^4	9.3×10^4
Chlorella	Pi/Na	4.9×10^7	4×10^5	3.6×10^5	2.8×10^4	2.2×10^4
Chlorogonium	Na	1.7×10^6	2×10^3	nd	nd	nd
Cryptomonas	Na	5×10^4	1.3×10^5	4×10^3	nd	nd
Ankistrodesmus	Na	6.6×10^6	nd	1.4×10^4	nd	1×10^3
Chrysochromulina	Na	1.3×10^5	1.6×10^5	4.3×10^4	1.3×10^4	2.4×10^4
Ochromonas	Na	nd	nd	2×10^3	nd	1×10^3
Unindent. flag.	Na	nd	4×10^3	nd	1.8×10^4	5×10^2
Ankyra	Na/Ne	nd	5×10^3	1×10^3	nd	nd
Phormidium	Na/Ne	nd	nd	2×10^3	nd	1×10^3
Chlamydomonas	Ne	5.2×10^6	2×10^4	nd	nd	nd
Total cell numbers		2×10^8	2.2×10^6	5.8×10^5	1×10^5	1.4×10^5

[a] Cell concentrations are given as cells ml^{-1}, *nd* indicates not detected. Size categories: *Ne* net (> 20 μm); *Na* nano (2–20 μm); *Pi* pico (< 2 μm). Values given are means of two samples taken on 21/1/88 and 14/2/88 (n = 3) for each sample).

Table 4. Maximum rates of photosynthesis observed in phytoplankton size fractions from each of the five study lakes, incubated in Heywood Lake on 15/2/88

Lake	Size fraction[a]			Total
	net (> 20 μm)	nano (2–20 μm)	pico (< 2 μm)	
Khyber	0.4 ± 0.1 (5)	4.4 ± 0.4 (57)	2.9 ± 0.6 (38)	7.7 ± 0.9
Moss	0.3 ± 0.1 (5)	3.2 ± 0.1 (58)	2.0 ± 0.2 (36)	5.5 ± 0.4
Sombre	0.3 ± 0.1 (3)	7.6 ± 0.7 (72)	2.7 ± 0.4 (25)	10.6 ± 1.4
Heywood	5.2 ± 0.6 (5)	52.5 ± 4.2 (55)	37.5 ± 7.2 (39)	95.2 ± 9.7
Wallows	134 ± 27 (2)	995 ± 40 (17)	2800 ± 610 (47)	5929 ± 650

[a] Rates are means of 4 replicates ± SE (mg carbon fixed $m^{-3} h^{-1}$). Values in parentheses are % total uptake.

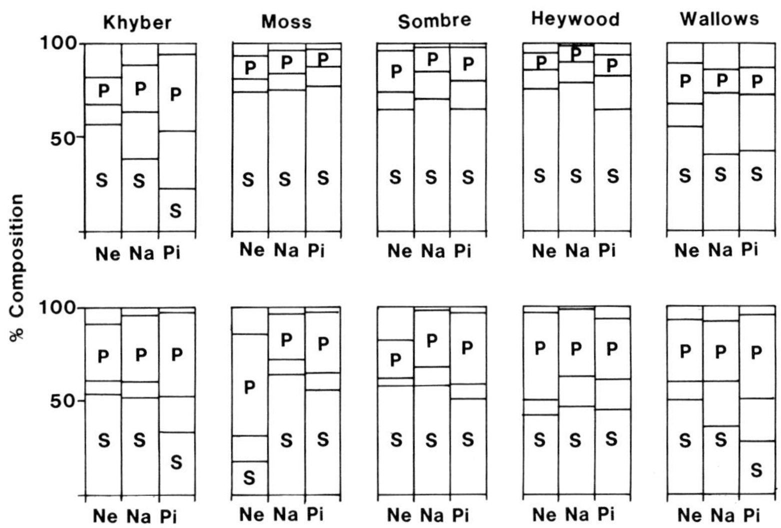

Fig. 4. Size fractionation of newly fixed photosynthate into four intracellular pools from data in Fig. 3. *Ne* net (> 20 mm); *Na* nano (2–20 μm); and *Pi* pico (< 2 μm). Intracellular components are (from bottom to top): *S* polysaccharide; lipid; *P* protein; *at top* metabolites

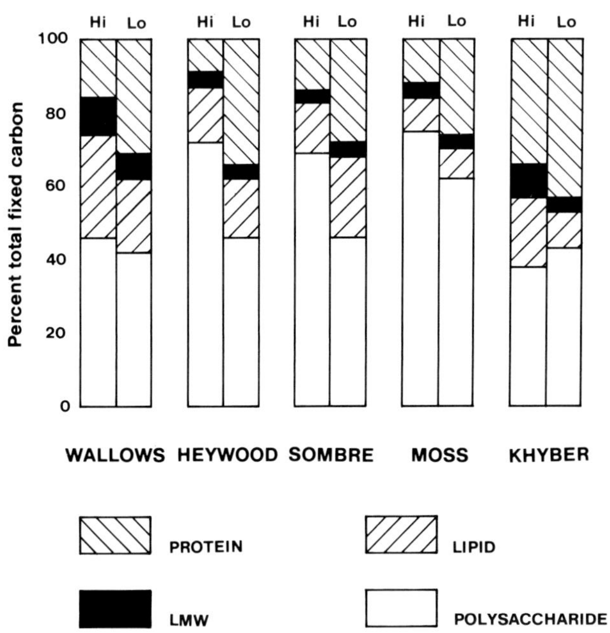

Fig. 5. Fractionation of newly fixed photosynthate into four intracellular pools. *Hi* and *Lo* refer to photon flux density at which samples were incubated (as Fig. 3)

the four intracellular pools examined (Figs. 4, 5). Greater proportions entered the protein pools of all fractions at low radiation flux than at the higher flux, this being reflected in lower proportions of fixed carbon in storage products (lipid and polysaccharide) under light-limiting conditions. Differences between the size fractions within and between lakes were slight, though there was a suggestion of higher specific protein synthesis rates in Khyber lake nanoplankton at high radiation flux (Fig. 4). Total phytoplankton photosynthate partitioning showed similar patterns to individual size fractions, with highest proportion of protein again in Khyber Lake (Fig. 5).

The disproportionately high contribution of nanoplankton to total carbon fixation rate seen in the laboratory incubations was evident in the field experiments, but only for The Wallows and Heywood Lake (Table 5). Very high chlorophyll-specific rates of photosynthesis (assimilation numbers) were recorded for the nanoplankton from the former, while nanoplankton rates in Heywood Lake were similar to those in Khyber Lake and only slightly higher than those in Sombre Lake. Moss Lake tended to have the lowest

Table 5. Chlorophyll-a specific rates of carbon fixation in phytoplankton size fractions from the five study lakes, incubated at a range of depths in Heywood Lake on 15/2/88

Lake	Depth (m)	Size fraction[a]			
		Net (> 20 μm)	Nano (2–20 μm)	Pico (< 2 μm)	Total
Wallows	0	1.2	3.2	0.8	1.7
	1	1.8	5.5	1.2	2.1
	2	3.0	11.5	7.3	8.6
	4	0.8	2.9	1.6	2.1
Heywood	0	4.4	1.9	0.5	1.1
	1	11.4	2.3	1.3	1.8
	2	8.7	2.4	1.1	1.7
	4	0.4	0.2	0.3	0.3
Sombre	0	0.3	0.6	1.1	0.6
	1	2.2	2.0	1.9	1.8
	2	1.6	1.5	1.1	1.3
	4	0.3	0.1	0.2	0.1
Moss	0	0.4	1.2	1.7	1.2
	1	0.7	1.2	1.7	1.2
	2	0.2	1.1	0.8	1.0
	4	0.1	0.1	0.1	0.1
Khyber	0	0.7	2.8	1.4	2.0
	1	1.2	2.3	2.6	2.3
	2	1.2	2.1	1.8	1.9
	4	0.0	0.2	0.0	0.1

[a] Rates are means of 4 replicates, units mg C mg^{-1} chlorophyll-a m^{-3}.

assimilation numbers for all size categories. The depth of maximum photosynthesis varied, being 2 m in The Wallows, 0–1 m in all others. In all cases, photosynthesis was undetectable at 6 m.

Although quantitative sampling of the lake benthos was not undertaken, scuba observations of the vegetation in the lake troughs showed that considerable differences existed between the five sites. No benthic vegetation was found on the floor of Heywood Lake or in The Wallows. In contrast, Moss Lake had a complex phytobenthos covering most of the lake floor below 5 m. Two communities dominate, one comprising aquatic mosses (mostly *Drepanocladus* and *Calliergon* spp.), with fronds up to 30 cm long and with a wide range of epiphytic algae (fully described by Priddle and Dartnall 1978). The other is a mat based on the cyanophyte *Tolypothrix*, which again has a host of diatom epiphytes (see Priddle and Belcher 1982; Oppenheim 1988 for more details).

The phytobenthos of Sombre Lake is less developed than Moss Lake, comprising mostly an incomplete mat of *Tolypothrix* with a few clumps of moss. Below 9 m, the upper limit of the winter oxycline, there is no permanent vegetation. The perennial mats in Sombre Lake are also characteristically overgrown in summer with dense, loosely attached growths of filamentous chlorophytes, mostly species of *Spirogyra* and *Mougeotia*, which were largely absent from Moss Lake.

The youngest of the lakes, Khyber, does not have perennial phytobenthos, but there is profuse growth of the annual filamentous alga (*Mougeotia*) on all of the exposed rock surfaces. There is little vegetation on the ice floor of the lake, but where sediment has collected in the scalloping on this ice, a variety of algae are found, particularly spores of the snow alga *Chlamydomonas nivalis* and species of *Cylindrocystis*, *Mesotaenium* and *Raphidonema*.

5 Discussion

The differences between the five lakes and their biota are profound. The sites range from a grossly enriched, highly productive pond, with vegetation dominated by phytoplankton (The Wallows), to an oligotrophic lake, with very clear water and very little phytoplankton, but a well-developed phytobenthos comprising perennial algae, cyanophytes and mosses (Moss Lake). The sites also vary in size, from small lakes, with high catchment area:lake volume ratio (Khyber Lake), to relatively large bodies with much smaller ratios (Moss Lake).

The lakes with highest nutrient concentration lie in the most highly vegetated and animal influenced catchments. The presence of birds and seals have a direct influence on the amount of N and P entering catchments (Allen et al. 1967) and thus reaching the lakes (Hawes 1983).

The stimulating effect of increased P loading to biomass of lake phytoplankton has been widely documented and there are numerous examples of predictive models linking the two (e.g. Dillon and Rigler 1974; Hickman 1980). Increased nutrient loading is also commonly accompanied by a change in species composition, as was observed in this study. Cold, oligotrophic alpine lakes are commonly dominated by small flagellates (Pechlaner 1971; Maeda and Ichimura 1973; Tilzer 1973) similar to those found in Moss and Sombre lakes. Increasing P loading, especially where this is accompanied by a fall in N:P ratio, often leads to an increased abundance of cyanophytes and increasing domination of water bodies by a few taxa (Margalef 1964; Reynolds 1984) as was found in Heywood Lake and The Wallows.

The unusually high nutrient concentrations in Khyber Lake reflect the high concentrations in Signy Island snowmelt (Allen et al. 1967; Hawes 1983, 1988). With a low volume and large catchment area this lake will have a rapid flowthrough of water, in contrast e.g. to Moss Lake, which will have much longer hydraulic residence times. Rapid flowthrough in Khyber Lake may result in relatively high rates of nutrient supply, despite low concentrations. In another Signy Island lake, Spirogyra Lake, a similar phenomenon occurs and results in a dense growth of benthic filamentous algae, capable of stripping nutrients from the overlying water (Hawes 1988). A similar effect may have resulted in the growth of *Mougeotia* seen in Khyber, and to a lesser extent Sombre Lakes.

Higher rates of photosynthesis in the nutrient-rich than the nutrient-poor lakes reflects both the increased biomass and, particularly in The Wallows, a very high chlorophyll-specific rate. Such high rates have been reported previously for Signy Island (Light et al. 1981; Hawes 1983). These high values may be related to the nutrient status of The Wallows phytoplankton; high assimilation numbers have previously been related to high internal nutrient concentrations (Senft 1978).

In small lakes like Khyber, with a potentially high rate of flow through, the phytoplankton population may be limited to a density which can maintain a high enough growth rate to persist. The high proportion of newly fixed carbon which enters the protein pool in Khyber Lake, compared to the other lakes, can be interpreted as indicating high specific growth rates. Studies elsewhere have shown that the amount of fixed carbon which enters the protein pool bears a close relationship to growth rate, while polysaccharides accumulate when growth rates are low compared to photosynthetic rate (Morris et al. 1974; Morris 1981; Li and Harrison 1982; Lancelot et al. 1986). All cell size categories studied may therefore be maintaining a higher specific growth rate in Khyber Lake than in most of the other water bodies.

The proportions of 20–40% fixed carbon as proteins and 40–80% as storage products are otherwise remarkably similar in all lakes despite the range of conditions under which populations developed. In North American lakes, increases in the proportions of storage materials (mostly lipid) have been observed under conditions of nutrient limitation (Groegger and Kimmel 1988). The proportions seen in this study appear to be typical of phytoplankton from a wide range of environments (e.g. Morris 1981; Li and Harrison 1982; Lancelot et al. 1986), including continental Antarctic lakes (Priscu et al. 1987). The increase in relative carbon flow into protein at low radiation flux is also commonly observed and results from saturation of protein synthesis at lower radiation fluxes than photosynthesis (Morris 1981).

Although The Wallows is such a shallow water body, the high concentration of clorophyll-a must make it optically deep. Using the relationship between chlorophyll-a concentration and the vertical extinction coefficient established by Talling (1957), an extinction coefficient of 5 can be calculated for this water body. Extinction of light by other suspended and dissolved material will increase this value and extinction of surface radiation to 1% will occur at a depth of less than 40 cm. Self-shading may therefore be an important consideration in this lake where nutrients are unlikely to be limiting growth, particularly in precluding the development of phytobenthos. A similar effect may be operating in Heywood Lake. The absence of benthic plants from Heywood Lake and The Wallows may, however, also be related to the effects of winter deoxygenation. Eutrophic Antarctic lakes frequently undergo winter anoxia, as a result of overwinter breakdown of organic material entering the system during summer (e.g. Ellis-Evans 1985).

The vegetation of maritime Antarctic lakes is related to the degree of enrichment of their catchments by marine animals. Changes in utilistion of the catchments by marine animals or other perturbations of nutrient supply are likely to result in changes in the lake biota. In the five lakes studied here phytoplankton abundance and rate of photosynthesis increase with increasing nutrient concentration, until densities where self-shading may occur are attained. In all lakes, small cyanophytes and chlorophytes were numerically dominant and picoplankton (< 2 μm) comprised most suspended chlorophyll-a. In eutrophic lakes, large flagellated algae occurred, whereas in oligotrophic lakes nanoplanktonic flagellates (2–20 μm) were common. There was little change in the partitioning of photosynthate between protein and storage substances between lakes. Phytobenthos only developed in the clear, nutrient-poor lakes.

Acknowledgements. Thanks are due to the British Antarctic Survey for providing facilities for this work and to all members of Signy Island station for assistance with fieldwork. Rupert Wedgewood obtained the under-ice oxygen profiles. Drs. J. C. Ellis-Evans, H. Burton and D. Adamson commented on the manuscript.

References

Allen SE, Grimshaw HM, Holdgate MW (1967) Factors affecting the availability of plant nutrients on an Antarctic island. J Ecol 55:381–396

Bourrelly (1972) Les algues d'eau douce. Initiation à la systémique I. Les Algues vertes (2nd edn). N. Boubée, Paris

Chaney AL, Marbach EP (1962) Modified reagents for the determination of urea and ammonia. Clin Chem 8:71–76

Dillon PJ, Rigler FH (1974) The phosphorus-chlorophyll relationship in lakes. Limnol Oceanogr 19:767–773

Ellis-Evans JC (1985) Decomposition processes in maritime Antarctic lakes. In: Seigfried WR, Laws RM, Condy PA (eds) Antarctic nutrient cycles and food webs. Proc 4th Symp Antarct Biol, Wilderness, Sept 1983. Springer Berlin Heidelberg New York, pp 253–260

Eisenreich SJ, Bannerman RT, Armstrong DE (1975) A simplified phosphorus analysis technique. Env Letters 9:43–53

Groeger AW, Kimmel BL (1988) Photosynthetic carbon metabolism by phytoplankton in a nitrogen-limited reservoir. Can J Fish Aquat Sci 45:720–730

Hawes I (1983) Nutrients and their effects on phytoplankton populations in lakes on Signy Island, Antarctica. Polar Biol 2:115–126

Hawes I (1988) The seasonal dynamics of *Spirogyra* in a shallow, maritime Antarctic lake. Polar Biol 8:429–437

Heywood RB, Dartnall HJG, Priddle J (1980) Characteristics and classification of the lakes of Signy Island, South Orkney Islands, Antarctica. Freshwater Biol 10:47–59

Hickman M (1980) Phosphorus chlorophyll and eutrophic lakes. Arch Hydrobiol 88:137–145

Lancelot C, Mathot S, Owens NJP (1986) Modelling protein synthesis, a step to an accurate estimate of net primary production: *Phaeocystis pouchetii* colonies in Belgian coastal waters. Mar Ecol Prog Ser 32:193–202

Laws RM (1956) The elephant seal (*Mirounga leonina* Linn.) II: General, social and reproductive behaviour. FID Surv Sci Rep no 68:124 pp

Li WKW, Harrison WG (1982) Carbon flow into the end products of photosynthesis in short and long incubations of a natural phytoplankton population. Mar Biol 72:175–182

Light JJ, Ellis-Evans JC, Priddle J (1981) Phytoplankton ecology in an Antarctic lake. Freshwater Biol 11:11–16

Mackereth JFH, Heron J, Talling JF (1978) Water analysis: some revised methods for limnologists. FBA Sci Publ no 36. Ambleside, Freshwater Biological Association

Maeda O, Ichimura S (1973) On the high density of a phytoplankton population found in a lake under ice. Int Rev Gesamten Hydrobiol 58:673–685

Margalef R (1964) Correspondence between the classic types of lakes and the structural and dynamic properties of their populations. Verh Int Ver Limnol 15:169–175

Marker AFH, Nusch EA, Rai H, Reimann B (1980) The measurement of photosynthetic pigments in freshwaters and standardisation of methods: conclusions and recommendations. Ergeb Limnol 14:91–106

Marker AFH, Jinks S (1982) The spectrophotometric analysis of chlorophyll-a and phaeopigments in acetone, ethanol and methanol. Arch Hydrobiol 16:3–17

Morris I (1981) Photosynthetic products, physiological state and phytoplankton growth. In: Platt T (ed) Physiological bases of phytoplankton ecology. Can Bull Fish Aquat Sci 210:83–102

Morris I, Glover HE, Yentsch CS (1974) Products of photosynthesis by marine phytoplankton: the effect of environmental factors on the relative rates of protein synthesis. Mar Biol 27:1–9

Pechlaner R (1971) Factors that control the production rate and biomass of phytoplankton in high-mountain lakes. Mitt Int Ver Theor Angew Limnol 18:125–145

Priddle J (1980) The production ecology of benthic plants in some Antarctic lakes I. In situ studies. J Ecol 68:141–153

Priddle J, Belcher JH (1982) An annotated list of benthic algae (excluding diatoms) from freshwater lakes on Signy Island. Br Antarct Surv Bull 57:41–53

Priddle J, Dartnall HJG (1978) The biology of an Antarctic aquatic moss community. Freshwater Biol 8:469–480

Priddle J, Heywood RB (1980) Evolution of Antarctic lake ecosystems. Biol J Linn Soc 14:51–66

Priscu JC, Priscu LR, Vincent WF, Howard-Williams C (1987) Photosynthate partitioning by microplankton in permanently ice-covered Antarctic desert lakes. Limnol Oceanogr 32:260–270

Reynolds CS (1984) The ecology of freshwater phytoplankton. Cambridge University Press, Cambridge

Senft WH (1978) Dependence of light-saturated rates of algal photosynthesis on intracellular concentrations of phosphorus. Limnol Oceanogr 23:709–718

Smith RIL (1988) Destruction of Antarctic terrestrial ecosystems by a rapidly increasing fur seal population. Biol Conserv 45:55–72

Talling JF (1957) The phytoplankton population as a compound photosynthetic system. New Phytol 56:133–149

Talling JF (1973) The application of some electrochemical methods to the measurement of photosynthesis and respiration in freshwater. Freshwater Biol 3:335–362

Tilzer M (1973) Diurnal periodicity of the phytoplankton assemblage of a high mountain lake. Limnol Oceanogr 18:15–30

Vollenweider RA (1969) A manual on methods for measuring primary production in aquatic environments. IBP handbok No 12, Blackwell, Oxford

A Preliminary Study of Benthic Diatoms in Contrasting Lake Environments

D. R. Oppenheim [1]

Summary. Eleven of the 17 freshwater lakes from Signy Island, South Orkney Islands, Antarctica were sampled by snorkel divers. A list of diatom taxa identified from benthic samples is presented. Most freshwater taxa listed were rare and only a few taxa were observed abundantly in many lakes. Assemblage composition varied between three broad lake categories: (1) proglacial; (2) oligotrophic; and (3) mesotrophic lakes. Assemblage composition also changed with lake depth in the larger lakes. A second, more detailed study was undertaken on an oligotrophic and a mesotrophic lake. No seasonal trends could be identified. The limitation of the methods used are discussed, and spatial and temporal variability in Antarctic lakes is considered.

1 Introduction

Large numbers of inland water bodies occur in Antarctica, particularly in ice-free areas and it would appear that benthic vegetation is a common feature of such systems (Priddle 1985). The freshwater algae have been described from South Georgia (Carlson 1913), South Orkneys (Fritsch 1912), the Antarctic Peninsula (Corte 1962; Parker et al. 1972), and many areas of the Antarctic Continent (Baker 1967; Opalinski 1972a, b; Wharton et al. 1983). Most work on freshwater Antarctic diatoms has focued on taxonomy (Reinsch 1890, Fukushima 1965, 1966; Kobayashi 1963a, b, 1965).

The present chapter represents some of the preliminary results of an ecological study of benthic diatoms inhabiting several freshwater lakes at Signy Island (60° 43′ S, 45° 38′ W). Initially a qualitative examination of 54 samples from 11 of the 17 different lakes was undertaken to identify the species present and to give a rough estimate of the dominant taxa present within the lakes examined. The assemblage structure of the epiphytic diatoms associated with filamentous mat-forming algae was then studied in oligotrophic Sombre Lake and mesotrophic Light Lake in more detail. Three sites in each lake were sampled at 4–6 week intervals from January 1986 to March 1987. The results presented show a preliminary list of the taxa identified, how composition within the mats of the lakes differd, and seasonal changes within both lakes are considered.

2 Materials and Methods

2.1 Study Area

Information about the main lakes of Signy Island is given in Heywood et al. (1980). The Khyber lakes have emerged with ice recession over the past 5–10 years and are considered to be proglacial. Such lakes vary in size according to levels of meltwater, but are of uniformly low nutrient content within the water column (Priddle and Heywood 1980). Sombre Lake and Moss Lake are typical oligotrophic systems with well-established cyanobacterial mats and benthic moss communities, while Heywood Lake and Bothy Lake represent nutrient enriched mesotrophic lakes with more developed phytoplankton communities. The 11 lakes sampled represent a wide range of environmental conditions, but all are ice-covered for at least 8 months each year. The location of all lakes is shown in Fig. 1a.

Sombre Lake and Light Lake (Fig. 1b, c) were chosen to represent two contrasting (oligotrophic and mesotrophic respectively) lakes not only for their differing water chemistry, but also due to their differences in surface area and bathymetry.

2.2 Collection of Mats for Initial Qualitative Analysis

Snorkel divers sampled the 11 lakes taking grab samples over a period of 3 months (February to April 1985) when the lakes were ice-free. All samples con-

[1] British Antarctic Survey, High Cross, Madingley Road, Cambridge CB3 0ET, UK.

Antarctic Ecosystems. Ecological Change and Conservation
(ed. by K. R. Kerry and G. Hempel)
© Springer-Verlag Berlin Heidelberg 1990

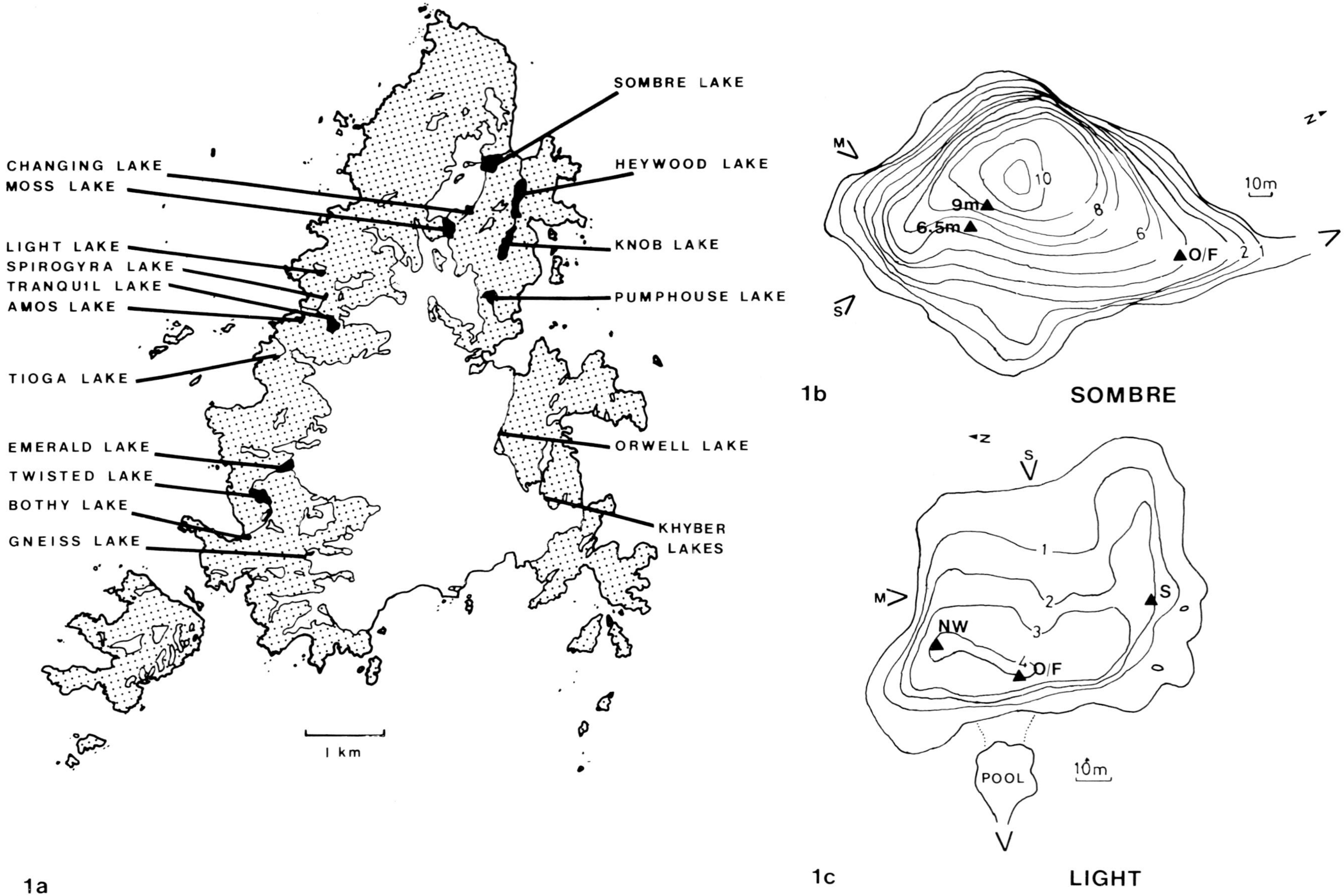

Fig. 1. **a** Map of Signy Island showing location of the 11 lakes studied. Bathymetric maps of **b** Sombre Lake, **c** Light Lake. *Arrows* indicate direction of water flow; *S* = Inflow stream; *M* = meltwater direct from snow slope; *O/F* = outflow site; *S* = south site; *NW* = northwest site

tained algal mats except the proglacial lakes (which were purely sediment). These were stored at − 20 °C and forwarded to the United Kingdom (UK) for subsequent analysis.

2.3 Collection of Mats from Sombre Lake and Light Lake

Six permanent sites were marked (three in each lake) using a surface rope. All algal mats were sampled from the surface using a modified form of the corer described in Ellis-Evans (1982), extruded and frozen at − 20 °C for transport to the UK.

2.4 Sampling for Diatom Enumeration

To obtain a representative sample from each mat, five subsamples were taken using a 0.5 cm diameter minicorer. These were then placed together in concentrated sulphuric acid, and oxidised using the potassium permanganate procedure described by Hendey (1964). The acid-cleaned material was mounted in Naphrax (R.I. 1.74) and 500 valves counted on each slide at 1000 × magnification using a Wild M20 microscope. A variety of methods have been attempted, but only Hendey's (1964) method achieved complete oxidation of organic debris. All untreated samples were examined to check for damage before counting. Further work on comparative sampling strategies has been undertaken and is discussed elsewhere (Oppenheim and Ellis-Evans 1989).

3 Results

A preliminary list of taxa identified is presented in Table 1. The list is deceptively small since many taxa have been omitted because further taxonomic work is required, particularly within the genera: *Achnanthes, Cymbella, Gomphonema* and *Navicula*. Considerable confusion exists for example in the clear identification of *Eunotia exigua* (Bréb. ex Kütz.) Rabenh. and *E. paludosa* Grun. as a whole range of intermediate morphological forms were observed between these two species. Several of the taxa identified are believed to be contaminants from the marine environment, but were only observed rarely in a few samples.

Most of the freshwater taxa listed were rare and occurred in low numbers in all the samples studied. Most of the taxa listed in Table 1 were found in all samples in all the lakes. However, when the relative abundance of taxa within each sample was examined, it was possible to identify distinguishing features in assemblage composition. Table 2 lists the dominant taxa identified from the different lakes studied. The 11 lakes examined were divided into three broad categories, although some differences in dominants found in the lakes within each category were observed.

The proglacial lakes were very distinctive from the other systems. The taxa identified, such as the *Eunotia* spp., were either not observed in the other lakes or, as in the case of *Achnanthes incognita* var., were only rarely observed in other lakes. The oligotrophic lakes were typically colonised by *Achnanthes minutissima*, and to a lesser degree by *Cymbella minuta*, while in the mesotrophic lakes *Gomphonema parvulum* and *Navicula veneta* were both very abudant. Another characteristic feature of the nutrient-enriched lakes was the greater abundance and diversity of *Nitzschia* spp. The dominant taxa observed in Changing Lake did not fit into any of the three main categories, but represented a mixture of taxa found in both the oligotrophic and mesotrophic lakes.

Certain features were common to all the lakes examined, most notably the diversity and abundance of small elliptic *Achnanthes* spp. While many *Achnanthes* spp. were present, different species were found within different lakes. Some taxa were common in all the lakes with no particular pattern in distribution e.g. *Synedra rumpens* var. *familiaris* and *Pinnularia abaujensis*. Other taxa were observed in high numbers of specific localities which could not be correlated to the nutrient status of the lake, viz. *Stauroneis anceps* var. *hyalina*, and *St. anceps* var. *linearis* which were always observed in the same assemblages.

In all samples examined mats taken from the same depth within the same lakes had the same composition of co-dominant taxa, varying only slightly in the proportional representation of individual taxa. Greater differences in assemblage composition were observed from samples taken at varying depths within a lake. The greatest differences in floristic composition occurred between lakes. Another distinctive feature of all benthic samples studied was that no species typical of open water or littoral phytoplankton populations, such as *Cyclotella, Asterionella, Stephanodiscus* or *Tabellaria,* were observed in any of the samples studied.

3.1 Quantitative Studies in Assemblage Composition of Sombre Lake and Light Lake

Regular sampling at the six permanently marked sites in the two lakes did not show distinct seasonal changes

Table 1. A preliminary checklist of species, subspecies and forms identified from Signy Island

Achnanthes
- *abundans* Manguin in Bourelly and Manguin
- *brevipes* Ag.
- *brevipes* var. *arctica* Kobayashi
- *delicatula* var. *australis* Manguin in Bourelly and Manguin
- *germainii* Manguin in Bourelly and Manguin
- *incognita* Krasske
- *kryophiloides* Foged
- *lanceolata* (Bréb. ex Kütz.) Grun. in Cleve et Grun.
- *lapponica* var. *ninckei* (Guerm. et Manguin) Reim. in Patrick and Reim.
- *linearis?*
- *manguinii* var. *elliptica* Manguin in Bourelly and Manguin
- *minutissima* Kütz.
- *pinnata* Hust.?
  - var. *japonica* Hust.?

Amphora
- *veneta* Kütz.
  - var. *capitata* E. Y. Haworth

Caloneis
- *bacillum* (Grun.) Cleve
- *ventricosa* var. *truncatula* (Grun.) Meister

Cocconeis
- *disculoides* Hust.[a]
- *disculus* (Schum.) Cleve[a]
- *fasciolata* (Ehrenb.) Brown[a]
- *utermoelii* Hust.

Cymbella
- *aubetii* Manguin in Bourelly and Manguin
- *minuta* Hilse ex Rabenh.
- *ventricosa* var. *obtusa* (Grun. in Schmidt) Cleve

Eunotia
- *exigua* (Bréb. ex Kütz.) Rabenh.
- *fallax* A. Cleve
- *paludosa* Grun.
- *praerupta* var. *inflata* Grun. in V.H.

Fragilaria
- *construens* var. *subsalina* Hust.
- var. *venter* (Ehrenb.) Grun.
- *elliptica* (Schum.) Haworth
- *intermedia* (Grun.) Grun. in V.H.
- *pinnata* Ehrenb.
  - var. *accuminata* A. Mayer
  - var. *lancettula* (Schum.) Hust. in A. Schmidt
- *vaucheriae* var. *longissima* Manguin in Bourelly and Manguin
- *virescens* var. *exigua* Grun. in V.H.

Gomphonema
- *angustatum* (Kütz.) Rabenh.
- *kamtschaticum* Grun. var.[a]
- *littorale* Hendey[a]
- *parvulum* (Kütz.) Kütz.

Hantzschia
- *amphioxys* var. *recta* O. Müller
  - f. *capitata* Kobayashi

Navicula
- *ammophila* var. *degenerans* Grun.
- *bicephala* Hust.
- *capitata* var. *hungarica* (Grun.) Ross
- *cuspidata* (Kütz.) Kütz.
- *dicephala* f. *australis* Manguin in Bourelly and Manguin
- *difficillima* Hust.
- *evanida* Hust.?
- *fragilarioides* Krasske
- *indifferens* Hust.
- *muralis* Grun. in V.H.
- *muticopsis* V.H.
  - f. *capitata* Carlson
- *pseudoingrata* Manguin var.
- *pseudomuralis* Hust. in A. Schmidt
- *pseudoscutiformis* Hust. in A. Schmidt
- *schönfeldii* Hust. (non Bourelly and Manguin)
- *simplex* Krasske
- *veneta* (Kütz.) Lange-Bertalot

Nitzschia
- *angulata* Hasle[a]
- *curta* (V.H.) Hasle[a]
- *cylindrus* (Grun. ex Cleve) Hasle[a]
- *frustulum* var. *perminuta* Grun.
- *gracilis* Hantzch.

Opephora
- *martyi* Hérib.

Pinnularia
- *abaujensis* (Pant.) Ross
- *biceps* f. *petersenii* Ross
- *borealis* Ehrenb.
- *cardinalis* (Ehrenb.) W. Sm
- *globiceps* Gregory
  - var. *krookei* (Grun.) Cleve
- *ignobilis* Krasske
- *krasskei* var. *ventricosa* Hust.
- *lapponica* Hust.
- *microstauron* var. *elongata* Bourelly and Manguin
- *molaris* (Grun.) Cleve
- *norrlandica* Cleve-Euler
- *viridis* var. *lata* Frenguelli

Rhopalodia
- *gibberula* var. *vanheurckii* O. Müller[a]

Stauroneis
- *anceps* var. *hyalina* M. Perag. et Brun. in Hérib.
  - var. *linearis* Rabenh.
  - f. *gracilis* Rabenh.

Surirella
- *ovata* var. *pinnata* (W. Sm.) Brun.

Synedra
- *rumpens* Kütz.
  - var. *familiaris* (Kütz.) Hust.

Thalassiosira
- *decipiens* (Grun.) Jørg.[a]

[a] Probable marine contaminants.

Table 2. Dominant taxa identified from the 11 lakes examined

Proglacial Khyber Lakes (Sediments)		
K1	K2	K4
Eunotia exigua *E. Paludosa* *Navicula* sp. *Achnanthes incognita*	*Eunotia exigua* *E. paludosa* *Pinnularia ignobilis*	*Nitzschia gracilis* *Pinnularia abaujensis*
Oligotrophic lakes (algal mats)		
Sombre Lake (7 m)	Emerald Lake (7 m)	Pumphouse Lake (2–3 m)
Synedra rumpens var. *familiaris* *Achnanthes minutissima* *Cymbella minuta* *Gomphonema angustatum*	*Navicula* sp. *Achnanthes minutissima* *Nitzschia* spp. *Gomphonema angustatum* *Navicula veneta* *Amphora veneta* *Stauroneis anceps* f. *gracilis*	*Gomphonema parvulum* *Achnanthes pinnata* var. *japonica?* *Nitzschia frustulum* var. *perminuta* *Achnanthes minutissima* *Synedra rumpens* var. *familiaris*
Sombre Lake (9–10 m)	Emerald Lake (13 m)	Moss Lake (10 m)
Stauroneis anceps var. *hyalina* *S. anceps* var. *linearis* *Navicula pseudoingrata*	*Synedra rumpens* var. *familiaris* *Achnanthes minutissima* *Gomphonema angustaum* *Caloneis bacillum* *Stauroneis anceps* f. *gracilis*	*Achnanthes minutissima* *A. linearis?* *Nitzschia frustulum* var. *perminuta* *Achnanthes* sp. *Synedra rumpens* var. *familiaris*
Mesotrophic lakes (algal mats)		
Light Lake (3 m)	Bothy Lake (3 m)	Twisted Lake (4 m)
Nitschia frustulum var. *perminuta* *Navicula veneta* *Stauroneis anceps* var. *hyalina* *Nitzschia gracilis* *Synedra rumpens* var. *familiaris* *Amphora veneta* var. *capitata* *Achnanthes pinnata?* *Gomphonema parvulum*	*Nitzschia gracilis* *Gomphonema parvulum* *Achnanthes krophiloides* *Nitzschia* spp. *Nitzschia frustulum* var. *perminuta* *Fragilaria intermedia*	*Nitzschia* spp. *Fragilaria virescens* var. *exigua* *Navicula veneta* *Gomphonema parvulum* *Gomphonema angustatum* *Amphora veneta* var. *capitata* *Nitzschia frustulum* var. *perminuta*
Heywood Lake (5 m)		
Achnanthes pinnata var. *japonica?* *A. abundans* *Gomphonema parvulum* *Achnanthes krophiloides* *Fragilaria elliptica*		

(Fig. 2). At the south site and to a lesser degree the north-west site in Light Lake, the results indicated that *Nitzschia frustulum* var *perminuta* and a *Nitzschia sp.* were more abundant during periods of open water, while other taxa such as *S. rumpens* var. *familaris* and *Fragilaria* spp. (an aggregate largely composed of *F. pinnata* var *accuminata*) reached maximum numbers during periods of ice cover. However, this seasonal pattern was not detected at any of the other sites. No seasonal trends were observed in counts recorded from Sombre Lake. Instead, the assemblage composition appeared to be a continuous presence of several co-dominant taxa present at differing levels of abundance. Certain taxa did, however, vary in number seasonally. *St. anceps* var. *hyalina* was recorded in highest numbers in the austral winter months at the 9-m site, while *Fragilaria virescens* var. *exigua* was recorded in highest numbers during periods of open water in the summer.

4 Discussion

Komarek and Ruzicka (1966) found diatoms to be only minor contributors to the algal assemblages found near Novolazerevskaya station which is located on the Antarctic continent. Epiphytic diatoms within

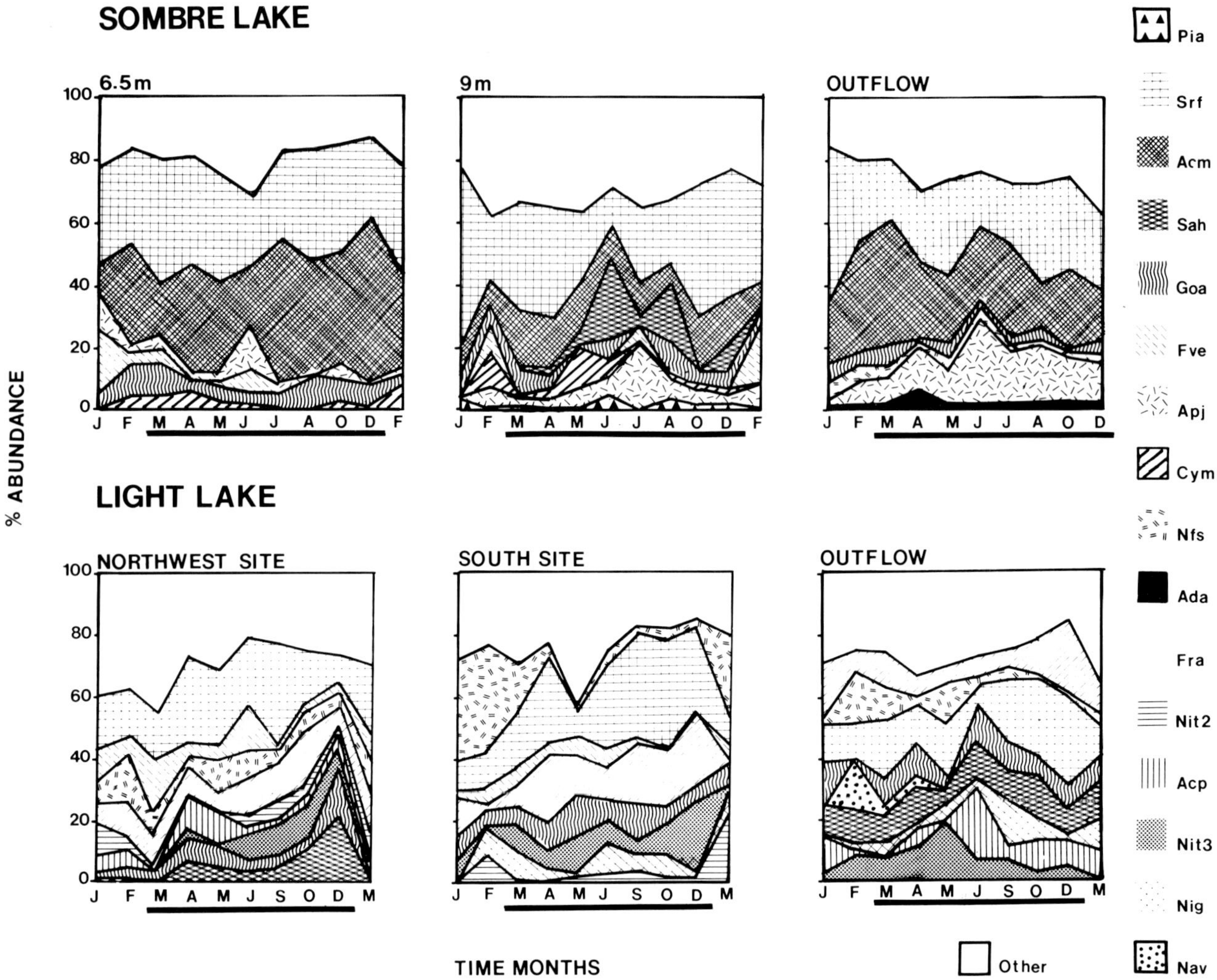

Fig. 2. Seasonal changes recorded for the six permanently marked sites in Sombre Lake and Light Lake. **Ada** = *Achnanthes delicatula* var. *australis*; **Acm** = *Achnanthes minutissima*; **Apj** = *Achnanthes pinnata* var. *japonica?*; **Cym** = *Cymbella minuta;* **Fra** = *Fragilaria* spp., *virescens* var. *exigua*; **Goa** = *Gomphonema angustatum;* **Nav** = *Naviula veneta;* **Nfs** = *Nitzschia frustulum* var. *perminuta;* **Nig** = *Nitzschia gracilis*; **Nit2** = *Nitzschia* sp.2; **Nit3** = *Nitzschia* sp.3; **Pia** = *Pinnularia abaujensis*; **Sah** = *Stauroneis anceps* var. *hyalina*; **Srf** = *Synedra rumpens* var. *familiaris*. Bar shows period of ice cover

the algal mats on Signy Island are extremely abundant, a finding in agreement with Opalinski (1972b) from continental sites. Some of the diatom assemblages from other Antarctic regions are very different from the taxa reported in this study. The pioneer taxa *Hantzschia amphyoxis* f. *capitata* and *Pinnularia borealis* reported from the Schirmacher Oasis ponds (Aleshinskaya and Bardin 1966) were rarely recorded at Signy Island. Similarly, the diatom assemblage associated with sea spray, reported by Karasawa and Fukushima (1977), was not observed on Signy, although *Navicula muticopsis* (a dominant member of this assemblage) and other closely allied varieties were recorded in low numbers.

Lavrenko (1965) reported a cosmopolitan diatom flora from a lake near Novolazerevskaya station. The Signy Island flora is also largely cosmopolitan. Many of the less commonly reported taxa identified from Signy Island have been previously observed on the sub-Antarctic Kerguelen Islands (Bourrelly and Manguin 1954; Le Cohu and Maillard 1983).

The initial qualitative sampling of algal mats from the 11 lakes gave some indication of the diversity of the epiphytic assemblages present. Dominant algal groups vary in lakes of differing nutrient status (Happey-Wood and Priddle 1984), and the classification of lakes according to floristic composition has long been recognised as a useful tool in diatom studies (Jørgensen 1948; Round 1957). The presence of abundant populations of *Nitzschia* spp. in more nutrient-enriched waters has been observed (Lange-Bertalot 1979), and the abundance of *Eunotia* in sediments and under more acidic conditions has also been reported elsewhere (Jørgensen 1948). However, the *Eunotia* spp. identified in the proglacial lakes have also been reported to grow on mosses (Petersen 1950).

The high numbers of *A. minutissima* in the oligotrophic lakes are difficult to explain, as this species is widely reported from temperate and high latitudes, growing on *Equisetum fluviatile* in Finland (Kairesalo 1984), in sediments (Hickman 1974), as an important member of assemblages growing on mosses (Round 1984) and in soils (Behre and Schwabe 1970). The species is clearly capable of growing under a much wider range of conditions than the limited distribution observed on Signy.

The absence of any typical planktonic freshwater diatoms from any of the samples prompts the question of the role of phytoplankton input. *Stephanodiscus astraea* var. *intermedia* Fricke and *Tabellaria fenestrata* (Lyngb.) Kütz. have been identified from Lake Glubokoye (Lavrenko 1965), a continental Antarctic lake which is much deeper than any of the Signy lakes and has far less snow cover than maritime Antarctic lakes (Ellis-Evans pers. commun.). Clark and Runnels (1975) concluded that no taxonomically defined diatom phytoplankton establishes in small ponds. The same species occur in the phytoplankton and epiphyton but exist as separate populations. In contrast, Brown and Austin (1973) reported an interaction between the two and could relate percentage abundance of common species in a shallow eutrophic lake. If there is an interaction between phytoplankton and the benthic diatoms in the Signy Island lakes, then the phytoplankton may be made up of benthic species. Further study of this interaction in the Signy Island lakes is required.

In studies from more temperate latitudes seasonal fluctuations of epiphytic algal populations is well established in terms of changes in algal biomass (Kairesalo 1984), seasonal abundance (Gumtow 1955), species composition (Jones and Mayer 1983) and photosynthetic response (Jones and Adams 1982). The sampling technique used in the present study unfortunately does not permit the distinction between living and dead material. This limitation masks any seasonal variation of the actively growing epiphytic diatom populations present, and may explain the lack of seasonal variation shown in Fig. 2. Using artificial substrata would be an alternative means of sampling the growing population without incurring the penalty of significant input of dead or senescent components of the epiphytic assemblages. However, the use of artificial substrata has its limitations (Wetzel 1964) and

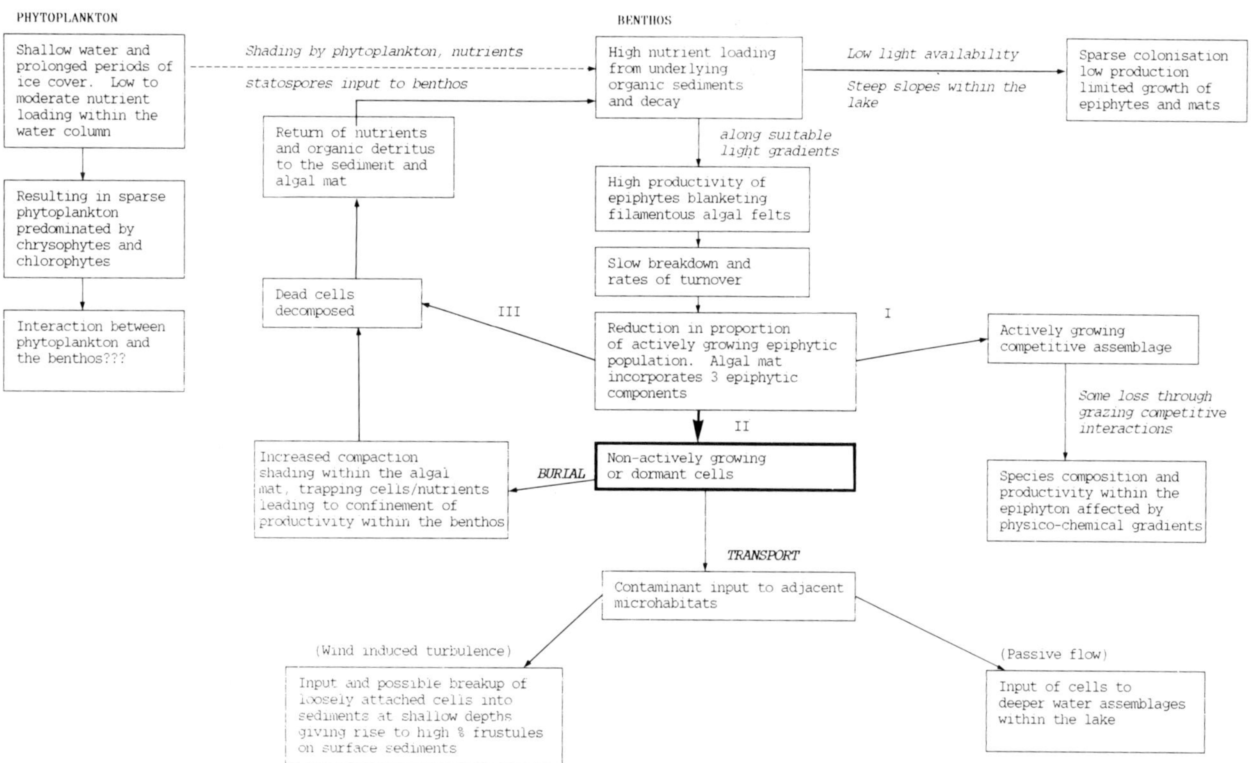

Fig. 3. Diagramitc flow chart of loss and growth processes leading to an abundance of senescent epiphytic cells incorporated into the filamentous algal mats

has been shown to inadequately reflect the true diversity present on natural substrates (Cattaneo and Kalff 1978). A more detailed quantitative study of the composition and morphology of the algal mats is needed to provide more information necessary for better interpretation.

Given that high levels of nutrients are available in the underlying sediments within the lakes (Ellis-Evans and Lemon 1989), and rates of breakdown and turnover are slow (Ellis-Evans 1982), then the sediment environment must have a significant effect on the overlying perennial algal mats. Input from the phytoplankton in the more oligotrophic lake is dominated by cryptophytes during ice-free periods and is sparse (Hawes 1983). A hypothesis encompassing the interactions of loss and growth processes for the diatom communities operating in oligotrophic lakes is proposed in Fig. 3. The most significant impact of such conditions might be the accumulation of a large inactive population within the perennial algal mat complex. If a large senescent population is present, then any seasonal changes in the actively growing component will be hidden when an acid-cleaning preparation is used where these components cannot be distinguished. The fate of the senescent population in terms of burial and transport will also affect the composition of algal mats in surrounding microhabitats and thus any ecological interpretation. Further studies of the distribution of phaeopigments within the lake basin might aid analysis and test the hypothesis proposed.

The abundance, composition and diversity of the epiphytic diatoms identified from Signy Island are not unlike communities found in more temperate systems, and the preliminary results indicate many aspects open to further investigation. The extent of variation within a lake must be studied in greater detail not only in terms of species composition, but also with respect to cell viability, and chlorophyll degradation. A variety of sampling methods must be used if the ecology of such a complex habitat is to be understood.

Acknowledgements. Thanks are extended to N. Rose for collection of material at Signy Island, and to Dr. J. C. Ellis-Evans for helpful comments on the manuscript. I am grateful to the British Antarctic Survey and the School of Biological Sciences at Portsmouth Polytechnic for providing facilities for completing this study.

References

Aleshinskaya AV, Bardin VI (1966) Diatomaceous flora of the Schirmacher ponds. Inf Byull Sov Antarkt Eksped 5:432–433

Baker AN (1967) Algae from Lake Miers, a solar-heated Antarctic lake. NZ J Bot 5:453–468

Behre K, Schwabe GH (1970) Auf Surtsey-Island in Sommer 1968 nachgewiesene nicht marine Algen. Schr Naturwiss Ver Schleswig-Holstein: 31–100

Bourelly P, Manguin E (1954) Contribution a la flore Algale d'eau douce des Iles Kerguelen. Mem Inst Rech Sci Madagascar Ser B 5:7–58

Brown SD, Austin AP (1973) Diatom succession and interaction in littoral periphyton and plankton. Hydrobiolgia 43:333–356

Carlson GWF (1913) Süsswasser-Algen aus der Antarktis Süd-Georgian und der Falkland Inseln. Wiss Ergebn der Schwedischen Sudpolor Exped 1901–1903, 4/14:1–94

Cattaneo A, Kalff J (1978) Seasonal changes in the epiphyte community of natural and artificial macrophytes in Lake Memphremagog (Que. & VT). Hydrobiologia 60:135–144

Clarke WJ, Runnels WC (1975) Diatoms in pond plankton: relationships to epiphytic and epipelic populations. Verh Int Limnol 19:2722–2728

Corte A (1962) Freshwater algae from semifrozen lakes at Hope Bay Antarctic Peninsula. Contrib Inst Antarct Argent 69:1–38 (in Spanish)

Ellis-Evans JC (1982) Seasonal microbial activity in Antarctic freshwater lake sediments. Polar Biol 1:129–140

Ellis-Evans JC, Lemon ECG (1989) Some aspects of iron cycling in maritime Antarctic lakes. Hydrobiolgia 172:149–164

Fritsch FE (1912) Freshwater algae of the South Orkneys. Report of the scientific results of the voyage of Scotia (1902–1904) 3:95–134

Fukushima H (1965) Preliminary report on diatoms from South Georgia. Antarct Rec 24:18–30

Fukushima H (1966) A brief note on the diatom flora of Antarctic inland waters. Proc Symp Pacific Antarctic Science. Tokyo JARE Sci Rep Spec Iss No 1, Nagata T (ed). Dept. of Polar Research, Nat Sci Museum Tokyo, Ueno Park Japan, pp 253–264

Gumtow RB (1955) An investigation of the periphyton in a riffle of the West Gallatin river, Montana. Trans Am Micro Soc 74:278–292

Happey-Wood CM, Priddle J (1984) The ecology of epipelic algae of 5 welsh lakes with special reference to Volvocalean green flagellates (Chlorophyceae). J Phycol 20:109–124

Hawes I (1983) Nutrients and their effects on phytoplankton populations in lakes on Signy Island, Antarctica. Polar Biol 2:115–126

Hendey NI (1964) An introductory account of the smaller algae of British coastal waters Part V. Bacillariophyceae diatoms. Min Ag Fish Inv Ser 4 (Lond) 317 pp

Heywood RB, Dartnall HJG, Priddle J (1980) Characteristics and classification of the lakes of Signy Island, South Orkney Islands, Antarctica. Freshwater Biol 10:47–59

Hickman M (1974) The epipelic diatom flora of a small lake on Baffin Island, Northwest Territories, Canada. Arch Protistenkd 116:270–279

Jones RC, Adams MS (1982) Seasonal variations in photosynthetic response of algae epiphytic on *Myriophyllum spicatum*. Aquat Bot 13:317–330

Jones RC, Mayer KB (1983) Seasonal changes in the taxonomic composition of epiphytic algal communities in Lake Wingra, Wisconsin, U.S.A. In: Wetzel RG (ed) Periphyton of freshwater ecosystems. Development in hydrobiology. Dr W Junk Publ, The Hague, pp 11–16

Jørgensen EG (1948) Diatom communities in some Danish Lakes and Ponds. Det Kongl Dansk Videns Sels Biol Skrift 5:73 pp

Kairesalo T (1984) The seasonal succession of epiphytic communities within an *Equisetum fluviatile* L. stand in Lake

Pääjärvi, Southern Finland. Int Rev Gesamten Hydrobiol 69:475–505

Karasawa S, Fukushima H (1977) Diatom flora and environmental factors in some freshwater ponds of East Ongul Island. Antarct Rec 59:46–53

Kobayashi T (1963a) Variability of *Hantzschia amphioxys* (Ehrenb.) Grun. var. *recta* O. Müller. Antarct Rec 17:59–63

Kobayashi T (1963b) Variations on some pennate diatoms from Antarctica (Parts I–V) Sci Rep Ser E, No 18, JARE 1956–1962, pp 1–20

Kobayashi T (1965) Variations on some pennate diatoms from Antarctica 2. (Parts VI–XI). Sci Rep Ser E, No 24, JARE 1956–1962, pp 1–23

Komarek J, Ruzicka J (1966) Freshwater algae from a lake in proximity of the Novolazarevskaya station Antarctica. Preslia 38:237–244

Lange-Bertalot H (1979) Toleranzgrenzen und Populationsdynamik benthischer Diatomeen. Arch Hydrobiol 56:184–219

Lavrenko GY (1965) Algae of a lake near Novolazarevskaya station. Inf Byull Sov Antarkt Eksped 56:57–61

Le Cohu R, Maillard R (1983) Les diatomées monoraphidées des île Kerguelen. Ann Limnol 19:143–167

Opalinski KW (1972a) Freshwater fauna and flora in Haswell Island (Queen Mary Land, eastern Antarctica). Pol Arch Hydrobiol 19:377–382

Opalinski KW (1972b) Flora and fauna in freshwater bodies of the Thala Hills oasis (Enderby Land, eastern Antarctica). Pol Arch Hydrobiol 19:383–398

Oppenheim DR, Ellis-Evans JC (1989) Depth-related changes in a maritime Antarctic lake. Polar Biol 9:525–532

Parker BC, Samsel G, Prescott B (1972) Freshwater algae of the Antarctic Peninsula 1. Systematics and ecology in the U.S. Palmer Station Area. In: Llano GA (ed). Antarctic terrestrial biology. American Geophysical Union, Antarct Res Ser 20:69–81

Petersen JB (1950) Observations on some small species of *Eunotia*. Dan Bot Ark 14:1–19

Priddle J (1985) Terrestrial habitats – inland waters. In: Bonner WN, Walton DWH (eds) Antarctica key environments. Pergamon press, Oxford, pp 118–132

Priddle J, Heywood RB (1980) Evolution of Antarctic lake ecosystems. Biol J Linn Soc 14:51–66

Reinsch PF (1890) Die Süsswasseralgen flora von Süd-Georgien. Die Internationale Polarforschung 1882–1883. Dtsch Exped Ergeb, Bd 2, Berlin, pp 329–365

Round FE (1957) The distribution of Bacillariophyceae on some littoral sediments of the English Lake district. Oikos 8:16–37

Round FE (1984) The ecology of algae. Cambridge Univ Press Cambridge, 653 pp

Wetzel RG (1964) A comparative study of the primary productivity of higher aquatic plants, periphyton, and phytoplankton in a large shallow lake. Int Rev Gesamten Hydrobiol 49:1–61

Wharton RA, Parker BC, Simmons GM (1983) Distribution, species composition and morphology of algal mats in Antarctic dry valley lakes. Phycologia 22:355–365

Seasonal Changes in Sea Ice Zones and off South Georgia

Seasonal and Regional Variation in the Pelagial and its Relationship to the Life History Cycle of Krill

V. SMETACEK, R. SCHAREK, and E.-M. NÖTHIG [1]

Summary. The earlier concept of Antarctic pelagic seasonality has changed drastically. We now know that the characteristic pelagic community resembles the oligotrophic communities of warm, nutrient-depleted waters. Results of recent sediment trap moorings show that the Antarctic oceanic pelagial behaves as a highly efficient retention system as losses due to sinking particles are exceptionally low. We show that the distinction into "new" and "regenerating" type pelagic systems only applies to restricted regions experiencing sizeable blooms where spring sedimentation pulses have been recorded. Apparently, community biomass is built up by channelling of nitrate into the recycling pool whereby balance between auto- and heterotrophs must be maintained, presumably within time scales of weeks. Interannual variability is likely to be much less pronounced in this type of buffered pelagic system.

In vast areas of the Southern Ocean, algal biomass is low but that of grazers comparatively high. We argue that Antarctic zooplankton, particularly the dominant copepods, are efficiently geared to the pelagic system overlying the deep ocean and maintain sizeable, stable stocks that undergo minor winter decline. The same is true for krill (*Euphausia superba*) but this animal is large enough to exploit the Antarctic pelagial in an unique way. It retreats to the ice undersurface during the long winter and feeds with high efficiency on plankton concentrations following ice melt. Apparently, krill has geared its life cycle to oceanic circulation patterns, including that of the ice, in a way that maximizes seasonal and regional exploitation of food resources. Adaptation to the rugged ice undersurface is probably the most important factor enabling maintenance of an uniquely large monospecific stock of planktivores in a low productive ocean. The concept of ice as a hostile habitat must be revised; rather than posing a problem for survival, it provides the answer to the age-old riddle of high animal biomass in an icy environment.

[1] Alfred-Wegener-Institut für Polar- und Meeresforschung, Columbusstrasse, D-2850 Bremerhaven, FRG.

1 Introduction

The Antarctic growth season has long been regarded as a brief but productive period characterized by large diatoms efficiently grazed by a short food chain of krill and whales. This view has changed radically since the turn of this decade (Olson 1980; Hewes et al. 1985; Hempel 1985). It is now accepted that the ice-free period is not exceptionally productive and that the sparse, flagellate-dominated community type is more characteristic than phytoplankton blooms of large species which only occur in restricted areas (El-Sayed 1984; Sakshaug and Holm-Hansen 1984; Heywood and Priddle 1987). However, the reasons for the low productivity are not obvious, as, in contrast to other oligotrophic waters, all macronutrients (N, P, Si) are available in excess. Currently, the deep mixed layer of circumpolar waters is widely considered responsible for low algal concentrations throughout most of the year (Hart 1934, 1942; El-Sayed and Mandelli 1965; Hasle 1969; Sakshaug and Holm-Hansen 1984; Hayes et al. 1984). The role of other potential controlling factors suggested by Hart (1934), viz. micronutrients (e.g. iron) and heavy grazing pressure is not yet clear. Indeed, new evidence indicates that iron availability can control oceanic phytoplankton growth (Martin and Fitzwater 1988). Further, if Antarctic waters are as oligotrophic as now believed, grazing pressure must be higher than previously assumed, as the stock size of metazoan grazers has not been revised downward; on the contrary, in the new conceptual picture the previously neglected herbivorous protozoans assume an important role (Hewes et al. 1985). Whatever the factor suppressing phytoplankton blooms, a consensus has emerged that small-sized algae with low biomass accumulation rate characterize the Antarctic pelagial (Sakshaug and Holm-Hansen 1984). The term "Pelagial", common in the German literature is used here in the sense of "pelagic realm" and comprises organisms and environment including oceanic structures and ice.

In this review, we examine the emerging picture of seasonality in the pelagic environment of the southern Antarctic – the region covered seasonally by sea ice with particular consideration of the Atlantic sector.

Antarctic Ecosystems. Ecological Change and Conservation
(ed. by K. R. Kerry and G. Hempel)
© Springer-Verlag Berlin Heidelberg 1990

We point out the implications of recent sediment trap studies from this area for our understanding of pelagic system functioning and vertical flux. We argue that some of the unusual features of the Antarctic pelagial, i.e. "oligotrophic communities in a sea of nitrate", supporting an exceptionally large grazer population (Sakshaug and Holm-Hansen 1984; Hewes et al. 1985), appear paradoxical against the background of the North Atlantic paradigm in which the spring bloom plays a pivotal role. The conceptual framework of this North Atlantic paradigm (Smetacek 1984) envisions a dormant winter stage disrupted by the spring bloom which grows in accordance with the Sverdrup (1953) model and is terminated by nitrate depletion. Thereafter, an oligotrophic community based on ammonia cycling develops which diminishes in size as essential elements are lost via sinking particles or conversion into dissolved organic matter. Input of new nutrients to this system results in nutrient competition and rapid conversion to biomass. The situation in the Antarctic is quite different as outlined below.

Much thought has been expended on explaining why Antarctic phytoplankton do not utilize all available nutrients. The validity of this question is based upon the implicit assumption that resources should be utilized to their utmost. Accepting the Antarctic situation as given, one might as well enquire why North Atlantic phytoplankton, by exhausting their nutrients, are unable to make full use of another basic resource, viz. solar energy, which is "wasted" by the sparse oligotrophic summer community.

2 Antarctic Pelagic Seasonality

Blooms worthy of the term (> 4 mg Chl. a m^{-3}; Sakshaug and Holm-Hansen 1984) occur in connection with the retreating ice edge but they tend to be restricted to specific geographic areas (Smith and Nelson 1986; von Bodungen et al. 1986; Heywood and Priddle 1987). A girdle of enhanced biomass present along the ice edge of the Weddell Sea has been revealed by satellite observations (Sullivan et al. 1988). However, these blooms are narrow patches that are prominent only along the Peninsula tip up to the South Orkneys; further to the east, chlorophyll concentrations in these patches tend to be low (< 4 mg Chl. a m^{-3}) which actually accentuates the degree of impoverishment of the surrounding circumpolar waters. Over vast regions of the Antarctic, chlorophyll concentrations under ice-free conditions rarely rise above 0.3 mg Chl. a m^{-3} (Sakshaug and Holm-Hansen 1984). Under the ice cover, water column values are lower by an order of magnitude (Schnack-Schiel 1987).

It now appears that the characteristic Antarctic pelagic community bears great resemblance to the oligotrophic communities of warm, stratified, nutrient-poor waters. Features shared by these two communities are: dominant phytoplankton represented by nanoflagellates and small pennate diatoms (Sakshaug and Holm-Hansen 1984); a comparatively large population of heterotrophic nano-, dinoflagellates and ciliates (Hewes et al. 1985); the bulk of primary production based on ammonia (Holm-Hansen 1985); a large (more than 75% of biomass) particulate detritus load (Schnack et al. 1985); substantial bacterial activity (Hodson et al. 1981; Bölter and Dawson 1982); heavy grazing pressure exerted by metazoa and copepods in particular (Schnack et al. 1985).

Another property of pelagic systems now receiving much attention is the vertical particle flux. Pelagic systems are commonly classified according to the criteria of "new" (nitrate-based) and "regenerated" (ammonia-based) production, whereby the former sets the upper limit to vertical flux from that system (Eppley and Peterson 1979). Sediment trap studies have shown that flux rates from new production systems (= blooms) are much higher than from those dominated by regenerated production. Peinert et al. (1989) introduce the terms "loss" and "retention" type systems based on the magnitude and composition of vertical flux. They identify two basic loss type systems: nitrogen-depleted phytoplankton blooms and those grazed heavily by large zooplankters (euphausiids, salps); sinking particles comprise cells and phytodetritus in the former case and faeces in the latter. Retention type systems are those heavily grazed by copepods as faeces are recycled in the surface layer. This latter system corresponds to the regenerating system. We next review recent information on vertical flux in the Antarctic within this framework.

3 Vertical Flux

Short-term sediment trap deployments at shallow depths in various coastal sites indicate that loss rates are generally low, ca. 1% of the suspended particle load per day (Schnack 1985; von Bodungen et al. 1987, 1988). With few exceptions, higher rates were found only when euphausiid faeces dominated trap collections. The exceptions recorded mass fallout of diatom cells from blooms that developed over shallow coastal water or under land-fast ice (von Bodungen et al. 1986; Matsuda et al. 1987; Fukuchi et al. 1988).

Moored multisample sediment traps provide continuous information on seasonality and magnitude of vertical flux. A 3-year deployment at 1588 m depth in

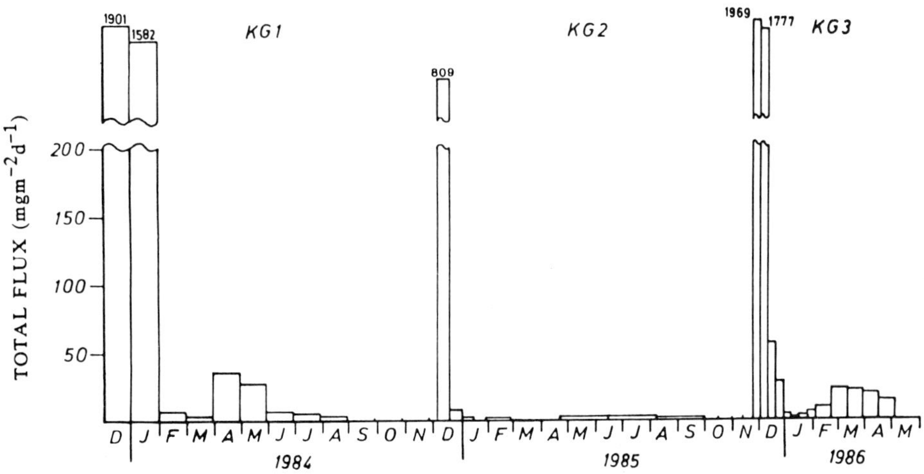

Fig. 1. Sedimentation rates (dry wt) measured with a multisample sediment trap moored at 1588 m depth for a 3-year period (*KG1–3*) in the Bransfield Strait. Note interannual variability in the restricted, sharp spring pulse of the 3 years (Wefer 1989)

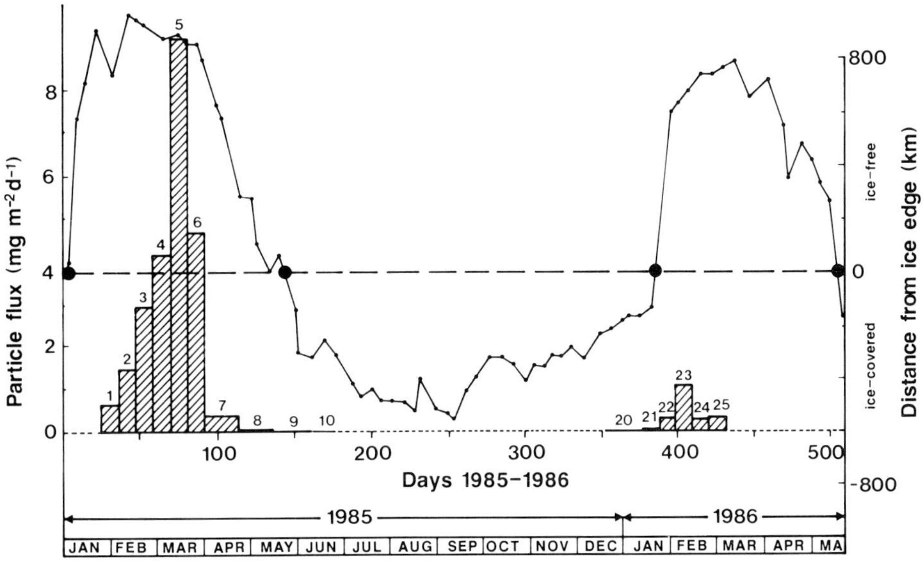

Fig. 2. Sedimentation rates (dry wt) measured with a multisample sediment trap positioned at 863 m and moored at 3880 m water depth in the Weddell Sea (62° 26 S; 34° 45 W). Note different scales on ordinate to Fig. 1. Total flux from January 1985 to January 1986 was 0.37 $g\,m^{-2}$. Superimposed is the shortest distance (km) from the approximate ice edge to the sediment trap site (Fischer et al. 1988)

the Bransfield Strait (2000 m water depth) yielded annual rates of 108, 12 and 36 g dry weight $m^{-2}\,yr^{-1}$ of which about 95% settled out in November/December, mostly comprising euphausiid faeces (Wefer 1989). The annual patterns recorded here are very similar to those obtained from various sites in the North Atlantic (Wefer 1989) except that the spring pulse is more pronounced and summer and winter rates are very low (Fig. 1).

A similar trap deployed at 863 m depth in the open northern Weddell Sea (3880 m water depth; Fischer et al. 1988) provided a radically different picture (Fig. 2). The annual sedimentation rate was only 1% in terms of dry weight of the average Bransfield Strait value. Further, there was no sign of a sedimentation pulse in spring: the bulk of the material arrived during the summer months with a peak in February. Krill faeces or remnants were rare and most of the material appeared to represent faeces of smaller animals. The abrupt decline to very low levels during the winter indicates that summer flux comprised rapidly sinking particles; a "background flux" of slowly sinking particles was absent as this would have prolonged the sedimentation signal into the winter.

The trap results have interesting implications for the structure and productivity of the overlying pelagic systems. They confirm that the Bransfield Strait is much more productive than the northern Weddell Sea, which is, however, much more representative of the Antarctic pelagial. However, the difference in productivity is not directly proportional to the difference in sedimentation. Thus, topography of the Bransfield Strait site could have exaggerated sedimentation rates. However, the sharpness of the pulse indicates a minor role of resuspension, that, were it of importance, would "smear" the signal into the summer. The difference between the two sites lies clearly in the absence of a spring pulse in the Weddell Sea trap. Summer and winter sedimentation rates are low in both regions.

The spring pulse in the Bransfield Strait is indicative of a typical bloom situation. Indeed, high chlorophyll concentrations (up to 10 mg Chl. a m^{-3}) due to large diatoms and *Phaeocystis* in the trap area during November/December have been recorded continuously for many years (von Bodungen 1986; von Bodungen et al. 1986, 1987). We conclude tht the spring bloom in the Bransfield Strait, even in the absence of nutrient exhaustion, conforms to the "loss type system" of Peinert et al. (1989). With the exception of inshore waters, where diatoms can settle out directly, vertical flux is controlled by euphausiid grazers.

A spring maximum of large diatoms should also occur in the northern Weddell Sea in the wake of the retreating ice edge (Smith and Nelson 1986; Sullivan et al. 1988). Indeed, at the time of deployment of the Weddell Sea trap (15 January 1985) maximum chlorophyll concentrations in the 28 m mixed layer were 2.4 mg Chl. a m^{-3} and diatoms dominated the biomass. Particulate organic carbon (POC) in the mixed layer amounted to 10.6 g C m^{-2}, which, at C:Chl a ratio of 171, indicates a large detrital component. This is a large stock of POC by any standards but it left a practically negligible signal in the trap, indicating that the material must have been recycled in the water column above the trap.

Jennings et al. (1984) estimated production based on nitrate uptake between winter and late summer in the northern Weddell Sea to be in the order of 20 g C m^{-2}. Particulate organic carbon collected by the trap was 20 mg^{-2} yr^{-1}, i.e. 0.1% of the "new" production sank out of the water column. This percentage, as also the absolute value (Fischer et al. 1988), is the lowest recorded in the world ocean to date. For instance, annual sedimentation rates in the oligotrophic Sargasso Sea, considered very much a regenerating system, albeit with a weak spring bloom, are higher by a factor of 40 (Deuser et al. 1981); the pulse from the spring bloom alone contributes one-third to total sedimentation.

Thus, the communities present throughout the year in the Weddell Sea site and during the summer months in the Bransfield Strait area represent highly retentive systems as defined by Peinert et al. (1989). We suggest that the sediment trap data provide indirect evidence for heavy copepod grazing, accompanied by faeces disintegration, probably within the mixed surface layer. The high detritus levels in the mixed layer over the Weddell Sea trap support this hypothesis.

4 Antarctic Regenerating System

Nitrate exhaustion by the preceding bloom is implicitly assumed to be the triggering factor leading to establishment of the regenerating system, i.e. autotrophs are dependent on release of reduced nitrogen by the heterotrophs. In the Weddell Sea and, by implication, in open Antarctic waters in general, the regenerating community establishes itself in the presence of abundant nitrate. Indeed, various studies have shown that this nitrate is also utilised (between 20 and 50%, Holm-Hansen 1985), although attention has focussed on the fact that ammonia provides the bulk of nitrogen requirements (Olson 1980; Koike et al. 1986). Collos and Slawyk (1986) even suggest that "nitrate can account for most of the primary production in offshore areas during the austral summer." The point we are making here, however, is that "new" production can be channelled directly and efficiently into a system that, by appearance, is a classical regenerating community and also behaves as a retention system. In other words, the nitrate exhaustion by the spring bloom is not necessary for the establishment of a regenerating community.

In the past, balancing pelagic nitrogen budgets has proved very difficult as particulate nitrogen pools (biomass and detritus) are invariably much smaller than the nitrate taken up by the community (Peinert 1986; von Bodungen et al. 1988). These authors assume that the "missing" nitrate is lost during episodic sinking events. Such events should have been recorded by the moored traps. The nitrogen pool that has eluded accurate measurement so far is the dissolved organic nitrogen (DON) pool. The recent introduction of a new method indicates serious underestimation of this pool in the past (Sugimura and Suzuki 1988). We now speculate that DON is the repository of this "missing" nitrogen and that, like the particulate detritus pool, it is an integral part of the regenerating community. The source of dissolved matter can well be zooplankton feeding (Jumars et al. 1989) or microbial populations (Brophy and Carlson 1989).

The efficiency of this system is indicated by the exceptional clarity of winter water: Secchi disc readings (> 60 m) from late winter in the Weddell and Ross Seas are the deepest recorded from the world ocean (Gieskes et al. 1987; Barry 1988). As Secchi disc depths in summer range between 7 and 18 m (von Bodungen et al. 1988), substantial organic pools must occur in these waters during the summer that are remineralised before the end of winter. This conclusion is supported by the surprisingly constant and horizontally homogeneous winter nitrate concentrations in the Weddell Sea (Jennings et al. 1984; Schnack-Schiel 1987). It appears unlikely that significant yearly flux losses to deeper waters would be so exactly compensated by vertical admixture of water from the warm deep water to the surface winter water.

Winter Weddell Sea plankton consists of an active regenerating community functioning at very great dilution (Scharek in prep.). The few winter observations of pelagic communities from other sites indicate that they are flagellate-dominated (Braarud 1935; Takahashi et al. 1978). Thus, a viable winter regenerating community is not unique to the Antarctic but is likely to be the rule. In the North Atlantic and shelf waters worldwide, the development of this community through the spring is obscured by the high biomass built up by the bloom. After its passage, the regenerating system appears to sustain itself by utilising ammonia regenerated from organic matter "left behind" because the previous bloom would have indeed exhausted nitrate. However, as shown above, this is not a prerequisite for the establishment of regenerating systems per se which can also build up community biomass by utilising nitrate. This community type can no longer be categorised as either a "new" or "regenerating" system as it combines characteristics of both. The term "retention" system is thus more appropriate.

Evidence from sediment traps now show that the regenerating community is highly efficient, even at great dilution and low temperature, and succeeds in recycling summer-produced organic matter more or less quantitatively back to the constituent nutrients. Its retention capacity is higher than widely imagined as the organism density at which this community can function appears to be very low. This conclusion was confirmed by observations of the community in water underlying the Ross Ice Shelf where apparently viable auto- and heterotrophs were encountered in spite of prolonged darkness (Azam et al. 1979).

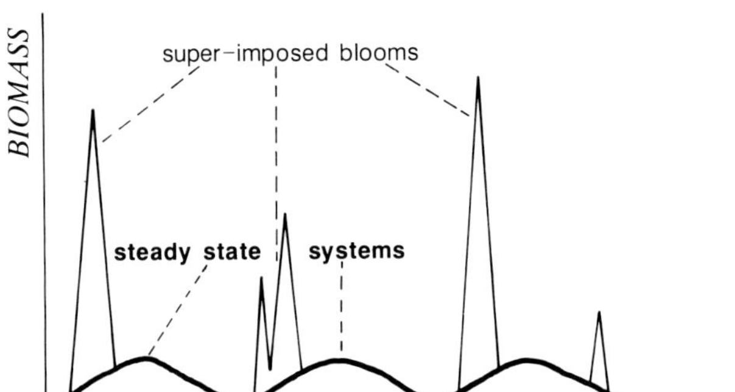

Fig. 3. A hypothetical scheme demonstrating the relationship between seasonality of the microbial network (regenerating system, *thick line*) with superimposed "blooms" (*thin lines*). The latter exhibit more variability as they are induced by weather conditions operating on a scale of weeks. They are followed by a correspondingly variable sedimentation pulse (*arrows*)

The "regenerating system" should now be regarded as the characteristic system of the pelagial with blooms, if they occur, representing a transitory event superimposed on the basal state (Yentsch et al. 1977). Accordingly, interannual variation of the regenerating system will be much less than in the case of blooms (Fig. 3). A similar situation prevails in the Sub-Arctic Pacific Gyre where, in the presence of abundant macronutrients, chlorophyll concentrations rarely rise above 0.5 mg Chl. *a* m^{-3} (Miller et al. 1984). Not surprisingly, this North Pacific Gyre has been accorded special status in textbooks as it does not experience a spring bloom (Parsons et al. 1984). It should be mentioned that surface macronutrient concentrations (with the exception of silicate) are high right up to the Sub-Tropical Convergence (41° S), indicating that the same situation exists in the broad Sub-Antarctic belt as in the Weddell Sea (Holm-Hansen 1985).

One of the terms in current usage generally considered synonomous with a regenerating system is the "microbial loop" implicitly attached to a food *chain*. Based on the above considerations, we suggest that the term "microbial network" as the basis of the pelagic food *web* (Pomeroy 1974) is much more appropriate. The concept of "network" confers a three-dimensionality to visualised thought that is more commensurate with the actual spatial situation. It is a useful visual framework of development of "natural history thinking" as described by Smetacek and Pollehne (1986). The concept of "network" as opposed to the two-dimensional "loop" will hence reduce the importance of the spring bloom in our thinking and render the question of "link" or "sink" redundant as already suggested by Sherr and Sherr (1988). Accepting the regenerating community as the "normal" pelagic system does not explain the absence of a bloom, merely independence from it.

5 Zooplankton

How do the metazoan grazers fit into the microbial network? The presence of huge stocks of herbivorous euphausiids is possibly the most characteristic feature of the Antarctic pelagial (Hempel 1985). However, sizeable stocks of copepods are also present (Smith and Schnack-Schiel in press). We shall first deal with the smaller metazooplankton before taking up the case of the krill.

A remarkable feature of Antarctic meso- and macrozooplankton is that biomass in the upper 1000 m is reported to be constant throughout the year (Foxton 1956, 1964; Hopkins 1971; Atkinson and

Peck 1988). Although all major zooplankton groups are represented in Antarctic waters, the bulk of the stock is contributed by a few large copepod species that overwinter in deep layers in a dormant state. In the high Antarctic, there is normally only one generation per year (Smith and Schnack-Schiel in press) and the size attained by this population during the reproductive season must be equivalent to that surviving the winter. In our opinion, this constancy is surprising because it indicates that the herbivore community is at steady state throughout the year.

The size of a given animal population is regulated by intrinsic (K-strategists) and/or extrinsic (r-strategists) factors. Considering the extreme seasonality of extrinsic factors in the Antarctic we must conclude that the herbivores here are controlled primarily by intrinsic factors (Fransz 1988). The mechanisms are obviously not "classic" K-strategy, with high juvenile survival rates and with density control of reproductive rate. Rather, there must be other mechanisms, possibly peculiar to copepods – this highly successful group of marine herbivores – that regulate stock size in relation to the external environment. Smetacek and Pollehne (1986) have suggested hypothetical mechanisms to explain this. Applying their conclusions to the Antarctic situation we in turn conclude that the copepod population is an intrinsic part of the microbial network, not only feeding on it as a "top" predator but helping to maintain it by adding detritus in the form of faeces and microzooplankton in the form of nauplii. Copepod faeces are part of the recycling pool. The salient features of this relationship are depicted in Fig. 4.

It has recently been shown that many of the large dinoflagellates are herbivores and can even tackle long diatom chains (Jacobson and Anderson 1986). These organisms, together with other known protozoan herbivores (many ciliates, heliozoans, radiolarians and foraminifers) are common in Antarctic waters (von Bröckel 1981; Hewes et al. 1985; Brandini and Kutner 1987; Nöthig 1988; Scharek in prep.) and can on occasion be abundant in restricted habitats (Spindler and Dieckmann 1986). The herbivorous protozoa must exert significant grazing pressure on the phytoplankton. However, they are in turn grazed by copepods that have been shown to selectively feed on them in the presence of low phytoplankton biomass (Stegmann and Peinert 1984; Nöthig 1986; Gifford and Dagg 1987). Hence, copepod biomass in a given region is only marginally dependent on the size of the spring bloom, and, being geared to the microbial network, should also exhibit little interannual variation.

6 The Role of Krill

Blooms are, by nature, transitory events. Herbivores adapted to feeding on blooms must hence either be geared to the time scales of the bloom in a given water body (i.e. they must give it time to develop and yet enter it before it crashes) or they must have the motile

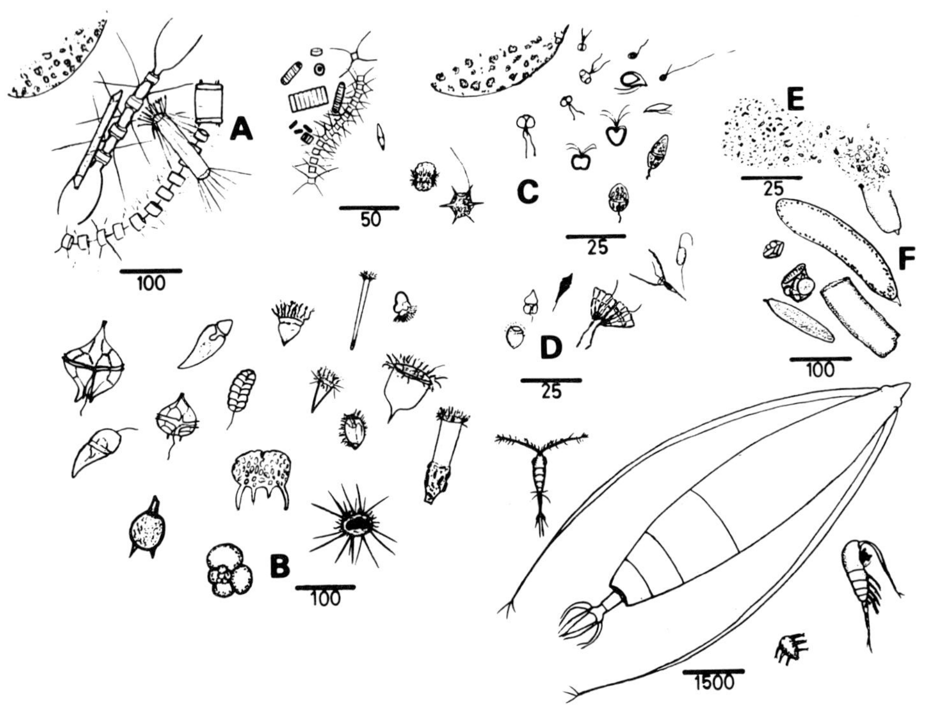

Fig. 4. The microbial network together with metazoans comprises the pelagic food web: Typical bloom diatoms and *Phaeocystis* (**A**) are fed on by herbivorous protozoans (**B**); small diatoms and flagellates (**C**) are eaten by other types of phagotrophic protozoans (**D**) that can also collect bacteria (**E**). Copepods feed on all these groups. Bacterial substrate is provided by dissolved organic carbon (released potentially by all the organisms) and particulate detritus emanating from faeces (**F**) particularly those of copepods. The borders between the above groups are not to be regarded as rigid because, for instance, small diatoms can also form blooms. Numbers under bars represent microns

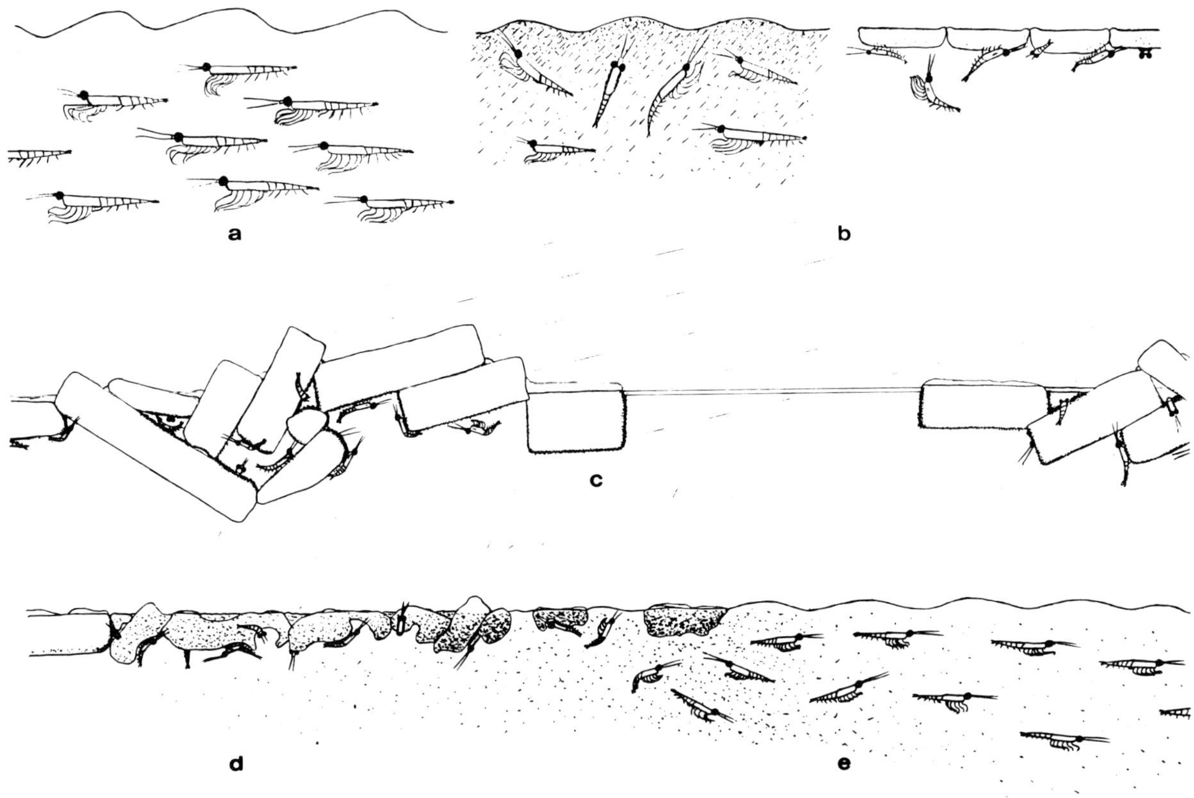

Fig. 5. Hypothetical annual cycle of krill. The diagrams are either drawn from direct observation (**a, c, d**) or document processes that inevitably occur, assuming that krill prefer the ice cover to the open water (**b, e**). **a** Compact swarms forage for food in the open water in the absence of ice. **b** During establishment of ice cover, frazil ice (*dashes*) consolidates to pancake ice collecting algal cells (*dots*) in the process. These algae can be efficiently utilised by adult and juvenile krill attracted to the ice. **c** Krill congregate under ridges because of protection against predators and improved food supply. Productivity of ice algae under ridges is greater than under flat ice because of increased surface area and light-trapping properties of the ridge which functions as a light dome. **d** Increased algal production in rotting ice provides a rich food source prior to water column blooming. **e** Melting ice can stabilise a shallow surface layer in which blooms rapidly develop. Krill switch from scraping ice algae to filtering phytoplankton

ability to follow blooms spatially. Krill apparently belongs to the latter category: sediment trap collections from the Peninsula area discussed above show that vertical flux emanating from bloom situations is comprised mainly of krill faeces, hence they are able to locate and intensively graze on blooms. The question then arising is how krill survive the non-bloom months as they lack a dormant stage and do not lay down substantial energy reserves. Their ability to starve for months with body and gonad regression observed in aquaria by Ikeda and Dixon (1982) is unlikely to be the sole survival mechanism. This long-standing enigma can now be resolved based on increasing evidence that krill thrive on the undersurface of the ice. Observations made around the continent (Marr 1962; Spiridonov et al. 1985; Kawaguchi et al. 1986; O'Brien 1988; Daly and Macauly 1988; Stretch et al. 1988) and in the Weddell Sea from 58°–76° S (Marschall 1988) show that they actually dwell on its undersurface. Krill distribution is restricted primarily to the region covered by ice in winter (Marr 1962); we must, therefore, assume that they are adapted to surviving under ice-covered waters. One can re-interpret known features of their distribution, behaviour and morphology in this light. Some of these, such as the ability to escape predators by darting into holes in the ice cover and to scrape and "suck" algae from the ice have been listed by Marschall (1988). There is now little reason to doubt that the bulk of the krill population moves into the ice habitat upon its formation.

We next postulate a life history pattern of krill for the Weddell Sea that is embedded in the seasonal dynamics of the physical environment (Fig. 5) and which seeks to explain why large stocks of this animal can be maintained in an apparently low-productive oceanic region. The biology of krill during the summer season is reasonably well known. Compact mobile swarms rove the open water in search of food in a patchy environment. They feed heavily on phytoplankton in blooms but are also suggested to utilise the microbial network and metazooplankton by feeding on protozoans (Hewes et al. 1985) and copepods (Boyd et al. 1984; Price et al. 1988).

Although krill are efficient collectors of suspended particles, feeding in this mode requires a minium particle density (Boyd et al. 1984). In the winter water column, particle concentrations are far below the krill feeding threshold. However, the ice/water interface is a much richer food source. Algal cells accumulate here by two mechanisms: collection ("harvesting") of algae from the water column by frazil ice (Garrison et al. 1983) and growth within and on the ice. Frazil ice, which dominates the Antarctic ice cover, is porous in its early stages and honey-combed with brine pockets and channels after consolidation. It provides a more favourable environment for algal growth than compact columnar ice which is more common in the Arctic (Weeks and Ackley 1982).

During autumnal ice formation, the phytoplankton concentrated in frazil ice provides easily accessible food that can be grazed at little energy expenditure by krill. The ice environment permits a much higher efficiency of light utilisation by its associated flora than in the case of phytoplankton suspended in the deep mixed layer below it. Visual observations from ship deck and by a TV camera mounted on a remote operated vehicle (ROV) conducted during the "Polarstern" cruise indicated that pressure ridges in the ice cover supported much higher algal biomass than adjoining stretches of undisturbed flat ice. These observations, made over wide stretches of the Weddell Sea (Marschall 1988), run counter to expectation but can be explained by the following factors, keeping in mind that albedo by the snow cover is the major source of light attenuation (Maykut 1985):

1. The roof-like piles of broken floes act as light domes for the low winter sun (Fig. 5). Reduced snow cover on oblique and vertical floe faces will act as "skylights". The light dome and skylight effects were particularly conspicuous in the "view from below" obtained with the ROV.

2. Surface area for growth of attached algae within ridges is substantially increased due both to more extensive floe surfaces and to significant enhancement of microenvironment provided by submerged snow. Wet snow loses its high albedo but retains its interstitial spaces within which the algae can grow.

3. Wind-induced movement of the ice cover results in both ridge and lead (stretches of open water) formation; ridged floes on the borders of leads will be exposed to high light levels resulting in enhanced growth. New ice forming in the leads will also support rapid algal growth.

We conclude that productivity within ridges is highly patchy and hence difficult to quantify but, on an areal basis, can well be higher than under flat ice.

Spring melting of the ice cover commences after the equinox and extends well into December. The albedo of the wettened snow cover declines dramatically. Further, melting ice greatly increases its porosity and hence the substrate area for algae. For these reasons, algal growth rates prevailing in melting ice will be at their annual maximum. Visual observations of rotting ice fields coloured brown by algae support these conclusions. Obviously, this environment represents extremely rich grazing grounds for krill which again could be confirmed by inspection with the ROV. Ice melt can result in stabilisation of a low-salinity surface layer in which the released algae can continue their rapid growth (Smith and Nelson 1986). The effect is particularly marked in topographically protected areas such as the Antarctic Peninsula region (von Bodungen et al. 1986). The accompanying krill population, now divested of their winter habitat, must merely switch from scraping algal-rich ice to filtering the phytoplankton bloom in the water, a transition which again occurs with a minimum of energy expenditure.

It is no coincidence that the largest krill stocks are encountered in the most productive waters along and eastward of the Antarctic Peninsula but also along the northern Ross Sea (Marr 1962). Extensive and hence energy-consuming migration by krill must not necessarily be invoked to explain this distribution pattern. Rather, the reasons will become apparent from a consideration of the large-scale hydrography/topography of the areas. Priddle et al. (1988) point out that high krill biomass concentrations occur where the Antarctic Coastal Current interacts with the Circumpolar Current along northwestern Gyre rims. To explain possible underlying causes we focus on the Peninsula region. Northward moving surface currents from the Bellingshausen and Weddell Seas converge, together with their ice cover, along the tip of the Peninsula and give rise to the Weddell-Scotia Confluence (Patterson and Sievers 1980). During spring melt, this substantial input of low-salinity water together with algal cells released from the ice induces large phytoplankton blooms into which the streams of melting ice deposit their overwintering krill populations. Hence this area not only receives but can also support large krill stocks with high transfer efficiency of phytoplankton production to krill. Copepod grazing pressure on blooms in this area is reported to be negligible (Schnack et al. 1985), presumably because their life-cycle strategy involving a deep-living, overwintering stage is disrupted by the hydrography/topography around the Peninsula.

Not surprisingly, the Peninsula area is also the major spawning site of krill (Marr 1962). Larvae develop to late furcilia and small juveniles by autumn (Marr 1962), and must overwinter in this stage. Juvenile krill but no adults have been found in high numbers at the ice edge (Mackintosh 1973); reports by divers indicate that they have essentially the same behaviour in the ice habitat (Garrison et al. 1986; Daly and Macaulay 1988; Hamner et al. 1989) as the adult krill observed in the southern Weddell Sea (Marschall 1988). Observations indicate that regional segregation of juveniles and adults occurs which would prevent food competition in this habitat. The northern edges of the ice receive more light during winter than in the south and will hence be more productive. Further, new ice is constantly being formed which permits faster algal growth than old, snow-covered ice. Thus, juvenile krill overwinter in a more favourable environment than the adults which migrate south in autumn (Marschall 1988). Adult krill can survive for months without food by utilising body tissue (Ikeda and Dixon 1982). This will enable them to overcome the months when ice algal growth will be severely retarded.

The winter ice habitat must be quite crowded, as a simple calculation shows: the number of individuals comprising the current range of stock estimates from 200–600 million tons (El-Sayed 1988) ranges between 2 and 6 × 10^{14} (based on a weight of 1 g/individual); homogeneous distribution under the total area covered by ice at its maximal extent (20 × 10^6 km^2; Maykut 1985) yields a krill density ranging between 10 and 30 individuals m^{-2}. However, the actual ice surface area will be greater by about one-third due to ridging of the cover. Krill congregations within ridges were indeed of this order (Marschall 1988). Experimental observations of krill grazing rate on algal-coated glass plates showed that an individual can sweep an area of 100 cm^2 clean of algae in about 5 min (Marschall 1988). Thus, the grazing pressure in the ice habitat must be very high indeed.

A surprising aspect of the Antarctic ice habitat is the paucity of other metazoan grazers such as copepods and amphipods which are common under Arctic sea ice (Carey 1985). As krill are also voracious predators it is possible that they have eliminated earlier competitors during their evolutionary invasion of the ice habitat. Cannibalism is commonly observed amongst krill in aquaria (McWinnie et al. 1979; Ishii et al. 1985), but the mode of prey capture and ingestion would be difficult in the water column. Hence, there is a tendency to regard it is an artifact induced by captivity. The confined spaces between ridged floes that were often occupied by dense krill aggregations resembled very much an aquarium habitat. Although not yet observed in nature, cannibalism, particularly if triggered by food shortage as reported from captive animals (Marschall pers. commun.), would ensure survival of vigorous individuals and conservation of biomass within the population. We suggest that cannibalism can have important implications as a self-regulatory mechanism gearing grazing pressure to available food supply and maintaining the population at maximum sustainable levels.

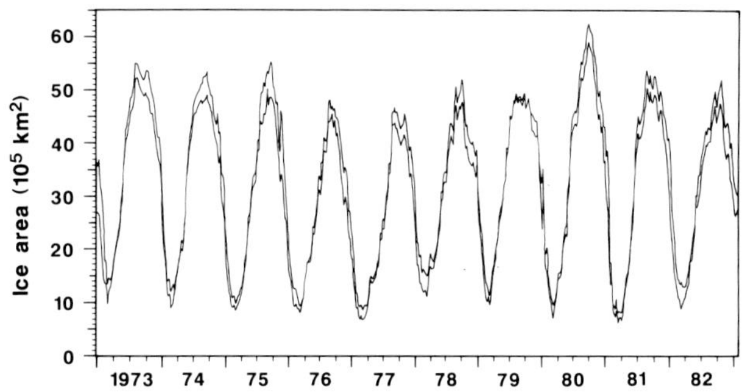

Fig. 6. Time series of total ice coverage and extent in the Weddell Sea for the years 1973–1982 (Sea Ice Climatic Atlas, Anonymous 1985)

Considering the close gearing of krill stocks to the seasonal ice cover, it is reasonable to expect that the stock size and recruitment will be affected by interannual variation in its extent and duration. There is indeed circumstantial evidence for such a relationship. The year 1981 has gone down in the history of krill biology as the year of the "super swarm" (Holm-Hansen and Foster 1981; Macaulay 1981; Brinton et al. 1981; Holm-Hansen and Huntley 1984). Breeding success in general was also exceptionally high as compared to other years (Hempel 1983). A glance at Fig. 6 shows that extent of ice cover in the previous winter stands out well above that of the other 10 years recorded. Obviously, more data are required before causal relationships of this type can be formulated, here, we merely wish to draw attention to this coincidence. Several possible mechanisms might underlie such a relationship: (1) large recruitment because of good juvenile food supply in a more extensive ice edge zone; (2) lower adult mortality because of more extensive ridging and hence lesser incidence of cannibalism; (3) better spring food supply because of more extensive ice-edge zone and/or more pronounced meltwater layers inducing larger blooms. Another unknown factor is the percentage of the population that remains under the multiyear ice fields of the western Weddell Sea. Monitoring krill under persistent ice in summer is an important future task.

7 Conclusions

It will now be apparent why such a large stock of a highly versatile planktivore can be maintained in a low productive environment. The adaptation of krill to both the ice and water column permits channelling of food produced in these differing habitats to the same animal, whose life-cycle strategy maximises ecological transfer efficiency. Because of the ridges, the ice cover provides agile, krill-sized animals more secure shelter from predators than the sea bottom. The earlier debate on whether krill is an "animal of the pack" or of the open water (Marr 1962) can now be resolved in this light. Until recently the winter ice cover was regarded as hostile to metazoans; now it is obvious that sea ice not only provides a substrate to some algae but is the segregated "kindergarten" and refuge of the key Antarctic herbivore. We should now regard sea ice as the "substrate" that enables such a proliferation of animal biomass in the Antarctic.

As bloom situations are transient events it follows that their occurrence will exhibit greatest interannual variability and accordingly also impact upon vertical flux. Although krill reproduction is likely to be geared to food produced by these blooms, its stock size is likely to be regulated more by conditions (food and predators) prevailing outside the spring, i.e. in the summer water column and under the winter ice cover. Hence, a "good bloom year" (e.g. 1983/84 of Fig. 1) could result in prolific breeding but not necessarily impact upon next year's stock. There is as yet no information on the effect of krill grazing on the microbial network. If krill indeed feed significantly on copepods this will release grazing pressure on the protistan community and result in a buildup of their biomass (Smetacek 1981). The latter would then be an alternative food source for krill (Hewes et al. 1985). Versatile feeding behaviour such as that of krill would hence be most advantageous in this type of pelagic community.

The fact that so many long-lived vertebrates have geared their life cycles to that of krill, which itself can live at least up to 6 years (Berman et al. 1989), suggests comparatively little interannual variability. We consider it unlikely that cyclic fluctuations such as those of the El Niño phenomenon will impact upon the Antarctic biota in a similarly drastic fashion. We argue that this is due to the comparatively minor interannual variation exhibited by the ice cover (Fig. 6) – the environment which we postulate is the main factor maintaining the size of the krill stock. However, yearly variation in distribution and abundance does occur, albeit not of the boom and bust type as in the Peru upwelling area. Priddle et al. (1988) suggest such variation to be due to "redistribution of krill by physical forces" rather than mass mortality. The northward extension of the ice cover during glacial periods will merely carry the krill population with it; as long as seasonal retreat of the ice cover is of the same extent as today, krill stock size need not have differed much between glacial and warm periods. Obviously, we are treading dangerously thin ice at this early stage of our knowledge regarding variability of the Antarctic pelagial. Our intention here is to stimulate research along the lines of these speculative conclusions.

References

Anonymous (1985) Sea ice climatic atlas: vol I Antarctic. Naval Oceanography Command Detachment, Asheville, 131 pp

Atkinson A, Peck JM (1988) A summer-winter comparison of zooplankton in the oceanic area around South Georgia. Polar Biol 8:463–473

Azam F, Beers JR, Campbell L, Carlucci AF, Holm-Hansen O, Reid FMH, Karl DM (1979) Occurrence and metabolic activity of organisms under the Ross Ice Shelf, Antarctica, at Station J9. Science 203:451–453

Barry JP (1988) Hydrographic patterns in McMurdo Sound, Antarctica and their relationship to local benthic communities. Polar Biol 8:377–391

Berman MS, McVey A, Ettershank G (1989) Age determination of Antarctic krill using fluorescence and image analysis of size. Polar Biol 9:267–271

Bodungen B von (1986) Phytoplankton growth and krill grazing during spring in the Bransfield Strait, Antarctica — implications from sediment trap collections. Polar Biol 6:153–160

Bodungen B von, Fischer G, Nöthig E-M, Wefer G (1987) Sedimentation of krill faeces during spring development of phytoplankton in Bransfield Strait, Antarctica. In: Degens ET, Honjo S, Izdar E (eds) Particle flux in the ocean. Mitteilungen aus dem Geologisch-Palaeontologischen Institut der Universität Hamburg, SCOPE UNEP, Sonderband 62:243–257

Bodungen B von, Nöthig E-M, Sui Q (1988) New production of phytoplankton and sedimentation during summer 1985 in the southeastern Weddell-Sea. J Comp Biochem Physiol 90B:475–487

Bodungen B von, Smetacek VS, Tilzer MM, Zeitzschel B (1986) Primary production and sedimentation during spring in the Antarctic Peninsula region. Deep-Sea Res 33:177–194

Bölter J, Dawson R (1982) Heterotrophic utilization of biochemical compounds in Antarctic waters. Neth J Sea Res 16:315–332

Boyd CM, Heyraud M, Boyd CN (1984) Feeding of the Antarctic krill *Euphausia superba*. J Crustacean Biol 4, Spec No 1:123–141

Braarud T (1935) The "Øst" expedition to the Denmark Strait 1929: 2. The phytoplankton and its conditions of growth. Hvalradets Sgkr 10:1–171

Brandini FP, Kutner MBB (1987) Phytoplankton and nutrient distributions off the northern South Shetland Islands (summer 1984 – BIOMASS/SIBEX). La Mer 25:93–103

Brinton E, Shulenberger E, Wormuth J, Antezana T (1981) Net sampling of plankton and krill in the Scotia Sea, January–March 1981. Antarct J U S 16:160–161

Bröckel K von (1981) The importance of nanoplankton within the pelagic Antarctic ecosystem. Kieler Meeresforsch 5:61–67

Brophy JE, Carlson DJ (1989) Production of biologically refractory dissolved organic carbon by natural seawater microbial populations. Deep-Sea Res 36:497–507

Carey AG (1985) Marine ice fauna: Arctic. In: Horner RA (ed) Sea ice biota. CRC Press, Inc, Florida, pp 173–190

Collos I, Slawyk G (1986) 14-C and 15-N uptake by marine phytoplankton, IV. Uptake ratios and the contribution of nitrate to the productivity of Antarctic waters (Indian Ocean sector). Deep-Sea Res 33:1039–1051

Daly KL, Macaulay MC (1988) Abundance and distribution of krill in the ice edge zone of the Weddell Sea, austral spring 1983. Deep-Sea Res 35:21–42

Deuser WG, Ross EH, Anderson RF (1981) Seasonality in the supply of sediment to the deep Sargasso Sea and implications for the rapid transfer of matter to the deep ocean. Deep-Sea Res 28:495–505

El-Sayed SZ (1984) Productivity of the Antarctic waters – a reappraisal. In: Holm-Hansen O, Bolis L, Gilles R (eds) Marine phytoplankton and productivity. Lecture notes on coastal and estuarine studies, vol 8, Springer, Berlin Heidelberg New York, pp 19–34

El-Sayed SZ (1988) The BIOMASS program. Oceanus 31:75–79

El-Sayed SZ, Mandelli EF (1965) Primary production and standing crop of phytoplankton in the Weddell Sea and Drake Passage. In: Llano GA (ed) Biology of the Antarctic seas II. Antarct Res Ser 5:87–106

Eppley RW, Peterson BJ (1979) Particulate organic matter flux and planktonic new production in the deep ocean. Nature 282:677–680

Fischer G, Fütterer D, Gersone R, Honjo S, Ostermann D, Wefer G (1988) Seasonal variability of particle flux in the Weddell Sea and its relation to ice cover. Nature 335:426–428

Foxton P (1956) The distribution of the standing crop of zooplankton in the Southern Ocean. Discovery Rep 28:191–236

Foxton P (1964) Seasonal variations in the plankton of Antarctic waters. In: Carrick R, Holgate MW, Prevost J (eds) Biologie Antarctique. Hermann, Paris, pp 311–318

Fransz HG (1988) Vernal abundance, structure and development of epipelagic copepod populations of the eastern Weddell Sea (Antarctica). Polar Biol 9:107–114

Fukuchi M, Hattori H, Sasaki H, Hoshiai T (1988) A phytoplankton bloom and associated processes observed with a long-term moored system in Antarctic waters. Mar Ecol Prog Ser 45:279–288

Garrison DL, Ackley SF, Buck KR (1983) A physical mechanism for establishing algal populations in frazil ice. Nature 306:363–365

Garrison DL, Sullivan CW, Ackley SF (1986) Sea ice microbial communities in Antarctica. Bioscience 36:243–250

Gieskes WWC, Veth C, Woehrmann A, Graefe M (1987) Secchi disc visibility world record shattered. EOS Transac 9:123

Gifford DM, Dagg MJ (1987) Consumption of ciliate microzooplankton by *Neocalanus plumchrus* in the subarctic North Pacific Ocean. (Abstr) Eos 68:1705

Hamner WM, Hamner PP, Obst BS, Carleton JH (1989) Field observations on the ontogeny of schooling of *Euphausia superba* furciliae and its relationship to ice in Antarctic waters. Limnol Oceanogr 34:451–456

Hart TJ (1934) On the phytoplankton of the southwest Atlantic and the Bellinghausen Sea. Discovery Rep 8:1–268

Hart TJ (1942) Phytoplankton periodicity in Antarctic surface waters. Discovery Rep 21:261–356

Hasle GR (1969) An analysis of the phytoplankton of the Pacific Southern Ocean: abundance, composition and distribution during the Brategg Expedition, 1947–48. Hvalradets Sgkr 52:1–168

Hayes PK, Whitacker TM, Fogg GE (1984) The distribution and nutrient status of phytoplankton in the Southern Ocean between 20° and 70° W. Polar Biol 3:153–165

Hempel G (1985) Antarctic marine food webs. In: Siegfried WR, Condy PR, Laws RM (eds) Antarctic nutrient cycles and food webs. Springer, Berlin Heidelberg New York, pp 266–270

Hempel I (1983) Studies in eggs and larvae of *Euphausia superba* and *Euphausia crystallorophias* in the Atlantic sector of the Southern Ocean. Rep Pol Res (Suppl) 4:30–46

Hewes CD, Holm-Hansen O, Sakshaug E (1985) Alternate carbon pathways at lower trophic levels in the Antarctic food web. In: Siegfried WR, Condy PR, Laws RM (eds) Antarctic nutrient cycles and food webs. Springer, Berlin Heidelberg New York, pp 277–283

Heywood RB, Priddle J (1987) Retention of phytoplankton by an eddy. Cont Shelf Res 7:937–955

Hodson RE, Azam F, Carlucci AF, Fuhrmann JA, Karl DM, Holm-Hansen O (1981) Microbial uptake of dissolved organic matter in McMurdo Sound, Antarctica. Mar Biol 61:89–94

Holm-Hansen O (1985) Nutrient cycles in Antarctic marine ecosystems. In: Siegfried WR, Condy PR, Laws RM (eds) Antarctic nutrient cycles and food webs. Springer, Berlin Heidelberg New York, pp 6–10

Holm-Hansen O, Foster TFD (1981) A multidisciplinary study of the eastern Scotia Sea. Antarct J U S 16:159–160

Holm-Hansen O, Huntley M (1984) Feeding requirements of krill in relation to food sources. J Crustacean Biol 4, Spec No 1:156–173

Hopkins TL (1971) Zooplankton standing crop in the Pacific sector of the Antarctic. In: Llano GA, Wallen IE (eds) Biology of the Antarctic Seas IV. American Geophysical Union, Washington DC, pp 347–362

Ikeda T, Dixon P (1982) Body shrinkage as a possible overwintering mechanism of the Antarctic krill, *Euphausia superba* Dana. J Exp Mar Biol Ecol 62:143–151

Ishii H, Omori M, Murano M (1985) Feeding behaviour of the Antarctic krill, *Euphausia superba* Dana I. Reaction to size and concentration of food particles. Trans Tokyo Univ Fish 6:117–124

Jacobson DM, Anderson JM (1986) Thecate heterotrophic dinoflagellates feeding behaviour and mechanisms. J Phycol 22:249–258

Jennings JC Jr, Gordon LJ, Nelson DM (1984) Nutrient depletion indicates high primary productivity in the Weddell Sea. Nature 309:51–53

Jumars PA, Penry DL, Baross JA, Perry MJ, Frost BW (1989) Closing the microbial loop: dissolved carbon pathway to heterotrophic bacteria from incomplete ingestion, digestion and absorption in animals. Deep-Sea Res 36:483–495

Kawaguchi K, Ishikawa S, Matsuda O (1986) The overwintering strategy of Antarctic krill (*Euphausia superba* Dana) under the coastal fast ice off the Ongul Islands in Lützow-Holm Bay, Antarctica. Mem Natl Inst Polar Res, Spec Iss 44:67–85

Koike I, Holm-Hansen O, Biggs DC (1986) Inorganic nitrogen metabolism by Antarctic phytoplankton with special reference to ammonium cycling. Mar Ecol Prog Ser 30:105–116

Macaulay MC (1981) Distribution and abundance of krill in the Scotia Sea as observed acoustically, 1981. Antarct J U S 16:166–167

Mackintosh NA (1973) Distribution of post-larval krill in the Antarctic. Discovery Rep 36:95–156

Marr JWS (1962) The natural history and geography of the Antarctic krill (*Euphausia superba* Dana). Discovery Rep 32:33–464

Marschall H-P (1988) The overwintering strategy of Antarctic krill under the pack-ice of the Weddell Sea. Polar Biol 9:129–135
Martin JH, Fitzwater SE (1988) Iron deficiency limits phytoplankton growth in the north-east Pacific subarctic. Nature 331:341–343
Matsuda O, Ishikawa S, Kawaguchi K (1987) Seasonal variation of downward flux of particulate organic matter under the Antarctic fast ice. Proc NIPR Symp Polar Biol 1:23–34
Maykut GA (1985) The ice environment. In: Horner RA (ed) Sea ice biota. CRC Press, Inc, Florida, pp 21–82
McWhinnie MA, Denys CJ, Parkin R, Parkin K (1979) Biological investigations of *Euphausia superba* (krill). Antarct J U S 141:163–164
Miller CB, Frost BW, Batchelder HP, Clemons MJ, Conway RE (1984) Life histories of large, grazing copepods in a subarctic ocean gyre: *Neocalanus plumchrus, Neocalanus cristatus,* and *Eucalanus bungii* in the northeast Pacific. Prog Oceanogr 13:201–243
Nöthig E-M (1986) Tank experiments with natural plankton populations: the influence of light and grazing pressure on pelagic system structure. Ophelia (Suppl) 4:179–190
Nöthig E-M (1988) On the ecology of the phytoplankton in the southeastern Weddell Sea (Antarctica) in January/February 1985. Rep Pol Res 53:118
O'Brien DP (1988) Direct observations of the behaviour of *Euphausia superba* and *Euphausia crystallorophias* (Crustacea: Euphausiacea) under pack-ice during the Antarctic spring of 1985. J Crustacean Biol 7:437–448
Olson RJ (1980) Nitrate and ammonium uptake in Antarctic waters. Limnol Oceanogr 25:1064–1074
Parsons TR, Takahashi M, Hargrave B (1984) Biological oceanographic processes. 3rd edn, Pergamon Press, Oxford, 332 pp
Patterson SL, Sievers HA (1980) The Weddell-Scotia confluence. J Phys Ocean 10:1584–1610
Peinert R (1986) Production, grazing and sedimentation in the Norwegian coastal current. In: Skreslet S (ed) The role of freshwater outflow in coastal marine ecosystems. Springer, Berlin Heidelberg New York, pp 361–374
Peinert R, Bodungen B von, Smetacek V (1989) Food web structure and loss rates: In: Berger WH, Smetacek VS, Wefer G (eds) Productivity of the oceans: present and past, Dahlem Konferenzen. John Wiley, Chichester, pp 35–48
Pomeroy LR (1974) The ocean's food web, a changing paradigm. Bioscience 24:499–504
Price HJ, Boyd KR, Boyd CM (1988) Omnivorous feeding behaviour of the Antarctic krill *Euphausia superba*. Mar Biol 97:67–77
Priddle J, Croxall JP, Everson I, Heywood RB, Murphy EJ, Prince PA, Sear CB (1988) Large-scale fluctuations in distribution and abundance of krill – a discussion of possible causes. In: Sarhage D (ed) Antarctic Ocean and resources variability. Springer, Berlin Heidelberg New York, pp 169–182
Sakshaug E, Holm-Hansen O (1984) Factors governing pelagic production in polar oceans. In: Holm-Hansen O, Bolis L, Gilles R (eds) Marine phytoplankton and productivity. Lecture notes on coastal and estuarine studies, vol 8. Springer, Berlin Heidelberg New York, pp 1–18
Schnack SB (1985) A note on the sedimentation of particulate matter in Antarctic waters during summer. Meeresforsch 30:306–315
Schnack-Schiel S (ed) (1987) The winter expedition of RV "Polarstern" to the Antarctic (ANT V/1–3). Rep Pol Res 39:259 pp
Schnack SB, Smetacek V, Bodungen B von, Stegmann P (1985) Utilization of phytoplankton by copepods in Antarctic waters during spring. In: Gray J, Christensen ME (eds) Proc 18th Eur Mar Biol Symp. John Wiley & Sons, Chichester, pp 225–232
Sherr E, Sherr B (1988) Role of microbes in pelagic food webs: a revised concept. Limnol Oceanogr 33:1225–1227
Smetacek V (1981) The annual cycle of protozooplankton in the Kiel Bight. Mar Biol 63:1–11
Smetacek V (1984) The supply of food to the benthos. In: Fasham MJ (ed) Flows of energy and materials in marine ecosystems: theory and practice. Plenum Press, New York, pp 517–548
Smetacek V, Pollehne F (1986) Nutrient cycling between water column and sediment: a reappraisal of the conceptual framework. Ophelia 26:401–428
Smith SL, Schnack-Schiel SB Polar zooplankton. In: Smith WO (ed) Polar oceanography. Academic Press, New York (in press)
Smith WO, Nelson DM (1986) Importance of ice edge phytoplankton production in the Southern Ocean. Bioscience 36:251–257
Spindler M, Dieckmann GS (1986) Distribution and abundance of planktonic foraminifera *Neogloboquadrina pachyderma* in sea ice of the Weddell Sea (Antarctica). Polar Biol 5:185–191
Spiridonov VA, Gruzov EN, Pushkin AF (1985) Investigations of schools of Antarctic *Euphausia superba* (Crustacea, Euphausiacea) under the ice. Zool Zh 64:1655–1660
Stegmann P, Peinert R (1984) Interrelationships between herbivorous zooplankton and phytoplankton and their effect on production and sedimentation of organic matter in Kiel Bight. Limnologica 15:487–495
Stretch JJ, Hamner PP, Hamner WM, Michel WC, Cook J, Sullivan CW (1988) Foraging behaviour of krill *Euphausia superba* on sea ice microalgae. Mar Ecol Prog Ser 44:131–139
Sugimura Y, Suzuki Y (1988): A high temperature catalytic oxidation method of non-volatile dissolved organic carbon in seawater by direct injection of liquid sample. Mar Chem 24:105–131
Sullivan CW, McClain CR, Comiso JC, Smith WO Jr (1988) Phytoplankton Standing crops within an Antarctic ice edge assessed by satellite remote sensing. J Geophys Res 93:12487–12498
Sverdrup HU (1953) On conditions for the vernal blooming of phytoplankton. J Cons Int Explor Mer 18:287–295
Takahashi M, Barwell-Clarke J, Whitney F, Koeller P (1978) Winter condition of marine plankton populations in Saanich Inlet, B C, Canada. I Phytoplankton and its surrounding environment. J Exp Mar Biol Ecol 31:283–301
Weeks WF, Ackley SF (1982) The growth, structure, and properties of sea ice. CRREL Monograph 82-1:130 pp
Wefer G (1989) Particle flux in the ocean: Present and Past In Berger WH, Smetacek VS, Wefer G – pp. 139–154. Dahlem Konferenzen, Chichester: Wiley
Yentsch DM, Yentsch ChS, Strube LR (1977) Variations in ammonium enhancement, an indication of nitrogen deficiency in New England coastal phytoplankton populations. J Mar Res 35:537–555

Primary Production and Consumption in McMurdo Sound, Antarctica

G. A. Knox [1]

Summary. This chapter provides a first synthesis of the data on primary production and consumption by invertebrate consumers in McMurdo Sound, Antarctica. The McMurdo Sound environment is briefly described, covering aspects such as bathymetry, sea ice development and breakout, and the hydrological regime. Based on available information an interpretation of the circulation patterns within the Sound and beneath the McMurdo Ice Shelf is given. A simple model of primary production and consumption is presented and the annual production cycle is discussed. In the eastern Sound the annual net production of the phytoplankton is estimated at 43.9 $g\,C\,m^{-2}$. Of this the annual bloom of *Phaeocystis* advected under the sea ice from the ice edge to the north contributes 16.1 $g\,C\,m^{-2}$. Annual sea ice microalgal net production is estimated at 11.6 $g\,C\,m^{-2}$. Overall the sea ice microalgae contribute up to one-third of the annual microalgal production. In the shallow waters of the eastern Sound the benthic microalgal production is estimated at 26.0 $g\,C\,m^{-2}$. Production of the phytoplankton and the benthic microalgae are much lower in the western Sound. Of the annual sea ice microalgal production at least 70% is released to the water column on the melting of the ice. The water column heterotrophs and the zooplankton consume no more than 3% of the annual phytoplankton production. At least 80% of this production sediments to the bottom. The impacts of perturbations, such as variations in the timing and extent of the sea ice breakout, the depth of snow cover on the sea ice, and changes in the current systems under the influence of external driving forces are briefly discussed. The main points to emerge are: (1) the McMurdo Sound primary production is high when compared with that of the open Southern Ocean, but it is comparable to that of the ice edge zone; (2) the important contribution made to primary production by the annual *Phaeocystis* bloom; (3) the major contribution which is made by the benthic microalgae in shallow inshore waters, a contribution that has been largely ignored in studies to date; (4) the high degree of natural variability within the system; (5) the potential for change in the various environmental variables that influence the system; and (6) the deficiencies in the data base and the need for integrated year-round studies of primary production and consumption.

1 Introduction

Since the International Geophysical Year (1956) there has been a considerable amount of research on varied aspects of the marine ecosystem of McMurdo Sound to date, however, there has been no attempt to integrate them into an overall view of the patterns of primary production and consumption. The objectives of this review are: (1) to develop a simple model of primary production and consumption; (2) to estimate the seasonal and annual primary production; (3) to estimate as far as possible the seasonal and annual consumption and the fate of the primary production; (4) to assess the relative importance of model forcing functions and environmental variables; (5) to evaluate the impact of natural perturbations on primary production and consumption; and (6) to highlight deficiencies in the data base and point out future research needs.

The theme of this Symposium is "Ecological change and the conservation of Antarctic ecosystems". While the data base for this work does not include long-term monitoring studies of primary production and consumption, there is nevertheless information obtained over many years and at different seasons suggestive of comparatively large-scale year-to-year variation. For example the pattern of ice breakout in McMurdo Sound is highly variable, as is the depth of the snow cover on the sea ice. Such variations can have a considerable impact on primary production, resulting in substantial differences from one season to the next. It is also possible that climate change involving increasing mean air and sea temperatures and changes in near-shore circulation patterns could result in permanent changes in the pattern of primary production. In

[1] Department of Zoology, University of Canterbury, Christchurch, New Zealand.

Antarctic Ecosystems. Ecological Change and Conservation
(ed. by K. R. Kerry and G. Hempel)
© Springer-Verlag Berlin Heidelberg 1990

order to predict future trends it is essential that we have an understanding of the various components of the production cycle in McMurdo Sound and of the variables that influence it. This work is a first attempt at such a synthesis.

2 The McMurdo Sound Environment

McMurdo Sound lies at the south-western corner of the Ross Sea, bounded by Victoria Land on the west and Ross Island on the east (Fig. 1). To the south is the McMurdo Ice Shelf, which is an easterly extension of the Ross Ice Shelf. The average width of the Sound is about 58 km.

2.1 Bathymetry

Figure 1 depicts the bathymetry of the Sound to the north of the McMurdo Ice Shelf. Of the total area of 5820 km^{-2} only 5.2% lies below 100 m, while 19% lies between 100 and 200 m. There are approximately equal areas between 200 and 500 m and over 500 m.

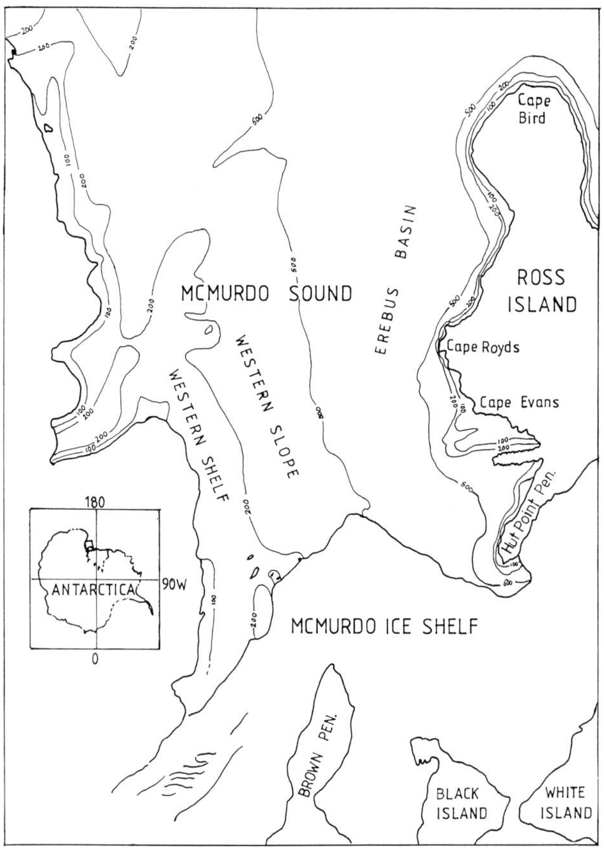

Fig. 1. Map of McMurdo Sound showing the bathymetry

On the eastern side of the outer sound the shelf is narrow and descends steeply into the Erebus Basin. The western shelf is broader and the slope to the Erebus Basin is more gradual.

2.2 Sea Ice

Dayton and Oliver (1977) examined aerial photographs taken of McMurdo Sound during the months of October to January over the years 1956 to 1979. These showed a predictable pattern of breakout of the annual sea ice very similar to that observed by the early Shackleton and Scott expeditions. The photographs documented very few breakouts in West Sound by February when aerial surveys stopped for the season. In the West Sound from Garwood Valley northward 2- to 3-year ice floes were observed, suggesting occasional breakup after February with different amounts of ice remaining in the area. The extent of the ice breakout to the south varies considerably from year to year (Heine 1963), and sea ice may remain at a variable distance from the McMurdo Ice Shelf for 2 or more years. The maximum breakout occurs during the last week of February and the first week of March with refreezing occurring by late March. However, the newly frozen ice may break up, drift out of the Sound and the surface water may refreeze several times before forming fast ice in July. The extent and timing of the breakout is determined by the strength of the northerly swells and southeasterly gales.

2.4 Circulation Patterns

Circulation patterns in McMurdo Sound are complex and variable (see reviews by Heath 1977; Raytheon 1983; Lewis and Perkin 1985). Based on observations carried out until 1975 (Gilmour et al. 1960, 1962; Tressler and Ommundsen 1962; Littlepage 1965; Heath 1971 a, b), Heath (1977) proposed a circulation pattern that has in large part been confirmed by subsequent studies. All of these studies identified a strong diurnal tidal influence on the current patterns, with the currents being strongest when the tidal range is greatest.

In the outer Sound a strong southerly current, the Cape Royds Current, passes round Cape Bird towards Cape Royds where it swings across the sound to join a northward flowing current on the west side of the sound. Estimates of the net movement of surface currents in the East Sound from north to south vary from 2.5 km day^{-1} (Barry and Dayton 1988), to 6.5 (Gilmour et al. 1962), 10.3 (Littlepage 1965) and 4–12 km (Palmisano et al. 1986), with a mean of 6.8 km.

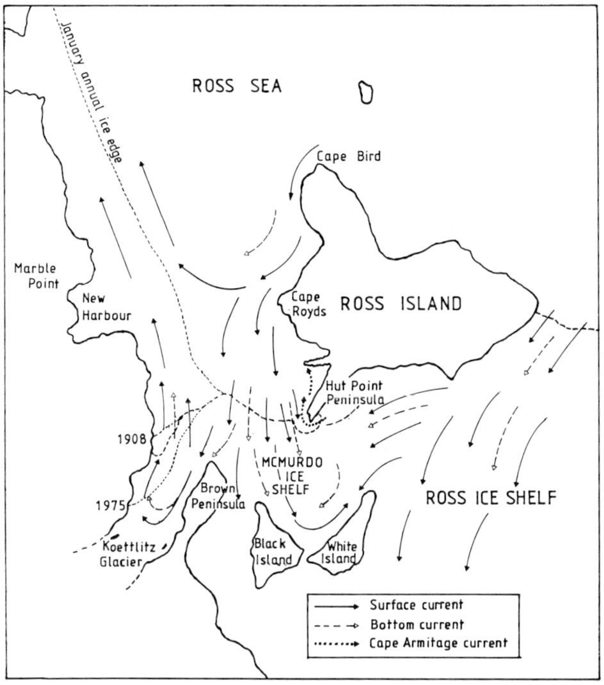

Fig. 2. Hypothesized circulation pattern in McMurdo sound and beneath the McMurdo Ice Shelf (after Knox 1985)

These show considerable variation indicative of either seasonal or year-to-year variations. The majority of studies have confirmed a northerly flow from under the shelf on the west side of the sound. Further north this flow from under the shelf joins the northward flow of water that originated from the Cape Bird Current.

The most controversial aspect of the current flows is a northward flowing current on the east side of the sound, the Cape Armitage Current, which has been identified by a number of investigators (Littlepage 1965; Evans 1965; Car and Codspoti 1968; Heath 1971a). Measurements made closer than about 2 km from the coast near Cape Armitage at the end of Hutt Point Peninsula indicated a flow out from under the shelf. Barry and Dayton (1988), on the other hand, considered that the flow along Cape Armitage is to the south. Heath (1971a), on the basis of hydrological measurements, considered that the northerly outflow near Ross Island probably originated from water flowing under the Ross Ice Shelf from the Ross Sea to the west of Ross Island. Figure 2 depicts a hypothetical circulation pattern based on the information discussed above.

3 A Production and Consumption Model

Figure 3 is a conceptual model incorporating the major components of production and consumption in McMurdo Sound. The dissolved and particulate organic matter pathways have not been included in the model and a number of minor pathways have been omitted. It is divided into three levels: (1) primary producers; (2) invertebrate consumers; and (3) vertebrate consumers. Four major (sea ice microalgae, phytoplankton, benthic microalgae and benthic macroalgae) and two minor (snow and melt pool microalgae and epiphytic microalgae) primary producers are shown. The invertebrates include the sea ice community consumers, the cryopelagic community consumers, the planktonic heterotrophs, the euphausiids, the non-euphausiid zooplankton, the benthic epifauna and the benthic infauna, while the vertebrate consumers include fishes, birds, seals and whales. Also depicted in the model are the pathways of import to and export from the McMurdo Sound marine system.

Table 1 gives an interaction matrix of the model forcing functions and environmental variables that influence the primary production processes. The penetration of solar radiation into the water column is determined by season, wind, sea ice cover and the depth of snow on the sea ice. Sea ice formation is determined by season, solar radiation, wind, water circulation, sea temperature and salinity. Sea ice ablation and breakout are determined by the same combination of factors as for its formation but in addition is influenced by the amount of snow cover. It can thus be seen that the primary production process is determined by a complex interaction of the variables listed in the matrix. Overall seasonality is the dominant influence with its impact on the light regime and weather, especially the wind strength and the prevalence and intensity of storms. While a number of studies have explored the impact of some of these variables on aspects of primary production, e.g. the impact of snow cover on ice algal and phytoplankton production (Palmisano et al. 1985a; Grossi et al. 1987), the complexity of their interactions has yet to be elucidated. This will be considered further in Section 6.

4 Primary Production

The major contributors to primary production are the phytoplankton, the sea ice microalgae and, in shallow water areas, the benthic microalgae. There are adequate data available to assess the annual contributions of these components. Other contributions come from the epiphytic microalgae and the benthic macroalgae, however, quantitative data are lacking A minor contribution is also made by the snow algae which may develop in association with the snow on the sea ice and the McMurdo Ice Shelf and the microalgae that grow

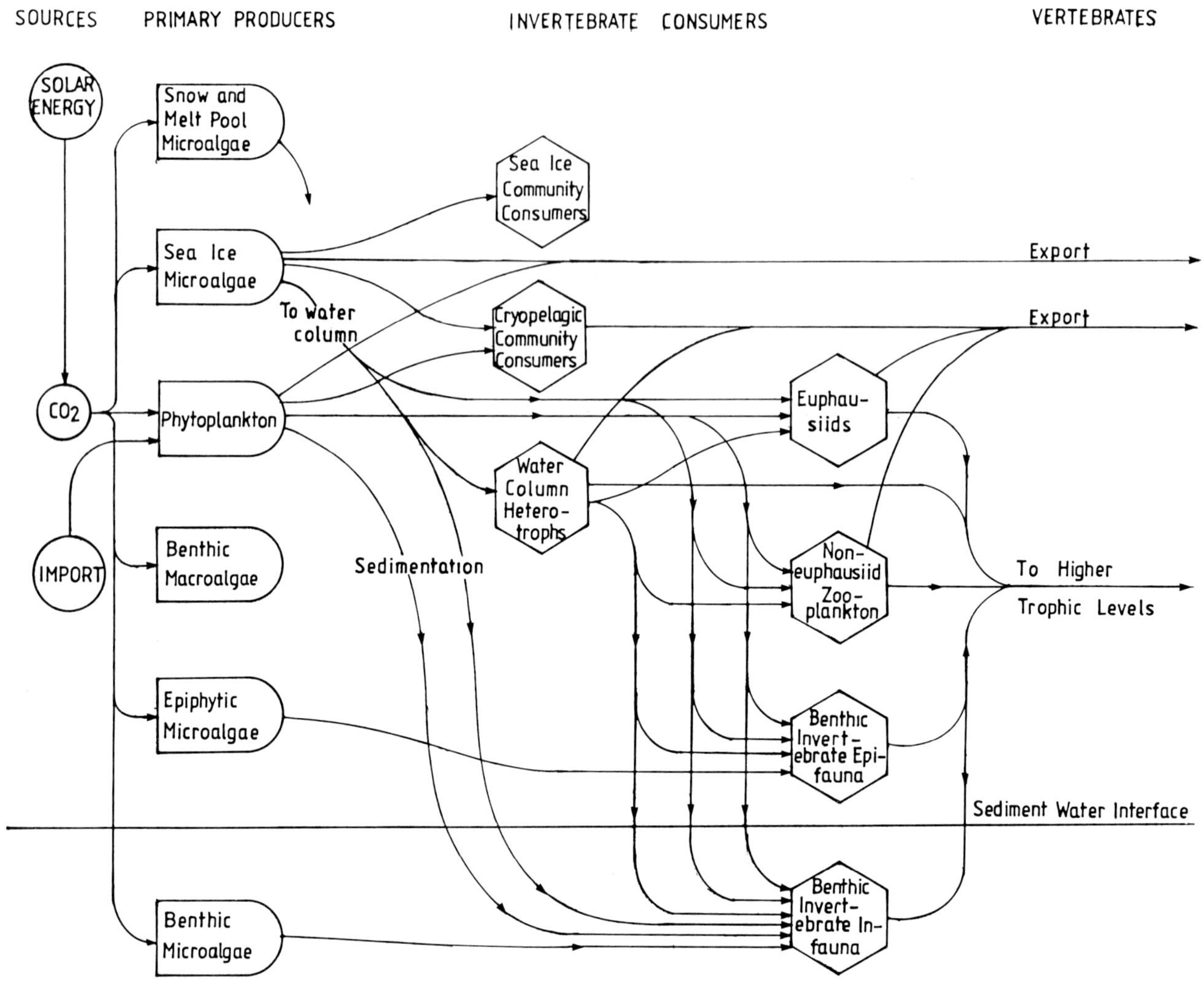

Fig. 3. A simple model of primary production and consumption in McMurdo Sound

in melt pools on the shelf. The various components of primary production will be considered in turn. A range of estimates of standing stocks and primary production are listed in Table 2.

4.1 Sea Ice Microalgae

The annual sea ice provides a growth substratum and refugium for a complex sea ice microbial community (SIMCO) consisting primarily of microalgae, bacteria and protozoa. In the inner McMurdo Sound near Cape Armitage diatoms dominate the SIMCO biomass, but other algae including *Phaeocystis pouchetii* and various small (10 μ) flagellates are also found (Palmisano and Sullivan 1983). Since the early 1980s US scientists have undertaken an integrated comprehensive programme on the SIMCOs in the McMurdo Sound region (e.g. Palmisano and Sullivan 1983, 1985; Grossi et al. 1984; Palmisano et al. 1985a, b, 1987; Sullivan and Palmisano 1984; Grossi and Sullivan 1985; Sullivan et al. 1985; Kottmier et al. 1987a, b). These investigators have recorded sea ice microalgal chlorophyll *a* concentrations averaging 131 ($\pm$ 121) mg Chl *a* m^{-2}, with a maximum of 309 mg. During the ice algal bloom, concentrations of chlorophyll *a* in the bottom 20 cm of the sea ice averaged 565 mg m^{-2}, 2000 times greater than the under ice phytoplankton at 1 m depths.

While Grossi (1985) estimated a production of 17.7 g C m^{-2} for the spring bloom of sea ice microalgae, other studies have given lower estimates. Based upon the average biomass during the spring bloom (131 mg Chl *a* m^{-2}) Palmisano and Sullivan (1983) estimated that annually the sea ice microalgae could contribute at least 4.1 g C m^{-2} to the benthic and pe-

Table 1. Matrix of model forcing functions and environmental variables influencing primary production

	Solar radiation	Tides	Wind	Water circulation	Sea temperature	Salinity	Nutrients	Sea ice formation	Sea ice melting and breakout	Sea ice snow cover	Consumption	Import	Export
Solar radiation			×					×	×	×			
Tides			×	×		×						×	×
Wind	×	×		×				×	×	×		×	×
Water circulation		×	×		×	×	×	×	×			×	×
Sea temperature				×		×		×	×				
Salinity		×		×	×			×	×				
Nutrients				×				×	×		×	×	×
Sea ice formation	×		×	×	×	×	×						
Sea ice melting and breakout	×		×	×	×	×	×			×		×	×
Sea ice snow cover	×		×						×				
Consumption							×					×	×
Import		×	×	×			×		×		×		
Export		×	×	×			×		×		×		

lagic consumers. They point out that this is clearly and underestimate because: (1) the spring ice algal bloom continued beyond their study; (2) they did not include losses to grazers; and (3) autumn growth may contribute further production.

Taking the above into account gross production is estimated at 16.9 $g\,C\,m^{-1}\,yr^{-1}$, respiration at 5.07 g (30% of gross production; Day et al. 1973) and net production at 11.83 g. It is probable that at least 70% of the production is released to the water column following the ablation and melting of the ice bottom layer in late December–January.

4.2 Phytoplankton

Chlorophyll *a* levels in the water column under the sea ice in the winter months are negligible, or very low, and remain so as the sea ice microalgae build up in the spring and reduce light penetration. Levels ranging from 0–0.09 mg Chl $a\,m^{-2}$ were recorded by Dayton et al. (1986) from a range of sites in the inner Sound. In mid-December, however, measured levels rise to 1.28 mg Chl $a\,m^{-2}$ as phytoplankton from the open waters in the Ross Sea is advected beneath the fast sea ice. This bloom is composed almost entirely of a single species, the colonial pyrmenesiophyte *Phaeocystis pouchetii*. With the advent of the bloom visibility for diving is lowered from about 300 to less than 1 m. During the 1984–85 spring and summer, Rivkin and Voytek (1987) found that *Phaeocystis* first appeared at the ice edge which lay between Cape Bird and Cape Royds on about 10 December, and then sequentially appeared at each of several sites along a north-south transect across the sea ice, arriving at Cape Armitage on 25–28 December.

Palmisano et al. (1986) found that *in situ* production beneath 1.8 m of annual ice at Cape Armitage averaged 2.8 mg C · Chl $a \cdot h^{-1}$. Assuming an euphotic zone under the sea ice of about 3 m (a conservative depth), it was estimated that the primary production would be 24 mg $C\,m^{-2}\,h^{-1}$. Bloom duration was estimated to be 4 weeks (again a conservative estimate). On this basis, the seasonal production of the early summer bloom of *Phaeocystis* in the eastern Sound was estimated at 16.1 $g\,C\,m^{-2}$.

The *Phaeocystis* bloom is followed later in the season by a second bloom composed mainly of diatoms. Smith and Nelson (1985a, b, 1986), Nelson and Smith (1986) and Wilson et al. (1986) determined the phytoplankton biomass near a receding ice edge off the Victoria Land coast just to the north of McMurdo Sound (between 77° and 78° S), over the period 26 January to 2 February. The measured chlorophyll *a* concentrations were much higher than those previously measured, averaging 4.08 ± 1.46 mg Chl $a\,m^{-3}$ at the depth of the chlorophyll maximum in the water column, and 128.2 ± 91.7 m^{-2} when integrated from the surface to 150 m. This compares with a mean value of 6.67 mg Chl $a\,m^{-2}$ for the Ross Sea region (Holm-

Table 2. Data on standing stocks and primary production for the microalgal producers

Parameter	Season	Units	East Sound	West Sound	Reference
Phytoplankton					
Chlorophyll *a*	Spring-summer	mg m^{-3}	0.69 (SD 0.87)	0.13 (SD 0.13)	Dayton et al. (1986)
Chlorophyll *a*					
Phaeocystis bloom	December	μg l^{-1}	12.0	–	Palmisano et al. (1986)
Diatom flux		No m^{-2} day^{-1}	390.0	90.0	Levanter and Dunbar (1986)
Productivity	Spring-summer	mg C m^{-2} h^{-1}	1.1	0.0	Bunt (1964)
Productivity					
pre-*Phaeocystis* bloom	Early summer	mg C m^{-2} h^{-1}	0.006 (max)	–	Kottmeier et al. (1984)
early-*Phaeocystis* bloom	Early summer	mg C m^{-2} h^{-1}	0.404 (max)	–	
Productivity					
Phaeocysits bloom	December	mg C m^{-2}h^{-1}	24.0	–	Palmisano et al. (1986)
Productivity	Summer	g C m^{-2} day^{-1}	2.0	0.4–1.0	Barry (1988)
Sea ice microalgae					
Chlorophyll *a*	Early summer	mg m^{-2}	520–1076	–	Bunt and Lee (1970)
Chlorophyll *a*	December	mg m^{-2}	309 (max)	–	Palmisano and Sullivan (1983)
Chlorophyll *a*	Early summer				Kottmeier et al. (1984)
	(a) Snow-free area		49.0		
	(b) Snow-covered area		8.0		
Chlorophyll *a*	December	mg m^{-2}	150–170	–	Palmisano et al. (1988)
Chlorophyll *a*	December	mg m^{-2}	93.1 (SD 95.3)	188.0 (SD 1.71)	Palmisano and Sullivan (1983)
Productivity	Early Summer	mg C m^{-2} h^{-1}			Kottmeier et al. (1984)
	(a) Snow-free area		22.0	–	
	(b) Snow-covered area		0.24	–	
Productivity	Spring	mg C m^{-2} h^{-1}	0.7–7.8	–	Grossi et al. (1987)
Productivity	Bloom duration	g C m^{-2} y^{-1}	17.7	–	Grossi (1985)
Productivity	Bloom duration	g C m^{-2} y^{-1}	4.1	–	Palmisano and Sullivan (1983)
Benthic microalgae					
Chlorophyll *a*	Summer	mg m^{-2}	294–863	18.3–145	Dayton et al. (1986)
	Winter		47.3–265		
Productivity	Summer	mg C · mg Chl *a* h^{-1}	0.01	–	Palmisano et al. (1985b)
Productivity	Summer	mg C m^{-2} h^{-1}	51.3–95.22	–	Dayton et al. (1986)

Hansen et al. 1977), average values of ca. 1.65 $mg\,m^{-2}$ in the southern Weddell Sea and 0.1 $mg\,m^{-2}$ in the northern Weddell Sea (El-Sayed and Taguchi 1981), and an overall mean for the Pacific sector of 0.26 $mg\,m^{-3}$ and 1.22 $mg\,m^{-2}$ (surface and integrated euphotic zone values) (El-Sayed and Turner 1977). The phytoplankton bloom, which was dominated by a single diatom, *Nitzschia curta* (considered to be an ice algal species), extended 200 km from the dense pack and appeared to be associated with a stable surface layer at the receding edge of the pack ice. The mean productivity of the bloom was estimated at 962 $mg\,C\,m^{-2}\,day^{-1}$ (Wilson et al. 1986).

Smith and Nelson (1986) calculated the production of the bloom by combining data on the bloom's spatial extent, its measured productivity, and its duration based on rates of ice retreat. They calculated that at one location the ice edge bloom produced 45.6 g $C\,m^{-2}$. Data are not available for the open waters of McMurdo Sound in January–February to determine the extent to which the bloom penetrates into the Sound. For the purpose of calculating the total phytoplankton production in McMurdo Sound, the January–February bloom is estimated at half the value given by Smith and Nelson (1986) for the Ross Sea waters to the north of the Sound.

The total annual phytoplankton production for eastern McMurdo Sound can be calculated as 16.1 $g\,C\,m^{-2}$ for the *Phaeocystis* bloom, plus 22.8 g $C\,m^{-2}$ for the later diatom bloom, plus 5 $g\,C\,m^{-2}$ for production at other times, making a total of 43.9 g $C\,m^{-2}$, which is more than three times that calculated by Holm-Hansen et al. (1977) for the open waters of the Southern Ocean (Table 3).

4.3 Benthic Microalgae

There is only limited data on the distribution, biomass and productivity of the benthic microalgae in McMurdo sound, and indeed in Antarctica as a whole. Dayton et al. (1986) measured benthic chlorophyll *a* and phaeopigments at several sites along the east and west sides of the sound. The benthic microalgae that occur here are remarkably shade-adapted. Palmisano et al. (1985b) found that they were light-saturated at only 11 $\mu E\,m^{-2}\,s^{-1}$, putting them among the most shade-adapted microalgae reported. Summer standing stocks of chlorophyll *a* in the East Sound ranged from 294 to 863 mg Chl $a\,m^{-2}$, while winter standing stocks ranged from 47.3 to 265 mg Chl $a\,m^{-2}$. West Sound summer standing stocks ranged from only 18.3 to 145 mg Chl $a\,m^{-2}$.

The data are insufficient to estimate the annual production of the benthic microalgae. However, if the mean maximum standing stock of chlorophyll *a* is about 500 $mg\,m^{-2}$ (the approximate mean of the measured November and December values) in the East Sound, this represents a standing stock of 13.5 mg $C\,m^{-2}$ (C:Chl *a* ratio of 38; Sullivan et al. 1982). In the West Sound a mean standing stock of 60 mg Chl $a\,m^{-2}$ represents a standing stock of 1.3 mg $C\,m^{-2}$. These estimates, however, take no account of the consumption of the benthic microalgae by the benthic invertebrate epifauna and infauna or of algal senesence. This could be considerable and the actual production could be very much greater than the mean maximum standing stock. While the measured summer productivity rates were low, benthic microalgal productivity may be considerably higher prior to the

Table 3. Primary production in McMurdo Sound[a]

Compartment	Annual net production $g\,C\,m^{-1}\,yr^{-1}$ (Eastern Sound)	Total annual production $t\,C\,yr^{-1}$ (for the whole Sound)
Snow and melt pool microalgae	Not known	Not known
Sea ice microalgae	11.83	500–700 000
Phytoplankton		
Phaeocystis	16.10	
Other species	27.8	
Total	43.9	1.5–2.25 million
Benthic macroalgae	Not known	Not known
Epiphytic microalgae	Not known	Not known
Benthic microalgae		
East Sound	26.0	25–30 000
West Sound	3.0	2– 3 500

[a] Assuming that phytoplankton production in the Western Sound is half of that in the Eastern Sound.

ice algal bloom which begins in late October and greatly reduces the photosynthetically available radiation (PAR) available to the benthic microalgae, and subsequent to the ice breakout and melting in late January and February.

4.4 Other Primary Producers

Howard-Williams et al. (1988) have shown that there is a considerable microalgal production (diatoms and cyanobacteria) in the anastomosing system of lakes, pools and streams on more than 1500 km of the McMurdo Ice Shelf. In the summer of 1982–83 there was a large runoff of meltwater from the shelf which covered the inshore sea ice to a depth of 1 m for some considerable distance from the edge of the ice shelf. However, the proportion of the annual production on the shelf that finds its way into the inshore marine system via runoff is unknown.

There are no measures of the standing stock or of the productivity of the populations of benthic macroalgae in McMurdo Sound, especially of the red algae *Iridaea skottsbergii* and *Phyllophora antarctica* which are common along the East Sound (Dayton et al. 1986). Nor has there been any estimates of the standing stock of the epiphytic microalgae which grow on the stems and fronds of the macroalgae and on the abundant sessile benthic invertebrate macrofauna growing in the shallow waters of the sound. These could contribute significantly to the overall primary production in shallow waters below 100 m. However, as these waters comprise only 5.2% of the total area of the sound, the contribution of the benthic macroalgae and the epiphytic microalgae to the overall primary production in the Sound would be minor.

4.5 The Seasonal Production Cycle

Figure 4 illustrates an attempt to depict the annual production cycle in McMurdo Sound. The vertical scale in this figure is not to be taken as representative of the comparative contributions of the various components to the total annual primary production. The phytoplankton cycle is now well established with diatoms dominating early in the season (mid-November to mid-December), followed by the *Phaeocystis* bloom from mid-December to early January and the subsequent diatom bloom in late January–February. The sea ice microalgae production commences in the spring, reaching a peak in mid-December and from mid-January the production is released to the water column with the melting of the ice. There may be a further smaller ice microalgal production in the autumn after the sea ice has reformed. The production curve for the benthic microalgae is somewhat speculative. It is hypothesized that a spring peak of production occurs in early November before the growth of the sea ice microalgae reduces the light available to the algae. A second peak is hypothesized after the breakout and melting of the sea ice in February–March.

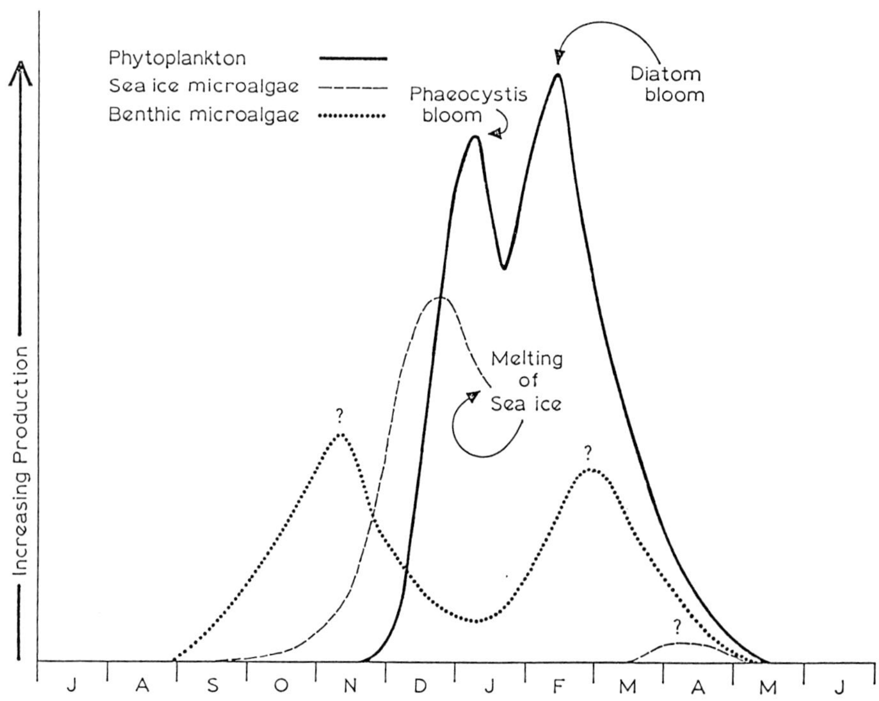

Fig. 4. The seasonal cycle of microalgal production

Table 3 gives estimates of the annual net primary production in eastern McMurdo Sound. The major producer is the phytoplankton with an annual net production of 43.9 g C m^{-2} yr^{-1}. The sea ice microalgae contribute an annual net production of 11.83 g C m^{-2} yr^{-1}, or 26.8% of that of the phytoplankton net production. While in the East Sound the benthic microalgae possibly contribute 25 g C m^{-2} yr^{-1} in the area under 100 m, this is only 1.8% of the total area of the Sound. Thus while the production may be significant to the benthic community there, in the context of the sound as a whole, its contribution to the overall primary production is minor. The sea ice microalgae and the phytoplankton in eastern McMurdo Sound together have an annual production of 55.66 g C m^{-2} yr^{-1} which is a very high value relative to the open waters of the Ross Sea and other parts of the Southern Ocean. Even if this estimate were halved the microalgal production in the eastern Sound would still have a value that would rank the area as one of the most productive in the Southern Ocean. A first-order estimate for the total Sound is also given in Table 3. From this it can be seen that the sea ice microalgae account for about one-third of the total production.

5 Fate of the Primary Production

5.1 Consumption, Decomposition and Sedimentation

To date there are insufficient data to accurately estimate the quantities of primary production that are directly consumed by the various primary consumers. These include the sea ice community heterotrophs, the cryopelagic community, the water column heterotrophs, the zooplankton and the benthic invertebrates.

Sea Ice Community Heterotrophs. These comprise heterotrophic flagellates and dinoflagellates, protozoa, small metazoans, such as harpacticoid copepods, and the nauplii and larval stages of copepods which occur as adults in the water column. These consumers constitute what has been called the sea ice *sympagic* community (Whitaker 1977; Carey 1985). They probably consume no more than 2 or 3% of the ice algal production.

The Cryopelagic Community. Associated with the under surface of the sea ice is the cryopelagic community composed of adult and larval stages of invertebrates and some fish species. Prominent among the algal grazers in this community are the amphipods *Paramoira walkeri*, *Orchomene plebs* and *Eusirus antarcticus*, adult and larval *Euphausia crystallorophias* and copepods. These species could potentially consume up to 15% of the ice algal production (Kottmeier et al. 1984; Hopkins 1987). Together the sea ice community heterothrophs and the cryopelagic community probably consume no more than 10–15% of the ice microalgal production. Thus, after taking decomposition into account, at least 70% of the microalgal production is released into the water column on the melting of the sea ice. Since a proportion of the sea ice is transported from the Sound after the breakup, an unknown proportion that will vary from year-to-year will be lost from the Sound.

Water Column Heterotrophs. To date there have been no studies on the ecological role of the water column heterotrophs (miscellaneous flagellates, some dinoflagellates, choanoflagellates, ciliates and tintinnids) in McMurdo Sound. Elsewhere in Antarctica they have been shown to be an important component of the nanoplankton (Buck and Garrison 1983; Heinbokel and Coats 1984, 1986; Garrison et al. 1984; Garrison and Buck 1985; Garrison and VanScoy 1985). These heterotrophs, especially the ciliates and choanoflagellates, can consume up to half the amount of phytoplankton consumed by the zooplankton and also play an important role in maintaining the bacterial biomass at relatively constant levels (Anderson and Fenchel 1985; Marchant 1985). They are an important link in the transfer of carbon from the extracellular carbon secreted by the phytoplankton via the bacteria they consume to higher trophic levels (Hewes et al. 1985).

Zooplankton. There have been a number of studies on the zooplankton of McMurdo Sound (Hicks 1974; Knox 1986; Forster 1987; Hopkins 1987; Knox et al. in press; Knox in prep.). The zooplankton is dominated by copepods and pteropods. Other species include ostracods, euphausiids, mysids, amphipods, chaetognaths, pelagic polychaetes, the larvae of benthic polychaetes and larval and post-larval stages of fishes, especially the mid-water fish *Pleuragramma antarcticum*. In contrast to more northerly Antarctic localities, e.g. that studied by Hopkins (1985) at Croker Passage on the Antarctic Peninsula, euphausiids are not as abundant and salps are rare. While the euphausiid *Thysanoessa macrura* occurs sparsely early in the spring, the principal euphausiid throughout the summer is *Euphausia crystallorophias*.

Summer zooplankton wet weights in southern McMurdo Sound have been estimated to range from 0.073 to 5.0 mg m^{-2} (Hicks 1974; Forster 1987), while in the northern part of McMurdo Sound in February,

Hopkins (1987) estimated zooplankton dry weights at 1.5 to 3.5 $g\,m^{-2}$. Hopkins investigated the trophic structure of the mid-water ecosystem by examining the diet of 35 species of zooplankton and micronekton. At the carbon levels recorded by Smith and Nelson (1985a) off Victoria Land (191 $\mu g\,C\,l^{-1}$ or 28.6 $g\,C\,m^{-2}$) the dominant copepod particle grazers would ingest up to 35% of their body weight in terms of carbon per day (Hopkins 1987). The mean biomass of the grazing copepods estimated by Hopkins (1987) at 2.45 g dry wt. m^{-2} or 0.6 $g\,C\,m^{-2}$, could potentially remove 0.21 g of particulate carbon $m^{-2}\,day^{-1}$, or 0.7% of the standing crop. If the remaining grazers removed a similar amount, then the total zooplankton consumption would be roughly 2% of the phytoplankton biomass. While the protists and other unicellular heterotrophs consume an as yet undetermined amount of phytoplankton, the total phytoplankton consumption is not likely to exceed 3% of the phytoplankton biomass. Thus at least 97% of the standing crop remains unconsumed by grazers.

Benthic Invertebrates. While there is sufficient information available to attempt to evaluate the role of the shallow water benthos in the McMurdo Sound ecosystem (e.g. Bullivant 1967; Dayton et al. 1969, 1970, 1974; Lowry 1976; Knox and Lowry 1977; Dayton and Oliver 1977), a detailed analysis is beyond the scope of this chapter. The epifaunal suspension feeders in depths above 200 m could consume as much as 10% of the phytoplankton as it sediments to the bottom, as well as a proportion of the sea ice microalgae released to the water column. A smaller amount of the phytoplankton and sea ice microalgae would be consumed by the infaunal suspension feeders. Nevertheless, the major proportion of the phytoplankton net production, up to 80%, is not consumed and is either decomposed in the water column or sedimented to the bottom. This estimate is in agreement with that of other workers in the region (Smith and Nelson 1986; Hopkins 1987). The proportion of the the benthic microalgal production that is consumed by the macrobenthic invertebrate deposit feeders, the benthic meiofauna and microfauna is not known.

5.2 Import–Export

The major input of organic matter is via the Cape Bird Current on the east side of the Sound. Other inputs are via the Cape Armitage Current from the Ross Sea to the east of Ross Island and the return flow from under the McMurdo Ice shelf on the west side of the Sound. The major exports are the input from the Sound to beneath the McMurdo Ice Shelf in the centre of the Sound (Knox 1986 and in prep.) and the northward flowing current along the Victoria Land coast on the west side of the Sound. Other contributions to the import–export balance come from the seasonal migrations of some species, especially fishes, seals, whales and birds. The data are inadequate for evaluation of the export–import equation.

6 Discussion

The main points that emerge from this analysis are: (1) the McMurdo Sound primary production is high when compared with that of the open waters of the Southern Ocean, but it is comparable to that of the ice edge zone; (2) the importance of the contribution made to primary production by the annual *Phaeocystis* bloom; (3) the major contribution which is made by the benthic microalgae in shallow inshore waters, a contribution that has largely been ignored in studies to date; (4) the high degree of natural variability in the system; (5) the potential for change in the various environmental variables that influence the system; and (6) the deficiencies in the data base and the need for integrated year-round studies of primary production and consumption.

6.1 Forcing Functions, Environmental Variables and Perturbations

As can be seen from Table 2 there are considerable differences in the various estimates of the standing stock and the primary production of the principal primary producers. In part this is a reflection of the time of the year when the observations were made, but also probably reflects year-to-year variations. Such variations would be due to changes in the forcing functions and environmental variables listed in Table 1.

Water circulation is one of the dominant factors influencing regional production within the Sound (Barry 1986; Barry and Dayton 1988). Mean currents in McMurdo Sound are the result of the additive effects of various driving forces including local and large-scale wind-driven circulation, tidal rectification, and pressure gradient forces arising from variations in density. Barry and Dayton (1988) contrasted their 1984 results with those of Lewis and Perkin (1985) in 1982 when northerly input from beneath the shelf dominated the southern Sound. They speculated that the differences were due to changes in the driving forces of the East Wind drift which flows west along the Ross Ice Shelf barrier and contributes strongly to the net transport into McMurdo Sound through the southern extension (the Cape Bird Current) sweeping into the Sound. It is probable that global atmospheric anomolies such as ENSO (El Nino Southern Oscillation) are linked to these differences (Barry and Dayton 1988), as they have been linked by numerous authors to global scale variability in atmospheric pressure and associated zonal winds (Rogers and VanLoon 1982). Barry and Dayton (1989) have demonstrated that the extent of fast ice from McMurdo Station north during the spring-summer channel breakout is greater in ENSO years (Seymour et al. 1984) than in non-ENSO years.

Early in the spring phytoplankton standing stocks and primary production are low due to the sea ice cover, which reduces the light penetration into the water column to very low levels. Where there is extensive snow cover light levels are further reduced. Increases in the phytoplankton standing stock and primary production in the late spring-early summer are due to the transport beneath the sea ice of the *Phaeocystis* bloom originating at the ice edge to the north of the Sound. Assuming a southward drift of 3.0 cm s^{-1} (Barry and Dayton 1988), transit of this bloom from Cape Bird to near McMurdo Station would take approximately 30 days. While the occurrence of the advected bloom is a regular annual feature, the timing of its arrival in the southern part of the Sound may differ by up to at least 3 weeks (pers. observ.). This difference is probably due in large part to the varying distance from year to year of the ice edge to the north of McMurdo Sound. The greater the distance, the longer is the time taken for the bloom to reach the southern Sound. An additional environmental variable is the strength of the Cape Bird Current flow into the Sound. As noted previously this may vary considerably. The *Phaeocystis* bloom is in turn followed by a diatom bloom again originating in processes occurring at the ice edge to the north. As the sea ice melts and breakup occurs the ice algae are released into the water column, and the increased water column stability from the meltwater input (El-Sayed and Mandelli 1965; El-Syed and Taguchi 1981; Smith and Nelson 1986) would provide conditions suitable for enhanced phytoplankton production within the Sound. Meltwater from the McMurdo Ice Shelf and the surrounding glaciers would also contribute to the meltwater input. Unfortunately, we know very little about the phytoplankton production in the open waters of the Sound following the sea ice breakout.

As detailed above phytoplankton import into the Sound via the inflowing current systems contributes substantially to the phytoplankton production. This is balanced to some extent by export to beneath the McMurdo Ice Shelf in the centre of the Sound (Knox in prep.). However, the magnitude of this export has yet to be quantified. Further export occurs in the northerly flowing current on the western side along the Victoria Land coast.

Nutrient levels are usually high and may not be of critical importance to phytoplankton production (El-Sayed and Mandelli 1965; Holm-Hansen 1985). However, data obtained by Barry (1988) in late January and early February indicated that nutrients ($NO_2 \cdot PO_4$) were low following the seasonal phytoplankton bloom, and were near zero at isolated locations. Depletion was greatest near the surface where water column stability was high. However, in areas where glacial meltwater is significant it can be enriched in ammonia (El-Sayed and Taguchi 1981; Horrigan 1981) and this may partially offset any nutrient limitation.

Grossi et al (1987) found that the timing of the development of the sea ice microalgal communities was strongly affected by surface snow cover. In 1982, in eastern McMurdo Sound, snow-free areas had an accelerated ice microalgal bloom that revealed a peak of chlorphyll *a* in early November, while levels of chlorophyll *a* in communities with more than 25 cm of surface snow continued to increase into December (Palmisano and Sullivan 1985). Palmisano et al. (1985a) conducted an in situ light perturbation experiment in 1981 during the spring ice microalgal bloom. A 100 m^{-2} quadrat was covered with snow to a depth of 70 cm reducing underwater radiance by 97% relative to a control quadrat with 7 cm of natural snow. As can be seen from the data in Table 2 the chlorophyll *a* at the termination of the 5-week experiment was six times greater in the control, while primary production, which was scarcely detectable in the snow-covered quadrat, was high (22 mg C m^{-2} h^{-1}) in the control. Thus the degree of snow cover can have a profound impact not only on the water column phytoplankton production but also on the sea ice microalgal production.

Thus the major natural perturbations, resulting in substantial variability of production within the Sound, are changes in the current systems strongly influenced by external driving forces, the timing and extent of the sea ice breakout, coupled with the extent and timing of the breakup of the sea ice to the north of McMurdo Sound, and the extent and depth of snow cover on the sea ice. These parameters will need to be monitored in order to further understand the natural variability of the system and to identify any long-term ecological changes.

6.2 Data Deficiencies and Future Research

The data presented in this chapter have highlighted some major deficiencies in the data base. This is partly due to the fact that the great bulk of the research was carried out over the summer period from mid-October to mid-January. This was largely dictated by logistics and the fact that work to date was dependent on the fast ice for access to the water column beneath. A very limited amount of research has been carried out in the winter and practically none in February to mid-March when there is open water in the Sound. It is essential that research on the production cycle be carried out during this period. There is also a great need for continuous year-round observations, whereby the major producers, the phytoplankton, the sea ice microalgae and the benthic microalgae should be studied together over the annual cycle. Information is also needed on the role of the epiphytic microalgae and the benthic macroalgae. Other gaps include data on the biomass and feeding of the water column heterotrophs, the consumption of the benthic microalgae and the sedimentation of the phytoplankton and the sea ice microalgae.

References

Anderson P, Fenchel T (1985) Bacterivory by microheterotroph flagellates in a sea water sample. Limnol Oceanogr 30:198–202

Barry JP (1988) Hydrographic patterns in McMurdo Sound, Antarctica and their relationship to local benthic communities. Polar Biol 8:377–391

Barry JP, Dayton PK (1988) Current patterns in McMurdo Sound, Antarctica and their relationship to local biotic communities. Polar Biol 8:367–376

Buck KR, Garrison DL (1983) Protists from the ice-edge region of the Weddell Sea. Deep-Sea Res 30:1261–1277

Bullivant JS (1967) The fauna of the Ross Sea. Part 5. Ecology of the Ross Sea benthos. N Z Oceanogr Inst Mem 32:49–75

Bunt JS (1964) Primary productivity under sea ice in Antarctic waters 1. Concentrations and photosynthetic activities of microalgae in the waters of McMurdo Sound, Antarctica. Antarct Res Ser 1:13–26

Bunt JS, Lee CC (1970) Seasonal primary production in Antarctic sea ice at McMurdo sound, Antarctica. J Mar Res 28:304–320

Car M, Codspoti LA (1968) Oceanographic cruise summary, Ross Sea Antarctica, February 1968. Informal Rep No 68–64, Naval Hydrographic Office, Washington, DC 12 pp

Carey AG (1985) Marine ice fauna: Arctic. In: Horner RA (ed) Sea ice biota. CRC Press, Inc, Boca Raton, pp 173–190

Day JW Jur, Smith WG, Wagner PR, Stowe WC (1973) Community structure and carbon budget of a salt marsh and shallow bay estuarine system in Louisiana. Publ No LSU-SG-72-04 Centre for Wetland Resources, Louisiana State University, Baton Rouge

Dayton PK, Robilliard GA, DeVries AL (1969) Anchor ice formation in McMurdo Sound, Antarctica and its biological effects. Science 163:273–274

Dayton PK, Oliver JS (1977) Antarctic soft-bottom benthos in oligotrophic and eutrophic communities. Science 197:55–58

Dayton PK, Robilliard GA, Paine RT (1970) Benthic faunal zonation as a result of anchor ice at McMurdo Sound. In: Holdgate MW (ed) Antarctic ecology, vol 1. Academic Press, London, pp 244–258

Dayton PK, Robillard GA, Paine RT, Dayton LB (1974) Biological accomodation in the benthic community at McMurdo Sound, Antarctica. Ecol Monogr 44:105–128

Dayton PK, Watson D, Palmisano A, Barry JP, Oliver JS, Rivera D (1986) Distribution patterns of benthic microalgal standing stock at McMurdo Sound, Antarctica. Polar Biol 6:207–213

El-Sayed SZ, Mandelli EF (1965) Primary production and standing crop of phytoplankton in the Weddell Sea and Drake Passage. Antarct Res Ser 5:87–106

El-Sayed SZ, Taguchi S (1981) Primary production and standing crop of phytoplankton along the ice-edge in the Weddell Sea. Deep-Sea Res 28A:1017–1032

El-Sayed SZ, Turner JT (1977) Productivity of the Antarctic and tropical subtropical regions: a comparative study. In: Dunbar MJ (ed) Polar oceans. Arctic Institute of North America, Calgary, pp 463–504

Evans RH (1965) Current observations and lead line soundings in McMurdo Sound. Antarctica. U.S. Naval Hydrographic Office. IMR N. 0-18-65 (unpubl)

Forster BA (1987) Composition and abundance of zooplankton under the spring sea-ice of McMurdo Sound, Antarctica. Polar Biol 8:41–81

Garrison DL, Buck KR (1985) Sea-ice algal communities in the Weddell Sea: species composition in ice and plankton assemblages. In: Grey JS, Christiansen ME (eds) Marine biology of polar regions and effects of stress on marine organisms. John Wiley, New York, pp 103–121

Garrison DL, VanScoy K (1985) Wilkes Land Expedition 1985: biological observations in the ice edge zone. Antarct J US 19:123–124

Garrison DL, Buck KR, Silver MW (1984) Microheterotrophs in the ice-edge zone. Antarct J US 19(5):109–111

Gilmour AE, MacDonald WJP, VanderHoeven FG (1960) Ocean currents in McMurdo Sound. Nature: 4740:867

Gilmour AE, MacDonald WPJ, VanderHoeven FG (1962) Winter measurements of sea currents in McMurdo Sound. N Z J Geol Geophys 5:778–789

Grossi SM (1985) Response of a sea ice microbial community to a gradient in under-ice irradiance. PhD Thesis, University of Southern California, Los Angeles

Grossi SM, Sullivan CW (1985) Sea ice microbial communities. V. The vertical zonation of the diatoms in an Antarctic fast ice community. J Phycol 21:401–409

Grossi SM, Kottmeier ST, Sullivan CW (1984) Sea ice microbial communities. III. Seasonal abundance of microalgae and associated bacteria, McMurdo Sound, Antarctica. Microbiol Ecol 10:231–242

Grossi SM, Kottmeier ST, Moe RL, Taylor GT, Sullivan CW (1987) Sea ice microbial communities. VI. Growth and primary production in bottom ice under graded snow conditions. Mar Ecol Progr Ser 35:153–164

Heath RA (1971 a) Circulation and hydrology under the seasonal ice in McMurdo Sound, Antarctica. N Z J Mar Feshwater Res 5:479–515

Heath RA (1971 b) Tidal constants for McMurdo Sound, Antarctica. N Z J Mar Freshwater Res 5:376–380

Heath RA (1977) Circulation across the ice shelf edge in McMurdo Sound, Antarctica. In: Dunbar MJ (ed) Polar oceans. Arctic Research Institute, Calgary, pp 129–149

Heinbokel JF, Coats DW (1985) Ciliates and nanoplankton in Arthur Harbour, December 1984 and January 1985. Antarct J US 19:135–136

Heinbokel JF, Coats DW (1986) Patterns of tintinnine abundance and reproduction near the edge of the seasonal pack ice in the Weddell Sea, November, 1983. Mar Ecol Progr Ser 33:71–80

Heine AJ (1963) Ice breakout around the southern end of Ross Island, Antarctica. N Z J Geol Geophys 6:395–401

Hewes CD, Holm-Hansen O, Sakshaug E (1985) Alternate carbon pathways in the lower trophic levels in the Antarctic food web. In: Siegfried WR, Condy PR, Laws RM (eds) Antarctic nutrient cycles and food webs. Springer, Berlin Heidelberg New York, pp 277–283

Hicks GRF (1974) Variation in zooplankton biomass with hydrological regime beneath the seasonal ice, McMurdo Sound, Antarctica. N Z J Mar Freshwater Res 8:67–77

Holm-Hansen O (1985) Nutrient cycles in Antarctic marine ecosystems. In: Siegfried WR, Condy PR, Laws RM (eds) Antarctic nutrient cycles and food webs. Springer, Berlin Heidelberg New York, pp 6–10

Holm-Hansen O, El-Sayed SZ, Franceschini GA, Cuhel GA, Cuhel RL (1977) Primary production and the factors controlling phytoplankton growth in the Southern Ocean. In: Llano GA (ed) Adaptations within the Antarctic ecosystem. Gulf Publ Co, Texas, pp 11–73

Hopkins TL (1985) The zooplankton community of Croker Passage, Antarctic Peninsula. Polar Biol 4:161–170

Hopkins TL (1987) Midwater food web in McMurdo Sound, Ross Sea, Antarctica. Mar Biol 96:93–106

Horrigan SG (1981) Primary production under the Ross Ice Shelf, Antarctica. Limnol Oceanogr 26:378–382

Howard-Williams C, Pridmore RD, Broady RD, Vincent WF (1988) Paper presented at the Fifth SCAR Symposium on Antarctic Biology, Hobart, Australia, 29 Aug–3 Sept, 1988

Knox GA (1986) Recent New Zealand marine research in the Ross Sea Sector of Antarctica. Mem Nat Inst Polar Res (Spec Iss) 40:345–363

Knox GA Summer plankton beneath the sea ice at three sites along the edge of the McMurdo Ice Shelf (in prep)

Knox GA, Lowry JK (1977) A comparison between the benthos of the Southern Ocean and the North Polar Ocean, with special reference to the Amphipoda and Polychaeta. In: Dunbar MJ (ed) Polar oceans. Arctic Institute of North America, Calgary, pp 423–462

Knox GA, Waghorn EH, Ensor PH, The summer plankton beneath the McMurdo Ice Shelf at White Island, McMurdo Sound, Antarctica (in press)

Kottmeier ST, Grossi SM, Sullivan CW (1987a) Sea ice microbial communities. VI. Growth and primary production in bottom ice under graded snow cover. Mar Ecol Progr Ser 35:153–164

Kottmeier ST, Grossi SM, Sullivan CW (1987b) Sea ice microbial communities. VIII. Bacterial production in annual sea ice of McMurdo Sound, Antarctica. Mar Ecol Progr Ser 35:175–186

Kottmeier ST, Muscat AM, Craft LL, Kastendiek JE, Sullivan CW (1984) Ecology of sea-ice microbial communities in McMurdo Sound, Antarctica, in 1983. Antarct J US 19:129–131

Leventer A, Dunbar RB (1986) Diatom flux in McMurdo Sound, Antarctica. Mar Micropaleontol 12:49–64

Lewis EL, Perkin RG (1985) The winter oceanography of McMurdo Sound, Antarctica. In: Jacobs SS (ed) Oceanology of the Antarctic Continental Shelf. Antarct Res Ser 43:145–166

Littlepage JL (1965) Oceanographic investigations in McMurdo Sound, Antarctica. In: Llano G (ed) Biology of the Antarctic Seas. Antarct Res Ser 2:1–37

Lowry JK (1976) Studies on the macrobenthos of the Southern Ocean. PhD Thesis. University of Canterbury, Christchurch, New Zealand

Marchant HJ (1985) Choanoflagellates in the Antarctic marine food chain. In: Siegfried WR, Condy PR, Laws RM (eds) Antarctic nutrient cycles and food webs. Springer, Berlin Heidelberg New York, pp 271–276

Nelson DM, Smith WO Jr (1986) Phytoplankton bloom dynamics of the western Ross Sea ice edge. II. Mesoscale cycling of nitrogen and silicon. Deep-Sea Res 33:1389–1412

Palmisano AC, Sullivan CW (1983) Sea ice microbial communities (SIMCO). I. Distribution, abundance, and primary production of ice microalgae in McMurdo Sound, Antarctica in 1980. Polar Biol 2:171–177

Palmisano AC, Sullivan CW (1985) Physiological response of microalgae in the ice platelet layer to ambient light conditions. In: Siegfried R, Condy PR, Laws RM (eds) Antarctic nutrient cycles and food webs. Springer, Berlin Heidelberg New York, pp 84–88

Palmisano AC, Kottmeier ST, Moe RL, Sullivan CW (1985a) Sea ice microbial communities. IV. The effects of light perturbation on microalgae at the ice-water interface in McMurdo Sound, Antarctica. Mar Ecol Progr Ser 21:37–45

Palmisano AC, SooHoo JB, Sullivan CW (1985b) Photosynthesis-irradiance relationships in sea ice microalgae from McMurdo Sound, Antarctica. J Phycol 21:341–346

Palmisano AC, SooHoo JB, SooHoo SL, Kottmeier ST, Craft LL, Sullivan CW (1986) Photoadaptation in *Phaeocystis pouchetii* advected beneath annual sea ice in McMurdo Sound, Antarctica. J Plankton Res 8:891–906

Palmisano AC, SooHoo JB, Moe RL, Sullivan CW (1987) Sea ice microbial communities. VII. Changes in the under-ice spectral irradiance during the development of Antarctic sea ice microbial communities. Mar Ecol Progr Ser 35:156–173

Palmisano AC, Lizotte MP, Smith GA, Nichols PD, White DC, Sullivan CW (1988) Changes in photosynthetic carbon assimilation in Antarctic sea ice diatoms during spring bloom variation. J Exp Mar Biol Ecol 116:1–13

Raytheon Service Company (1983) Report on the McMurdo Sound water quality study. Unpubl Rep to the US Office of Polar Programs

Rivkin RB, Voytek MA (1987) Photoadaptations of photosynthesis and carbon metabolism by phytoplankton from McMurdo Sound, Antarctica. I. Species specific and community responses to reduced irradiation. Limnol Oceanogr 32:249–259

Rogers JC, VanLoon H (1979) Spatial variability of sea levels, pressure and 500 mb height anomalies over the Southern Hemisphere. Monthly Weather Rev 110:1375–1392

Seymour RJ, Strange RR III, Cayan DR, Nathan RA (1984) Influence of El Ninos on Californias's wave climate. In: Edge BL (ed) Nineteenth Coastal Engineering Conference, Proc Int Conf, September 3–7 1984. I. American Society of Civil Engineers New York, pp 577–592

Smith WO Jr, Nelson DM (1985a) Phytoplankton biomass near a receding ice edge on the Ross Sea. In: Siegfried WR, Condy PR, Laws RM (eds) Antarctic nutrient cycles and food webs. Springer, Berlin Heidelberg New York, pp 70–77

Smith WO Jr, Nelson DM (1985b) Phytoplankton bloom produced by a receding ice edge in the Ross Sea: spatial coherence with the density field. Science 227:163–166

Smith WO Jr, Nelson DM (1986) The importance of ice edge phytoplankton blooms in the Southern Ocean. Bioscience 36:251–257

Sullivan CW, Palmisano AC (1984) Sea ice microbial community: distribution, abundance and diversity of ice bacteria in McMurdo Sound, Antarctica. Appl Environ Microbiol 47:788–795

Sullivan CW, Palmisano AC, Kottmeier S, Moe R (1982) Development of the sea-ice microbial community (SIMCO) in McMurdo Sound, Antarctica. Antarct J US 17:155–157

Sullivan CW, Palmisano AC, Kottmeier S, Grossi SM, Moe R (1985) The influence of light on growth and development of sea ice microbial community in McMurdo Sound. In: Siegfried R, Condy PR, Laws RM (eds) Antarctic nutrient cycles and food webs. Springer, Berlin Heidelberg New York, pp 84–88

Tressler WL, Ommundsen AM (1962) Seasonal oceanographic studies in McMurdo Sound, Antarctica. USN Hydrographic Office Technical Report 125

Whitaker TM (1977) Sea ice habitats of Signey island (South Orkneys) and their primary productivity. In: Llano GA (ed) Adaptations within Antarctic ecosystems. Gulf Publ Co, Houston, pp 75–82

Wilson DL, Smith WO Jr, Nelson DM (1986) Phytoplankton bloom dynamics of the western Ross Sea ice edge – I. Primary productivity and species specific production. Deep-Sea Res 33:1375–1387

Seasonal and Geographic Variations in Sea Ice Community Structure of the Weddell Sea, Antarctica

M. Spindler, G. S. Dieckmann and M. A. Lange[1]

Summary. Seasonal and spatial variation in sea ice organisms were studied in the Weddell Sea region during the austral summer 1985 and winter and spring 1986. Sea ice cores were collected from 22 stations. Winter values for chlorophyll *a* measured per 10 cm segment of drilled ice core ranged between 0 and 49 $\mu g\,l^{-1}$ (mean 3.4) during July to September and between 0 and 450 $\mu g\,l^{-1}$ (mean 7.6) during October to December, while mean summer values were considerably higher (mean 34.5) with a maximum of 2220 $\mu g\,l^{-1}$. Increasing numbers of sea ice organisms were also observed in sea ice cores from stations reoccupied after several weeks in spring. Samples taken later in season (November compared to October) revealed a higher biomass and more organisms. Algal biomass and absolute numbers of sea ice organisms including foraminifers, copepods and ciliates decreased along a transect from the ice edge towards the continent and along the eastern Weddell Sea coast from northeast to southwest. However, the distribution of organisms was patchy and replicate cores taken only a few centimeters apart contained numbers of organisms which differed considerably. This is partly due to different ice formation processes which show both small and mesoscale variation and may account for the observed abundances of ice organisms.

1 Introduction

Sea ice from the Arctic and Antarctic contains an internal system of pores, brine channels and brine pockets which serve as a habitat for a variety of organisms, both plants and animals. This system has been known for more than a century (Ehrenberg 1841; Hooker 1847) but detailed studies on species diversity, physiology and trophic activity have only begun recently (for a review, see Horner 1985a). Dominant groups in the sea ice system include bacteria, diatoms, flagellates, fungi, foraminifers, ciliates, nematodes, polychaetes, turbellarians, copepods and various larvae of different taxonomic origin. However, studies on the community of Antarctic sea ice have mainly concentrated on the investigation of ice algae and bacteria and relatively little attention has been paid to the faunal components (Carey 1985).

Most observations from the Antarctic sea ice community have been made near land-based stations (for reviews, see Horner 1985a, b; Garrison et al. 1986) thus limiting the information to sea ice near the continent. The majority of the studies have been qualitative and semi-quantitative and only recently quantitative results on the sea ice community have been published (e.g. Garrison et al. 1986; Garrison and Buck 1985, 1986). Since sea ice formation close to the ice shelf edge differs significantly from those processes observed in the central Weddell Sea (Lange et al. 1989) it is likely that species diversity and abundance will be different also. In addition, sea ice community investigations in the past were more or less restricted to summer months. During winter the pack ice zone has been mostly inaccessible because strong ice-breaking research vessels were not available. Winter data on the Antarctic sea ice community are thus limited to sea ice off the Antarctic Peninsula or to sea ice close to overwintering stations on the perimeters of the continent (Bunt and Lee 1970; Hoshiai 1985; Hoshiai and Tanimura 1986; Kottmeier and Sullivan 1987; Watanabe and Satoh 1987). The Winter Weddell Sea Project 1986 (WWSP '86) divided into leg 1 from July to September (ANT V/2) and leg 2 from October to December 1986 (ANT V/3) offered the opportunity to study sea ice organisms within the large pack ice zone during both winter and spring.

2 Material and Methods

Sea ice cores were collected from stations within the Weddell Sea pack ice zone and along the eastern Weddell Sea coast during the ANT III/3, ANT V/2 and ANT V/3 expeditions of RV "Polarstern" in the austral summer 1985, and winter and spring 1986 [see Spindler and Dieckmann (1986) for ANT III/3 sta-

[1] Alfred-Wegener-Institut für Polar- und Meeresforschung, Columbusstrasse, D-2850 Bremerhaven, FRG.

Antarctic Ecosystems. Ecological Change and Conservation
(ed. by K. R. Kerry and G. Hempel)
© Springer-Verlag Berlin Heidelberg 1990

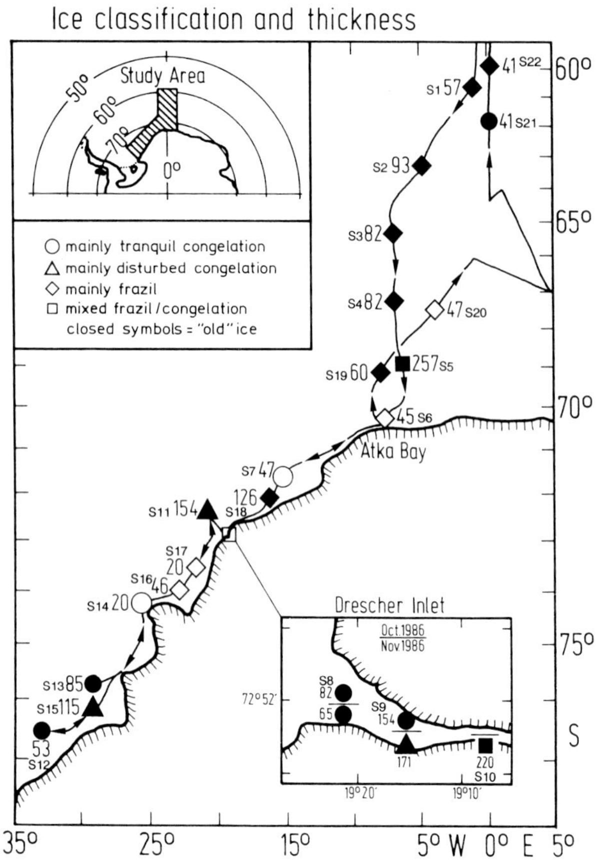

Fig. 1. Cruise track of Polarstern's ANT V/3 expedition in the Weddell Sea and along its ice shelf coast with ice stations 1 to 22. Ice classification is indicated by *symbols*. *Numbers* indicate ice thickness in cm

tions; Lange et al. (1989) for ANT V/2; and Fig. 1 and Spindler et al. (1987) for ANT V/3 stations]. Several closely spaced cores were collected at each station, using modified SIPRE ice corers (7.5 and 10 cm diameter). From each station one core was stored at − 30 °C for reference purposes. Additional 7.5 cm diameter cores were cut into 10-cm intervals and thawed in 3 l of seawater filtered through 0.2 μm mesh on board the ship. This method assured that during thawing organisms living in the ice in the highly saline brine pockets and brine channels were not subjected to low salinities which could have been lethal to them. Ciliates and copepods were individually picked and counted from these samples, since ciliates may disintegrate in a directly melted treatment and for copepods it was possible to discriminate between live and dead specimens. Numbers given for copepods are those of living individuals. A similar method of melting the ice in large volumes was described recently by Garrison and Buck (1986).

One 10-cm-diameter core was returned to the ship's cold laboratory where a longitudinal section of 1–2 cm thickness was taken to investigate the ice texture. Sections of various length but uniform textural class were selected, thawed in the dark, and volume and salinity recorded (for further details, see Lange 1988). Foraminifers were picked from the samples before 50-ml aliquots were removed for chlorophyll determinations, after the method of Evans and O'Reilly (1983).

Ice was classified according to its texture. Frazil ice is granular in structure with isometric grains which are formed in the upper part of the water column. Congelation ice shows a columnar structure with elongated parallel grains. Eicken and Lange (1989) differentiated congelation ice into tranquil congelation ice and disturbed congelation ice. Tranquil congelation ice is formed at the ice-water interface under stable conditions (hydrodynamically smooth regime) resulting in formation of ice of columnar texture. Disturbed congelation ice results in the formation of intermediate columnar-granular ice under unstable conditions (hydrodynamically rough regime) at the ice-water interface. As a consequence of the two hydrographically different regimes the grains of disturbed congelation ice are less elongated and of generally smaller size than seen in the columnar ice of tranquil congelation.

The sea ice cores were grouped into two categories, "young" and "old" ice, although nearly all cores were obtained from first-year ice with the possible exception of a few cores from Drescher Inlet (Stations 9, 10) The estimated age of the sampled sea ice floes was based mainly on ice thickness as compared to surrounding floes, surface characteristics (pressure ridges, rafting) and the nature of the snow cover. The terms "young" and "old" are thus relative and do not imply an exact age definition in days, weeks or months.

For comparison of foraminiferal abundances in the ice with the water column, vertical Multinet hauls using 100 μm mesh size and Apstein net hauls using 20 μm mesh size were carried out from 60 m depth layer to surface.

3 Results

Ice classification, approximate age and thickness of the ice cover during ANT V/3 are shown in Fig. 1.

Chlorophyll *a* data from the sea ice cores collected during different times of the year from the Weddell Sea (Table 1) clearly indicate that there is a marked increase in chlorophyll *a* from winter to summer. During the winter season highest chlorophyll *a* values measured for a 10-cm segment of a sea ice core reached 48 $\mu g\,l^{-1}$ while those during spring increased to 450 $\mu m\,l^{-1}$ and peaked at 2220 $\mu g\,l^{-1}$ during sum-

Table 1. Chlorophyll *a* values from sea ice cores collected during different seasons from the Weddell Sea

Season	Months	Expedition	No. of cores analyzed	Chlorophyll *a* ($\mu g\, l^{-1}$) range	mean
Winter	Jul.–Sep.	ANT V/2	31	0–48.7	2.1
Spring	Oct.–Dec.	ANT V/3	26	0–450	7.6
Summer	Jan.–Mar.	ANT III/3	13	0–2220	34.5

mer. The mean values for the complete cores increased correspondingly.

This increase in chlorophyll is followed by an increase in numbers of heterotrophic consumers. The planktonic foraminifer *Neogloboquadrina pachyderma*, which is known to be a member of the Antarctic sea ice community (Lipps and Krebs 1974; Spindler and Dieckmann 1986) generally occurred in higher numbers in sea ice cores collected at geographically identical stations during the austral summer 1985 compared to spring values from 1986 (Table 2). While ice cores close to Atka Bay contained no foraminifers in October 1986, an average of 30 individuals l^{-1} were encountered in three cores sampled in January and February 1985 (Table 2; Station 6). Similar results were found in sea ice cores southwest of Atka Bay and in those from Drescher Inlet. However, in Drescher Inlet the innermost station (Station 10) is the only exception with higher foraminiferal numbers in November 1986 compared to the February 1985 sample (26 to 16 individuals l^{-1}).

Two stations in the Drescher Inlet (Stations 8, 9) which were reoccupied within a 5-week interval indicated a similar increase in numbers of foraminifers later in the season (Table 2).

Summer and spring abundances for other heterotrophs (e.g. ciliates and copepods) were difficult to compare due to a lack of sufficient samples from the same location at different seasons, and low organism numbers in comparable cores. Sea ice cores from four reoccupied stations (Stations 8, 9, 10, 18) were analyzed for copepods. However, in most cases numbers of copepods per litre melted sea ice were low at those stations, except for one from the Drescher Inlet. Here a total of 293 copepods or 36 specimens per liter (mainly copepodites and adults of the harpacticoid *Drescheriella glacialis*, and two calanoid tisbeids) were found in Feburary 1985 (Table 2; Station 9).

Heterotrophic consumers such as foraminifers, ciliates and copepods were also present in sea ice of the Weddell Sea during winter and spring, sometimes even with quite high numbers of individuals. These organisms were found within the ice as members of the interior and the interstitial assemblage (definitions after Horner et al. 1988) where they feed on autotrophic

Table 2. Numbers of foraminifers, copepods and ciliates in sea ice cores collected at Stations 1 to 22 from the Weddell Sea during ANT III/3 and ANT V/3

Station No.[a]	Date	Foraminifera No./l	Foraminifera Total	Copepods No./l	Copepods Total	Ciliates No./l	Ciliates Total
S 01	05. 10. 86	262	930	10	16	92	140
S 02	07. 10. 86	136	726	0	0	5	20
S 03	08. 10. 86	58	285	0.6	2	2	5
S 04	09. 10. 86	20	99	0	0	2	5
S 05	10. 10. 86	16	249	–	–	–	–
S 06 a	19. 01. 85	30	307	–	–	–	–
S 06 b	12. 10. 86	0	0	0	0	0	0
S 07	13. 10. 86	0	0	0.6	1	0	0
S 08 a	25. 01. 85	8	57	1.5	11	–	–
S 08 b	15. 10. 86	0	0	0	0	0	0
S 08 c	17. 11. 86	0.4	1	3	13	35	83
S 09 a	31. 01. 85	20	176	36	293	14	116
S 09 b	15. 10. 86	2.5	23	–	–	–	–
S 09 c	18. 11. 86	5	48	1.2	8	14	97
S 10 a	13. 02. 85	16	139	0.5	4	–	–
S 10 b	20. 11. 86	26	322	0.9	8	10	90
S 11	17. 10. 86	0.1	1	1	5	2	10
S 12	27. 10. 86	0	0	0	0	0	0
S 13	30. 10. 86	0.2	1	1.5	5	21	67
S 14	01. 11. 86	0	0	0	0	0	0
S 15	10. 11. 86	0	0	15	65	66	295
S 16	14. 11. 86	0	0	1	2	0	0
S 17	15. 11. 86	0	0	0	0	1	1
S 18 a	28. 01. 85	41	319	0.4	3	–	–
S 18 b	26. 11. 86	2	10	0	0	9	40
S 19	29. 11. 86	134	396	0	0	13	30
S 20	30. 11. 86	0	0	2	3	41	60
S 21	04. 12. 86	20	38	0	0	83	130
S 22	05. 12. 86	2125	1944	62	94	16	25

[a] Repeated occupation of station is indicated by letters a–c.

diatoms and flagellates, and bacteria. With an increase in temperature and light towards the austral summer, algal standing stock (expressed as the amount of chlorophyll *a*), and numbers and production of bacteria increase (Kottmeier and Sullivan 1987). This also corresponds to an increase in heterotrophic consumers.

Geographically, there is a decrease in foraminiferal number within ice cores from the ice edge zone towards the ice shelf coast. Numbers of *Neogloboquadrina pachyderma* diminish steadily from 262 individuals l^{-1} at the outermost station to lower numbers at stations closer to the ice shelf near Atka Bay (Table 2; Stations 1–6). On the way out of the ice (Stations 19–22), however, this was not observed, although the outermost station (Station 22) again showed the highest numbers of foraminifers. Foraminifers in sea ice along the ice shelf coast are scarce or completely absent (e.g., Stations 11–17). Higher numbers were, however, encountered in ice cores from the Drescher Inlet.

Similar distributional patterns were obtained for copepods and ciliates (Table 2) with highest abun-

Table 3. Distribution of foraminifers, copepods and ciliates in sea ice cores collected from the Weddell Sea during ANT V/3[a]

Ice type	Foraminifers			Copepods			Ciliates		
	Mean	Range	No specimen (%)	Mean	Range	No specimen (%)	Mean	Range	No specimen (%)
Young	0	–	100	0.6	0–2	50	7	0–41	67
Old	156	0–2125	17	6	0–62	44	23	0–92	13

[a] Given are mean and range of numbers of individuals per core, and the percentage of cores without any specimen. For all organism groups 6 "young" ice cores were investigated, for "old" 18 for foraminifers and 16 for copepods and ciliates respectively. For explanation of "young" and "old" ice consult text.

dances close to the ice edge on the way in and out, and smaller numbers along the ice shelf coast. The sea ice from the Drescher Inlet again showed higher densities at least for ciliates. The second most southerly station (Station 15) was exceptional in containing larger numbers of both copepods and ciliates (15 and 66 l^{-1}).

These distributional patterns of the organisms within the ice are related to sea ice properties such as age and texture (see Fig. 1), and to the distributional patterns of the organisms in the water column.

It is evident from the data summarized in Table 3 that young sea ice contains no foraminifers and less copepods and ciliates than older ice. For example, the three ice cores southwest of Drescher Inlet (Stations 14, 16, 17) were obtained from thin sea ice floes (thicknesses ranged between 20 and 46 cm), which were recently formed in an open lead. Organism numbers in this type of ice were low. This is also true for mean and maximum numbers and for the percentage of cores without individuals of the respective group. This young ice was formed under late winter or early spring conditions in the inner Weddell Sea, the respective organisms were either absent from the water column or less abundant than earlier in season when the surrounding thicker ice was formed. This is true for the planktonic foraminifer *Neogloboquadrina pachyderma* (unpubl. results) and for copepods in general (Hempel et al. 1987).

In addition to age, the ice texture influenced species abundance. The ice textures of 45 cores obtained during ANT V/2 were analyzed and grouped into five different classes according to Lange et al. (1988). The number of foraminifers varied considerably within ice cores for each class (Fig. 2). Ice cores which consisted of predominantly congelation ice (> 80%) contained no foraminifers. Highest foraminiferal abundances with a mean of 170 individuals l^{-1} were encountered in sea ice cores mainly composed of frazil ice and intermediate numbers in sea ice with lower percentages of each textural ice type (Fig. 2). During ANT V/3 an additional type of congelation ice was found, which was termed "disturbed" congelation. This type of ice was found primarily close to the ice shelf edge. Sea ice composed mainly of disturbed congelation ice

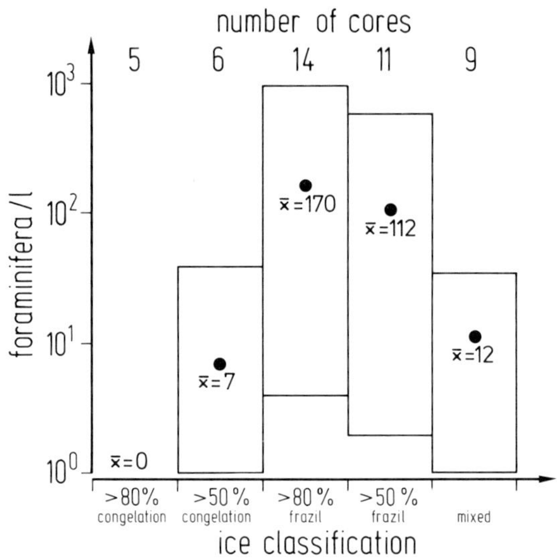

Fig. 2. Distribution of the foraminifer *Neogloboquadrina pachyderma* in relation to sea ice texture. The mean values are indicated by *dots*, minimum and maximum numbers of foraminifers per liter by the column extension

was much richer in organisms compared to sea ice of "tranquil" congelation ice. The second most-southerly station (Station 15) already mentioned above, had exceptionally high numbers of ciliates and copepods and consisted mainly of disturbed congelation ice. The two additional stations with this type of ice (Stations 9b, 11) also contained foraminifers, copepods and ciliates (Fig. 1, Table 2).

Geographical differences of abundances of organisms enclosed into the sea ice system depended also on the abundance within the water column at the time and location of sea ice formation. Foraminifers were extremely abundant in surface waters close to the ice edge (Fig. 3A) which in both polar regions is known to be an extremely productive area with abundances for phyto- and zooplankton exceeding those of adjacent waters (e.g. Schandelmeier and Alexander 1981; Rey and Loeng 1985; Smith and Nelson 1985; Spies 1987). In this region we collected up to 4500 individuals m^{-3} within the upper 60 m. Further within the pack ice zone only 58 foraminifers m^{-3} were collected

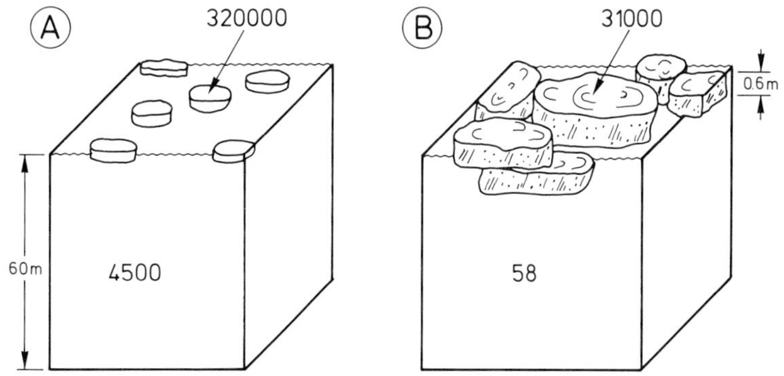

Fig. 3. Abundance of formainifers m^{-3} in 2- to 3-day-old pancake ice at the ice edge zone (**A**) and within the older pack (**B**) compared with numbers of foraminifers within the upper 60 m of the water column at those stations

within the uppermost 60 m water underlying a 60-cm-thick sea ice cover (Fig. 3 B). Accordingly, numbers of *Neogloboquadrina pachyderma* were high in 2- to 3-day-old pancake ice at the ice edge and lower in older pack ice.

4 Discussion

The amount of chlorophyll *a* from sea ice cores from the Weddell Sea was used in this study as a measure of the amount of algae in sea ice. Despite the limitations of this method (Apollonio 1961; Falkowski et al. 1985; Sullivan et al. 1985; Grossi et al. 1987) our data show that the autotrophic algae in the sea ice increased from winter through spring to summer as did the total numbers of heterotrophic consumers. Some of the consumers in the ice do not depend on autotrophic flagellates and diatoms as a food source but on bacteria (e.g. some ciliate species). Kottmeier et al. (1987) reported on the development of bacterial number, biomass and production from October to December 1982. Numbers and biomass increased by a factor of about 10 in the sea ice of McMurdo Sound during that period and bacterial cell production increased by more than three orders of magnitude. Food organisms available for heterotrophic consumers are thus present during the winter and the amount and production increase during the course of the year. With more abundant food, the number of meiofaunal components of the sea ice community also rises. In contrast to these findings, Hoshiai and Tanimura (1986) who studied the sea ice faunal development at Syowa Station found that here copepods appeared continuously from March to late November but disappeared in December. Highest abundances of copepods in the ice were encountered in mid-winter. Repeated observations in later years showed large fluctuations in abundance but still highest numbers during the winter months (Hoshiai and Tanimura 1986).

Fluctuations in numbers of sea ice organisms are often encountered and thus comparisons of the sea ice community both on a geographical and a seasonal scale are difficult to interpret. These changing abundances are mainly due to different ice properties such as age and structure, regionally different abundances of organisms in the water column which will eventually be enclosed into the developing sea ice, and regionally different ice-forming processes which result in different types of ice.

We have shown that the relative age of the sea ice has a strong influence on the abundance of foraminifers, copepods and ciliates within the ice. Newly forming sea ice can only enclose organisms present in the water column at the time of ice formation. During winter open water is generally created by divergent ice flow driven by surface wind. Smaller and sometimes large leads and coastal polynyas develop and again refreeze at low temperatures. Thus young ice is formed at times when the water column is almost devoid of phyto- and zooplankton. Even during spring the water close to the ice shelf did not contain significant amounts of plankton which resulted in Secchi disc visibilities between 70 and 80 m (Gieskes et al. 1987).

Ice type is important also for species distribution and abundance. Sea ice mainly composed of congelation ice contains less organisms than frazil ice. This difference between the types may be explained by the following facts. Tranquil congelation ice is the result of slow ice growth underneath an existing ice cover while frazil ice crystals accumulate in the open water column. Here, they either form into a solid ice sheet or, in the presence of a wave field, into pancake ice (Weeks and Ackley 1982; Lange et al. 1989). Frazil ice crystals are generally formed under turbulent conditions and are considered to "harvest" or "scavenge" organisms into the forming sea ice (Ackley 1982; Garrison et al. 1983). Algal cells from the water column are thus physically concentrated during frazil ice formation. An alternate mechanism for enriching biological material during frazil ice formation was recently proposed by Ackley et al. (1987). These authors suggest that a propagating wave field may concentrate organisms into a layer of ice crystals. During congelation ice formation, harvesting, scavenging or enrichment by a wave field do not take place.

Another difference between both types of ice is the different porosity. Frazil ice with more irregularly shaped crystals offers more space to be utilized by organisms for settlement.

Since distribution and abundance of sea ice organisms depend on several parameters of the ice, including age and structure, comparisons of sea ice communities of different geographical stations or different seasons therefore have to include structural analysis of the sea ice. Even the use of replicate cores to determine different parameters in the ice may be problematic. Differences in ice texture may occur even on small scales. Spindler and Dieckmann (1986) demonstrated that foraminiferal numbers in replicate cores taken less than 30 cm apart, varied by one order of magnitude.

On a geographical scale, difficulties in evaluation of abundances of organisms arise by little knowledge of distributional patterns of organisms in the water column, which may become part of the sea ice community, and different ice formation processes. Lange et al. (1989) describe for the central Weddell Sea a process of pancake ice formation, their thickening due to rafting and the creation of new open water areas, which again give rise to frazil/pancake formation ("pancake cycle"). This results in floes of less than 1 m thickness consisting primarily of frazil ice. Additional types of ice were found close to the ice shelf. These were large (up to 15 cm long but 1 to 2 mm thick) platelets, which probably were formed in deeper waters (Dieckmann et al. 1986; Eicken and Lange 1989). We observed up to 9 m of loosely aggregated ice platelets underneath a consolidated ice sheet. In several cases these platelets can become partly incorporated into the existing ice sheet. In this layer space between crystals is even larger and additional organisms may become part of the sea ice community. Eicken and Lange (1989) further differentiated congelation ice into tranquil and disturbed congelation ice. The latter type was frequently encountered along the eastern Weddell Sea coast during ANT V/3. Organisms were more abundant in disturbed than in tranquil congelation ice.

Thus comparisons of sea ice communities should always be made only after careful consideration of the history of the ice. This can in fact be determined by the crystal structure and in relation to probable site of formation (e.g. open water, coastal area). In addition, sea ice structure analysis preferentially has to be carried out on part of the same ice core from which other parameters such as distribution of organisms have been recorded.

Acknowledgements. We wish to thank all who assisted us in the coring during both cruises ANT III/3 and ANT V/3, especially R. Steinmetz. Thanks are due to the crews of the RV "Polarstern" and the helicopters. We express our gratitude to H. Eicken and U. Vogel for analyzing the ice texture, K. Beyer for help in the laboratory, and G. Dansauer for drawing the figures. The manuscript was critically read and improved by K. Kerry. This is Contribution No. 160 of the Alfred Wegener Institute for Polar and Marine Research.

References

Ackley SF (1982) Ice scavenging and nucleation: two mechanisms for incorporation of algae into newly-formed sea ice. Eos 63:54

Ackley SF, Dieckmann G, Shen H (1987) Algal and foram incorporation into new sea ice. Eos 68:1736

Apollonio S (1961) The chlorophyll content of Arctic sea-ice. Arctic 14:197–200

Bunt JS, Lee CC (1970) Seasonal primary production in Antarctic sea ice at McMurdo Sound in 1967. J Mar Res 28:304–320

Carey AG (1985) Marine ice fauna: Arctic. In: Horner RA (ed) Sea ice biota. CRC Press, Inc. Boca Raton, pp 173–190

Dieckmann G, Rohardt G, Hellmer H, Kipfstuhl J (1986) The occurrence of ice platelets at 250 m depth near the Filchner Ice Shelf and its significance for sea ice biology. Deep-Sea Res 33:141–148

Ehrenberg CG (1841) Ein Nachtrag zu dem Vortrage über Verbreitung und Einfluß des mikroskopischen Lebens in Süd- und Nord-Amerika. Monatsber Dtsch Akad Wiss Berlin 1841:202–207

Eicken H, Lange MA (1989) Development and properties of sea ice in the coastal regime of the southeastern Weddell Sea. J Geophys Res 94:8193–8206

Evans CA, O'Reilly JE (1983) A handbook for the measurement of chlorophyll *a* in netplankton and nanoplankton. BIOMASS Handbook No. 9. SCAR/SCOR/IABO/ACMRR Group of specialists on Living Resources of the Southern Ocean 43 ff

Falkowski PG, Dubinsky Z, Wyman K (1985) Growth-irradiance relationships in phytoplankton. Limnol Oceanogr 30:311–321

Garrison DL, Buck KR (1985) Sea-ice algal communities in the Weddell Sea: species composition in ice and plankton assemblages. In: Gray JS, Christiansen ME (eds) Marine biology of polar regions and effects of stress on marine organisms. John Wiley & Sons, New York, pp 103–121

Garrison DL, Buck KR (1986) Organism losses during ice melting: a serious bias in sea ice community studies. Polar Biol 6:237–239

Garrison DL, Ackley SF, Buck KR (1983) A physical mechanism for establishing algal populations in frazil ice. Nature 306:363–365

Garrison DL, Sullivan CW, Ackley SF (1986) Sea ice microbial communities in Antarctica. Bioscience 36:243–250

Gieskes WWC, Elbrächter M, Rabsch U, Scharek R, Schaumann K, Smetacek V, Veth C (1987) Phytoplankton and heterotrophic microorganisms in the water column. Ber Polarforsch 39:190–196

Grossi SM, Kottmeier ST, Moe RL, Taylor GT, Sullivan CW (1987) Sea ice microbial communities. VI. Growth and primary production in bottom ice under graded snow cover. Mar Ecol Prog Ser 35:153–164

Hempel I, Fransz G, Hagen W, Marschall HP, Marschall S, Mizdalski E, Schiel S (1987) Zooplankton. Ber Polarforsch 39:197–205

Hooker JD (1847) The botany of the Antarctic voyage of HM Discovery ships Erebus and Terror in the years 1835–1843. 1. Flora Antarctica. Reprinted 1963. J Cramer, Weinheim, pp 503–519

Horner RA (1985a) History of ice algal investigations. In: Horner RA (ed) Sea ice biota. CRC Press, Boca Raton, pp 1–19

Horner RA (1985b) Ecology of sea ice mircroalgae. In: Horner RA (ed) Sea ice biota. CRC Press, Boca Raton, pp 1–19

Horner RA, Syvertsen EE, Thomas DP, Lange C (1988) Proposed terminology and reporting units for sea ice algal assemblages. Polar Biol 8:249–253

Hoshiai T (1985) Autumnal proliferation of ice-algae in Antarctic sea-ice. In: Siegfried WR, Condy PR, Laws RM (eds) Antarctic nutrient cycles and food webs. Springer, Berlin Heidelberg New York, pp 89–92

Hoshiai T, Tanimura A (1986) Sea ice meiofauna at Syowa Station, Antarctica. Mem Natl Inst Polar Res, Spec Iss 44:118–124

Kottmeier ST, Sullivan CW (1987) Late winter primary production and bacterial production in sea ice and seawater west of the Antarctic Peninsula. Mar Ecol Prog Ser 36:287–298

Kottmeier ST, Grossi SM, Sullivan CW (1987) Sea ice microbial communities VIII. Bacterial production in annual sea ice of McMurdo Sound, Antarctica. Mar Ecol Prog Ser 35:175–186

Lange MA (1988) Basic properties of Antarctic sea ice as revealed by textural analysis of ice cores. Ann Glaciol 10:95–101

Lange MA, Ackley SF, Wadhams P, Dieckmann GS, Eicken H (1989) Development of sea ice in the Weddell Sea, Antarctica. Ann Glaciol 12:92–96

Lipps JH, Krebs WN (1974) Planktonic foraminifera associated with Antarctic sea ice. J Foraminiferal Res 4:80–85

Rey F, Loeng H (1985) The influence of ice and hydrographic conditions on the development of phytoplankton in the Barents Sea. In: Gray JS, Christiansen ME (eds) Marine biology of polar regions and effects of stress on marine organisms. John Wiley & Son, Chichester, pp 49–63

Schandelmeier L, Alexander V (1981) An analysis of the influence of ice on spring phytoplankton population structure in the southeast Bering Sea. Limnol Oceanogr 26:935–943

Smith WO, Nelson DM (1985) Phytoplankton bloom produced by a receding ice edge in the Ross Sea: spatial coherence with the density field. Science 227:163–166

Spies A (1987) Phytoplankton in the marginal ice zone of the Greenland Sea during summer, 1984 Polar Biol 7:195–205

Spinder M, Dieckmann GS (1986) Distribution and abundance of the planktic foraminifer *Neogloboquadrina pachyderma* in sea ice of the Weddell Sea (Antarctica). Polar Biol 5:185–191

Spindler M, Bartsch A, Beyer K, Eicken H, Elbrächter M, Schaumann K, Steinmetz R (1987) Biology and structure of sea ice. Ber Polarforsch 39:152–156

Sullivan CW, Palmisano AC, Kottmeier S, Grossi SM, Moe R (1985) The influence of light on growth and development of the sea-ice microbial community of McMurdo Sound. In: Siegfried WR, Condy PR, Laws RM (eds) Antarctic nutrient cycles and food webs. Springer, Berlin Heidelberg New York, pp 78–83

Watanabe K, Satoh H (1987) Seasonal variations of ice algal standing crop near Syowa Station, East Antarctica, in 1983/84. Bull Plankton Soc Jpn 34:143–164

Weeks WF, Ackley SF (1982) The growth, structure, and properties of sea ice. Monogr 82-1, Cold Regions Research and Engineering Laboratory, Hannover

Seasonal Variation in Ice Algal Assemblages in the Fast Ice Near Syowa Station in 1983/84

K. WATANABE[1], H. SATOH[2] and T. HOSHIAI[1]

Summary. Ice algal assemblages in the fast ice near Syowa Station were investigated from March 1983 to January 1984. Peaks in the standing crop (chlorophyll *a*) occurred in April–June and October–November; the largest (125 mg chlorophyll *a* m^{-2}) occurred in mid-November at a site with moderate snow cover. Chlorophyll was most highly concentrated in the bottom of the ice, where the diatoms *Amphiprora kufferathii, Berkeleya rutilans, Nitzschia lecointei, N. stellata, N. turgiduloides* and *Pleurosigma directum* were dominant. The chlorophyll concentration within the ice developed mainly in May with little change during the following months at heavily snow-covered sites. Centric diatoms, e.g. *Chaetoceros* spp., *Eucampia antarctica* and *Rhizosolenia alata* were major components in the interior assemblage. The chlorophyll concentration increased in the upper consolidated snow layer during October–January. The dominant species there, *Tropidoneis* sp., was a minor component of the interior and bottom ice layers. These results suggest that the interior ice algal assemblage was formed by mechanical inclusion of microalgae at the time of sea ice formation, and did not grow in the ice. However, the bottom assemblage grew when irradiance was high enough. The surface assemblage, which was presumed to be inoculated by an interior assemblage through vertical channels, grew in spring.

1 Introduction

The sea ice of the Southern Ocean covers approximately 20×10^6 km^2 in September when it is at its maximum and 4×10^6 km^2 in March when it is at its minimum extent (Zwally et al. 1983). Although many varieties of ice algal assemblages have been reported from Antarctic waters (Horner 1985), their occurrence within the seasonal cycle of sea ice formation to disintegration is not clear. These processes are presumably closely related to environmental conditions. It is important for an understanding of the dynamics of assemblages to monitor their formation, development and disintegration processes from freezing to melting of the sea ice. However, long-term observations of ice algal assemblages at a fixed location covering that period are limited (Hoshiai 1969, 1981; Bunt and Lee 1970; Whitaker 1977), and very little is known about species succession of the ice algal assemblages (Hoshiai and Kato 1962; Krebs et al. 1987).

Here we report the seasonal variation of chlorophyll *a* standing crop in ice and diatom species composition of the assemblages. The development of the assemblages is proposed on the basis of their species composition.

2 Materials and Methods

Ice cores were collected from March 1983 to January 1984 with a SIPRE type auger (7.6 cm in diameter) at four locations, Stations I, III, V and X (Fig. 1), with differing ice and snow conditions in the fast ice area near Syowa Station (69°00′S, 39°35′E). A scale on a pole fixed into ice, marking the sampling sites was used to measure the depth of snow and the thickness of the sea ice. Three ice cores were collected for each sampling, two were used for the measurement of the chlorophyll *a* concentration and one for the examination of species composition. Ice cores were sectioned and melted at room temperature in the laboratory at Syowa Station. For chlorophyll *a* and phaeopigment measurements, meltwater samples were processed by the fluorometric method of Strickland and Parsons (1972), as modified by Aruga (1979) using a Hitachi model 650-40 spectrofluorometer. The average of the chlorophyll *a* concentrations from two ice cores was used. Meltwater was fixed with a formalin and acetic acid (1:1) mixture (Hasle 1978) for the examination of species composition of diatoms and other robust algae.

Species were identified after cleaning with distilled water or acid (Simonsen 1974), air drying and ion sputtering with gold using a JEOL T-100 and a JSM-

[1] National Institute of Polar Research, 9-10 Kaga 1-chome, Itabashi-ku, Tokyo 173, Japan.
[2] Tokyo University of Fisheries, 5-7 Konan 4-chome, Minato-ku, Tokyo 108, Japan.

Antarctic Ecosystems. Ecological Change and Conservation (ed. by K. R. Kerry and G. Hempel)
© Springer-Verlag Berlin Heidelberg 1990

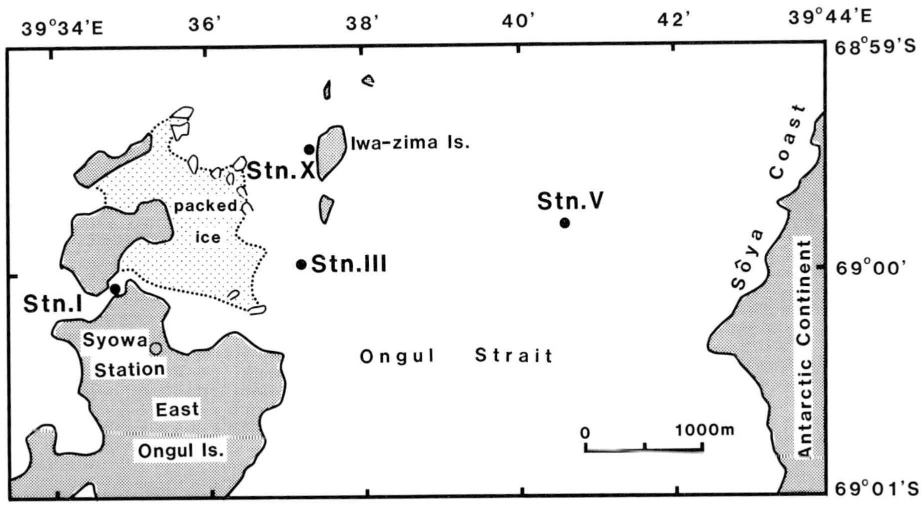

Fig. 1. Location of study area and sampling sites

35 scanning electron microscope. Intact diatom cells in water mounts were examined with a Nikon Diaphoto-TMD phase contrast inverted microscope. Species compositions were compared using the percent similarity index (PSI) of Whittaker (1952) and a dendrogram drawn according to the Mountford method (Davis 1963).

3 Results and Discussion

3.1 Seasonal Variation in Chlorophyl *a* Standing Crop in Ice

During this study, the ice sheet broke into pieces and flowed away from the coast of Ongul Islands on May 3 as a result of swell and currents. New ice formed at the four sampling sites. A white granular ice layer originating from frazil ice developed, ca. 20 cm at Station I and 70 cm in thickness, at Station III. The maximum thickness of sea ice (93–120 cm) was reached between mid-August and late October at the four stations. The maximum snow depths recorded were 6 cm at Station X on November 28, 34 cm at Station I on November 27, 30 cm at Station V on September 2, and 76 cm at Station III on November 14.

As the light extinction coefficient for snow is much greater than that for sea ice, the snow was the principal determinant of irradiance within the ice. The difference of the in situ irradiance in the sea ice habitats resulting from the depth of the snow cover seems to account for the differences of the seasonal variation in chlorophyll *a* standing crop in ice at Stations I, III, V and X (Watanabe and Satoh 1987).

Peaks of standing crop occurred in April–June and October–December at Stations I, V and X, and were similar to results obtained previously from the Syowa area (Hoshiai 1969, 1981, 1985). Spring-summer peak values ranging from 20.9–125 mg chlorophyll a m^{-2} were 3–20 times greater than the autumn values. The maximum value of 125 mg chlorophyll a m^{-2} was reached on November 18, 1983 at Station I. The values of 6.22–8.88 mg chlorophyll a m^{-2} for the autumn peak were lower than those reported by Hoshiai (1969, 1981) and probably due to the short time for the development of the ice algal assemblage in the new ice following the outflow of the ice sheet in early May.

Microalgal flocs were found incorporated into newly formed sea ice from late March to early May at Stations I, III and X. As particulate matter in the water column is known to be incorporated into sea ice by physical mechanisms of scavenging and nucleation (Weeks and Ackley 1982; Garrison et al. 1983), we presume that not all of the autumnal increase of standing crop in the ice was a result of ice algal growth. On the other hand, the spring increase in standing crop is considered to be a result of ice algal growth alone. Since there was no growth of the sea ice at this time, there was no opportunity for the incorporation of microalgae from the water column into the sea ice. The annual production of ice algae therefore is equal to the sum of the spring and autumn increases of ice algal standing crop minus the sum of the algae incorporated from the water column into the newly forming ice during autumn and winter. The annual production was therefore estimated to be 124–130, 36.1–42.9 and 19.1–25.3 mg chlorophyll a m^{-2} year^{-1} at Stations I, V and X respectively. Applying a carbon/chlorophyll a ratio of 26.3, measured from ice algal samples collected from the bottom of sea ice in early November near Station I (Satoh and Watanabe 1986), these values can be converted to 3.26–3.42, 0.95–1.13 and 0.50–0.97 g C m^{-2} $year^{-1}$, respectively.

3.2 Vertical Distribution of Chlorophyll *a* in Ice

The new part of the multi-year ice which formed between late March and early April at Station I contained small brown flocs of microalgae. The bottom layer of the multi-year ice turned brown by April 23. The chlorophyll *a* concentration of the bottom 3-cm layer was 56.9 mg m^{-3}. After the outflow of sea ice in early May, small brown masses of microalgae were again found in the newly formed sea ice layer until mid-May. Brown discoloration of ice ca. 1 cm thick was observed a few centimeters above the bottom between late May and early June. A maximum chlorophyll *a* concentration in the bottom 5-cm ice layer was 40.4 mg m^{-3} on May 20. The bottom chlorophyll-rich layer appeared to shift downward when the ice was growing fast until mid-June. From late June to July, this shift stopped and the chlorophyll-rich layer was left 5–10 cm above the bottom. This occurred presumably because the incident irradiance decreased to a level insufficient for the growth and the subsequent shift of the ice algae, while the sea ice continued to develop under lower air temperature conditions. The spring increase of chlorophyll *a* in the ice, causing discoloration, occurred on August 28 in the bottom of the ice layer. It reached a maximum of 2980 mg chlorophyll *a* m^{-3} for the bottom 4-cm layer on November 18. After that, it declined rapidly due to the melting away of the bottom layer of the sea ice.

At Station III, intensive bottom discoloration was observed in mid- and late April. An autumn maximum of 247 mg chlorophyll *a* m^{-3} was reached on April 17. The bottom few centimeters of the newly formed sea ice appeared brown in late May. This colored layer then shifted downward with the growth of sea ice from late May to early July as was observed at Station I. The highest chlorophyll *a* concentration in the ice in autumn, 96.5 mg m^{-3}, was recorded for the bottom 4-cm layer on May 28. The interior layer of new ice (20–70 cm deep) appeared yellow to pale brown with the chlorophyll *a* concentration ranging between 16.6–40.4 mg m^{-3} on June 2; it remained high, i.e. 5.61–41.8 mg m^{-3}, until December 29. Due to heavy snow and developing vertical brine channels, and seawater flooded snow, a consolidated snow (snow ice) layer was formed in mid-October. In this consolidated snow layer, yellowish green to pale brown discoloration was observed on December 3. The maximum chlorophyll *a* concentration was 17.9 mg m^{-3} for the uppermost 10 cm of the snow ice layer on December 18. These three layers of discoloration were observed in an ice core taken on December 18, although the bottom one was weak.

At Station I, the bottom assemblage contributed the most to the chlorophyll *a* standing crop in ice. At the time of the maximum standing crop in mid-November, 96% of the crop was concentrated in the bottom 9 cm of the ice layer. The interior assemblage dominated the standing crop at Station III after the formation of new ice in May. Chlorophyll *a* standing crop in the interior layer ranged between 8.44–16.4 mg m^{-2} and accounted for 64–90% of the total crop in the ice at Station III between June 2 and January 12. The surface assemblage reached a maximum of 3.82 mg chlorophyll *a* m^{-2} on December 18, which amounted to 27% of the total chlorophyll *a* crop at Station III.

3.3 Species Composition of Ice Algal Assemblages

Species identified in the ice algal samples examined are listed in Table 1. Because of the ice melting process and the fixative which was used, the fragile species like flagellates and cilliates were not preserved in the samples (Garrison and Buck 1986). Species composition of the bottom assemblage at Station I, and the interior and surface ones at Station III are shown in Figs. 2–4, together with total diatom abundance and chlorophyll *a* concentration in the ice layer examined. *Nitzschia* species which formed a ribbon-shaped colony were divided into two groups: (a) species 1 group where the valve costae were coarse enough to be seen in a girdle view under the light microscope at a magnification of 200 ×, and (b) species 2 group which was hardly visible at a magnification of 400 × (Watanabe 1988). Scanning electron microscopy revealed that the *Nitzschia* species 1 group included *N. obliquecostata* and *N. sublineata*; the species 2 group included *N. curta* and *N. cylindrus*.

3.4 Bottom Assemblage at Station I

Seasonal variation in species composition, total diatom cell numbers and chlorophyll *a* concentration of the bottom ice layer (ca. 5 cm thick) at Station I are shown in Fig. 2. More than 98% of the diatoms consisted of pennate species except on May 10, when centric species, mainly *Rhizosolenia alata*, composed 8.8% of the assemblage. In March and April, *Nitzschia lecointei* increased in the bottom layer and became abundant (25–75%). In mid- and late April,

Table 1. Algal species found from sea ice near Syowa Station

DIATOMS

Centrales

Actinocyclus actinochilus (Ehrenberg) Simonsen
Asteromphalus sp.
Chaetoceros bulbosum (Ehrenberg) Heiden
C. criophilum Castracane
C. dichaeta Ehrenberg
C. gausii Heiden et Kolbe
C. neoglacile Van Landingham
C. spp.
Corethron criophilum Castracane
Coscinodiscus asteromphalus Ehrenberg
Dactyliosolen antarcticus Castracane
Eucampia antarctica (Casracane) Mangin
Odontella litigiosa (Van Heurck) Hoban
O. weissflogii (Janish) Grunow
Porosira pseudodenticulata (Hustedt) Langerheim
Rhizosolenia alata Brightwell
R. sima Castracane
R. bifrons Jouse
Stellarima microtrias (Karsten) Hasle et Sims
Thalassiosira australis Peragallo
T. tumida (Janish) Hasle
T. spp.

Pennales

Amphiprora kufferathii Mangin
A. oestrupii Van Heurck
Berkeleya rutilans (Trentepohl) Grunow
Gomphonema spp.
Haslea trompii (Cleve) Simonsen
Licmophora luxuriosa Heiden et Kolbe
Navicula directa (Wm. Smith) Ralfs in Pritchard
N. glaciei Van Heurck
N. grevillei (Agardh) Heiberg
N. neglecta Kutzing
N. spp.
Nitzschia closterum (Ehrenberg) Wm. Smith
N. curta (Van Heurck) Hasle
N. cylindrus (Grunow) Hasle
N. lecointei Van Heurck
N. obliquecostata (Van Heurck) Hasle
N. stellata Manguin
N. sublineata Hasle
N. turgiduloides Hasle
N. spp.
Pinnularia quadratarea (Schmidt) Cleve
Pleurosigma directum Grunow
Stauroneis spp.
Synedra spp.
Thalassionema sp.
Thalassiothrix longissima Cleve et Grunow
Tropidoneis antarctica (Grunow) Cleve
T. belgicae (Van Heurck) Heiden et Kolbe
T. spp.

DINOFLAGELLATES

Dinophysis spp.
Peridinium spp.
Prorocentrum antarcticum (Hada) Balech

CHRYSOPHYTES

Distephanus speculum (Ehrenberg) Haeckel

PRYMNESIOPHYTES

Phaeocystis pouchetii (Hariot) Lagerheim

when discoloration in the bottom was observed, *N. stellata* (49%) and *Nitzschia* species 2 group (43%) appeared as major components. After the outflow of sea ice in early May, the species composition changed. *N. turgiduloides* was the dominant species (38–74%) and *Amphiprora kufferathii, N. lecointei, Nitzschia* species 1 group and *Pleurosigma directum* appeared also. From late August to mid-November, when the ice algal standing crop increased greatly, *Amphiprora kufferathii, Nitzschia lecointei* and *N. turgiduloides* became abundant and comprised 25–58, 4.2–35 and 2.3–28% of the cell numbers, respectively. This indicates that these species grew rapidly in the bottom ice layer as major components. These species could have stayed in the bottom layer until late summer, when the layer melted, as was suggested by the species composition in the bottom ice layer in March and April 1983. Therefore some major species in the bottom ice layer may remain for all or most of the year, thus providing the basis for recruitment.

3.5 Interior Assemblage at Station III

The species composition in the interior ice layer (35–40 cm deep) at Station III is shown in Fig. 3. This pale yellow layer was formed before May 24 and was composed of granular ice which originated from frazil ice. Centric species, e.g. *Actinocyclus actinochilus, Chaetoceros* spp., *Eucampia antarctica* and *Rhizosolenia alata* contributed 43–67% of the diatom abundance. The species composition did not change greatly until mid-October. Taking this into consideration along with the fact that the chlorophyll *a* concentration in the interior ice layer remained relatively constant, it is suggested that the interior assemblage was formed when the granular ice layer was established and that it did not grow in the ice. After mid-October, *Nitzschia lecointei* and other pennate species increased in abundance and seemed to migrate from other ice assemblages.

3.6 Surface Assemblage at Station III

Consolidated snow (snow ice) was collected and examined from October 10. A strong discoloration was observed in early and mid-December. The species composition in the uppermost snow ice layer (10–18 cm thick) at Station III is shown in Fig. 4. Pennate species consisted of more than 95% of the diatoms except for the sample collected on December 18, in which unidentified centric species (*Thalassiosira* spp.) and *Rhizosolenia alata* contributed 11 and 5.1%, respectively. At the beginning of the snow ice formation, unidentified pennate species, *Tropidoneis* sp. dominat-

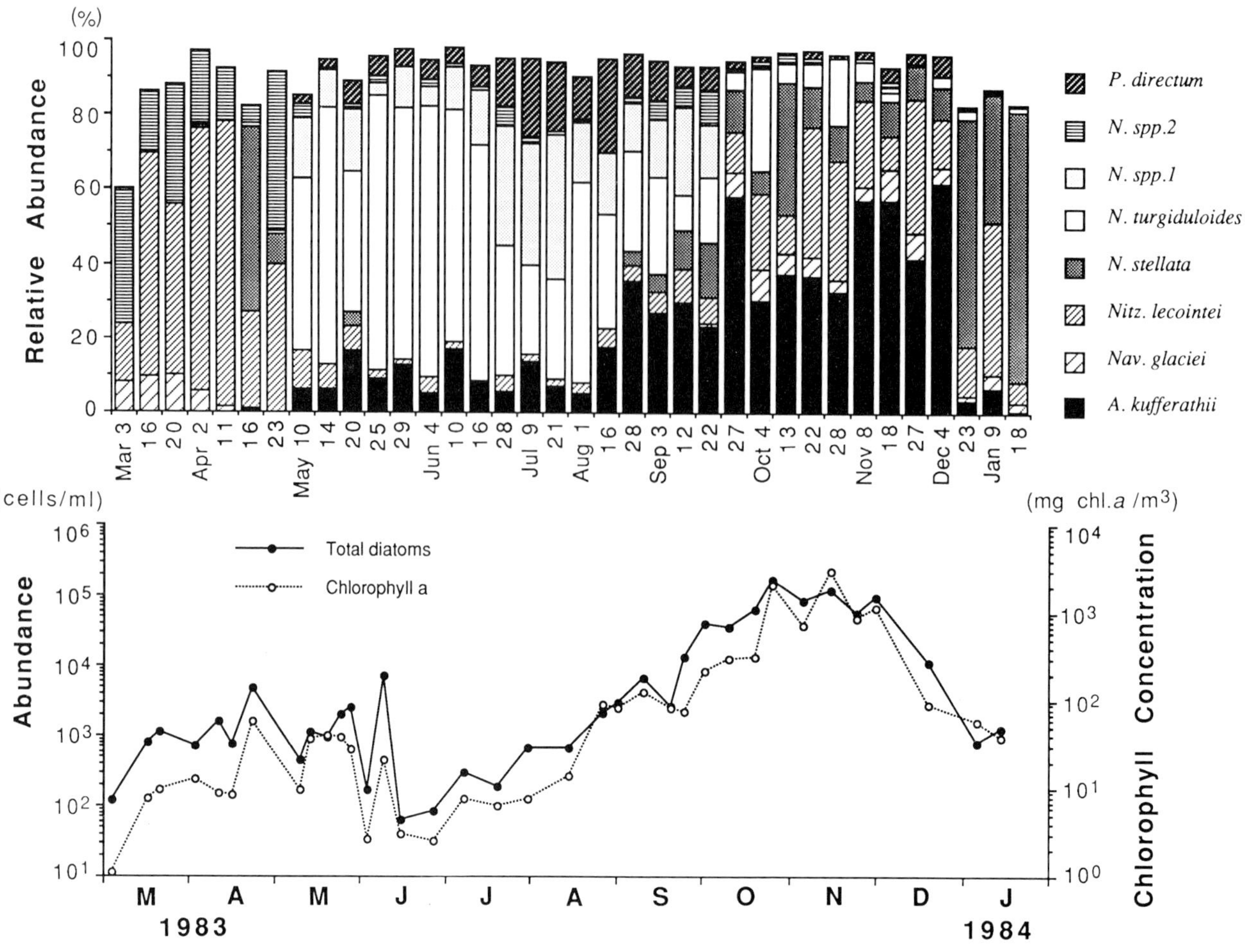

Fig. 2. Seasonal variation in relative abundance of major diatom species, chlorophyll *a* concentration and total cell numbers in the bottom ca. 5-cm layer at Station I

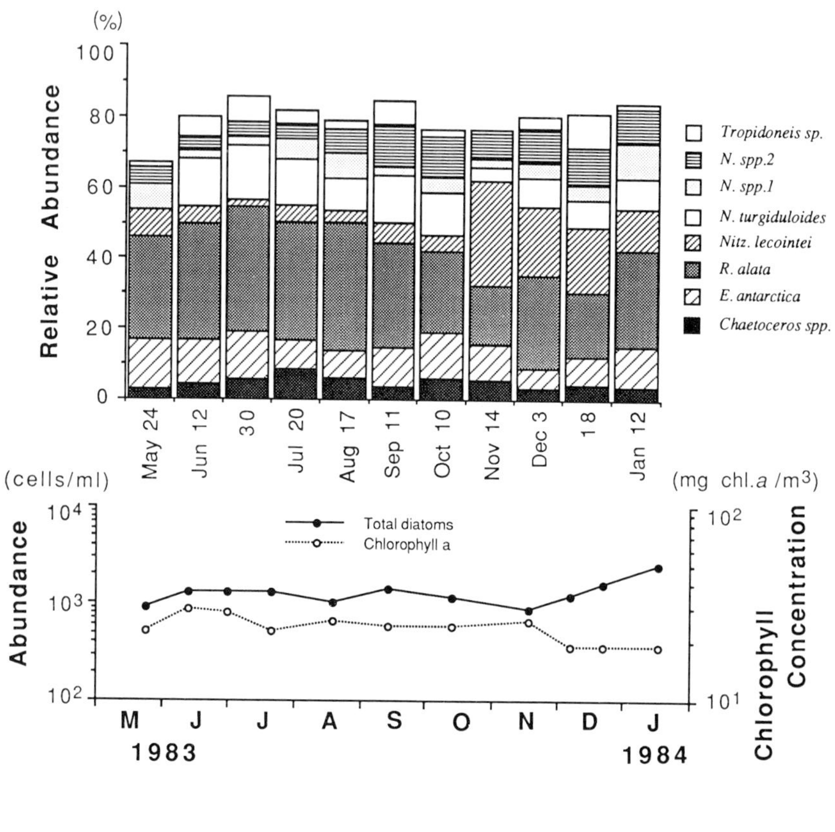

Fig. 3. Seasonal variation in the relative abundance of major diatom species, chlorophyll *a* concentration and total cell numbers in the interior 35–40 cm deep layer of sea ice at Station III

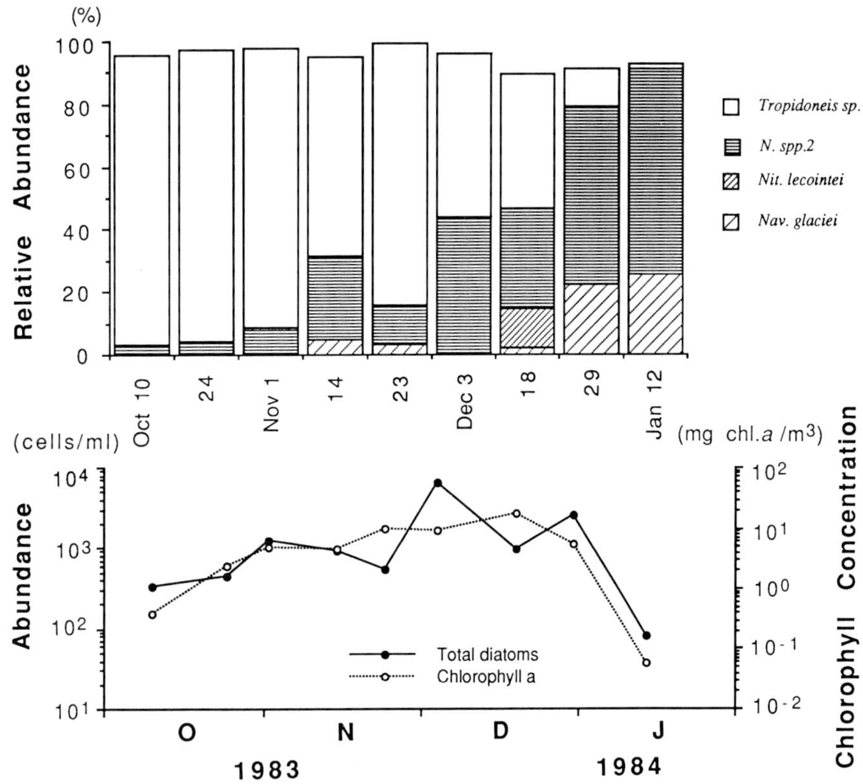

Fig. 4. Seasonal variation in relative abundance of major diatom species, chlorophyll *a* concentration and total cell numbers in the upper 10–18 cm of the snow ice layer at Station III

ed with *Nitzschia closterium, N. lecointei, N. turgiduloides,* and the *Nitzschia* species 2 group appeared. These species were observed both in the interior and the bottom ice layers. However, the relative abundance of the species in the surface snow ice was more similar to the interior one than the bottom one at the same location. This suggests that the surface assemblage was inoculated by the interior ice assemblage presumably through vertical channels in the sea ice, which developed in spring. Relative abundance of *Tropidoneis* sp. decreased and *Nitzschia* species 2 and *Navicula glaciei* increased gradually to become the dominant species in mid-January.

3.7 Comparison of Species Composition Between the Bottom, Interior and Surface Assemblages at Station III

Species compositions of assemblages in fixed layers of the same ice core, collected at Station III, were compared and dendrograms were drawn on the basis of the percentage similarity index (PSI), from October 10, 1983 to January 12, 1984 (Fig. 5). From October 10 to December 18, surface layers (S1, uppermost 10–18 cm snow layer; S2, lowest 10 cm snow ice layer) and interior layers (I1, 15–20 cm deep layer; I2, 35–40 cm deep layer) contained a similar composition of species. The bottom ca. 5-cm layer had the most different composition from that of the others. The PSI between the surface and the interior assemblages increased from 15.4 to 51.1% during that period. These indicate that species contained in the interior layers like *Nitzschia lecointei* and *Nitzschia* species 2 group (Fig. 3) had moved upward to the surface layers and had grown to become major components (Fig. 4). They were presumed to be transported through the vertical channels in the sea ice, which were found from October onwards. On January 12, the lower surface layer and the interior layers became similar in species composition, showing the increase of vertical transport through the channels.

4 Conclusions

1. The seasonal variations of the standing crop of ice algae in the bottom layer in the fast ice area near Syowa had two peaks in autumn and spring-summer. Dominant diatom species changed progressively from (1) *Nitzschia lecointei, N. stellata* and *Nitzschia* species 2 group, to (2) *N. turgiduloides* and *Nitzschia* species 1 group, to (3) *Amphiprora kufferathii* and *N. lecointei,* and to (4) *N. stellata. Amphiprora kufferathii, N. lecointei* and *N. turgiduloides* contributed most to the spring-summer increase of the standing crop. These

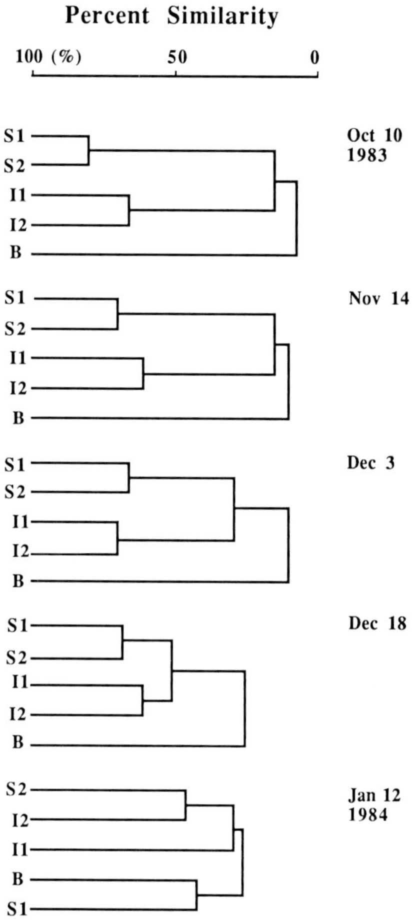

Fig. 5. Dendrograms showing the percent similarity of diatom assemblages from the upper 10–18 cm (*S1*) and the lowest 10 cm (*S2*) of snow ice, 15–20 cm (*I1*) and 35–40 cm (*I2*) of the interior layer and bottom ca. 5 cm (*B*) of sea ice at Station III

species presumably stay in the sea ice throughout the year for recruitment.

2. The interior assemblages were formed at the time of sea ice formation but did not grow there. The species composition did not change greatly with *Rhizosolenia alata, Eucampia antarctica, Nitzschia turgiduloides* and *N. lecointei* as major diatom components.

3. The surface assemblage developed in spring and summer and was dominated by *Tropidoneis* sp., *Nitzschia* species 2 group and *N. lecointei*. The similarity of the composition between the surface and the interior layers suggests that the surface assemblage was inoculated by the interior assemblage through vertical channels in ice.

Acknowledgements. We would like to thank Prof. S. Mae (University of Hokkaido), the leader of the 24th Japanese Antarctic Research Expedition, and the members of the party, especially Drs. H. Kanda (NIPR) and E. Takahashi (University of Kobe) for their cooperation in the fieldwork in Antarctica. We are grateful to the referees for improving the manuscript.

References

Aruga Y (1979) Fluorometric determination of chlorophyll *a* and phaeopigments. In: Nishizawa K, Ichimura M (eds) Methods in phycological studies. Kyoritsu Shuppan, Tokyo, pp 400–401 (in Japanese)

Bunt JS, Lee CC (1970) Seasonal primary production in antarctic sea ice at McMurdo Sound in 1967. J Mar Res 28:304–320

Davis BNK (1963) A study of micro-arthropod communities in mineral soils near Corby, Northants. J Anim Ecol 32:49–71

Garrison DL, Buck KR (1986) Organism losses during ice melting: a serious bias in sea ice community studies. Polar Biol 6:237–239

Garrison DL, Ackley SF, Buck KR (1983) A physical mechanism for establishing ice algal populations in frazil ice. Nature (Lon) 306:363–365

Hasle GR (1978) Some specific preparations, diatoms. In: Sounia A (ed) Phytoplankton manual. UNESCO Press, Paris, pp 136–142

Horner RA (1985) Ecology of sea ice microalgae. In: Horner RA (ed) Sea ice biota. CRC Press, Boca Raton, pp 83–103

Hoshiai T (1969) Ecological observations of the colored layer of the primary production in the Antarctic Ocean. Antarct Rec 27:99–101 (in Japanese with English Abstr)

Hoshiai T (1981) Proliferation of ice algae in the Syowa Station area, Antarctica. Mem Natl Inst Polar Res Ser E 34:1–12

Hoshiai T (1985) Autumn proliferation of ice-algae in Antarctic sea-ice. In: Siegfried WR, Condy PR, Laws RM (eds) Antarctic nutrient cycles and food webs. Springer, Berlin Heidelberg New York, pp 89–92

Hoshiai T, Kato M (1962) Ecological notes on the diatom community of the sea ice in Antarctica. Bull Mar Biol Syn Asamushi 10:221–230

Krebs WN, Lipps JH, Burckle LH (1987) Ice diatom floras, Arthur Harbor, Antarctica. Polar Biol 7:163–171

Satoh H, Watanabe K (1986) Photosynthetic nature of ice-algae under fast ice near Syowa Station, Antarctica. Mem Natl Inst Polar Res Spec Iss 44:34–42

Simonsen R (1974) The diatom plankton of the Indian Ocean Expedition of RV "Meteor" 1965–65. Meteor Forschungsergeb (D Biol) 19:1–66

Strickland JDH, Parsons TR (1972) A practical handbook of sea water analysis. Bull Fish Res Boar Can 167:1–310

Watanabe K (1988) Sub-ice microalgal strands in the Antarctic coastal fast ice area near Syowa Station. Jpn J Phycol 36:221–229

Watanabe K, Satoh H (1987) Seasonal variations of ice algal standing crop near Syowa Station, East Antarctica, in 1983/84. Bull Plankton Soc Jpn 34:143–164

Weeks WF, Ackley SF (1982) The growth, structure and properties of sea ice. US Army CRREL, Hanover (CRREL Rep, 82-1)

Whitaker TM (1977) Sea ice habitats of Signy Island (South Orkneys) and their primary productivity. In: Llano GA (ed) Adaptations within Antarctic ecosystems. Gulf Publ, Houston, pp 75–82

Whittaker RH (1952) A summer study of foliage insect communities in the Great Smokey Mountains. Ecol Monogr 22:1–44

Zwally HJ, Comiso JC, Parkinson CL, Campbell WJ, Carsey FD, Gloersen P (1983) Antarctic sea ice, 1973–1976: satellite passive-microwave observations. NASA, Washington (NASA SP, 459)

Seasonal Variation of Particulate Organic Matter Under the Antarctic Fast Ice and its Importance to Benthic Life

O. MATSUDA [1], S. ISHIKAWA [2] and K. KAWAGUCHI [3]

Summary. Qualitative and quantitative analyses were carried out on organic matter in suspended particulate matter of seawater for a period of 1 year under the Antarctic fast ice near Syowa Station in Lüzow-Holm Bay during 1984/85. Marked seasonal variation of suspended organic matter, in particular chlorophyll *a* was observed. The concentration of suspended organic matter was generally high in summer and low in winter. The maximum value of chlorophyll *a* standing stock through the water column was 25 mg m^{-2} and the minimum was 0.4 mg m^{-2}. Maximum standing stock of particulate organic carbon was 9076 mg m^{-2} and the minimum was 1632 mg m^{-2}. However, the magnitude of the variation was low compared with the downward flux of particulate organic matter already reported for the same station. Suspended particles were rich in fresh algae and nutrients only during summer due to proliferation of phytoplankton and ice algae. Results of the present investigation suggest that feacal pellets produced by zooplankton and ice algal flocs detached from the bottom of fast ice in summer are responsible for the energy supply to benthic life under ice.

1 Introduction

It is known that the biomass of marine benthos is generally high in some Antarctic coastal regions (White 1984). Although Knox (1970) demonstrated the importance of primary production in the water column to the underlying benthos, the process of food supply to the benthic community has not been extensively studied. There are very few publications on suspended organic matter under fast ice in winter and on carbon flux to the sediment in the Antarctic Sea (Clark 1985).

The primary production by phytoplankton and ice algae is confined only to summertime under the fast ice in the Antarctic region (Bunt and Lee 1970; Fukuchi et al. 1985), since underwater irradiance is insufficient in the other seasons (Sullivan et al. 1985). Despite the absence of primary production under the ice during long and dark winter conditions, the animals are able to survive. Therefore, the availability of an energy source for the organisms under the ice during winter is an important aspect to be investigated in the Antarctic marine ecosystem. From this point of view, it is suggested that the suspended particulate matter in seawater under ice could play a vital role in the food supply to benthic life. However, difficult access to ice-covered regions allowed only few investigations of the ecosystem under ice and these are mainly concerned with the summer season (Bunt 1960; Fukuchi et al. 1985; Satoh et al. 1986).

In the present study, qualitative and quantitative surveys of organic matter were performed on suspended particles in the water column under the Antarctic fast ice between 18 February 1984 and 4 January 1985 during the 25th Japanese Antarctic Research Expedition (JARE 25). Seston weight (dry weight of suspended matter), concentration of chlorophyll *a* (chl. *a*), pheopigments, particulate organic carbon (POC), and particulate organic nitrogen (PON) were determined as well as nutrient concentration in seawater and environmental factors. As qualitative measures, variation of POC/dry weight ratio, POC/chl. *a* ratio, and pigment ratio (% of chl. *a*/chl. *a* + pheopigments) were investigated. Special emphasis was also given to clarify the fate of algal organic products in the ecosystem under the Antarctic fast ice.

The authors have already reported on the seasonal variation of downward flux of particulate organic matter investigated at the same station during the same expedition (Matsuda et al. 1987). During a field survey both studies were conducted simultaneously, whereas comparisons are made in the present study between suspended particles and sinking particles trapped in the sediment trap.

[1] Faculty of Applied Biological Science, Hiroshima University, Saijo, Higashi-Hiroshima 724, Japan.
[2] Department of Biology, Faculty of Science, Kochi University, 5-1, Akebonocho 2-chome, Kochi 780, Japan.
[3] Otsuchi Marine Research Center, Ocean Research Institute, University of Tokyo, Akahama, Otsuchi, Iwate 028-11, Japan.

Antarctic Ecosystems. Ecological Change and Conservation
(ed. by K. R. Kerry and G. Hempel)
© Springer-Verlag Berlin Heidelberg 1990

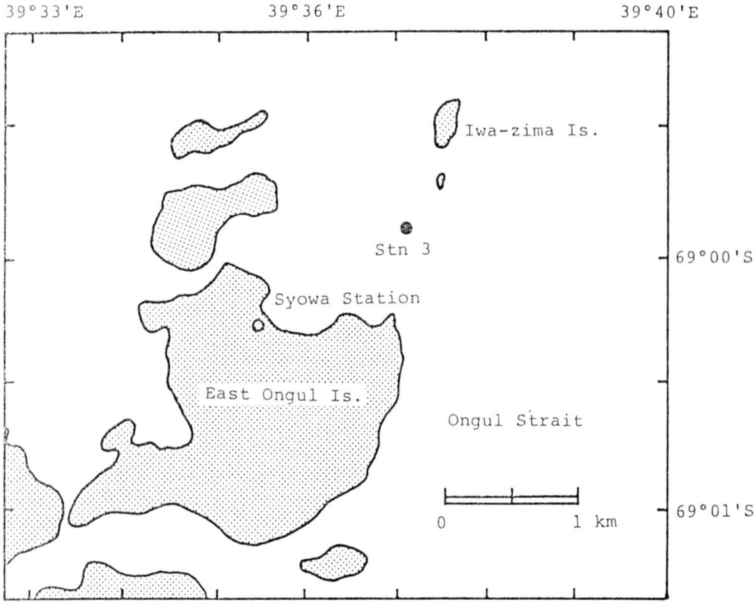

Fig. 1. Location of a sampling station in the Kita-no-ura cove, Lützow-Holm Bay

2 Materials and Methods

Field surveys were carried out at St. 3 in the Kita-no-ura Cove off East Ongul Island in Lützow-Holm Bay (Fig. 1) at intervals throughout the year. At the station, an ice hole of 1 × 1 m was established and access to the water was maintained during the study period. Seawater samples were collected using 6-litre Van Dorn bottles from depths of 0 (water surface), 2.5, 5, 10, 15, 25 and 35 m. Temperature and salinity profiles were obtained using a CTD monitor (FIC Model AFC-III). Dissolved oxygen, pH and nutrients in seawater were measured following the procedure described already by Fukuchi et al. (1985).

Measurements of dry weight, POC, PON, chl. *a* and pheopigments were made on the suspended particulate matter. Determinations of POC and PON were made by the CHN analyzer (Yanagimoto MT-3). Chlorophyll *a* and pheopigments were determined spectrophotometrically by the method of Lorenzen (1967).

3 Results and Discussion

3.1 Seasonal Variation of Environmental Factors

Solar radiation measured at Syowa Station fluctuated markedly from nearly zero to ca. 1000 MJ m^{-2} $month^{-1}$ (Fig. 2). Although water temperature

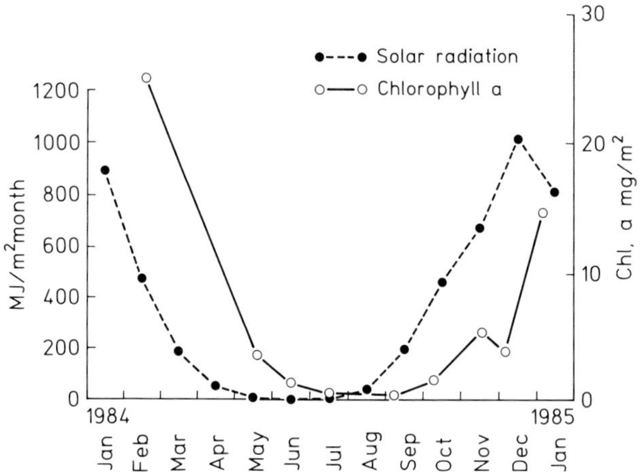

Fig. 2. Seasonal variation of solar radiation at Syowa Station and that of the integrated chlorophyll *a* value through the water column observed under the Antarctic fast ice

ranged from − 1.91° to − 0.35 °C, it was usually below − 1.5 °C. Salinity was usually higher than 33.33‰ throughout the water column but it was very low in the surface waters during January and February due to melting ice. The concentration of dissolved oxygen was uniform (mostly 7 to 8 ml l^{-1}) and exceeded 9 ml l^{-1} at the surface in January and February.

Nutrient concentrations were relatively constant throughout the year, but the concentrations decreased in the surface water in January and February. For example, phosphate concentrations ranged from 1.33 to 2.25 μg-at l^{-1} except in January and February when 0.58 and 0.22 μg-at l^{-1} respectively were measured at the surface. The concentrations of silicate and nitrate also showned a marked decrease only in January and February at the surface. However, judging from the nutrient concentration measured in summer, these primary nutrients presumably did not play a limiting role in the primary production of the area investigated.

3.2 Seasonal Variation of Suspended Particulate Matter in Seawater

The chl. *a* concentration in seawater showed an extreme seasonal variation. It was highest (2.02 $\mu g\,l^{-1}$) in February at 5 m depth and lowest (0.01 $\mu g\,l^{-1}$) from July to September through the entire water column. As shown in Fig. 2, integrated chl. *a* values ($mg\,m^{-2}$) through the water column also showed remarkable variation. Values were highest (25 $mg\,m^{-2}$) in February, and sharply decreased towards May to reach a minimum in September. There has been a gradual increase since then, and the maximum was reached in summer. The seasonal variation of integrated chl. *a* values followed basically a similar pattern to the solar radiation, although the phase was approximately 2 months behind.

The seasonal variation of seston weight integrated over the top 35 m in terms of dry weight ranged from 99 to 431 $g\,m^{-2}$ (Fig. 3). It is interesting to note that a high seston weight of ca. 300 $g\,m^{-2}$ was observed in midwinter (June and July), when primary production completely ceased and the integrated chl. *a* values were also low (0.34–0.63 $mg\,m^{-2}$).

The seasonal variation of integrated POC through the water column showed two peaks in the year (Fig. 4). The highest value appeared in January and

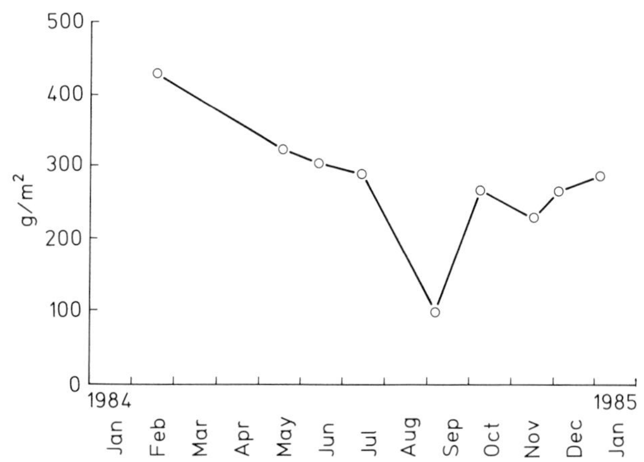

Fig. 3. Seasonal variation of the integrated seston weight through the water column observed under the Antarctic fast ice

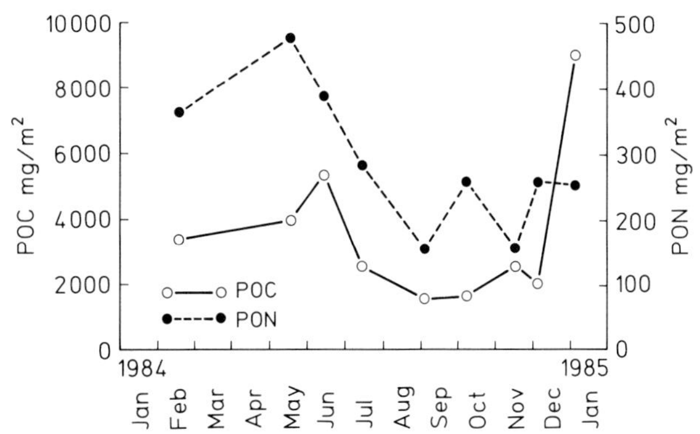

Fig. 4. Seasonal variation of the integrated POC (particulate organic carbon) and PON (particulate organic nitrogen) through the water column observed under the Antarctic fast ice

the second highest value in June. The pattern of this seasonal variation differs greatly from that of integrated chl. *a*, where a lower value was recorded in June as previously described. Therefore, it is strongly suggested that suspended particles in midwinter are composed of a relatively highly organic but chl. *a*-depleted substance.

3.3 Comparison of Suspended and Sinking Particles

According to the previous study (Matsuda et al. 1987), the downward flux of chl. *a* showed a marked seasonal variation compared with POC and PON flux. Variation of chl. *a* flux was also higher relative to the variation of the integrated chl. *a* value in the water column. High chl. *a* flux in February suggested the contribution of algal flocs which were detached from the bottom of fast ice and sank to the seafloor. Higher POC flux (max: 136 $mgC\,m^{-2}\,day^{-1}$) was also measured from January through March in 1984. A flux depletion occurred between March and May, and the flux was lowest (approximately 1.5 $mgC\,m^{-2}\,day^{-1}$) from June through September, and then gradual recovery was observed. The pattern of this variation differed from that of integrated POC through the water column.

The POC/dry weight ratio (%) of suspended particles was usually lower than 2%. A slightly higher ratio, indicating the higher content of organic matter, was observed only at the surface in January, presumably due to the suspension of detached ice algae and growth of phytoplankton. The POC/dry weight ratio of sinking particles increased in summer and decreased in winter from ca. 18 to 1% (Matsuda et al. 1987), which is higher than the ratio of suspended particles.

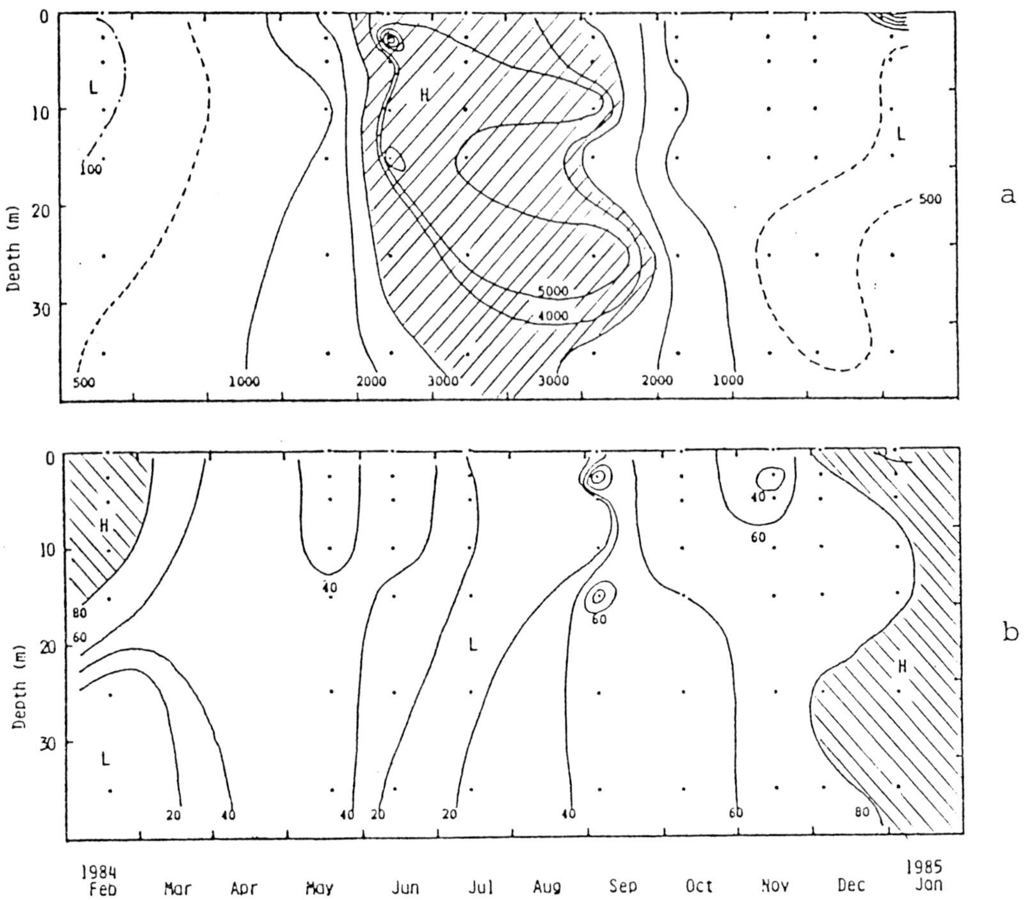

Fig. 5. Seasonal variation of POC/chl. *a* ratio (**a**) and pigment ratio (**b**) of suspended particles in seawater observed under the Antarctic fast ice

The POC/chl. *a* ratio of sinking particles was also reported to show marked seasonal variation, from 30 to 790 (Matsuda et al. 1987). On the other hand, the same ratio of suspended particles was generally much higher (70 to 8000) than the ratio of sinking particles. The ratio was extremely high (3000 to 5000) at all depths from June to September (Fig. 5). Since detrital material had a higher POC/chl. *a* ratio than fresh algae, lower ratios of sinking particles from November through February suggest that they contain numerous fresh algae. However, the higher ratio of suspended matter in winter suggests that they consist of chlorophyll-poor organic matter (Fig. 5).

The pigment ratio indicates the relative freshness of the algal component. Higher ratios ($> 90\%$) of suspended particles at the upper layers in January and February (Fig. 5) indicate that microalgae in seawater are fresh. In constrast, lower ratios ($< 30\%$) in July indicate the relative dominancy of deteriorated or inactive microalgae. Moreover, high pigment ratios were occasionally observed from September to December, which might be caused by the sampling of detached ice algae. The fact that there were may faecal pellets of zooplankton in trapped particles (Matsuda et al. 1987) supports the hypothesis that chlorophyll-rich suspended particles are ingested, digested and egested in the form of faecal pellets by zooplankton.

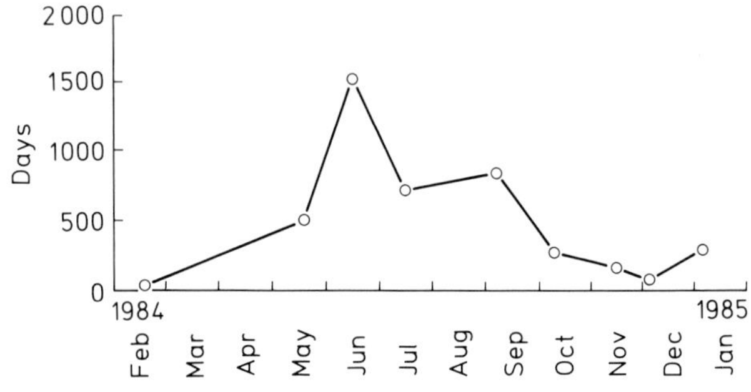

Fig. 6. Seasonal variation of the apparent residence time of particulate organic matter under the Antarctic fast ice

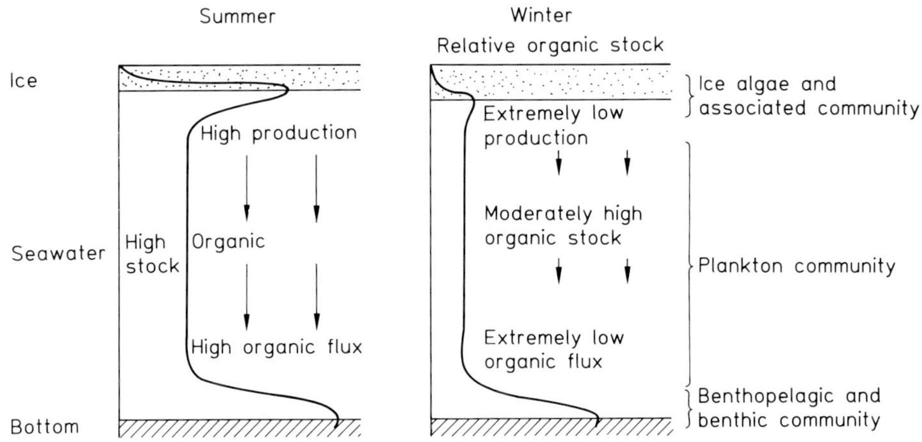

Fig. 7. Schematic model of an ecosystem under the Antarctic coastal fast ice

3.4 Apparent Residence Time and Food Supply to Benthic Life

In order to understand the quantitative relationship between integrated POC through the water column and downward flux of POC, the apparent residence time of POC was calculated by dividing the former by the latter. It was shortest (36 days) in February and longest in June (1536 days; Fig. 6). The shorter residence time in summer suggests the active vertical transport of POC, presumably through feacal pellet production by zooplankton and partly through direct sedimentation of ice algal flocs.

The results of the present investigation can be summarized in two ways. Firstly, the seasonal variation of the standing stock of suspended particles is smaller than that of organic flux. Secondly, suspended particles differ from sinking particles in quality, the benthic community depends more strongly on organic flux as energy source than on the standing stock of suspended particles. Thus the food supply to the benthos is concentrated in summer under the Antarctic fast ice.

Hoshiai et al. (1987) reported the importance of ice algae as food source of an ice-associated copepod such as *Paralabidocera antarctica* in the same area. Also, Kawaguchi et al. (1986) observed benthopelagic life of *Euphausia superba* in winter at the same site by the use of light traps. Therefore, the food source of an under ice ecosystem primarily depends on the algal growth, including both ice algae and phytoplankton and secondly on the transportation of these organic products onto the seafloor. Thus, the following hypothetical model of an under ice ecosystem can be constructed (Fig. 7). In summer, the rate of algal growth and the consumption of algae by zooplankton are remarkably high, which in turn increases the supply of organic material to benthic life. The vertical profile of the relative abundance of organic matter shows bimodal distribution. Namely the standing stock is higher at both ends of the water column and lower in the water column per se. In winter, the algal standing stock is extremely low and hence the vertical organic flux decreases to a minimum. However, at the sea bottom, a relatively abundant benthic community is present. The mechanism of utilization of sedimented organic matter by the benthic community during long winters is not clarified in the present study. The recycling of feacal pellets through the microbial food chain is one possible mechanism, but further study is needed.

Acknowledgements. The authors would like to express their gratitude to Prof. T. Hirasawa of the National Institute of Polar Research, a leader of the JARE 25, for his helpful arrangements and encouragement. Thanks are due to all of the wintering members of JARE 25 for their kind cooperation in the severe field. Thanks are also extended to Dr. S. Uye of Hiroshima University for critical reading and for the valuable comments on the manuscript.

References

Bunt JS (1960) Introductory studies: hydrology and plankton. Mawson, June 1956–February 1957. ANARE Rep. 56:1–135

Bunt JS, Lee CC (1970) Seasonal primary production within and beneath Antarctic sea ice. J Mar Res 28:304–320

Clarke A (1985) Energy flow in the Southern Ocean food web. In: Siegfried WR, Condy PR, Laws RM (eds) Antarctic nutrient cycles and food webs. Springer, Berlin Heidelberg New York, pp 571–580

Fukuchi M, Tanimura A, Ohtsuka H (1985) Marine biological and oceanographical investigations in Lützow-Holm Bay, Antarctica. In: Siegfried WR, Condy PR, Laws RM (eds)

Antarctic nutrient cycles and food webs. Springer, Berlin Heidelberg New York, pp 52–59

Hoshiai T, Tanimura A, Watanabe K (1987) Ice algae as food on an Antarctic ice-associated copepod, *Paralabidocera Antarctica* (IC Thompson). Proc NIPR Symp, Polar Biol 1:105–111

Kawaguchi K, Matsuda O, Ishikawa S, Naito Y (1986) A light trap to collect krill and other micronectonic and planktonic animals under the Antarctic coastal fast ice. Polar Biol 6:37–42

Knox GA (1970) Antarctic marine ecosystems. In: Holdgate MW (ed) Antarctic ecology, vol 1. Academic Press, London, pp 69–96

Lorenzen CJ (1967) Determination of chlorophyll and pheopigments; spectrophotometer equations. Limnol Oceanogr 12:343–346

Matsuda O, Ishikawa S, Kawaguchi K (1987) Seasonal variation of downward flux of particulate organic matter under the Antarctic fast ice. Proc NIPR Symp, Polar Biol 1:23–34

Satoh H, Watanabe K, Kanda H, Takahashi E (1986) Seasonal change of chlorophyll *a* standing stocks and oceanographic conditions under fast ice near Syowa Station, Antarctica, in 1983/84. Antarct Rec 30:19–32

Sullivan CW, Palmisano AC, Kottmeir S, McGrathgrossi S, Moe R (1985) The influence of light on growth and development of the sea-ice microbial community in McMurdo Sound. In: Siegfried WR, Condy PR, Laws RM (eds) Antarctic nutrient cycles and food webs. Springer, Berlin Heidelberg New York, pp 78–83

White MG (1984) Marine benthos. In: Laws RM (ed) Antarctic ecology. Academic Press, London, vol 2, 421 p

Seasonal Patterns of Ichthyoplankton Distribution and Abundance in the Southern Weddell Sea

G. HUBOLD [1]

Summary. The distribution and abundance of fish larvae were analysed in early spring (Oct.–Nov. 1986) and summer (Jan.–Feb. 1985) during two cruises of the RV "Polarstern" to the southern Weddell Sea. In October, mostly post-larval and juvenile notothenioids were collected. A few larval specimens were represented by *Muraenolepis microps* and *Macrourus* sp. Larvae of *Pleuragramma antarcticum* and two channichthyids hatched near Vestkapp (73° S, 19° W) by mid-November. In January, post-larval *Pleuragramma antarcticum* dominated the ichthyoplankton (98% by numbers). The abundance of this species decreased by a factor of 5 between January and February, while its distribution extended far beyond the continental shelf. Other notothenioids and oceanic *Electrona* sp. and *Notolepis* sp. appeared later in the season, and were equally abundant in both months. Of a total of 23 ichthyoplankton species, no more than 16 occurred together in each month. Species were ranked according to neritic or oceanic occurrence. Spatial and temporal distribution patterns of the most abundant taxa were interpreted as adaptations in the life history to extend the geographical basis of the larval drift beyond the limited continental shelf area.

1 Introduction

Over most of the oceanic extension of the Southern Ocean, the pelagic fish community is composed of mesopelagic species and their larvae. Antarctic fishes of the endemic suborder Notothenioidei are mostly neritic (Andriashev 1965). Their larvae, post-larvae, and juveniles are found regularly in surface zooplankton samples taken in the vicinity of islands and close to the continent, and do not occur in great numbers beyond the shelf, possibly due to competition with mesopelagic species (e.g. DeWitt 1970; Keller 1983; Slosarczyk and Rembiszewski 1982; Hubold 1984; Kellermann and Kock 1984; Williams 1985; Kellermann 1986).

Many larval notothenioids depend on small zooplankton as first food (Kellermann 1986; North and White 1987). Plankton peak production in Antarctic waters is confined to a relatively short season. Thus, two important factors limit the distribution and abundance of notothenioids: the extension of shallow shelf areas, and seasonality of food for pelagic stages.

Towards higher Antarctic latitudes, plankton peak production becomes even more restricted to a shorter period, later in the season (Hart 1942; Voronina 1970). Except for the shelves of the Ross Sea and Weddell Sea, shallower areas around the continent are scarce, and often interrupted by deep trenches and innershelf depressions, which allow mesopelagic species to occur close to the coast. Nevertheless, considerable numbers of pelagic stages of notothenioids are found at high latitudes, including the only known fully pelagic planktivorous species of the Antarctic, *Pleuragramma antarcticum* (DeWitt 1970; Hubold 1984, 1985; Williams 1985). Special adaptations in the life histories of these species should be expected to explain the relative abundance of ichthyoplankton and nekton in the high Antarctic.

Due to difficult access to the "permanent pack ice zone" (Hempel 1985), systematic investigations on the fish fauna of the Antarctic continental shelf are scarce. During two expeditions with the German research vessel FS "Polarstern" we had the opportunity to investigate zooplankton and fish in the southern Weddell Sea over an extended period in summer and, for the first time ever, to obtain samples from the coastal polynya and pack ice in late winter. Seasonal patterns of larval fish distribution and abundance from these two cruises are presented in this chapter.

2 Methods

Based on previous knowledge of large-scale zooplankton and fish distribution in the Weddell Sea (Keller

[1] Institute for Polar Ecology, University of Kiel, Olshausenstr. 40, 2300 Kiel 1, FRG.

Antarctic Ecosystems. Ecological Change and Conservation
(ed. by K. R. Kerry and G. Hempel)
© Springer-Verlag Berlin Heidelberg 1990

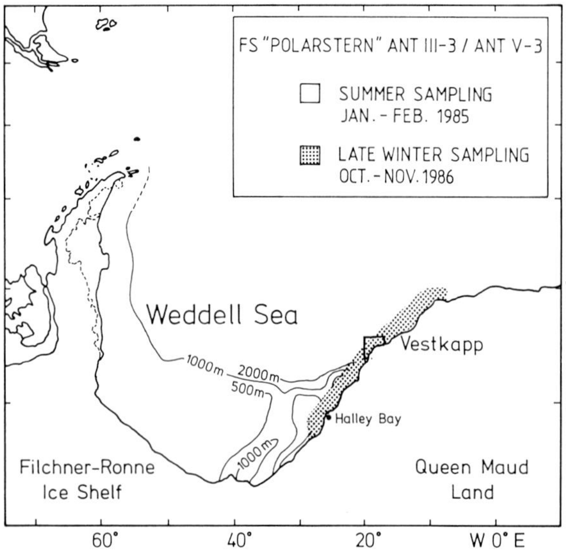

Fig. 1. Investigation area in the south-eastern Weddell Sea

1983; Hempel 1983; Hubold 1984, 1985), mesoscale sampling was conducted in summer 1985 in a target area at Vestkapp (73° S, 19° W). A 50 × 50 mile station grid in the ice-free coastal polynya was sampled twice in the course of 5 weeks (Fig. 1). The location and size of this "box" was chosen to cover the entire seaward extension of the Weddell Sea coastal current, to sample shelf, slope and deepsea areas, and to include the peculiar topographic situation of a deep trench under a floating ice shelf (water depth > 2000 m at the "coast") within one sampling zone, which would also allow for fast ("quasi-synoptic") repeated sampling of the grid (see Fig. 3b, c). A detailed bathymetry of the south-eastern Weddell Sea is given by Johnson et al. (1981).

In summer 1985, standard oblique (300/500 μ) bongo net hauls in the upper 200 m represented the principal sampling method used, and in addition RMT-m hauls were made to 300 m depth. Larvae from Multinet (Weikert and John 1981) hauls down to 1000 m were provided by S. Schiel, Bremerhaven. Twenty-four standard (bongo net) sampling stations were distributed on four legs perpendicular to the coast. The grid was sampled within 1 week (January 23–Feb. 1, 1985) and repeated 3 weeks later (February 12–21). Some additional stations were made after the repetition of the standard grid in the same area.

In October 1986, for the first time, a research vessel penetrated the Weddell Sea pack ice in late winter to conduct biological research in the coastal polynya. Samples were taken between Atka Bay (8° W) and Halley Bay (29° W; Fig. 1). The main sampling effort was concentrated in the Vestkapp area. Due to heavy pack ice, vertical bongo net hauls between 200 or 500 m and surface were made mainly through ice holes broken by the ship. Open leads and the coastal polynya were used for oblique tows between 200 or 500 m and surface with RMT and bongo nets.

Sampling effort (as total numbers of hauls and filtered water volume, mean bottom depths, and distances-to-coast of sampling stations) is shown in Table 1 for the months October and November 1986, and January and February 1985 separately. Total effort was 196 hauls, filtering 825 000 m^3 of water.

Sampling stations were distributed almost identically in October and November, and in January and February, respectively. Sampling during the summer, however, extended farther offshore than winter sampling, as shown by mean bottom depth and distance to coast (Table 1). These differences bias the distribution patterns between seasons, whereas comparisons of months within seasons are more reliable.

Since the two cruises were not made in the same year, the spring data from 1986 could only be compared to the summer data from 1985. Interannual variability in the ichthyoplankton of the Weddell Sea is unknown. Summer ichthyoplankton collections from other cruises in the same area (Keller 1983; Hubold 1984, 1985), however, contained comparable numbers and species of ichthyoplankton. Thus, as a first approach, similarity of the ichthyoplankton in the two years will be assumed. The present observations in late winter and spring are the only information available in this part of the Antarctic.

Methods and effort adopted during the two cruises are not fully comparable. The Multinet caught mainly yolk-sac larvae in November, but otherwise contained

Table 1. Fishing effort as numbers of hauls and volume (m^3) filtered by net types and months[a]

Month/ year	Bo (n)	Vol (m^3)	RMT (n)	Vol (m^3)	MN (n)	Vol (m^3)	Total (m^3)	Mean (max.) bottom depth (m)	Mean (max.) distance from coast(nm)
Oct 1986	17	5918	8	103460	9	1650	111028	1236 (3680)	11 (38)
Nov 1986	44	18160	18	456748	16	2873	477780	1311 (3406)	11 (36)
Jan 1985	24	33071	5	66049	15	1883	97469	2518 (4182)	26 (61)
Feb 1985	24	37858	4	98289	12	2075	138222	2505 (4168)	26 (61)

[a] Bo = bongo net; RMT = Rectangular Midwater Trawl; MN = Multinet.

few specimens; RMT nets were used without flow meters in October–November sampling, and thus cannot be quantitatively compared with summer samples. For quantitative considerations, only bongo net hauls were used. Quantitative aspects therefore refer to larvae and post-larvae, but not to juvenile stages because the latter were able to avoid the smaller plankton nets (Kellermann 1986).

Plankton samples were sorted on board the ship, and larvae were measured fresh at in situ water temperature immediately after capture. Large catches of *Pleuragramma antarcticum* were subsampled, and 30–100 individuals were measured. Size measurements (to the lower mm) refer to notochord length in larvae and standard length (SL) in post-larvae and juveniles, respectively.

Water temperatures higher than 0 °C led to rapid deterioration (shrinkage) of early larvae in the catch. Sizes of post-larval *Pleuragramma antarcticum* decreased by 20% within 35 min if not kept at low temperatures.

Fish larvae were preserved in 4% formaldehyde or 96% alcohol. Formaldehyde reasonably preserved the size at the moment of fixation; shrinkage due to preservation accounted for approximately 10% within the following 17 months. Some taxa were measured fresh (Oct.–Nov. samples) or preserved in formaldehyde (Jan.–Feb. samples), so that a 10% size correction may be considered in the summer material.

3 Results

3.1 Ichthyoplankton Abundance

Early life stages of fish were collected in all investigated months. Forty-four individuals were collected by all types of equipment in October, and 3145 in November. January samples contained 4797, and February samples 1367 specimens (Fig. 2a). These specimens were from 15–16 species in November, January, and February, and from 7 species in October (Fig. 2b). The differences are only partly due to different fishing methods, as shown by the standardized abundance estimates: Mean abundances of ichthyoplankton from positive hauls (bongo nets) were 6 and 227 larvae/1000 m^3 in October and November, and 130 and 26 larvae/1000 m^3 in January and February, respectively (Fig. 2c). Median abundances showed the same trend (Fig. 2c). Both in November and January, locally high larval concentrations were found as shown by maximum abundances (Fig. 2d). Frequency of occurrence, as the percentage of positive bongo net hauls, ranged from 12% in October to 92% in January (Fig. 2e).

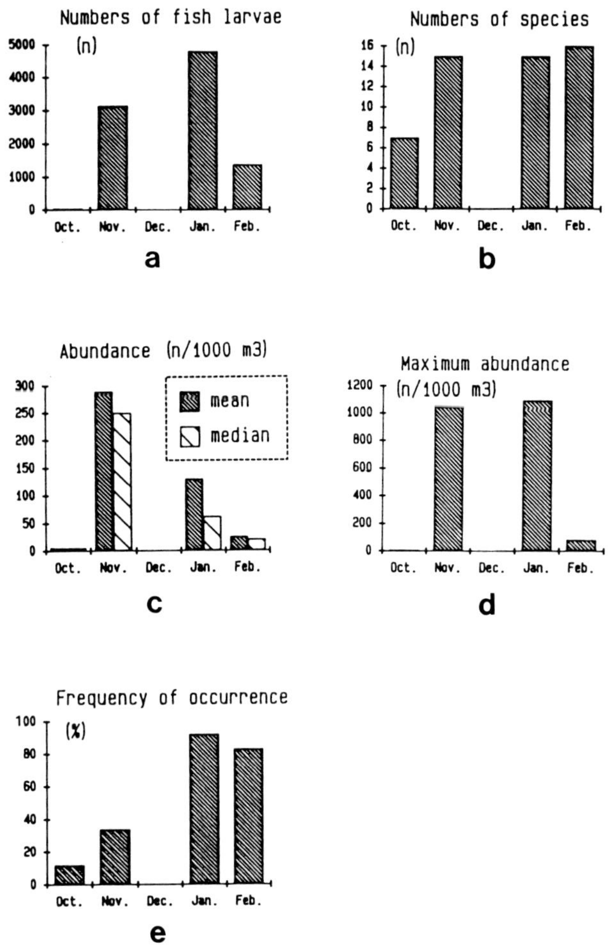

Fig. 2. Indices of ichthyoplankton abundance by month. **a** Numbers of fish larvae and juveniles collected by all nets; **b** numbers of fish species from all nets; **c** mean and median standardized abundance of ichthyoplankton in bongo net hauls; **d** maximum values of ichthyoplankton abundance from bongo net hauls; **e** frequency of occurrence as percentage of bongo net hauls containing early life stages of fish

3.2 Ichthyoplankton Composition

During the two expeditions, 23 species of larval, post-larval and juvenile fish were collected. Eighteen of these were notothenioids (Table 2). Pelagic fish eggs were not found in plankton samples (maximum sampling depth 1000 m), however, the stomachs of adult *Trematomus centronotus* and *Trematomus scotti*, two benthos feeding fish collected at Vestkapp, contained several hundreds of fish eggs in October and November 1986. The eggs contained embryos at an advanced stage of development; they could be identified as *Pleuragramma antarcticum* (egg diameter 1.95 mm), and a channichthyid, probably *Chionodraco myersi* (egg diamter 3.1 mm).

Table 2. Numbers, percentages and mean standard lengths (SL, in mm) of ichthyoplankton collected in the south-eastern Weddell Sea in different months by RMT, bongo, and Multinet (J: juveniles)

Taxon	October			November			January			February		
	n	SL	(%)	n	SL	(%)	n	SL	(%)	n	SL	(%)
Nototheniidae												
Pleuragramma antarcticum J	27	38	(61)	133	38	(4)						
Pleuragramma antarcticum				2949	9	(89)	4656	15	(97)	1217	19	(89)
Trematomus centronotus	9	18	(21)	1	18	(< 1)						
Trematomus scotti							12	11	(< 1)	17	15	(1)
Trematomus lepidorhinus							3	17	(< 1)	5	19	(< 1)
Trematomus eulepidotus							2	33	(< 1)	6	50	(< 1)
Pagothenia brachysoma							1	30	(< 1)			
Aethotaxis mitopteryx				8	15	(< 1)	3	20	(< 1)			
Channichthyidae												
Chaenodraco wilsoni	1	40	(2)	2	40	(< 1)						
Pagetopsis maculatus J	3	40	(7)									
Pagetopsis maculatus				1	12	(< 1)	7	18	(< 1)	3	20	(< 1)
Pagetopsis macropterus				2	36	(< 1)						
Cryodraco antarcticus				4	30	(< 1)	1	45	(< 1)			
Chionodraco myersi J				1	58	(< 1)						
Chionodraco myersi				33	19	(1)	8	28	(< 1)	9	30	(< 1)
Dacodraco hunteri							9	20	(< 1)	7	22	(< 1)
Bathydraconidae												
Gerlachea australis	1	26	(2)	3	44	(< 1)						
Racovitzia glacialis				1	16	(< 1)	8	20	(< 1)	4	22	(< 1)
Prionodraco evansii				1	13	(< 1)	1	15	(< 1)	1	26	(< 1)
Gymnodraco acuticeps										1	41	(< 1)
Artedidraconidae												
Artedidraconidae spp.				2	10	(< 1)				2	17	(< 1)
Macrouridae												
Macrouridae sp. J				1	50	(< 1)						
Macrouridae sp.	1	10	(2)	1	10	(< 1)				2	13	(< 1)
Muraenolepidae												
Muraenolepis microps	2	7	(5)							1	10	(< 1)
Bathylagidae												
Bathylagus sp.				1	10	(< 1)	11	15	(< 1)	14	16	(1)
Paralepididae												
Notolepis sp.				1	38	(< 1)	66	30	(1)	66	35	(5)
Myctophidae												
Electrona sp.							9	14	(< 1)	12	16	(< 1)
Total specimens (taxa)	44 (7)			3145 (15)			4797 (15)			1367 (16)		

In spite of considerable sampling effort (Table 1) in early spring 1986 (October 21–31) only 44 specimens of larval and juvenile fish of 7 species were collected (Table 2). Mesopelagic taxa were not found, although samples were taken also over depths of more than 3600 m. One Macrourid (*Macrourus holotrachys*?) and two *Muraenolepis microps* were collected as yolk-sac larvae, probably hatched in October. Five notothenioid species were advanced post-larvae (*Trematomus centronotus, Gerlachea australis*), or juveniles (*Pleuragramma antarcticum, Chaenodraco wilsoni* and *Pagetopsis maculatus*). *Pleuragramma antarcticum* juveniles of a mean length of 38 mm dominated the ichthyoplankton by numbers. Other nototheniids accounted for 21% and channichthyids for 9% of the collected specimens (Table 2).

In November, 3145 specimens of 15 species were collected (Table 2). Post-larvae and juveniles of *Pleuragramma antarcticum, Trematomus centronotus, Chaenodraco wilsoni, Pagetopsis macropterus, Cryo-*

draco antarcticus, Chionodraco myersi, Gerlachea australis, Macrourus holotrachys (?), and *Notolepis* sp. were found.

Yolk-sac larvae of *Pagetopsis maculatus, Aethotaxis mitopteryx,* and *Chionodraco myersi* of 12, 15 and 19 mm, respectively, appeared from early November on. Identification of early larvae of *Chionodraco* spp. is still unclear (Kellermann pers. commun.), but the regular occurrence of adult *Chionodraco myersi* in the area (Ekau et al. 1987), and the sizes of identified post-larval *Chionodraco myersi* in January and February, indicate hatching by this species off Vestkapp in November.

From mid-November onwards, yolk-sac larvae of *Pleuragramma antarcticum* were collected. Larval lengths were 7–12 mm. Early stages of one species of an unidentified artedidraconid (10 mm) and older larvae of the bathydraconids *Gerlachea australis, Racovitzia glacialis* and *Prionodraco evansii*, and mesopelagic *Bathylagus* sp. also occurred. Larval *Chionodraco myersi* and *Pleuragramma antarcticum* occurred first at depths below 500 m in samples collected with the Multinet and appeared in the surface layers a few days later. Hatching at depth and successive ascent to the surface can be inferred from this observation.

In January, 15 fish species were present. Species composition differed from November samples. Myctophids *Electrona* sp. and five additional notothenioid species appeared: larval *Trematomus scotti* (11 mm) and *Trematomus lepidorhinus* (17 mm); and post-larvae of *Trematomus eulepidotus, Pagothenia brachysoma* and *Dacodraco hunteri* of 33, 30, and 20 mm, respectively. Post-larval *Pleuragramma antarcticum* had a mean length of 15 mm in January, and *Aethotaxis mitopteryx* were 20 mm long. One juvenile *Cryodraco antarcticus* was collected. Post-larval *Pagetopsis maculatus* and *Chionodraco myersi* measured 18 and 28 mm. The bathydraconids, *Prionodraco evansii* and *Racovitzia glacialis,* were found. Mesopelagic species *Electrona* sp., *Bathylagus* sp. and *Notolepis* sp., although collected in considerable numbers, accounted for less than 2% of the ichthyoplankton (Table 2). Sixteen species were represented in samples collected during February (Table 2). *Pleuragramma antarcticum* post-larvae had a mean length of 19 mm. Three *Trematomus* species were collected. Channichthyids were *Pagetopsis maculatus, Chionodraco myersi* and *Dacodraco hunteri.* The bathydraconid *Gymnodraco acuticeps* occurred only in this month in the catches. Macrouridae and Muraenolepidae were found in low numbers, and the percentage of mesopelagic species was relatively high (7%) due to lower numbers of *Pleuragramma antarcticum* in February.

Pleuragramma antarcticum dominated the samples in all months investigated. Percentage share of the species varied between 61% (juveniles) in October, 98% (juveniles and yolk-sac larvae) in November and 97% (post-larvae) in January. In February, the share of *Pleuragramma antarcticum* decreased to 89% (Table 2).

3.3 Distribution and Abundance of *Pleuragramma antarcticum*

Mesoscale horizontal distribution and abundance of *Pleuragramma antarcticum* in the Vestkapp area is shown in Fig. 3a–c. In October and November, the occurrence of juveniles was limited to a few specimens collected over the shelf and slope (Fig. 3a). The majority of the juveniles, however, were caught on the shelf to the south of Halley Bay (Fig. 1). Recently hatched larvae occurred from mid-November 1986 in a small area to the south-west of Vestkapp over bottom depths of 1000–2000 m in great abundances (Fig. 3a).

In January 1985, post-larvae occurred at high density of more than 50 individuals m^{-2} over the trench, and further offshore over the slope of the continental shelf off Vestkapp (Fig. 3b). Abundances were high to the south-west of the cape, and decreased towards the north-east and open sea.

Abundances in February were only 20% of January values, and post-larvae were found mostly in the south-westerly part of the sampled area (Fig. 3c). To the north-east of Vestkapp, the plankton contained generally less than 5 larvae m^{-2}, whereas to the south-west densities were 5–10 larvae m^{-2} or more.

Abundances of all other taxa never exceeded 1–4 larvae m^{-2}.

3.4 Distribution Patterns

Due to the typical continental Antarctic shelf topography with "innershelf depressions" (Andriashev 1965) and trenches such as the one under Vestkapp, great water depths can be found in the immediate vicinity of the coast. Distribution patterns may thus be influenced by both bottom depth and distance from the coast. Separated for each month, and as a mean over all months, median bottom depths and distances to the coast of larval fish occurrences are shown in Table 3.

Notothenioid larvae (median numbers of collected larvae) occurred 1 to 26 nautical miles (nm) from the coastline over water depths from 270 to 3000 m. Macrouridae, *Muraenolepis microps, Bathylagus* sp., *Notolepis* sp., and *Electrona* sp. occurred over mini-

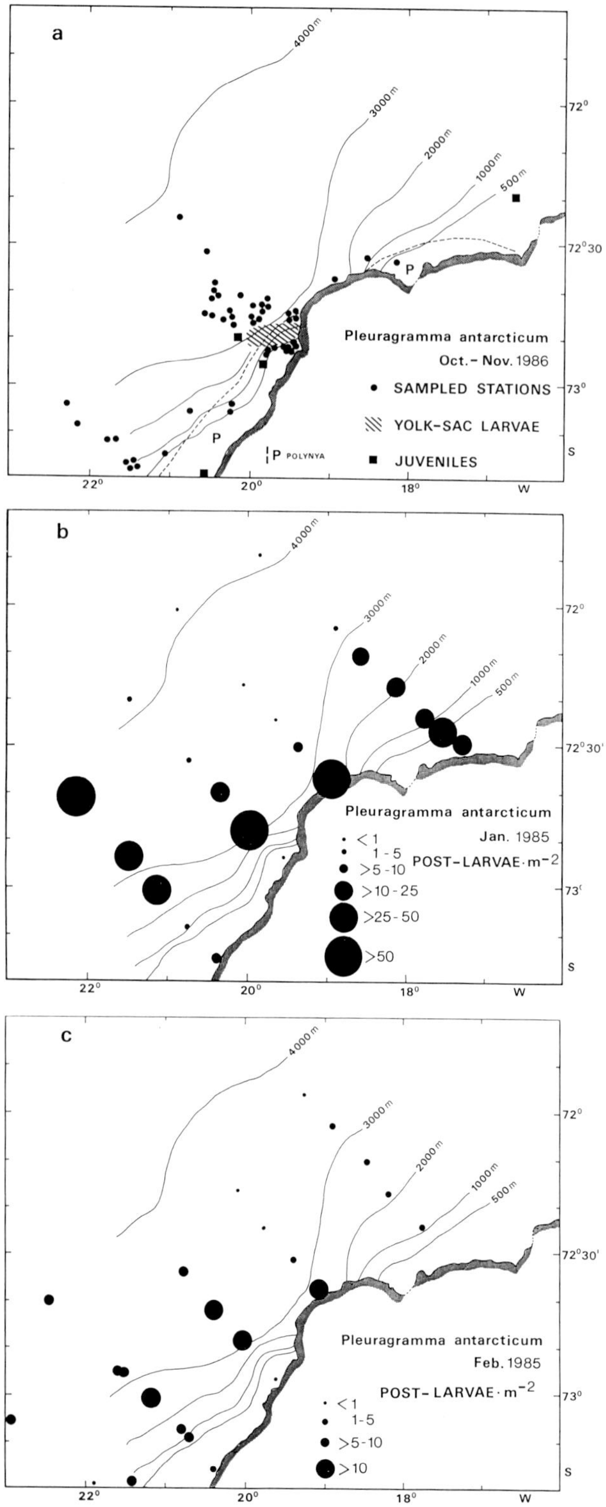

Fig. 3. Horizontal distribution and abundances of *Pleuragramma antarcticum* larvae from bongo net catches at Vestkapp. **a** Occurrence of yolk-sac larvae and juveniles in early spring. **b** Standardized abundances of post-larvae in January. **c** Standardized abundances of post-larvae in February

mum depths of 1700 to more than 3000 m, within 6 to 41 nm from the coast, respectively. Mean monthly depth and distance of occurrence of all taxa increased from October to November 1986 from 900 to 1200 m and from 9 to 12 miles, and from January to February 1985 from 1700 to 1900 m and 13 to 21 miles, respectively.

Ranking the species by an index of occurrence as a product of depth and distance from the coast divided by 1000, an ordination of neritic to oceanic species was obtained (Fig. 4). With the exception of Macrouridae (index 26; rank 16) and *Bathylagus* sp. (index 41; rank 20) all non-notothenioid fish ranked higher than notothenioids. Juveniles of *Pleuragramma antarcticum* and *Chionodraco myersi* showed a more neritic distribution than post-larvae of the same species (Table 3).

The distribution of single taxa differed between months: Juvenile *Pleuragramma antarcticum* mainly occurred over wide (5–19 nm) and shallow (400–550 m) parts of the shelf in October and November. Yolk-sac larvae of the same species occurred closer to the coast (3 nm) but over greater depth (1400 m) in November, i.e. over the slope of the Vestkapp trench. Later in the season, larvae were found over greater depths (2700 m) and farther offshore (27 nm).

Cumulative percentages of larval and juvenile *Pleuragramma antarcticum* over increasing bottom depths showed that 90% of the juveniles collected in October and November occurred over bottom depths less than 500 m (mean 432 m; Fig. 5). Larvae in the November collections occurred only rarely over shallow water. Only 10% of yolk-sac stages were collected over depths shallower than 1000 m; the bulk of larvae occurred over the slope of the continental shelf over a narrow depth range between 1000 and 1600 m (mean 1397 m; Table 3). In January, only 20% of post-larvae were found over depths shallower than 2000 m; 60% were distributed over depths between 2000 and 2500 m, and some 20% were found beyond that depth. Mean depth was 2332 m. In February, a further displacement towards greater depth was found, and only 5% of *Pleuragramma antarcticum* post-larvae occurred over the shelf. A significant portion of the post-larvae were collected over 1500 to 2300 m; more than 50% occurred beyond 3600 m depth. Mean depth of occurrence was 2697 m.

Distribution analysis by month for other species is difficult due to the low frequency of occurrence. Other notothenioids such as *Trematomus eulepidotus* and *Pagetopsis maculatus* were also found in greater distance from the coast in summer than in spring, whereas *Trematomus scotti, Trematomus lepidorhinus* and *Racovitzia glacialis* occurred over shallower depth at greater distance from the coast in February than in

Table 3. Bottom depth (m) and distance from coast (nm) for ichthyoplankton taxa in the Weddell Sea (boldface: yolk-sac larvae; J: juveniles)

Taxon	Median bottom depth (m)				Mean of months	Median distance to coast (nm)				Mean of months
	Oct	Nov	Jan	Feb		Oct	Nov	Jan	Feb	
Nototheniidae										
Pleuragramma antarcticum J	548	409	–	–	479	5	19	–	–	12
Pleuragramma antarcticum	–	**1397**	2332	2697	1242	–	**3**	16	27	15
Trematomus centronotus	493	847	–	–	670	2	3	–	–	3
Trematomus scotti	–	–	**2497**	1000	1749	–	–	**3**	10	7
Trematomus lepidorhinus	–	–	2497	1000	1749	–	–	7	20	14
Trematomus eulepidotus	–	–	421	2116	1269	–	–	4	26	15
Pagothenia brachysoma	–	–	900	–	900	–	–	5	–	5
Aethotaxis mitopteryx	–	**1231**	640	–	936	–	**1**	8	–	5
Channichthyidae										
Chaenodraco wilsoni	278	367	–	–	323	8	23	–	–	16
Pagetopsis maculatus	267	**1138**	2020	2640	1932	8	1	22	19	14
Pagetopsis macropterus	–	414	–	–	414	–	18	–	–	18
Cryodraco antarcticus	–	671	859	–	765	–	8	10	–	9
Chionodraco myersi J	–	359	–	–	359	–	20	–	–	20
Chionodraco myersi	–	**1033**	2961	2684	2226	–	**14**	25	24	21
Dacodraco hunteri	–	–	859	421	640	–	–	10	4	7
Bathydraconidae										
Gerlachea australis	389	480	–	–	435	10	3	–	–	7
Racovitzia glacialis	–	**2100**	2311	995	1802	–	**24**	16	24	21
Prionodraco evansii	–	**1600**	270	423	764	–	**2**	2	3	2
Gymnodraco acuticeps	–	–	–	2760	2760	–	–	–	15	15
Artedidraconidae										
Artedidraconidae spp.	–	**308**	–	1050	679	–	**20**	–	9	15
Macrouridae										
Macrouridae sp.	1675	**2635**	–	1675	1995	6	**19**	–	14	13
Muraenolepidae										
Muraenolepis microps	**2590**	–	–	3451	3020	**15**	–	–	41	28
Bathylagidae										
Bathylagus sp.	–	**2900**	2020	2255	2391	–	**13**	13	24	17
Paralepididae										
Notolepis sp.	–	2890	2742	3047	2893	–	13	27	31	24
Myctophidae										
Electrona sp.	–	–	2548	3400	2974	–	–	25	41	33
Monthly mean	891	1222	1705	1859	1402	9	12	13	21	14

January samples (Table 3). The non-notothenioid taxa in general showed a more pronounced deep-water distribution: except for larval macrourids, mean occurrence of these species was beyond 2000 m bottom depth.

4 Discussion

In the present ichthyoplankton collections from the southern Weddell Sea, 23 taxa occurred over the entire summer period between October and February; in individual months up to 16 species were found. Only about ten larval and juvenile species were found in Prydz Bay, a similar Antarctic shelf sea, in January–March (Williams 1985). Unfortunately, no further information on species composition of larval fish communities of continental Antarctic shelves, e.g. Ross Sea, is available at present. From the composition of the ichthyoplankton of the south-eastern Weddell Sea it may be concluded that species diversity on high Antarctic shelves is as high as that at lower latitudes,

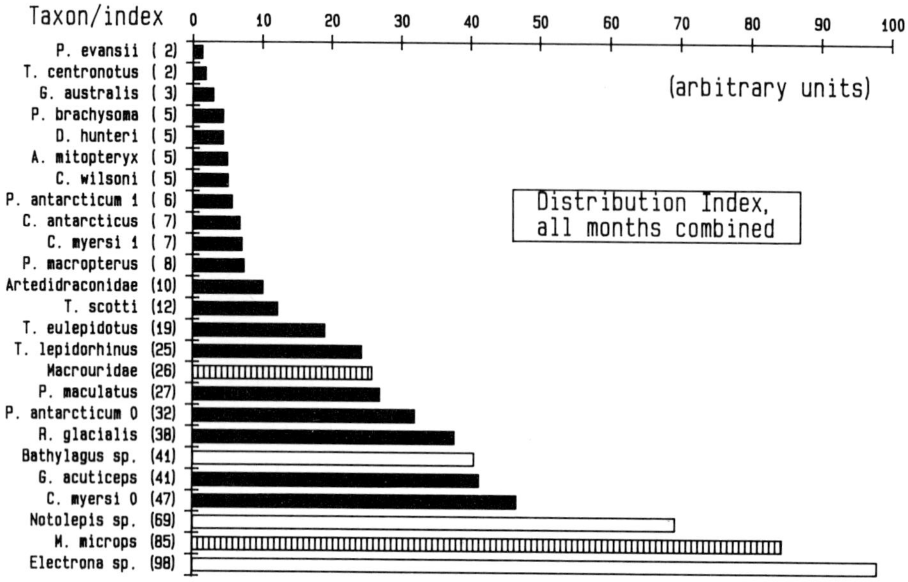

Fig. 4. Index of distribution of ichthyoplankton taxa. Mean depth of occurrence (m) × distance from coast (nm) divided by 1000 (*1* = juveniles)

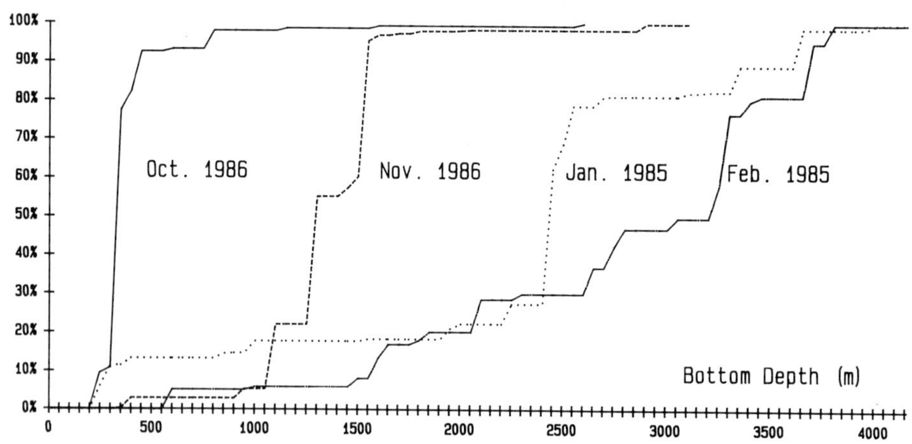

Fig. 5. Cumulative percentages of collected larvae, post-larvae, and juveniles of *Pleuragramma antarcticum* over increasing bottom depths at Vestkapp. October 1986: only juveniles; November 1986: juveniles and yolk-sac larvae; Jan. and Feb. 1985: post-larvae

where 14 to 20 fish species were regularly present (Kellermann and Kock 1984; Slosarczyk 1986; Kellermann 1986). White and North (1987) reported a total of 29 species, including several oceanic species, in collections made around the South Shetland Islands. Species composition and genera of the Notothenioidei on the Weddell Sea shelf, however, differed considerably from the Antarctic Peninsula and South Shetland samples, whereas the oceanic myctophids and paralepidids were represented by the same genera. Bathylagids occurred more frequently in the Weddell Sea than in the north. In the extensive ichthyoplankton collections analyzed by Kellermann (1986), only seven notothenioid species were in common with the present Weddell Sea samples. In both areas, *Pleuragramma antarcticum* was abundant in all months. Except for the myctophid, *Krefftichthys anderssoni*, all species collected by Williams (1985) in the Prydz Bay area were also found in the Weddell Sea, indicating similarity of the two regions with respect to composition of reproducing fish species. Differences in species diversity may be attributed to lower sampling effort in the Prydz Bay, so that rare species were not collected.

In the Weddell Sea, *Pleuragramma antarcticum* dominated numerically by two orders of magnitude (Table 2). Second in abundance was *Notolepis* sp. The other mesopelagic species were also amongst the more abundant species.

As expected, the distribution of different species in relation to bottom depth and distance from the coast showed a more neritic occurrence of most of the notothenioids than for most of the mesopelagic species (Fig. 4). There were, however, interesting exceptions to this pattern, e.g. mesopelagic *Bathylagus* sp. ranked lower (index 41) than notothenioids *Gymnodraco acuticeps* and *Chionodraco myersi* (indices 41 and 47, respectively). Post-larval *Pleuragramma antarcticum* (index 32) showed a more oceanic and deep-water occurrence than juveniles of the same species (index 06). Thus, oceanic and neritic species do not seem to be completely separated geographically during the larval phase, as was stated for the mid-water fauna of the Ross Sea (DeWitt 1970). Due to the very narrow continental shelf in the study area, mesopelagic species occur close to the coast, and notothenioids occur over oceanic depths.

The first information about seasonal differences in distribution and abundance of larval fish in the Weddell Sea can be derived from the present data, when comparing October–November 1986 and January–February 1985 samples. Although collected in different years, the different values of distribution and abundance indicate a consistent pattern of seasonal development of the ichthyoplankton community (Figs. 2a–e, 3a–c). Larvae of most taxa did not hatch before November or December; in October very low numbers of mainly juvenile fish were found to dominate the ichthyoplankton. Overall lower larval abundances observed in February were due to lower numbers of *Pleuragramma antarcticum*, whereas numbers of some other taxa increased in catches during February. Extrapolating from the present data suggests that the highest ichthyoplankton abundances in the area can be expected in December; however, samples from this month are not yet available.

Seasonality in ichthyoplankton abundance parallels the production cycle in phyto- and zooplankton. Phytoplankton concentrations were extremely low in October (Gieskes et al. 1987). From November on, a gradual increase of phytoplankton was recorded (Elbrächter et al., 1987). In spite of low phytoplankton stocks, cyclopid copepods were actively reproducing, and approximately 800 individuals m^{-3} of the species *Oithona similis* and *Oncaea curvata* were present in the surface zooplankton in October, anticipating the phytoplankton spring bloom (Fransz 1988). *Pleuragramma antarcticum* larvae are known to feed on the cyclopids (Kellermann 1986), and thus are able to use this early food resource. Channichthyids *Chionodraco myersi* were observed to hatch in November, in the vicinity of *Pleuragramma antarcticum*; post-larvae and juveniles of these species prey on *Pleuragramma antarcticum* (Hubold and Ekau 1989). Most of the other nototheniids were collected in January and February only, and may have hatched during December. Small *Trematomus scotti* in January and February samples indicate that this species hatches even later. *Trematomus scotti, Trematomus lepidorhinus* and *Trematomus eulepidotus* were collected in higher numbers in February and appear to occupy a late niche in the summer plankton of the Weddell Sea.

Confined to a relatively narrow continental shelf and a short productive season, summer development of high latitude Antarctic ichthyoplankton follows a pronounced seasonal pattern, which is, however, not simply described by maximum abundance of all species at the time of maximum zooplankton production. The ichthyoplankton community consists of more than 23 species, of which no more than 16 co-occur in individual months. Temporal and spatial partitioning of the habitat by the different larval fish taxa takes place. Older post-larvae and juveniles of the notothenioids show a rather restricted neritic distribution, whereas post-larvae can be widely dispersed. Ontogenetic coastward migration of juveniles may be inferred from this distribution pattern.

Species with widely dispersed post-larvae such as *Pleuragramma antarcticum* and *Chionodraco myersi* are the most abundant species on the Weddell Sea shelf (Ekau et al. 1987), showing a possible relation between stock size and larval drift (*sensu* Cushing 1975) or dimension of the respective retention area (*sensu* Sinclair 1988). Early stages of some notothenioids and mesopelagic species occur in the same geographical area. They are, however, temporally separated by differences in their seasonal cycles. *Pleuragramma antarcticum* appears early in the season, whereas myctophids and paralepidids, and a few other notothenioids gain relative influence in the ichthyoplankton by late summer.

To understand these spatial and temporal interactions between ichthyoplankton species, especially for the future purpose of modelling ecosystem dynamics and monitoring of man-made changes in these systems, further detailed studies with high spatial and temporal resolution of the seasonal development of pelagic communities will be required.

Acknowledgements. Special thanks go to Adi Kellermann, Bremerhaven, who kindly identified most of the ichthyoplankton taxa in the present collection, and to Sigrid Schiel, who provided additional material from Multinet samples. The plankton sampling programme was undertaken from RV "Polarstern" in cooperation with W. Hagen, I. Hempel, E. Mizdalski, and S. Schiel. Thanks are due to all colleagues on the research vessel, in the Alfred Wegener Institute, Bremerhaven and in the Institute for Polar Ecology, Kiel, who contributed to the collection and processing of the material, and to Martin White, Cambridge, who provided helpful advice for the finalization of the manuscript.

References

Andriashev AP (1965) A general review of the Antarctic fish fauna. In: Mieghem J van, Oye P van (eds) Biogeography and ecology in Antarctica. Monogr Biol, vol XV. Junk Publ, The Hague, pp 491–550

Cushing DH (1975) Marine ecology and fisheries. Cambridge Univ Press, Cambridge, 278 pp

DeWitt HH (1970) The character of the midwater fish fauna of the Ross Sea, Antarctica. In: Holdgate MW (ed) Antarctic ecology, vol I. Academic Press, London, pp 305–314

Ekau W, Hubold G, Wöhrmann A (1987) Fish and fish larvae. In: Schnack-Schiel SB (ed) The winter expedition of RV "Polarstern" to the Antarctic (ANT V/1–3). Rep Pol Res 39:210–218

Elbrächter M, Gieskes WWC, Rabsch U, Scharek R, Schaumann K, Smetacek V, Veth C (1987) Phytoplankton and heterotrophic microorganisms in the water column. In: Schnack-Schiel SB (ed) The winter expedition of RV "Polarstern" to the Antarctic. Rep Pol Res 39:190–196

Fransz G (1988) Vernal abundance, structure and development epipelagic copepod populations of the eastern Weddell Sea (Antarctica). Polar Biol 9:107–114

Gieskes WWC, Veth C, Wöhrmann A, Gräfe M (1987) Secchi disc visibility world record shattered. EOS 68(9):123

Hart TJ (1942) Phytoplankton periodicity in Antarctic surface waters. Discovery Rep 21:261–356

Hempel G (1983) Die Expedition ANTARKTIS-I mit FS "Polarstern" 1982/83. Rep Pol Res 14:1–141

Hempel G (1985) Antarctic marine food webs. In: Siegfried WR, Condy PR, Laws RM (eds) Antarctic nutrient cycles and food webs. Proc 4th SCAR Symp Antarct Biol. Springer Berlin Heidelberg New York, pp 266–270

Hubold G (1984) Spatial distribution of *Pleuragramma antarcticum* (Pisces: Nototheniidae) near the Filchner and Larsen Ice Shelves (Weddell Sea/Antarctica). Polar Biol 3:231–236

Hubold G (1985) On the early life history of the high Antarctic silverfish *Pleuragramma antarcticum*. In: Siegfried WR, Condy PR, Laws RM (eds) Antarctic nutrient cycles and food webs. Proc 4th SCAR Symp Antarct Biol. Springer Berlin Heidelberg New York, pp 445–451

Hubold G, Ekau W (1990) Trophic interactions of post-larval and juvenile notothenioids in the Weddell Sea, Antarctica. Polar Biol (in press)

Johnson GL, Vanney JR, Elverhoi A, LaBreque JL (1981) Morphology of the Weddell Sea and southwest Indian Ocean. Dtsch Hydrogr Z 34(6):263–272

Keller R (1983) Contributions to the early life history of *Pleuragramma antarcticum* Boul. 1902 (Pisces, Nototheniidae) in the Weddell Sea. Meeresforschung 30:10–24

Kellermann A (1986) Zur Biologie der Jugendstadien der Notothenioidei (Pisces) an der Antarktischen Halbinsel. Ber Polarforsch 31:149 pp

Kellermann A, Kock KH (1984) Postlarval and juvenile notothenioids (Pisces, Perciformes) in the southern Scotia Sea and northern Weddell Sea during FIBEX 1981. Meeresforschung 30:82–93

North AW, White MG (1987) Reproductive strategies of Antarctic fish. Proc V Congr Eur Ichthyol, Stockholm 1985, pp 381–390

Sinclair M (1988) Marine populations. An essay on population regulation and speciation. Books in Recruitment Fishery Oceanography. Wash Sea Grant Prog, Univ Wash Press, Seattle, 252 pp

Slosarczyk W (1986) Attempts at a quantitative estimate by trawl sampling of distribution of post-larval and juvenile notothenioids (Pisces, perciformes) in relation to environmental conditions in the Antarctic Peninsula region during SIBEX 1983–84. Mem Natl Inst Polar Res Spec Iss 40:299–315

Slosarczyk W, Rembiszewski JM (1982) The occurrence of juvenile Notothenioidei (Pisces) within krill concentrations in the region of the Bransfield Strait and the southern Drake Passage. Pol Polar Res 3:299–312

Voronina NM (1970) Seasonal cycles of some common Antarctic copepod species. In: Holdgate MW (ed) Antarctic ecology, vol 1. Academic Press, London New York, pp 162–172

Weikert H, John H-Ch (1981) Experiences with a modified Be multiple opening-closing plankton net. J Plankton Res 3(2):167–176

White M, North A (1987) Postlarval Notothenioidei and midwater fish collected during the SIBEX cruise by British Antarctic Survey, 1985. Proc V Congr Eur Ichthyol, Stockholm 1985, pp 405–411

Williams R (1985) The potential impact of a krill fishery upon pelagic fish in the Prydz Bay area of Antarctica. Polar Biol 5:1–4

The Distribution of Zooplankton in Relation to the South Georgia Shelf in Summer and Winter

A. ATKINSON and J. M. PECK [1]

Summary. Zooplankton was sampled with RMT (1 + 8) nets on a grid of stations surrounding South Georgia in early summer, 1981/82 and in winter, 1983. This chapter compares the summer and winter distributions of zooplankton in the shelf and oceanic portions of the survey area. During both summer and winter, mesopelagic species and salps were rare over the shelf. Neritic species were present at both times of year. In winter, the species which undergo seasonal vertical migrations tended to occur in greater abundance over the shelf than in the oceanic surface layer. This was thought to be due to the shelf having prevented the descent of these animals in the preceding autumn. These distribution patterns all suggest a relatively slow interchange of plankton between the oceanic and shelf systems, which may cause relict distributions of species over the South Georgia shelf.

1 Introduction

Antarctic zooplankton species have frequently been studied in oceanic areas, but less is known of their life cycles and distributions in shelf regions. Island shelves are known to have a marked influence on oceanic current streams (Allanson et al. 1985; Deacon 1983) and comprise a large part of the foraging ranges of many of Antarctica's land-based predators. These areas therefore deserve a more detailed appraisal.

An intensive study of zooplankton in the South Georgia area was made by Hardy and Gunther (1935), but at that time there was insufficient knowledge of the ecology of individual species for their distributions to be fitted into a coherent pattern. The British Antarctic Survey sampled zooplankton around South Georgia in early summer 1981/82 and during winter 1983. The mesoscale survey grids covered shelf, slope and the oceanic areas around the island. Within the oceanic portions of these survey grids, Atkinson and Peck (1988) compared the species composition and vertical distribution of the zooplankton. The two surveys also provide a good opportunity to investigate the influence of the South Georgia shelf on the distribution of zooplankton in summer and winter. This is the aim of the present work.

[1] British Antarctic Survey, High Cross, Madingley Road, Cambridge, CB3 0ET, UK.

2 Materials and Methods

The summer and winter surveys around South Georgia involved sampling at stations along a series of parallel transects. The stations were at 30 nautical mile intervals and formed a rectangular grid around the island (Fig. 1). The early summer survey (24 November to 19 December 1981) sampled at 63 stations, whereas the winter survey (27 July–20 August 1983) was curtailed by a medical emergency and only 42 of the summer stations were sampled. Most of these stations were also displaced slightly from their summer positions for ease of navigation and logistical reasons.

At each station, zooplankton was sampled using a Rectangular Midwater Trawl (RMT 1 + 8; see Baker et al. 1973). The RMT 1 net (mesh size 330 μm, nominal mouth area 1 m^2) is attached to the top of the larger RMT 8 net (codend mesh size 4.5 mm, nominal mouth area 8 m^2). The nets were generally towed obliquely downwards at a speed of 2 knots, and were opened and closed acoustically. The comparable depth horizons sampled during the two surveys are shown in Fig. 1. Where several depth horizons were fished (Fig. 1), a multiple RMT system (Roe and Shale 1979) was used. At some of the stations sited over the shelf, the water depth was insufficient to allow sampling to 250 m. In these cases the nets were fished to within 10 and 30 m of the seabed. The mean depth reached by the shelf hauls was 177 m. At eight of the shelf stations during the winter survey, the second net pair was opened as the first was closed at its maximum depth. These nets were then fished obliquely upwards through the water column. Thus two consecutive net samples were obtained from the same depth horizon. Six of the summer survey RMT 1 samples and three of the winter RMT 1 samples could not be analysed be-

Antarctic Ecosystems. Ecological Change and Conservation
(ed. by K. R. Kerry and G. Hempel)
© Springer-Verlag Berlin Heidelberg 1990

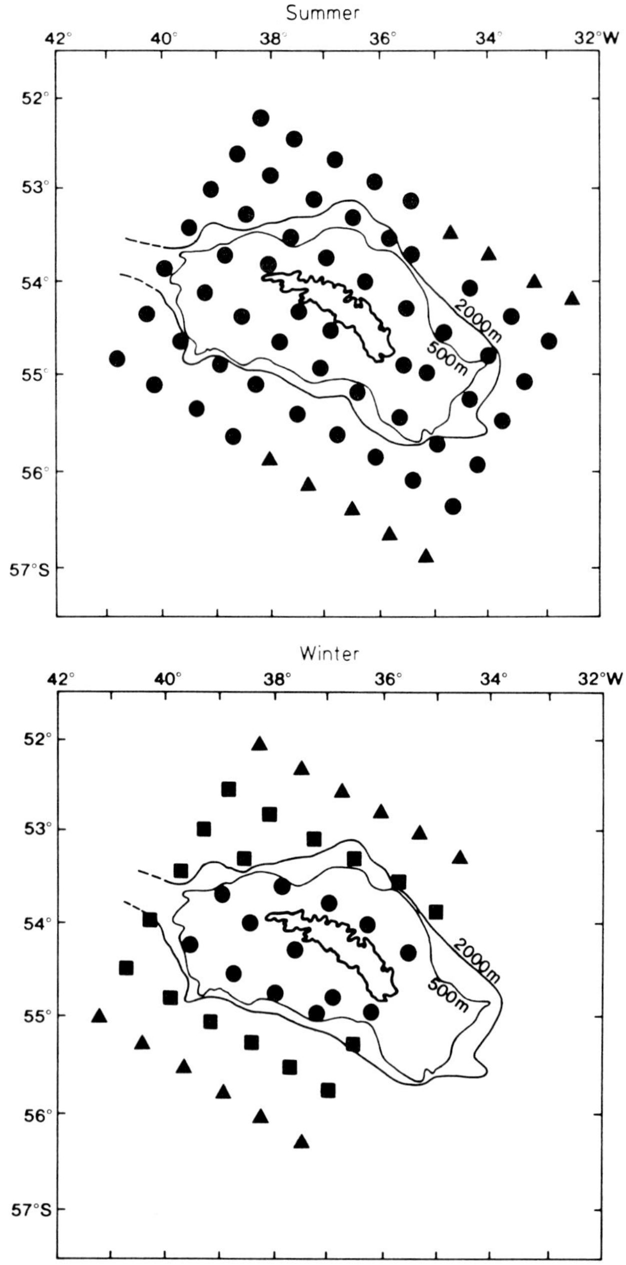

Fig. 1. The stations occupied during the summer and winter South Georgia surveys. Comparable depth horizons sampled during the two surveys are shown. The shelf break and oceanic areas are indicated by the 500 and 2000 m isobaths. ●, Shelf haul (see text) or 10–250 m oceanic haul; ■, 10–250 plus 250–500 m; ▲, 10–250, 250–500, 500–1000 m

cause of sampling problems. These included clogging of the nets with phytoplankton or krill, premature closing of the net, and loss or spillage of the collected sample. The duration of the shelf hauls, the 10–250 m hauls and the 250–500 m haul was generally 30 min, and that of the 500–1000 m hauls was approximately 60 min. A flowmeter was mounted on the net monitor, which enabled volume filtered to be estimated from equations of Pommeranz et al. (1983). Filtration was assumed to be 100% efficient. Net sampling took place as soon as the ship came onto station, irrespective of time of day. The longer day length in summer resulted in a larger proportion (75%) of summer net samples being taken in daylight, compared with 34% in winter. At all stations, salinity and temperature profiles were obtained using a Plessey 9041 STD-SV telemetering unit and data logger. This instrument was calibrated using reversing thermometers and a "Guildline Autosal" salinometer. Water samples were collected on each profile using a rosette of Niskin bottles attached to the STD unit.

The RMT 8 samples were sorted and processed on board, but all RMT 1 samples were fixed and then preserved in 4% buffered formalin for laboratory analysis. This analysis, described by Atkinson and Peck (1988), was carried out in a conventional manner, and involved firstly the removal of macrozooplankton from the whole samples. Smaller and rarer species or developmental stages were then sorted from aliquots obtained using a Folsom plankton splitter. The calculated abundance of each taxon or species (per 1000 m^3) refers to all identifiable maturity stages encountered, with the exception of the euphausiids, whose developmental stages are not treated here.

3 Results

Before regional comparisons were made, the RMT 1 and RMT 8 nets were compared by analysing the distribution patterns of species caught in both nets. Their distributions tended to be very similar but abundance and biomass estimates were invariably higher from the RMT 1 catches. The finer mesh RMT 1 nets also caught a greater number of species. The following results therefore refer only to catches from the RMT 1 nets, with the exception of *Euphausia superba* and *E. triacantha*. RMT 8 data were used for these two large and mobile species.

3.1 Distribution of Zooplankton Biomass

The displacement volumes of zooplankton (ml 1000 m^{-3}) caught in the RMT 1 nets over the shelf and in the top 250 m of the oceanic layer are shown in Fig. 2. Within this layer, zooplankton biomass was generally higher in summer than in winter. During the summer survey, the distribution of biomass appeared to be rather more patchy and the stations with high biomass (mainly over the E shelf and NE area) were attributable to large catches of krill (*Euphausia super-*

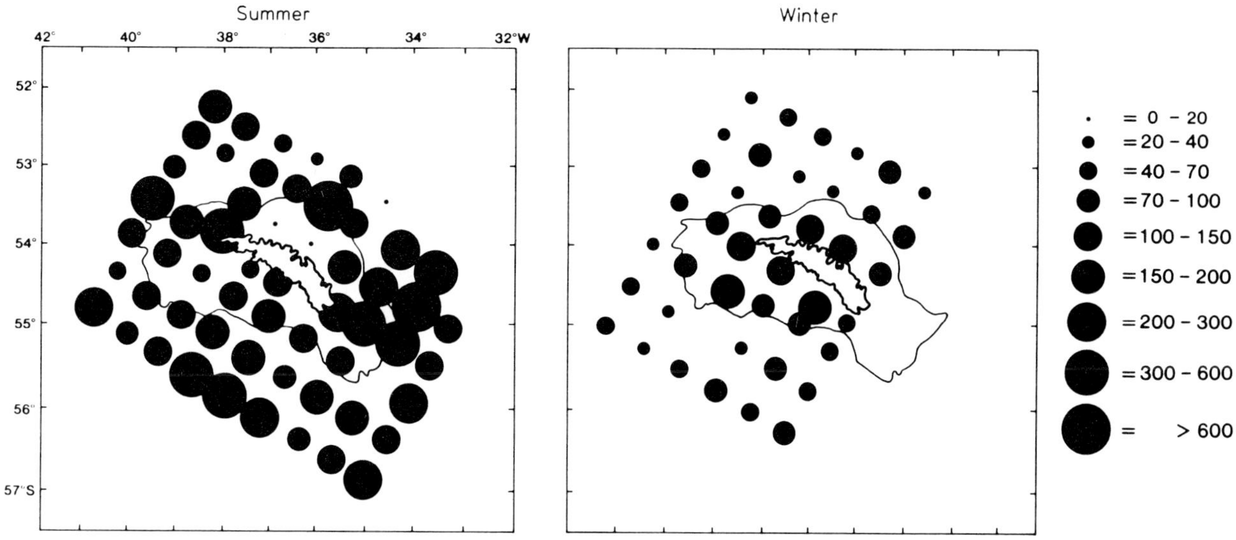

Fig. 2. Total zooplankton biomass (ml 1000 m^{-3}) from the RMT 1 net in shelf and oceanic 10–250 m hauls. The 500-m isobath marks the position of the shelf break. At the eight winter shelf stations where a second net was fished (see Sect. 2) mean values (ml 1000 m^{-3}) are presented

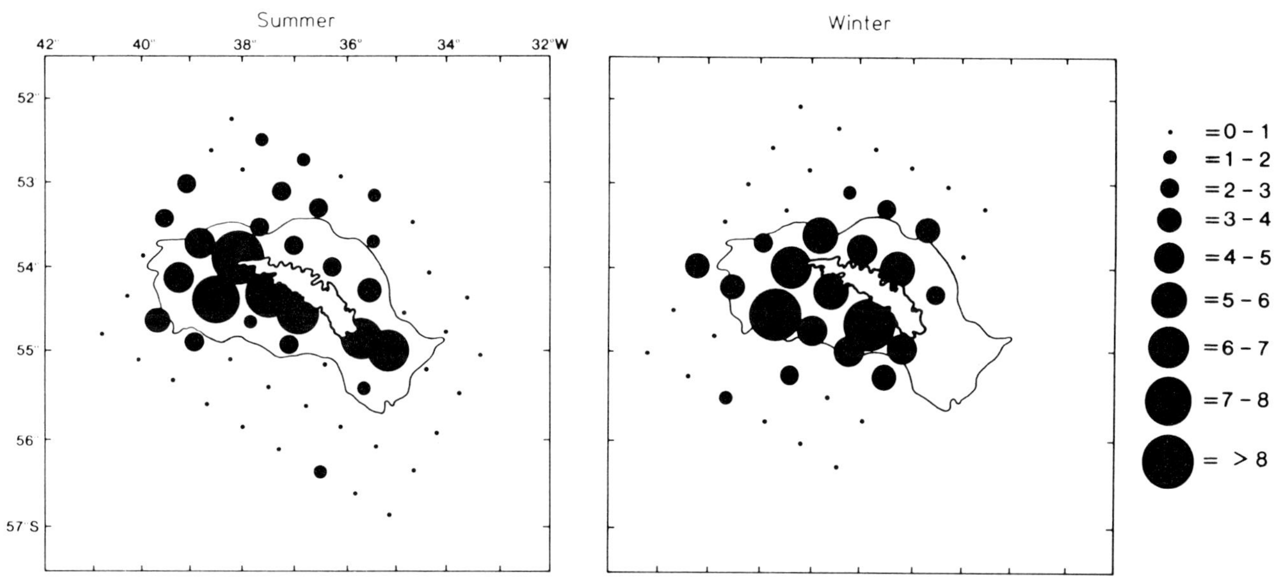

Fig. 3. Distribution of the three neritic species throughout the shelf and oceanic 10–250 m layer. Abundances are expressed as the sum of the logged (abundance + 1) of *Drepanopus forcipatus*, *Antarctomysis maxima* and *Notocrangon* spp.

ba). Although zooplankton biomass was low at several of the summer shelf stations, there was, in general, no noticeable onshelf/offshelf difference in biomass in summer. During the winter survey, however, biomass within the surface layers tended to be highest over the shelf (Fig. 2).

Regional comparisons of biomass within the deeper oceanic layers is not possible because stratified sampling was only undertaken at nine of the summer stations (Fig. 1). However, mean profiles of zooplankton biomass during the two surveys are presented in Atkinson and Peck (1988). These show that most of the summer biomass was concentrated in the top 250 m layer, whereas in winter it was found mainly below 250 m.

3.2 Distribution of Neritic Species

Three neritic species were encountered during both surveys and their pooled distributions within the surface layer are shown in Fig. 3. The commonest was

Drepanopus forcipatus, a holoplanktonic copepod (family Clausocalanidae) which was patchy in occurrence but extremely abundant at several shelf stations. The maximum recorded abundance of this species was 1060 individuals m^{-3}. This species was more widespread over the shelf and strayed further into oceanic waters than the other two neritic species. At several winter survey stations to the SW of the shelf, *Drepanopus forcipatus* occurred in the 250–500 m layer as well as in the surface layer.

Antarctomysis maxima is a benthopelagic mysid which was restricted to the broader southern shelf of the island during the summer survey. *Notocrangon* spp. is a meroplanktonic decapod, whose larvae occurred in low abundance at several shelf stations.

3.3 Comparison of Zooplankton Abundance in Shelf and Oceanic Areas

Distribution maps were drawn for all of the other species encountered during the two surveys. Analysis of these indicated that many of the species tended either to be concentrated over the shelf or excluded from shelf waters. These onshelf/offshelf differences were compared for each species by calculating the ratio of mean abundance over the shelf to that in the sampled water column of the surrounding ocean. Information on smaller scale variability is inevitably lost when abundances are averaged over large areas, but this method reduces experimental error and is the most reliable in highlighting regional differences.

The onshelf/offshelf abundance ratios of the species ranged widely, but tended to be related to their vertical distribution within the oceanic layer. The more abundant species were therefore categorised into "seasonal migrants", "epipelagic, non-seasonal migrants" and "mesopelagic species". These groupings of species were made by Atkinson and Peck (1988) from their oceanic vertical distributions during the two surveys. The seasonal migrants inhabited the surface layer (10–250 m) during the summer survey but were found mainly in the deeper strata in winter. The epipelagic, non-seasonal migrants also inhabited the surface layers in summer but showed little evidence for a winter descent. Mesopelagic species occurred predominantly below 250 m at both times of year, but the limitations of sampling made it difficult to deduce whether they made seasonal vertical migrations. Table 1 presents the median ratio of shelf to oceanic abundance for each group of species.

In neither the shelf area nor the oceanic area was the entire water column sampled. Nevertheless, Table 1 indicates that in both summer and winter mesopelagic species were greatly depleted in shelf waters. Although the seasonal migrants and the epipelagic, non-seasonal migrants also generally occurred in decreased abundance in the sampled water column over the shelf, they were not depleted over the shelf to nearly the same extent as the mesopelagic inhabitants.

Table 1. Ratios of mean abundance (under 1 m^2 of the sampled water column) in the shelf area to that in the oceanic area. The median ratios of three groups of species are presented, with their interquartile ranges in parentheses. These species groups were constructed according to the vertical distributions of the species within the oceanic water column (see text). Members of each group are listed. RMT 1 data is used for the majority of species, but RMT 8 data is used for larger and scarcer species (indicated by "8")

Species group	Median ratio of abundance (Interquartile ranges are in parentheses)	
	Summer	Winter
Seasonal migrants	0.497 (0.420–0.695)	0.360 (0.120–0.684)
Epipelagic, non-seasonal migrants	0.491 (0.173–2.41)	0.467 (0.301–1.80)
Mesopelagic species	0.0316 (0.008–0.182)	0.0319 (0–0.334)

Seasonal migrants	Epipelagic, non-seasonal migrants	Mesopelagic species
Calanus simillimus	*Clausocalanus laticeps*	*Metridia lucens*
Calanoides acutus	*Euaetideus australis*	*M. gerlachei*
Rhincalanus gigas	*Racovitzanus antarcticus*	*M. curticauda*
Eucalanus longiceps	*Haloptilus oxycephalus*	*Pleuromamma robusta*
Ctenocalanus spp.	*Vanadis antarctica*	*Gaidius antarcticus*
Euchaeta antarctica	Ostracoda	*Lucicutia ovalis*
Chaetognatha	*Themisto gaudichaudii*	*Heterorhabdus* spp.
Salpa spp.	*Euphausia superba* (8)	*Oncaea antarctica*
	Euphausia frigida	*Pelagobia longicirrata*
		Gymnosomata
		Euphausia triacantha (8)

A further trend was apparent when the mean abundance of each species was compared between the shelf and the surface layer (10–250 m) of the surrounding ocean. Table 2 presents the median ratios for the three groups of species. The range of ratios within each group is large, but it is evident that in winter especially the seasonal migrants were relatively more abundant over the shelf than the epipelagic, non-seasonal migrants or the mesopelagic species.

Table 2. Ratios of mean abundance (No. 1000 m^{-3}) over the shelf to that in oceanic 10–250 m layer. The median ratios of the three groups of species are presented. Interquartile ranges are in parentheses

Species group	Median ratio of abundance in shelf to that in the oceanic 10–250 m layer	
	Summer	Winter
Seasonal migrants	0.750 (0.583–1.39)	2.28 (0.458–15.1)
Epipelagic, non-seasonal migrants	0.521 (0.127–1.03)	0.95 (0.558–3.24)
Mesopelagic species	0.241 (0.027–0.641)	0.728 (0.0556–1.43)

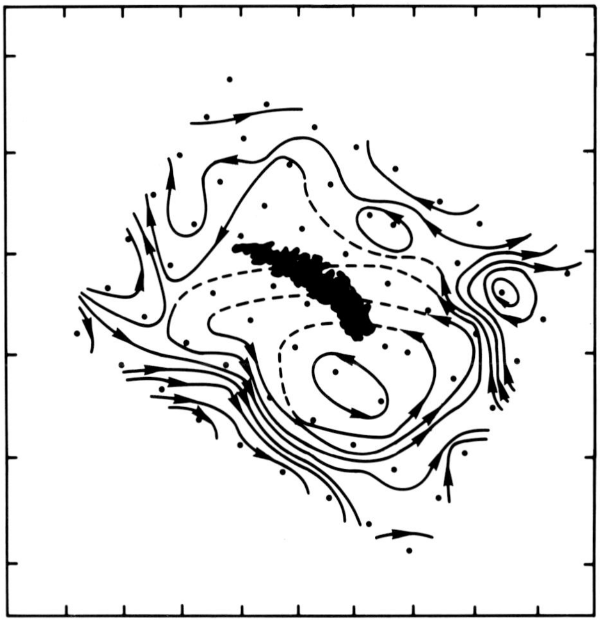

Fig. 4. Geopotential topography of the summer survey grid (Priddle et al. 1986). The isolines, which indicate direction of current flow, were plotted from the 1000 d bar reference level. The closeness of their spacing is a measure of relative current velocity. A similar geostrophic current pattern was obtained for the winter survey (R. B. Heywood unpubl. data)

4 Discussion

The presence and distribution of neritic species give some indication of water movement around the island. The occurrence of these species to the edge of the shelf would indicate a relatively slow rate of exchange of water between the Antarctic Circumpolar Current and the shelf. Transport within the Antarctic Circumpolar Current can be relatively fast, with mean surface velocities of 28 and 40 cm s^{-1} respectively within the Antarctic Zone and Polar Front in the Atlantic sector (Hofmann 1985). It would appear that this surface current stream must have been partially deflected around the edge of the shelf rather than flow directly over it.

Hulsemann (1985) concluded that the population of *Drepanopus forcipatus* around South Georgia was a self-maintaining one, disjunct from the main population on the continental shelf of South America and the Falkland Islands. Occurrences of small numbers of this species within the oceanic area around the island are therefore likely to have resulted from currents having swept the animals off the shelf. Studies of macroscale circulation in the Atlantic Sector have shown that the Antarctic Circumpolar Current approaches South Georgia from the SW and is deflected eastward and then northwards by the South Georgia shelf (Gordon and Goldberg 1970). This sets up an anticlockwise circulation pattern around the island, which was confirmed during the two surveys by plots of geopotential topography (Fig. 4) and by the orientation of the surface isotherms (Atkinson 1989; Heywood et al. 1985). From this pattern, one would expect that neritic species would tend to be swept downstream (i.e. to the north) of the shelf. This is indeed noticeable during the summer survey (Fig. 3). However, at both times of year neritic species were present in small numbers to the SW of the shelf, apparently upstream of the island. It is possible that these animals had been swept off the shelf by currents in the western arm of the gyre to the SE of the island (Fig. 4). However, strong winds are frequent at these latitudes, and wind-induced surface currents are also likely to be important transport mechanisms. The strengths and directions of these variable currents have not yet been assessed.

These results are in contrast to the situation found by Boden and Parker (1986) around the Prince Edward Islands (47° S, 38° E). Despite an eddy system between these islands there was no evidence for the existence of a true neritic community. The South Georgia shelf is much larger and more extensive than that surrounding the Prince Edward and several other of the smaller sub-Antarctic islands, and consequently the neritic species appear to be better established.

The distribution of oceanic species relative to the South Georgia shelf provides further evidence for a low rate of water exchange across the shelf break. Hopkins et al. (1981) showed that the penetration of oceanic species onto the Florida shelf was closely related to their vertical distribution in the open ocean. Epipelagic species extended further over the shelf than deeper living species. We also found epipelagic species to be relatively more abundant over the shelf than mesopelagic inhabitants (Tables 1, 2). Clearly the limited water depth over many parts of the shelf is unsuited to mesopelagic species.

This situation is rather more complicated in Antarctic waters because many summer inhabitants of the surface layer overwinter at depth. Within the oceanic area, these seasonal migrants occurred at similar winter depths to the permanent inhabitants of the mesopelagic zone (Atkinson and Peck 1988). However, in winter they tended to occur in much greater abundance over the shelf than the mesopelagic inhabitants (Tables 1, 2).

The most likely explanation for this is that in the preceding autumn the seabed impeded the seasonal vertical migration of individuals situated over the shelf. Thus, they remained in the shallow water over the shelf throughout the winter, whilst individuals in the oceanic region overwintered mainly below 250 m and were rare in the surface layers. This explanation is consistent with a comparatively slow rate of water exchange between the oceanic surface water and the strata inhabited by those species on the shelf. If there was a greater rate of flushing, these overwintering seasonal migrants would have been swept off the shelf and replaced by oceanic surface waters containing few of these species.

The numerous physical and biotic factors dictating the distribution of a species relative to the South Georgia shelf act in concert, and it may be expected that there are several exceptions to a simple classification of species into neritic/oceanic and seasonal/non-seasonal migrants. *Salpa* spp., for example, is a seasonal migrant which was conspicuous by its near absence over the shelf in both summer and winter. It is possible that diurnal vertical migration influenced the distribution of this species. From the summer and winter data sets, *Salpa* spp. (as well as two other mesopelagic metridiid copepods) showed statistical evidence for a nighttime migration out of the top 250 m layer during both surveys (A. Atkinson, unpubl. data). Hardy and Gunther (1935) and Piatkowski (1985) also found that salps underwent diurnal vertical migrations, and noted their scarcity in shelf areas. Possibly the limited water depth over the shelf is unsuitable for salps because their diurnal migratory behaviour is restricted.

Onshelf/offshelf differences in mean abundance can also result partly from reproductive activity. The population of *Euchaeta* spp. for example, was dominated by large numbers of early copepodids over the shelf in winter (Ward and Wood 1988) and Ward and Robins (1987) noted a winter spawning period for *Euchaeta antarctica* in fjords around South Georgia. Also, spawning of *Calanoides acutus* and *Rhincalanus gigas* was found to have been delayed over the shelf during the summer survey (Atkinson 1989). This resulted in low abundance of their summer generations over the shelf compared to the oceanic surface layer.

Euphausia superba has a potential life span of 4–7 years (e.g. Ikeda and Thomas 1987) and reproductive behaviour probably does not account for onshelf/offshelf differences in abundance of this species. During the summer survey, krill were found in abundance in both the shelf and the oceanic regions. However, in winter, they were practically absent from the oceanic area, and were only encountered at several inner shelf stations. Priddle et al. (1988) and Atkinson and Peck (1988) hypothesised that the disappearance of this Antarctic species from the oceanic region in the 1983 winter was due to the southward extension of the Polar Front which resulted in a greater influence of sub-Antarctic water. Possibly, individuals were present over the shelf earlier in the winter survey season, but the movement of the Polar Front then supplied South Georgia with water containing very few krill. The limited interchange of shelf and oceanic water then enabled this micronektonic species to maintain its distribution in pockets over the shelf. Boden and Parker (1986) suggested that the co-occurrence of warm and cold water species around the Prince Edward Islands may have been due to a complex circulation pattern which included transport of core rings. The presence of *Euphausia superba* on the South Georgia shelf in winter could therefore support Boden and Parker's (1986) hypothesis that "alien" species may be relicts of previous hydrographic regimes.

In conclusion, the major factors which influence zooplankton abundance over the South Georgia shelf are the depth range of each species and the extent of their seasonal vertical migration. Other interrelated and contributing factors include shelf modifications to reproductive cycles and possibly diurnal vertical migration. The circulation pattern maintains neritic populations over the shelf and may have also given rise to a relict winter distribution of krill around the island.

Acknowledgements. We would like to thank those aboard RRS "John Biscoe" who helped to collect and process the RMT samples. Peter Ward was involved in sorting and curation of the material and made helpful comments on an early draft of this manuscript. We are also grateful to Alistair Murray who gave statistical advice, and to Julie Thomson who typed the manuscript.

References

Allanson BR, Boden P, Parker L, Rae CD (1985) A contribution to the oceanology of the Prince Edward Islands. In: Siegfried WR, Condy PR, Laws RM (eds) Antarctic nutrient cycles and food webs. Proc 4th SCAR Symp Antarct Biol. Springer, Berlin Heidelberg New York, pp 35–40

Atkinson A, (1989) Distribution of six major copepod species around South Georgia in early summer. Polar Biol (9:353–363)

Atkinson A, Peck JM (1988) A summer-winter comparison of zooplankton in the oceanic area around South Georgia. Polar Biol 8:463–473

Baker A de C, Clarke MR, Harris MJ (1973) The NIO combination net (RMT 1 + 8) and further developments of rectangular midwater trawls. J Mar Biol Assoc UK 53:167–184

Boden BP, Parker LD (1986) The plankton of the Prince Edward Islands. Polar Biol 5:81–93

Deacon GR (1983) Kerguelen, Antarctic and subantarctic. Deep-Sea Res 30:77–81

Gordon AL, Goldberg RD (1970) Circumpolar characteristics of Antarctic waters. Antarctic Map Folio Ser 13:10 pp

Hardy AC, Gunther ER (1935) The plankton of the South Georgia whaling grounds and adjacent waters, 1926–1927. Discovery Rep 11:1–465

Heywood RB, Everson I, Priddle J (1985) The absence of krill from the South Georgia zone, winter 1983. Deep-Sea Res 32:369–378

Hofmann EE (1985) The large-scale horizontal structure of the Antarctic Circumpolar current from FGGE drifters. J Geophys Res 90:7177–7187

Hopkins TL, Milliken DM, Bell LM, McMichael EJ, Hefferman JJ, Cano RV (1981) The landward distribution of oceanic plankton and micronekton over the west Florida continental shelf as related to their vertical distributions. J Plankton Res 3:645–658

Hulsemann K (1985) Two species of *Drepanopus* Brady (Copepoda, Calanoida) with discrete ranges in the Southern Hemisphere. J Plankton Res 7:909–925

Ikeda T, Thomas PG (1987) Longevity of Antarctic krill (*Euphausia superba* Dana) based on a laboratory experiment. Proc NIPR Symp Polar Biol 1:56–62

Piatkowski U (1985) Distribution, abundance and diurnal migration of macrozooplankton in Antarctic surface waters. Meeresforschung 30:264–279

Pommeranz T, Herrmann C, Kuhn C (1983) Mouth angles of the Rectangular Midwater Trawl (RMT 1 + 8) during paying out and hauling. Meeresforschung 29:267–274

Priddle J, Heywood RB, Iheriot E (1986) Some environmental factors influencing phytoplankton in the Southern Ocean around South Georgia. Polar Biol 5:65–79

Priddle J, Croxall JP, Everson I, Heywood RB, Murphy EJ, Prince PA, Sear CB (1988) Large-scale fluctuations in distribution and abundance of krill – a discussion of possible causes. In: Sahrhage D (ed) Antarctic ocean and resources variability. Springer-Verlag, Berlin, pp 169–182

Roe HSJ, Shale DM (1979) A new multiple rectangular midwater trawl (RMT 1 + 8) and some modifications to the Institute of Oceanographic Science's RMT 1 + 8. Mar Biol 50:283–288

Ward P, Robins DB (1987) The reproductive biology of *Euchaeta antarctica* Giesbrecht (Copepoda:Calanoida) at South Georgia. J Exp Mar Biol Ecol 108:127–145

Ward P, Wood AG (1988) The distribution of the Euchaetidae (Copepoda:Calanoida) around South Georgia. Polar Biol 9:45–52

Ecological and Population Changes in Sea Birds and Mammals

Ross Sea Emperor Penguin Breeding Populations Estimated by Aerial Photography

G. L. KOOYMAN[1] and J. L. MULLINS[2]

Summary. In December 1983 aerial photographs of all emperor penguin breeding colonies along the Ross Sea coast from Cape Roget south to Beaufort Island were taken for the purpose of censusing this species. Cape Crozier, the most southerly colony, was counted from the ground. Chicks and adults could not be distinguished from the photographs. They were differentiated by using chick:adult ratios of 7 chicks to 1 adult (± 2.14 standard deviation, n = 11) obtained from ground counts on 12 to 14 December 1986 at Cape Washington. Of the six colonies, the smallest was Cape Crozier (78 chicks) and the largest was Coulman Island (estimated 22137 chicks). Except for Cape Crozier and Beaufort Island, for which the most recent published censuses were in 1977, the other colonies had not been counted since their first and last census in 1968 or earlier. There are differences between some of the earlier and the 1983 censuses. The estimated total of breeding birds in 1983 was about 120000. A common characteristic of all colonies within the Ross Sea is their proximity to open water.

1 Introduction

Since the discovery of the first emperor penguin colony at Cape Crozier in 1902, there have been sporadic searches for, and cursory counts of other emperor penguin breeding sites in the Ross Sea. With the exception of Cape Crozier, all other colonies were discovered between 1956 and 1965 (Table 1). There are now six known breeding sites within the Ross Sea (Fig. 1), excepting a reportedly large colony at Cape Colbeck (Harrington 1959).

The most closely followed colony has been Cape Crozier. The size of this colony has fluctuated greatly, and since the turn of the century it has nearly disappeared at least twice; in 1911 and again in 1976 when only 100 and 127 adults were present (Table 2). Whether other colonies have fluctuated greatly over the years is unknown.

To assess the variability in these colonies and their usefulness as environmental monitors, a census every year would be ideal, but, considering logistic constraints, one every 3 to 5 years may be adequate if the

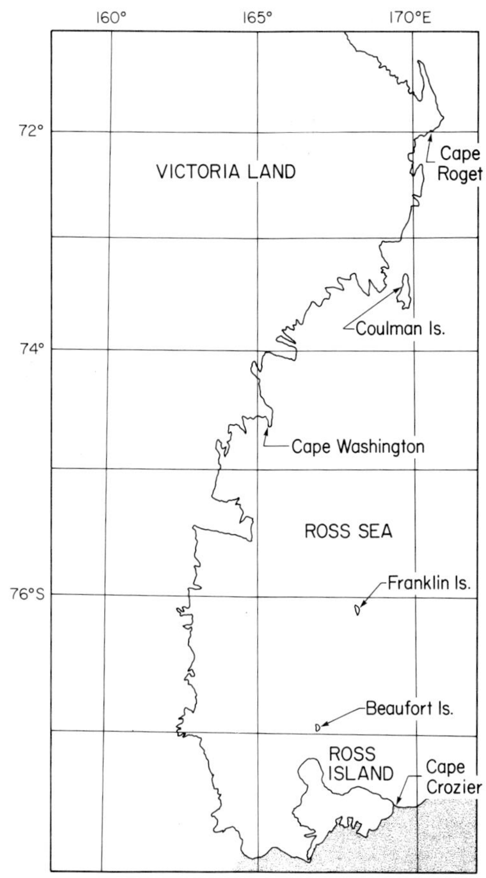

Fig. 1. Distribution map of the known Ross Sea emperor penguin colonies. Note that there are two major year-round research stations on Ross Island (United States and New Zealand) and another summer station about 50 km south of Cape Washington (Italy). There are also a small year-round station on Ross Island (Greenpeace) and at present an inactive station 40 km south of Cape Washington (West Germany). The *stippled area* south of Ross Island is the Ross Ice Shelf

[1] Physiological Research Laboratory, Scripps Institution of Oceanography, La Jolla, CA 92093, USA.
[2] Geological Survey, U.S. Department of the Interior, Reston, VA 22092, USA.

Antarctic Ecosystems. Ecological Change and Conservation (ed. by K. R. Kerry and G. Hempel)
© Springer-Verlag Berlin Heidelberg 1990

Table 1. Censuses prior to 1983 of all known emperor penguin colonies in the western Ross Sea except for Cape crozier

Locality	Count					
	Chicks	Adults	Ratio	Census date	Discovery date	Source
Cape Roget 170° 31′ E; 71° 59′ S	11 700	10 000	1.2	9 Nov 1964	1964	Cranfield (1966)
Coulman Is. 169° 38′ E; 73° 20′ S	21 000	20 000	1.1	7 Nov 1964	1958	Cranfield (1966)
Cape Washington 165° 22′ E; 74° 39′ S	4–6 000	4–6 000	1.0	29 Oct 1965	1965	Cranfield (1966)
	7 600[a]			14 Nov 1968		Stirling and Greenwood (1970)
Franklin Is. 168° 15′ E; 76° 07′ S	1 700	2 000	0.9	19 Nov 1964	1964	Stonehouse (1969)
Beaufort Is. 167° 03′ E; 76° 56′ S	1 069	707	1.5	6 Dec 1962	1956	Stonehouse (1966)
	572	1 000	0.6	2 Nov 1976		Todd (1980)[b]
	764	737	1.0	25 Nov 1977		Todd (1980)

[a] Undifferentiated. [b] Also reported 1115 carcasses.

colonies do not exhibit the wide population variations of Cape Crozier. It is timely to begin these censuses since levels of human activity in the Ross Sea are increasing with the continued establishment of new bases, free-lance expeditionary activity, and the potential for commercial exploitation by tourist, minerals, and fishing industries. With this in mind, the purpose of this chapter is to report on: (1) a census obtained from all colonies in a single year and at the same time in the breeding cycle (Dec 1983); (2) the results of a simultaneous aerial and ground count; (3) the variation in chick/adult ratios; and (4) to note the effect of aerial overflights as observed from the ground in order to recommend a safe census flight altitude.

2 Methods

All colonies, except Cape Crozier, were photographed on 9 and 14 December 1983 from an LC-130 aircraft at an altitude of 500 m. A Swiss RC-8, large-format aerial mapping camera with a 152-mm lens was used to take vertical, black and white photographs (Fig. 2). On 6 November 1986, aerial photographs of the Cape Washington colony were taken with similar equipment, but at an altitude of 1500 m. On 13 November 1987 oblique aerial photographs were taken from a twin otter aircraft at 500 m with a Hasselblad C camera and 80-mm Planar f 2.8 lens. High oblique aerial photographs of the Cape Crozier colony were taken on 16 October 1986 from a UHIN helicopter at an altitude of 1000 m with a Hasselblad C camera and 150-mm Sonnar f 3.5 mm lens. All photographs were taken during clear weather when both individual birds and their shadows were easily resolved on the photographs.

On 6 November 1986, aerial photographs and a ground count were obtained of one group of approximately 1000 birds at the Cape Washington colony in order to compare the two census techniques. The ground count was done by three independent counters, stationed at widely separated locations around the perimeter of the group. The ground counts were averaged and the mean compared with the counts from aerial photographs of the same group. Ground counts of chicks of the entire Cape Washington colony

Table 2. Chronological census of the Cape Crozier (169° 25′ E; 77° 30′ S) emperor penguin colony (ground counts) (After Todd 1980; data from 1980 to 1983, D. Ainley pers. commun.; datum of 1986 from aerial photographs, G. L. Kooyman)

Date	Count		Ratio (chicks/adults)
	Chicks	Adults	
Oct 1902	30	400	0.075
Sep 1903	150	1000	0.15
Nov 1903	30	400	0.075
Jul 1911	25– 50	100	0.25–0.5
Sep 1957	100–270	700–800	0.143–0.338
Nov 1958	523	618	0.846
Oct 1961	1325	2000 +	0.633
Dec 1962	1250	1000–1200	1.25–1.042
Nov 1963	885	1050–1300	0.843–0.681
Oct 1967	0	500 +	–
Oct 1968	651	1000	0.651
Oct 1969	680	1300	0.523
Nov 1974	249	600 +	0.415
Oct 1975	94	274	0.343
Oct 1976	46	127	0.362
Nov 1977	40	204	0.196
		Average	0.15
8 Dec 1980	11	?[a]	–
15 Dec 1981	65	?[a]	–
19 Dec 1982	109	1[a]	–
16 Dec 1983	78	15[a]	5.2
16 Oct 1986		477[b]	–

[a] D. A. Ainley (pers. commun.).
[b] Adults, aerial census this report; chicks would not be visible this time of year.

Fig. 2. An example of the aerial photographs used to count the emperor penguins

were made on 29 November 1986. The colony was naturally divided into about 15 separate groups, each of which was counted independently by two people. If they disagreed by $>10\%$, the count was repeated.

Chick/adult ratios of the Cape Washington colony were determined at various times throughout the season by one of four investigators. Critical to this report, or any report on counts where adults and chicks cannot be distinguished, were the ratios determined between 12 to 14 December. The average of 11 counts was 7.0 ± 2.14 standard deviation chicks/adults and this factor was used to calculate the number of chicks in all colonies for which aerial photographs were obtained.

3 Results and Discussion

Few census reports of emperor penguin colonies have been published. The last reports for the Ross Sea colonies were from Beaufort Island and Cape Crozier in 1977 (Todd 1980). Since the 1960s none of the larger colonies has been counted, and the colonies at Cape Roget and Coulman Island have been counted only once (Table 1), nearly 20 years prior to our survey in 1983. These rather dated censuses formed the basis for assessing the Ross Sea populations in reports as recent as 1984 (Wilson and Taylor 1984; Harper et al. 1984).

The only colony in the Ross Sea for which serial counts have been made on a long-term basis is Cape Crozier. The longest gap since the first count in 1902 was 46 years between 1911 and the beginning of the modern era of Antarctic research in 1957. The last published count was in 1977 (Todd 1980), but more recent censuses are reported here (D. Ainley, pers. commun.; see Table 2). It is unfortunate that this emphasis has occurred at Cape Crozier because, as mentioned by Todd (1980), and as indicated by our comparative observations at Cape Washington, the Cape Crozier colony is atypical. The 26 adults found killed by avalanches from the nearby cliffs at Cape Crozier in 1976 when compared to the 6 dead adults at Cape Washington in 1986 (Kooyman pers. observ.) or the 4 per year from 1952–1970 in a colony of 12 000 to 6000 birds near Dumont d'Urville (Jouventin 1975) indicates an unusually high adult mortality (Todd 1980). Furthermore, the chick/adult ratio in October and November (Table 2) as compared to Cape Washington is skewed greatly toward adults. The average ratio at Cape Crozier is 0.15 compared to a ratio of 3 for November at Cape Washington. At Cape Washington, the ratio increases in December (Fig. 3). The variation in the ratio in late December at Cape Washington is due largely to the egress of chicks each day as fledging progresses, and the small number of adults for which a small number of birds arriving or departing cause the ratio to rise or decline greatly.

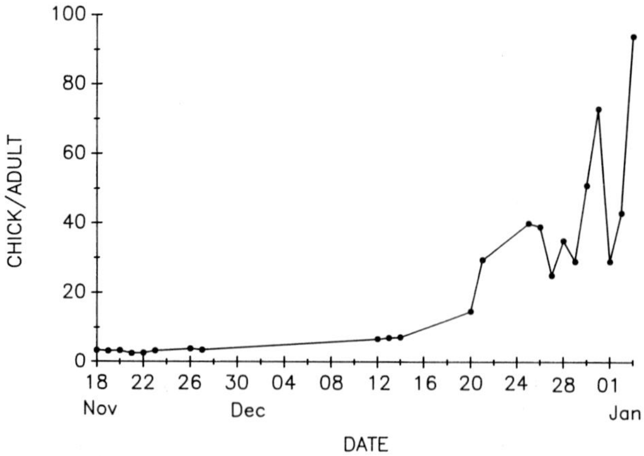

Fig. 3. The changing ratio of chicks to adults at Cape Washington in 1986. After 20 December the ratio fluctuates greatly because of the egress of large numbers of chicks as they fledge, the large ratio of chicks to adults, and the uneven arrival and departure of adults

Table 3. Aerial censuses on 9 and 14 Dec 1983, for all colonies except Cape Crozier in which a ground count was made on 16 Dec 1983. No distinction can be made between chicks and adults from the aerial photographs

Locality	Count	Chick Estimate 7 chicks/ 1 adult = 8 total Factor = 0.875[a]
Cape Roget	4402	3851
Coulman Is.	25 299	22 137
Cape Washington	18 725	16 384 19 364[b]
Franklin Is.	5702	4989
Beaufort Is.	204	179
Cape Crozier[c]	78	78
Note: 1983 Chick total		50 598[d]
~ 15% Mortality	(*Chick count*) 0.85	59 527
~ 1% Egg loss		595
Grand total		60 122
Breeding adults		120 244

[a] Based on ratios obtained on 12 to 14 Dec 1986.
[b] Ground count of chicks on 29 Nov 1986. In addition, 1850 dead chicks were counted.
[c] Cape Crozier from the 1983 ground count.
[d] 1986 count for Cape Washington used in sum.

We presume that the chick/adult ratio at Cape Washington is more typical of a healthy, stable colony than that at Cape Crozier for the following reason. The Cape Washington colony is fragmented into numerous groups of 1000 to 2000 chicks. In three groups where the chick/adult ratio was less than one, many of the adults behaved like inexperienced breeders, that is: (1) there was a higher chick mortality than in most of the other groups (Kooyman unpub. observ.); (2) the chicks were much smaller (younger?); (3) adults were frequently fighting over chicks; and (4) none of the chicks entered crèche until much later than in the other groups. The chick/adult ratio of the inexperienced groups appears to be more like that at Cape Crozier. If the similarity is real, then it may be that young birds tend to collect at Cape Crozier, moving north with more experience in subsequent years. This is a highly speculative proposal and it is not the only explanation why the colony population fluctuates widely (Table 2), and why none of the birds banded at Cape Crozier in the 1960s were ever sighted (D. Ainley pers. commun.).

Aerial survey results are presented in Table 3. Coulman Island is the largest emperor penguin colony known. Based on our Cape Washington chick/adult ratio, we estimate that there were 22 137 chicks that were close to fledging at Coulman Island at the time of the aerial survey. Cape Washington is a close second in size, with about 25% (1983 estimate) fewer chicks (Table 3). However, until more data on chick/adult ratios are obtained, the above projections and the others in Table 3 must be tentative. A modest error in the ratios of early December can cause a large error in the estimate of chicks.

The 1983 census at Cape Washington is considerably more than the previous estimates of about 5000 adults in 1965 and 7600 in 1968 (Table 1). In both of these previous estimates the observations were made during an aerial passage at some distance from the colony. In reviewing some of Stirling's photographs (pers. commun.), the guano stain on the ice and distribution of the birds look somewhat similar to that in 1986 when one of us (GLK) camped at the site for 2.5 months, and also to that noted on 13 November 1987 when aerial observations were made (Fig. 4). We suspect the previous counts were likely to have been underestimates, and that the colony has not doubled or tripled since the 1960s. However, it is worth noting that in 1986 about 3300 more chicks were counted than in 1983, about a 21% increase.

Emperor penguins are ideal subjects for a vertical aerial photographic census because they are large and cast a clear shadow on the sea ice (Fig. 2). A comparison of aerial and ground counts on 6 November 1986 show that from about 1000 birds the 2 differ by 4%. For the three ground surveyors, individual replicates were within 5%, and between counters they differed by 18%. At this time of year and with the particular group counted the chicks make the counts difficult. Some are under adults and cannot be counted, while others are in crèches but mobile. Our opinion is that

Fig. 4. Aerial photographs of the Cape Washington colony. The upper photograph (**A**) was taken on 14 Nov 1968. The *curved arrows* mark the boundary to the arcing dispersal of clusters of birds. The *arrow* to the *far right* points to a large group beyond the edge of the print. The lower photograph (**B**) was taken on 13 Nov 1987 at 500 m altitude

the most accurate counts can be made from aerial photographs taken in December. The photographs freeze the birds in place, and the count can be rechecked. Two disadvantages are that birds in shadows cannot be counted and overflights may disturb the birds.

We have found few reports on aircraft disturbance to emperor penguins. The impact of aircraft overflights depends on the time in the breeding cycle. Problems could be minor late in the season, or major and decrease breeding success if done early in the season when adults may be flushed from eggs or chicks. Budd

Fig. 5. Satellite image of the western Ross Sea taken on 16 Nov 1986. Note the large polynya next to the Drygalski Ice Tongue (*curved arrow*). Cape Washington is 90 km to the north (*straight arrow*). Photograph courtesy of US Navy Meteorological Service at McMurdo Station

(1962) mentions the early season effects of low level overflights by a Beaver aircraft at 150 m. The circling aircraft caused incubating birds to begin to march in line away from the aircraft. Such a response shows that 150 m is an unsatisfactory altitude, at least if the aircraft orbits the birds.

During the 6 November 1986 overflight of an LC 130, we noted the reaction of the birds to the aircraft as it passed over at 1500 m. We had noted earlier that a graded stress response of the birds to any external disturbance was: (1) neck craning; (2) flapping of the wings while standing in place; (3) walking away from the source of disturbance; (4) tobogganing away from the disturbance powered by feet only; and (5) tobogganing at high speed and powered by both feet and wings (panic). During the overflight, much head craning and wing flapping occurred. No further aerial passes at lower altitudes where made.

In late season when chicks are not protected by their parents, flights should be made at various altitudes to determine a safe height that will not stampede the birds. Such an assessment should be made from the ground, because it is unlikely that those in the plane will be able to note the disturbance during the pass, or appreciate the full degree of disturbance. Based on our observation a single pass by an LC-130 at 500 m is probably safe once the chicks are in crèche stage.

The ice conditions around Cape Washington seem to be one of reliable fast ice in the immediate vicinity of the colony with broken pack ice or open water a short distance away. This seems typical of all the colony locations within the Ross Sea, but not those near the Australian research stations of Davis and Mawson (Budd 1961). Of six colonies described by Budd, all were from 25 to 115 km from the ice edge and the largest colony of about 12000 pairs at Amanda Bay was farthest. At some of these sites the birds had access to water through seal breathing holes and cracks around stranded icebergs. Many birds were diving and feeding at these holes. These conditions are in marked contrast to the open water or loose pack ice, even in winter, at most emperor penguin colonies in the Ross Sea.

The open water is possibly due to several factors, but one of the major causes at Cape Washington is well documented. As early as 1914 Priestley described the fierce katabatic winds of winter that created a great polynya near the Drygalski Ice Tongue (Fig. 5). The details of the airflow off the polar cap and its effect on the sea surface in its path have been discussed in detail (Kurtz and Bromwich 1985). In contrast to these fierce conditions only about 40 km from the Cape, the colony at the Cape is situated in an area of light winds where much snow is deposited.

A liability of nearby open water is the risk of early fast-ice breakup. Most vulnerable are the colonies at Beaufort and Franklin Islands which are found on narrow ice ledges next to shore. Another unusual hazard is the unstable cliffs adjacent to the Cape Crozier colony. Falling rocks have probably been a major factor in mortality at the colony since the birds were forced to move close to the cliffs after the offshore rifts in the shelf disappeared in the late 1960s (Todd 1980).

4 Conclusions

The Ross Sea emperor penguins can be accurately censused by vertical aerial photographs. From a priori reasoning, it would seem that emperor penguins could be useful indicator species in assisting the assessment of an alteration in environmental productivity of some of the coastal areas of the Ross Sea. (1) Emperors are one of the top predators in the environment, and variations in oceanic productivity should be indicated by changes in the breeding populations and breeding success. (2) The colonies and the birds are easily distinguished on sea ice. (3) The breeding sites for Ross Sea colonies are on annual sea ice which clears each year, removing dead birds and eggs from the previous year, enabling adult and chick mortality as well as egg loss to be assessed for a current year. (4) The phases of the breeding cycle are reliable enough that an aerial census may be selected to include only adults or chicks. The best time of year is mid-December before the chicks disperse and when most birds in the colonies are chicks. By mid-December the census also serves as an estimate of fledging success. (5) The colonies are distributed throughout the western Ross Sea thereby permitting local variations in breeding success to be compared for an overall assessment of productivity in much of this region.

It is important to begin these surveys soon to serve as a baseline for future environmental studies, before any impact from human activity occurs. This type of census is proposed not because it will yield the best data set, but because it may be the most feasible study presently available for assessing environmental fluctuations on a fairly large scale compared to the enormous effort necessary for more detailed studies that might give breeding success or assessment of some of the other parameters proposed by Croxall et al. (1988) in their analysis of monitoring studies to determine reproductive performance. For example, if it is assumed that the number of adults attempting to breed each year is relatively constant – not unreasonable in a long-lived species – then several years of fledging counts will give a rough idea of the variability in breeding success. It will also help to more clearly expose the type of detailed studies that should be pursued, such as counts of breeding adults.

Finally, based on calculations presented in the note to Table 3, the total breeding population in the Ross Sea is about 60000 pairs. This assumes a chick mortality of 15% and egg loss of 1% in 1983. Both assumptions are based on data obtained in 1986 from the Cape Washington Colony.

Acknowledgements. Supported by NSF DPP 83-16963. We wish to thank D. Croll, S. Smith, and S. Stone who assisted in the ground counts at Cape Washington, and D. Ainley who provided the Cape Crozier data for 1980–1983. The U.S. Navy VXE-6 provided air support and Eric Chiang of NSF provided much help and advice on operations.

References

Budd GM (1961) The biotopes of emperor penguin rookeries. Emu 61:161–178

Budd GM (1962) Population studies in rookeries of the emperor penguin *Aptenodytes forsteri*. Proc Zool Soc Lond 139:365–388

Cranfield HJ (1966) Emperor penguin rookeries of Victoria Land. Antarctic 4:365–366

Croxall JP, McCann TS, Prince PA, Rothery P (1988) Reproductive performance of seabirds and seals at South Georgia and Signy Island, South Orkney Islands, 1976–1987: implications for Southern Ocean monitoring studies. In: Sahrhage D (ed) Antarctic Ocean and resources variability. Springer, Berlin Heidelberg New York, pp 261–285

Harper PC, Knox GA, Spurr EB, Taylor RH, Wilson GJ, Young EC (1984) The status and conservation of birds in the Ross Sea sector of Antarctica. In: Croxall JP, Evans PGH, Schreiber WH (eds) Status and conservation of the world's seabirds. ICBP Tech Publ No 2, pp 593–608

Harrington HJ (1959) Narrative of a visit to the newly discovered emperor penguin rookery at Coulman, Island, Ross Sea, Antarctica. Notornis VIII:127–132

Jouventin P (1975) Mortality parameters in emperor penguin *Aptenodytes forsteri*. In: Stonehouse B (ed) The biology of penguins. MacMillan, London, pp 435–446

Kurtz DD, Bromwich DH (1985) A recurring, atmospherically forced polynya in Terra Nova Bay. In: Jacobs SS (ed) Oceanology of the Antarctic Continental Shelf. Antarct Res Ser 43:177–201

Priestley RE (1914) Antarctic adventure. Bell and Cockburn, Toronto, 372 pp

Stirling I, Greenwood DJ (1970) The emperor penguin colony at Cape Washington in the Western Ross Sea, Antarctica. Notornis 17(4):277–279

Stonehouse B (1966) Emperor penguin colony at Beaufort Island, Ross Sea, Antarctica. Nature 210(5039):925–926

Stonehouse B (1969) Emperor penguins *Aptenodytes forsteri* at Franklin Island, Ross Sea, Antarctica. Ibis (Short Commun) 111:627–628

Todd FS (1980) Factors influencing emperor penguin mortality at Cape Crozier and Beaufort Island, Antarctica. Gerfaut 70:37–49

Wilson GJ, Taylor RH (1984) Distribution and abundance of penguins in the Ross Sea sector of Antarctica. N Z Antarct Rec 6(1):1–7

The Effect of Disturbance on the Heart Rate and Behaviour of Adélie Penguins (*Pygoscelis adeliae*) During the Breeding Season

B. CULIK [1], D. ADELUNG [1] and A. J. WOAKES [2]

Summary. The behaviour and heart rate of nine incubating Adélie penguins (*Pygoscelis adeliae*) and one chick at crèche age were recorded in a breeding colony at Esperanza Station, Antarctica, during natural activity and disturbance. Heart rates were obtained by safety-pin electrode ECG and implantable radiotransmitters. In adults, minimal heart rate at rest was 86 bpm (beats per minute), rising to a maximum of 127 bpm when the birds were standing. Heart rates reflecting the degree of disturbance rose from resting levels to 110, 118, 127 and 145 when the stimulus was sheathbill, congener, human and helicopter, respectively. In the chick, normal heart rates were higher by 140 bpm than in adults (2.2 kg as opposed to 4.5 kg). Heart rates during natural activity ranged from 220 (sleep) to 287 bpm (food run). Disturbance resulted in heart rate increasing from resting levels to 240 (congener), 260 (approaching helicopter), 273 (helicopter overhead) to 310 bpm (capture by a human). Helicopters even when 1500 m away caused scattering of the birds and other panic reactions. A hand-reared chick in an enclosure did not show any fearful reaction (behaviour, heart rate) to helicopter approach whilst the observer was standing close by which indicates habituation to human disturbance. We propose that wild Adélie penguins, although often seemingly unconcerned, react strongly to man-made stimuli during the breeding season, resulting in reduced fledging and hatching success. Disturbance should be kept to a minimum.

1 Introduction

In the Antarctic, ice-free areas near the coast, suitable for penguin colonies or research bases are rather rare. The resulting conflict between a base and breeding colonies in the vicinity can be very direct, e.g. at Pointe Geologie, where an airfield runs directly through an Adélie penguin (*Pygoscelis adeliae*) colony (P. Jouventin pers. commun.), at Cape Halett, where a penguin colony was leveled to build a station (Wilson et al. 1988) or at Hope Bay where escaped sledge dogs kill dozens of penguins in the breeding colony before they can be caught again (Romero, pers. commun.). Frequently, helicopters and airplanes are used to supply the bases flying low over adjacent breeding colonies before landing, thereby scattering scores of penguins, breaking the breeding routine and exposing eggs and chicks to predation by skuas (Sladen and Leresche 1970; Thompson 1977). On the ground, all types of vehicles as well as individual people or groups of tourists repeatedly cause disturbance, subjecting the birds to unnatural stress.

According to Perry (1973), the more immediate response to an unusual stimulus is the "fight or flight" syndrome, with adrenaline and noradrenaline having a direct effect on heart rate. A considerable increase in heart rate in otherwise undisturbed birds may result from the approach of a potential predator. This has been observed in Herring gulls (*Larus argentatus*) (Kanwisher et al. 1978) and diving ducks (*Aythia ferina* and *A. fuligula*) (Butler and Woakes 1979). Ball and Amlaner (1980) believe that these "emotional" heart rate responses are adaptively significant and that they provide information on the bird's assessment of the danger to which it is exposed during approach. Stout and Schwab (1980) monitored heart rate of gulls (*Larus delawarensis*) to test the relative effectiveness of dispersal-inducing stimuli of airfields.

We report here on heart rate of incubating Adélie penguins during the breeding season, and on the variations in heart rate due to natural or man-made interference. Observations on a wild chick in the breeding colony are compared to those obtained from a hand-raised chick in an enclosure. Our aim is to investigate the relationship between changes in heart rate and the degree of disturbance as a better means to assess interference than behavioural observations alone in often seemingly undisturbed individuals.

[1] Institut für Meereskunde, Abteilung Meereszoologie, Düsternbrooker Weg 20, D-2300 Kiel 1, FRG.
[2] School of Biological Sciences, University of Birmingham, P.O. Box 363, Birmingham B15 2TT, UK.

Antarctic Ecosystems. Ecological Change and Conservation
(ed. by K. R. Kerry and G. Hempel)
© Springer-Verlag Berlin Heidelberg 1990

2 Materials and Methods

Nine incubating Adélie penguins (4–5 kg) in a small colony (approx. 50 pairs) located on Seal Point, Esperanza Bay (63.5° S, 57° W), Antarctic peninsula, were observed from a small, solid hiding placed 3 m from the nearest nest. Between November 27 and December 16, 1987, we conducted five experiments, recording heart rate and activity of one incubating penguin at a time for periods of 24–27 h. In four of these experiments, we also recorded the activity of eight other incubating birds (controls). Further data were acquired during 12 short-term observations lasting from 2–6 h.

The electrocardiogram (ECG) recorder, described in Heise (1989), was placed 0.5 m from the bird in an insulated and heated housing (20 × 20 × 20 cm). Two gold-plated safety pins were inserted into the penguin's skin a few centimetres posterior to each flipper. The safety pins were connected to the ECG recorder via crocodile clamps and a short (0.5 m) coaxial cable to avoid interference. ECG output was fed to a Linseis plotter and recorded by counting pen deflection for 60 s every 5 min. To check the validity of counts, the signal was plotted for 30 s every 15 min. Several experiments had to be ended because of a short circuit between the ECG electrodes due to rain.

The complete process of equipping the penguin with the electrodes and cables lasted approximately 10 min. Subsequently, the bird was released in the vicinity of its nest where incubation of the eggs was resumed within a few seconds. Heart rates were high after manipulation of the bird, decreasing to a steady level within 60 to 90 min. ECG-fitted and control birds stood up 1.47 (n = 4, SD = 0.40) and 1.19 (n = 31, SD = 0.47) times per hour respectively during each 24-h period. The difference is not significant ($p = 0.27$) and the behaviour of ECG-fitted penguins did not differ from that of surrounding birds. We never observed any penguin trying to remove the safety pins or the cables, or attempting to flee from the nest.

The activity of the ECG-fitted penguins was recorded at all times. Rated activities were: rest (lying flat on the nest and incubating motionless), rest with head movement, standing (including standing still as well as short activity bursts such as rearranging eggs, preening, picking at stones) and disturbance (sheathbill, penguin, human or helicopter approach). The large helicopters used in logistics at Esperanza station were of the type "Sea King" (Agusta). Their operation was typical for base provisioning.

Telemetric heart rate transmitters were implanted into one wild chick from the colony and one hand-reared, caged chick. At the time of the operation both chicks weighed 1.7 kg and their weight had increased to 2.2 kg when the experiments were begun. The caged chick was obtained from a nearby colony at crèche age (2 weeks) when thermally stable and weighing 1.2 kg. Twice daily the chick was hand-fed deboned fish, duck food pellets and vegetable oil (16:3:1) in gelatine, and kept together with others (n = 12) in a 36 m^2 cage (Adelung and Culik 1988). The chick was weighed every third day and started to chase the parent for food after 5 days (cf. Culik 1987). The transmitters were implanted into the body cavity under halothane and local xylocaine anaesthesia and both electrodes were placed adjacent to the heart. The operation lasted approximately 30 min. During suture the halothane anaesthesia was reduced from 2 to 0% and the chicks recovered to consciousness within approximately 5 min. After injection of Binotal antibiotic and 24-h recovery in an insulated shelter, both chicks were released into the cage and the colony respectively. The behaviour of both chicks, 1 day after the operation, did not differ from the behaviour observed previously. In experiments using a treadmill within a respiration chamber CO_2 and O_2 data as well as walking performance of implanted chicks did not differ from controls (n = 4 in each group, $p > 0.05$) (Culik et al. 1990). Both chicks fed and grew normally. After 3 weeks, the transmitters were surgically removed and the birds also recovered well from the second operation.

The telemetry ECG signal was received up to a distance of 60 m, varying with topography and posture of the birds. The signal was received with a commercial Air Band radio and stored on a Sony Professional quartz locked stereo cassette tape recorder. Simultaneously, behaviour observations were recorded orally on the same tape, making use of stereo separation. In the laboratory, the ECG signal was decoded and recorded on a Linseis plotter, together with the observations (cf. Woakes and Butler 1986; Butler and Woakes 1984).

The behaviour of the transmitter-equipped colony chick was observed from the blind mentioned above. We were able to differentiate between natural activities: sleep, rest (lying on the ground), stand, walk and run, and disturbance: by adult congeners, approaching helicopters, overhead helicopters and humans during capture (including weighing) and release of the bird. To avoid the Farraday effect of the cage, the activity of the caged chick was observed from within the enclosure during helicopter operation, the observer standing 1 m from the chick. The chick was tame and showed no fear of the observer.

Several Adélie penguin colonies and their access routes to the sea via snow fields and rocky slopes were observed from a 45-m hill behind the base, simultaneously with helicopter operations between the base and the supply ship.

3 Results

3.1 Incubating Adults

The heart rates of incubating adults (Fig. 1) resting quietly on their nest averaged 86 beats per minute (bpm) at zero wind speed (cf. Culik et al. 1989). When head movements were observed during incubation, heart rates rose to a mean of 99 bpm. The highest heart rates were recorded in the brief (60 ± 25 s, $n = 11$) instances when the birds were standing (approx. 1.5 times/h), with 127 bpm. In resting birds, disturbance by sheathbills (*Chionis alba*) landing or walking next to the incubating penguin caused heart rates to increase from 86 to 110 bpm. Congeners running noisily through the colony and perhaps pursued by another penguin, when approaching within reach of the incubating bird, caused its heart rate to rise to 118 bpm. Humans approaching from a distance of 30 m had a marked effect on heart rate, signaling to the observer in the blind the arrival of his relief. Although the birds were seemingly undisturbed and continued incubation (minimum approach distance 3 m), heart rates rose from resting rates to the same value found in standing penguins. In one bird, heart rate was measured while a helicopter flew over the colony (altitude 20 m). The mean of 145 bpm ranged from 139 to 150 and was the highest value we ever recorded in an incubating penguin.

3.2 Colony Chicks at Crèche Age

Without the constraints of incubation, the chick was able to move about freely and more activities could be

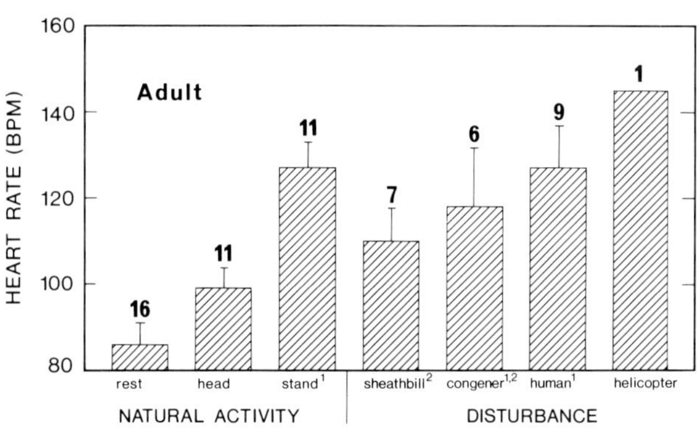

Fig. 1. Mean heart rates of adult incubating Adélie penguins recorded with safety-pin electrodes during natural behaviour and disturbance during rest. *Vertical bars* indicate 95% confidence intervals. *Numbers* above the *vertical bar* give individual recordings. *Superscripts* [1,2,3] indicate no statistical difference between categories with same number ($p = 0.05$, Duncan 1970)

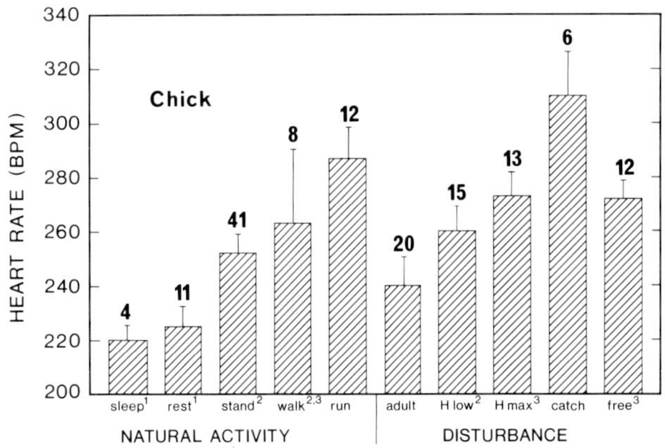

Fig. 2. Mean heart rate of one wild chick (crèche age) in the colony recorded by telemetry during natural behaviour and disturbance by a variety of stimuli. *H low* = Helicopter approaching; *H max* = helicopter overhead

observed than in the adults. Heart rates in the smaller chick were also generally higher by approximately 140 bpm than in the incubating adults (Fig. 2). The lowest heart rates (220 bpm) were recorded when the bird was sleeping, followed by resting (225 bpm). Standing led to 252 bpm, whereas walking slowly with pauses every few steps resulted in an average heart rate of 262 bpm. In the rare instances when the chick was running after the newly returned parent, begging for food, the heart rate rose to a maximum of 287 bpm. Disturbance of the resting chick by an adult, i.e. harassment or trampling, caused a heart rate of 240 bpm. Perception of an approaching helicopter led to 260 bpm, and caused the resting chick to stand up. When the helicopter flew over the colony at less than 50 m height, the heart rate rose to 273 bpm just prior to running. Much higher heart rates were observed when the chick was captured and weighed in a bag (310 bpm). After the release of the bird, heart rates decreased to 272 bpm and did not reach resting values within 30 min.

3.3 Comparison of Colony and Hand-Reared Chicks

Both chicks were observed simultaneously during the same sequence of helicopter approaches (Fig. 3). The hand-reared chick stood most of the time prior to disturbance, and only walked a few steps. Its heart rate remained within 230–250 bpm, but dropped shortly before helicopter approach to 220 bpm. Although the helicopter flew right over the cage, the heart rate of the hand-reared chick did not increase significantly, and the observer standing among the

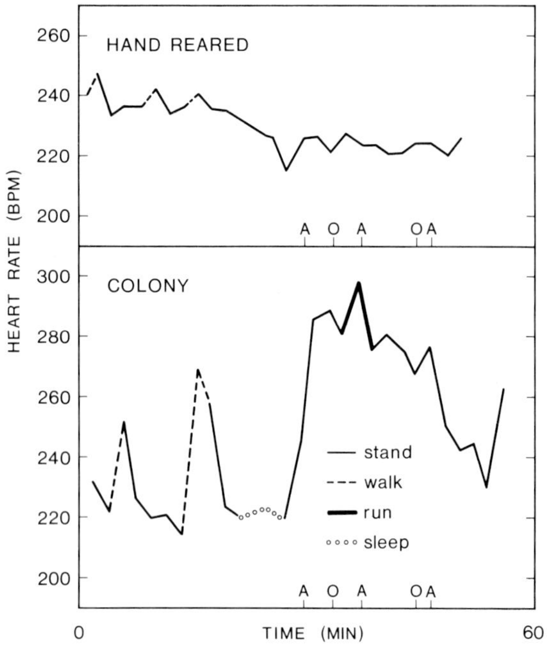

Fig. 3. Comparison of the heart rate of a hand-reared, caged chick and a wild chick from the colony measured by telemetry during the same sequence of helicopter approaches. *A* approaching helicopter; *O* helicopter overhead. The hand-reared chick did not show any behavioural or heart rate response to the helicopter

caged chicks did not notice any fearful behaviour. The colony chick, on the other hand, was much more active prior to helicopter operations. During walking, heart rate rose sharply to approximately 260 bpm and decreased to resting levels within a few minutes (230 bpm). Immediately prior to helicopter approach, the chick was sleeping (220 bpm). When the helicopter was only 200 m away, the chick stood up, its heart rate rising sharply to 280 bpm. By that time the adults in the colony had begun to run in panic, whereas the chicks did not move until the helicopter was almost overhead.

3.4 Behaviour of Adults

Early in the breeding season, when adult Adélie penguins were incubating, helicopter approach to within 20 m would in most cases not lead to desertion of the nest, although heart rate rose to 145 bpm on that occasion. However, later in the season, when the chicks had formed crèches, adults and chicks fled from the colony when the approaching helicopter was still 300 m away.

Prior to helicopter operation, Adélie penguins were observed to descend from their colony towards the sea, crossing a 100 × 150 m snow field (cf. Wilson et al. 1989). From their location, the birds could not see the supply ship which was approximately 1500 m away. As soon as the helicopter started from the deck and rose to sufficient height to be seen from the penguins, the birds rushed back uphill to their colonies, in an attempt to increase the distance to the approaching helicopter. Although the helicopter never came closer to that particular snow field than 400 m, and never actually flew over these colonies, the scene was repeated with every flight.

4 Discussion

Most incubating Adélie penguins have a high affinity for their nest and would not desert their eggs even when attacked by a dog (Romero pers. commun.). Disturbance by sheathbills, congeners and humans, in most cases only resulted in increased awareness, documented by head movements, and rarely led to a threatening behaviour or even escape of the bird. Heart rate, however, was found to increase very early, often before a behavioural response could be observed. Although seemingly docile, incubating Adélie penguins will react to disturbance as demonstrated by an increase in heart rate. The strongest stimuli were approaching humans and helicopters.

There is evidence that the "looming quality" of an approaching object provides a strong fear-provoking stimulus in several species (Jones et al. 1981). A rapidly inflating balloon caused alarm and panic reactions in adult domestic fowl (Duncan and Filshie 1980). The heart rate of mallards (*Anas platyrhynchos*) increased sharply when a human was approaching, and reached a higher value at a particular distance than when the person was receding (Heise 1989). As pointed out by Ball and Amlaner (1980), the approach speed results in changes in the response, a faster speed leading to an earlier, more drawn out and perhaps more extreme reaction. The approach speed and concurrent size increase of an approaching object perceived by a penguin ranges from sheathbill < congener < human < helicopter. This pattern was matched by the observed heart rate, indicating the degree of stress caused by these different factors. Capture and handling of a chick led to the highest heart rate recorded, demonstrating the intense "stress" induced during handling. Even 30 min after release of the chick, heart rate remained as high as during helicopter operation.

Kanwisher et al. (1978) found in Herring gulls that without outward indication of alarm, heart rate increased suddenly with the slightest noise from the blind. Heart rate in a seemingly unconcerned bird approached by a child rose from 160 to 440 bpm, returning to 220 bpm as the child backed off. In a study of short-term stress in domestic fowl, Duncan and Filshie (1980) noted that docile birds were as frightened as flighty birds when comparing heart rates, although the behavioural response was different.

Habituation to a frightening stimulus like pyrotecnics or distress calls emitted from approaching vehicles has been observed in gulls (*Larus delawarensis*) (Stout and Schwab 1980). However, the stimulus was used for dispersal from an airfield several times per day. Jones and Faure (1981) reported that regular handling reduced fear of humans in domestic chicks, while the responses of adult domestic hens were very resistant to modification. In incubating Adélie penguins, and in chicks at crèche age, habituation to human disturbance is unlikely. Helicopter disturbance is sporadic at most bases. Visits to the colony by tourists of personnel are not sufficiantly regular and frequent to lead to habituation, and even scientists will check chick weight maximally once daily (Davis and McCaffrey 1986; Lishman 1985) or only every third day (Volkman and Trivelpiece 1984).

Hughes and Black (1976) report that regularly handled hens were not only fearless of humans, but also of inanimate objects. Jones and Faure (1981) attribute this finding to the fact that the human observer was standing within 2 m of the fearless hen during the experiment. During helicopter operation, one of us was standing quietly next to the telemetry-equipped penguin chick, and neither behaviour or heart rate (HR) indicated any reaction to the stimulus. In the colony, however, the adult Adélie penguins engaged in panic runs before the maximal approach of the helicopter, and with some time delay were followed in their reaction by the chicks. We believe that Adélie penguin chicks learn to some extent the reaction to a stimulus from the adults, or from what they consider to be the adult (cf. Spurr 1975). This observation supports our view that hand-reared penguin chicks will be less fearful and stressed under laboratory conditions than wild animals, provided that the observer remains nearby.

The body temperature of Adélie and Emperor penguins rose by 1 °C and stayed high for several hours after handling of the birds and rose by 0.4 °C in an Adélie penguin when a congener was captured several metres away (Boyd and Sladen 1971). Independently from our heart rate measurements, these findings show that Adélie penguins, although they may seem unconcerned, respond very strongly to a human stimulus. Physiological parameters like body temperature and heart rate allow quantification of the degree of the response. Human approach and handling were the strongest stimuli, and these findings may lead to the reconsideration of which human activities in a breeding colony are strictly necessary for scientific reasons, and how much disturbance can be accepted. Helicopter approach, a strong visual and auditory stimulus, had a very high impact on heart rate and also triggered escape behaviour in non-incubating Adélie penguins and chicks at crèche age.

Sladen and Leresche (1970) observed that helicopters flying below 600 m strongly affect the behaviour of incubating Adélie penguins and that landing within 180 m caused 50–80% of the birds of flee from their territories. Later in the season we observed panic runs and escape reactions as far as 1500 m away from the flying helicopter, in areas which the helicopter never approached by more than 400 m. These observations should be considered when the minimum allowable approach distance of a helicopter to a breeding colony is being discussed. The 200-m limit accepted today, in our view is too low, although we are aware of the fact that small, relatively quiet helicopters may cause less disturbance than the large Sea Kings.

Physiological parameters as well as behavioural observations such as the fight or flight response indicate fear reactions and help single out quickly the stimulus most likely to have a strong impact on Adélie penguins and perhaps other species in Antarctica. If human interference is not controlled in the light of these findings, we have to expect in the short term reduced hatching or fledging success in the disturbed breeding colony, followed by an overall decline in the breeding population.

Although the Adélie penguins used in this study seemed to behave normally during and after termination of the experiments, their breeding success, i.e. hatching of the eggs, was very low. This may have been due to the fact that their nests were at the edge of the colony (Davis and McCaffrey 1986), thus resulting in stronger exposure to meteorological conditions and predation by skuas. It is more likely, however, that human interference from experiments and provisioning of the base were the cause. This view is supported by Lishman (1985), who reported that his daily visits to a colony, which included adult and egg measurement, had a considerable negative impact on the breeding success of the birds as opposed to the non-visited colonies nearby. Thompson (1977) attributed the sharp decline in penguin numbers at Cape Royds between 1955 and 1963 "to the nearly constant interference on fine days by visitors on foot and more significantly to helicopter flying low over the colony". However, since the restriction of human activities in that area, the population has recovered.

Acknowledgements. We would like to thank the Instituto Antarctico Argentino for the invitation to the 1987–88 expedition to Esperanza station, and the personnel of Esperanza, the Armada Argentina and P. Pfähler of DeTeWe Argentina for logistic support. This work was funded by a grant of the Deutsche Forschungsgemeinschaft MZ AD 24/11. Drs. H. Parish and R. J. Cucchiara aided in the implantation of the HR transmitter. N. R. "Coco" Coria and Hugo Spairani were of invaluable help during the expedition. We would like to thank Drs. J. Croxall, M. Tilzer, R. P. Wilson and E. C. Young for encouragement and support.

References

Adelung D, Culik B (1988) A method to obtain hand-reared Adélie penguins for physiological experiments. 1st Int Conf on Penguins, Dunedin, New Zealand

Ball NJ, Amlaner CJ Jr (1980) Changing heart rates of Herring gulls when approached by humans. In: Amlaner CJ, Macdonald DW (eds) A handbook on biotelemetry and radio tracking. Pergamon Press, Oxford, pp 589–594

Boyd JC, Sladen WJL (1971) Telemetry studies of the internal body temperatures of Adélie end Emperor penguins at Cape Crozier, Ross Island, Antarctica. Auk 88:366–380

Butler PJ, Woakes AJ (1979) Changes in heart rate and respiratory frequency during natural behaviour in ducks, with particular reference to diving. J Exp Biol 79:283–300

Butler PJ, Woakes AJ (1984) Heart rate and aerobic metabolism in Humboldt penguins *Spheniscus humboldtii* during voluntary dives. J Exp Biol 108:419–428

Culik B (1987) Fluoride turnover in Adélie penguins and other bird species. Polar Biol 7:179–187

Culik B, Adelung D, Heise M, Wilson RP, Coria NR, Spairani H (1989) In situ heart rate and activity of incubating Adélie penguins (*Pygoscelis adeliae*). Polar Biol 9:365–370

Culik B, Woakes AJ, Adelung D, Wilson RP, Coria NR, Spairani HJ (1990) Energy requirements of Adélie penguin (*Pygoscelis adeliae*) chicks. J Comp Physiol B (in press)

Davis LS, McCaffrey FT (1986) Survival analysis of eggs and chicks of Adélie Penguins (*Pygoscelis adeliae*). Auk 103:379–388

Duncan DB (1970) Multiple comparison methods for comparing regression coefficients. Biometrics 26:141–143

Duncan IJH, Filshie JH (1980) The use of radio telemetry devices to measure temperature and heart rate in domestic fowl. In: Amlaner CJ, Macdonald DW (eds) A handbook on biotelemetry and radio tracking. Pergamon Press, Oxford, pp 579–588

Heise M (1989) Human-induced tachycardia in wild and tame Mallard (*Anas platyrhynchos*). Comp Biochem Physiol 92A:125–128

Hughes BO, Black AJ (1976) The influence of handling on egg production, egg shell quality and avoidance behaviour of hens. Br Poult Sci 17:135–144

Jones RB, Duncan IJH, Hughes BO (1981) The assessment of fear in domestic hens exposed to a looming human stimulus. Behav Processes 6:121–134

Jones RB, Faure JM (1981) The effects of regular handling on fear responses in the domestic chick. Behav Processes 6:135–144

Kanwisher JW, Williams TC, Teal JM, Lawson KO (1978) Radiotelemetry of heart rates from free ranging gulls. Auk 95:288–293

Lishman GS (1985) the comparative breeding biology of Adélie and Chinstrap penguins *Pygoscelis adeliae* and *P. antarctica* at Signy Island, South Orkney Islands. Ibis 127:84–99

Perry G (1973) Can the physiologist measure stress? New Sci 60:175–177

Sladen WJL, Leresche RE (1970) New and developing techniques in Antarctic ornithology. In: Holdgate WM (ed) Antarctic ecology, vol 1. Academic Press, London, pp 585–596

Spurr EB (1975) Behaviour of the Adélie penguin chick. Condor 77:272–280

Stout JF, Schwab ER (1980) Telemetry of heart rate as a measure of the effectiveness of dispersal inducing stimuli in seagulls. In: Amlaner CJ, Macdonald DW (eds) A handbook on biotelemetry and radio tracking. Pergamon Press, Oxford, pp 603–610

Thompson RB (1977) Effects of human disturbance on an Adélie penguin rookery and measures of control. In: Llano GA (ed) Adaptations within Antarctic ecosystems. Smithsonian Inst. Washington, DC, pp 1177–1180

Volkman NJ, Trivelpiece W (1984) Growth in pygoscelid penguin chicks. J Zool Lond 191:521–530

Wilson GJ, Taylor RH, Barton KJ (1988) Long-term impact of humans on Adélie penguins at Cape Halett. 5th SCAR Symposium on Antarctic Biology, Hobart, Australia

Wilson RP, Culik B, Adelung D, Coria R, Spairani H (1990) To slide or stride; when do Adélie penguins toboggan? Can J Zoo (in press)

Woakes AJ, Butler PJ (1986) Respiratory, circulatory and metabolic adjustments during swimming in the diving duck *Aythia fuligula*. J Exp Biol 120:215–231

The Impact of Man on Adélie Penguins at Cape Hallett, Antarctica

K-J. Wilson [1], R. H. Taylor [2] and K. J. Barton [2]

Summary. The joint US-NZ base at Cape Hallett, Antarctica, was established in December 1956 on a stony spit occupied by breeding Adélie penguins (*Pygoscelis adeliae*). Over the 16 years the base was occupied it expanded to cover 4.4 ha, roads were constructed, habitat modified and scientific studies conducted. The colony declined from 62 900 pairs by 1959 to 37 000 pairs in 1968, then may have increased to 50 000 pairs in 1972. We censused the colony annually since 1981 and made ground observations in 1983. These show that, while the population has increased to about its original size (over 66 000 pairs in 1987), few areas modified by man have been fully recolonized.

During human occupation subcolonies close to the station and roads decreased markedly but have now recovered. The greatest declines were in isolated subcolonies subject to intensive banding and study, some of which declined to extinction and few of these have been re-inhabited.

During construction roads and the station environs were bulldozed flat so destroying the mounds on which the penguins nested. Even in the absence of people such areas were recolonized only where mounds remained, or where they were constructed when the base was demolished between 1984 and 1988.

1 Introduction

The Adélie penguin (*Pygoscelis adeliae*) colony at Cape Hallett (72° 19′ S, 170° 12′ E) was discovered in 1895 by a Norwegian whaling expedition (Bull 1896). The penguins nest on Seabee Spit which extends into Moubray Bay from the base of cliffs 300 m high. The spit is hook shaped, generally less than 5 m above sea level, about 1200 m long and between 130 and 575 m wide. Its area is about 41.1 ha of which 24.3 ha were occupied by nesting penguins before 1957 (Reid 1964).

On the spit, penguins nest on well-drained ridges and hummocks where snow does not accumulate, shunning low-lying areas between ridges and flat, featureless terrain below the cliffs. There are also large subcolonies on scree fans at the base of the cliffs where there is irregular, sloping terrain where neither snow nor meltwater accumulate.

Cape Hallett was next visited on 11 February 1956 by a US Navy reconnaissance party searching the Northern Victoria Land Coast for a site to build a research station. Seabee Spit was chosen, and on 29 December 1956 personnel arrived to build the station (Reid 1964). Nesting penguins occupied all suitable building sites, so over the next few days 7580 penguins, including 3318 chicks, were evicted from the 0.83 ha required for the station (Eklund 1959, 1961; Reid 1964). The requisitioned area was bulldozed flat before the buildings were constructed.

Over the next few years the station expanded. By 1960 the buildings occupied 1.8 ha and their associated roads, dumps, fuel caches and aerials a further 2.8 ha, reducing the area available to penguins from 24.3 to 19.8 ha. A total of 4.4 ha had been bulldozed flat, destroying the mounds favoured by penguins (Fig. 1).

Beyond the base environs, fuel, supplies and rubbish were cached at numerous sites, occupying space usually used by nesting penguins. Approximately 1250 m of road was built, about 550 m of which went through areas occupied by penguins (map in Reid 1964). Ridges were flattened, leaving four colonies ending abruptly at the road edge.

In 1959 much of the equipment scattered over the spit was consolidated around the station, rubbish was bulldozed from the dump onto the sea ice and large objects relocated so snowdrifts that formed downwind of them did not cover penguin subcolonies.

The base was an overwinter facility until 1964, and continued as a summer-only station (November to February) until it was abandoned on 17 February 1973. Between 1984 and 1988 the base was demolished and rubbish, aerials and fuel drums removed from the spit.

[1] Entomology Department, Lincoln University, Canterbury, New Zealand.
[2] Ecology Division, D.S.I.R., Private Bag, Nelson, New Zealand.

Antarctic Ecosystems. Ecological Change and Conservation
(ed. by K. R. Kerry and G. Hempel)
© Springer-Verlag Berlin Heidelberg 1990

Fig. 1. Adélie penguins nesting on hummocked terrain, even in the presence of trash, but avoiding the smooth compacted habitat about the base

While the station was occupied the number of penguins and skuas (*Catharacta maccormicki*) breeding at Cape Hallett had declined due to both disturbance and loss of breeding habitat (Reid 1964; Fredrickson 1971). It was generally assumed that penguin numbers would increase as they recolonized areas now vacated by man, but Reid (1964) suggested that with the ridges gone they would never again be sites where penguins would breed with much success.

This chapter summarizes the recovery of the breeding population since 1973, the recolonization of previously abandoned nesting areas, and recent attempts to modify the habitat to encourage recolonization by penguins.

2 Methods

The information presented in this study comes from several sources. A map of the colony made in 1962 (Reid 1964) and our ground observations of 1983 provided baselines against which change could be measured. Counts and photographs from the period of human occupation and our aerial photographs taken in the 1980s allowed us to follow population changes at both colony and subcolony levels. The methods by which our data were obtained are described below.

2.1 Aerial Photographs

Since 1981 the colony has been censused annually from aerial photographs taken during an extensive survey (Taylor et al. in press), sometime between 29 November and 8 December each year, coinciding with the early incubation period when fewest non-incubating birds are ashore (Taylor 1962; Reid 1964).

Hercules C-130 aircraft flew at 300–400 m above sea level and oblique photographs were taken through cockpit windows or open paratroop doors. Large-format (6 × 7) and 35-mm cameras with various lenses and monochrome film were used (Taylor and Wilson 1985; Taylor et al. in press). Sections of these photographs were enlarged enabling counts to be made of the occupied nests in each subcolony, and in the entire colony each year. Randomly selected subsamples were checked by three additional counters, and 95% confidence limits calculated for each season.

2.2 Ground Observations

In January and February 1983 GJW noted the location of nesting penguins in respect to buildings, roads, dumps, fuel drums and aerial masts. At each object the presence or absence of nesting penguins, plus the condition of habitat, whether it was modified or not, raised or flat, smooth or rough, was recorded. The area around the base and its main aerial fields and dump were mapped in sufficient detail to reveal the loss of subcolonies or formation of new subcolonies, and any gross changes in subcolony size or shape since the previous map was made in 1962 (Reid 1964).

2.3 Artifical Nest Mounds

To see if penguins would colonize highly modified areas if suitable terrain was reformed, an area of flat, compacted ground was dug up in February 1983 to form a ridge 5 × 3 m in area and about 1 m high. The ridge had an irregular surface and was equidistant between two colonies, about 10 m from each. Larger mounds were built with a bulldozer in February 1986 and in November 1987, by base clean-up parties. In 1986 two smooth surface mounds were constructed, one in line with, and one across the prevailing wind. In 1987 both smooth and rough, circular and elongated mounds were built.

Table 1. Numbers of breeding pairs of Adélie penguins at Cape Hallett[a]

Year	Number breeding pairs	Type of count	Reference
1959	62900	Ground	Reid (1964)
1960	61955	Ground	Reid (1964)
1962	55579	Ground	Choate (1967)
1963	50200	Ground	Cranfield et al. (undated)
1964	52430	Ground	Cranfield (1966)
1965	over 53000	Ground estimate	F. C. Kinsky (1966, unpubl.)
1966	46242	Ground	Choate (1967)
1967	43000	Ground	Tenaza (1971)
1968	37000	Ground	K. E. Westerskov (1969 unpubl.)
1970	48045	Ground	L. H. Fredrickson (pers. commun. 1983)[b]
1970	45006	Aerial estimate	L. H. Fredrickson (pers. commun. 1983)[b]
1971	47442	Aerial estimate	L. H. Fredrickson (pers. commun. 1983)[b]
1972	50156	Aerial estimate	L. H. Fredrickson (pers. commun. 1983)[b]
1981	37573	Aerial	This study
1982	43822	Aerial	This study
1983	51874	Aerial	This study
1984	58084	Aerial	This study
1985	60984	Aerial	This study
1986	58243	Aerial	This study
1987	66319	Aerial	This study

[a] The 1970–1972 counts were made sometime in November; all other counts were made between mid-November and early December.
[b] However, Fredrickson (1971) and Baker (1973) indicate that numbers continued to decline over this period.

3 Results

3.1 Population Changes 1959–1987

The first accurate census of 62900 breeding pairs was made in late November 1959 (Reid 1964). Subsequently the rookery was censused between mid-November and early December most years and numbers declined steadily to 37000 pairs in 1968 (K. E. Westerskov unpubl.; Table 1).

In 1970 disturbance to penguins was more tightly controlled (Fredrickson 1971). In November 1970 Fredrickson made a ground count of 48045 breeding pairs. He also obtained aerial photographs of the colony, probably about mid-November from 1970 to 1972, and the subsequent estimates suggest an increase in nesting pairs from 45006 in 1970 to 50156 in 1972 (L. H. Fredrickson pers. commun. 1983, 1988). However, Baker (1973), who reported on a six-summer study at Cape Hallett concluding in 1972/73, said that the population had been in continuous decline since the station was founded.

Counts made from aerial photographs each year since 1981 show an increase from 37500 breeding pairs in 1981 to 66000 in 1987 (Table 1). The increase was initially linear (Fig. 2), but counts made since 1984 suggest that the population may be stabilizing although a slow upward trend continues. Since 1984 the estimates range from 58000 (1984) to 66000 (1987) but show extensive overlap between their 95% confidence limits (Fig. 2).

3.2 1983 Inspection

In 1983 the surface of the base environs and roads (Fig. 1) remained compacted, devoid of any surface irregularities. No recolonization had taken place on the bulldozed 4.4 ha, but some ungroomed ground adjacent to the base and an area of rubble piled up during road construction had been colonized by nesting penguins.

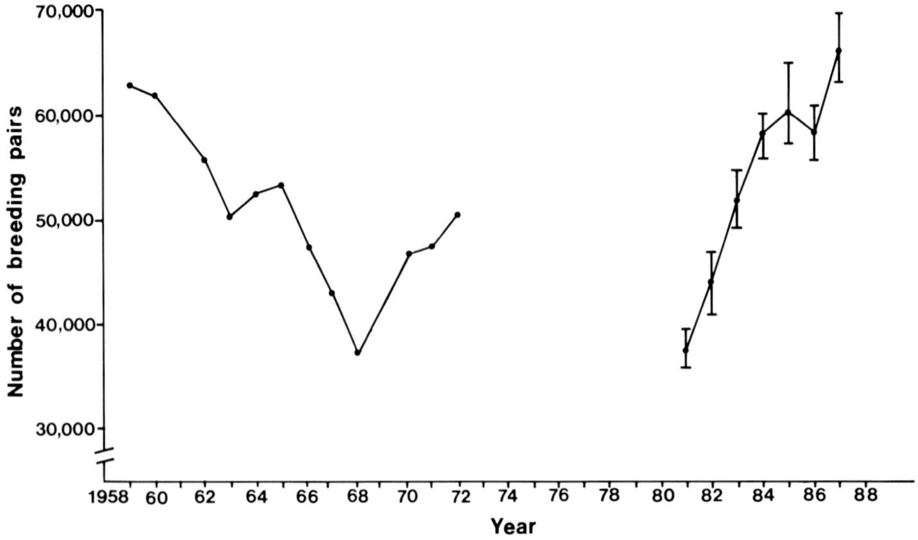

Fig. 2. Number of breeding pairs of Adélie penguins at Cape Hallett, 1959–1968, 1981–1987

The rubbish dump was also set in the penguin colony. Though the ridges were flattened, the surface was left rough and in 1983 penguins were nesting on some ground shown as dump in 1962 (Reid 1964).

Numerous radio aerials were erected at widely scattered locations. The ground beneath three conical monopole antennae erected in the penguin colony had been partially flattened, but up to eight rounded mounds utilized by penguins remained beneath each mast, the radiating ground wires being laid over these features. In 1983 all but one of the subcolonies present beneath these masts in 1962 (Reid 1964) were still occupied, although some further nesting habitats had been lost and one new subcolony (213 chicks) formed.

Where other aerials and power poles had been erected, rubble had been heaped up to support the masts and anchor their guys, so creating new mounds, generally 1–1.5 m high and 1.5–5 m diameter. In 1983 penguins were nesting on at least 18 such mounds, 10 of which are not shown on the 1962 map.

Snowdrifts form downwind of large objects and some of these had obliterated penguin subcolonies. A large snowdrift north of the communications building shown in the 1970, 1971, 1981 and 1982 aerial photographs was present throughout our visit in 1983 and covered two subcolonies shown on the 1962 map. Early in the 1970s, fuel drums, trailers, and other large objects were lined up 0–10 m from a group of penguin subcolonies, causing snowdrifts to cover up to a third of four large subcolonies.

The impact on penguins of smaller items was negligible. Penguins nested beside 19 of 25 solitary oil drums on well-drained, elevated ground with up to six nests by some drums. Other rubbish including wooden boxes, lumber, scrap metal, wire and a truck battery were found in subcolonies (Fig. 1).

Three small subcolonies had formed where abandoned objects had created well-drained, elevated terrain.

3.3 Recolonization and Population Growth in Areas Modified by Man

To investigate the long-term influence of the station and roads on the penguin population, we compared on the 1962 map and later aerial photos, a group of subcolonies in the vicinity of the station, with subcolonies near the road, and with a control group where the erection of radio masts was the only habitat modification and disturbance had been light. An aerial photograph taken in November 1970 allowed us to determine the existence or absence of many of these subcolonies, but detail was not sufficient for counts to be made.

Station Environs. On undulating ground adjacent to the biology laboratory and communications buildings, some changes in penguin numbers and distribution occurred between 1962 and 1981. The population declined by 20%, two subcolonies had been abandoned (at least one due to bulldozing) and seven new subcolonies formed, none of which appear on the 1970 photograph (Table 2). However, after 1981, the numbers of penguins increased rapidly and an additional eight new subcolonies were established.

Throughout our visit in January–February 1983, a snowbank extended north from the communications building, covering four subcolonies present in 1962.

Between 1970 and 1983, three subcolonies formed around objects left by man, one on a piece of twisted

Table 2. Changing numbers of breeding pairs of Adélie penguins in three study areas on Seabee Spit with differing levels of human disturbance

	1962	*1981*	*1984*	*1987*
Station environs				
Number of subcolonies	24	29	31	37
Total number of breeding pairs	2301	1839	3082	3965
Road Area				
Number of subcolonies	17	19	21	21
Total number of breeding pairs	1929	1567	3501	3196
Control – little human disturbance				
Number of subcolonies	35	48	51	53
Total number of breeding pairs	2654	2726	4433	3981

metal decking, and two on debris heaped around the base of poles.

By December 1984 the communications building had been removed, leaving heaps of scree mixed with snow, and the subcolony closest to the building was temporarily abandoned. To the south of the station one small subcolony had formed where drums of trash had been removed since 1983. There was a general increase in the size of most nearby subcolonies.

In 1987 the overall numbers in this area had increased again (Table 2). The snow on the communications building site had melted leaving the debris piles lower and penguins nested on one mound. Two new colonies (19 and 14 nests) had established close by, the snowbank was gone and in its place was a new subcolony (48 nests). The subcolony abandoned in 1984 had been reoccupied. The trash-site group had grown from 7 to 134 nests.

Road Area. Between 1962 and 1981 numbers nesting near the road had declined by 19%. By 1981 one subcolony had been abandoned but there were three new subcolonies not shown on the 1962 map. One of these had formed where rubble had been pushed aside while constructing the road. Two subcolonies had expanded but few other changes were apparent.

By 1984 the number of occupied nests had doubled but only two new subcolonies (six and seven nests) had formed. By 1987 numbers in this study area had declined overall, though a few subcolonies had increased in size.

Control Area. Numbers in the control area increased by 3% between 1962 and 1981, one subcolony had been abandoned and 14 new ones formed, six of which appear on the 1970 photograph. Three new subcolonies were formed between 1981 and 1984, and two more by 1987, but all these remained small (1 – 57 nests in 1987). Two of the subcolonies first noticed in 1970 had formed on debris heaped up around the base of aerials or their guys; one further such group appeared by 1981 and two more by 1984.

Numbers increased in this zone from 1981 to 1984, but declined slightly by 1987 (Table 2).

3.4 Effect of Disturbance

At Cape Hallett all types of disruptive activities occurred. Helicopters were present only while ships were close by, usually late in the breeding season. Vehicles passed along the roads most days. There was a constant hum of machinery and frequent foot traffic of sightseers, photographers and scientists. Disturbance was not evenly spread over the colony, being most intense close to the station or at subcolonies used by scientists, and least at remote areas.

Up to 1962/63 about 85 subcolonies, scattered over the entire spit, had been used for banding or scientific observations (Reid 1964). Between 1969/70 and 1972/73, 361 adults plus 15217 chicks were banded, mainly close to the station, in at least 25 subcolonies (Baker 1973).

We have identified some of these subcolonies on our aerial photographs and tested the long-term impact that scientific study has had on penguin populations. As we have subcolony by subcolony counts only for

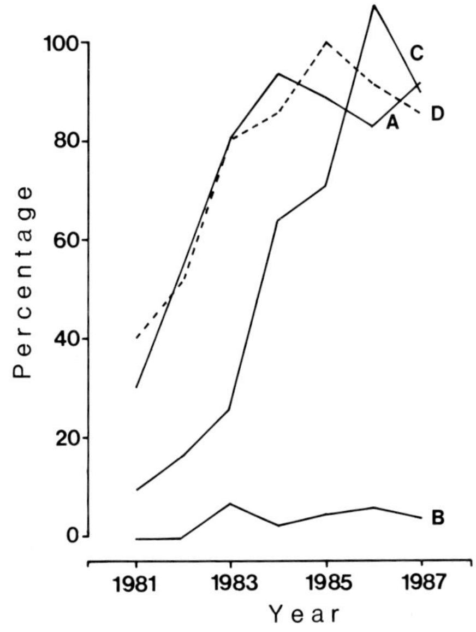

Fig. 3. Changing numbers of breeding pairs of Adélie penguins in four study areas remote from Hallett Station with differing levels of scientific disturbance, expressed as a percentage of numbers present in November 1962. *A* Pair banding 1961/62 in central subcolonies; *B* pair banding 1961/62 in peripheral subcolonies; *C* annual banding 1958/59–1964/65; *D* control group in central subcolonies

1962/63 it is this season we are comparing with our 1981–1987 data.

Each year from 1981 to 1987, we have identified on our photos 15 of the 68 subcolonies used for pair banding in 1961/62. All 15 are at the eastern end of Seabee Spit, remote from Hallett Station. Of these, one had been abandoned, and one other supported only a single nest during the summer following banding (Reid 1964). Six were central subcolonies (located within 10 m of an occupied subcolony; Fig. 3 A), and four of these were occupied during the 1980s. The population of this group declined by 70% between 1962 and 1981, but by 1984 had recovered to about 90% of the 1962 levels. The other nine were peripheral subcolonies (10–100 m from occupied subcolonies in 1981; Fig. 3 B). None was occupied in 1981 and only one has been recolonized since. The total population of this group is still less than 10% of the 1962 level.

Seven other subcolonies (Fig. 3 C) in which birds were banded and closely observed over seven seasons (1958/59–1964/65) declined by more than 90% during that period (Cranfield et al. undated; Reid 1968). Six of these subcolonies were small (4–46 pairs in 1962) and peripheral, and had been abandoned by 1970. In 1981 only one was occupied, but one other has been recolonized since 1983, and one to five pairs have nested in two other subcolonies in some years. The seventh subcolony was large (498 pairs in 1962), adjacent to other subcolonies and showed a 26% decline from 1962 to 1981. The total population has now recovered to the 1962 level but is still less than 40% of what it was in 1959 (Fig. 4).

We chose 15 subcolonies close to the above groups at which no study had taken place (Fig. 3 D). All were extant in both 1962 and throughout the 1980's. The total population of these colonies declined by 40% between 1962 and 1981, but by 1985 numbers were back to 1962 levels.

3.5 Artifical Nest Mounds

The number of birds nesting on the artificial mounds have been counted using the aerial photographs each year. Birds first nested on the mound built in February 1983, two seasons after construction, and numbers had increased to 18 nests in 1987 and 1988 (Table 3). There were eight chicks on this mound on 2 February 1986.

Birds first nested on the two mounds built in February 1986 in 1988 (two and three nests respectively) by which time erosion had made the originally smooth, compacted surfaces rough. Penguins nested on the two

Table 3. Number of Adélie penguins nesting on the artificial mound built in February 1983

Date	No. occupied nests	Total birds
2 Dec. 1983	0	2
7 Dec. 1984	4	7
4 Dec. 1985	9	13
11 Dec. 1986	16	18
8 Dec. 1987	18	20
4 Dec. 1988	18	19

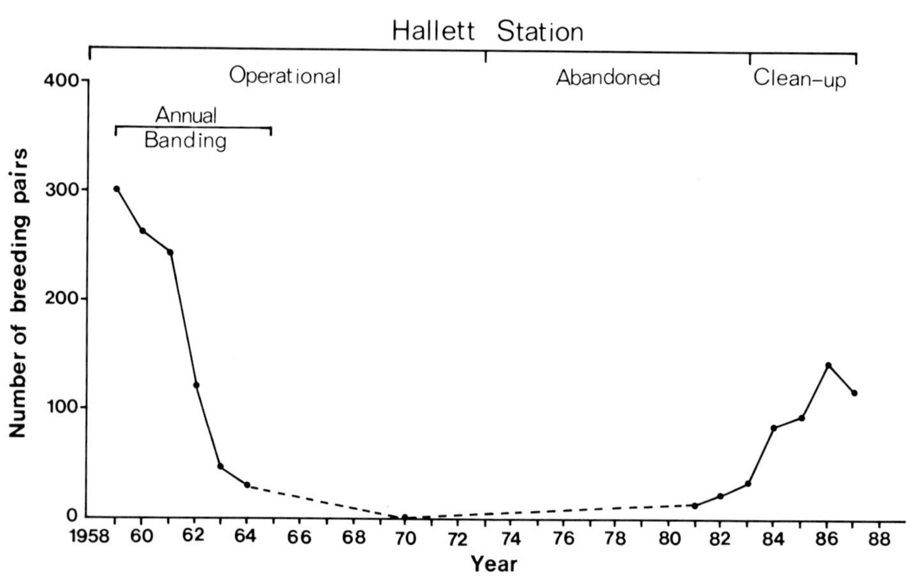

Fig. 4. Decline in numbers of breeding pairs of Adélie penguins in six subcolonies subjected to annual banding and study between 1959 and 1965 (Cranfield et al. undated) and their subsequent recovery

rough surfaced mounds built in November 1987 the next season (23 and 9 nests), but shunned the similarly sized smooth mounds. Another mound, half-smooth half-rough, had nine pairs nesting on it in 1988, all of which were on the rough section.

4 Discussion

The number of Adélie penguins that nested at Cape Hallett before the station was constructed is not known, as neither of the early estimates by Austin (1957) or Eklund (1961) is reliable.

The total number of birds nesting at Cape Hallett declined markedly between 1959 (62 900 occupied nests) and 1968 (37 000). When we began our counts in 1981 there were again about 37 000 occupied nests. Subsequently, they have increased steadily until 1984, thereafter fluctuating about the pre-decline level (Fig. 2). The story is complicated by counts made between 1970 and 1972 (L. H. Fredrickson pers. commun. 1983). He suggested that numbers had begun increasing by 1970, although in a published account (Fredrickson 1971) he stated: "Counts ... in 1970 verified their further decline". Baker (1973) suggested that the decline continued right up to 1972/73. The 1970–1972 counts were probably made about mid-November, when more nests are occupied than during the standard late November to early December census period (Taylor 1962), thus they need not indicate a real population increase.

There is insufficient information to determine whether similar population changes occurred in other Northern Victoria Land colonies during the 1960s. The only contemporary estimates were made by Cranfield (1966) in November 1964. His rough ground estimates for Cape Cotter, Coulman Island and Possession Islands, all within 140 km of Cape Hallett, are (like the 1964 Hallett estimate) similar to aerial photograph counts of the same colonies in the early 1980's (Taylor et al. in press). At Cape Adare, 130 km north of Hallett, numbers may have declined between January 1961 [when Reid (1962) estimated that there were between 370 500 and 394 500 chicks] and January 1981 (when GJW estimated that there were only 285 750). Reid (1962) thought this represented 289 400 breeding pairs, compared to the 257 000 and 282 000 counted from aerial surveys in 1983 and 1986 (Taylor et al. in press).

There is mounting evidence to show that Adélie penguin populations undergo long-term natural fluctuations (Taylor et al. in press; Wilson in press). Regardless, at Cape Hallett [as at Cape Royds where Adélies were also subject to repeated human disturbance (Thomson 1977)], the decline while the base was operational was primarily due to land acquisition by man, disturbance and scientific study. As expected, the greatest upheaval for the penguins occurred within the station environs where nesting habitat was flattened, buildings were erected, subcolonies covered by snowdrifts and where there was continual human disturbance. Less predictably, the impact was equally drastic in roaded areas away from the immediate vicinity of the base. Penguin numbers showed a much smaller decline in areas remote from the base, where there was little human disturbance.

Since the base was abandoned, few areas with altered topography have been recolonized by nesting penguins, as predicted by Reid (1964). Penguins nest only where mounds or ridges provide ground where neither snow nor meltwater accumulate. These mounds may be natural or man-made. The penguins have shown remarkable adaptability in nesting on a variety of artificial mounds, such as metal sheeting, stacks of fuel hoses, and roadside tailings. Elevated, snow-free, well-drained terrain is the common factor, but our experimental nest mounds indicate that a rough surface with freely available stones (for nest building) is also a necessary attribute.

Scattered oil drums and other rubbish did not deter penguins from nesting close to them. However, large objects such as fuel tanks and buildings create downwind snowdrifts, and in the early 1980s several such drifts completely covered some subcolonies that had existed in 1962.

Other than habitat destruction, the most effective and long-lasting disturbance appears to be banding, coupled with intensive study, particularly in peripheral subcolonies (Fig. 3). Cranfield et al. (undated) divided the colony into nine zones and followed population changes in each from 1960/61 to 1964/65. They found that changes in numbers of breeding pairs were closely correlated to the degree of disturbance. While the colony as a whole declined by 13.5% over this period, numbers actually increased between 11.9 and 57.5% in the three least disturbed zones. In the four zones close to the base or roads numbers decreased between 16.8 and 40.2%. The decline was greatest (89.5%) in seven subcolonies subjected to handling, banding and observations by scientists (Fig. 4; see also Reid 1968). Even banding of adults without subsequent study caused a 33% decline in some subcolonies, possibly by increasing the mortality of the banded birds (Cranfield et al. undated). At Cape Crozier Adélie penguins banded as chicks showed a 28% higher mortality than unbanded birds in their first year, but only a 3% higher mortality in subsequent years (Ainley et al. 1983).

Once a decline is precipitated it appears to continue even after disturbance is curtailed, suggesting that the decline is largely due to pre-breeders being dissuaded from settling there, rather than scaring off established breeders. Ainley et al. (1983) also suggested that human presence lowered recruitment of young birds. They found that subcolonies closest to their hut declined by 90% over 11 years, while those 800–1000 m distant declined by 33%. They noted that neither nest desertion nor egg or chick mortality differed between disturbed and undisturbed nests.

It was 12–14 years after Hallett Station was abandoned by man that penguins reached their pre-decline level. At Cape Royds the number of breeding pairs began to increase as soon as human disturbance was reduced (Thomson 1977), but again it took 12–14 years to attain the pre-decline population (Harper et al. 1984).

Once areas were abandoned by nesting penguins, recolonization occurred only at suitable sites close to existing subcolonies. Only 3 of 14 abandoned peripheral subcolonies (more than 10 m from occupied subcolonies) had been recolonized by 1987.

Partly as a result of our visit in 1983, a clean-up of seabee spit and the demolition of the station were begun in 1984/85 and completed in 1987/88. In the area of the base, the numbers of breeding pairs increased by 28% between 1984 and 1987 as subcolonies grew and new subcolonies established on rough surfaced mounds left after the removal of buildings and trash. By comparison over the same period numbers in the entire rookery increased by only 14%, and those in a zone around the base, where the removal of aerials and scattered drums was the only habitat change, declined by 10%. Further changes will no doubt occur and we plan to continue monitoring these.

Acknowledgements. We are indebted to the RNZAF for making our aerial censuses possible and to the US Coast Guard for transport to Cape Hallett in 1983 and 1986. We are grateful to the Antarctic Division, DSIR for their logistic support. Jonathan Pascoe kindly provided assistance in the field. We thank members of the 1986 and 1987/88 base clean-up parties for constructing nesting habitat at our request.

Peter Wilson, Bruce Thomas, Brian Reid and Prof. L. H. Fredrickson kindly allowed us access to unpublished information. We thank the following for their help in the preparation of this paper; Jocelyn Tilley, drafting, Peter Alspach, statistical advice, Tim Fitzgerald, photography and counting, Sue Trotter, typing and Dr P. R. Wilson, Dr R. M. Emberson and Tony Pritchard for comments on the manuscript.

References

Ainley DG, Le Resche RE, Sladen WJL (1983) Breeding biology of the Adélie penguin. University of California Press, Berkeley

Austin OL (1957) Notes on banding birds in Antarctica and on the Adélie penguin colonies of the Ross Sea. Bird-Banding 28:1–26

Baker JR (1973) Penguin and skua studies at Hallett Station. Antarct J US 8:200–201

Bull HJ (1896) The cruise of the 'Antarctic' to the south polar regions. Edward Arnold, London

Choate TS (1967) Skua and penguin research at Cape Hallett, 1966–7. Antarctic 4:637–640

Cranfield HJ (1966) Adélie penguin rookeries in the Cape Hallett region. Antarctic 4:315–317

Cranfield HJ, Kinsky FC, Reid BE (undated) The effect of banding and continued observation on an Adélie population. Unpubl rep, Antarctic Division Library, Christchurch, NZ

Eklund CR (1959) Antarctic ornithological studies during the IGY. Bird-Banding 30:114–118

Eklund CR (1961) Distribution and life history studies of the south polar skua. Bird-Banding 32:187–223

Fredrickson LH (1971) Environmental awareness at Hallett Station. Antarct J US 6:57

Harper PC, Knox GA, Spurr EB, Taylor RH, Wilson GJ, Young EC (1984) The status and conservation of birds in the Ross Sea sector of Antarctica. In: Croxall JP, Evans PGH, Schreiber RW (eds) Status and conservation of the worlds seabirds. ICBP Tech Publ No 2, pp 593–608

Reid BE (1962) An assessment of the size of the Cape Adare Adélie penguin rookery and skuary – with notes on petrels. Notornis 10:98–111

Reid BE (1964) The Cape Hallett Adélie penguin rookery – its size, composition, and structure. Rec Dom Mus (Wellington) 5:11–37 (with accompanying map)

Reid BE (1968) An interpretation of the age structure and breeding status of an Adélie penguin population. Notornis 15:193–197

Taylor RH (1962) The Adélie penguin *Pygoscelis adeliae* at Cape Royds. Ibis 104:176–204

Taylor RH, Wilson PR (1985) Adélie penguin rookeries at Coulman Island, western Ross Sea, Antarctica. Notornis 32:101–107

Taylor RH, Wilson PR, Thomas BW Status and trends of Adélie penguin populations in the Ross Sea region. Pol Rec (in press)

Tenaza R (1971) Behaviour and nesting success relative to nest location in Adélie penguins (*Pygoscelis adeliae*). Condor 73:81–92

Thomson RB (1977) Effects of human disturbance on an Adélie penguin rookery and measures of control. In: Llano GA (ed) Adaptations within Antarctic ecosystems: Proceedings of the Third SCAR Symposium on Antarctic Biology. Smithsonian Institution, Washington DC, pp 1177–1180

Wilson K-J. Fluctuations in the Adélie penguin (*Pygoscelis adeliae*) population at Cape Bird, Antarctica. Pol Rec (in press)

Adélie and Chinstrap Penguins: Their Potential as Monitors of the Southern Ocean Marine Ecosystem

W. Z. TRIVELPIECE, S. G. TRIVELPIECE, G. R. GEUPEL, J. KJELMYR, and N. J. VOLKMAN [1]

Summary. Analysis of annual variability in the parameters recommended by CCAMLR for ecosystem monitoring revealed considerable variation among the parameters for which we have sufficient data to make an assessment. Interpretation of the observed variability was greatly enhanced when several parameters were considered simultaneously. In concert, the parameters should be useful for assessing perturbations in the Antarctic ecosystem. The choice of Adélie and chinstrap penguins as key indicator species seems an excellent one, given their very different responses to the highly variable conditions in the Antarctic Peninsula region, their species-specific breeding patterns, and the asynchrony between their respective breeding seasons. If a time series of data becomes available, it may be possible to differentiate between fishery-induced and natural perturbations in the ecosystem.

1 Introduction

A primary goal of the US participation in the Commission for the Conservation of Antarctic Marine Living Resources (CCAMLR) is to provide information needed for US policy regarding the conservation and management of the marine living resources in the oceans surrounding Antarctica. To fulfil this goal an internationally coordinated, directed research program was recently established to detect, monitor, and predict the effects of fishing and associated activities on target and dependent marine species and the ecosystem of which they are a part. The major focus of this effort is to obtain an accurate measurement of the interannual variation in population parameters of the major taxonomic groups of the Antarctic marine ecosystem; including both the potential targets of fisheries (e.g., krill, *Euphausia superba*, and organisms dependent on these prey, e.g., birds, seals, and whales).

Among the most accessible members of this Antarctic community are those species which, while they depend on marine organisms for their food and spend most of their lives in the marine environment, must come ashore to breed. Seabirds are the largest and most diverse group exhibiting these attributes, and it has been suggested that measurements of aspects of the breeding biology of selected, easily studied predators may furnish good indices of the state of other, less accessible, krill-dependent predator populations (Everson 1977; Croxall and Prince 1979; SCAR 1979). Furthermore, monitoring seabirds, which are potential competitors with the developing Antarctic krill fishery, may help substantiate data collected aboard research vessels, on the abundance and composition of the krill stocks themselves.

A meeting of the *ad hoc* Working Group on Ecosystem Monitoring (CCAMLR 1985) identified the seabird species and biological parameters that would likely provide the most sensitive indices to ecosystem changes potentially wrought by the development of a large-scale fishery for krill. The Antarctic seabird community is comprised of approximately 45 species with penguins accounting for 60–70% of the total avian biomass of the Southern Ocean ecosystem (prevost 1981). Recognizing the significant ecological role of penguins in the Antarctic, the Working Group for the CCAMLR Ecosystem Monitoring Program identified three penguin species, the Adélie (*Pygoscelis adeliae*), chinstrap (*P. antarctica*), and macaroni (*Eudyptes chrysolophus*), and several of their life history parameters as key components in an ecological monitoring program designed to assess the potential impacts by commercial fisheries on ecosystem processes (CCAMLR 1985, 1988).

The parameters currently recommended for penguin monitoring studies are: breeding population size, breeding success, incubation shift durations, diet characteristics, foraging trip duration, chick fledging weight, demography, and adult arrival weight. Also identified were a series of land-based study sites where seabird monitoring should be carried out. The US sites include Palmer Station, Admiralty Bay, and Seal Island (Fig. 1) and are all located in an integrated study area along the western side of the Antarctic Peninsula, a region where a major krill fishery is likely to evolve.

[1] Point Reyes Bird Observatory, 4990 Shoreline Highway, Stinson Beach, CA 94970, USA.

Antarctic Ecosystems. Ecological Change and Conservation
(ed. by K. R. Kerry and G. Hempel)
© Springer-Verlag Berlin Heidelberg 1990

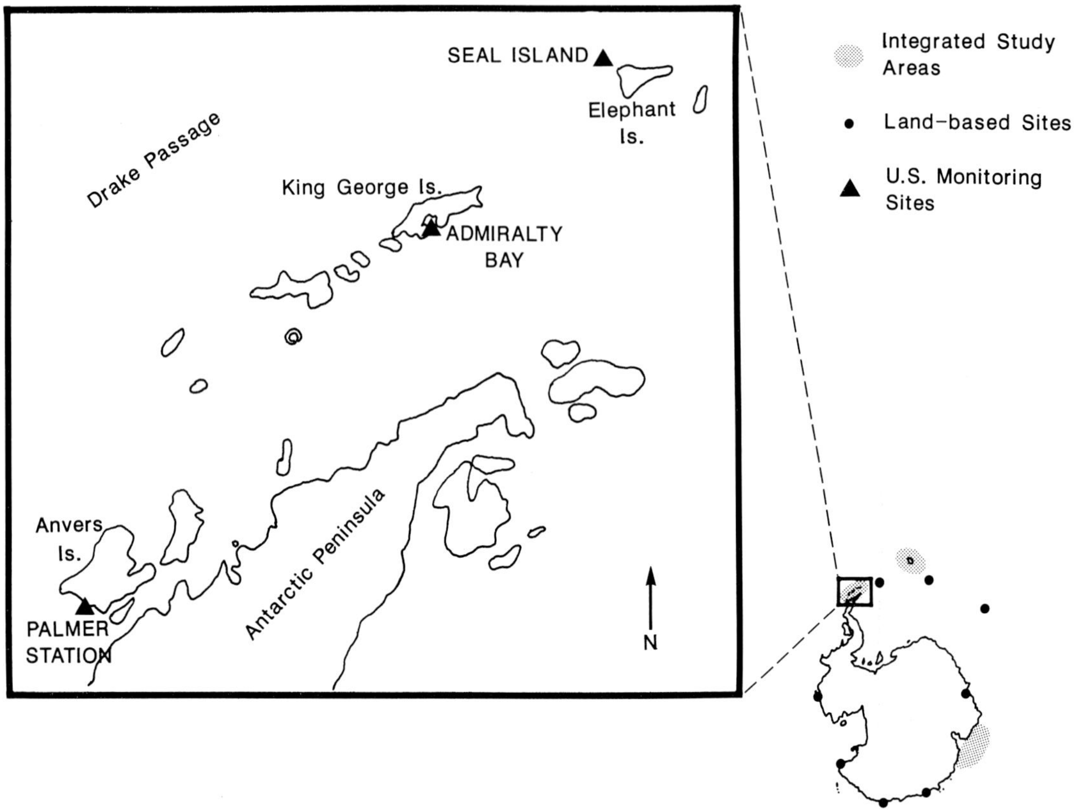

Fig. 1. The location of the US CCAMLR predator monitoring sites in the integrated study area of the Antarctic Peninsula region

The breeding biology and feeding ecology of Adélie and chinstrap penguins have been the focal point of a decade-long study at Admiralty Bay, King George Island, South Shetland Islands, Antarctica (62° 10′ S, 58° 30′ W) (Fig. 2 a, b). This report summarizes data currently available from this site with regard to the indicator parameters recommended for predator monitoring studies (CCAMLR 1988). It should be noted, however, that most of these data were collected prior to the development of the ecosystem predator monitoring concept, as part of research projects with different scientific aims. The methods used are similar but not wholly in accordance with those recommended under the CCAMLR ecosystem monitoring program (CCAMLR 1988).

The main objectives of this chapter are (1) to describe the interannual variation found within the various parameters and (2) to examine this variation among the parameters within and between seasons to evaluate their potential for monitoring purposes.

2 Methods and Results

2.1 Breeding Population Size

Penguins occupying nests with eggs were counted 1 week following the peak of egg laying for each species; early November for Adélies, early December for chinstraps. In the initial census year (1977), Adélie penguins were counted in all small breeding groups of less than 350 nests (n = 21); in larger breeding goups (n = 2), numbers were estimated by measuring the breeding group's area and applying nest-density figures derived from the smaller breeding groups (1.13 pairs m^{-2}; Trivelpiece and Volkman 1979). The smaller Adélie breeding groups contained approximately 30% of the colony population. Population data for 1981 to 1986 were estimated by recounting these 21 small breeding groups. The resultant numbers were then divided by the 1977 count of the 21 breeding groups, and this ratio was multiplied by the 1977

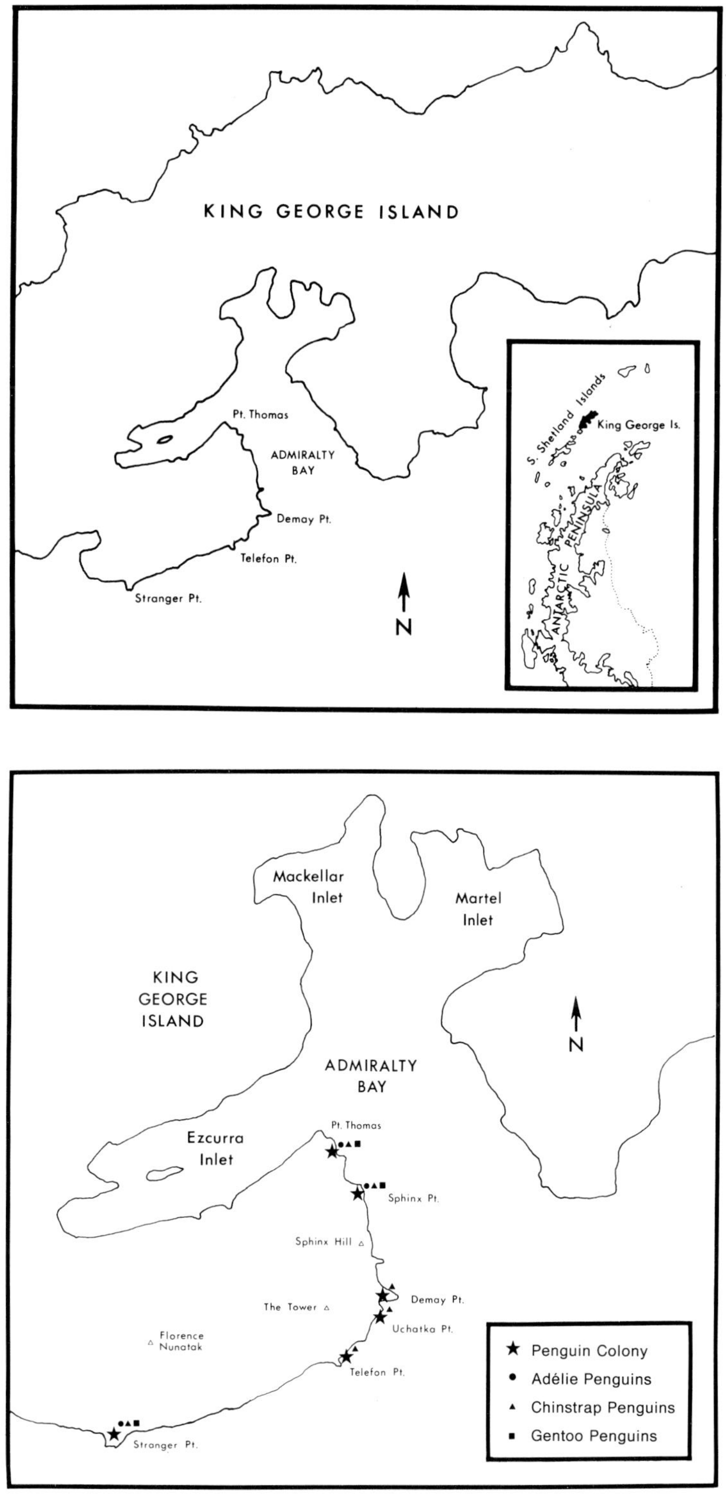

Fig. 2a. The location of the Admiralty Bay monitoring site, King George Island, South Shetland Islands, Antarctica. **b** The species composition of the penguin colonies in Admiralty Bay region. Our research was conducted in the Point Thomas and Sphinx Point colonies

Table 1. Number of breeding pairs of Adélie and chinstrap penguins in the Sphinx colony, Admiralty Bay, 1977–1986[a]

Census year	Adélie penguin	Annual change (%)	Chinstrap penguin	Annual change (%)
1977	7735	–	290	–
1979[b]	6728	– 6.5	349	+ 10.2
1981	9272	+ 18.9	240	– 15.6
1982	7316	– 21.1	280	+ 16.7
1984	7673	+ 2.4	259	– 3.8
1985	6870	– 10.5	282	+ 8.9
1986	7844	+ 14.2	218	– 22.7

[a] Calculations of percent annual change measure the direction (increasing or decreasing) and magnitude of change in Adélie and chinstrap breeding numbers compared to the most recent census data (e.g., the preceding year's census). Whenever more than 1 year elapsed between successive counts, the resultant % change in population numbers was divided by the number of years since the previous count to give an average % annual change over the intervening (2-year) period:

$$\text{percent annual change} = \frac{(\text{population year i} + 1 - \text{population year i})/\text{population year i}}{\text{number of years between censuses.}}$$

[b] Data from Jablonski (1984).

colony population estimate of 7735 pairs to obtain an estimate in each subsequent year. Data from 1979 are from a complete census by Jablonski (*1984 pers. commun.*) which entailed aerial photography of the two large breeding groups and counts of the remaining breeding groups. All chinstrap population estimates are direct counts of the few breeding pairs at this site.

The size of the Adélie and chinstrap breeding populations is known for 7 years between 1977 and 1986 (Table 1). The data presented are only from the Sphinx colony which is protected from disturbance due to its distance and isolation from Arctowski Station (3 km). It is also known that no birds have been collected at this site.

Annual variation in the size of the Adélie and chinstrap breeding populations was large, exceeding 20% in some years. Interestingly, years of increasing Adélie numbers coincided with years of decreasing chinstrap numbers.

2.2 Breeding Success

The breeding success of several hundred pairs of Adélie and chinstrap penguins was determined annually for seasons 1977/78, 1981/82, 1982/83, and 1984/85–1986/87. Pairs were randomly selected for study upon their arrival at the colony, in October (Adélies) or November (chinstraps). Nest sites were marked and drawn on a colony map and one or both members of the pair were banded with metal flipper bands. Nest sites were visited daily until chicks entered crèches at approximately 3 weeks of age.

Adélie breeding success was fairly constant during the study period, except for the significantly lower breeding success during the 1982/83 season (Table 2). Breeding success was highest in the 1986/87 season.

Chinstrap penguins experienced significantly lower breeding success in 1982/83 and 1986/87. During this latter year, success was actually lower than the reproductive data indicated and was comparable to the 1982/83 season when total population success was calculated. Twenty-two percent of the banded chinstraps with previous breeding experience did not attempt egg laying in 1986/87 although they returned to their nest sites in November as usual. If these nonbreeding birds are taken into account in the calculation of breeding success, the actual number of chicks fledged per potential pair was 0.65, close to the 1982/83 success rate of 0.53 and well below other years (Table 3). Particularly high reproductive success occurred in 1984/85 and 1985/86.

Table 2. Reproductive success of Adélie penguins ($\bar{x} \pm$ SE)

Year	No. breeding pairs	Eggs laid per pair	Chicks hatched per pair	Chicks fledged per pair
1977/78	550	1.84 ± 0.02	1.18 ± 0.12	0.98 ± 0.12
1981/82	809	1.94 ± 0.02	1.39 ± 0.04	1.02 ± 0.04
1982/83	289	1.88 ± 0.02	0.99 ± 0.06	0.67 ± 0.06[a]
1984/85	254	1.91 ± 0.02	1.33 ± 0.08	1.12 ± 0.08
1985/86	166	1.91 ± 0.03	1.20 ± 0.14	0.97 ± 0.11
1986/87	207	1.98 ± 0.01	1.48 ± 0.06	1.17 ± 0.10
MEAN	379	1.91 ± 0.02	1.26 ± 0.08	0.89 ± 0.08

[a] Significantly different from all other years; $P < 0.01$, ANOVA and Duncan's New Multiple Range Tests (DNMRT).

Table 3. Reproductive success of chinstrap penguins ($\bar{x} \pm$ SE)

Year	No. breeding pairs	Eggs laid pe pair	Chicks hatched per pair	Chicks fledged per pair
1977/78	330	1.87 ± 0.02	1.10 ± 0.05	1.06 ± 0.04
1981/82	317	1.88 ± 0.02	1.07 ± 0.07	0.98 ± 0.06
1982/83	161	1.86 ± 0.05	0.62 ± 0.10	0.53 ± 0.08[a]
1984/85	200	1.89 ± 0.04	1.37 ± 0.13	1.13 ± 0.13
1985/86	360	1.92 ± 0.03	1.35 ± 0.13	1.16 ± 0.13
1986/87	260	1.83 ± 0.04	0.94 ± 0.16	0.83 ± 0.09[a]
MEAN	271	1.88 ± 0.03	1.08 ± 0.11	0.95 ± 0.09

[a] Significantly different from all other years, $P < 0.01$, ANOVA and DNMRT.

2.3 Incubation Shifts

The nest attendance patterns of Adélie and chinstrap penguins during the incubation period were determined in 1981/82 (Adélie only) and during the four seasons, 1982/83, 1984/85–1986/87 for both species. One or both members of all pairs were flipper-banded during the courtship period and pairs were sexed by copulation and/or tread marks on the females' back (Sladen 1958; Ainley et al. 1983). Pairs were visited daily until chicks hatched. Data collection for Adélies began on the day the clutch was completed (day = 0). as both members of an Adélie pair remain on the nesting territory throughout the courtship and egg-laying period. Following clutch completion, Adélie males normally incubate first while females go to sea to feed. Reverse incubation routines (females incubating first) were rare, characteristically observed among younger, inexperienced breeders, and usually resulted in nest failure. Reverse incubations were not included in the sample.

Data were collected on chinstrap pairs from the day the first egg was laid because males often depart for sea before completion of the clutch. In determining the duration of chinstrap incubation shifts, we used the clutch completion date as day zero. However, the total time spent feeding at sea by Chinstrap males that left between eggs, accounts for all the days they were gone, including any days prior to clutch completion.

Data analysis was only performed on pairs that laid two-egg clutches and successfully hatched both chicks. Whenever both members of a pair were found at the nest during daily checks, each received credit for half a day. Parameters examined included: length of the first incubation shift of males and females following the courtship fasting period, the combined total first at-sea foraging period for males and females, the total number of days spent incubating by each sex, and the number of nest relief shifts during incubation.

Male Adélie penguins incubated more total days than females in all years (19.2 vs 14.8, respectively); however, there were no differences within males or females between years over the five study seasons (Table 4). Data collected on the remaining Adélie incubation variables showed that patterns in the 1982/83 season were significantly different from those in other years. The mean first incubation shift was longest for males and females in 1982/83 (a mean 13.5 vs 11.9 for males; 12.1 days vs 9.4 for females in all other years). The total male plus female first at-sea foraging period was also longer in 1982/83 (25.4 days) than in all other years ($\bar{x}$ = 21.1 days). Finally, the mean number of

Table 4. Nest attendance patterns of Adélie penguins during the incubation period ($\bar{x} \pm$ SE)

Year	n$^{t\,a}$	Male		Female		Total days 1st foraging trip by M + F[a]	Number of nest reliefs
		1st shift	Total days	1st shift	Total days		
1981/82	108	12.8 ± 0.3	19.8 ± 0.2	9.4 ± 0.2	13.1 ± 0.2	21.7 ± 0.4	4.3 ± 0.1
1982/83	52	13.5[b] ± 0.3	19.6 ± 0.3	12.1[b] ± 0.3	14.2 ± 0.3	25.4[b] ± 0.5	3.4[b] ± 0.2
1984/85	65	11.5 ± 0.3	18.8 ± 0.3	9.7 ± 0.4	14.0 ± 0.3	21.1 ± 0.5	5.5 ± 0.2
1985/86	31	11.1 ± 0.5	18.5 ± 0.4	8.6 ± 0.6	13.9 ± 0.4	19.6 ± 0.8	6.7 ± 0.4
1986/87	149	12.2 ± 0.2	19.1 ± 0.2	10.4 ± 0.4	13.7 ± 0.2	22.2 ± 0.3	4.1 ± 0.1

[a] n^t Number of pairs that laid two eggs and successfully hatched both; M = Male, F = female.
[b] Significantly different from all other years; $P < 0.01$, ANOVA and DNMRT.

Table 5. Nest attendance patterns of chinstrap penguins during the incubation period ($\bar{x} \pm$ SE)

Year	n$^{t\,a}$	Male		Female		Total days 1st foraging trip by M + F[a]	Number of nest reliefs
		1st shift	Total days	1st shift	Total days		
1982/83	31	7.9[b] ± 0.6	17.3 ± 0.4	5.0[b] ± 0.5	15.4 ± 0.4	14.0[b] ± 0.8	7.0[b] ± 0.5
1984/85	41	4.7 ± 0.5	17.3 ± 0.4	2.8 ± 0.3	15.1 ± 0.4	7.7 ± 0.7	10.9 ± 0.6
1985/86	45	4.1 ± 0.4	16.9 ± 0.4	3.3 ± 0.3	16.0 ± 0.4	8.1 ± 0.5	10.9 ± 0.4
1986/87	19	8.5[b] ± 0.6	17.4 ± 0.6	6.2[b] ± 0.7	16.0 ± 0.6	16.1[b] ± 0.8	8.1[b] ± 0.5

[a] n^t Number of pairs that laid two eggs and successfully hatched both; M = males, F = female. Males incubated first in 48 of 136 pairs (35%), females in 88 of 136 pairs (65%) during the 4 years.
[b] Denotes no significant difference between 1982/83 and 1986/87 or between 1984/85 and 1985/86 for any variable. However, the former two seasons are significantly different from the latter two seasons for all variables except total days spent incubating by males and by females; $P < 0.01$, ANOVA and DNMRT.

nest reliefs in 1982/83 (3.4) was fewer than in all other years.

Male chinstraps spent a mean 17.3 days on the nest during incubation, which was significantly more days than the mean 15.5 spent by females. However, like Adélies, there were no differences among males or females between years for total incubation time. Chinstrap penguins exhibited similar patterns in the 1982/83 and 1986/87 seasons, which were markedly different from those in 1984/85 and 1985/86 seasons. First incubation shifts during the 1982/83 and 1986/87 seasons averaged longer for males (8.3 vs 4.4 days) and females (5.6 vs 3 days); first at-sea foraging trips were longer (15.4 vs 8.5 days for males and females combined); and nest reliefs during the incubation period were fewer (7.6 vs 10.9, respectively; Table 5).

2.4 Diets

Prey characteristics in the diets of Adélie and chinstrap penguins are available for six seasons, 1977/78–1982/83. Data from the first (Volkman et al. 1980) and final two seasons (Volkman et al. in prep.) are from our study; the 1978/79–1980/81 data are from Jablonski (1985). Adult stomach samples were collected during the chick-rearing period: December to January for Adélies and January to February for chinstraps. Adults were captured upon returning from the sea but prior to feeding their chicks. Stomach contents were acquired by killing birds (1977/78), using emetics (1978/79 to 1980/81) or stomach pumping (1981/82 and 1982/83; Wilson 1984). Prey were sorted into three categories, krill, fish, and other, and weighed to the nearest gram.

Adélies relied almost exclusively on krill (*E. superba*) in all years except 1980/81, when many amphipods were taken. The mean composition of the Adélie diet over the 6 year period was 95.4% krill, 1.6% fish, and 3% other prey, primarily amphipods (Table 6).

Chinstraps ate predominantly krill, except in the 1980/81 season when fish was the major dietary item brought to chicks. Chinstraps also captured significant numbers of amphipods in 1978/79 and 1980/81. The mean composition of the chinstrap diet over the 6-year period was 83.6% krill, 11.1% fish, and 5.3% other, primarily amphipods (Table 7).

The amount of food brought back to chicks by foraging parents varied between a mean 419–654 g for Adélies and 402–498 g for Chinstraps during the 6-year period, 1977/78–1982/83 (Table 8). Adélie penguins were captured on the beach in 1977/78; thus, both breeders and nonbreeders were likely to be included in the stomach sampling, which may account for the lighter food loads in that year. In all other years (1978/79–1982/83), penguins were captured upon arrival at their nest sites, thereby assuring us they were returning to feed chicks.

There were no interannual differences in the amount of food brought to chicks by Adélie or chinstrap penguins. However, Adélies returned with significantly greater amounts of food for their young in every season where we were certain that only successful adults were being sampled (x = 605 vs 460 g for Adélies and chinstraps, respectively, 1978 to 1982).

There were no differences in the mean size of krill eaten by Adélies and chinstraps in any year (Table 9). The sizes of individual krill are from a subsample of 50

Table 6. Percent wet weight of food items in Adélie penguin stomach samples, 1977/78–1982/83

Season	n	Krill (%)	Fish (%)	Other (%)
1977/78	13	100.0	0.0	0.0
1978/79[a]	24	98.2	1.3	0.5
1979/80[a]	19	95.1	3.4	1.5
1980/81[a]	54	79.8	4.3	15.9
1981/82	14	99.6	0.4	0.0
1982/83	14	99.6	0.4	0.0
MEAN	23	95.4	1.6	3.0

[a] From Jablonski (1985).

Table 7. Percent wet weight of food items in chinstrap penguin stomach samples, 1977/78–1982/83

Year	n	Krill (%)	Fish (%)	Other (%)
1977/78	18	99.7	0.3	0.0
1978/79[a]	13	80.1	0.0	19.9
1979/80[a]	20	97.3	0.0	2.7
1980/81[a]	51	25.7	64.7	9.6
1981/82	15	98.7	1.3	0.0
1982/83	15	99.8	0.2	0.0
MEAN	22	83.6	11.1	5.3

[a] From Jablonski (1985).

Table 8. Mean (± SE) food load size (g) brought to Adélie and chinstrap chicks by foraging parents

Year	Adélie penguins		Chinstrap penguins	
	n	Load weight	n	Load weight
1977/78[a]	13	419 ± 63.3	18	468 ± 61.5
1978/79[b]	24	600	13	498
1979/80[b]	19	654	20	474
1980/81[b]	54	599	51	470
1981/82	14	596 ± 41.1	15	402 ± 49.7
1982/83	14	577 ± 50.6	15	454 ± 48.7
MEAN[t]	25	605	23	460

[a] Means calculated excluding 1977/78 when sampling may have included nonbreeding birds.
[b] From Jablonski (1985); no data available for ± SE.

Table 9. Comparisons of length (mm) of euphausiids consumed by Adélie and chinstrap penguins; mean (± SE)[a]

Species	n	1977/78 Length	n	1981/82 Length	n	1982/83 Length
Adélie penguin	13	41.8 (0.3)	14	43.2 (0.2)	14	39.8 (0.2)
Chinstrap penguin	18	42.5 (0.2)	15	43.8 (0.2)	15	40.3 (0.2)

[a] No significant differences between species.

animals selected randomly from the fresh portion of each stomach sample. Krill lengths were measured from the eye to the tip of the telson.

2.5 Foraging Trip Durations

The duration of foraging trips of banded pairs of Adélie and chinstrap penguins was recorded during 1981–82 and 1982–83. We defined a foraging trip as the time between the departure of a penguin from its nest site and its return to the nest from the sea. We stood near the study colonies and watched penguin pairs continuously for 12 to 15 h day^{-1} during 1981–82, and made periodic checks of nests between observation periods. Data from nests where we were uncertain whether or not a relief had taken place during our absence were not used. We observed nests 24 h day^{-1} from wooden huts on the edge of colonies during 1982–83. Observations were conducted during the time each species was rearing 1–2-week-old chicks: 8–20 December for Adélies, 7–15 January for chinstraps.

Table 10. Mean (± SE) foraging trip duration (h) for Adélie and Chinstrap penguins during their respective early chick guard stages

Species	1981–82 n^a	1981–82 h	1982–83 n^a	1982–83 h	F^b	P^b
Adélie penguin	19	22.0 (0.9)	18	26.7 (1.2)	9.93	0.004
Chinstrap penguin	22	16.2 (0.8)	25	17.1 (1.6)	0.28	NS
F		28.6		21.1		
P		< 0.0001		< 0.0001		

[a] n = Number of pairs.
[b] F, F-test; P, probability.

Adélie foraging trips were significantly longer than those of chinstraps in both years ($\bar{x}$ = 24.4 and 16.7 h, respectively), and Adélies spent a significantly longer time at sea in 1982/83 than in 1981/82 (i.e., total time away from nest minus time spent on land). There were no between-year differences in duration of chinstrap feeding trips (Table 10).

2.6 Fledging Weights

Data on the weights of Adélie and chinstrap fledglings are available for 7 and 6 years, respectively 1977/78–1982/83, plus 1986/87 for Adélies (Figs. 3, 4). Weights recorded in 1977/78, 1981/82, and 1982/83 were from

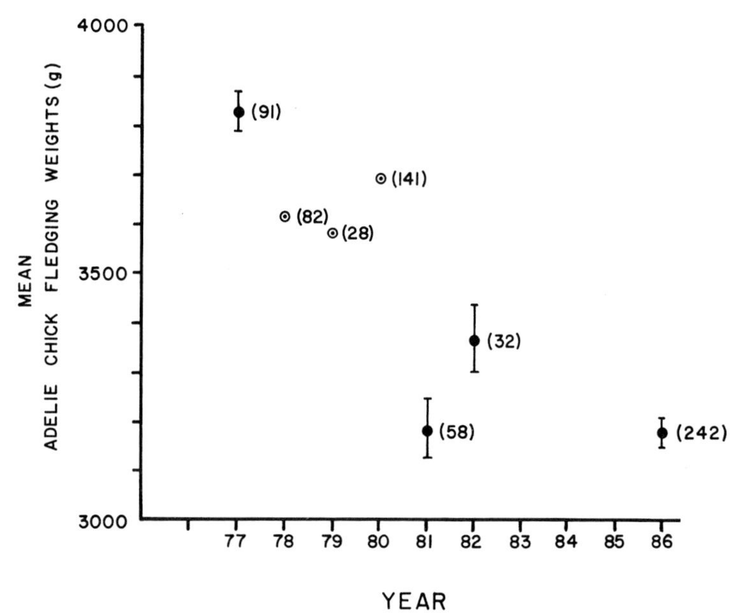

Fig. 3. Adélie chick fledging weights ($\bar{x}$ ± SE) for the period 1977/78 and 1986/87. *Solid circles* are data from our studies; *open-dot circles* are from Jablonski (1985). Sample sizes in *parentheses*

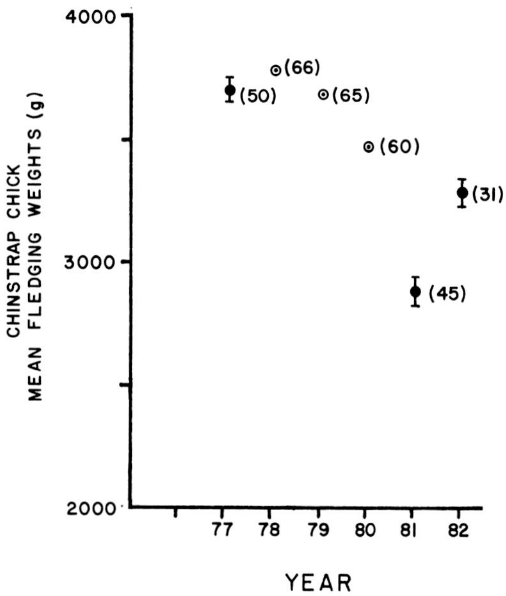

Fig. 4. Chinstrap chick fledging weights ($\bar{x} \pm SE$) for the period 1977/78 to 1982/83. *Solid circles* are data from our studies; *open-dot circles* are from Jablonski (1985). Sample sizes in *parentheses*

our comparative study of penguin chick growth. Chicks were weighed every 3 days from hatching to fledging, defined as the last day a chick was located in or near its natal colony.

Weights for the 3 years, 1978/79–1980/81, were from a feeding study conducted by Jablonski (1985); figures for chicks from single- and two-chick broods for these years were averaged. Weights were obtained in the breeding colonies during the final stage of chick moult (Jablonski 1985), and are comparable to those gathered in the growth study. Data from 1986/87 for Adélies are the mean weights of chicks captured on the colony beaches as they were about to depart for sea. Thus, assuming that these chicks may not have fed for a few days, we might expect these figures to be lower than those collected by the other methods. Chinstrap weights were unavailable that year due to our early departure date.

There were no significant differences in the weights of chicks from single- or two-chick broods during our study. The mean age of fledging for Adélie and chinstrap chicks was 50–55 days and did not differ statistically.

2.7 Demography

Demographic data have been collected since 1984/85 on known-age chick cohorts banded since 1980 (chinstraps) or 1981 (Adélies). It is too early to assess the full potential of including demographic data in monitoring studies; however, indications are that this may be one of the most sensitive indices of environmental variability. The percent of young (3- to 5-year-old) birds that attempted breeding between 1984 and 1986 varied considerably and may serve as an indicator of the winter/spring feeding conditions. Significantly fewer 3-year-old Adélies attempted breeding in 1984/85 than in all other years; while significantly fewer younger chinstraps attempted breeding in 1986/87 (Table 11).

This parameter may serve as an indicator of the general breeding condition of Adélie and chinstrap penguins at the beginning of each season and may aid in interpreting the results of other parameters, such as incubation shift durations, that are themselves influenced by both body condition and spring food supply.

3 Discussion

The census data for Adélie and chinstrap penguins collected in 7 of 10 years reveals two facts pertinent to monitoring studies. First, the year-to-year variation in numbers of breeding birds was large for both the Adélie and chinstrap populations, with changes of over 20% between some years (Table 1). This large annual variation suggests that interpreting population trends from spot censuses (i.e., counts taken at 5- or 10-year intervals) may give misleading impressions of the actual status of a penguin population. For instance, if we examine the Adélie data in Table 1, we find that during the 10-year period 1977 to 1986, the Adélie population was stable (n = 7735 pairs to 7844 pairs; 0.1% annual

Table 11. Percent of Adélie and chinstrap age groups that attempted breeding in the 1984/85 to 1986/87 austral summers following different overwinter conditions

Year	Environmental conditions preceding winter-spring in South Shetland Islands, Antarctica[a]	Adélie age cohorts			Chinstrap age cohorts			
		3	4	5	3	4	5	6
1984/85	Mild temp., little pack ice	1.8	–	–	22.2	46.7	–	
1985/86	Mild temp., little pack ice	9.2	26.5	–	20.2	56.3	75.0	
1986/87	Cold temp., heavy pack ice	9.9	55.6	81.0	2.9	23.5	28.6	100.0

[a] Temperature and ice information from overwinter personnel at nearby Polish Arctowski Station.

Table 12. Relationship between actual breeding success and breeding success estimated from chick counts in chinstrap penguins, 1984/85–1986/87

Baseline census year	Breeding population size	Breeding success estimates[a] 1984/85	1985/86	1986/87
1977	290	1.03	1.13	0.69
1979	349	0.86	0.94	0.57
1981	240	1.25	1.36	0.83
1982	280	1.07	1.17	0.71
Actual breeding success		1.13	1.16	0.83

[a] 1. Estimated breeding success: the number of chicks fledged in a given year divided by the breeding population size from a previous baseline census year (e.g. 1977, 1979, 1981, 1982).

$$\text{Est. breeding success} = \frac{\text{No. chicks fledged year i + 1, or i + 2, or i + 3, etc.}}{\text{pop. size year i (baseline census year)}}$$

2. The number of chicks fledged per breeding pair determined by following 200, 360, and 260 chinstrap pairs daily from egg laying to chick fledging in 1984/85, 1985/86, and 1986/87, respectively.

change). If 1979 is taken as the baseline year (n = 6728 pairs) the 7-year interval between 1979 and 1986 computes to a +2.4% annual increase in the Adélie population. However, if 1981 becomes the baseline year (n = 9272 pairs), we calculate a − 3.1% annual decrease in the Adélie population. Two to 3% annual changes in long-lived seabird populations could be significant if they are real. The only way to assess these data are by measurements of the annual variability in the populations over many years.

Second, the magnitude of change in breeding population size was similar but opposite in direction for the two species in all years surveyed to date (Table 1). Thus, large annual increases in the Adélie breeding population occurred coincident with large decreases in the chinstrap breeding population and vice versa. This pattern indicates that the two congeners are affected differentially by winter conditions, perhaps through their choice of habitat or diet. Recent information gathered during two Antarctic winter cruises in 1988 as part of NSF's AMERIEZ program has provided the first data that explain the differences in overwinter survival between Adélie and chinstrap penguins. Adélies were found in very large concentrations in the Weddell Sea pack ice during the July cruise (Fraser et al. in prep.); however, no chinstraps were observed associated with the pack ice in this region. Chinstraps, but not Adélies, were found wintering pelagically well to the north of the pack ice, near the Antarctic convergence during the August AMERIEZ cruise (Fraser et al. in prep.). This strongly suggests that differing winter habitats are the primary, ultimate cause of the differential overwinter survival we have reported in our Adélie and chinstrap populations; although variations in prey type or availability between the pack ice and pelagic communities are likely proximate factors as well.

Breeding success (chicks fledged per pair) varied little over the 6 years studied, except for the low breeding success in the 1982/83 season for both species, and in 1986/87 for chinstraps. Among the reproductive variables, the percent of chicks hatched and chicks fledged, but not clutch size, showed appreciable variation. The variability and correspondence in breeding success and the variability in breeding numbers suggest that these are parameters suitable for the monitoring program.

Unfortunately, we had no way to evaluate chick counts, a parameter being considered by CCAMLR for inclusion in monitoring programs as an index to colony reproductive status (i.e., population + success). Variations in the number of chicks produced in a colony may provide an index to long-term trends (5–20 years). However, using chick counts to estimate breeding success in a given year will only be possible if data are also available on the breeding population size in the monitored colonies for that same year. The large, annual variation in breeding numbers of Adélie and chinstrap penguin populations (Table 1) will greatly bias any breeding success estimate based on an earlier year's census of the breeding population (see Table 1). For example, depending on the census year chosen as the baseline for reproductive calculations (1977, 1979, 1981, or 1982), breeding success estimates vary from − 24% to + 11% of the 1984/85 actual mean success, − 19% to + 17% of the 1985/86 mean success, and as low as − 31% below the actual 1986/87 breeding success (Table 12). Breeding success can only be determined by following a sample of selected nests throughout a complete breeding season.

The juxtaposition of highly variable breeding numbers against less variable breeding success is most likely the result of interactions between the quality of potential breeders and the environment. Breeder quality is a function of age, experience, and genetic makeup. The environment affects a bird mainly through physiological fitness (i.e., fat reserves). Most reproductive failures occur during the incubation period and are the proximate result of poorly timed nest reliefs. Ultimately, nest relief timing is a function of foraging success and the rate of physiological recovery following fasting. Penguins that begin breeding in a poor physiological state are less likely to successfully complete the prolonged fast during the courtship and incubation periods. Younger, less experienced birds are more likely to be in a poor physiological state. We hypothesize that following winters of low prey availability (a function of prey distribution and numbers, and ice conditions), the composition of the breeding population is largely composed of older, experienced birds, and the number of young birds is significantly reduced (Table 11). If physiologically fit, these older birds should fare well. Alternatively, we hypothesize that following winters of high prey availability, older birds are more likely to be in good physiological condition and they are unlikely to have difficulty fasting during the courtship and incubation periods. However, the number of younger, inexperienced penguins making their first (and usually unsuccessful; Ainley et al. 1983) breeding attempt is greatly increased (Table 11). The inclusion of the younger cohorts may dilute the average breeding success of the population.

The interaction of the breeding condition and the age composition of the breeding population in any given breeding season tends to dampen the overall variability in breeding success and is a primary reason for the relatively stable number of chicks fledged per pair in 4 or 5 of the 6 study years (Tables 2, 3).

Data on the diets of Adélie and chinstrap penguins were confined to respective chick-rearing periods. The diets of the two species were similar overall, with a mean 95.8 and 83.6%, respectively, of the diet being krill. Adélies fed exclusively on krill in 5 of the 6 years, while chinstraps were exclusive krill eaters during 4 of these same seasons. The 1980/81 season showed the most divergence in prey type. Adélies ate appreciable amounts of amphipods (~ 16% by wet wt) and chinstraps consumed primarily fish (65% by wet wt) and ate significant amounts of amphipods (~ 10% by wet wt). The remainder of the diet was krill. Chinstraps also fed heavily on amphipods during the 1978/79 year when these crustaceans comprised nearly 20% of their diet.

Information on Adélie and chinstrap diets from 1978/79 to 1980/81 was collected at our study site by Jablonski (1985) during a period when we were not present. Thus, we have no ancillary data on breeding success or other related parameters that might help interpret his diet data. However, we do have data on Adélie and chinstrap penguin diets during 1982/83 when reproduction was poor; both species relied entirely on krill, just as in the 1977/78 and 1981/82 years of high success. Furthermore, the food load sizes brought to chicks in 1982/83 were not different from the earlier years (Table 8) and the fledging weights of chicks in 1982/83 were 200 g heavier than in 1981/82 (Figs. 3, 4). Krill, although plentiful in the January 1983 sampling period, may have been less available earlier in the breeding season.

The probability of different feeding conditions early vs late in the 1982/83 season is further supported by the data on foraging trip durations. The length of foraging trips made by Adélies in 1982/83 was significantly longer than trips at this same time of year in 1981/82 (Table 10). There were, however, no differences in foraging trip durations between the years for chinstraps (Table 10). In that Adélies breed 1 month earlier than chinstraps at our Admiralty Bay site, this suggests that feeding conditions may have improved significantly between the mid-December Adélie and mid-January chinstrap data collection times in 1982/83. The reproductive data tend to support this conclusion. Whereas the earlier breeding Adélies lost 47% of their eggs and 33% of their chicks during the 1982/83 season, the later breeding chinstraps lost only 15% of their chicks; however, chinstraps experienced very heavy egg losses with 67% of the eggs laid failing to hatch. The considerably higher egg losses found among chinstraps in 1982/83 than among Adélies is also consistent with poor spring and early summer feeding conditions. Adélies have significantly longer incubation shifts than chinstraps (Tables 4, 5), which implies both more time and a greater potential foraging range available to locate food in sparse years; hence, Adélies may have better chances of being successful, other factors being equal.

Nest attendance patterns also indicated the anomalous breeding success in 1982/83. The mean length of the first trip to sea (in days) by male and female Adélie penguins was significantly longer in 1982/83 than in any other year (Table 4). Accordingly, the number of nest exchanges by pair members during incubation were significantly fewer than in any other year (Table 4). This indicates that food was less available in the early phase of the 1982/83 breeding cycle. Chinstraps show a similar pattern for 1982/83 (Table 5).

Additionally, the 1986/87 season was comparable to that of 1982/83 for chinstraps, and very different from either the 1984/85 or 1985/86 years (Table 5). The differences between chinstraps and Adélies in 1986/87 are probably best explained by the apparent difference in physiological condition among individuals of each species at the start of the breeding season. Indicators are that chinstraps began the 1986/87 year in very poor condition. This is clear from three separate pieces of evidence: (1) The 1986/87 chinstrap breeding population was 23% lower than in 1985/86 (Table 1); (2) 22% of banded chinstraps with previous breeding experience did not attempt to breed in 1986/87; and (3) the percent of younger, known-age birds attempting to breed in 1986/87 was reduced by 50 to 90% of the 1984/85 or 1985/86 levels for 3-, 4-, and 5-year-olds (Table 11). On the other hand, the Adélie breeding population was 14% higher in 1986/87 than in 1985/86, and twice as many younger-aged birds attempted to breed compared to the 1985/86 season (Table 11). Spring feeding conditions may have been low during 1986/87, a factor that would have greatly exacerbated the already hard-pressed Chinstraps while presenting less of a problem to the Adélie penguins that arrived in excellent condition.

The analysis of the life history parameters of Adélie and chinstrap penguins undertaken here clearly shows that interpretation of results is greatly enhanced when data are collected simultaneously on several variables. Population size is highly variable and data on this parameter should be collected annually. Breeding success appears to be less variable but provides clear indications of major perturbations, as seen in the 1982/83 season. Chick counts at selected colonies, a parameter currently substituting for breeding success at sites where less than full season efforts are undertaken, may provide information on long-term trends in colony status. However, breeding success cannot be inferred from chick counts unless breeding numbers are also known from the same year (Table 12).

Incubation shift durations provide insights into the body condition of breeders in the early spring. Along with populations size (and arrival weights, a parameter we could not evaluate), incubation patterns provide an indication of early-season feeding conditions.

Information on penguin diets are good indicators of the food resources in summer, and prey switching between years may be a key indicator of food resource levels. To date we have no corroborating breeding success data during the years when prey switching occurred (1978/79, 1980/81), so this possibility remains to be evaluated. However, diet studies without ancillary data on fledging weights, foraging trip duration, and other factors such as breeding success, can be misleading. This is clearly seen in the 1982/83 season when krill was the sole dietary item of Adélies and chinstraps, food load sizes brought to chicks were comparable to other years and chick-fledging weights were higher than in several years, yet breeding success was significantly reduced.

Where full-season efforts are undertaken at CCAMLR monitoring sites, the inclusion of population data collection among the parameters to be monitored would be highly desirable. Demographic data (i.e., overwinter survival and the proportion of young birds breeding) varied consistently with other parameters, and likely are more precise indicators. Annual variations in overwinter survival and in the number of younger-age birds that attempt breeding (Table 11) have already provided us with strong evidence that resources available during the preceding winter period are highly variable, a key to understanding the data collected subsequently during the breeding season. Additional insights into the interaction of predators will be garnered when known-age birds are used for studying any of the other monitoring parameters (e.g., breeding success, incubation routines, foraging abilities, diet studies). Designing monitoring studies that include demographic data collection will greatly strengthen our overall interpretive ability.

The life history parameters designated as ecosystem monitoring indicators by CCAMLR show varying degrees of sensitivity to environmental changes. As such they can be valuable tools in assessing perturbations to the Antarctic ecosystem. In assessisng the parameters identified by CCAMLR as key components for predator monitoring studies, one thing, above all else, is clear. Collecting data on several parameters simultaneously is required to detect and interpret annual differences in food availability and to ascribe significant portions of this variation to differences in overwinter, spring and summer conditions. Without the inclusion of data on population size, breeding success, incubation patterns, and demography (which requires a full-season effort), data from only the January–February summer months cannot be related to conditions earlier, and vice versa. In that a krill fishery will operate for 5 to 6 months annually, any potential effects on the food web will not be detectable through a monitoring program, unless we conduct the research throughout the austral summer season.

Given the support required to finance full-field seasons at CCAMLR monitoring sites, we feel that penguins and other seabirds will provide meaningful data on the impact of the developing krill fishery on other dependent predator populations and, in concert with other directed research components, help substantiate data on the abundance and composition of the krill stocks themselves.

Acknowledgements. We wish to thank the Polish Academy of Sciences and members of Arctowski Station for their hospitality during our 1977 to 1986 research in Admiralty Bay. We also thank the Captains and crew members of the R/V *Hero* and R/V *Polar Duke* for assistance and transportation, and USARP/USAP personnel for their logistical support. We thank S. Ware and K. Young for field assistance and Prof. K. Jazdzewski and his colleagues at Lodz University, Lodz, Poland, for help in identifying the food items in the penguins' diets. Finally, we thank D. G. Ainley, D. Adelung, and K. R. Kerry for their helpful comments on and discussions of this research. This study was conducted under NSF-Division of Polar Programs grants to the authors; numbers 18-17205 and 83-14667. This is contribution number 440 of the Point Reyes Bird Observatory.

References

Ainley DG, LeResche RE, Sladen WJL (1983) Breeding biology of the Adélie penguin. University of California Press, Los Angeles

CCAMLR (1985) Report of the fourth meeting of the Scientific Committee, Annex 7. Rep Ad hoc Working Group on Ecosystem Monitoring. SC-CAMLR-IV, Hobart, Australia

CCAMLR (1988) Standard methods for monitoring parameters for species. CCAMLR Ecosystem Monitoring Program, Hobart, Australia

Croxall JP, Prince PA (1979) Antarctic seabird and seal monitoring studies. Polar Rec 19:573–595

Everson I (1977) The living resources of the Southern Ocean. FAO GLO/SO/77/a, Rome

Jablonski B (1984) Distribution and numbers of penguins in the region of King George Island, South Shetland Islands in the breeding season 1980/81. Pol Pol Res 5:17–30

Jablonski B (1985) The diet of penguins on King George Island, South Shetland Islands. Acta Zool Cracov 29:117–186

Prevost J (1981) Population, biomass and energy requirements of Antarctic birds. In: El-Sayed SZ (ed) Biomass, vol II. Scott Pol Res Inst, Cambridge, p 125

SCAR (1979) Fifteenth meeting of SCAR, Chamonix, 16–26 May 1978 Appendix A. Working Group in Biology. Polar Rec 19:304–312

Sladen WJL (1958) The pygoscelid penguins. I. Methods of study. II. The Adélie penguin (*P. adeliae*). Sci Rep Falkld Is Depend Surv 17:1–97

Trivelpiece WZ, Volkman NJ (1979) Nest site competition between Adélie and Chinstrap penguins: an ecological interpretation. Auk 96:675–681

Volkman NJ, Presler P, Trivelpiece WZ (1980) Diets of pygoscelid penguins at King George Island, Antarctica. Condor 82:373–378

Wilson RP (1984) An improved stomach pump for penguins and other seabirds. J Field Ornithol 55:109–112

Foraging Patterns of Adélie Penguins During the Incubation Period

L. S. Davis[1] and G. D. Miller[1,2]

Summary. We report on the duration of foraging trips taken by Adélie penguins *Pygoscelis adeliae* during the incubation period at Cape Bird, Ross Island, Antarctica. In years when birds foraged longest, males apparently adjusted the duration of the Second Foraging Trip relative to the time available until hatching. In seasons where the sum of the first two foraging trips was well short of the incubation period, males showed no evidence of adjustment; however, females taking the Third Foraging Trip did. Cues for adjustment of foraging trips were not derived from eggs. Neither did the physiological condition (i.e. time spent fasting before foraging) of the birds affect the duration of foraging trips. It seems that time until chicks hatch is the crucial determinant of time spent foraging. If this is so, and food is not limiting, then attempts to relate foraging trip times to krill stocks may be ill-founded.

1 Introduction

Much of the rationale for using Adélie penguins (*Pygoscelis adeliae*) to monitor Antarctic krill stocks (*Euphausia superba* and *E. crystallorophias*) is predicated on an assumption that aspects of their reproductive performance will reflect changes in the distribution and abundance of krill (Croxall et al. 1988). For this to hold, krill, the principal component of the diet of Adélie penguins (Emison 1968; Lishman 1985), must in some way be a limiting resource that directly influences the breeding performance or behaviour of Adélie penguins (i.e. a perturbation in krill stocks will produce a perturbation in indices of penguin reproductive performance).

Durations of foraging trips, both during incubation (equivalent to measuring incubation spell length) and chick rearing, are purported to be amongst the most sensitive indices to changes in krill abundance (Croxall et al. 1988, Table 7). We know surprisingly little, however, about factors that influence breeding performance in Adélie penguins; especially the extent to which the abundance of krill may be important.

In the Northern Rookery at Cape Bird, Ross Island, two principal causes account for over 80% of all breeding failures (Davis and McCaffrey 1986):

(1) South Polar skuas (*Catharacta maccormicki*) are the main cause of chick mortality, and can also be responsible for a large proportion of egg losses, particularly from penguins in peripheral nests (Davis 1982a; Davis and McCaffrey 1986). The risk of predation from skuas remains relatively constant throughout incubation and the guard stage (it is minimal for older chicks) (Davis and McCaffrey 1986), and from season to season (Davis and Miller unpubl. data).

(2) Failure to co-ordinate nest relief is the other major cause of breeding failure and occurs when a bird spends too long foraging at sea. This can have two consequences: either its mate will desert or its chicks will starve (Davis 1982b). In contrast to losses from skuas, the risk from this form of loss is concentrated around critical periods during the breeding season (Davis and McCaffrey 1986), and losses are highly variable between seasons (Spurr 1975; Yeates 1975; Ainley et al. 1983). Over one-third of all nests may be affected (Davis 1982b).

As desertions and starvations result from foraging trips that are too long, they are more likely to be related to changes in the Antarctic marine ecosystem than is predation by skuas. Here we present results from observations and experiments carried out at Cape Bird on factors involved in the failure of Adélie penguins to co-ordinate their foraging trips, and we comment on the apparent importance of krill as a limiting resource for the penguins.

2 Background: Patterns of Foraging at Cape Bird

Adélie penguins are biparental incubators. One bird must stay on the nest to protect the eggs from cold and

[1] Department of Zoology, University of Otago, P.O. Box 56, Dunedin, New Zealand.
[2] Department of Biology, University of New Mexico, Albuquerque, New Mexico 87131, USA.

Antarctic Ecosystems. Ecological Change and Conservation
(ed. by K. R. Kerry and G. Hempel)
© Springer-Verlag Berlin Heidelberg 1990

skuas, while the other goes to sea to feed. Immediately after egg laying, with occasional exceptions, the female goes to sea for about 2 weeks: the First Foraging Trip (FFT). The male takes the Second Foraging Trip (SFT) and is away for about 2 weeks also. Incubation takes around 34 days, and usually the female takes a Third Foraging Trip (TFT), lasting 1–7 days, just before the chicks hatch (Davis 1988). Chicks must be fed regularly, and parents shorten their foraging trips to about 1 day after their chicks have hatched.

An Adélie pair must co-ordinate two, and possibly three, nest reliefs between the completion of laying and hatching. The foraging bird must balance its need to procure sufficient energy against the risks of staying away too long. We can identify two distinct periods and types of risk associated with nest reliefs during the incubation period.

2.1 The Nest Relief Between FFT and SFT

The risk if the female stays away too long is that the male will desert. Including an average of 12 days spent in courthsip (Davis and Speirs in press), males must fast for about 1 month. In a sample of 130 nests, 17% of males deserted after their partners had been away an average of 6 days longer than the mean FFT (Davis 1982b). Most females eventually return to the nest (Davis 1988).

2.2 Nest Reliefs Before Hatching

Here the risk is not so much that the incubating partner will desert (it has recently replenished its fat reserves), but that the chicks will hatch and starve to death. Unfed chicks die about 6.5 days after hatching (Reid and Bailey 1966). The greatest risk of starvation for chicks up to 40-days-old occurs when they are 6–8 days, because their foraging parent does not return to the nest once they have hatched (Davis 1982b). Starvation is unlikely after chicks reach the crèche stage at about 3 weeks; suggesting that, even if their mates are late or do not return from foraging, single parents can cope with feeding chicks once emancipated from guard duties (Davis and McCaffrey 1986). Furthermore, the risk of starvation is the same for chicks in one- and two-chick broods (Davis and McCaffrey 1986). This suggests that while starvation is a major cause of chick loss, food per se may not limiting.

3 Methods

Adélie penguins were studied over four breeding seasons at the Northern Rookery, Cape Bird, Ross Island, Antarctica (77° 13′ 10″ S, 166° 28′ 30″ E): 1977/78, 1984/85, 1986/87 and 1987/88. [References to colonies, or groups of colonies (blocks), within the rookery follow the conventions of Knox and Ensor (1978)].

3.1 Observations

The durations of foraging trips (FFT, SFT, and TFT) were recorded for pairs in H-Block (1977/78) or colony C2 (1984/85–1987/88) by inspecting all nests at least once per day. Birds were banded for individual identification using numbered aluminium alloy flipper bands, and these could be read at a distance of 15 m using 10 × 50 binoculars. All residents of colony C2 were banded, and at least one member of each pair monitored in H-Block. Banding was carried out before the start of laying, and was the only time the birds were handled.

3.2 Experiment 1

In 1986/87, clutches were transferred between ten early and ten late nests in three colonies of B-block (three nests were subsequently lost to predation by skuas) to test whether penguins derived cues to determine the incubation period (and, hence, the duration of foraging trips) from the eggs themselves. This altered the apparent incubation periods from the host parents' perspective: eggs hatched 26 days after laying was completed in the "late" host nests, and after 36 days in the "early" host nests (Miller and Davis unpubl. data).

3.3 Experiment 2

To test the effect on foraging trip times of increased energy expenditure prior to foraging, 11 females were induced to lay a third egg (usual clutch size is two) by removing their first egg soon after laying. Compared to a control group of 24 pairs, this added an average of 4 days spent at the nest fasting before the experimental females could leave to forage (Miller and Davis unpubl. data). During courtship and incubation, fasting males lose 52 g/day and fasting females lose 79 g/day (Green and Davis unpubl. data). In addition, an egg costs around 109 g of body weight to produce (Astheimer and Grau 1985). Given the average weights of males (5.0 kg, n = 37) and females (4.5 kg, n = 18) during the courtship period in 1986/87, the manipulations cost them 4 and 9% of their body weights, respectively.

3.4 Experiment 3

In 1987/88, after nine females had completed their clutches and left their colonies, they were captured and kept in pens for 4 days before being allowed to go to sea. The aim of this experiment was to test the effect of shortening the time available for foraging during the incubation period, while at the same time increasing the energy expenditure of the female. This experiment imposed similar physiological demands on the female as in Experiment 2 (without the production of another egg) but effectively reduced the time available for foraging during the incubation period by 4 days. Unless the female reduced her time at sea in response to the manipulation, the male would similarly be forced to fast longer yet face a reduction in the time available in which to complete the SFT.

It was predicted that if food is limiting, the birds should forage longer; whereas, if time available to the hatching of chicks limits the duration of foraging trips, the birds should forage for shorter durations than pairs in 76 unmanipulated control nests.

To determine whether the observations and experiments influenced reproductive success, each year a group of colonies was selected as a control group. The control colonies were visited only twice: once to count the number of occupied nests, and later, to count the number of chicks. The reproductive success of birds in the study colonies did not differ from that of the controls (Davis 1982 b; Davis 1988; Miller and Davis unpubl. data). Numbers of pairs breeding in C2 declined, however, because of a lack of recruitment by females (Davis and Speirs in press).

4 Results

4.1 Observations: Do Adélie Penguins Adjust Their Foraging Trips Near the End of the Incubation Period?

In the 1977/78 and 1984/85 breeding seasons, the time males took on the SFT was inversely related to the time they had fasted since the completion of laying (equivalent to the FFT) (Table 1). Males that fasted longest, foraged for the shortest time, suggesting that they were adjusting the duration of their SFTs in relation to the time left till hatching. Whereas in the 1986/87 and 1987/88 seasons the time males spent at sea on the SFT was not related to the duration of their partners' FFT (Table 1). In the latter two seasons, pairs completed their first two foraging trips in an average of only 26 days, so that males were likely to return from the SFT well before their chicks hatched.

By contrast, time available for females to take the TFT was always limited by the imminent hatching of their chicks. In all years, the duration of the TFT was inversely related to the time that had elapsed since the females laid their eggs (i.e. the sum of the FFT and SFT) (Table 1).

4.2 Experiment 1: Are Cues for Adjustment of Foraging Trips Derived from the Egg?

Foraging times were the same for host parents in "early" and "late" nests (Fig. 1 a). The birds did not adjust their foraging in response to cues derived from the eggs, implying that any assessment of the incubation period must be endogenously based.

Table 1. Means and standard deviations (ranges in parentheses) of durations of first (FFT), second (SFT) and third (TFT) foraging trips (in days), and correlations between FFT vs SFT (r^a) and FFT + SFT vs TFT (r^b), for pairs of Adélie penguins in each season[a]

Season	n	FFT	SFT	TFT	r^a	r^b
1977/78	79	16.5 ± 3.6 (9–25)	12.4 ± 2.9 (7–20)	3.6 ± 2.1 (1–9)	− 0.69***	− 0.46***
1984/85[b]	23	19.0 ± 3.5 (10.7–24.7)	12.6 ± 2.7 (6.5–16.1)	3.1 ± 1.7 (1.4–7.1)	− 0.55**	− 0.48*
1986/87	16	14.2 ± 2.1 (10.5–18.5)	12.0 ± 1.5 (9.5–15)	4.3 ± 1.8 (1.5–7)	− 0.07 ns	− 0.82***
1987/88	13	15.8 ± 2.8 (11.5–20.5)	9.7 ± 1.9 (7.5–13.5)	4.3 ± 2.7 (2–7)	0.04 ns	− 0.68*

[a] ns = Non-significant, * $P < 0.05$, ** $P < 0.01$, *** $P < 0.001$.
[b] From Davis (1988).

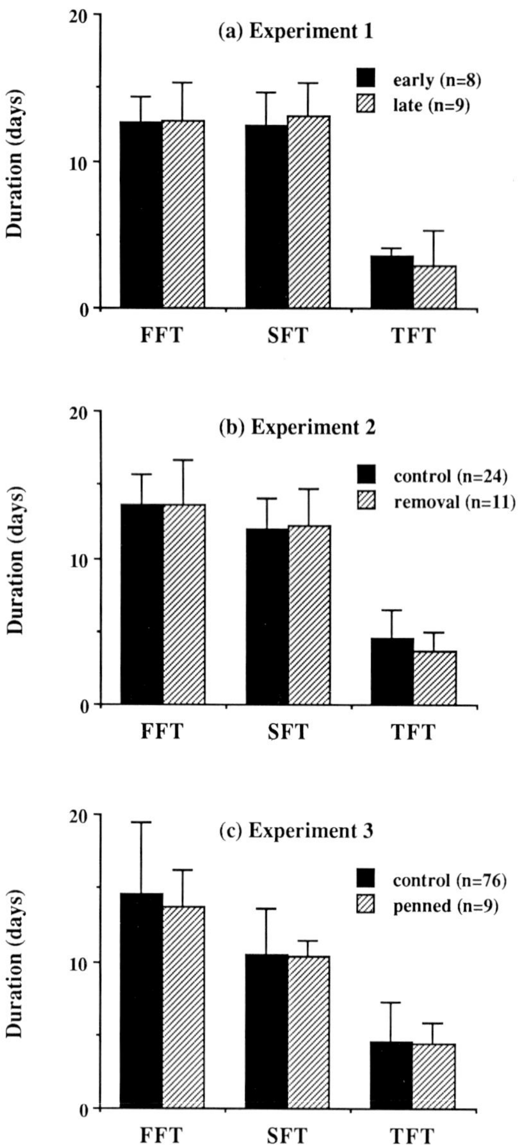

Fig. 1. Mean ($\pm$ SD) durations of the first (FFT), second (SFT) and third (TFT) foraging trips taken by Adélie penguins in **a** Experiment 1, **b** Experiment 2, and **c** Experiment 3. No comparisons within experiments were significant (i.e. $P > 0.05$, Student's *t*-test). (**a** and **b** from Miller and Davis unpubl. data)

4.3 Experiment 2: Do Energy/Nutrient Requirements Determine Time Away from the Nest?

The durations of the FFTs, SFTs and TFTs for the 11 removal pairs (i.e. those where the female had been induced to lay a third egg) were not statistically different from those of the control birds (Fig. 1 b). The penguins did not lengthen the duration of their foraging trips to compensate for poorer physiological conditions.

4.4 Experiment 3: Is Time or Food the Limiting Factor?

The foraging trips of the penned females and their mates were not statistically different from those of the controls (Fig. 1 c), although the mean FFT was nearly 1 day shorter for penned females. Despite increased time spent fasting before foraging, experimental birds did not increase their foraging time.

5 Discussion

During the incubation period, Adélie penguins foraging near Cape Bird feed almost exclusively on euphausiids (*E. crystallorophias*) (van Heezik 1988). Time spent foraging could reflect: (1) time needed to catch krill, which will be affected by their distribution and abundance, (2) time spent getting to and from foraging areas, which could be affected by ice conditions, or (3) some factor independent of foraging, such as reducing the likelihood of leopard seal predation at the ice edge (Ainley et al. 1983). The latter might explain why foraging trips during the incubation period are long, but it would not account for the seasonal variation that we observed in foraging times.

Our results demonstrate that Adélie penguins adjust the duration of foraging trips close to hatching, and that the duration of foraging trips is not affected by a bird's physiological condition (within the range tested). For this to have been so, the penguins must have accomodated the nutritional deficiencies brought about by our experimental manipulations in the time normally available for foraging. Radiotelemetry has shown that Adélie penguins foraging near Cape Bird spend about one-third of their time on ice floes (Davis et al. 1988). This suggests that food per se is not the limiting factor that determines the time away from the nest during the incubation period.

We suggest that variation in durations of foraging trips between seasons result from: (1) ice conditions (these are especially likely to affect the duration of the FFT), and (2) as a consequence of this, from adjustment of the SFT and TFT in relation to the time available to hatching of the chicks. In seasons with extensive fast-ice and late-ice breakout, first foraging trips are longer and desertions more prevalent (Ainley and LeResche 1973). At Signy Island, breeding success of Adélie penguins is correlated with the date of ice breakout: success is poorest in years where the ice persists longest, and this seems to be primarily due to its effect on time taken to find open water suitable for foraging (Croxall et al. 1988).

It is possible that the extent and persistence of ice cover is correlated with changes in krill abundance (Croxall et al. 1988), and that a major reduction in krill stocks could potentially increase foraging times independently of ice conditions. However, it must be concluded from the evidence available that food is not limiting during the incubation period, and that variations in foraging times do not reflect changes in krill stocks.

Foraging times during chick rearing may be more directly influenced by euphausiid abundance. But, the similar growth rates (Ainley and Schlatter 1972) and survival prospects (Davis and McCaffrey 1986) of chicks from one- and two-chick broods, and the paucity of starvations once both parents are emancipated from guard duties (Davis 1982b), do not suggest that krill stocks are limiting during the guard or crèche stage of chick rearing (Davis and McCaffrey 1986). Availability of euphausiids might be limiting when chicks are older than 40 days. However, if foraging times are to be used as indices of reproductive performance to monitor euphausiid stocks, then a link must be established between foraging times and krill abundance. This has not been done.

Acknowledgements. This research was supported in part by a UGC grant 86/012 (LSD) and an NSF Postdoctoral Fellowship INT 8603504 (GDM). We thank E. Speirs and R. Johnstone for field assistance. Logistic support was provided by the Antarctic Division, DSIR, and the US Navy VXE-6 squadron.

References

Ainley DG, LeResche RE (1973) The effects of weather and ice conditions on breeding in Adélie penguins. Condor 75:235–239

Ainley DG, Schlatter RP (1972) Chick raising ability in Adélie penguins. Auk 89:559–566

Ainley DG, LeResche RE, Sladen WJL (1983) Breeding biology of the Adélie penguin. University of California Press, Berkley

Astheimer LE, Grau CR (1985) The timing and energetic consequences of egg formation in the Adélie penguin. Condor 87:256–268

Croxall JP, McCann TS, Prince PA, Rothery P (1988) Reproductive performance of seabirds and seals at South Georgia and Signy Island, South Orkney Islands, 1976–1987: implications for Southern Ocean monitoring studies. In: Sahrhage D (ed) Antarctic Ocean and resources variability. Springer, Berlin Heidelberg New York, pp 261–285

Davis LS (1982a) Creching behaviour of Adélie penguin chicks (*Pygoscelis adeliae*). NZ J Zool 9:279–286

Davis LS (1982b) Timing of nest relief and its effect on breeding success in Adélie penguins (*Pygoscelis adeliae*). Condor 84:178–183

Davis LS (1988) Coordination of incubation routines and mate choice in Adélie penguins (*Pygoscelis adeliae*). Auk 105:428–432

Davis LS, McCaffrey FT (1986) Survival analysis of eggs and chicks of Adélie penguins (*Pygoscelis adeliae*). Auk 103:379–388

Davis LS, Speirs EAH (in press) Mate choice in penguins. In: Davis LS, Darby JT (eds) Penguin Biology. Academic Press, San Diego

Davis LS, Ward GD, Sadleir RMFS (1988) Foraging by Adélie penguins during the incubation period. Notornis 35:15–23

Emison WB (1968) Feeding preferences of the Adélie penguin at Cape Crozier, Ross Island. In: Austin OL Jr (ed) Antarctic bird studies. Antarct Res Ser, vol 12. American Geophysical Union, Washington DC, pp 191–212

Knox GA, Ensor PH (1978) University of Canterbury Antarctic Research Unit Expedition 16. University of Canterbury, Christchurch

Lishman GS (1985) The food and feeding ecology of Adélie penguins (*Pygoscelis adeliae*) and Chinstrap penguins (*P. antarctica*) at Signy Islands, South Orkney Islands. J Zool Lond (A) 205:245–263

Reid B, Bailey C (1966) The value of the yolk reserve in Adélie penguin chicks. Rec Dom Mus (Wellington) 5:185–193

Spurr EB (1975) Breeding of the Adélie penguin *Pygoscelis adeliae* at Cape Bird. Ibis 117:324–338

van Heezik Y (1988) Diet of Adélie penguins during the incubation period at Cape Bird, Ross Island, Antarctica. Notornis 35:23–27

Yeates GW (1975) Microclimate, climate and breeding success in Antarctic penguins. In: Stonehouse B (ed) The biology of penguins. Macmillan, London, pp 397–409

Long-Term Changes in Seabird and Seal Populations in the Southern Ocean

P. JOUVENTIN and H. WEIMERSKIRCH[1]

Summary. This chapter reports on trends in the breeding populations of selected species at Dumont D'Urville, Antarctica and at Crozet, Amsterdam and Kerguelen Islands (French Austral and Antarctic territories). Populations of 12 seabird and three seal species are reviewed and analyzed. While most of the penguin and fur seal populations are today increasing, several populations of albatrosses and petrels and elephant seals are dangerously declining. Most of the changes reported here are considered to have been induced by past and present human activities. The importance of monitoring and demographic studies of birds and seals for managing and protecting the living resources of the Southern Ocean is stressed.

1 Introduction

Various authors (Holdgate and Wace 1961; Carrick 1964; Holdgate 1970; Roberts 1977; Clark and Dingwall 1985) have drawn attention to the conservation values of Antarctic and Subantarctic islands. Although the exploitation of seals and birds is now minimal and the level of awareness of ecosystem management is higher than previous times, the increasing number of permanent stations and fishing activities consitute an increasing threat.

As the pressure upon marine resources accelerate, particularly with technological developments, the conservation and management of the Southern Ocean ecosystem becomes more critical than ever before. Since seals and birds depend on marine organisms for food, changes in the marine ecosystem will consequently have a direct impact on these predators. The importance and usefulness of birds and seals as indicators of environmental changes have been recognized in recent years with the development of international monitoring programmes, such as the International Survey of Antarctic Seabirds (ISAS), Biological Investigations of Marine Antarctic Systems and Stocks (BIOMASS) and Commission for the Conservation of Antarctic Marine Living Resources (CCAMLR), for selected predator species. Species were selected as they are the major consumers of marine resources, but other unselected species as petrels and albatrosses are also threatened directly by human activities. The monitoring of these species will be of major importance for conservation.

The purpose of this chapter is to provide a description of the French monitoring of seabirds and seals species at several localities in the Southern Ocean. Data concerning trends of 12 bird populations are presented.

2 Study Areas and Methods

Monitoring studies were conducted at four localities in the Southern Ocean.

1. In the Antarctic zone: at the Pointe Géologie Archipelago (66° 40′ S, 140° 01′ E), in Adélie Land.

2. In the Subantarctic zone: on Possession Island (46° 25′ S, 51° 45′ E), Crozet Islands (46° 25′ S, 51° 45′ E) and three localities on the Kerguelen Islands (49° S, 70° E): the Courbet Peninsula (albatrosses, penguins, seals), Mayes Island, Morbihan Gulf (petrels, penguins) and the Jeanne d'Arc Peninsula (albatrosses).

3. In the subtropical zone, on Amsterdam Island (37° 50′ S, 77° 30′ E).

Bird and mammal studies have been conducted over the last 30 years. A total of four species of pinnipeds and 24 species of birds, of which 23 are seabirds, are now being monitored on a regular basis (see Jouventin and Weimerskirch in press). Some 85 000 individuals have been banded. Field records are stored on computer and analyzed for estimation of demographic parameters, population fluctuations and modelling.

The size of populations refers for every species to annual populations and not total populations that are different in biennial breeding species like some species of albatrosses or King penguin (*Aptenodytes patagonicus*). For birds, population sizes correspond to the

[1] Centre d'Etudes Biologiques de Chizé, Centre National de la Recherche Scientifique, 79360 Beauvoir sur Niort, France.

Antarctic Ecosystems. Ecological Change and Conservation
(ed. by K. R. Kerry and G. Hempel)
© Springer-Verlag Berlin Heidelberg 1990

number of breeding pairs, i.e. generally the number of birds incubating an egg, or nests with eggs. Counts were always made just after the end of the laying period except for Macaroni penguins (*Eudyptes chrysolophus*), where counts were made during the brooding period. For petrels, albatrosses, gentoo penguins (*Pygoscelis papua*), number of pairs were counted individually and are probably accurate to better than ± 5%. For King, Adélie (*Pygoscelis adeliae*), and Macaroni penguins counts were made from photographs taken from helicopters or from promontories with a ± 5% accuracy. For Emperor penguins (*Aptenodytes forsteri*), count population estimates refer to the number of chicks fledged, added to the number of eggs and chicks lost. For fur seals and the southern elephant seals (*Mirounga leonina*), counts refer to the maximum number of adult females present on the beaches during the breeding cycle.

3 Results

3.1 Changes in the Sizes of Populations

Seals and Fur Seals. Southern elephant and both fur seal populations have followed opposite trends over the last 20 years (Table 1). On Amsterdam Island, Subantarctic fur seals (*Arctocephalus tropicalis*) have increased from 800 females in 1955/56 to 12975 females in 1981/82 (Paulian 1964; Segonzac 1972; Hes and Roux 1983). On the Crozet Islands, Subantarctic (*A. tropicalis*) and Antarctic fur seals (*A. gazella*) have both recolonized Possession Island only recently (Jouventin et al. 1982). The Subantarctic fur seal had reached a population of 100 females in 1980 (Jouventin et al. 1982), and their numbers have remained fairly stable (Table 1) probably because the single breeding site is fully occupied. Since a neighbouring beach is currently colonized by more than 100 non-breeding females and males, a further increase by breeding on this beach is likely to occur in the next few years. Antarctic fur seals reached a population of 80 females in 1987/88 (Table 1). At Kerguelen, this species is now recolonizing the mainland from the Nuageuses Islands on the north-west tip of the archipelago, where 2500 breeding females were found in 1984/85. Small isolated groups of females with pups are found in numerous zones around the mainland coast.

On Possession Island, the only regularly monitored island of the Crozet group, elephant seals showed a constant decline since at least 1970 (Barrat and Mougin 1978), the decline being more rapid since 1983. Between 1980 and 1988, the number of breeding females decreased from 1044 to 698. At Kerguelen, elephant seals declined drastically during the 1970s (Pascal 1982; Van Aarde 1980) and early 1980s.

Penguins. King penguins have increased dramatically on Crozet and Kerguelen Islands during the last three decades (Table 2). At Kerguelen, the monitored

Table 1. Annual numbers of breeding female elephant and fur seals

Species Locality	*Arctocephalus tropicalis*		*Arctocephalus gazella* Possession Island	*Mirounga leonina*	
	Possession Island (Crozet Is.)	Amsterdam Island		East Coast Possession Island (Crozet Is.)	Courbet Peninsula Kerguelen
Reference[a]	(1)	(2)	(1)	(3)	(4)
1955/1956		808			
1956/1957					73466
1958/1959					56041
1960/1961					61120
1966/1967				1050	
1969/1970		3370			
1970/1971					70945
1973/1974				642	
1976/1977				568	
1977/1978	(25)				46923
1978/1979	42		2		
1979/1980	61				47253
1980/1981	95		4	480	
1981/1982	100	12975	3		
1984/1985	104		19	443	
1986/1987			57	388	39704
1987/1988	113		80		

[a] (1) Jouventin et al. (1982); (2) Hes and Roux (1983); (3) Barrat and Mougin (1978); Guinet et al. (unpubl.); (4) Pascal (1982); Guinet et al. (unpubl.).

Table 2. Annual breeding populations of penguins

Locality	Adélie Land		Crozet Islands/Possession Island				Kerguelen Islands			
Species	Emperor penguin	Adélie penguin	King penguin			Gentoo penguin	King penguin			Macaroni penguin
Colony[a]			(1)	(2)	(3)		(4)	(5)	(6)	(7)
Reference[b]	(a)	(a)	(b)			(c)	(d)			(d)
1952	6081									
1958	6000	16220								
1960										
1961										
1962	5623		45000	4000						
1963	6236						12800	7400		69660
1964	6000									
1965	5015		43400							
1966	5800									
1967	5850			3500	5000					
1968	5370									
1969	6083									
1970	5500					2000				
1971	5750									
1972	6000									
1974	5000								3000	
1975	5900									
1976	4850									
1977	3500									
1978	4400									
1979	3400									
1980	3450									
1981	3350		32000	11700	14700					
1982	2550									
1983	2800									
1984	3050	27925								
1985	3270					1400	52414	36397	21491	82198
1986	2517		40500	21700	33000	1450				

[a] (1) Baie du Marin; (2) Petite Manchotière; (3) Jardin Japonais; (4) Ratmanoff; (5) Cape Digby; (6) Larose Bay; (7) Cape Cotter + Cape Cabane.
[b] (a) Thomas (1986); (b) Jouventin and Weimerskirch (in press); (c) Bost (unpubl.); (d) Weimerskirch et al. (1989).

colonies have grown by four- to seven-fold between 1963 and 1985 (Weimerskirch et al. 1989). On Possession Island of the Crozet group, two small colonies have appeared recently, numbering 100 and 280 pairs in 1987. Two other colonies have grown by a factor of 5 and 5.6 in 20 years but the large colony near the permanent base has declined slightly during the same period (Table 2). The only indication of change in the populations of Macaroni penguins comes from Kerguelen where a 1.2-fold increase has been observed in the number of birds at two colonies between 1963 and 1985 (Weimerskirch et al. 1989, Table 2). In Adélie Land, Adélie penguins have increased on all the islands of the Pointe Geologie Archipelago except the largest island where the station is baded (Table 2). There, the Emperor penguin colony has declined to half its size during the 1970s. Finally, on Possession Island the population of gentoo panguins declined from 2000 pairs in 1972 to 850 pairs in 1987 (Table 2). The Emperor penguin population breeding near the base on the Pointe Geologie Archipelago has declined from ca. 5500 pairs in 1962 to 2300 pairs in 1989 (Table 2). After a stable period from 1962 to 1975, the decline began abruptly in 1976 and continued with large annual variations (for a detailed analysis, see Jouventin and Weimerskirch in press).

Albatrosses and Petrels. Wandering albatrosses (*Diomedea exulans*) have decreased (Table 3) on the three monitored islands of the Crozet group (Weimerskirch and Jouventin 1987) and on all the monitored colonies of the Kerguelen archipelago (Weimerskirch et al. 1989). On Possession Island, the population of northern (*Macronectes halli*) and southern giant petrel (*Macronectes giganteus*), decreased during the same period between 1981/82 and 1985/86 but to a lesser extent (Table 3). At Kerguelen, a decrease has also been noted for the monitored colony of black-browed albatrosses (*Diomedea melanophrys*) (Table 2).

The small breeding population of Antarctic fulmars (*Fulmarus glacialoides*) in Adélie Land, has undergone sharp fluctuations over the last 25 years, with the

Table 3. Annual breeding populations of albatrosses and giant petrels

Species	Black-browed albatross	Wandering albatross			Northern giant petrel	Southern giant petrel
Locality		Crozet Islands		Kerguelen Part of Courbet Pen.	Crozet Islands Possession Is.	Crozet Islands Possession Is.
		Possession	Cochons			
Reference[a]	(1)	(2)	(2)	(1)		
1960		475				
1964			1 500			
1968		500				
1971				260		
1974			1 210			
1975		324				
1978	1 200					
1980					558	84
1981			1 060			
1982		256				
1985		262		114	388	64
1986	1 050					
1987	905					

[a] (1) Weimerskirch et al. (1989); (2) Weimerskirch and Jouventin (1987).

Table 4. Changes in the number of nests (eggs laid) of the Antarctic fulmar in Adélie Land

Year	Nests	Year	Nests	Year	Nests
1962	25	1972	39	1982	47
1963	26	1973	39	1983	51
1964	32	1974	34	1984	21
1965	31	1975	23	1985	21
1967	22	1976	32	1986	32
1968	28	1977	9	1987	19
1969	24	1978	20	1988	40
1971	45	1979	22		

number of eggs laid varying each year from 9 to 51 (Table 4).

4 Discussion

4.1 Changes in the Size of Populations

Three species of penguin and two species of fur seal have increased in numbers over the last 25 years. Both fur seals species and King penguin populations are probably still recovering from past exploitation (Rounsevell and Copson 1982; Weimerskirch et al. 1989; Croxall et al. 1988). This delayed recovery has probably been caused by several factors. Both fur seals were still being culled sporadically up to the first half of this century and may have been completely eradicated from the Crozet Islands where recolonization started only in the 1970s (Jouventin et al. 1982). For both fur seals as for King penguins, rates of increases from very low populations will be small to start at the beginning and gradually increase as the population increase. A minimum density of breeding fur seals is necessary to achieve the maximum rate of population growth (see Roux 1987 for *A. tropicalis*). Small colonies of King penguins can produce fledglings, and consequently recruits, only when they have reached a minimum size. Larger colonies have a higher survival of chicks, and in colonies smaller than 100–200 pairs, all the chicks are killed by predators (unpubl. data).

Macaroni penguins have probably suffered less from exploitation than King penguins and fur seals, and Adélie penguins have never been exploited. The increase of Macaroni populations on Kerguelen and Adélie penguins on Adélie Land is probably a consequence of the same factor that has caused increases of Adélie and chinstrap penguins elsewhere in the Southern Ocean (Croxall et al. 1981, 1988). This factor is generally supposed to be a reduction in the numbers of whales making increased quantities of food available (particularly Antarctic Krill *Euphausia superba*) to other species. This supplementary food resource may also have accelerated the recovery of populations of King penguin and both fur seals, which now display extremely high rates of increase.

Although the overall populations of Adélie and King penguins have increased, certain colonies have declined after the settlement of the permanent base. The population of Adélie penguins has doubled everywhere on the Pointe Géologie Archipelago of Adélie Land, but it has decreased slightly on the island where the station is based due to human impact (Thomas

1986). On Possession Island, the only colony of King penguins to show a slight decrease is the one near the station where landing installations have been constructed on the breeding grounds. The construction of a road through the colony in 1982 has been successfully compensated by the occupation of an adjacent and similar area that was immediately colonized by the evicted birds.

Although many Procellariiforms and perhaps southern elephant seals could have taken advantage of the postulated increase in marine resources as have most penguins, their populations have been constantly diminishing over the last two decades. Weimerskirch and Jouventin (1987) have analyzed the general decrease in the wandering albatross population at the Crozet Islands, showing that it resulted from an abnormally high mortality of adults, particularly females, and suggested that fishing activities were the major cause of the population's decline. Similar conclusions were obtained for the black-browed albatross population at Kerguelen (Weimerskirch et al. 1987) and possibly the southern giant petrels, which suffer a high mortality from fishing operations, and from deliberate shooting at sea.

The southern elephant seals have declined at Crozet and Kerguelen Islands, also at Macquarie (Hindell and Burton 1987) and Marion Islands (Condy 1979), but not in South Georgia where the population has remained stable over the same period (McCann and Rothery 1988). The Possession Island population of southern elephant seals declined continually since 1966, while the Kerguelen population was stable between 1984 and 1988 following a reduction during the 1970s and early 1980s. These differing trends of elephant seal populations show that factors acting on the populations must be analyzed locally. At Crozet, the decrease of the population is suspected to result, at least in part, from predation by killer whales. For Kerguelen Island, Pascal (1982) has connected the decline of the population with a period of heavy fishing on the continental shelf, but further data on diet, population trends and fishing effort are required to corroborate this suggestion.

Finally, some populations of birds appear to be subject to important year to year oscillations in the number of breeders. The number of southern fulmars breeding in the small colony of Adélie Land appear to be related to the amount of snowfall on the colony in spring, because a heavy cover prevents a large proportion of pairs from breeding. Gentoo penguins show large changes in the number of pairs breeding each year, possibly in relation to the availability of food resources (Croxall et al. 1988).

The low breeding rate of seabirds prevents these species from compensating for even minor decreases in adult survival that is generally at a higher rate (Croxall 1982; Jouventin and Weimerskirch in press). The populations of these species, particularly those displaying high adult survival, are consequently very sensitive to small increases in adult mortality. Species most at risk are the small Procellariiforms which are subject to predation by introduced mammalian predators (cats and rats) on the breeding grounds, and the large Procellariiforms which are subject to accidental mortality in fishing material or deliberate killing. The increase in the number of King penguins probably results from a consistently high survival of adult and immature birds alike, even though breeding success and breeding frequency are still relatively low.

5 Conclusion

Seabirds and seals provide convenient links between the marine environment and the terrestrial habitats where studies can be more easily (and less expensively) carried out. It is thus not surprising that past, present and particularly future international programmes (ISAS, BIOMASS, CCAMLR) have promoted studies of the biology of Antarctic seabirds and seals, to use them as indicators of change in the marine environment. More cooperative programmes must consequently be developed between marine biologists and ornithologists (and seal biologists).

The need for management of the Southern Ocean has caused the different nations involved in these programmes to concentrate their studies on the major predator species in terms of biomass, i.e. penguins and seals. The results of our long-term studies on the population trends and the demographic parameters show that apart from the strictly applied and economic aspects, and considering that Procellariiform species are threatened by human activities, attention has to concentrate also on species that are less important in terms of biomass, but equally important in terms of conservation. Because of the development of human activities on the breeding grounds and in the feeding zones of Southern Ocean seabirds and seals, an increased conservation effort is urgently required for the different species and their habitats.

Acknowledgements. We are greatly indebted to the Administration of "Terres Australes et Antarctiques Françaises" for their financial and logistical support of the long-term monitoring programe in its entirety. We thank everyone who has been involved in collecting the field data. We are grateful to L. Ruchon for drawing the figures, to D. Besson for help in processing the data and for typing the text, and to P. Doncaster for improving the English.

References

Barrat A, Mougin JL (1978) L'Eléphant de mer *Mirounga leonina* de l'île de la Possession, Archipel Crozet. Mammalia 42:143–174

Carrick R (1964) Problems of conservation in and around the Southern Ocean. In: Carrick R, Holdgate M, Prevost J (eds) Antarctic biology. Herman, Paris, pp 589–598

Clark MR, Dingwall PR (1985) Conservation of islands in the Southern Ocean. IUCN Publ, Cambridge Univ Press, Cambridge

Condy PR (1979) Annual cycle of the southern elephant seals *Mirounga leonina* (L) on the Prince Edwards Islands. S Afr J Antarct Res 8:42–48

Croxall JP, Rootes OM, Price RA (1981) Increases in penguin populations at Signy Island, South Orkney Islands. Br Antarct Surv Bull 54:47–56

Croxall JP, McCann TS, Prince PA, Rothery P (1988) Reproductive performance of seabirds and seals at South Georgia and Signy Island, South Orkney Islands, 1976–1986: implication for Southern Ocean monitoring studies. In: Sahrage D (ed) Antarctic Ocean and resources variability. Springer, Berlin Heidelberg New York, pp 261–285

Hes AD, Roux JP (1983) Population increase in the subantarctic fur seal *Arctocephalus tropicalis* at Amsterdam Island. S Afr J Antarct Res 13:29–34

Hindell MA, Burton HR (1987) Past and present status of the southern elephant seal (*Mirounga leqnina*) at Macquarie Island. J Zool Lond 213:365–380

Holdgate MW (1967) The antarctic ecosystem. Phil Trans R Soc Lond B252:363–383

Holdgate MW (1970) Conservation in the Antarctic. In: Holdgate MW (ed) Antarctic ecology, vol 2. Academic Press, New York, pp 924–945

Holdgate MW, Wace NM (1961) The influence of man on the floras and faunas of southern islands. Polar Rec 10:475–493

Jouventin P, Weimerskirch H (in press) Analysis of past and present changes in the number of seabirds of French Antarctic and Subantarctic territories and its implication for the management of southern populations. Bird Population studies: their relevance to conservation and management. Oxford, Blackwell

Jouventin P, Stahl JC, Weimerskirch H (1982) La recolonisation des îles Crozet par les otaries (*Arctocephalus tropicalis* et *A. gazella*). Mammalia 46:505–514

McCann TS, Rothery P (1988) Population size and status of the southern elephant seal (*Mirounga leonina*) at South Georgia, 1951–1985. Polar Biol 8:305–309

Pascal M (1982) Evolution numérique de la population d'éléphants de mer (*Mirounga leonina* L.) de l'archipel des Kerguelen au cours des 30 dernières années. Com Nat Fr Rech Antarct 51:119–135

Paulian P (1964) Contribution à l'étude de l'Otarie de l'île Amsterdam. Mammalia 28:3–146

Roberts BB (1977) Conservation in the antarctic. Phil Trans R Soc Lond B279:97–104

Rounsevell De Copson GR (1982) Growth rate and recovery of a King penguin *Aptenodytes patagonicus* population after exploitation. Aust Wildl Res 9:519–525

Roux JP (1987) Recolonization process in the subantarctic fur seal *Arctocephalus tropicalis,* on Amsterdam Island. NOAA Techn Rep 51:189–194

Segonzac M (1972) Données récentes sur la faune des îles Saint-Paul et Nouvelle Amsterdam. L'Oiseau et la RFO 42:3–68

Thomas T (1986) L'effectif des oiseaux nicheurs de l'archipel de Pointe Géologie (Terre Adélie) et son évolution au cours des trente dernières années. L'Oiseau et la RFO 56:349–368

Van Aarde RJ (1980) Fluctuation in the population of southern elephant seal *Mirounga leonina* (L.) at Kerguelen. S Afr J Zool 15:99–106

Weimerskirch H, Jouventin P (1987) Population dynamics of the wandering albatross (*Diomedea exulans*) of the Crozet Islands: causes and consequences of the population decline. Oikos 49:315–322

Weimerskirch H, Clobert J, Jouventin P (1987) Survival in five southern albatrosses and its relations with their life history. J Anim Ecol 56:1043–1056

Weimerskirch H, Zotier R, Jouventin P (1989) The avifauna of the Kerguelen Islands. Emu 89:15–29

Annual Fluctuations in Productivity and Breeding Success of Adélie Penguins and Fulmarine Petrels in Prydz Bay, East Antarctica

M. D. Whitehead [1, 2], G. W. Johnstone [3] and H. R. Burton [1]

Summary. Annual chick productivity and breeding success, recorded at four Adélie penguin, *Pygoscelis adeliae*, colonies at Magnetic Island in eastern Prydz Bay, are presented for the seven breeding seasons 1981/82 to 1987/88. The adult breeding population remained relatively stable during the first 4 years of the study, and increased in the last 2 years. Substantial annual variation in breeding success occurred over the study period, ranging between an estimated 0.69 and 1.33 chicks surviving until late crèche stage per nest for seasons 1985/86 and 1982/83 respectively. Annual patterns of chick productivity in southern fulmar, *Fulmarus glacialoides*, and Antarctic petrel, *Thalassoica antarctica*, populations within Prydz Bay were synchronous with those of Adélie penguins. In the years of highest and lowest reproductive performance, prey abundance within the likely foraging areas was correspondingly high and low. Reproductive performance was greatest in years when fast-ice breakout occurred before the end of December (1981/82, 1982/83, 1986/87 and 1987/88) and lowest when breakout was after (1983/84, 1984/85 and 1985/86) and pack-ice cover persisted within the foraging range of the birds during the chick-rearing period.

1 Introduction

Understanding the factors responsible for any observed variability in population size or reproductive performance is valuable for predicting the likely effect on the species concerned of some perturbation within the environment, either natural or anthropogenic.

Such knowledge is vital to comprehend the influence of change in the marine environment of Antarctic nesting seabirds. The Commission for the Conservation of Antarctic Marine Living Resources (CCAMLR) is responsible for gathering such information for management purposes, particularly in response to the developing krill fishery in Antarctic waters. It is with this application in mind that the CCAMLR Ecosystem Monitoring Program (CEMP) was established (CCAMLR 1986).

Though data of this kind have important ecological and management implications, few long-term records of annual fluctuation in Antarctic seabird population size or reproductive performance exist. The only previous accounts of breeding success fluctuation in Antarctic seabirds spanning more than a 2- to 3-year period are those of Croxall et al. (1989), centred at South Georgia and Signy Island populations, and Trivelpiece and Trivelpiece (this vol.) in the South Shetland Islands. Even less often has concurrent information been collected concerning environmental factors that may be determining any changes. This can be attributed to logistic difficulties and the remoteness of this environment.

The present study monitored the reproductive performance of three abundantly nesting seabirds in the Prydz Bay region (Adélie penguin, *Pygoscelis adeliae*; Antarctic petrel, *Thalassoica antarctica* and southern fulmar, *Fulmarus glacialoides*) in east Antarctica over a 7-year period, to determine the extent of natural fluctuations of this parameter. The study attempts to interpret the observed fluctuations with reference to prey abundance, ice and weather conditions.

2 Methods

2.1 Adélie Penguin

The Adélie penguin colony at Magnetic Island (68° 33′ S, 77° 54′ E), lying 3 km offshore of the Vestfold Hills in Prydz Bay (Fig. 3) was monitored during the breeding seasons between 1981/82 and 1987/88.

The date that Adélie penguins first began arriving ashore to commence nesting in each of the seasons, was recorded, except in 1985/86.

Birds were captured and weighed as they travelled across the sea ice toward their breeding sites on the

[1] Australian Antarctic Division, Channel Highway, Kingston, Tasmania 7050, Australia.
[2] Present address: Department of Botany and Zoology, Monash University, Clayton, Victoria 3168, Australia.
[3] Deceased.

Antarctic Ecosystems. Ecological Change and Conservation (ed. by K. R. Kerry and G. Hempel)
© Springer-Verlag Berlin Heidelberg 1990

island. As the average arrival weight declines as the occupation period progresses (Ainley and Emison 1972), valid comparisons of annual arrival weights can only be made if the data are collected at the same stage of each occupation period. We weighed incoming birds 3 days after the first adults arrived between 1983/84 and 1987/88, except in 1984/85 when birds were not weighed, and 1985/86 when the date of the first arrival was not recorded. Birds arriving in 1985/86 were weighed on 17 October 1985. It is unlikely that birds began arriving atypically early in this season as fast ice was more extensive than usual. Penney (1968) and Ainley and LeResche (1973) showed that in years of greater than average ice extent initial arrival was delayed by up to 2 weeks.

Four Adélie penguin colonies were selected as sites for monitoring inter-season nesting population size and breeding success. They were chosen on the basis of being representative of the typical size range of colonies present and came from a range of locations on Magnetic Island. Discrete colonies were selected to facilitate accurate counting of adults and chicks.

The numbers of nesting pairs were counted at each of the colonies in the first week of December, at the conclusion of the egg-laying period in each season. In the 1985/86 season, however, counts were not made until late December. Chicks surviving in each study colony until the late crèche stage of development were counted in late January of each study season as a measure of breeding success.

2.2 Antarctic Petrel and Southern Fulmar

One Antarctic petrel and one southern fulmar colony were monitored at each of three islands (Hop, Filla and Buchan) in the Rauer Islands (68° 51′ S, 77° 50′ E), 20 km south of Davis (Fig. 3), in the summers of 1983/84 to 1987/88. These study colonies were described by Green and Johnstone (1986). The numbers of Antarctic petrel and southern fulmar chicks surviving in each of the colonies until the late nestling stage of development were counted in early February and early March respectively. Due to the relatively difficult access to these islands, it was not always possible to visit early in the season to determine the number of birds attempting to nest.

2.3 Krill Abundance

As the euphausiids *Euphausia crystallorophias* and *E. superba* are major components of Prydz Bay Adélie penguin diet (Puddicombe and Johnstone 1988; Green and Johnstone 1988), we examined their abundance within the Prydz Bay region from data collected during marine science cruises in the summers of 1982/83 and 1984/85 (see Williams et al. 1986; Ikeda et al. 1986), and our own sampling program in the 1985/86 summer. All samples were made with a Rectangular Midwater Trawl net (Baker et al. 1973) of 8 m^2 mouth area (RMT-8) and 4.5-mm mesh size, towed at a speed between 1.8 and 3.0 knots.

During the Australian Antarctic Division's marine science cruise in the 1982/83 season, zooplankton were sampled at 15 sites over the Prydz Bay continental shelf in mid-December 1982. The net was towed horizontally at selected depths between the surface and 105 m. The composition of these samples was reported by Williams et al. (1986).

During the Australian Antarctic Division's cruise in the 1984/85 season, three sites in the same region were sampled in mid-January 1985. The net was towed obliquely over an integrated depth range between the surface and 260 m. The composition of these samples was reported by Ikeda et al. (1986). The neritic distribution and abundance of *E. crystallorophias*, based on the samples from these two cruises, was described by Thomas and Green (1988).

As part of the 1985/86 ornithology program, zooplankton samples were sampled at 18 sites over the Prydz Bay shelf region in late January 1986. Stations along two transect lines, west and north-west of Magnetic Island were sampled. The net was towed obliquely between the surface and 100 m.

The volume of water filtered by each tow was calculated from the speed and duration of the tow, and abundance of euphausiids expressed as number per 1000 m^3.

The biases introduced to abundance estimates by the two differing strategies of oblique and horizontal tows were considered by Thomas and Green (1988), who believed that comparisons of relative abundance were still valid.

The samples were not taken at the same stage of each season but comparisons between years are still valid if a relatively constant biomass over one season (Heywood et al. 1985) is assumed.

2.4 Ice and Weather Conditions

Data on the extent and coverage of ice within Prydz Bay throughout the breeding seasons of this study were obtained from NOAA satellite pictures of ice conditions (Northern Ice Limit 1981–1988). Dates of fast-ice breakout near Davis Station (68° 35′ S, 77° 58′ E) (3 km inshore of the Magnetic Island study site) were obtained from sea-ice observation records made weekly over the study period (I. Allison, pers. commun.). Ice breakout at the sea-ice observation site usually occurred between 2 and 3 weeks following breakout along the seaward shore of Magnetic Island. These dates, however, do give an indication of years of early and late breakout.

Meteorological information was obtained from the records kept by the Bureau of Meteorology for Davis Station.

3 Results

3.1 Adélie Penguin

The first Adélie penguins were seen arriving at the Vestfold Hills breeding colonies between October 7 and 16 over this study period (Table 1). In 1982/83 and 1987/88 the initial arrival dates were earlier than in other years.

Mean arrival weight was lower in 1985/86 than any of the other years for which it was recorded (Table 1). Although analysis of variance revealed no significant difference between mean annual arrival weights ($F = 2.65$, $0.05 < p < 0.10$); considered independently, arrival weight was significantly lower in 1985/86 than in 1986/87 ($t = 2.589$, $p < 0.02$).

In the first 4 years of the study, the number of nesting pairs did not differ markedly (Table 2). The lower number of occupied nests in 1985/86 compared with the earlier counts in other years may represent either a real decline in the number of pairs which nested in 1985/86, a high mortality rate during the incubation and early chick-rearing period, or a combination of these processes. In other years, the number of nests remaining in late December was never as low (Table 2). The 348 nests remaining in late December 1985 represented a reduction of 27% of the number of nests remaining in late December averaged over the three seasons 1981/82–1983/84. The number of birds nesting in the 1986/87 and 1987/88 summers was greater in each of the study colonies than previously recorded (Table 2). The number of birds attempting to nest in all study colonies in the 1987/88 summer was 10.7% greater than numbers nesting in 1981/82, the first season of the study.

The number of chicks surviving to late crèche stage at each of the study colonies varied largely as a result of the fluctuations in breeding success (Fig. 1); but the increased number of birds nesting in 1986/87 and 1987/88 (Table 2), combined with slightly increased breeding success, was responsible for the greatly increased number of chicks surviving to late crèche stage over the last 2 years of the study (Fig. 1).

The breeding success of Adélie penguins averaged for the four Magnetic Island study colonies was highest in 1982/83, the only year when breeding success exceeded one chick surviving per nest. Breeding success was low in 1983/84, 1984/85 and 1985/86. In 1981/82, 1986/87 and 1987/88 breeding success was close to average for the study period. The fluctuations in success were generally consistent between the study colonies (Fig. 1). As the number of pairs nesting was not counted in early December 1985, the breeding success could not be precisely determined for this season. The points shown in Fig. 1 for 1985/86 represent the breeding success if the number of pairs attempting to nest had been the same as the number of pairs still occupying nests when counted in late December. The data from previous years (Table 2) indicate that some nests would already have failed by this stage, so the points in Fig. 1 for 1985/86 breeding success are conservatively low estimates of nest failure. If the number of pairs initially attempting to nest had been the same as the average number attempting to nest for the previous four breeding seasons, the success would have been only 0.45 chicks per nest.

Table 1. Adélie penguin arrival date, mean arrival weight (± 1 SD) (n = sample size), and ice breakout date at the Vestfold Hills, 1981/82 to 1987/88 (NR = not recorded)

Season	First arrival	Arrival weight ± 1 SD (kg)	Sample size (n)	Ice breakout date
1981/82	14 Oct	NR	–	15 Dec 1981
1982/83	7 Oct	NR	–	31 Dec 1982
1983/84	14 Oct	6.25 ± 0.55	100	7 Feb 1984
1984/85	16 Oct	NR	–	9 Jan 1985
1985/86	NR	6.09 ± 0.49	69	21 Jan 1986
1986/87	14 Oct	6.30 ± 0.56	100	25 Dec 1986
1987/88	9 Oct	6.39 ± 0.46	10	27 Dec 1987

Table 2. Numbers of Adélie penguins nesting at each of the four Magnetic Island study colonies, and the total for all colonies combined, in early December and in late December (in parentheses), between 1981/82 and 1987/88 (NR = not recorded)

Season	Numbers nesting				
	Colony 1	Colony 2	Colony 3	Colony 4	Colonies 1–4
1981/82	193 (170)	88 (79)	103 (96)	159 (149)	543 (494)
1982/83	186 (163)	81 (78)	114 (110)	131 (119)	512 (470)
1983/84	189 (170)	77 (80)	119 (109)	138 (114)	523 (473)
1984/85	192 (NR)	89 (NR)	107 (NR)	130 (NR)	518 (NR)
1985/86	NR (141)	NR (59)	NR (72)	NR (76)	NR (348)
1986/87	218 (210)	92 (83)	123 (120)	155 (150)	588 (563)
1987/88	232 (222)	89 (78)	118 (115)	162 (160)	601 (575)

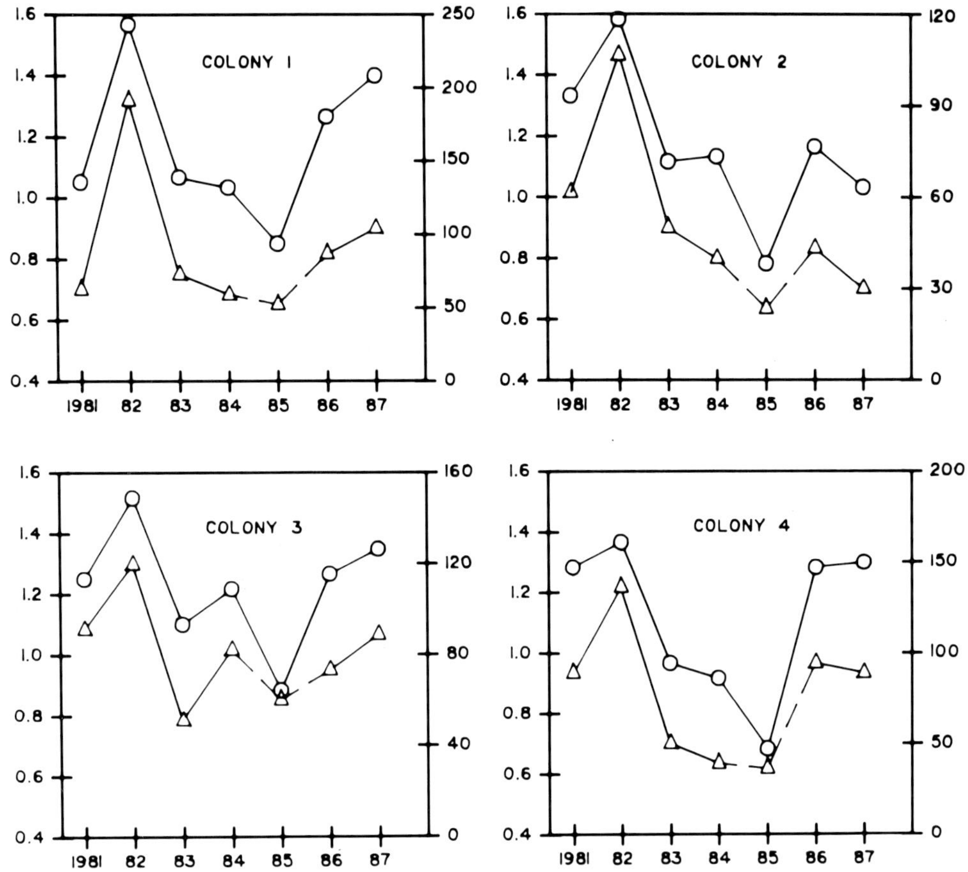

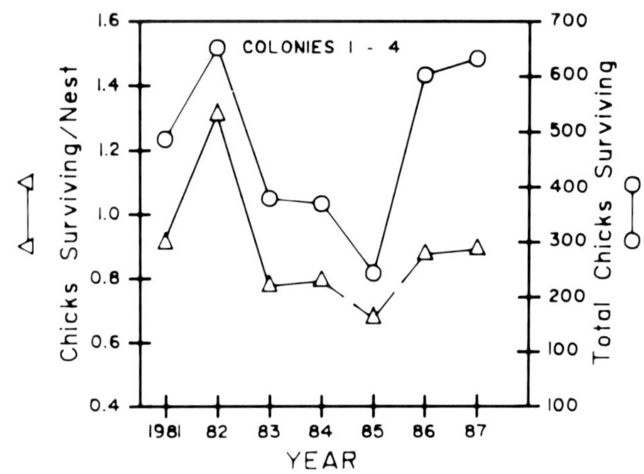

Fig. 1. Total number of Adélie penguin chicks surviving, and number surviving per nest (breeding success) to late crèche stage at each of the four Magnetic Island study colonies, and all colonies combined, 1981/82 to 1987/1988. Seasons indicated by the year in which they commenced. *Dashed line* indicates maximum possible 1985/86 breeding success

3.2 Antarctic Petrel and Southern Fulmar

In the Rauer Island study colonies, 1985/86 was also a season of reduced productivity for Antarctic petrels and southern fulmars. More chicks survived to the late nestling stage in 1986/87 and 1987/88 than in 1983/84 and 1984/85, a situation similar to that for Adélie penguins at Magnetic Island (Fig. 2). Antarctic petrel and southern fulmar chick productivity in 1985/86 averaged over the three study sites was 57.3 and 88.0% lower, respectively, than in 1984/85; compared to the 1986/87 season it was 70.8 and 89.9% lower respectively. These data suggest that Antarctic petrels were less affected by conditions resulting in the reduced productivity than southern fulmars. Adélie penguin chick production in 1985/86 was 40.5 and 57.4% low-

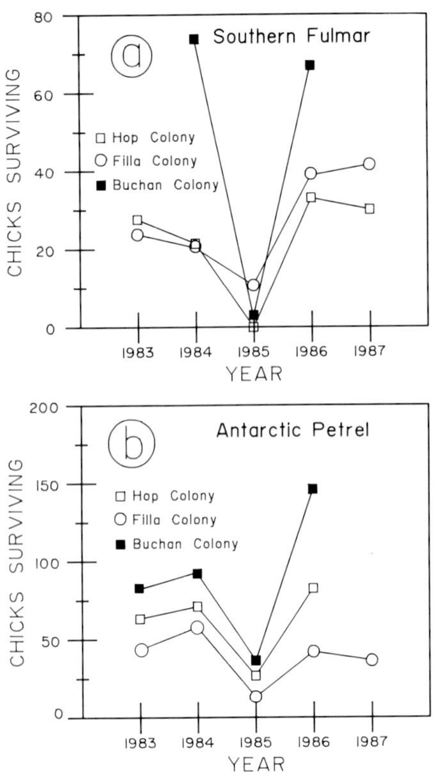

Fig. 2. Number of chicks surviving to late nestling age at the Hop, Filla and Buchan Island study colonies, in the Rauer Island Group, 1983/84 to 1987/88 of **a** Southern fulmars and **b** Antarctic petrels. Seasons indicated by the year in which they commenced

er than in the 1984/85 and 1986/87 seasons respectively, suggesting that they were less susceptible than either Antarctic petrels or southern fulmars to the prevailing conditions.

3.3 Krill Abundance

In January 1985 and January 1986 fewer euphausiids were captured over the Prydz Bay continental shelf than in December 1982 (Figs. 3, 4). Though geographic patchiness and rapid temporal fluctuation are well-known problems in adequately assessing euphausiid abundance (Siegel 1985), the fact that no patches of dense euphausiid aggregation were sampled in the 1984/85 or 1985/86 cruises, but were sampled regularly on the 1982/83 cruise, does suggest a real reduction in their abundance within the Prydz Bay shelf region during the period of the two latter cruises.

The areas of greatest *E. superba* and *E. crystallorophias* density sampled in 1982/83 were two orders of magnitude more dense than any areas sampled on the 1984/85 and 1985/86 cruises (Figs. 3, 4) and may

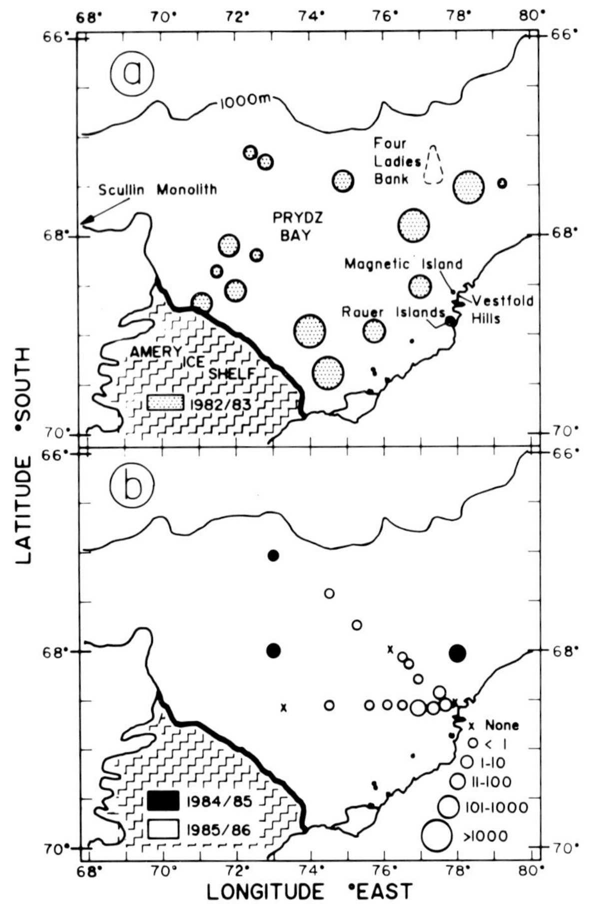

Fig. 3. Abundance of *E. crystallorophias* (No. 1 000 m^{-3}) at Prydz Bay sampling sites during **a** the 1982/83 research cruise, and **b** the 1984/85 and 1985/86 research cruises

Table 3. Average density of *E. crystallorophias* and *E. superba* sampled over the Prydz Bay continental shelf (± 1 SD) on research cruises during the 1982/83, 1984/85 and 1985/86 summers. Sample size (n) indicates number of stations sampled

Season	Sampling dates	n	*E. crystallorophias* density ± 1 SD (No. 1 000 m^{-3})	*E. superba* density ± 1 SD (No. 1 000 m^{-3})
1982/83	8–16 Dec	15	654.11 ± 980.87	230.16 ± 447.14
1984/85	13–14 Jan	3	2.01 ± 1.92	13.35 ± 13.17
1985/86	23–26 Jan	18	1.61 ± 3.58	0.01 ± 0.01

have represented the existence of a number of large swarms within the area at this time.

The average density of both species sampled on each of the cruises is indicated in Table 3. Despite the enormous variation between sampling sites, indicated by the large standard deviations, the average density of *E. crystallorophias* sampled was significantly lower in 1985/86 than it was in 1982/83 (Tukey test; $q = 4.13$, $0.01 < p < 0.025$). Although the mean den-

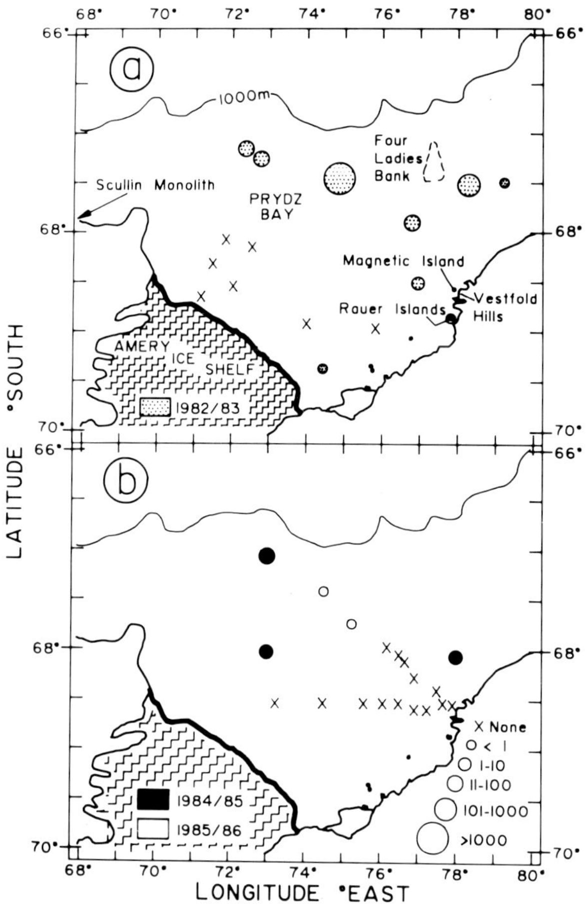

Fig. 4. Abundance of *E. superba* (No. 1 000 m^{-3}) at Prydz Bay sampling sites during **a** the 1982/83 research cruise, and **b** the 1984/85 and 1985/86 research cruises

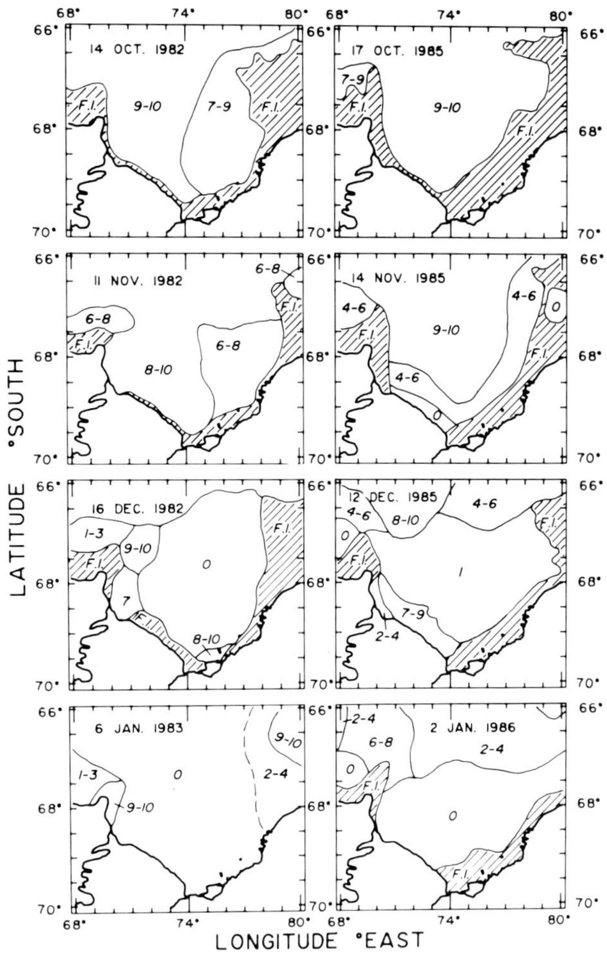

Fig. 5. Fast-ice extent (*F.I.*) and pack-ice concentration (indicated in tenths) in the Prydz Bay region in mid-October, mid-November, mid-December and early January during the 1982/83 and 1985/86 summers

sity was also low in 1984/85, the sample size was small and the difference between it and the high mean density of 1982/83 was not significant (Tukey test; q = 2.28, $0.20 < p < 0.50$). Trends in *E. superba* average density were the same as for *E. crystallorophias* (Table 3), however, the difference between years was not significant (ANOVA; F = 2.69, $0.05 < p < 0.10$).

3.4 Ice and Weather Conditions

Persistence of fast ice in the inshore areas of Prydz Bay varied in the study period by up to 7 weeks. In 1981/82, 1982/83, 1986/87 and 1987/88 the fast ice had broken out from the observation site at Davis by the end of December. In the other years, breakout occurred between the second week of January and the first week of February (Table 1). The years of later fast ice breakout coincided with the years of lowest reproductive performance in Adélie penguins, Antarctic petrels and southern fulmars.

The pack-ice conditions in Prydz Bay are characterized by the recurring development of a large polynya during early summer (Streten and Pike 1984), while the ocean to the north is still extensively covered by pack ice. The concentration and extent of pack ice within the region did, however, vary between the summers of this study. NOAA satellite pictures (Northern Ice Limit) indicated that the years of late fast-ice breakout coincided with the persistence of pack ice within the Prydz Bay region. In years of early breakout, the pack-ice edge zone had usually retreated south to join the Prydz Bay polynya by early January. In other years the pack ice persisted into January and the polynya remained until late January.

The ice conditions within Prydz Bay during the two summers of most differing seabird reproductive performances are indicated in Fig. 5. Fast ice was much more extensive in 1985/86 and persisted well into Jan-

Table 4. Number of days of strong wind (> 11.3 ms^{-1}) and mean monthly wind speed (ms^{-1}) (in parentheses) per month, between October and January, 1981/82 to 1987/88

Season	Month			
	Oct	Nov	Dec	Jan
1981/82	2 (3.4)	21 (11.1)	10 (9.3)	7 (11.8)
1982/83	13 (10.0)	11 (9.8)	3 (7.2)	5 (9.3)
1983/84	8 (9.3)	7 (16.7)	9 (10.0)	4 (9.7)
1984/85	7 (7.8)	14 (12.5)	17 (12.1)	6 (9.4)
1985/86	11 (10.5)	6 (11.0)	13 (11.5)	2 (8.0)
1986/87	17 (11.2)	1 (7.2)	7 (9.0)	6 (8.7)
1987/88	8 (7.2)	17 (11.3)	14 (11.3)	6 (9.9)

uary, whereas in 1982/83 it broke out in December. In mid-October 1985, the fast-ice belt extended for about 60 km from the coast, but at the same time in 1982 it extended for only 12 km. Pack ice within Prydz Bay persisted longer into the 1985/86 season, with large areas of pack ice to eight-tenths concentration still enclosing the Prydz Bay polynya in early January.

Ainley and LeResche (1973) described a strong relationship between persistence of ice and wind conditions, finding that in stormy seasons ice cover over the Ross Sea adjacent to Cape Crozier, Ross Island (77° 20′ S, 169° 15′ E) was generally reduced. Not one of the summers during this study was so obviously atypical in terms of local wind conditions to allow comment on the influence of wind on the persistence of ice cover (Table 4). None of the years had consistently high or low average wind speeds or numbers of days of strong wind.

4 Discussion

4.1 Seabird Population Parameters, Food Availability and Ice Conditions

High Adélie penguin breeding success when *E. crystallorophias* was abundant is understandable considering the importance of this species in their diet within Prydz Bay (Puddicombe and Johnstone 1988) and elsewhere (Emison 1968). The lower number of chicks raised in each of the Magnetic Island study colonies in 1983/84 and 1984/85 compared to the two previous seasons has been attributed to less frequent chick feeding, as a result of adult Adélie penguins having to swim further to feed on *E. superba* in response to a shortage of *E. crystallorophias* over the shelf region (Thomas and Green 1988; Green and Johnstone 1988). This hypothesis was supported by the fact that in the two seasons of lowered chick productivity, *E. superba* was the dominant euphausiid in the diet, whereas the more neritic *E. crystallorophias* predominated in the earlier season of greatest chick productivity. Our data, indicating very low *E. crystallorophias* abundance in 1985/86 and related low chick productivity, are consistent with the view that high *E. crystallorophias* availability is necessary for high Adélie penguin breeding success.

The reduced initial arrival weight of Adélie penguins in 1985/86 suggests that there was some process operating prior to the breeding season or as birds were coming ashore. This weight may have been lost due to the longer than usual walking time necessary to cross the extensive fast-ice belt of that year to their breeding sites. A shortage of food prior to the breeding season may also have contributed to the lower than average arrival weights.

Ainley and LeResche (1973) and Lishman (1985) attributed reduced breeding success of Adélie penguins at Ross Island in 1968/69, and at Signy Island in 1980/81 respectively, to the increased energetic cost of travelling across sea ice in years of persistent ice. This travel deprived the birds of fat stores that would otherwise have been available for the reproductive effort. The lack of energy reserves resulted in a reduced proportion of the population attempting to breed, increased nest desertions during incubation, and reduced frequency of chick feeding, resulting in increased egg and chick mortality.

The correlation between Antarctic petrel, southern fulmar, and Adélie penguin reproductive performance reported in this study suggests the influence of ice cover on energetic cost of travelling is not solely responsible for the annual variations in chick productivity. Antarctic petrels and southern fulmars, being aerial seabirds, regularly fly hundreds of kilometres to foraging areas (Ainley et al. 1984), possibly beyond the extent of concentrated pack ice, even in 1985/86. Nesting colonies of Antarctic petrels exist in the Theron Mountains (79° 05′ S, 28° 15′ W) and the Muhlig-Hofmannfjella (72° 00′ S, 5° 20′ E), up to 250 km inland of the coastline (Brook and Beck 1972), so the increased distance to foraging areas imposed by a year of unusually persistent ice may not be a major limitation on the breeding performance of this species at the coastal Rauer Island colonies.

Alonso et al. (1987) reported a greatly reduced reproductive performance of Antarctic petrels at Scullin Monolith (67° 47′ S, 66° 42′ E) in the 1985/86 season, with only 2% of pairs that attempted to nest, successfully raising young. At Scullin Monolith in 1984/85 and 1986/87, 35 and 77% of pairs, respectively, were successful. This pattern is consistent with all trends observed in this study. Scullin Monolith is approximately 480 km west of the Rauer colonies (Fig. 3), indicating that the phenomena occurred across a greater area than the immediate Prydz Bay region.

In the 1987/88 season of average reproductive performance, southern fulmars and Antarctic petrels breeding at the Rauer Islands ate no *E. crystallorophias,* but relied on *E. superba* as the euphausiid component of their diet. Southern fulmar diet consisted of 64% *Pleuragramma antarcticum*, a pelagic fish, and 36% *E. superba* by weight and Antarctic petrel diet 77% *P. antarcticum* and 22% *E. superba* by weight, though differences were not significant (J. P. Y. Arnould and M. D. Whitehead, unpubl. data).

The correlated trends in reproductive performance of these two species of surface-nesting petrel and Adélie penguins suggests that the petrel food sources were also less available in 1985/86 than in other years. Fish, including *P. antarcticum*, were less important in the diet of Adélie penguins in 1983/84 and 1984/85 than in 1982/83 (Green and Johnstone 1988). *P. antarcticum* abundance in the Prydz Bay region may have been reduced as a consequence of the demonstrated reduction in *E. crystallorophias*, as this fish has been captured associated with krill swarms in Prydz Bay (Williams 1985). If so, given the importance of *P. antarcticum* in the diet of both petrel species, their reduced reproductive performance in these 2 years is not surprising. This circumstantial evidence indicates that the suppression of marine resources during the 1985/86 summer was probably a general reduction in food availability, rather than a scarcity of only *E. crystallorophias*. In discussing the distribution and abundance of *E. superba* in oceanic waters of the Prydz Bay region, Hosie et al. (1988) reported that the mean biomass of *E. superba* was higher in samples from the 1982/83 season than in samples taken in the 1983/84 and 1984/85 seasons. Our data on *E. superba* abundance over the continental shelf agree with these trends, and indicate that 1985/86 was also a season of low *E. superba* abundance. Based on these data, Antarctic petrel and southern fulmar reproductive performance would probably also have been higher during the 1982/83 season. This information further supports the view that the reduction in marine resources in 1985/86 was a general pehnomenon responsible for the poor breeding performances of all species studied.

Reduced abundance of *E. superba* during the 1985/86 season may also have contributed to the reduced arrival weight of Adélie penguins, as they probably rely on *E. superba* in the winter and early spring when they are feeding beyond the continental shelf. Adélie penguins ate *E. superba* almost exclusively in continental slope areas of the Ross Sea, but fed on *E. crystallorophias* over the continental shelf (Ainley et al. 1984). Adélie Penguins in the Prydz Bay region similarly rely largely on *E. crystallorophias* during the summer months when they are foraging over continental shelf waters (Puddicombe and Johnstone 1988).

Coincident trends in reproductive performance between different Antarctic seabird species has also been reported by Croxall et al. (1989). They reported poor breeding success at South Georgia of black-browed albatross, *Diomedea melanophrys*, macaroni penguin, *Eudyptes chrysolophus*, and gentoo penguin, *P. papua*, in 1977/78 and again in 1983/84 along with grey-headed albatross, *D. chrysostoma*. The 1982/83 season was one of poor breeding success for both Adélie penguins and chinstrap penguins, *P. antarctica*, at Signy Island. As the trends were not synchronous between these two sites, it is not surprising that they do not coincide with the patterns we have described. Krill availability was low around South Georgia in the austral winter of 1983 (Heywood et al. 1985), prior to the poor reproductive performance of many of the seabirds the following summer (Croxall et al. 1989).

4.2 Ocean/Ice/Atmosphere Interactions and Food Availability

We have observed the relationship between seabird reproductive performance and food availability, and food availability and sea-ice conditions. Though a causal relationship between early ice dispersal and high food availability during the birds breeding season is uncertain, ice conditions have been linked with altered vertical distribution of prey species compared to their open water distribution (Ainley et al. 1986) and horizontal distribution of prey in open water has altered in response to movements of proximate ice (Daly and Macaulay 1988). In any year ice conditions would be expected to influence seabird food availability.

Reduced availability of *E. superba* over the Prydz Bay continental shelf waters when pack ice persisted in more northern areas, such as in 1985/86, may be a reflection of a general response by euphausiids to follow retreating ice. Kanda et al. (1982) observed large swarms of *E. superba* moving southwards in early summer in areas of open water. It does not, however, explain the reduction in availability of *E. crystallorophias* in coastal waters, which is the usual habitat for this species (Thomas and Green 1988), or lowered abundance of *E. superba* in more offshore waters.

Krill stocks foraged upon by Adélie penguins and other breeding seabirds within Prydz Bay lie within the body of water whose movement is determined by the action of the Prydz Bay Gyre. First suggested by Grigor'yev (1967), this is a cyclonic circulation system, travelling in a westward direction along the coast between approximately 87 and 65° E, where it is deflected north to approximately 66° S, then travelling eastward before turning south again at about 87° E (Smith et al. 1984). Though the krill population within this gyre are not confined to it (MacDonald et al. 1986), their distribution at any time is probably influenced by its action.

Deacon (1976) indicated the importance of the similar Weddell Sea circulation system to the distribution and dispersal of krill within it, and Streten and Wells (1977) highlighted the influence of the westerly atmospheric circulation on the maintenance and variability of these systems.

A change in the balance between the coastal easterly and offshore westerly components of the airflow that drive the Prydz Bay circulation may alter the net effect of atmospheric forcing to the degree that usually occurring oceanic fronts and areas of upwelling and high productivity are interrupted.

The influence on oceanic productivity and seabird reproductive performance, of changing wind environments, associated with the El Niño Southern Oscillation (ENSO) (in the Pacific Ocean), and anomalous climatic events off the southern African coast has been well documented (e.g. Schreiber and Schreiber 1984; LaCock 1986).

The importance of the westerly wind belt of the southern hemisphere, operating via its influence on the west-wind drift, on timing of the spring bloom and stocks of commercial marine species in the waters of southern Australia, has recently been shown (Harris et al. 1988). Although our data on the coastal easterly wind regime over the summer periods at Davis Station did not indicate any year of atypical wind strengths, the patterns of the west-wind drift system must be considered before the influence of climatic features on the variation in food availability and seabird reproductive performance can be rejected. These changing ocean-atmosphere interactions, most directly evidenced in anomalous surface water temperatures (usually a warming), probably operate in the Antarctic. Measurements have been made of the sea surface temperature in the Prydz Bay region at a number of sites between the 1980/81 and 1987/88 summers (e.g. Kerry et al. 1987a, b), however, variability was so great between sites due to the localized cooling influence of pack-ice cover that it was not possible to easily detect annual anomalies in sea surface temperature. The geographic extent of the reduced productivity in 1985/86 suggests that a meso-scale process (Hunt and Schneider 1987) such as a widespread climatic anomaly was operating, as opposed to only a local change in conditions.

5 Conclusions

The fluctuations in Adélie penguin, southern fulmar and Antarctic petrel reproductive performances reported here were a product of variable food supply, probably reflecting variability within the realm of the ice-ocean-atmosphere interaction.

Some of the difficulty in separating the effects of physical conditions and variable food supply (Croxall et al. 1989) have been overcome by comparing three seabird species at two locations (Scullin Monolith and Prydz Bay sites) with concomitant data on abundances of dominant food items. Our results suggest that monitoring each of these species would be useful in detecting changes in the marine environment, this being the main aim of the CEMP (CCAMLR Ecosystem Monitoring Program; CCAMLR 1986).

The influence of naturally varying environmental parameters on the reproductive performance of Antarctic seabirds is probably of sufficient magnitude to obscure the effects of human-induced change over the short term. Consequently, irregular monitoring will not be sufficient to provide definite conclusions regarding impact of human activities.

Longer-term integrated studies of avian population parameters (see CCAMLR 1986), zooplankton distribution and abundance, weather conditions, and fluctuations of oceanographic features will be necessary before the physical processes ultimately determining seabird food availability and thus, reproductive performance can be thoroughly resolved. Maybe then, the operation of the "complex interaction between the spectrum of (natural) perturbations and the spectrum of biological responses" (Harris 1985) will be understood to a degree enabling recognition of human-induced change.

Acknowledgements. These data were collected by the Australian Antarctic Division biologists stationed at Davis between 1981 and 1986, prior to the arrival of MDW. We thank R. Puddicombe (1982/83 and 1983/84), K. Green (1983/84 and 1984/85) and N. Gales (1985/86) for making their data freely available. The help given by all expeditioners at Davis over this period, particularly by R. Besso, J. Kirkwood, P. Osborne, L. Parr, J. Arnould and T. Howard, who all assisted with counting and weighing, is very much appreciated. We are grateful to J. Kirkwood for conducting the zooplankton sampling in January 1986. I. Allison kindly allowed us access to his data on dates of ice breakout. J. Cox drafted the figures for this paper. Prof. J. M. Cullen, K. Green, S. Nicol, R. Williams and E. Woehler made valuable comments on the manuscript.

References

Ainley DG, Emison WB (1972) Sexual size dimorphism in Adélie penguins. Ibis 114:267–271

Ainley DG, LeResche RE (1973) The effects of weather and ice conditions on breeding in Adélie penguins. Condor 75:235–239

Ainley DG, O'Connor EF, Boekelheide RJ (1984) The marine ecology of birds in the Ross Sea, Antarctica. American Ornithologists' Union Ornithological Monograph Number 32, Washington DC

Ainley DG, Fraser WR, Sullivan CW, Torres JJ, Hopkins TL, Smith WO (1986) Antarctic mesopelagic micronekton: evidence from seabirds that pack ice affects community structure. Science 232:847–849

Alonso JC, Johnstone GW, Hindell M, Osborne P, Guard R (1987) Las aves del monolito Scullin, Antartida oriental (67° 47′ S, 66° 42′ E). In: Castellvi J (ed) Actas del segundo symposium Espanol de estudios Antarticos. Consejo Superior de Investigaciones Cientificas, Madrid, pp 375–386

Baker A de C, Clarke MR, Harris MJ (1973) The NIO combination net (RMT 1 + 8) and further developments of rectangular midwater trawls. J Mar Biol Assoc UK 53:167–184

Brook D, Beck JR (1972) Antarctic petrels, snow petrels and south polar skuas breeding in the Theron Mountains. Br Antarct Surv Bull 27:131–137

CCAMLR (1986) Report of the Working Group for the CCAMLR Ecosystem Monitoring Program (Annex 6). In: Report of the Fifth meeting of the Scientific Committee. SC-CAMLR-V, Hobart, Australia, pp 143–190

Croxall JP, McCann TS, Prince PA, Rothery P (1989) Variation in reproductive performance of seabirds and seals at South Georgia and Signy Island, South Orkney Islands, 1976–1987: implications for Southern Ocean monitoring studies. In: Sahrhage D (ed) Antarctic ocean and resources variability. Springer, Berlin Heidelberg New York, pp 261–285

Daly KL, Macaulay MC (1988) Abundance and distribution of krill in the ice edge zone of the Weddell Sea, austral spring 1983. Deep Sea Res 35:21–41

Deacon GER (1976) The cyclonic circulation in the Weddell Sea. Deep Sea Res 23:125–126

Emison WB (1968) Food preferences of the Adélie penguin at Cape Crozier, Ross Island. In: Austin OL Jr (ed) Antarctic research series, vol 12. American Geophysical Union, Washington DC, pp 191–212

Green K, Johnstone GW (1986) Breeding distribution and abundance of surface-nesting petrels in the Rauer Islands, East Antarctica. ANARE Res Notes No 35, 54 pp

Green K, Johnstone GW (1988) Changes in the diet of Adélie penguins breeding in East Antarctica. Aust Wildl Res 15:103–110

Grigor'yev YA (1967) Circulation of the surface waters in Prydz Bay. Sov Antarct Exped 7:74–76

Harris GP (1985) The answer lies in the nesting behaviour. Freshwater Biol 15:375–380

Harris GP, Davies P, Nunez M, Meyers G (1988) Interannual variability in climate and fisheries in Tasmania. Nature 333:754–757

Heywood RB, Everson I, Priddle J (1985) The absence of krill from the South Georgia zone, winter 1983. Deep Sea Res 32:369–378

Hosie GW, Ikeda T, Stolp M (1988) Distribution, abundance and population structure of the Antarctic Krill (*Euphausia superba* Dana) in the Prydz Bay region, Antarctica. Polar Biol 8:213–224

Hunt GL Jr, Schneider DC (1987) Scale-dependent processes in the physical and biological environment of marine birds. In: Croxall JP (ed) Seabirds: feeding ecology and role in marine ecosystems. Cambridge University Press, Cambridge, pp 7–41

Ikeda T, Hosie G, Stolp M (1986) SIBEX II cruise krill/zooplankton data. ANARE Res Notes No 32, 70 pp

Kanda K, Takagi K, Seki Y (1982) Movement of the larger swarms of Antarctic krill *Euphausia superba* population off Enderby Land during 1976–1977 season. J Tokyo Univ Fish 68:25–42

Kerry KR, Woehler EJ, Robb MS (1987a) Oceanographic data: Prydz Bay region – SIBEX II, MV Nella Dan, January 1985. ANARE Res Notes No 53, 122 pp

Kerry KR, Woehler EJ, Wright S, Dong Z (1987b) Oceanographic data: Prydz Bay region – FIBEX, MV Nella Dan, January–March 1981. ANARE Res Notes No 49, 114 pp

LaCock GD (1986) The southern oscillation, environmental anomalies, and mortality of two southern African seabirds. Clim Change 8:173–184

Lishman GS (1985) The comparative breeding biology of Adélie and chinstrap penguins *Pygoscelis adeliae* and *P. antarctica* at Signy Island, South Orkney Islands. Ibis 127:84–99

MacDonald CM, Williams R, Adams M (1986) Genetic variation and population structure of krill (*Euphausia superba* Dana) from the Prydz Bay region of Antarctic waters. Pol Biol 6:233–236

Northern Ice Limit (1981–1988) Prepared by US Navy-NOAA Joint Ice Center. Suitland

Penney RL (1968) Territorial and social behaviour in the Adélie penguin. In: Austin OL Jr (ed) Antarctic research series, vol 12. American Geophysical Union, Washington DC, pp 83–131

Puddicombe RA, Johnstone GW (1988) The breeding season diet of Adélie penguins at the Vestfold Hills, East Antarctica. In: Ferris JM, Burton HR, Johnstone GW, Bayly IAE (eds) Biology of the Vestfold Hills, Antarctica: Proc Symp, Hobart, August 1984. Developments in hydrobiology. Kluwer Academic Publishers, Dordrecht, pp 239–253

Schreiber RW, Schreiber EA (1984) Central Pacific seabirds and the El Niño Southern Oscillation: 1982 to 1983 perspectives. Science 225:713–716

Siegel V (1985) The distribution pattern of krill, *Euphausia superba,* west of the Antarctic Peninsula in February 1982. Meeresforschung 30:292–305

Smith NR, Dong Z, Kerry KR, Wright S (1984) Water masses and circulation in the region of Prydz Bay, Antarctica. Deep Sea Res 31:1121–1147

Streten NA, Pike DJ (1984) Some observations of the sea-ice in the southwest Indian Ocean. Aust Met Mag 32:195–206

Streten NA, Wells NC (1977) Atmospheric and oceanic circulation at high latitudes of the South Atlantic and the southwest Indian Oceans. Deep Sea Res 24:791–793

Thomas PD, Green K (1988) Distribution of *Euphausia crystallorophias* within Prydz Bay and its importance to the inshore marine ecosystem. Pol Biol 8:327–331

Williams R (1985) The potential impact of a krill fishery upon pelagic fish in the Prydz Bay area of Antarctica. Pol Biol 5:1–4

Williams R, Kirkwood JM, O'Sullivan DB (1986) ADBEX I cruise zooplankton data. ANARE Res Notes No 31, 107 pp

Human Impacts and Ecological Constraints on Skuas

A. D. HEMMINGS [1]

Summary. Skuas are amongst the most-studied animals in the Antarctic and subantarctic regions. This report documents human impacts across the range of natural environments occupied by skuas in the Southern Ocean. Increases in their populations are shown in response to station garbage, introduction of exotic prey species and as a result of human perturbations of the marine environment. Population reductions are being effected by direct destruction, the exclusion of birds from breeding grounds as a result of construction, agriculture or other human activity, or indirectly through reductions in local prey availability. These gross effects are compounded by the isolated nature of many of the ecosystems, which may recover only slowly or stabilise in different forms altogether.

1 Introduction

Two species of skua occur in the Antarctic region: the subantarctic skua, *Catharacta lonnbergi* of the Antarctic Peninsula, Scotia Arc Islands and subantarctic and the Antarctic skua *Catharacta maccormicki* of the Antarctic Continent, Antarctic Peninsula, South Shetlands and South Orkney Islands. This chapter reviews the nature of human impacts upon these skuas across their natural range.

Skuas are amongst the most-studied animals in the Southern Ocean (Furness 1987) and have been investigated to some extent across virtually their entire breeding range from Continental Antarctica, through the Peninsula, subantarctic and temperate oceanic islands. These environments range from the severe and markedly seasonal, where breeding success is low, to near-temperate regions where breeding success is high and populations may be sedentary (Hemmings in press). Given their wide range, they experience a number of human impacts which are, I suggest, of some use as indicators of human impacts.

Moreover, we have a substantial body of information on skuas available (which is not necessarily the case with other species) from which to make first-order assessments of the type and intensity of human impacts on the genus. Known human impacts on skuas are listed in Table 1.

2 The Types of Impacts

2.1 Killing and Maiming of Skuas

Currently, this is generally not a serious problem, although skuas were killed by individuals on the early Antarctic expeditions (Wilson 1907). However, Johnston (1971) included wing breakages in Antarctic skuas at Hallett in his assessment of mortality due to human activities and, during the first Chinese expedition to King George Island, it is reported that skuas were killed (McDonald 1985, Barsch D pers. commun.). Skuas are shot by farmers on Pitt Island in the Chathams group despite legal protection, on islands in the Straits of Magellan (Parmelee and MacDonald 1975) and also in the Falklands and Tristan da Cunha (Richardson 1984). This destruction is largely confined to the northerly populations outside the Antarctic region.

Moynihan (1962) reported egging of Chilean skua colonies and presumably this has occurred elsewhere on those few islands which have been long occupied by humans. Deliberate destruction of eggs has occurred in the Chathams where farming interests view the skua as a pest. The removal of some eggs of the Antarctic skua in the vicinity of the Italian Terra Nova Station to facilitate work around the station (Greenpeace 1988) seems to have been done with some sensitivity – the eggs were apparently placed under other skuas – but is, nonetheless, disruptive of the local population.

[1] Department of Zoology and Antarctic Policy Group, Environmental Science, University of Auckland, Private Bag, Auckland, New Zealand.

Antarctic Ecosystems. Ecological Change and Conservation
(ed. by K. R. Kerry and G. Hempel)
© Springer-Verlag Berlin Heidelberg 1990

Table 1. Known human impacts on skuas

Effects	Cause	Mortality (adults)	Level of impact[a]	Location affected	Species	Reference
Direct:						
Killing/maiming	Personnel	Yes	Low (L)	Hallett Station	*maccormicki*	Johnston (1971)
Killing/maiming	Personnel	Yes	Low (L)	King George Is	*lonnbergi* *maccormicki*	Barsch pers. commun. McDonald (1985)
Killing/maiming	Personnel	Yes	Low (L)	Chatham Is.	*lonnbergi*	pers. observ.
Killing/maiming	Personnel	Yes	Low (L)	Straits of Magellan	(*chilensis*)[b]	Parmelee and MacDonald (1975)
Killing/maiming	Personnel	Yes	High (L)	Tristan da Cunha	*lonnbergi*	Richardson (1984)
Egging	Locals	No	?	Straits of Magellan	(*chilensis*)[b]	Moynihan (1962)
Incidental:						
Egg moving	Personnel	No	?	Terra Nova Station	*maccormicki*	Greenpeace (1988)
Poisoning (lead batteries)	Personnel	Yes	Low (L)	Hallett Station	*maccormicki*	Johnston (1971)
Poisoning (baits for other species)	Personnel	Yes	Low (L)	Codfish Is (NZ)	*lonnbergi*	A Cox pers. commun.
Choking (corn cobs/batteries)	Personnel	Yes	Low (L)	Hallett Station	*maccormicki*	Johnston (1971)
Entanglement in fishing line	Fishermen	Yes	Low	Chatham Is.	*lonnbergi*	pers. observ.
Striking aerials	Personnel	Yes	Low (W)	Antarctic/Subantarctic Stations	*maccormicki* *lonnbergi*	pers. observ. Morvan et al. (1967)
Fowl cholera	Natural or introduced?	Yes	High (L)	Palmer Station	*lonnbergi*	Parmelee et al. (1979)
Fowl cholera	Natural or introduced?	Yes	High (L)	King George Is	*lonnbergi*	Trivelpiece et al. (1981)
Oiling	Shipping[c]	?	High	Palmer Station	*lonnbergi* *maccormicki*	NSF (1989)
Habitat modification (construction)	Personnel	No	High (L)	Dumont d'Urville Station	*maccormicki*	Jouventin et al. (1984)
Disturbance of food chains (construction)	Personnel	No	Low (L)	Hallett Station	*maccormicki*	Johnston (1971)
Fisheries activities?	Commercial operators	No	High	Signy Station	*lonnbergi* *maccromicki*	Hemmings (1984) unpub. data: this study
Additon of food sources (wastes)	Personnel	No	Low (L)	Dumont d'Urville Station	*maccormicki*	Morvan et al. (1967) Jouventin and Guillotin (1979)
Addition of food sources (wastes and exotic introductions)	Personnel	No	Low (W)	Antarctic/Subantarctic Stations	*maccormicki* *lonnbergi*	Trivelpiece and Volkman (1982) Hall (1900) Jones and Skira (1973) Sinclair (1980) Williams (1980) Stahl and Mougin (1986) Hunter (this Vol.)

[a] Low = affecting only small part of population; high = affecting large part of population; L = localised effect; W = more widespread; ? = unknown effect.
[b] Species in parentheses outside Antarctic regions.
[c] Bahia Paraiso sinking oil spillage

2.2 Indirect Killing of Skuas Through Accidental Poisoning, Entanglement in Fishing Nets and Lines or Aerial Arrays at Stations, or as a Result of Introduced Avian Diseases

Poisoning of skuas has been recorded by F. C. Kinsky (in Johnston 1971), as a result of ingesting parts of lead battery plates. A consistent feature of abandoned bases in Antarctica seems to be the presence of such batteries (personal experience in Antarctic Peninsula and Greenpeace 1988) and this may be a widespread problem, although probably a 'low-level' one in terms of the numbers of animals involved. Poisoning, whether from batteries or other noxious substances, is an impact that is likely to affect skuas, kelp gulls (*Larus dominicanus*) and sheathbills (*Chionis alba*) particularly as these birds are likely to fossick through human debris ashore.

Skuas have also died as a result of mechanical obstruction caused by trying to eat strange objects discarded by humans. Johnston's (1971) referral to dead skuas with their crops ruptured by corn cobs is the only account in the scientific literature, although there are a number of anecdotal accounts of a similar vein.

Enormous quantities of fishing line are lost or discarded in fishing operations, and skuas as opportunistic scavengers are attracted to objects attached to these. On South East Island in the Chathams group there were 39 breeding territories holding 92 adult skuas in 1986/87. Eleven birds present in 1985/86 were missing in 1986/87. Two of the 11 (from the same territory) were found dead, entangled in separate fishing lines on the same stretch of coastline.

Aerial arrays at stations present a consistent low-level threat to any birds transiting the area, although there are few reliable figures for mortality. During 2.5 years at Signy Island I witnessed three events – all with skuas panicked by helicopters – where the birds struck the arrays or became entangled with them and on another occasion found a recently dead bird with a broken wing beneath the aerials. Morvan et al. (1967) reported two separate events of Antarctic skuas striking aerials at Dumont d'Urville.

An outbreak of fowl cholera at Palmer Station in 1978/79 may have caused the deaths of several dozen brown skuas, including breeding birds, but did not affect the more numerous Antarctic skuas (Parmelee et al. 1979). Over a 9-year period there were between six and nine pairs of subantarctic skuas each year (Parmelee and Pietz 1987), so this represents a large impact on that population. Similarly, 90% of a banded subantarctic skua population in Admiralty Bay at King George Island disappeared between 1977/78 and 1980/81 (Trivelpiece et al. 1981). These may have been natural outbreaks, but the absence of records elsewhere must leave open the possibility (no more than that) of accidental introduction through human agency. Spoiled avian food products (eggs, egg powder and poultry) present a risk of introducing exotic avian disorders when discarded in the Antarctic and such materials have been encountered near stations.

Dumped food wastes, including eggs, at the Spanish station King Juan Carlos I (Greenpeace 1988) may have been scavenged by skuas, gulls and sheathbills. Another potential vector for avian disease are pigeons taken to the Great Wall Station in apparent breach of the 1964 Agreed Measures for the Conservation of Antarctic Flora and Fauna. Some of these pigeons were killed by skuas and gulls (Greenpeace 1988).

During February and March 1989, oil pollution resulted from the grounding of the Argentine vessel *Bahia Paraiso* impacting particularly severely on local skua populations. Adult skuas have been oiled and fledglings deserted by their parents or poisoned by the oil. At Anvers Island, almost all the young of the 1988/89 breeding season around Palmer Station appear to have perished as a result (NSF 1989).

2.3 Habitat Destruction (Breeding and Feeding Grounds Ashore) Through Base Construction, Vehicle Activity and Other Human Disturbance

The two best documented cases of habitat destruction associated with base activities are at Hallett (Johnston 1971) and Dumont d'Urville (Jouventin et al. 1984) stations, although the problem is not confined to these localities. With Hallett, the decline of the Antarctic skua was approximately coincident with a decline in the Adélie penguin population and that case is therefore considered under the effects of disturbance of existing food chains.

At Point Geologie the construction of an airstrip by France has resulted in the disturbance of six pairs of Antarctic skuas (Jouventin et al. 1984) from a total of 37 pairs in the entire Point Geologie archipelago (Thomas 1986). Since probably half the breeding population of skuas in the whole of Terre Adélie is located at Point Geologie (Thomas 1986), any impact there could be of considerable importance. Quite apart from the direct effects during airstrip construction, an operational airstrip will increase the number of people at the station and the amount of activity there (Thaler 1984). Measures to prevent bird-strike during air operations may have to be taken, and skuas may present serious problems in this respect. The two most likely effects of increased human activity at Dumont d'Urville, namely reduced avian populations and increased human garbage, will have opposing effects on the skua population and the outcome is far from clear at present.

That skua breeding territories and densities may be maintained despite long-term proximity to human activity has been amply shown at Cape Bird (Young, this Vol.). However, whereas the territorial pattern may not have changed appreciably over 22 seasons, caution should be exercised in interpreting this as evidence of no human impact. A population which is "acclimatised" to human activity is still a perturbed population and the behavioural and reproductive consequences of this need to be addressed further.

2.4 Disturbance of Components of Existing Food Chains

Johnston's examination of the decline of skua numbers at Hallett Station (Johnston 1971) over a 9-year period showed it to be mainly due to the abandonment of the colony by breeding pairs of skuas. There was a 54% reduction in breeding pairs over this period, from 181 pairs in 1960/61 to 98 pairs in 1968/69, coincident with the period of human habitation and a decline in the Adélie penguin population.

Johnston was not sure whether the decline in skua numbers was due to direct human disturbance or the reduction of the penguin population (itself a result of human activity).

At Signy Island the skua populations have shown a substantial increase over the past 20 years. The subantarctic skua population has doubled (from around 70 breeding pairs in 1958/59 to 143 breeding pairs in 1982/83, Table 2) and a small population of Antarctic skuas has established itself (Hemmings 1984). Whereas many scientific stations have attracted skuas by their dumping of human food wastes, this does not appear to have been the case at Signy. The station area is divided between two skua territories, and has been for the past 20 years. Largely as a result of this, few other skuas are found in the vicinity of the station and such food scraps, as are discharged into the sea, are monopolised by the pair of skuas which occupy each of these territories. Undoubtedly, the location and operation of the station since its establishment in 1947 have affected the use of the area by skuas, however, there is no evidence that it has facilitated the increase in the skua population seen at Signy (Hemmings unpub. data), contrary to the suggestion in Ainley et al. (1986).

Table 2. Number of breeding pairs of subantarctic skuas and rate of increase at Signy Island 1958/59 to 182/83

Year	Total No. breeding pairs	No. breeding pairs by region					
		1	2	3	4	5	6
1958/59	70[a]	7	5	15	15[b]	6	22
1960/61	63	8	4	13	11	6	21
1961/62	77	9	5	16	9	9	29
1962/63	64	9	5	14	6	10	20
1964/65	94[c]	13	6	21	–	13	29
1982/83	143	21	11	28	27	21	35
Increase (%) 1958/59–1982/83	104	200	180	87	80	250	59

[a] Calculated using estimate in Region 4 (see footnote b).
[b] Estimated by comparing a partial count of 4 pairs in part of Region 4 in 1958/59 with the same part in subsequent years.
[c] Burton (1968) suggested that the total breeding population was about 110 pairs, but had firm evidence for only 81 pairs. I think it unlikely that uncensused Region 4 would have had around 30 pairs in 1964/65. Even assuming a doubling of the population in Region 4 since 1962/63, the total population for that year would be only 94 pairs.

Work by Croxall et al. (1981) on the penguin populations of Signy has shown that Adélie penguins have increased at a rate of 3–4% per annum and chinstrap penguins at a rate of 7–10% per annum over a period of 20–30 years up to 1978/79 and attributed this substantial increase to the enhanced food availability following human reduction of whale populations in Antarctica.

Although skuas at Signy do not feed exclusively on penguins, they are probably an important source of food during the summer by virtue of their numbers alone and therefore some increase in skua numbers might have been expected as a result of the increased penguin numbers. Signy is roughly triangular, with penguin colonies at the apices. The three regions within which these colonies occur (Regions 3, 4 and 6 in Table 2) have shown the smallest increases in skua numbers since 1958/59 (87, 80 and 59% respectively), although these are still substantial changes. In the other three areas, where skuas feed largely on petrels, increases of 180–250% have occurred. There are no reliable data for changes in the petrel population of Signy, but this substantial increase in skua numbers may point to such an increase, although it might also be due, in part, to any increase directly attributable to the penguin increase. Examination of the unpublished reports and maps recording the skua populations since 1958 (held in the Archives of the British Antarctic Survey) has shown that the increased population has been accomodated by increased breeding density, rather than by occupation of new areas on the island.

Very few skuas have nested above 100 m a.s.l., even in areas which are invariably ice-free during the breeding season; in 1982/83 only 4 of 143 pairs nested above 100 m. Taking the 150-m contour line as the cut-off point, this increase in the skua population has therefore occurred in a summer ice-free land area of just 9.3 ha.

The colonisation of the South Orkneys by the Antarctic skua during the late 1970s may also reflect human influence on the Antarctic marine environment. Since 1947 substantial studies of a great many species have been conducted on, and around, Signy Island. In none of these studies did the pelagic nototheniid fish *Pleuragramma antarcticum* figure as a prey item despite its frequent detection in studies at Peninsula and Continental localities. Yet, Antarctic skuas at Signy subsist almost entirely on this fish, as do many skuas at Continental sites (Young 1963). It is therefore possible that the skuas' recent arrival and breeding in the South Orkneys may be due to increased availability of the krill eating *Pleuragramma*.

2.5 Additions to Food Sources Available for Skuas, Such as: Station Garbage, Fishing Wastes and the Introduction of Stock and Pest Species

Dumps, which include human food wastes, are often found at Antarctic stations and skuas are often in attendance (pers. obser. and Greenpeace 1988). The changes in skua numbers at Point Geologie appear to be largely due to the Dumont d'Urville station. Of the various human activities which have affected the skua population there, the additional feeding opportunities presented by wastes from the base appear particularly important, and were remarked upon by Morvan et al. (1967). Between 1965/66 and 1976/77 the skua population increased from around 80–90 to 320 and the number of breeding pairs from 29 to 40 (Jouventin and Guillotin 1979). Thereafter the population levelled off; in 1981/82 there were 39 breeding pairs (Thomas 1986), representing an increase of 1.7% per annum since 1964/65. During the most rapid period of increase (1970–1972) most pairs reared two chicks (Jouventin and Guillotin 1979), a rare event with Continental colonies but reported from the Signy Island population (Hemmings 1984). Jouventin and Guillotin attributed this increase to the availability of station garbage and showed that the skuas' previous tight synchronisation with the breeding cycle of Adélie penguins had broken down over the same period. More recent work reports a reduction in the number of skua breeding pairs between 1981/82 and 1984/85 by 9.4% per annum to the 1965/66 figure of 29 (Thomas 1986). Why the skua population should have shown this recent decline is, as yet, unclear, the more so as the local Adélie population has increased. Since increased human activity at Point Geologie associated with the airstrip construction, and use, is likely to adversely affect breeding skuas, this decline warrants further attention.

On King George Island Trivelpiece and Volkman (1982) found one pair of Antarctic skuas nesting 25 m from a dump at Arctowski station and defending the dump. This pair commenced egg laying some 3 weeks before the next pair of Antarctic skuas.

At Marion Island, with a subantarctic skua population of about 400, during the summer of 1978/79 between 20 and 50 birds regularly fed on station wastes and petrels dazzled by station lights were collected by patrolling skuas (Sinclair 1980). The availability of food scraps from the station allowed 10 to 25 skuas to overwinter on the island, whereas most skuas were absent from May to August (Williams 1980).

Several authors have referred to the attraction of skuas to the wastes from whaling operations in the subantarctic. Lonnberg (1906), for example, noted skuas scavenging around seal and whale carcasses at the beginning of shore-based operations on South Georgia, and Stonehouse (1956) suggested that this may have helped to maintain a larger skua population than the island could otherwise support. There are no data for any increases in skua populations that could be used to assess this statement, but given the continuance of whaling operations over some 50 years (at South Georgia) it would seem a plausible claim.

Most of the subantarctic and temperate islands have one or (usually) more introduced mammals on them. Skuas, as the largest predators ashore, are able to exploit these animals as food sources, either killing the smaller species (mice, rats, rabbits) or scavenging upon the carcasses of the larger species (sheep, moufflon, reindeer, cattle). Skuas may also benefit from the greater accessibility of some prey species following modifications to the vegetation cover or trampling effected by these introductions. Their eggs and young, in particular, are at risk from predation by cats and rats. Skuas may also face competition from cats for avian prey (Hunter, this Vol.).

On those islands where they have been introduced, exotic mammals may constitute quite a large proportion of the skuas' diets. In the Crozet group, Stahl and Mougin (1986) found 30% of skua pellets were of black rat (*Rattus rattus*) remains and 67% avian remains on Ile de la Possession, whereas on Ile de l'Est, where rats do not occur, 95% of pellets were of avian remains, and most of the remainder were rabbit, which are not found on Possession.

Rabbits are taken by skuas on Kerguelen (Hall 1900) and Macquarie Islands, where their effects on populations of the subantarctic skua have been particularly well documented. Rabbits have been a particularly important food source for plateau nesting skuas on Macquarie Island, since these skuas do not have the alternate natural prey available to the coastal birds (Jones and Skira 1979). Young rabbits are also available during the period when the skuas are themselves rearing young (Johnston 1973). Jones and Skira predicted that any reduction in rabbit numbers would reduce the numbers of skuas on the plateau, but not on the terrace of the island, and this has proved to be the case, following the introduction of myxomatosis in 1978. Between 1974 and 1983 the number of skuas nesting on the plateau decreased by 54% and the number breeding along the coastal terrace increased by 17% (Skira 1984). It has been suggested that skuas have been forced to concentrate their territories closer to penguin colonies (Brothers and Copson 1988). Skira suggests that because of introduced predators (cats, rats, mice, wekas) with some burrowing petrel species confined to offshore stacks in low numbers, the extra predation of skuas following the reduction in rabbits may be a significant check on the petrel population. Other workers have suggested that the number of skuas will decrease with control or eradication of the rabbits and that this will facilitate greater breeding success by the burrow nesting petrels (Rounsevell and Brothers 1984); and presumably they will recolonise the main island. Whichever scenario proves correct (they could both be of course, if Skira is describing the short term and Rounsevell and Brothers the long term), it is clear that rabbits have had a considerable effect on Macquarie's skuas.

3 The Ecological Context

The foregoing catalogue of human impacts on skuas must be considered in the context of our knowledge of their ecology and behaviour. Skuas are effectively the top avian predators in the Antarctic and subantarctic, and, in common with most such animals, few in number and therefore vulnerable to impacts which might at first glance appear quite trivial. Although they are long-lived birds, with a high annual survival as adults, their reproductive rate is not high, ranging from a mean 0.34 fledged chicks per breeding pair for Antarctic skuas at Cape Crozier over 8 years (Wood 1971) to 1.4 per breeding territory for subantarctic skuas in the Chatham Islands over 3 years (Hemmings 1989). Maintenance of skua populations seems dependent upon their high survivorship and stable pair bonds in many cases (Ainley et al. 1984). Any factors which tended to increase adult mortality could therefore have severe implications for local populations. Despite their substantial annual migrations, skuas in the Antarctic regions appear conservative in breeding site preferences and do not tend to stray far from their own natal grounds. Wood's finding (1971) that of 1800 Cape Crozier breeders banded only one was found anywhere else within 160 km of Cape Crozier, and a more recent estimate of the Crozier emigration rate at about 0.3% (Ainley et al. 1986) suggests that the effects of depletion or extirpation of a skua population in one area may not easily be overcome by immigration from undisturbed areas. Nonetheless, skuas can move considerable distances, as evinced by the finding of about a dozen breeder birds in the Ross Sea area which had been banded as chicks at Point Geologie (Ainley et al. 1984). Of course, whether a population can be re-established is only part of the issue; the changes wrought in the meantime might in any case be very severe and not reversible. Furthermore, these disjunct populations may be reproductively isolated and thus genetically distinct.

Most Antarctic skua pairs do not rear their second chick, so human activity, which enhances food availability, may effect a substantial increase in production merely by aiding the survival of this second chick. This has been reported where food wastes are available to skuas (Jouventin and Guillotin 1979). The extent to which the production rates of skua populations in the subantarctic are generally raised above the natural level by the availability of introduced prey has not been determined. In turn, any increasing skua population will put additional pressure upon its traditional prey – birds which may already be stressed by other human impacts. Such additive and synergistic effects are poorly studied to date.

The fundamental problem of human impact on skua populations is "competition" for access to the same ice-free "oases" on the coasts of Antarctica and to the, generally, small subantarctic islands around the Continent. This problem is well illustrated by the recent surveys of the skua populations of the Ross Sea which yielded an estimate of 4798 breeding pairs (Ainley et al. 1986). This survey documented changes in the skua populations, attributing this in part to human activity (primarily wastes from the stations in southern McMurdo Sound) and more than doubled the known breeding sites. Skuas were found nesting on most areas of flat coastal land in the Ross Sea region (Ainley et al. 1984).

Acknowledgements. This chapter draws on work supported by the British Antarctic Survey (1980–84), Royal Forest & Bird Protection Society QEII Scholarship and Auckland University Research Grant 449.226, to which bodies I extend my thanks. Attendance at the Fifth SCAR Symposium was made possible by Travel Grants from the Department of Zoology, University of Auckland; Royal Forest & Bird Protection Society Inc., Wellington and Royal Forest & Bird Protection Society, Central Auckland Branch. I am grateful to G. Hempel, two anonymous referees and especially Knowles Kerry for helpful comments on an earlier draft.

References

Ainley DG, Morrell SG, Wood RC (1986) South polar skua breeding colonies in the Ross Sea region, Antarctica. Notornis 33:155–163

Ainley DG, Ribic CA, Wood RC (1984) Population studies of the south polar skua. Antarct J US 14:167–168

Brothers NP, Copson GR (1988) Macquarie Island flora and fauna management – interpreting progress and predictions for the future. Pap Proc R Soc Tasmania 122:129–135

Burton RW (1968) Breeding biology of the brown skua, *Catharacta skua lonnbergi* (Mathews), at Signy Island, South Orkney Islands. Bull Br Antarct Surv 15:9–28

Croxall JP, Rootes DM, Price RA (1981) Increases in penguin populations at Signy Island, South Orkney Islands. Bull Br Antarct Surv 54:47–56

Furness RW (1987) The skuas. Poyser, Calton

Greenpeace (1988) Expedition report 1987–88 Greenpeace Antarctic Expedition. Stichting Greenpeace Council, Lewes

Hall R (1900) Field notes on the birds of Kerguelen Island. Ibis 6:1–34

Hemmings AD (1984) Aspects of the breeding biology of McCormick's skua *Catharacta maccormicki* at Signy Island, South Orkney Islands. Bull Br Antarct Surv 65:65–79

Hemmings AD (1989) Communally breeding skuas: breeding success of pairs, trios and groups of *Catharacta lonnbergi* on the Chatham Islands, New Zealand. J Zool Lond 218:393–405

Hemmings AD (1990) Winter territory occupation and behaviour of skuas at the Chatham Islands, New Zealand. Emu 90

Johnston BR (1971) Skua numbers and conservation problems at Cape Hallett, Antarctica. Nature 231:468

Johnston GC (1973) Predation by southern skua on rabbits on Macquarie Island. Emu 73:25–26

Jones E, Skira IJ (1979) Breeding distribution of the great skua at Macquarie Island in relation to numbers of rabbits. Emu 79:19–23

Jouventin P, Guillotin M (1979) Socio-ecologie du skua Antarctique a Point Geologie. La Terre et La Vie, Rev Ecol 33:109–127

Jouventin P, Stahl JC, Weimerskirch H, Mougin JL (1984) The seabirds of the French subantarctic islands and Adélie Land, their status and conservation. In: Croxall JP, Evans PGH, Schreiber RW (eds) Status and conservation of the world's seabirds. ICBP Technical Publication No 2, Cambridge, pp 609–625

Lonnberg F (1906) Contributions to the fauna of South Georgia. Kungl Svenska Vetenskapsakademiens Handlingar 40:58–62

McDonald H (1985) Chasing penguins. Far Eastern Economic Review 27 June:34

Morvan PL, Mougin JL, Prevost J (1967) Ecologic du skua Antarctique (*Stercorarius skua maccormicki*) dans l'archipel de Point Geologie (Terre Adélie). L'Oiseau et RFO 37:193–220

Moynihan M (1962) The Chilean great skua. In: Hostile and sexual behaviour patterns of South American and Pacific Laridae. Behaviour (Suppl 8), Brill, Leiden

NSF (National Science Foundation) (1989) Press release 89-16. 11 March

Parmelee DF, MacDonald SD (1975) Recent observations on the birds of Isla Contramaestre and Isla Magdalena, Straits of Magellan. Condor 77:218–219

Parmelee DF, Maxson SJ, Bernstein NP (1979) Fowl cholera outbreak among brown skuas at Palmer Station. Antarct J US 14:168–169

Parmelee DF, Pietz PJ (1987) Philopatry, mate and nest-site fidelity in the brown skuas of Anvers Island, Antarctica. Condor 89:916–919

Richardson ME (1984) Aspects of the ornithology of the Tristan da Cunha group. Cormorant 12:123–199

Rounsevell DE, Brothers NP (1984) The status and conservation of seabirds at Macquarie Island. In: Croxall JP, Evans PGH, Schreiber RW (eds) Status and conservation of the world's seabirds. ICBP Technical Publication No 2, Cambridge, pp 587–592

Sinclair JC (1980) Sub-antarctic skua, *Catharacta antarctica,* predation techniques on land and at sea. Cormorant 8:3–6

Skira J (1984) Breeding distribution of the brown skua on Macquarie Island. Emu 84:248–249

Stahl JC, Mougin JL (1986) La segregation alimentaire chez le skua subantarctique *Stercorarius skua lonnbergi* dans l'archipel Crozet. L'Oiseau RFO 56:193–208

Stonehouse B (1956) The brown skua *Catharacta skua lonnbergi* (Mathews) of South Georgia. Falkl Isl Depend Surv Sci Rep No 14

Thaler L (1984) Rapport du comite des sages concernant l'impact ecologique du project de piste d'atterrissage en Terre Adélie. Paris

Thomas T (1986) L'effectif des oiseaux nicheurs de l'archipel de Point Geologie (Terre Adélie) et son evolution au cours des trente dernieres annees. L'Oiseau RFO 56:349–368

Trivelpiece W, Volkman NJ (1982) Feeding strategies of sympatric south polar *Catharacta maccormicki* and brown skuas *Catharacta lonnbergi.* Ibis 124:50–54

Trivelpiece WZ, Butler RG, Volkman NJ (1981) Pygoscelid penguin research in Admiralty Bay. Antarct J US 16:150

Williams AJ (1980) Aspects of the breeding biology of the subantarctic skua at Marion Island. Ostrich 51:160–167

Wilson EA (1907) Aves. In: National Antarctic Expedition 1901–1904, Natural History, vol 2, Zoology (ii), Lond, pp 64–76

Wood RC (1971) Population dynamics of breeding south polar skuas of unknown age. Auk 88:805–814

Young EC (1963) Feeding habits of the south polar skua *Catharacta maccormicki.* Ibis 105:301–318

Long-Term Stability and Human Impact in Antarctic Skuas and Adélie Penguins

E. C. YOUNG [1]

Summary. This chapter assesses the impact of human activity associated with a field station on the long-term population stability of Adélie penguins (*Pygoscelis adeliae*) and Antarctic skuas (*Catharacta maccormicki*) at the Northern Colony, Cape Bird, Ross Island (77° 13′ S, 166° 28′ E). The field station has been occupied over 22 years at this site.

Human impact was assessed by comparing an area of the colony near the station with a second one at a distance which was much less disturbed. Skuas appeared largely unaffected by the station with breeding density and pattern of territories little changed from first years of research here. Breeding success and age structure of the populations in the two areas were similar. The skuas maintained their high tolerance of people intruding into the territory that had been developed in the first years of the station's operation. This tolerance has persisted even though nearly all the original breeding birds have been replaced. The stability recorded in the skuas contrasts with the local impact found on the breeding penguins. Breeding groups nearest the station declined in numbers even though the total population of the colony markedly increased.

1 Introduction

People have similar requirements to penguins and skuas for ice-free terrain near open water. Thus with man's expanding activities in the Antarctic the chances of causing significant impacts are increasing. Management and conservation will be increasingly concerned therefore with ensuring that birds and people can coexist successfully and that natural penguin and skua breeding conditions are retained. Past experience at a number of places has demonstrated that great care will be needed to achieve this (Stonehouse 1965; Johnstone 1971; Thomson 1977).

Only a small proportion of the coast of Antarctica is suitable for breeding Adélie penguins (*Pygoscelis adeliae*) and Antarctic skuas (*Catharacta maccormicki*). The Victoria Land coast of the Ross Sea, for example, has been estimated by Gregory et al. (1984) to be over 2190 km of which only 210 km are low rocky shores and beaches. On Ross Island a coastline of 320 km has less than 40 km free of ice cliffs, of which no more than 10 km is accessible to penguins.

A research program on penguins and skuas was initiated in November 1965 at the Northern Colony, Cape Bird, Ross Island (73° 13′ S, 166° 28′ E). In the following season a field hut was sited on a small bench adjacent to the penguin colony from which fieldwork has been carried out annually. From the outset rules were established to minimise both the handling and disturbance to the penguins and skuas being studied and in particular to minimise the predation by skuas on neighbouring skua eggs and chicks (Young 1970). Prior to the start of this research program, only brief human contact had been made with these birds. Caughley visited the region in 1955 (Caughley 1960) to count nesting penguins and Wood visited in 1964 for several days to band skuas.

Detailed records had been obtained in 1965 on the penguin groups and skua populations, territories and reaction to human intruders providing an opportunity to assess the impacts on these species of all the activities including research which had taken place from the field hut. A visit to the area therefore was made in the austral summer of 1986/87 to assess how the penguins and skuas had responded to the intervening 22 years of continuous research. This chapter reports the results of this study.

2 Methods

The field hut is located on a bench overlooking the penguins and skuas of H Block (see below). The position of the hut provides good access to the main penguin colony area without having to walk through concentrations of breeding birds along the beach. It was located also on the outer edge of the skua breeding

[1] Department of Zoology, University of Auckland, Auckland, New Zealand.

Antarctic Ecosystems. Ecological Change and Conservation
(ed. by K. R. Kerry and G. Hempel)
© Springer-Verlag Berlin Heidelberg 1990

area overlapping the penguin colony and was backed by steeply rising slopes shunned by skuas. The helicopter pad is located 200 m south of the penguin colony. This has been a well used station catering to a mixed range of research topics including ornithology, marine sciences and physiology. Research teams of four to ten people have been in residence for several months each summer.

2.1 Study Areas

Although most areas of this colony have been studied over the years, this study focused on only two. H Block was located close to the field station; EF Block was located at a distance. These represent extremes of human contact for the colony. The location of these areas and the location of the field station and the helicopter landing pad are shown in Fig. 1.

Penguin groups and their contiguous and overlapping skua territories were mapped in the 1965/66 season and again in the 1986/87 season. Counts of penguin nests were made in early December each year since 1965. Skua territories were delimited by observing contacts between neighbouring pairs on the ground. Apparent overlap of territories occurred only over nesting penguins where ground contact was precluded.

The response of breeding skuas to humans was assessed for skuas in H Block in mid-January (at the peak of hatching) in 1968 and 1987. At each observation a record was made of the response of the nesting pairs to a person walking at a steady pace between nests, stopping at each nest for a set interval of 30 s before moving on to the next. Specifically, the records were of the distance from the nest before the first alarm call and when the sitting bird left the nest, and of the type and intensity of the attack behaviour.

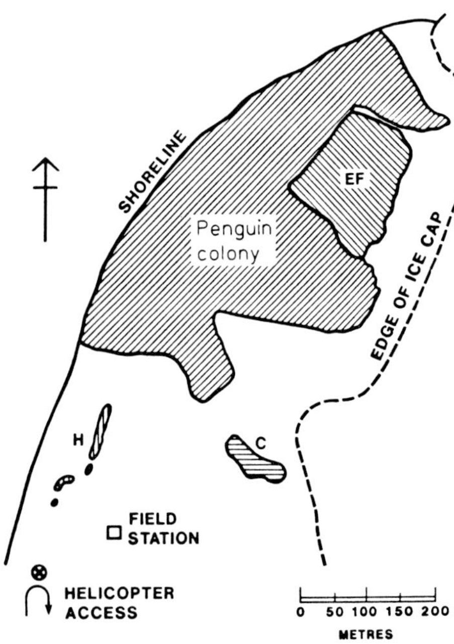

Fig. 1. The Northern penguin colony at Cape Bird, Ross Island, showing the location of the field station and the two areas, *H* and *EF Bocks*, used to measure human impact at this site

3 Results

Maps of the penguin breeding groups and overlapping skua territories are presented for H Block in Fig. 2 and for EF Block in Fig. 3. A comparison is made in each case between the 1965/66 season and the 1986/87 season.

3.1 Comparison of Skuas in H Block and EF Block

3.1.1 Density of Breeding Pairs

The density of pairs breeding within each area at two dates is shown in Table 1. Although the mean territory size has increased slightly in EF, the higher density of birds occupying H Block has reduced the mean area there.

3.1.2 Changes in the Pattern of Skua Territories

The patterns of territories illustrated in Figs 2 and 3 are a very accurate representation of the territories in late December and early January. The overall pattern is determined by nesting density, by the topography of the area and by the distribution and shape of the penguin groups. When possible, skuas nest on broken ground (to avoid wandering penguins and for shelter

Table 1. Density of breeding pairs of skuas in the two areas at the establishment of the field station and 22 years later

Year	Number of pairs	Mean territory size (m^2)
H Block (total area 22000 m^2)		
1965/66	17	1290
1986/87	20	1045
EF Block (total area 16800 m^2)		
1965/66	11	1527
1986/87	10	1680

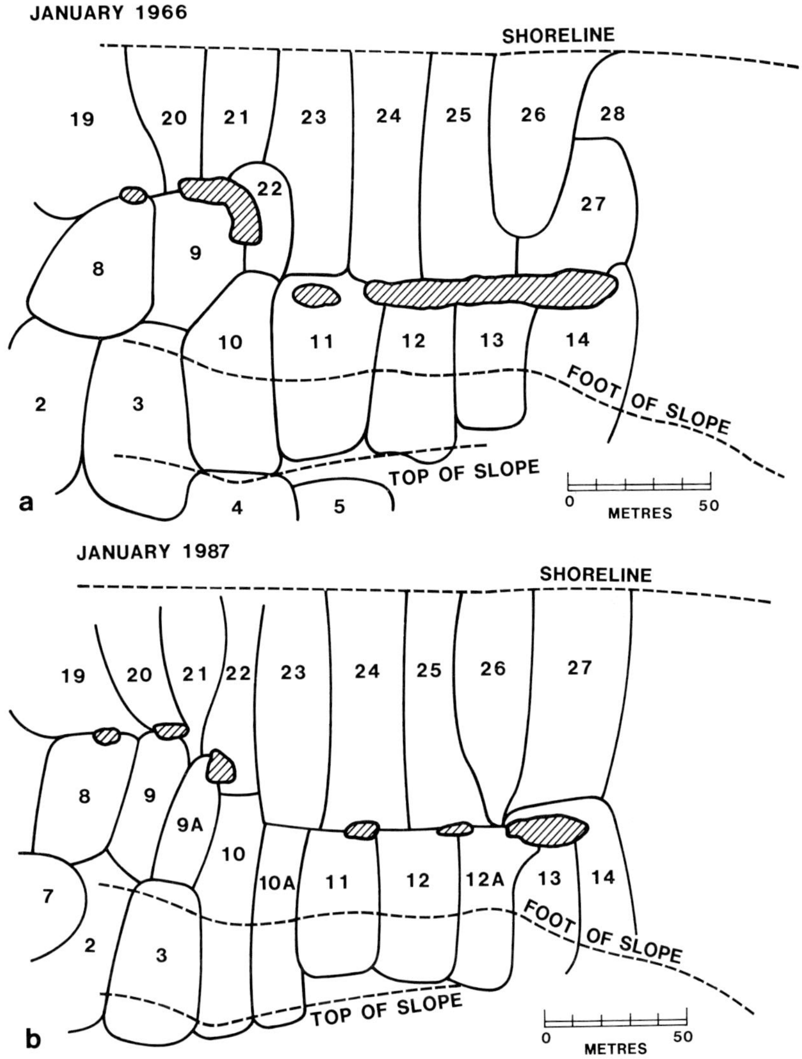

Fig. 2. Changes in the pattern of skua territories and Adélie penguin breeding groups in H Block, 1965/66 and 1986/87. Skua territories are *outlined* and numbered *2–28*; penguin groups are *cross-hatched*

from wind and blowing sand) and on the higher parts of the territory. Taken together these factors result in territories that extend from the slopes across the colony area. This pattern is clearly evident in the territories of pairs 152 and 153, for example, in EF Block and for pairs 10 to 14 in H Block. Nesting is difficult, often impossible, among closely spaced penguin groups.

In spite of the changes in the distribution and size of the penguin groups in H Block in 1987, the skua territories there have retained their original pattern. The pairs on the seaward and landward sides still meet at the marginally higher points of the beach ridge, even where these have lost breeding penguins. The pattern of breeding territories on the seaward side (pairs 19–27) has remained remarkably constant with time even though these territories lie across an essentially featureless beach. The numbers assigned to these territories in 1965 could be applied again in 1987 without modification.

A low-level banding effort that maintained a banded breeding population was continued at Cape Bird until 1980 and can be used to compare the ages of skuas breeding in the two areas. In 1987 the two areas contained identical proportions of older and younger skuas, represented by banded and unbanded birds. Twenty-two of the 42 birds in H Block and 16 of the 28 birds in EF Block were banded. There was no evidence that young birds had been deterred from settling in H Block.

3.1.3 Breeding Success to Chick Hatching

The records of eggs laid and surviving to hatch for the two areas have been compared for 4 of the first 5 years

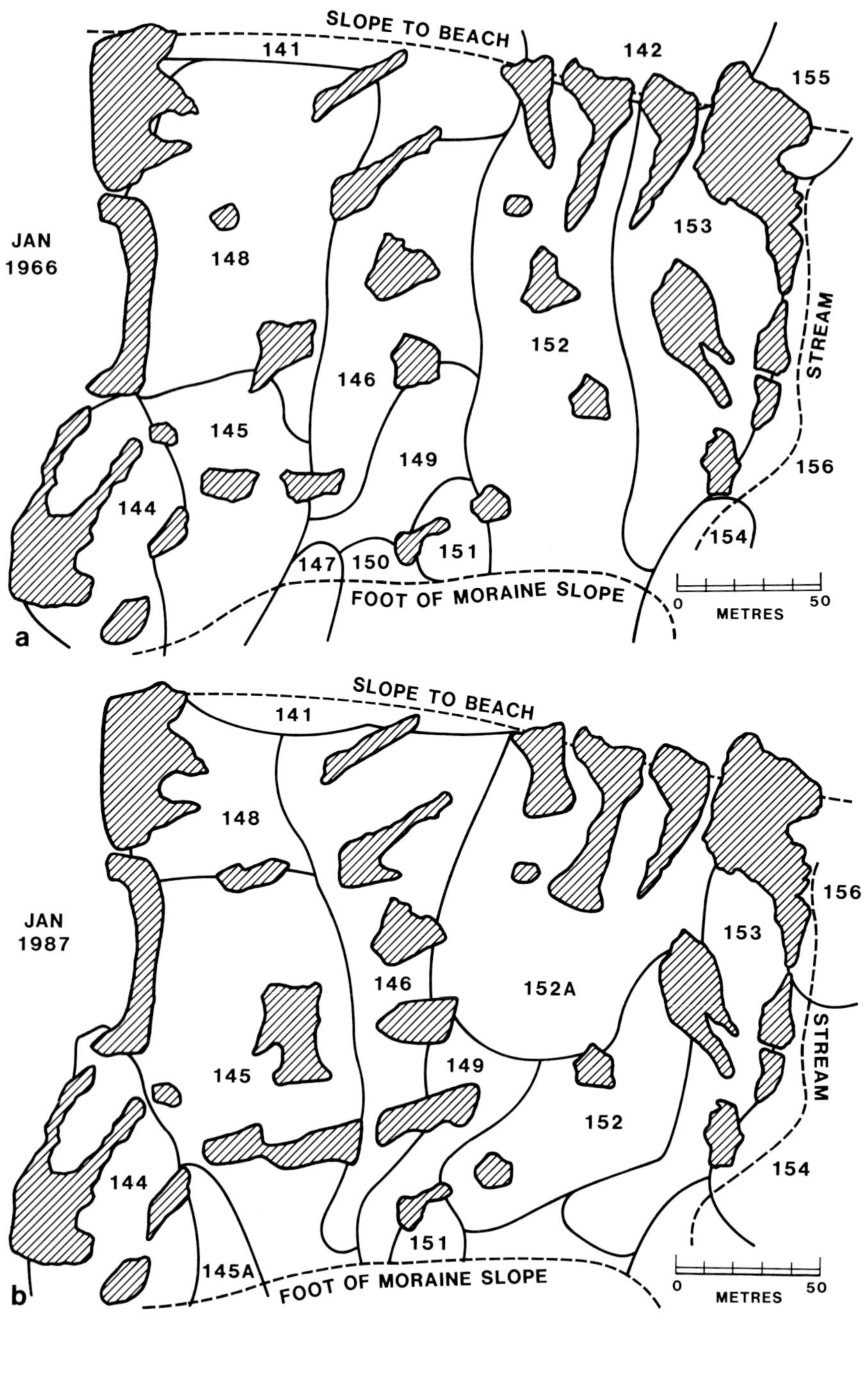

Fig. 3. Changes in the pattern of skua territories and Adélie penguin breeding groups in EF Block, 1965/66 and 1986/87. Skua territories are numbered *141*–*156*; penguin groups are *cross-hatched*. The increased numbers of breeding penguins in 1987 is indicated by the larger groups and the newly established group in skua territories *145, 146* and *149*

of the station's existence. These data were not collected in the second year. Overall, in these four years 59% of 153 eggs laid in H Block survived compared with 37% of 105 eggs in EF Block. In the first and last years hatching rates were similar. In the middle years H Block rates were significantly higher. According to these criteria the breeding success in H Block was even more successful than in the less disturbed EF Block.

3.1.4 Response of the Breeding Skuas

The skuas in this colony became progressively more tolerant of human behaviour during the first years of research. By the third season many skuas had given up flight attacks on people in their territory and a proportion remained tightly on their nests until approached very closely. Some needed to be lifted from the nest to

Table 2. The declining relative importance of H Block as a penguin breeding area in the Northern Colony at Cape Bird

Year[a]	Total colony nest number	Numbers of nests in H Block (total nests, %)
1965/66	25065	864 (3.4)
1966/67	24800	883 (3.6)
1967/68	21560	737 (3.4)
1968/69	14929	426 (2.8)
1969/70	19803	583 (2.9)
1984/85 a	28978	366 (1.2)
1985/86 b	35103	366 (1.0)
1986/87 c	37278	391 (1.0)

[a] From counts of nests in early to mid-December. The records for the last three seasons are from unpublished reports; a D. S. Horning (1985); b L. A. Logan, J. Fennell, R. Brown (1986); c G. J. Wilson (1988).

check the eggs and chicks (Young 1970). This tolerance to people is in strong contrast to the usually fierce habits of skuas in defence of the nest. Strong nest defence continued in the skuas at the middle and southern penguin colonies at Cape Bird, which were seldom visited by ornithologists, and was still evident there in 1987. Although the penguins were treated with similar care (almost all observations were made from outside the study breeding groups), there was no evidence during these first years of research that they were also becoming more tolerant of people.

Two levels of response can be compared from the records made in 1968 and 1987: the distance in metres from the nest when it was abandoned by the skua and the numbers of pairs carrying out swoop and soar attacks. Both are dependent on weather conditions. Under cold, windy conditions skuas sit more tightly on the nest and attack more vigorously when they leave it. On the occasions compared here the weather was warm and calm. Even though the experiment was more demanding of the skuas on the first occasion, with more time spent at the nest, this does not significantly affect the conclusion that these skuas have retained their tolerance of people; indeed, the evidence indicates that this was even higher in 1987 than earlier. It should be noted that only one skua remained of those worked with in 1965–1970; the rest were new birds. In 1968 skuas left the nest on average at a distance of 18.7 m and 6 of the 21 pairs carried out mild swoop and soar flights directed at the intruder. In 1987 a distance of 6.4 m on average was measured and only 1 of the 14 pairs tested in this year carried out a swoop and soar attack. In each year only one or two pairs were in flight at any time during these nest recording visits and all nests were protected again by skuas once the person had reached the next territory or was at most two territories away.

4 Discussion

4.1 Stability in Skua and Penguin Breeding Numbers and Distribution

Clear differences have been found in the comparison of long-term changes in skua and penguin numbers in the two areas examined to measure the impact of the station's location on this colony. No measurable impact on the breeding skua was detected by any of the methods used. Most significantly the density and pattern of skua territories in the area closest to the field station had been maintained. Skua breeding territories had also been maintained throughout the general body of the penguin colony. Chick hatching rates and recruitment of young birds, other indicators of human impact, were similar in the two areas compared. Overall, it may be concluded that the skuas here have adapted very well to the station's presence. The situation of the penguins is less satisfactory. The decline in numbers of breeding penguins in H Block is shown graphically in the maps of this area and is recorded clearly in the annual counts. In this instance the colony as a whole is not at risk, its numbers have been increasing rapidly, but the distribution of its breeding groups has been affected. It might be suggested that the falloff in the numbers of penguins in H Block has been caused by the depredations of the skuas there, but even higher concentrations of skuas occur around isolated small penguin groups at other colony locations and their numbers have not declined. C Block contains similarly isolated groups of penguins with numerous surrounding skua territories. It is the obvious site to distinguish the impacts of skuas and people on this colony. This area has, however, also been studied intensively by scientists and its penguin numbers have also declined.

Oelke (1978) has stressed the importance of maintaining the balance between penguins and skuas for the conservation of the Cape Crozier penguin breeding groups. Breeding groups reduced in size through human disturbance are unable to resist skua attacks. The situation in H Block is close to the critical level for survival of breeding groups of penguins. This decline in numbers in H Block is also of concern for a second reason. This area was a marvelous place for the direct observation of the interrelations of penguins and skuas, but this has now been disturbed, as evidenced by the altered ratio of the numbers of the two species.

4.2 Stability of the Pattern of Skua Territories in H Block

Although there is a general conformity in the pattern of skua territories throughout the colony over these years, this pattern is to some extent contrained by topography and a constant density of breeding birds. It would indeed be surprising if the pattern had changed grossly given these preconditions. In one area, however, the pattern has been remarkably constant. On the seaward side of the penguin groups in H Block there are nine territories whose margins have hardly varied during this long period. Moreover, the boundaries on the longest sides are not in any way linked to topography. The beach extending from the penguin groups towards the shore is almost featureless, a few stones larger than the others are the only obvious markers across an otherwise uniform reach of shingle and sand. There are no raised roosts. The inland edge of each of these territories is anchored by the penguin groups or by the ridges they once occupied. The other end extends across the broken ice of the sea margin in spring and to the sea edge in summer when the ice has broken away. The stability of pattern shown here is not a product of the crude location of the boundaries. On this beach the territorial boundaries can be mapped from an observation point to within less than 1 m. Nor is it dependent on the longevity of the original territorial birds; only one bird remains of the set occupying these territories in 1970. This stability has apparently arisen through the perfect replacement of pairs within former territories, as recorded in the pair-removal experiments carried out in this colony earlier (EC, Young 1972).

4.3 Maintenance of Tameness in Skuas

The most remarkable feature of the skuas at the Northern Colony in 1987 was their tolerance of people within territories. This has persisted, even increased, since 1970 although virtually all of the original birds of the study group of 1965–70 had been replaced. This tolerance provides a unique working environment for behavioural studies at this colony. Among other advantages is the ease with which bands can be read using field glasses on birds standing quietly on the territory. The birds do not need to be handled nor frightened and their acceptance of scientists and visitors ensures that minimal impact occurs to breeding behaviour. The greatest benefit lies, however, in the prospect that the pattern of skua territories within and around the penguin colony can be maintained indefinitely. This experience suggests that in any area where intensive human influence is expected there is much to be gained from spending the first few years of contact in taming the skuas.

References

Caughley G (1960) The Adélie penguins of Ross and Beaufort Islands. Rec Dom Mus (Wellington) 3:263–282

Gregory MR, Kirk RM, Mabin MCG (1984) Shore types of Victoria Land, Ross Dependency, Antarctica. N.Z. Antarct Rec 5:23–40

Johnston BR (1971) Skua numbers and conservation problems at Cape Hallett, Antarctica. Nature 231:468

Oelke H (1978) Natürliche oder anthropogene Populationsveränderungen von Adéliepinguinen (*Pygoscelis adeliae*) im Ross-Meer-Sektor der Antarktis. J Orn 119:1–13

Stonehouse B (1965) Animal conservation in Antarctica. N.Z. Sci Rev 23:3–7

Thomson RB (1977) Effects of human disturbance on an Adélie penguin rookery and measures of control. In LLano GA (ed) Adaptations within Antarctic ecosystems. Smithsonian Institution, Washington DC, pp 1177–1180

Young EC (1970) The techniques of skua-penguin study. In Holdgate MW (ed) Antarctic ecology, vol 1. Academic Press, London, pp 568–584

Young EC (1972) Territory establishment and stability in McCormick's skua. Ibis 114:234–244

Young P (1971) Penguin summer. Reed, Wellington

Influence of the Increasing Fur Seal Population on Wandering Albatrosses *Diomedea exulans* Breeding on Bird Island, South Georgia

J. P. Croxall, S. P. C. Pickering and P. Rothery [1]

Summary. Since 1961, the wandering albatross breeding population at Bird Island, South Georgia has decreased by 22% (1% per annum). Since 1960 Antarctic fur seals at South Georgia have increased by 17% per annum; Bird Island was the main focus of this increase. The seals have destroyed large areas of tussock grassland (the albatrosses breeding habitat) and frequent interactions with albatrosses are observed. We investigated the potential adverse effect of fur seals on wandering albatrosses by comparing breeding numbers, success and other aspects of population dynamics in areas with different abundances (including absence) of seals. Albatross breeding numbers have decreased more rapidly in areas with most seals. In contrast, albatross breeding success has increased over the last decade and is highest in areas with most seals. There is no evidence that seals have any adverse effect on established breeding albatrosses but juvenile recruits to the breeding population tend to avoid areas with abundant seals. Because young birds have lower reproductive success than older birds, this movement may create the increase in average breeding success of the (older) birds remaining in areas with numerous seals.

1 Introduction

That breeding populations of wandering albatrosses *Diomedea exulans* in the Southern Ocean are declining has been recognised fairly recently (Croxall 1979; Jouventin et al. 1984; Tomkins 1985). Croxall et al. (1984) and Jouventin et al. (1984) suggested that incidental mortality due to fishing activity away from the breeding grounds might be implicated in the South Georgia and Crozet Island population declines. Recent analysis of the Crozet data (Weimerskirch and Jouventin 1987) has reinforced this suggestion.

At South Georgia, however, the population of Antarctic fur seals *Arctocephalus gazella* increased from a few thousand in the 1960's to about 1 million nowadays at a rate of ca. 17% per annum (Payne 1977; Doidge and Croxall 1985). The wandering albatross breeding population at Bird Island, which holds 58% of the total South Georgia breeding population (Croxall 1979), has declined by about 22% since 1961, at an average rate of 1% per annum (Croxall et al. 1988a). Bird Island was the main nucleus for the expansion of the Antarctic fur seal population and some 180000–200000 seals occur there in summer. By 1982 seals had access to 60% of the 190 ha of tussock grassland (the habitat in which the albatrosses breed) and had caused significant damage to the tussock grass in about half of this (Croxall et al. 1984). Large areas of the coastal grasslands have now been destroyed by the seals and they are increasingly active throughout all but about 30% (from which they are absent) of Bird Island's grassland habitat (Bonner 1985).

In this chapter we investigate whether fur seals are having an adverse effect on breeding wandering albatrosses and so contributing to their population decline.

2 Methods

Full details of the field methods recording wandering albatross breeding population size and success, for adult survival and juvenile recruitment can be found in Croxall et al. (1988b). Briefly, all occupied nests are counted on 31 January (immediately after the end of egg laying) in each of 24 subdivisions covering the whole of Bird Island (see Hunter et al. 1982, Fig. 1). The original number of breeding pairs is estimated using the proportion of nests failing prior to this date in the detailed study population (10% of the island total), on Wanderer Ridge which is checked daily throughout the egg-laying period. Using the corrected counts as a basis, hatching success is taken to be the proportion of chicks surviving on 1 April (a census is taken on this day) and overall breeding success as the

[1] British Antarctic Survey, High Cross, Madingley Road, Cambridge, CB3 0ET, UK.

Antarctic Ecosystems. Ecological Change and Conservation
(ed. by K. R. Kerry and G. Hempel)
© Springer-Verlag Berlin Heidelberg 1990

proportion actually fledging. Breeding success data are available since 1976. In 1958, 1962, 1963 and every year since 1972 (except 1975 and 1979) most chicks on the island were ringed in October prior to fledging in December. Through standard recapture surveys all breeding birds of known age have been retrapped annually (except 1975) since 1972, providing estimates of adult survival and juvenile recruitment.

The present (1987–88) density of fur seals in each of the 24 census subdivisions of the island was assessed on a four-point scale by an independent observer with detailed familiarity with their distribution and abundance. The categories used were: 0 – seals absent; 1 – seals local (individuals, but rarely groups usually present); 2 – seals common (individuals always present, groups not infrequent); 3 – seals abundant. Changes in wandering albatross breeding population size and success were then compared (using linear regression of log-transformed counts) between areas grouped according to the fur seal abundance categories.

3 Results

Since 1977 the breeding population of wandering albatrosses at Bird Island has declined by 0.98% per annum ($r^2 = 0.46$; $P < 0.01$), or 14% overall. Populations have declined in all areas of the island but when subdivided in terms of seal abundance (Fig. 1a, Table 1) only for the areas with most seals is the decrease statistically significant. If we compare the rates of decrease in the different area categories (Table 1), all show a lower rate than the areas with abundant seals; the difference, however, is only significant for the comparison with category 1 areas.

Since 1977 the breeding success of the island's population has increased by 1% per annum ($r^2 = 0.46$, $P = 0.01$). When analysed in relation to seal abundance categories there are significant increases both for areas without seals and for those with most seals (Table 2). The trend for breeding success to increase with increasing number of seals is not significant but birds in the areas with most seals (category 3) have performed better then birds in areas with fewer or no seals in six of the nine years for which full data exist (Mann-Whitney U-test, $P < 0.05$). The average difference in breeding success between birds in category 3 areas and those elsewhere (5–6%; Table 2) is significant in two of three comparisons.

Circumstances which might account for the changes in breeding numbers include changes in any or all of adult survival, adult breeding frequency, recruitment to the breeding population and movements of birds

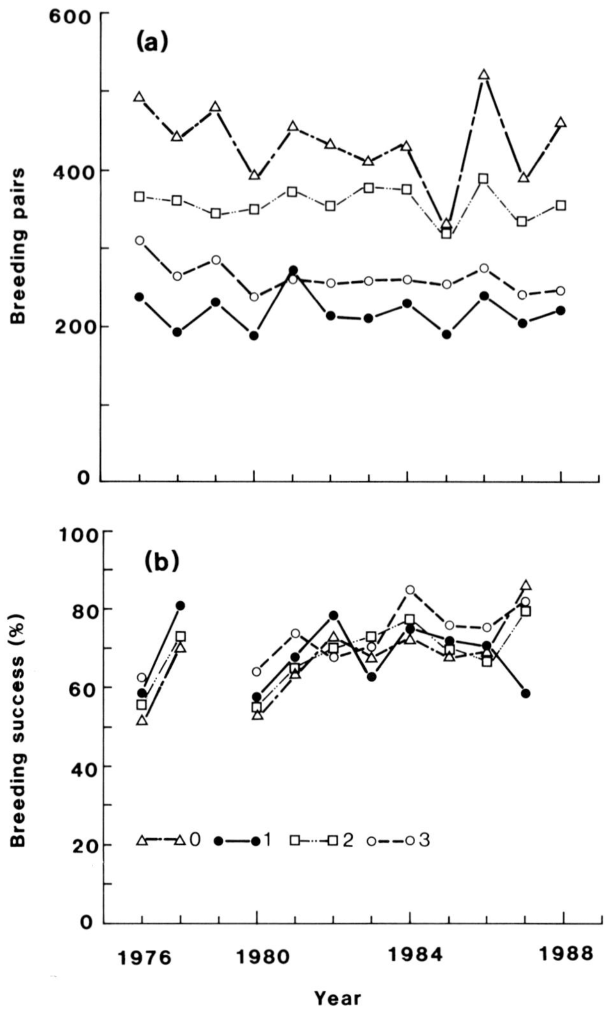

Fig. 1. Wandering Albatross **a** breeding population size (pairs) and **b** breeding success (chicks fledged from eggs laid), in relation to Antarctic fur seal abundance at Bird Island, South Georgia, 1977–1988. *0* Seals absent; *1* seals local; *2* seals common; *3* seals abundant

Table 1. Absolute and relative breeding population changes of wandering albatrosses at Bird Island, South Georgia 1977–1988 in relation to Antarctic fur seal abundance. Relative change compares changes in areas of different seal abundance relative to areas of category 3[a]

Seal abundance category	Absolute change (%)			Relative change (%)	
	Overall	Annual		Mean difference	SE
		Mean	SE		
0: Absent	– 7.3	– 0.83	1.11	– 0.37	0.88
1: Local	– 8.3	– 0.02	0.74	– 1.22*	0.42
2: Common	– 4.6	– 0.24	0.46	– 0.96	0.56
3: Abundant	– 20.5*	– 1.20*	0.53	–	–

[a] *Significant at $P < 0.05$ level.

Table 2. Breeding success (percent chicks fledged from eggs laid) of wandering albatrosses at Bird Island, South Georgia 1977–88 in relation to Antarctic fur seal abundance. Relative change is the difference in breeding success relative to areas of category 3[a]

Seal abundance category	Breeding success (%)					
	Overall		Realtive change		Annual change	
	Mean	SE	Mean	SE	Mean	SE
0: Absent	67.5	3.1	– 6.1*	2.2	2.1*	0.7
1: Local	68.6	2.7	– 6.2	2.9	– 0.05	0.9
2: Common	68.6	2.6	– 5.1*	1.5	1.5	0.7
3: Abundant	73.3	2.5	–	–	1.8**	0.5

[a] Differences significant at $P < 0.05$ = *; at $P < 0.01$ = **.

Table 3. Recruitment of wandering albatrosses in relation to Antarctic fur seal abundance

Seal abundance category	Albatross breeding population[a]	Cohort (%)				
		1975	1976	1977[b]	1978[b]	Total
0	731	14.3	11.6	7.1	5.3	38.4
1	354	9.8	7.1	6.8	3.4	27.1
2	614	13.6	10.9	5.7	4.2	34.5
3	473	11.2	7.6	2.7	3.1	24.7
1–3	1441	11.9	8.9	5.0	3.7	29.4

[a] Average number of birds over the 6-year period while these cohorts were being recruited.
[b] Recruitment to breeding population not yet complete.

between areas. There are no significant differences in either adult survival or breeding frequency in relation to seal abundance. Adult wandering albatrosses are extremely site faithful and we have no evidence of birds that have once bred moving their breeding site to areas with fewer fur seals. However, while the majority of chicks return to breed in their natal areas, a small proportion do nest away from the area from which they fledged.

The change in relative numbers between areas might, therefore, be due to discrimination by recruits in favour of areas with less disturbance by fur seals. Proportionately fewer birds from the 1975, 1976, 1977 and 1978 cohorts have been recruited in areas of highest seal abundance compared to areas where seals are absent (Table 3; Friedman test for differences across cohorts $\chi^2 = 9.3$, $P = 0.01$,). We can examine this in more detail using data from the 1975 and 1976 cohorts, which are the most complete (Table 4). These date indicate that, while the survival rate to first breeding attempt by chicks fledging in different areas is similar, there is a net increase in the number of birds returning to breed in areas where seals are absent and

Table 4. Recruitment: survival ratios (sample sizes in parentheses are of surviving birds) of wandering albatross chicks fledged from areas of different seal abundance. The ratio compares, for chicks fledging from each category of area, the number recruited to breed into each category of area with the total number surviving to breed. Ratios > 1 indicate disproportionate immigration; < 1 disproportionate emigration

Seal abundance category	Recruitment: survival ratio		
	1975 cohort	1976 cohort	Total
0	1.07 (98)	1.16 (73)	1.11 (171)
1	1.06 (33)	0.93 (27)	1.00 (60)
2	1.01 (83)	0.95 (60)	0.99 (143)
3	0.84 (63)	0.63 (53)	0.77 (116)
1–3	0.96 (179)	0.84 (140)	0.91 (319)

a net decrease in the number of birds returning to breed in areas of high seal abundance.

4 Discussion

The breeding population of wandering albatrosses at Bird Island is decreasing and this seems to be happening fastest in areas with most seals. There is no evidence that seals are adversely affecting breeding succcess or survival. However, a proportion of birds being recruited into the breeding population are avoiding areas where seals are abundant and are breeding in areas where seals are absent. This may be due to the disturbance of juvenile albatrosses by seals. Thus seals tend to move away from the beaches (where pups are born in December) as the summer season progresses. Maximum seal abundance on wandering albatross nesting areas usually occurs in February and March, which coincides with young birds visiting the island to establish potential nest sites, display with each other and from pair bonds (Pickering 1989). Seals have frequently been observed disturbing young birds performing courtship displays in areas where seals are abundant. It may be this disturbance, before birds have become attached to a breeding site, that is causing them to shift to areas with fewer seals.

This differential recruitment between areas may also account for the increased breeding success in areas where seals are abundant. The breeding success of young, inexperienced birds (< 12-year-old) is lower than older, more experienced birds (unpubl data). Thus areas with reduced recruitment will tend to have higher than average breeding success because there will be fewer of the young birds which have a higher failure rate.

The differential recruitment may be unrelated to seal abundance. Because the population is declining, nest sites left vacant on the higher ridges away from the lower flat areas may be preferentially selected by young birds being recruited to the breeding population. These nest sites may be selected not because of the absence of seals but for other reasons such as the ease of landing and takeoff. However, we have no evidence of competition for nest sites or that the number of birds breeding in any area is limited by the number of nest sites.

In conclusion we have no evidence that the massive increase in Antarctic fur seal numbers since 1960 has influenced the decline in the breeding population of wandering albatrosses on Bird Island. Nor do seals appear to have affected the breeding success of established breeders, even in areas of highest seal density. It is possible, however, that seals are causing a gradual redistribution of the albatross breeding population because young birds are more likely to start breeding in areas with fewer seals.

Acknowledgements. We thank C. D. Duck for help in the assessment of seal abundance, J. Thomson for typing the manuscript and. P. A. Prince and two anonymous reviewers for useful comments.

References

Bonner WN (1985) Impact of fur seals on the terrestrial environment at South Georgia. In: Siegfried WR, Condy PR, Laws RM (eds) Antarctic nutrient cycles and food webs. Springer, Berlin Heidelberg New York, pp 641–646

Croxall JP (1979) Distribution an population changes in the wandering albatross *Diomedea exulans* at South Georgia. Ardea 67:15–21

Croxall JP, Prince PA, Hunter I, McInnes SJ, Copestake PG (1984) The seabirds of the Antarctic Peninsula, islands of the Scotia Sea and Antarctic Continent between 80° W and 20° W: their status and conservation. In: Croxall JP, Evans PGH, Schreiber RW (eds) Status and conservation of the worlds seabirds. ICBP, Cambridge, pp 637–666

Croxall JP, McCann TS, Prince PA, Rothery P (1988a) Reproductive performance of seabirds and seals at South Georgia and Signy Island, South Orkney Islands, 1976–1987: implications for Southern Ocean monitoring studies. In: Sahrhage D (ed) Antarctic Ocean and resources variability Springer, Berlin Heidelberg New York, pp 261–285

Croxall JP, Prince PA, Boyd IL (1988b) Bird Island monitoring programme. British Antarctic Survey, Cambridge

Doidge DW, Croxall JP (1985) Diet and energy budget of the Antarctic fur seal *Arctocephalus gazella*. In: Siegfried WR, Condy PR, Laws RM (eds) Antarctic nutrient cycles and food webs. Springer, Berlin Heidelberg New York, pp 543–550

Hunter I, Croxall JP, Prince PA (1982) The distribution and abundance of burrowing seabirds (Procellariiformes) at Bird Island, South Georgia. I. Introduction and methods. Bull Br Antarct Surv 56:49–67

Jouventin P, Stahl JC, Weimerskirch H, Mougin JL (1984) The seabirds of the French sub-Antarctic islands and Adelie Land, their status and conservation. In: Croxall JP, Evans PGH, Schreiber RW (eds) Status and conservation of the world's seabirds. ICBP, Cambridge, pp 609–625

Payne MR (1977) Growth of a fur seal population. Phil Trans R Soc B 279:67–79

Pickering SPC (1989) Attendance patterns and behaviour in relation to experience and pair-bond formation in the wandering albatross *Diomedea exulans* at South Georgia. Ibis 131:183–195

Tomkins RJ (1985) Reproduction and mortality of wandering albatrosses on Macquarie Island. Emu 85:40–42

Weimerskirch H, Jouventin P (1987) Population dynamics of the wandering albatross, *Diomedea exulans*, of the Crozet Island: causes and consequences of the population decline. Oikos 49:315–322

Fluctuations in Breeding Populations of Elephant Seals *Mirounga leonina* at Stranger Point, King George Island[1] 1980–1988

D. F. VERGANI and Z. B. STANGANELLI[2]

Summary. The chapter presents data on annual fluctuations of the population of elephant seals at Stranger Point (King George Is.) between 1980–1988 (excluding 1981). A maximum of 825 breeding cows was recorded in 1980. A decline in the population was observed in 1980–1982 which was followed by annual increments until 1986. A further decrease was observed in 1987. The female component was mainly affected in 1980–1982 ($r = -0.575$) while the male component was affected in 1987 ($r = -0.463$). A possible influence of El Niño Southern Oscillation (ENSO) events upon the numerical changes in the studied population is discussed.

1 Introduction

The breeding distribution of elephant seals, *Mirounga leonina*, is circumpolar comprising three stocks: South Georgia, Kerguelen and Macquarie Is. (Laws 1960).

The current census information from all major elephant seal populations indicate that Kerguelen and Macquarie Is. stocks are declining while South Georgia populations appear to be stable (Hindell and Burton 1987).

Present research is directed at establishing whether the population decline is due to fishery pressures, natural environmental changes or factors intrinsic to the seal population. Several studies dealing with population changes of elephant seals have been undertaken (Van Aarde 1980; Bester 1982; Skinner and Van Aarde 1983; McCann 1985; Hindell and Burton 1987; Vergani et al. 1987).

Female elephant seals tend to return to the same breeding beaches every year (Nichols 1970). Changes in numbers from even a small number of beaches can therefore reflect changes involving entire populations (Hindell and Burton 1987). Regular counts of elephant seal numbers at selected beaches are thus a practical method of monitoring the numerical status of a population.

This chapter records the numerical changes in abundance of both males and females of *M. leonina* at Stranger Point, King George Is. 62° 14′ S–58° 30′ W. Environmental factors, either biotic or abiotic, which could have affected observed changes in abundance, are discussed. The population belongs to the South Georgia stock (Laws 1960) and occurs at the southern limit of the dispersion area of *M. leonina*. Data obtained include records from 1980 to 1988 (excluding 1981).

2 Methods

A study area of 5 km of beach was established at Stranger Point. It was bounded by Potter Cove to the southwest and by a glacier to the northeast, forming a discrete study area (Fig. 1).

The first population census for this study was made from October 1980 to February 1981. The 1981–1982 season was missed, but from 1982 counts throughout ensuing years were conducted. In this chapter we included only the data from breeding seasons.

During the periods of study prior to 1985, counts were carried out on a weekly basis. The interval between counts, however, was too great to establish clearly the dates of peak abundance. To determine the date of peak female haul-out with more precision, censuses in following seasons were conducted on alternate days. Finally, in 1987, the frequency of counts was applied only in the proximity of the predicted peak and weekly counts were made in the rest of the season. At each count the number of females, males and pups (dead and weaned) was recorded. Males were recorded as either associated or not with the harem. If associated, the social rank (beachmaster or challenger) was indicated.

The maximum number of females hauled out during each breeding season at Stranger Point was calculated from the sum of: number of females in the peak of hault-out + weaned pups and the accumulative number of dead pups on the date of the females peak

[1] Known in Argentina as Isla 25 da Mayo.

[2] Instituto Antártico Argentino, Cerrito 1248, 1010 Capital Federal, Argentina.

Antarctic Ecosystems. Ecological Change and Conservation
(ed. by K. R. Kerry and G. Hempel)
© Springer-Verlag Berlin Heidelberg 1990

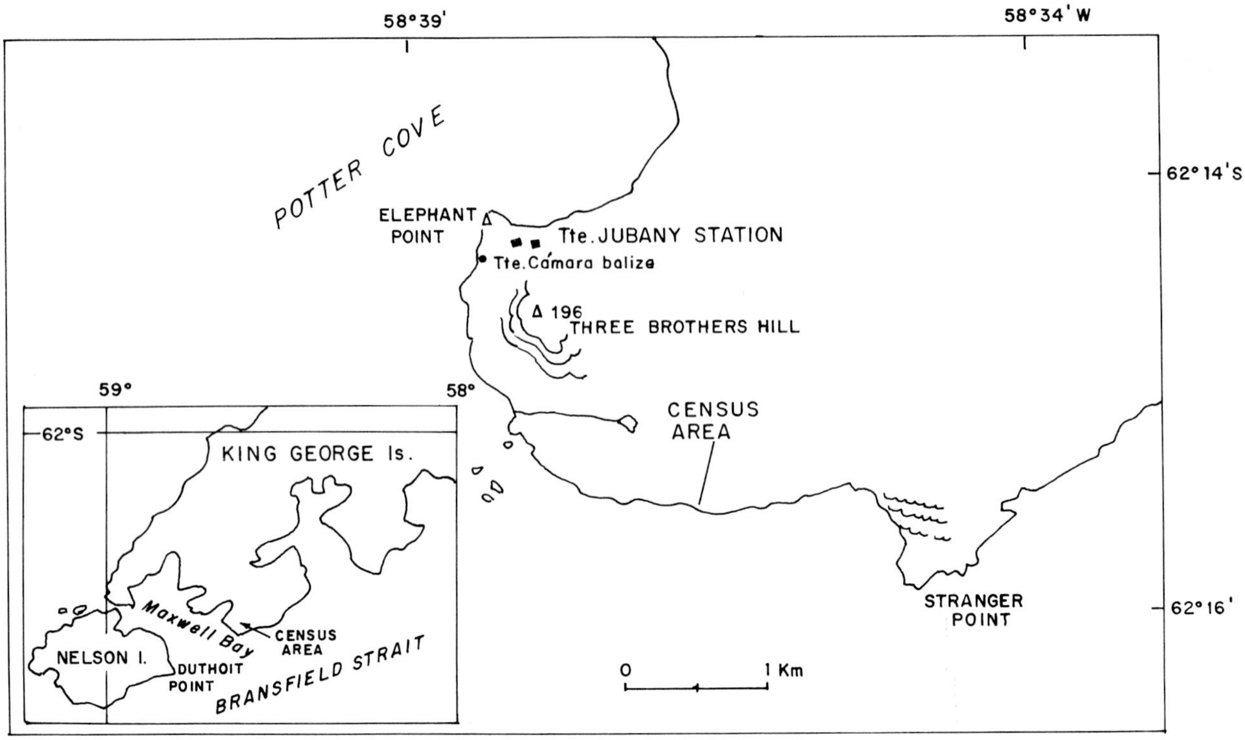

Fig. 1. Census area on King George Island

(McCann 1985). Estimates of intrinsic population growth rates were based on the exponential function $N_t = N_0 \exp(\pi t)$ (Caughley 1977; Skinner and Van Aarde 1983). The maximum number of males on land was established when the sum of challengers plus beachmasters was greatest and was a sum of all adult males ashore on that date (including those not associated with harems). The data were processed to determine dates of peak abundance in each season and to ascertain changes to population structure.

An analysis of adult sex ratio and the relationship of maximum number of bulls and females on land during the same year and the year t + 1 was made, as described by Skinner and Van Aarde (1983).

3 Results

The haul-out pattern of the breeding females in different years showed that cows start coming ashore in September (mean date 16, range 8 to 26 Sep., n = 8) and the peak number of cows ashore was in late October (mean date 26, range 21 to 31 Oct., n = 8). Males come ashore earlier than females but do not keep determined positions on the beach until females come into oestrus. The peak number of males was also in October with the same mean and range as for females.

Information on date of peak haul-out, estimation of total number of females, maximum number of males, sex ratio and pup mortality for each season is summarized in Table 1.

Maximum adult bull numbers decreased overall from 1980 to 1988 at a mean annual intrinsic rate of – 0.5%, due mostly to a crash in numbers in 1986–1987. During the same period, the female component increased at a rate of 0.4%. Declines in female (– 57.5%) and male components (– 11.2%) of the population were observed between the 1980–1982 period, followed by a mean annual increase of 5.2% in males and 11.3% in females between 1982–1986. A further decrease of 46.3% in males and 16.8% in females was observed between 1986–1987. In females, the decline continued with a drop of 6.6% from 1987 levels being recorded in 1988. Males showed an increase in the same period (34.5%). This information is summarized in Table 2.

The adult sex ratio changed significantly from year to year in the 1980–1988 period ($\chi^2 = 20.93$, contingency coefficient 0.061), but in the period 1982–1986 no significant change was observed ($\chi^2 = 3.81$, contingency coefficient 0.034). No correlation was observed between the maximum number of males and total number of cows hauled out during the same breeding season ($r_8 = 0.319$). However, excluding both periods of decrease there is a strong correlation

Table 1. Counts of elephant seals at Stranger Point, King George Island during the breeding seasons 1980, 1982–1988

Year	Male Maximum	Date Maximum	Beach-masters	Challengers	Adult sex ratio Male:Female
1980	94	28/10	22	22	1 :8.78
1982	84	03/11	10	20	1 :5.52
1983	93	31/10	11	18	1 :5.26
1984	93	27/10	17	34	1 :6.29
1985	95	30/10	18	29	1 :6.35
1986	108	28/10	14	41	1 :6.82
1987	68	30/10	18	18	1 :9.16
1988	96	25/10	16	27	1 :6.07

Year	Female Total	Date Maximum	Total dead pups	Pup mortality (%)
1980	825	28/10	22	2.7
1982	464	25/10	12	2.6
1983	489	31/10	07	1.4
1984	585	31/10	21	3.6
1985	603	20/10	27	4.3
1986	737	21/10	27	3.7
1987	623	30/10	19	3.0
1988	583	23/10	10	1.7

($r_5 = 0.933$, $P < 0.01$) between them. There is no correlation between the maximum number of bulls in year t and the total number of cows in the year t + 1 ($r_6 = 0.394$); this calculation was made between years in the 1982 to 1988 period.

Pup mortallity ranged between 1.4 and 4.3% (mean = 2.9%, SD = 1.0) of total pup births for each season.

4 Discussion

Published census information on South Shetland elephant seal population is scarce and dated. Laws (1953) estimated the 1951 breeding population of the entire achipelago at 250 animals. In 1976, a count at Stranger Point was made by Müller Schwarze et al. (1978). Though 212 females and 12 males were counted, the census was made on 12 October, 2 weeks before the mean date of maximum numbers ashore. This figure would therefore have underestimated the size of the breeding population. Müller Schwarze also made mention of a single harem (females, 8; bulls, 1) at Thomas Point, Admiralty Bay) King George Island.

A reconnaissance of the whole archipelago during the breeding season was logistically not possible in the present study. Occasional searches resulted in locating an additional small colony (females 106, bulls 21) at Duthoit Point (Nelson Is.). With present knowledge, the colony at Stranger Point is the largest in the South Shetland Islands.

Three possible hypotheses for the population decrease in 1980–1982 and 1986–1987 are postulated here:

1. Overfishing in the area;
2. Natural regulating factors;
3. Environmental changes.

1. Overfishing in the Area. Little is known about the diet of elephant seals. According to Laws (1977) and McCann (1985), elephant seals eat 75% squid and 25% fish. In contrast, N. Gales and H. Burton (1987 pers. commun.) suggest that elephant seals in the southern extremity of their range tend to feed mainly on fish. Gales and Burton use data from the study made by Lubimova (1985) on the diet and ecology of pelagic squid in the Southern Ocean to support their hypothesis. Lubimova concluded that the abundance of true mesopelagic and bathipelagic squid is very low in waters south of 60° S. However, the biomass of fish does increase greatly on the continental shelf that surrounds Antarctica. Changes in this prey sprecies could affect elephant seal populations.

Large-scale harvesting of finfish in the Antarctic Peninsula Region started in 1978–1979. It was reported for two successive seasons and for 1982–1983. After the 1978–1979 season, yields dropped substantially (CCAMLR 1987). The reduction in yield may in itself have been caused by fishing or perhaps some other factor which removed fish from the fishery area. Overfishing in the area might explain the decrease in elephant seals from 1980 to 1982 but not the decrease for 1986–1987 because there were no fishery activities over these years in the Antarctic Peninsula area (CCAMLR 1987).

2. Natural Regulating Factors. Skinner and Van Aarde (1983) postulated that the number of males could be

Table 2. Rates of Increase

Year	1980–82	1982–83	1983–84	1984–85	1985–86	1986–87	1987–88	Mean
n =	2	2	2	2	2	2	2	8
r Cows	– 0.575	0.052	0.179	0.030	0.201	– 0.168	– 0.066	0.004
r Bulls	– 0.112	0.102	0.000	0.021	0.128	– 0.463	0.345	– 0.005

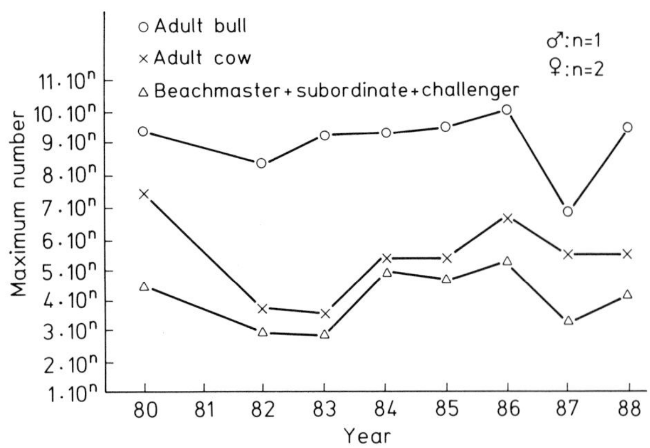

Fig. 2 Population trend in the reproductive component

too few to inseminate all females thereby resulting in a drop in pup production and hence population size. In support of their hypothesis, Skinner and Van Aarde reported that at Marion Is. between 1974 and 1982 the number of bulls in year t and the number of females in year t + 1 were related. When investigated in the present study, the number of bulls in year t was found to be more closely related to the number of cows in year t than the number of cows in year t + 1; at least in normal years. A number of males remain ashore but unassociated with breeding groups during the breeding seasons. When the number of females ashore increased, the proportion of males ashore which were not associated with breeding groups decreased (Fig. 2). Thus the number of males holding or challenging for harems corresponded with the number of the females ashore. Our observations cannot be explained by the hypothesis of Skinner and Van Aarde.

Condy (1977) hypothesized that increased killer whale, *Orcinus orca* predation could be responsible for the decrease in elephant seal populations. At Peninsula Valdes (Patagonia) killer whales tend to attack elephant seal pups near shore at the end of the breeding season. They have rarely been observed successfully attacking adults and only one adult bearing the scars of an obvious killer whale attack has been sighted at Stranger Point (Vergani 1985). In contrast, predation pressure by killer whales is much greater at Peninsula Valdes (pers. obser.) a site where the elephant seal population is increasing (Vergani et al. 1987). The hypothesis that killer whale predation is responsible for the decrease in elephant seal numbers at Stranger Point cannot be supported given the present information.

3. Environmental Changes. During 1982 and 1983 disturbance to the Southern Ocean environment was caused by an El Niño Southern Oscillation (ENSO). At South Georgia in 1984 changes in foraging behaviour of seabirds and seals could have reflected oceanographic perturbations initiated by ENSO events (Croxall et al. 1987). At Stranger Point the coast was free of ice one month earlier than other seasons (per. obser.) and there were harems in Potter Cove during 1982 and 1983 but neither before nor after these years. Many elephant seals were seen in the Palmer Station area during 1983. In addition, elephant seals pupped at Palmer Station for the first time in 1983. Four elephant seals were born between 15 October and 9 November 1983 (Heimark and Heimark 1986). It is possible that elephant seals moved southwards in response to changes in distribution of their prey. Le Boeuf et al. (1986) also suggested that during 1983 female northern elephant seals, *M. angustirostris* travelled further to obtain prey because of the effect of an intensive and widespread El Niño event in progress throughout much of the Eastern Pacific that caused a reduction in plankton, fish and squid.

Alterations to the marine environment caused by ENSO events could modify the feeding area and the distribution of elephant seal populations. With an enforced change in distributions, elephant seals may not have the time or navigational ability to return to traditional breeding sites. As a result, numbers at traditional breeding sites, such as Stranger Point, would appear to decrease as cows come ashore to pup at their nearest accessible beach. This hypothesis explains the anomalies of low breeding cows at Stranger Point in 1982–1983 and cows calving in Potter cove and in the vicinity of Palmer Station in 1983. Following this hypothesis several factors could have affected the continuation of low cow numbers at the Stranger Point site through the period 1983–1986. One theory suggested here is that cows could have been attempting to return to "ENSO enforced" rearing beaches. The breeding success of these cows would be hampered by ice conditions, cold and thus poor rates of pup raising and absence of bulls for insemination. The observed gradual annual increase in cow numbers at Stranger Point over the period 1983–1986 may be explained by either strayed cows returning to the traditional breeding colony, or perhaps it represented a genuine population increase. Information from 1986 indicates that another ENSO has occurred (Trillmich 1987; Boersma 1987). This could again have affected the Stranger Point elephant seal population, but it differed from the 1982 decrease in affecting principally the male component of the population.

Results suggest that male and female components of the population are affected independently. An explanation of this could be that their feeding grounds and times of foraging are different. The sexual dimorphism in the size of the adult elephant seals stresses the potential for distinct feeding abilities. Timing and geographic range of ENSO events may not be the same and could affect elephant seal population components in different ways.

It is our opinion that changes in elephant seal numbers at the Stranger Point colony between 1980–1988 are better explained by ENSO-related effects than by either other natural regulating factors or overfishing of prey species.

According to Croxall et al. (1987), the objective of ecosystem monitoring in Antarctica is to detect as quickly as possible changes that result from changes in food availability and, if possible to distinguish between responses due to natural events and those induced by commercial exploitation. Variation in elephant seal numbers would appear to be a useful tool in detecting changes in the Southern Ocean ecosystem at a regional level.

Acknowledgements. We wish to thank all those who took part in the counts; in particular P. Lumberg, D. Gaona, M. Alcalde, H. Spairani, A. Corvalan, R. Cassaux and A. Mazzota. C. Aguire helped in the translation and Dr. J. Schnack critically reviewed the manuscript.

We thank in particular K. Green, R. Kirkwood and the Australian Antarctic Division for discussion, encouragement and assistance in the presentation at the Symposium and the preparation of the final manuscript.

References

Bester MN (1982) An analysis of the southern elephant seal *Mirounga leonina* breeding population at Kerguelen. S Afr J Antarct Res 12:11–16

Boersma PD (1987) El Niño behind penguin deaths? Nature 327:96

CCAMLR (1987) Report of the ad hoc working group on fish stock assessment. In: Report of the Sixth Meeting of the Scientific Committee 6:185–234

Caughley G (1977) Analysis of vertebrate populations. John Willey & Sons, New York, p 232

Condy PR (1977) Annual cycle of the southern elephant seal *Mirounga leonina* at Marion Island. S Afr J Wildl Res 8:159–168

Croxall JP, McCann TS, Prince PA, Rothery P (1988) Variation in reproductive performance of seabirds and seals at South Georgia, 1976–1986: Implication for Southern Ocean monitoring studies. In: Sahage D (ed) Antarctic Ocean and Resources Variability. Springer-Verlag Berlin Heidelberg, pp 261–285

Hindell MA, Burton HR (1987) Past and present status of the southern elephant seal (*Mirounga leonina*) at Macquarie Island. J Zool Lond 213:365–380

Heimark RJ, Heimark GM (1986) Southern elephant seal pupping at Palmer Station, Antarctica. J Mammal 67:189–190

Laws RM (1953) The Seals of the Falkland Islands and Dependencies. Oryx 2:87–97

Laws RM (1960) The southern elephant seal (*Mirounga leonina* Linn) at South Georgia. Norsk Hvalfangst-Tidende 10:466–476, 11:520–542

Laws RM (1977) Seals and whales of the Southern Ocean. Phil Trans R Soc Lond B 279:81–96

Le Boeuf BJ, Costa DP, Huntley AC, Kooyman GL, Davis RW (1986) Pattern and depth of dives in northern elephant seals. *Mirounga angustirostris*. J Zool (Lond) 208:1–7

Lubimova TG (1985) Result of Soviet investigation of the distribution and ecology of pelagic squids (*Oegopsida*) in the Southern Ocean. SC-CCAMLR-IV/BG 18:79–112

McCann TS (1985) Size, status and demography of southern elephant seals (*Mirounga leonina*) populations. In: Ling JK, Bryden MM (eds) Studies of sea mammals in south latitudes. South Australia. Geographic Services Pty Ltd, pp 1–19

Müller Schwarze D, Waltz EC, Trivelpiece W, Volkman NJ (1978) Breeding status of southern elephant seals at King George Island. Antarct J US (13) 4:157–158

Nicholls DG (1970) Dispersal and dispersion in relation to the birth site of the southern elephant seal, *Mirounga leonina* (L.), of Macquarie Island. Mammalia 34:598–616

Skinner JD, Van Aarde RJ (1983) Observations on the trend of the breeding population of southern elephant seals, *Mirounga leonina*, at Marion Island. J Appl Ecol 20:707–712

Trillmich P (1987) Annual and seasonal variability and the maternal strategy of Galapagos fur seal: a long-term study. In: Abstracts of the Seventh Biennial Conference on the Biology of Marine Mammals, Miami Florida, p 71

Van Aarde RJ (1980) Fluctuations in the population of the southern elephant seals *Mirounga leonina* at Kerguelen Island. S Afr J Zool 15:99–106

Vergani DF (1985) Estudio comparativo de las poblaciones de Antártida y Patagonia del Elefante Marino del Sur *Mirounga leonina* y su Metodologia. DNA Buenos Aires Pub 15, p 94

Vergani DF, Lewis MN, Stanganelli ZB (1987) Observation on haul out patterns and trends of the breeding populations of southern elephant seal at Peninsula Valdes (Patagonia) and Stranger Point (25 de Mayo – King George Is.) SC-CAMLR-VI/BG/36:1–9

A Simulation of the Age Structure of Crabeater Seals in a Fluctuating Environment

J. W. Testa[1]

Summary. A population model of crabeater seals was constructed to assess the sensitivity of the age strucktureof crabeater seals to environmental fluctuations. The model incorporates density dependence in age-specific reproductive and survival schedules, with reproduction and juvenile survival being most sensitive to changing population size or carrying capacity. Environmental fluctuations were simulated by varying the populations's carrying capacity in four ways: (1) random fluctuations; (2) random walk; (3) sinusoidal oscillations with minor stochastic variation; and (4) recurrent episodes of good years with variable periodicity. High carrying capacity produced strong cohorts from pups of that year and yearlings. However, the magnitude of these strong cohorts was only moderately correlated with the magnitude of the carrying capacity in the year of birth. Reasons for this include the reduced juvenile survival following population growth in good years, and the effects of carrying capacity in previous years that determined population size. These simulations indicate that fluctuations in the environment for crabeater seals can be transmitted to the age structure, although some "noise" is inherent in the age structure signals.

1 Introduction

Apparently cyclic phenomena have been described in different population attributes of three species of Antarctic seals. Weddell seals (*Leptonychotes weddellii*) have undergone periodic declines in adult reproduction since 1971 (Testa 1987) and leopard seals (*Leptonyx hydrurga*) have oscillated in abundance at Maccquarie Island since the early 1950s (Rounsevell and Eberhardt 1980). The period of these fluctuations is close to 5 years. The longest continuous sequence of data is from the collections of crabeater seals (*Lobodon carcinophagus*) made near the Antarctic Peninsula (Laws 1984; Bengtson and Laws 1985). The age structure of these collections show strong cohorts at intervals of about 5 years. What are the possible relationships between these age structure anomalies and underlying ecological signals? Variations in size, availability or distribution of krill stocks, or in the extent and distribution of pack ice are environmental variables that could affect crabeater productivity and/or survival. An age-structured population model of crabeater seals was constructed that varied the carrying capacity to simulate a varying environment. By controlling the pattern of variation in this parameter one can examine some of the possible conditions that might produce cyclic patterns like those seen int he age structure of crabeater seals, and look at the quantitative relationship between the hypothetical environmental signal and cohort strenghts of crabeater seals.

2 Methods

An age-structured model was programmed in FORTRAN for a crabeater seal population that included 40 year classes. A variety of density-dependent (DD) functions have been proposed in single species population models (e.g., May et al. 1974) to vary population growth rates or vital parameters as a function of population size or density. Where such functions are used to vary survivorship and fecundity, such as in DD Leslie matrices (e.g., Usher 1972), these rates are usually varied between zero and some age-specific maximum in either a linear or nonlinear fashion. Eberhardt and Siniff (1977) and Fowler (1981, 1987) discussed reasons for believing that the vital rates of marine mammals behave in a nonlinear, DD manner. A nonlinear function was used to vary age-specific survival and fecundity schedules between hypothetical minimum and maximum rates (Fig. 1). The minimum rates would result in the population declining at 13% per year, while the maximum rates would result in population growth of 13% per year. As modeled, density dependence operates most strongly on the ages where maximum and minimum rates differ most. The following DD function was used to vary survival and repro-

[1] Institute of Marine Science, University of Alaska, Fairbanks, AK 99775-1080, USA.

Antarctic Ecosystems. Ecological Change and Conservation
(ed. by K. R. Kerry and G. Hempel)
© Springer-Verlag Berlin Heidelberg 1990

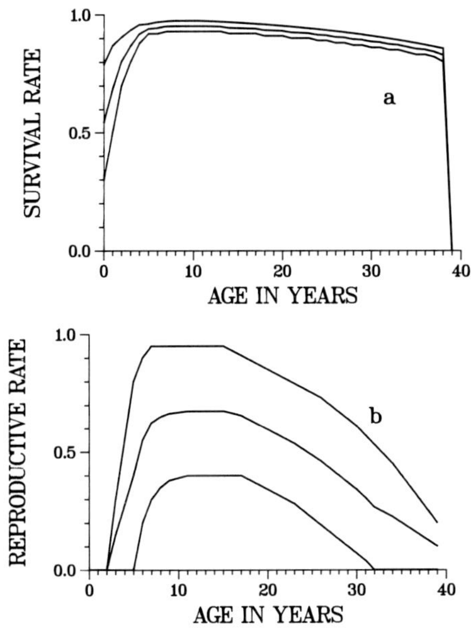

Fig. 1. Maximum, mean, and minimum age-specific schedules of **a** survivorship and **b** reproduction for the population model of crabeater seals

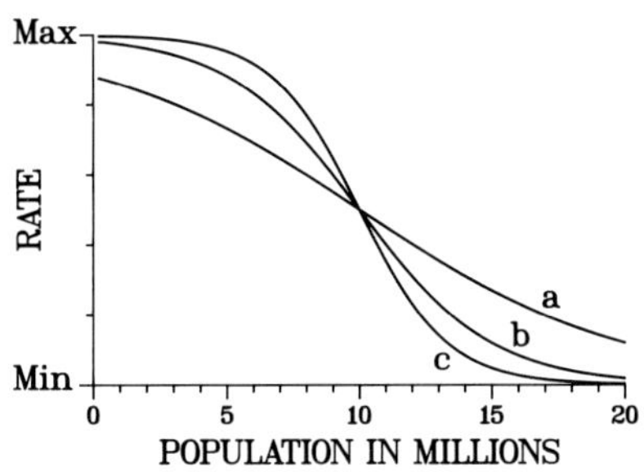

Fig. 2. Functions used to determine both survival and reproductive rates between the maximum and minimum values shown in Fig. 1 in the case of **a** weak, **b** moderate, and **c** strong density dependence. The inflection point (P 4) for all three curves is near the equilibrium, or carrying capacity of the environment. The simulations were conducted while varying the horizontal position of the inflection point (see text)

duction between the maximum and minimum schedules shown in Fig. 1:

$$\text{RATE} = \text{P1} - \left\{ \text{P2} / (1 + \text{EXP}\left[\text{P3} * \frac{(\text{N} - \text{P4})}{\text{P4}} \right] \right\},$$

where N is the population size; P 1 is the maximum possible rate; P 2 is the difference between the maximum and minimum rates; P 3 is a shape constant chosen from values less than −1.0; and P 4 is the population size at the inflection point of the function (see Fig. 2). When the maximum rates of population growth and decline are equal, as in the present case, P 4 corresponds to the carrying capacity, or equilibrium. For these simulations three shape parameters were chosen (P3 = −2.0, −4.0, and −6.0) to vary the slope of the DD function. These parameter values shape the function to simulate weak, moderate and strong density dependence over the range of population sizes to be simulated (see Fig. 2). The chosen schedules and parameters produce several desirable features for a simulated seal population: the maximum and minimum rates of population growth are "reasonable" for the range of population sizes encountered in the simulations, vital rates are most sensitive to population size near carrying capacity, and the most sensitive population parameters are juvenile survivorship and rates of reproduction (including age of first reproduction).

Environmental "signals" were simulated by varying the carrying capacity, or P 4, in several ways. An arbitrary starting population of approximately 10 million seals was chosen and P 4 was varied around that value. An endless variety of signals can be imagined, but of interest were signals that might produce cycles in cohort strengths such as those observed in collections of crabeater seals. Two types of random fluctuations and two cyclic patterns (with stochastic elements) were used to simulate environmental variation:

1. *Random* fluctuations chosen from a uniform distribution between 8 and 12 million (Fig. 3 a);
2. *Random walk* between 8 and 12 million with a step size chosen from a uniform distribution between −2 and 2 million (Fig. 3 b);
3. *Sinusoidal* variation between 8 and 12 million with random walk in time steps between −1 and 1 year (Fig. 3 c);
4. *Episodic* recurrence of peaks of 12 million and lows of 9.5 million with a mean period of 5 years, randomly varying between 3 and 7 years (Fig. 3 d).

Simulations were carried out for 115 iterations, with weak, moderate, and strong density dependence in each case. At each time step the population was subjected to age-specific mortality. Age-specific pupping rates were then applied to the new population to produce the newest cohort. Density dependence in survival and reproduction was based on the population size at the start of the time step. P 4 was envisioned here as an average carrying capacity during the previous year.

A cohort was identified by the year of its birth and the strength of a cohort was defined as the proportion of the total population that belonged to a particular age class, with analyses focused on cohort strengths of 0-, 5-, and 15-year-old seals. Comparisons of cohort strengths were made within particular age classes across 100 years. Three aspects were of interest: (1)

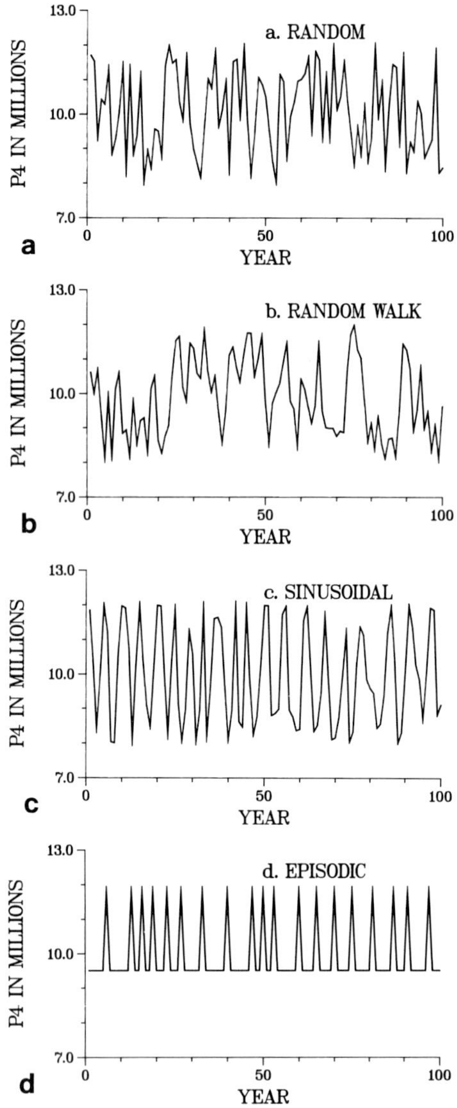

Fig. 3. Pattern of yearly variation used in simulations: **a** *random* fluctuations in carrying capacity (*P4*); **b** *random walk* of P4; **c** *sunusoidal* variation of P4 with random walk in time; **d** *episodic* recurrence of high P4 values at intervals of 3 to 7 years, chosen from a uniform, random distribution

the quantitative relationship between P4 (carrying capacity) proximate to the year of birth and cohort strengths observed in the subsequent age structure; (2) the qualitative faithfulness of spikes or troughs in cohort strength to similar features in the time series of P4; and (3) the importance of density dependence to the quantitative and qualitative signals produced in age structure.

The analyses consisted of (1) regressions of cohort strength on P4 for years including and immediately surrounding the year of birth; (2) comparisons of periods between peak years of P4 and periods between peak years in cohort strengths of 0-, 5-, and 15-year-old seals; and (3) comparisons of peak cohort years with peaks in carrying capacity to determine possible time lags. The cohort strength of age 0 seals was expected to be a good indication of the immediate effect of P4. Cohort strengths 5 and 15 years after a given P4 may or may not be a good quantitative representation of the quality of the environment for seals in their birth year, due to the effects of age-specific survival schedules and to variation in environment and population size during the intervening time. As the quantitative relationship between environmental quality and cohort strengths degrades, so too may the qualitative one. The analysis of these simulations was applied only at the population level, and ignored the problems of sampling design and choosing statistical models to detect patterns in the age structure.

3 Results

3.1 Quantitative Signals

Since the number of models is large, the tabular and graphical data were abbreviated to illustrate the emergent properties. The strength of a cohort is determined by both its initial size and subsequent survival. Figure 4a shows the regression coefficients (analogous to the slope of a simple regression) of cohort strength on P4 in 7 years surrounding and including the year of birth (year 0 plus and minus 1 to 3 years) for all of the models using moderate density dependence. The first important feature is the broad similarity between all models in the relative influence of carrying capacity in years surrounding the birth year. P4 in the years preceding birth of the cohort determines population size at time of birth, and the regression coefficients became strongly negative just prior to the birth year. As expected, strength of the cohort just born was strongly positively correlated with carrying capacity that year. The relative importance of carrying capacity each year diminished with succeeding years and increasing age of seals in that cohort. However, the relative importance of P4 in years surrounding birth of the cohort was maintained into the older age classes, as demonstrated by the similarity of the regression coefficients from the 5- and 15-year-old cohorts.

This regression analysis used several years to predict the strength of a cohort, but it is easier to imagine the effects of a single good year with, for example, very high food availability preceded and followed by years of average food availability. Two strong cohorts would be produced, corresponding with the pups and yearlings present that year. Earlier cohorts would also

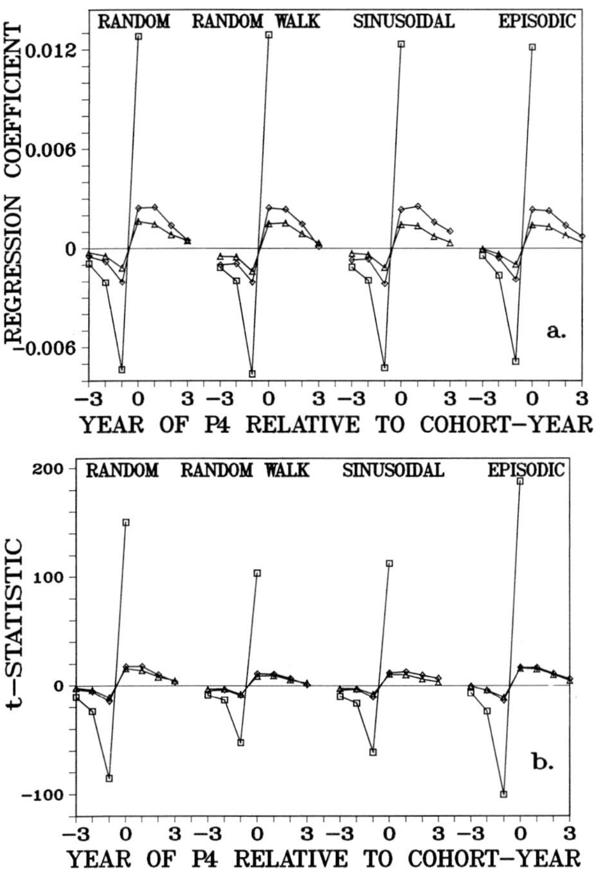

Fig. 4a. Regression coefficients and **b** associated t-statistics from the regression of cohort strengths, 0 (□), 5 (◇), and 15 (△) years after birth on carrying capacity (*P4*) in the 7 years including and surrounding the birth year. P4 was varied as described in section 2 with all four models models and moderate density dependence (P3 = −4.0) was operative

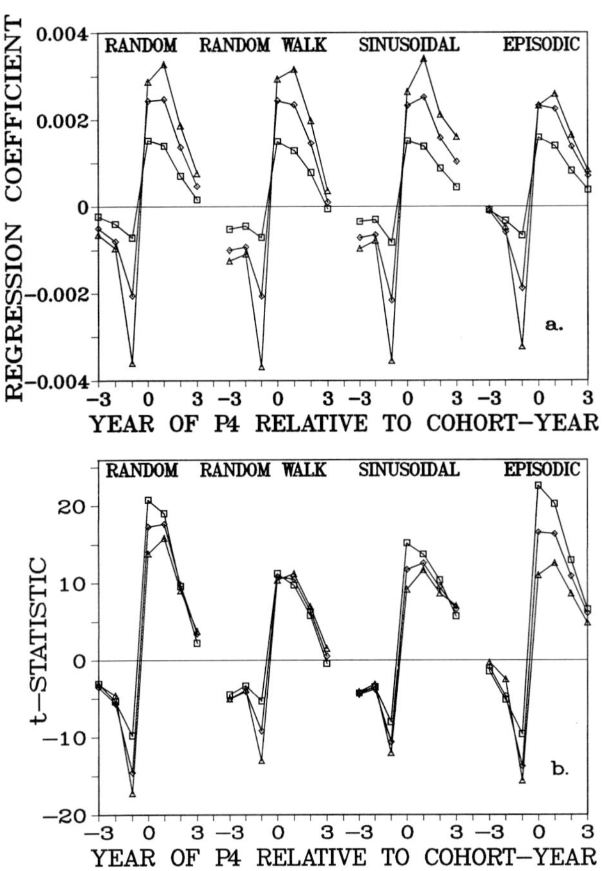

Fig. 5a. Regression coefficients and **b** associated t-statistics from the regression of cohort strength 5 years after birth on carrying capacity (*P4*) in the 7 years including and surrounding the birth year. P4 was varied as described in section 2 with all four models. Simulations were repeated with weak (△), moderate (◇), and strong (□) density dependence

be strengthened to the degree shown by the regression. If conditions returned to normal in the next year, the larger population size would be detrimental to pup production that year and lead to a weak cohort in the age structure.

Although regression coefficients were similar for all four models, the predictive power of P4 in those years did vary slightly among models. Both the random walk and Sinusoidal models gave lower t-statistics (Fig. 4b) for the regression coefficients. Random walk and sinusoidal models differed from the other two in that P4 was not independent between years, being positively correlated in adjacent years. If the abundance of a prey population depends on its previous population size, such dependence between years may slightly weaken the signal transmitted to the predator age structure.

To understand how the anomalies in age structure might be influenced by degree of density dependence, and since the patterns in Fig. 4 are similar for 5- and 15-year-old cohorts, the regression analysis was repeated for the four models using only the 5-year-old cohorts with three levels of density dependence. Increasing density dependence increased the regression coefficient in all cases (Fig. 5a), especially the negative coefficient of P4 in the year prior to birth. There was also a slight increase in the effect of carrying capacity in the year after birth relative to that in the birth year. This reflects the slightly greater sensitivity of first-year survival relative to reproduction when density changes in these models. The t-statistics (Fig. 5b) associated with the regression coefficients show that strong density dependence weakened the statistical relationship between carrying capacity and later cohort strengths. While virtually all of the t-statistics shown would test significant for the regression coefficient, strong density dependence made the relationship of carrying capacity to age structure less predictable. The second interesting feature in Fig. 5b is the slightly lower values from the random walk and sinusoidal models. The positive correlation in these two models between P4 in adjacent years may produce an interaction that makes cohort strenghts less predictable.

3.2 Qualitative Signals

The intervals between peaks of P4 and cohort strengths were examined as a measure of periodicity, and compared between P4 and cohort strengths of various ages (Table 1). This is a crude measure, and the differences in interpeak intervals usually resulted from differences in the number of small peaks present in the time series, rather than substantial differences in the larger scale anomalies. In both the random and random walk models, the period between peaks in age 0 cohorts was similar to that of P4. Among these cohort after 5 and 15 years the average interpeak interval was somewhat longer, but the variance also increased. This was the result of dampening of the signal and loss of some of the small peaks as the cohort aged. Smaller step sizes were tried in the random walk model and produced longer interpeak intervales in the older cohorts, but the variance in mean interval was always higher than that seen in the cyclic models of similar period.

Both the sinusoidal and the episodic models produced cycles in cohort strengths (Table 1). Interpeak intervals were of similar length and variability to those in the underlying P4 values. Increasing density dependence in the models decreased the interpeak intervals somewhat, especially in the episodic model. This was due to occasional small peaks in cohort strength that occurred among cohorts of very similar size, pointing to the inadequacy of using peaks in cohort strength without regard for magnitude. Visually examining the time-series data provided more useful observations.

Figure 6 shows the time series of P4 values and cohort strengths (offset by the age of the cohort) for the sinusoidal model given moderate density dependence. Peaks in cohort strengths among the age 5 and 15 categories often occurred for the cohorts born the year before the peak in carrying capacity. In the models based on cycling P4, this shift of peak P4 to the cohort born in the previous year was accentuated by

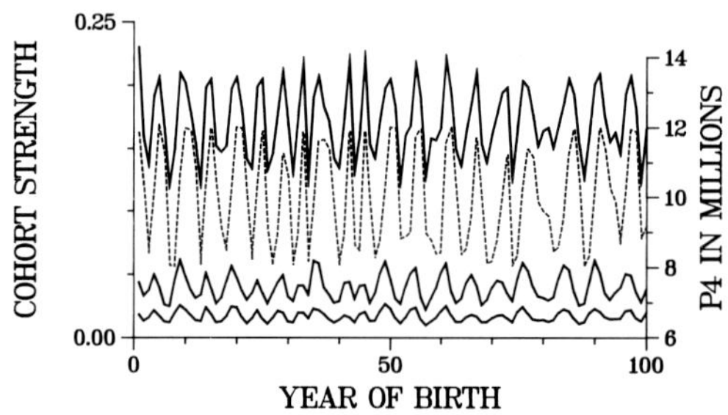

Fig. 6. Yearly variation in carrying capacity (P4, *dashed lines*), and cohort strengths (*solid lines*) after 0, 5, and 15 years (*top to bottom*) from the sinusoidal model with strong density dependence

Table 1. Means (with SD) of the intervals between peaks in P4 (carrying capacity) in the year of birth, and between peaks in cohort strengths of the same cohorts 0, 5 and 15 years later. Stochastic elements in each of the four models repeated while varying the strength of the density dependent (DD) function

	Years between peaks			
	P4	Age 0	Age 5	Age 15
Random model				
Weak DD	3.133 (1.479)	2.938 (1.014)	4.000 (2.147)	4.174 (2.269)
Moderate DD	3.133 (1.479)	2.938 (1.105)	3.692 (1.892)	4.000 (2.064)
Strong DD	3.133 (1.479)	2.848 (0.939)	3.429 (1.574)	3.692 (1.692)
Random walk model				
Weak DD	3.800 (1.472)	3.538 (1.555)	5.167 (2.093)	4.130 (2.380)
Moderate DD	3.800 (1.472)	3.276 (1.192)	4.650 (1.954)	4.130 (2.455)
Strong DD	3.800 (1.472)	3.276 (1.192)	4.429 (2.111)	4.130 (2.455)
Sinusoidal model				
Weak DD	5.111 (1.278)	4.842 (1.015)	5.052 (1.311)	4.800 (1.196)
Moderate DD	5.111 (1.278)	4.600 (1.095)	5.105 (1.243)	4.619 (1.160)
Strong DD	5.111 (1.278)	4.381 (1.203)	5.105 (1.100)	4.619 (1.117)
Episodic model				
Weak DD	5.056 (1.514)	5.056 (1.514)	5.000 (1.291)	4.700 (1.380)
Moderate DD	5.056 (1.514)	5.056 (1.514)	4.364 (1.529)	3.840 (1.344)
Strong DD	5.056 (1.514)	4.895 (1.522)	3.957 (1.552)	3.917 (1.316)

Table 2. Mean times (with SD) from peaks in carrying capacity to the birth year of the strong cohorts

	Time in years		
	Age 0	Age 5	Age 15
Sinusoidal model			
Weak DD	−0.21 (0.42)	−0.68 (0.48)	−0.61 (0.49)
Moderate DD	−0.21 (0.42)	−0.82 (0.56)	−0.74 (0.45)
Strong DD	−0.32 (0.48)	−0.95 (0.62)	−0.92 (0.42)
Episodic model			
Weak DD	0.00 (0.00)	−0.16 (0.37)	−0.16 (0.37)
Moderate DD	0.00 (0.00)	−0.16 (0.37)	−0.18 (0.38)
Strong DD	0.00 (0.00)	−0.66 (0.47)	−0.58 (0.51)

increasing the degree of density dependence (Table 2). When the peaks in P4 were extreme (i.e., the episodic model) this tendency was reduced somewhat, but strong density dependence would often cause the strongest signal from P4 to appear in the cohort born in the previous year, and occasionally that born 2 years previously.

4 Discussion

Parameter ranges and DD functions were chosen that seemed most reasonable for crabeater seals based on published rates of these and other Antarctic seals, but our knowledge of many aspects of seal life histories is limited. Some of the most significant areas of ignorance and the survival rates of juveniles, the size of DD responses in life-history parameter, and the stability of the underlying resource base or carrying capacity.

The results presented follow from the structure of the model. Tho proportion of pups was positively correlated to carrying capacity (P4) in the birth year, and negatively correlated to P4 values in the years prior to birth bue to their influence on population size. The clear signal that was produced in the initial age class was dampened most by DD juvenile survival before age 5, but the remaining signal was maintained at later ages under more constant adult survival rates. The variation in cohort strengths initially caused by varying P4 was accentuated by strong density dependence. In these models, DD juvenile survival accentuated the positive effects of P4 on cohorts born previously, especially yearlings. Given the model assumptions used here, identification of a strong cohort year could indicate a strong resource base in the year or year after birth (occasionally 2 years later) or a weak resource base and population decline prior to the birth of the strong cohort.

The sensitivity of crabeater age structure was analyzed here with regard to two unknowns: varation in the carrying capacity, presumably related to the resource base, and the degree of density dependence. To do this, assumptions were made on the relative importance of certain life-history parameters in their response to changing resources or population size. Some qualitative generalizations can be made on the effects of changing these assumptions. The most responsive DD factors were reproduction and juvenile survival. It is likely that greater DD variation in adult survival would dampen variations in cohort strengths by distributing the effects of P4 fluctuations on a cohort over more ages. Increasing the relative importance of juvenile survival in the model would increase the importance of P4 in the years immediately following the birth of the cohort. The reverse would be true if the relative importance of reproductive rates were greater than in these models. The strong negative correlation of cohort strength with P4 prior to birth, and the diminished importance of juvenile survival would lead to accentuated fluctuations in cohort strength and stronger positive correlation with P4 in the birth year.

Correctly interpreting the signal seen in crabeater seal age structure data depends not only on the statistical analysis relating the samples to the population, but on many of the assumptions included in these models. The limitations outlinded above did mask the transmission of environmental fluctuations to the age structure of the model populations. Though some caution is warranted, these models indicate that valuable information on the underlying environment for seal populations, and on DD population processes, may be found in age-structure data. However, the key features in the real environment for crabeater seals have not yet been measured on the temporal and physical scales relevant to our understanding of the population data from crabeater seals. There is good biological rationall for using population attributes of crabeater seals as indicators of krill abundance (Laws 1984; Bengtson and Laws 1985), but validation of these measures may take some time.

Acknowledgements. I wish to thank J. L. Bengtson, P. Boveng, and P. D. Shaughnessy for constructive review of the manuscript. Support was provided by National Science Foundation grant DPP-8 716 073.

References

Bengtson JL, Laws RM (1985) Trends in crabeater seal age at maturity: an insight into Antarcitc marine interactions? In: Siegfried WR, Condy PR, Laws RM (eds) Antarcitc nutrient cycles and food webs. Springer, Berlin Heidelberg New York, pp 669–675

Eberhardt LL, Siniff DB (1977) Population dynamics and marine mammal management policies. J Fish Res Board Can 34:183–190
Fowler CW (1981) Density dependence as related to life history strategy. Ecology 62:602–610
Fowler CW (1987) A review of density dependence in populations of large mammals. In: Genoways HH (ed) Current mammalogy, vol 1, Plenum Press, New York, pp 401–441
Laws RM (1984) Seals. In: Laws RM (ed) Antarctic ecology, vol 2. Academic Press, London, pp 621–715
May RM, Conway GR, Hassel MP, Southwood TRE (1974) Time-delays, density dependence and single-species oscillations. J Anim Ecol 43:747–770
Rounsevell D, Eberhard I (1980) Leopard seals, *Hydrurga leptonyx* (Pinnipedia) at Marcquarie Island from 1949 to 1979. Aust Wildl Res 7:403–415
Testa JW (1987) Long-term reproductive patterns and sighting bias in Weddell seals (*Leptonychotes weddelli*). Can J Zool 65:1091–1099
Usher MB (1972) Developments in the Leslie matrix model. In: Jeffers JNR (ed) Mathematical models in ecology. Blackwell Scientific Publications, Oxford, pp 29–60

Continental Estimates and Population Trends of Antarctic Ice Seals [1]

A. W. ERICKSON [2] and M. B. HANSON [2, 3]

Summary. Seal census data accrued during 1968–1983 were incorporated into a standard data format and revised density estimates were calculated. A key element in this reanalysis was time correction of the census counts to peak diurnal haul-out values using recently revised and evaluated correction equations (Erickson et al. 1989). This reanalysis generally lowered previously published density and population estimates.

Comparison of the corrected census data on crabeater seals taken in the western Weddell Sea (20–60° W) in 1968–69 and in 1983, and in the Pacific Ocean sector (90–160° E) in 1973–74 and in 1983, showed declines in population density values. The causes of these lower density estimates of crabeater seals in the Weddell Sea [from 11.38 seals per nm^2 (nautical miles squared) in 1968–69 to 4.28 in 1983] and in the Pacific Ocean sector (from 4.93 seals per nm^2 in 1973–74 to 1.95 seals in 1983) are unknown. There may have been a change in the distribution of seals, either because of a movement from one area to another, or because of a new pattern of local distribution. Alternatively, the abundance of seals in these two areas may have declined in recent years. There is no evidence of an acute die-off of crabeater seals, which suggests that the population declines, if real, have occurred over several years and may be food related. They may have resulted from increasing competition for food as the stocks of baleen whales recovered.

1 Continental Estimates and Population Trends of Antarctic Ice Seals

Four species of lobodontine seals have a circumpolar distribution within the pelagic pack ice of the Southern Ocean. They are the crabeater seal (*Lobodon carcinophagus*), the Weddell seal (*Leptonychotes weddellii*), the leopard seal (*Hydrurga leptonyx*), and the Ross seal (*Ommatophoca rossii*). The distributional occurrence of each species is markedly influenced seasonally by the annual expansion and contraction of the pack ice surrounding the continent. Within the ice pack, the crabeater seal and the leopard seal occur principally throughout the outer unconsolidated reaches of the pack, with highest densities occurring in areas of broken ice dominated by cake and small floe ice (Siniff et al. 1970). The Ross seal appears to prefer deeper pack ice areas with heavier concentrations of ice dominated by medium and larger floe ice (Skinner 1984). The Weddell seal is primarily associated with inshore fast ice, where if necessary it maintains gnawed access holes to the water. However, significant numbers of Weddell seals are found in unconsolidated pack ice, particularly in the deeper pack dominated by large ice floes.

This chapter presents an assimilative analysis of the large body of Antarctic seal census data developed by the senior author and his colleagues during 1968–1983. The results of these surveys have been reported, but a reworking of the data is appropriate at this time to re-order the accrued data into a common format to derive revised regional and continental population estimates and to discern possible population trends.

2 Methods and Data

The data used in this analysis were derived from nine separate strip censuses (four aerial, five shipboard) taken of pack ice areas surrounding the continent (Table 1). Each of the censuses was taken during the austral summer, using U. S. Coast Guard icebreakers and the vessels' onboard helicopters. The censuses were performed as reported by Gilbert and Erickson (1977). The strip widths for the shipboard counts were 1/8 nautical mile (nm) to either side of the survey vessel, and for the aerial censuses 1/4 nm to either side of the flight path. The transects were flown at elevations ranging from 200 to 450 ft. The shipboard counts were conducted by team members rotating on

[1] Contribution no. 796, School of Fisheries WH-10, University of Washington, Seattle, WA 98195, USA.

[2] Wildlife Science Group, Fisheries Research Institute, School of Fisheries WH-10, University of Washington, Seattle, Washington 98195, USA.

[3] Present address: NOAA, National Marine Fisheries Service P.O. Box 21668, Juneau, Alaska 99802, USA.

Antarctic Ecosystems. Ecological Change and Conservation
(ed. by K. R. Kerry and G. Hempel)
© Springer-Verlag Berlin Heidelberg 1990

Table 1. Species composition of lobodontine seals observed in six regions of Antarctic pelagic pack ice

Region	Data set[a]	Census		Number of seals counted				
		Method	Date	Crab-eater	Weddell	Leopard	Ross	Total
Amundsen and Bellingshausen	3, 4	Aerial	1/23–2/15/72	6118	181	285	109	6693
	3	Shipboard	1/23–2/15/72	1931	8	74	13	2026
Seas, 60° W–130° W			TOTALS	8049	189	359	122	8719
Western Ross Sea	3, 4	Aerial	2/6–2/14/72	717	4	12	2	735
Eastern Ross Sea	3, 5	Aerial	1/16–1/26/73	633	38	35	14	720
130° W–160° E			TOTALS	1350	42	47	16	1455
Southern Pacific	3, 5	Aerial	1/16–1/26/73	1438	34	110	44	1626
Ocean 90° E–160° E	6	Aerial	1/18–1/28/74	1682	183	104	100	2069
90° E–160° E	6	Shipboard	1/18–1/28/74	530	8	20	12	570
	7	Aerial	1/30/83	53	42	23	6	124
	7	Shipboard	1/24–2/2/83	109	3	15	5	132
			TOTALS	3812	270	272	167	4521
Southern Indian	7	Aerial	2/3–2/9/83	543	241	13	3	800
Ocean	7	Shipboard	2/3–2/11/83	119	14	3	8	144
20° E–90° E			TOTALS	662	255	16	11	944
Eastern Weddell Sea	7	Aerial	2/12–2/16/83	1102	23	38	24	1187
20° E–20° W	7	Shipboard	2/12–2/16/83	206	6	11	2	225
			TOTALS	1308	29	49	26	1412
Western Weddell Sea	1, 2	Shipboard	1/30–3/13/68	773	5	11	1	790
20° W–60° W	2	Shipboard	2/18–3/24/69	1130	10	22	3	1165
		Subtotal '68–'69		1903	15	33	4	1955
	7	Aerial	2/17–3/3/83	423	201	13	5	642
		Shipboard	2/17–3/3/83	1248	31	114	2	1395
		Subtotal Air-Ship '83		1671	232	127	7	2037
			TOTALS	3574	247	160	11	3992
		SHIPBOARD TOTALS		6046	85	270	46	6447
		AERIAL TOTALS		12709	947	633	307	14596
All regions		GRAND TOTALS		18755	1032	903	353	21043

[a] 1 = Siniff et al. (1970); 2 = Erickson et al. (1971); 3 = Erickson et al. (1972); 4 = Gilbert and Erickson (1977); 5 = Erickson et al. (1973); 6 = Erickson et al. (1974); 7 = Erickson et al. (1983).

2-hr schedules. The aerial counts were taken by the senior author with the sighting observations being relayed to a recorder. During both types of census, ice coverage and ice types were classified throughout the census together with observational conditions.

For analysis, the shipboard data were recorded in 15-min time blocks and the aerial observations were recorded in 3-min time blocks. Each time block represented a sample unit identifiable to time and location, measured as to area, and contained all observational data and related information. The accrued census data were standardized into a common data format, and prior to density calculations all counts were time-corrected to peak haul-out values (Erickson et al. 1989).

To examine the population status of Antarctic pack ice seals on a regional and continental basis, the seal census data were analyzed according to the five oceanographic ice regions of the Southern Ocean characterized on the basis of oceanographic and meteorological features (Zwally et al. 1983). The five regions were the Amundsen-Bellingshausen Seas (60–130° W), the Ross Sea (130° W–160° E), South Pacific Ocean (90°–160° E), South Indian Ocean (20°–90° E), and the Weddell Sea (20° E–60° W). The Weddell Sea in these analyses was further divided into an eastern and western sector at 20° W latitude.

3 Results

Tables 1 and 2 present the results of the several censuses taken of lobodontine seals in the five pelagic pack ice areas of Antarctica used in this analysis. In sum, censuses were conducted over 3476.8 nm^2 of pack ice, yielding a count of 21043 seals and an overall density of 6.05 seals per square nautical mile (nm^2).

Table 1 (continued)

Species composition (%)						
Including Weddell seals				Excluding Weddell seals		
Crab-eater	Weddell	Leopard	Ross	Crab-eater	Leopard	Ross
91.4	2.7	4.3	1.6	93.9	4.4	1.7
95.3	0.4	3.7	0.6	95.7	3.7	0.6
92.3	2.2	4.1	1.4	94.4	4.2	1.4
97.6	0.5	1.6	0.3	98.0	1.7	0.3
87.9	5.3	4.9	1.9	92.8	5.2	2.0
92.8	2.9	3.2	1.1	95.6	3.3	1.1
88.4	2.1	6.8	2.7	90.3	6.9	2.8
81.3	8.9	5.0	4.8	89.2	5.5	5.3
93.0	1.4	3.5	2.1	94.3	3.6	2.1
42.7	33.9	18.6	4.8	64.7	28.0	7.3
82.5	2.3	11.4	3.8	84.5	11.6	3.9
84.3	6.0	6.0	3.7	89.7	6.4	3.9
67.9	30.1	1.6	0.4	97.1	2.3	0.6
82.6	9.7	2.1	5.6	91.5	2.3	6.2
70.1	27.0	1.7	1.2	96.1	2.3	1.6
92.9	1.9	3.2	2.0	94.6	3.3	2.1
91.5	2.7	4.9	0.9	94.1	5.0	0.9
92.6	2.1	3.5	1.8	94.6	3.5	1.9
97.9	0.6	1.4	0.1	98.5	1.4	0.1
97.0	0.9	1.9	0.2	97.8	1.9	0.3
97.3	0.8	1.7	0.2	98.1	1.7	0.2
65.9	31.3	2.0	0.8	95.9	3.0	1.1
89.5	2.2	8.2	0.1	91.5	8.4	0.1
82.0	11.4	6.2	0.4	92.6	7.0	0.4
89.5	6.2	4.0	0.3	95.4	4.3	0.3
93.8	1.3	4.2	0.7	95.0	4.3	0.7
87.1	6.5	4.3	2.1	93.1	4.6	2.3
89.1	4.9	4.3	1.7	93.7	4.5	1.8

3.1 Species Composition

The species composition of the 21043 seals counted during the nine surveys broadly distributed in Antarctic pack ice was 89.1% crabeater seals, 4.9% Weddell seals, 4.3% leopard seals, and 1.7% Ross seals (Table 1). These figures are at some variance with the composite results of 14 surveys reported by other investigators (Table 3). The species composition of seals in those surveys was 79.7% crabeater, 9.2% Weddell, 6.4% leopard, and 5.1% Ross seals. The results of the two data sets are not directly comparable, however, because of the wide variability in the manner in which the censuses of the other investigators were taken (see references in Table 3).

As shown in Table 1, the species composition of seals varied greatly between individual surveys. In part, this variability is perhaps due to the small sample size of some of the surveys. However, a major factor affecting the species composition of the surveys was the highly variable occurrence of Weddell seals. The compositional occurrence of Weddell seals in the surveys ranged from 0.4% to 31.3% and averaged 4.9%. The variability of the species composition within the data sets for the remaining seal species narrowed markedly with the removal of Weddell seals (Table 1).

On average the compositional occurrence was higher for crabeater seals and lower for Weddell and Ross seals in the shipboard surveys than in the aerial surveys (Table 1). Since our shipboard surveys did not generally penetrate the deep inner ice pack to the same extent as the aerial surveys, it follows that this relationship should exist given that crabeater seals prefer dispersed ice types more prevalent at the periphery of the ice pack, while Weddell and Ross seals prefer the denser ice types more prevalent in the inner ice pack (Siniff et al. 1970; Gilbert and Erickson 1977).

The species composition of seals observed in these censuses was quite consistent between regions, except for the variability introduced by the Weddell seal component, particularly in the Pacific and Indian Ocean sectors (Table 1). Perhaps the most outstanding variable was the remarkably sparse compositional occurrence of Ross seals in the Weddell Sea.

3.2 Population Densities

Table 2 presents population densities of seals observed in the several surveys of Antarctic pelagic pack ice areas. For all regions and combining both shipboard and aerial survey results, composite corrected densities were as follows: crabeater seals, 6.62; Weddell seals, 0.39; leopard seals, 0.32; and Ross seals, 0.12 per square nautical mile.

Although the seal density data for individual censuses were highly variable, there were few profound differences between the regional composite densities of the several seal species (Table 2). For the crabeater seal, the highest composite density value recorded was 12.99 per square nautical mile for the eastern Weddell Sea in 1983, and the lowest composite density value recorded was 4.39 per square nautical mile for the Ross Sea in 1972–1973. The most noteworthy aspect of the census data developed for the crabeater seal was the marked difference between the results of the censuses taken in the western Weddell Sea in 1968 and 1969, when the mean crabeater composite density values for the 2 years averaged 11.38 seals per square nautical mile, and in 1983 when a composite aerial and shipboard census averaged 4.28 crabeater seals per square nautical mile (Table 2). An analysis of this difference and its possible significance is discussed below.

Table 2. Population densities of lobodontine seals observed in six regions of Antarctic pelagic pack ice

Region	Data set[a]	Census			Crabeater				Weddell			
		Method	Date month/ day/ year	Total area (nm^2)	Corrected[b]				Corrected[b]			
					No. obs.	No.	Dens. (nm^2)	Cor-rect. (%)	No. obs.	No.	Dens. (nm^2)	Cor-rect. (%)
Amundsen	3, 4	Aerial	1/23–2/15/72	1076.4	6 118	6449	5.99	5	181	188.1	0.175	4
and Bellings-hausen Seas,	3	Shipboard	1/23–2/15/72	184.4	1 931	2972	16.12	54	8	12.5	0.068	56
60° W–130° W			TOTALS	1260.8 (36.3%)	8 049	9421	7.47	17	189	200.6	0.159	6
West. Ross Sea	3, 4	Aerial	2/6–2/14/72	163.7	717	768	4.69	7	4	4.2	0.058	5
East. Ross Sea	3, 5	Aerial	1/16–1/26/73	164.2	633	672	4.09	6	38	40.5	0.247	7
130° W–160° E			TOTALS	327.9 (9.4%)	1 350	1440	4.39	7	42	44.7	0.136	6
Southern Pacific	3, 5	Aerial	1/16–1/26/73	452.0	1 438	1508	3.33	5	34	35.5	0.078	4
Ocean	6	Aerial	1/18–1/28/74	254.7	1 682	1,974	7.75	17	183	204.5	0.803	12
90° E–160° E		Subtotal '73–'74		706.7	3 120	3482	4.93	12	217	240.0	0.340	11
	6	Shipboard	1/18–1/28/74	50.3	530	1036	20.61	95	8	9.8	0.194	22
	7	Aerial	1/30/83	48.1	53	64	1.33	21	42	47.6	0.989	13
	7	Shipboard	1/24–2/2/83	50.1	109	128	2.55	17	3	3.3	0.067	10
		Subtotal Air-Ship '83		98.2	162	192	1.95	16	45	50.9	0.518	13
			TOTALS	855.2 (24.6%)	3 812	4710	5.51	24	270	300.7	0.352	11
Southern	7	Aerial	2/3–2/9/83	95.2	543	637	6.69	17	241	360.6	3.788	50
Indian Ocean	7	Shipboard	2/3–2/11/83	55.8	119	233	4.18	96	14	27.3	0.490	95
20° E–90° E			TOTALS	151.0 (4.3%)	662	870	5.76	31	255	387.9	2.568	52
Eastern	7	Aerial	2/12–2/16/83	90.9	1 102	1222	13.44	11	23	26.0	0.286	13
Weddell Sea	7	Shipboard	2/12–2/16/83	30.8	206	359	11.64	74	6	8.0	0.259	33
20° E–20° W			TOTALS	121.7 (3.5%)	1 308	1581	12.99	21	29	34.0	0.279	17
Western	1, 2	Shipboard	1/30–3/13/68	110.5	773	1145	10.38	48	5	8.3	0.075	66
Weddell Sea	2	Shipboard	2/18–3/24/69	132.7	1 130	1622	12.22	32	10	16.0	0.121	100
20° W–60° W		Subtotal '68–'69		243.2	1 903	2767	11.38	45	15	24.3	0.100	62
	7	Aerial	2/17–3/3/83	331.9	423	473	1.42	12	201	308.5	0.930	53
		Shipboard	2/17–3/3/83	185.1	1 248	1741	9.41	39	31	51.7	0.280	67
		Subtotal Air-Ship '83		517.0	1 671	2214	4.28	32	232	334.1	0.698	44
			TOTALS	760.2 (21.9%)	3 574	4981	6.55	39	247	358.4	0.471	45
		AERIAL GRAND TOTAL		2677.1	12 709	13 767	5.14	8	947	1215.5	0.456	28
		SHIPBOARD GRAND TOTAL		799.7	6 046	9236	11.55	53	85	136.9	0.171	61
All regions		GRAND TOTALS		3476.8 (100.0%)	18 755	23 003	6.62	23	1032	1352.4	0.389	31
Uncorrected densities							5.39				0.298	

[a] 1 = Siniff et al. (1970); 2 = Erickson et al. (1971); 3 = Erickson et al. (1972); 4 = Gilbert and Erickson (1977); 5 = Erickson et al. (1973); 6 = Erickson et al. (1974); 7 = Erickson et al. (1983).
[b] Observational densities corrected for diurnal haul-out (Erickson et al. 1989).

Table 2 (continued)

Leopard				Ross				Total All species	
Corrected [b]				Corrected [b]					
No. obs.	No.	Dens. (nm^2)	Correct. (%)	No. obs.	No.	Dens. (nm^2)	Correct. (%)	Uncorr.	Correct.
285	301.5	0.280	6	109	116.4	0.108	7		
74	131.8	0.715	78	13	15.6	0.085	20		
359	433.3	0.344	21	122	132.0	0.105	8	8 719	10 187
12	12.9	0.079	7	2	2.1	0.013	5		
35	37.1	0.226	6	14	14.9	0.091	6		
47	50.0	0.152	6	16	17.0	0.052	6	1 455	1 552
110	114.6	0.253	4	44	46.7	0.103	6		
104	121.6	0.478	17	100	134.2	0.527	34		
214	236.2	0.334	10	144	180.9	0.256	26		
20	28.3	0.563	41	12	15.7	0.313	31		
23	27.6	0.575	20	6	6.8	0.142	13		
15	18.9	0.377	26	5	6.0	0.120	20		
38	46.5	0.473	22	11	2.8	0.130	16		
272	311.0	0.364	14	167	209.4	0.245	25	4 521	5 531
13	16.5	0.174	27	3	9.3	0.098	210		
3	6.6	0.118	120	8	11.7	0.210	46		
16	23.1	0.153	44	11	21.0	0.139	91	944	1 302
38	43.6	0.479	15	24	25.5	0.292	6		
11	19.8	0.643	80	2	2.9	0.094	45		
49	63.4	0.521	29	26	28.4	0.233	9	1 412	1 707
11	15.0	0.136	36	1	1.0	0.009	4		
22	28.1	0.211	100	3	3.5	0.026	200		
33	43.1	0.177	31	4	4.5	0.018	12		
13	16.5	0.050	27	5	5.4	0.016	8		
114	180.3	0.974	58	2	2.4	0.013	20		
127	196.8	0.381	55	7	7.8	0.015	11		
160	239.9	0.316	50	11	12.3	0.016	12	3 992	5 591
633	691.9	0.259	9	307	361.3	0.135	18		
270	428.0	0.536	59	46	58.8	0.073	29		
903	1119.9	0.322 0.260	24	353	420.1	0.121 0.102	19	21 043	25 870

Of the seal density data developed for the four species of Antarctic pack ice seals, those for the Weddell seal were by far the most variable (Table 2). Composite density abundances of the Weddell seal ranged from a low of 0.16 seals per square nautical mile in the Amundsen-Bellinghausen Seas to a high of 2.57 seals per square nautical mile for the Indian Ocean sector. The composite mean density of Weddell seals for all areas combined was 0.39 per square nautical mile, and this value is probably a reasonably reliable expression of the abundance of this species in Antarctic pack ice generally.

The composite density abundance of leopard seals was 0.32 per square nautical mile, and ranged from a low of 0.15 per square nautical mile in the Ross Sea and the Indian Ocean, to a high of 0.52 per square nautical mile in the eastern Weddell Sea.

Only 353 Ross seals were tallied during the course of all surveys. The lowest density observed was 0.016 seals per square nautical mile in the western Weddell Sea, while the highest densities observed were 0.24 in the Pacific sector and 0.23 in the eastern Weddell Sea.

3.3 Population Estimates

Table 4 presents regional population estimates of crabeater seals derived from the several censuses conducted in the pelagic pack ice of each region. The estimates for each regional survey include a true estimate for the seaward or outer portion of the ice pack actually sampled and an inferred estimate for the shoreward or inner portion of the ice pack which was not sampled. The population estimates for the outer pack ice areas were derived by multiplying the mean density abundance of crabeater seals in each census by the proportion of the pack ice sampled (Table 4, column 4). This component of the population estimates for the several surveys represents real or hard estimates since it was derived from known census data.

For the uncensused inner pack ice area, populations were estimated by multiplying the amount of unconsolidated pack ice in the inner pack by the mean density abundance of crabeater seals observed in the inner censused portion of the outer pack. The rationale for this approach is that as a rule denser and more uniform ice types were characteristic of our surveys of the deeper pack and crabeater numbers tended to stabilize at low levels in this area (Gilbert and Erickson 1977). The presumption was that these conditions persisted throughout the deeper inner unconsolidated pack ice area.

The population estimates derived for the inner pack ice areas are thus inferred estimates, and accordingly the projected regional and continental population estimates, which include these derived estimates, must also be considered subjective. The population estimates of Table 4 marked with an asterisk denote those censuses which in our opinion are the most representative and valid.

Table 3. Summary of seal censuses in Antarctic pelagic pack ice reported by other researchers[a]

Reference	Census dates month/day/year	Method	Region	Number of seals				
				Crab.	Wedd.	Leop.	Ross	Total
			Ross Sea					
Ray (1970)	12/29/65–1/12/66	Aerial	165°–175° E	275	250	30	17	572
Stirling (1969)	Jan–Feb 67 and 68	Shipboard	163°–170° E	82	122	17	1	222
Eklund and Atwood (1962)	12/17/56–1/10/57	Shipboard	166°–177° E	785	0	50	11	846
	Totals			1142	372	97	29	1640
			Pacific and Indian oceans					
Prevost (1964) after Øritsland (1970)	Jan 1961	Shipboard	About 130° E	295	26	7	0	328
Eklund (1964) after Condy (1976)	1/25–28/57	Shipboard	105°–112° E	–	–	–	–	–
Zenkovich (1962) after Øritsland (1970)	12/7/57–2/13/58	Shipboard	65°–165° E	1000	117	59	5	1181
	Totals			1295	143	66	5	1509
			Eastern Weddell Sea					
Øritsland (1970)	8/25–10/31/64	Shipboard	37°–55° W	729	7	110	15	861
Bonner and Laws (1964), Laws (1964)	Summer 1950–52	Shipboard	20° W–10° E	92	0	8	13	113
		Subtotal,	10° E–55° W	821	7	118	28	974
Nel (1966)	12/63–12/64	Shipboard	1°–6° W	97	4	4	15	120
Hall-Martin (1974)	1/17–2/2/74	Shipboard	1°–6° W	133	1	2	40	176
Wilson (1975)	1/20–30/75	Shipboard	2°–4° W	238	16	3	132	389
Condy (1976)	1/19–2/14/76	Shipboard	3° E–9° W	287	29	8	47	371
Condy (1977)	1/20–2/8/77	Shipboard	9° E–3° W	344	5	5	38	392
	Subtotal, King Haakon Sea			1099	55	22	272	1448
	Totals, E. Weddell Sea			1920	62	140	300	2422
			Western Weddell Sea					
Aguayo and Torres (1967) after Øritsland (1970)	12/65–3/66	Shipboard	55°–65° W	737	19	83	0	839
Erickson et al. (1970)	1/1–24/70	Shipboard	61°–66° W	108	2	5	0	115
	Totals			845	21	88	0	954
	Grand totals, all regions***			5202	598	391	334	6525

[a] Modified from previous summaries of Øritsland (1970) and Condy (1976).
[b] *Calculated from means because of missing numbers; **calculated as mean of values because of missing area totals; ***grand total seal densities includes only censuses reporting amount of census area.

Assuming a steady population over the period of these surveys, these surveys suggest a minimal continental population of crabeater seals within the unconsolidated pack ice on the order of approximately 7 000 000 and distributed as follows: Amundsen-Bellinghausen Seas, 633 000; Ross Sea, 1 293 000; southern Pacific Ocean, 939 000; southern Indian Ocean, 746 000; eastern Weddell Sea, 806 400; and western Weddell Sea, 2 500 000 (Table 4).

Regional and continental population estimates for Weddell, Ross and leopard seals are presented in Table 5. Since the observed numbers of these species were few and the density values highly variable (Table 2), population estimates for Weddell, Ross, and leopard seals were calculated for only the key censuses denoted in Table 4. Assuming steady state abundance over the time spread of the censuses, the summed regional population estimates for these seal species indi-

Table 3 (continued)

Percent composition[b]							Total area (nm^2)	Seal densities (nm^2)			
Including Weddell seals				Excluding Weddell seals							
Crab.	Wedd.	Leop.	Ross	Crab.	Leop.	Ross		Crab.	Wedd.	Leop.	Ross
48.1	43.8	5.2	2.9	85.4	9.3	5.3	386.0	0.71	0.65	0.08	0.05
36.9	54.9	7.7	0.5	82.0	17.0	1.0	94.2	0.87	1.29	0.18	0.01
92.8	0.0	5.9	1.3	92.8	5.9	1.3	?	0.17	0.00	0.01	0.00
69.6	22.7	5.9	1.8	90.1	7.6	2.3		0.58**	0.65**	0.09**	0.02**
89.9	8.0	2.1	0.0	97.7	2.3	0.0	–	–	–	–	–
92.8	0.0	1.3	5.9	92.8	5.9	1.3	?	1.11	0.00	0.03	0.00
84.7	9.9	5.0	0.4	94.0	5.5	0.5	–	–	–	–	–
89.1*	6.0*	2.8*	2.1*	94.8*	4.6*	0.6*	–	1.11	0.00	0.03	0.00
84.7	0.8	12.8	1.7	85.4	12.9	1.7	–	–	–	–	–
81.4	0.0	7.1	11.5	81.4	7.1	11.5	–	–	–	–	–
84.3	0.7	12.1	2.9	84.9	12.2	2.9	–	–	–	–	–
80.8	3.3	3.3	12.5	83.6	3.4	12.9	–	–	–	–	–
75.5	0.6	1.1	22.7	76.0	1.2	22.8	54.1	2.46	0.02	0.04	0.74
58.5	4.0	0.7	32.4	63.8	0.8	35.4	45.4	5.24	0.35	0.07	2.91
77.4	7.8	2.1	12.7	83.9	2.3	13.8	56.1	5.12	0.52	0.14	0.84
87.7	1.3	1.3	9.7	88.9	1.3	9.8	83.7	4.11	0.06	0.06	0.45
75.9	3.8	1.5	18.8	78.9	1.6	19.5	239.3	4.19	0.21	0.08	1.07
79.3	2.6	5.8	12.3	81.4	5.9	12.7					
87.8	2.3	9.9	0.0	89.9	10.1	0.0	–	–	–	–	–
93.9	1.8	4.3	0.0	95.6	4.4	0.0	22.0	4.91	0.09	0.23	0.00
88.6	2.2	9.2	0.0	90.6	9.4	0.0					
79.7	9.2	6.0	5.1	87.8	6.6	5.6	741.5	1.98	0.57	0.09	0.37

cates a continental pelagic pack ice abundance for these species as follows: Weddell seal, 798 983; Ross seal, 131 362; and leopard seal, 296 454 (Table 5).

3.4 Population Trends

Of the censuses of seals taken during 1968–1983, only the censuses taken of the western Weddell Sea region in 1968–69 and 1983 and of the Pacific Ocean sector in 1973–1974 and 1983 provide data useful in assessing population trends (Table 2). In the case of crabeater seals, these comparative data sets suggest significant changes in the status of the stocks between the late 1960s-early 1970s and 1983, particularly in the Weddell Sea.

Figure 1 displays the densities of crabeater seals observed in the 1968, 1969, and 1983 shipboard censuses and in the 1983 aerial censuses of the western Weddell Sea. For the entire western Weddell Sea (left side of Fig. 1), these data show relatively comparable and high crabeater densities for the three shipboard censuses (approximately 10 nm^2) but a density of only

Table 4. Regional population estimates of crabeater seals in pelagic pack ice of the Southern Ocean

Region and census identification[a]			Total pack ice area (nm^2)[b]	Total area censused (nm^2)	Ice pack surveyed (%)	Mean seals nm^{-2}		Estimated populations		
						Outer pack[c]	Inner pack[d]	Outer pack[e]	Inner pack	Total estimate
Amundsen-Bellingshausen Seas	– Air '72		219 323	1077	46	5.99	0.24	604 323 *	28 424	632 747 *
	– Shipboard '72		219 323	184	20	16.12	8.35	707 097	1 465 078	2 072 178
W. Ross Sea	– Air '72		290 720	164	33	4.69	2.90	449 947 *	564 869	1 014 816
E. Ross Sea	– Air '73		72 680	164	90	4.09	1.46	267 535 *	10 611	278 146
	Total '72–'73		363 400	328				717 482 *	575 480	1 292 962 *
W. Pac. Ocean	– Air '73		67 155	452	47	3.33	0.61	634 004 *	21 711	655 715
E. Pac. Ocean	– Air '74		136 345	255	60	7.75	0.93	232 468 *	50 720	283 188
	Total '73–'74		203 500	707		4.93		866 472 *	72 431	938 903 *
Pacific Ocean	– Air '83		203 500	48	79	1.33	0.33	213 817 *	14 102	227 918 *
	– Shipboard '83		203 500	50	35	2.55	1.03	181 624	136 243	317 867
Indian Ocean	– Air '83		130 800	95	54	6.69	4.55	472 528 *	273 376	745 904 *
	– Shipboard '83		130 800	56	46	4.18	0.84	251 502	59 331	310 833
E. Weddell Sea	– Air '83		60 000	91	100	13.44	–	806 400 *	–	806 400 *
	– Shipboard '83		60 000	31	70	11.64	3.04	488 880	54 720	543 600
W. Weddell Sea	– Shipboard '68	(N. of 70° S)	153 000	63	72	6.86	5.08	755 698	217 627	973 325
		(S. of 70° S)	128 340	47	48	15.08	5.08	928 976	339 023	1 267 999
	Total '68		281 340	110		10.38		1 684 674 *	556 650	2 241 324 *
	– Shipboard '69	(N. of 70° S)	170 100	30	18	0.85	5.86	26 010	816 884	842 894
		(S. of 70° S)	220 500	103	29	15.96	5.86	1 020 562	917 412	1 939 994
	Total '69		390 600	133		12.22		1 046 572 *	1 734 296	2 780 868 *
	– Shipboard '83	(N. of 70° S)	155 700	85	31	18.25	52.89	880 873	5 682 213	6 563 086[f]
		(S. of 70° S)	200 700	100	20	1.89	4.21	75 865 *	675 958	751 823 *
	Total '83		356 400	185		9.41		956 738	6 358 171	7 314 909[f]
	– Aerial '83	(N. of 70° S)	155 700	0	0	–	–	–	–	–
		(S. of 70° S)	200 700	332	25	1.42	3.28	71 248 *	493 722	564 970 *
	Total '83		356 400	332		1.42				

[a] From Table 1.
[b] Calculated for the Amundsen, Bellingshausen and Weddell seas surveys, remainder averages from Zwally et al. (1983).
[c] From Table 2.
[d] Estimated from deepest pack ice census values.
[e] *Denotes most reliable censuses.
[f] Distorted estimate, explanation in text.

1.42 seals per square nautical mile for the 1983 aerial census.

Since the 1983 aerial and shipboard censuses were taken concurrently, the striking difference in the density abundances of crabeater seals for these two surveys would appear attributable to differences between the two types of census. Rather, the major difference appears certainly due to an artifact associated with the 1983 shipboard census. By chance, 923 (74%) of the 1248 total crabeater seals tallied in the 1983 western Weddell Sea shipboard survey (Table 2) resulted from censuses of only 38.2 nm^2 of area located between 62° and 64° S latitude near the tip of the Antarctic Peninsula. By comparison, only 325 crabeater seals were observed in 146.8 nm^2 of shipboard census distributed throughout the remaining portion of the western Weddell Sea. Since the total ice area north of 64° S latitude during the 1983 surveys was only 22 265 nm^2 or 6.2% of the total pack ice present in the western Weddell Sea, the sampling artifact associated with the small ice

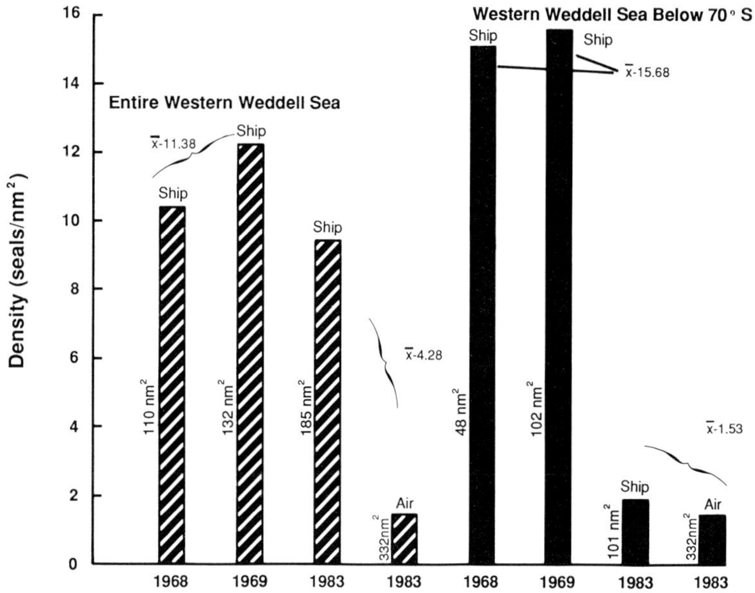

Fig. 1. Weddell Sea crabeater seal census densities 1968/69 vs 1983

Table 5. Population estimates of Weddell, Ross and leopard seals in Antarctic pelagic pack ice

Region and census identification[a]	Total pack ice area (nm^2)[b]	Densities (seals nm^{-2})[a]			Population estimates[c]		
		Weddell	Ross	Leopard	Weddell	Ross	Leopard
Amundsen-Bellingshausen Seas-Air '72	219 323	0.175	0.108	0.280	38 381	23 687	61 410
Ross Sea-Air '72 and '73	363 400	0.136	0.052	0.152	49 422	18 897	55 237
Pacific Ocean-Air '73 and '74	203 500	0.340	0.256	0.334	69 190	52 096	67 969
Indian Ocean-Air '83	130 800	3.788	0.098	0.174	495 470[d] (64 092)	12 818	22 759
E. Weddell Sea-Air '83	60 000	0.286	0.292	0.479	17 160	17 520	28 740
W. Weddell Sea-Shipboard '68	281 340	0.753	0.009	0.136	211 849	2 532	38 262
-Shipboard '69	390 600	0.120	0.026	0.211	46 872	10 156	82 417
x̄ '68–'69					129 360	6 344	60 339
Total for all areas					798 983 (367 605)	131 362	296 454

[a] From Table 2.
[b] From Table 4.
[c] Densities projected to total areas of pack ice.
[d] Improbable estimate, shipboard value in parentheses more probable.

area at the tip of the Antarctic Peninsula completely distorted the population estimates derived from the 1983 shipboard census.

With the elimination of the Antarctic Peninsula count, shipboard crabeater density values for the remainder of the western Weddell Sea averaged 2.21 per square nautical mile. This value approximated the 1.42 per square nautical mile density value of the 1983 aerial census of the western Weddell Sea.

Unfortunately, no aerial surveys were performed north of 70° S latitude in 1983 (Table 4). Thus, the only portion of the western Weddell Sea for which comparable density abundance data were derived for the 1968 and 1969 shipboard surveys, and for the 1983 shipboard surveys was for the area south of 70° S (Table 4). This comparison shows corrected density abundances of 15.08 and 15.96 crabeater seals per square nautical mile, respectively, for the 1968 and 1969 shipboard surveys, and densities of 1.89 and 1.42 per square nautical mile, respectively, for the 1983 shipboard and aerial surveys (Table 4; right side of Fig. 1). The similarity of the density values for the 1968 and 1969 surveys and for the 1983 shipboard and 1983 aerial surveys, and the marked disparity between the density values for the surveys for the two time periods (right side of Fig. 1), suggest a dramatic population decline of crabeater seals in the western Weddell Sea between 1968–69 and 1983.

A similar pattern emerges upon examination of the 1973–74 and 1983 census data available for the Pacific Ocean sector, although the limited amount of census data available for 1983 renders this comparison less meaningful (Tables 2, 4). The 1973–74 aerial censuses yielded crabeater densities of 4.93 seals per square nautical mile, as compared with a density of 1.95 seals per square nautical mile for the composite 1983 aerial-shipboard densities.

As regards population trends for the remaining species of Antarctic ice seals, the numbers of seals in the comparative surveys are too few and irregular to permit meaningful comparisons.

4 Discussion and Conclusions

These analyses indicate that the population sizes of Antarctic ice seals within the pelagic pack ice of the Southern Ocean are minimally on the order of 7 000 000 crabeater seals, 800 000 Weddell seals, 300 000 leopard seals, and 130 000 Ross seals. These population estimates should be considered minimal values since they do not include estimates for an unknown fraction of seals that did not haul out on the ice during the peak haul-out period, and thus were not included in the counts and in the estimates derived from the counts.

Two points of evidence suggest that appreciable numbers of seals do not haul out during the peak haul-out period. On March 10 and 11, 1969, Erickson et al. (1971) replicated a midday aerial census over a 26-nm^2 area in the Weddell Sea containing a dense population of crabeater seals, and obtained counts of 1173 seals on the first day and 574 seals on the second day, a 49% difference. Similarly, J. L. Bengtson (pers. commun. 1988) found irregular haul-out patterns among six crabeater seals fitted with time-depth recorders in the Weddell Sea in March 1986. During a total of 24 activity days (an activity day = 1 seal monitored for 1 day), these seals failed to haul out during 4 (17%) of the days.

These observations suggest that the population estimates for crabeater seals in pelagic pack ice should be expanded by at least 17% to account for the fraction of the seal population which did not haul out daily and thus would not have been counted in the censuses. The same factor would presumably apply to the other seal species.

Although the population estimates noted above are believed to represent reasonably reliable estimates of the ice seals present in the unconsolidated pack ice areas of the Southern Ocean, they should not be construed as representing population estimates of Antarctic ice seals for the entire Southern Ocean. There are no reliable estimates for these seal species in the ice-free areas of the Southern Ocean, but circumstantial evidence indicates that significant numbers of crabeater seals occur seaward from the ice edge (Bertram 1940; LaGrange 1962; A. W. Erickson, unpubl. data) in ice-free waters adjacent to the continent (Bonner and Laws 1964; Turbott 1952), and significant numbers of crabeater seals, leopard seals, and Weddell seals are also found in the vicinity of sub-Antarctic islands (Laws 1981; Hofman et al. 1977; Csordas 1963).

Taken together, these considerations lead to the logical conclusion that the minimal population estimate of 7 million crabeater seals in the pelagic pack ice is very conservative and it would be reasonable to estimate that the continental population size of this species is more on the order of 11–12 million.

The continental population estimates of crabeater seals derived in this work are somewhat lower than previous estimates, which suggested populations of between 15 and 30 million (Erickson et al. 1971; Gilbert and Erickson 1977; Laws 1984). The downscaling of the population numbers for most of the regional as well as the continental estimates in this work is largely attributable to improved analytical procedures, particularly the revised equation for correcting off-peak haul-out counts to peak haul-out values (Erickson et al. 1989). In the case of the Amundsen-Bellinghausen Seas crabeater population, the changed estimate from 1 193 367 (Gilbert and Erickson 1977) to 632 747 seals (Table 4) represents a 53 % reduction. Even larger corrections resulted for the 1968 and 1969 estimates of crabeater seals in the western Weddell Sea (Siniff et al. 1970; Erickson et al. 1971).

The population estimates for the crabeater seal discussed above were predicated on the assumption that the status of the crabeater seal stocks remained relatively unchanged during 1968–1983. The censuses taken in the Pacific Ocean in 1972 and 1973, and in the western Weddell Sea in 1968 and 1969, and repeated in both areas in 1983, suggest that the crabeater seal populations in these two regions may have declined dramatically. The logical questions are whether the indicated population declines are real, and if so, what was the cause of the decline.

As to the first question, the primary census data indicative of a population decline of crabeater seals between the early and late period surveys of the Weddell Sea and Pacific Ocean sector were the marked reductions in density abundances between the two survey periods. The obvious variables that would affect the density abundance estimates include the distributional characteristics of seals during the several censuses and the representativeness of our sampling efforts. Although neither factor is known, the differences between the density abundances noted between sampling periods appear to be real since the census data were derived by the same methodology and the same investigators. Furthermore, the ubiquitous occurrence of the crabeater seal in all of the censuses argues against sampling artifact. Nonetheless, the possibility exists that crabeater seals may have been dispersed differently during the 1983 surveys than during the earlier surveys. Two possibilities include dispersal into the relatively uncensused deep pack and into ice-free areas adjacent to the coast and seaward from the ice edge.

As possible causes that might explain a population decline of crabeater seals between 1968–70 and 1983, a number of speculations could be made. However, in the absence of evidence denoting a population die-off of crabeater seals, we are of the view that a food-based decline attributable to increased foraging competition affords a logical explanation.

Mackintosh (1970) has estimated that the Antarctic biomass of baleen whales, excluding the minke whale, progressively decreased from 22 million metric tons at the beginning of the 1920s to a low of two million metric tons in the mid-1960s. He concluded that between 100 and 150 million metric tons of krill formerly grazed by the large baleen whales became available to other foragers during this period. It logically follows that the obligate krill-feeding crabeater seal would have benefited from this largess and its populations would have burgeoned. By chance, our 1968 and 1969 seal surveys in the Weddell Sea occurred precisely at the nadir of the great whales' population decline, and the 1973 and 1974 surveys in the Pacific Ocean sector followed soon after. Thus, those surveys probably recorded near all-time peaks in the abundance of crabeater seals. Since that time, the great whale populations have had an opportunity to recover (Bryden et al., this vol.) and the reduced census densities encountered in the 1983 surveys possibly reflect this change.

A further possible competitive depressant on the crabeater population in the Weddell Sea since the 1968–69 surveys is the advent of commercial krilling, with total landings of krill now amounting to about 400 000 metric tons annually (CCAMLR 1987). Taken together, these factors provide a logical explanation for the apparent recent precipitous decline in the Weddell Sea crabeater population, and the same trend could logically be expected in other regions of the Southern Ocean. The obvious question is, how far will the crabeater population decline, and at what level might it stabilize?

As regards population trends for the remaining species of Antarctic ice seals, the numbers of seals counted in the comparative surveys are too few and irregular to permit meaningful comparisons.

References

Aguayo LA, Torres ND (1967 Observaciones sobre mamiferos marinas durante la Vigesima Comision Antartica Chilena. Primero censo de pinipedios en las Islas Shetland del Sur. Revta Biol Mar 13:1–57

Bertram GCL (1940) The biology of the Weddell and crabeater seals; with a study of the comparative behavior of the Pinnipedia. Brit Graham Land Exped, 1934–37, Brit Mus (Nat Hist), Sci Rep No. 1

Bonner WN, Laws RM (1964) Seals and sealing. In: Priestley R, Adie RJ, Robin G de Q (eds) Antarctic research. Butterworths London, pp 163–190

Bryden MM, Kirkwood GP, Slade RW This symposium

CCAMLR (1987) Report of the Sixth Meeting of the Scientific Committee SC-CAMLR-VI, Hoborth, Australia

Condy PR (1976) Results of the third seal survey in the King Haakon VII Sea, Antarctic. S Afr J Antarct Res 6:2–8

Condy PR (1977) Results of the fourth seal survey in the King Haakon VII Sea, Antarctica. S Afr J Antarct Res 7:10–13

Csordas SE (1963) Leopard seals in Macquarie Island. Victorian Nat 79:358–362

Eklund CR, Atwood EL (1962) A population study of Antarctic seals. J Mammal 43:229–238

Eklund CR (1964) Population studies of Antarctic seals and birds. In: Carrick R, Holdgate M, Prevost J (eds) Biologie Antarctique: Antarctic biology. Hermann, Paris, pp 421–432

Erickson AW, Bledsoe LJ, Hanson MB (1989) Bootstrap correction for diurnal activity cycle in census data for Antarctic seals. J Mar Mam Sci 5(1):29–56

Erickson AW, Denney RN, Brueggeman JJ, Sinha AA, Bryden MM, Otis J (1974) Seal and bird populations off Adelie, Clairie, and Banzare coasts. Antarctic J US 9:292–296

Erickson AW, Gilbert JR, Otis J (1973) Census of pelagic seals off the Oates and George V coasts, Antarctica. Antarctic J US 8:191–193

Erickson AW, Gilbert JR, Petrides GA, Oehlenschlager RJ, Sinha A, Otis J (1972) Population studies of seals, whales and birds in the pack ice of the Bellinghausen and Amundsen seas. Antarctic J 7:70–72

Erickson AW, Hanson MB, Kehoe DM (1983) Population densities of seals and whales observed during the 1983 circumnavigation of Antarctica by the USCGC POLAR STAR. Antarctic J US 18:163–166

Erickson AW, Hofman RJ, Thomas WL, Oehlenschlager RJ (1970) Seal survey in the South Shetland and South Orkney Islands. Ant J US 5:130–131

Erickson AW, Siniff DB, Cline DR, Hofman RJ (1971) Distributional ecology of Antarctic seals. In: Deacon G (ed) Symposium on Antarctic ice and water masses. Scient. Comm. on Antarctic Research, Cambridge, England pp 55–76

Gilbert JR, Erickson AW (1977) Distribution and abundance of seals in the pack ice of the Pacific sector of the southern ocean. In: Llano GA (ed) Adaptions within Antarctic ecosystems Proc. Third SCAR Symp Antarctic Biology, Smithsonian Institution, DC, pp 703–740

Hall-Martin AJ (1974) Observations on population density and species composition of seals in the King Haakon VII Sea, Antarctica. S Afr J Ant Res 4:34–39

Hofman RJ, Reichle RA, Siniff DB, Muller-Schwarse D (1977) The leopard seal (*Hydrurga leptonyx*) at Palmer Station, Antarctica. In: Llano GA (ed) Adaptions within Antarctic ecosystems Proc. Third SCAR Symp Antarctic Biology. Smithsonian Institution, Washington DC, pp 769–782

LaGrange JJ (1962) Notes on the birds and mammals on Marion Island and Antarctica (SANAE). J South Afr Biol Soc 3:27–84

Laws RM (1964) Comparative biology of Antarctic seals. In: Carrick R, Holdgate M, Prevost JP (eds) Biologie Antarctique: Antarctic biology. Hermann, Paris, pp 445–454

Laws RM (1981) Seal surveys, South Orkney Islands, 1971 and 1974. Brit Ant Surv Bull 54:136–139

Laws RM (1984) Seals. In: Antarctic ecology. Academic Press, London, pp 621–715.

Mackintosh NA (1970) Whales and krill in the twentieth century. In: Antarctic ecology, vol 1. Academic Press, London New York, pp 194–212

Nel JAJ (1966) On the behavior of the crabeater seal (*Lobodon carcinophagus*). Zool Afr 2:91–93

Øritsland T (1970) Sealing and seal research in the southwest Atlantic pack ice, September-October 1964. In: Holdgate MW (ed) Antarctic ecology, vol1. Academic Press, London, pp 367–376

Prevost J (1964) Observations complementaires sur les pinnipedes de l'Archipel de Pointe Geologie. Mammalia 28:351–358

Ray C (1970) Population ecology of Antarctic seals. In: Holdgate MW (ed) Antarctic ecology, vol 1. Academic Press, London, pp 398–414

Siniff DB, Cline DR, Erickson AW (1970) Population densities of seals in the Weddell Sea, Antarctica in 1968. In: Holdgate M (ed) Antarctic ecology, vol1. Academic Press, London, pp 377–397

Skinner JD (1984) Research on the Ross seal, *Ommatophoca rossii*, in the King Haakon VII Sea, Antarctica. S Afr J Sc 80:30–31

Stirling I (1969) Distribution and abundance of the Weddell seal in the Western Ross Sea, Antarctica. NZJ Mar Freshwat Res 3:191–200

Turbott EG (1952) Seals of the Southern Ocean. In: Simpson FA (ed) The Antarctic today, N Z Antarc. Soc, Wellington, NZ, pp 195–215

Wilson, VJ (1975) The second seal survey in the King Haakon VII Sea, Antarctica. S Afr J Ant Res 5:31–36

Zenkovich BA (1962) Sea mammals as observed by the round-the-world expedition of the Academy of Sciences of the USSR in 1957–58. Norsk Hvalfangsttid. 51:198–210

Zwally HJ, Comiso JC, Parkinson CL, Campbell WJ, Carsey FP, Gloersen P (1983) Antarctic sea ice, 1974–1976: satellite passive-microwave observations. NASA SP-459, National Aeronautics and Space Administration, Washington, DC, 206 pp

Abundance of Antarctic Fur Seals in the South Shetland Islands, Antarctica, During the 1986/87 Austral Summer

J. L. Bengtson [1], L. M. Ferm [1], T. J. Härkönen [2] and B. S. Stewart [3]

Summary. A survey of Antarctic fur seals (*Arctocephalus gazella*) in the South Shetland Islands, Antarctica, during the 1986/87 austral summer indicated that this species is continuing its population recovery and recolonization of rookery sites following 19th Century commercial exploitation. Eleven fur seal pupping sites were identified, some of which had not previously been reported. The largest pupping sites were at Telmo Island and Cape Shirreff, Livingston Island, and at the Seal Islands, Elephant Island. Total fur seal pup production in the South Shetland Islands in 1986/87 was estimated to be approximately 4000 individuals.

1 Introduction

Similar to the overexploitation of Antarctic fur seals at South Georgia in the late 1700s and early 1800s, Antarctic fur seals at the South Shetland Islands were subjected to an intensive commercial harvest and were nearly exterminated soon after the islands' discovery in 1819 (Stackpole 1955; Bertrand 1971). In the 1930s small numbers of fur seals (tens of individuals) were once again seen at Bird Island, South Georgia (Bonner 1968). Since 1956, when censuses began in the South Georgia vicinity, the number of pups born at Bird Island and South Georgia has increased to a population estimated in the early 1980s at just under 1 million (Payne 1977; Bonner 1981; Laws 1984). At present, the majority of Antarctic fur seal pups are born at Bird Island and the Willis Islands, South Georgia, although daughter rookeries have been reestablished during the past several decades throughout the Scotia Arc and elsewhere (Bonner 1968; Laws 1984). In February 1959, the first post-exploitation pups in the South Shetland Islands were observed by O'Gorman (1961) at Cape Shirreff, Livingston Island. During the 1965/66 austral summer, Aguayo and Torres (1967) surveyed Antarctic fur seals throughout most of the South Shetland Islands. The census described here also focused on the South Shetland Islands, with the objectives of documenting the recolonization and current distribution of Antarctic fur seal rookeries and estimating the current annual pup production throughout the archipelago.

2 Methods

From 23 December, 1986, to 2 January, 1987, helicopters from the USCGC *Glacier* were used to survey all coastlines of the Elephant Island group, including the Seal Islands, Cornwallis, Clarence, Gibbs, Aspland, Eadie, and O'Brien Islands (Fig. 1). Surveys were flown at approximately 100 m altitude. Surveys from the R/V *Prof. Siedlecki* were conducted from 25 January to 12 February, 1987, by navigating inflatable boats near the shoreline. All ice-free coasts of the South Shetland Islands between Low and Smith Islands (to the southwest) and King George Island (to the northeast) were examined in this fashion (Fig. 1). For both aerial and boat surveys, when fur seal pups or large groups of males were seen, observers landed to search for pups and to count all seals by walking through haul-out areas. Becauses these censuses were conducted from late December to mid-February, the social behavior and designation of some adult males changed during the course of surveys; therefore, for reporting purposes, all males older than pups are reported together (subadult, adult territorial, adult nonterritorial). All counts were restricted to seals hauled out on land. The peak of fur seal pupping in the South Shetland Islands is in early December, so all pups would have been born prior to the time of the surveys.

3 Results

A total of 18 881 Antarctic fur seals was counted during this census: 11 100 males (58.8 %), 3957 females

[1] National Marine Mammal Laboratory, Northwest and Alaska Fisheries Center, 7600 Sand Point Way N.E., Seattle, Washington 98115, USA.
[2] Tjärnö Marine Biological Institute, Pl. 2781, S45200, Strömstad, Sweden.
[3] Hubbs Marine Research Center, 1700 South Shores Road, San Diego, California 92109, USA.

Antarctic Ecosystems. Ecological Change and Conservation
(ed. by K. R. Kerry and G. Hempel)
© Springer-Verlag Berlin Heidelberg 1990

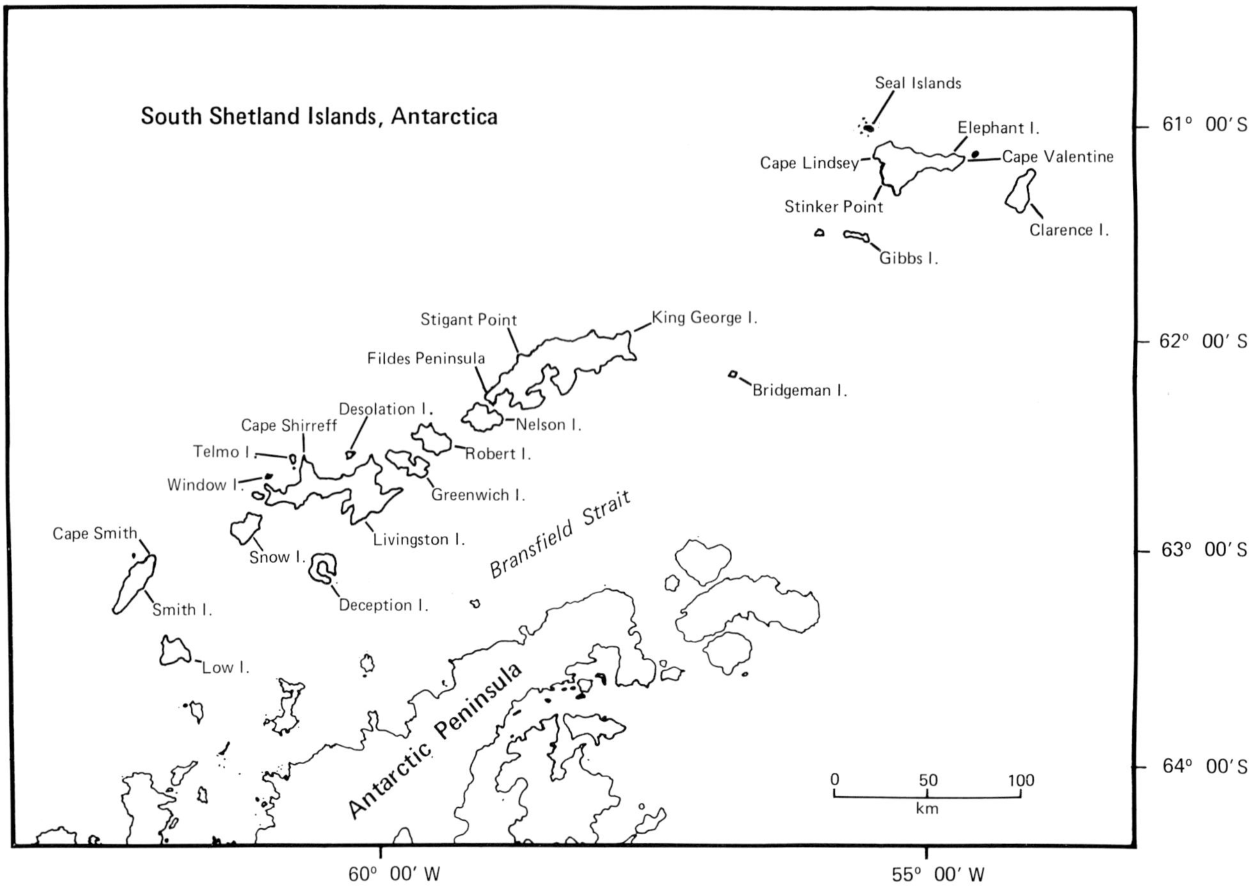

Fig. 1. Chart of the South Shetland Islands, Antarctica, where Antarctic fur seal abundance was surveyed from December, 1987, to February, 1988

(21%), and 3824 pups (20.2%) (Table 1). Eleven pupping sites were identified, some of which had not been reported previously. Numbers of Antarctic fur seal pups at various sites ranged between a single pup each at Fildes Peninsula (62° 11′ S, 58° 59.5′ W), Desolation Island (62° 29.5′ S, 60° 18′ W), and Cape Smith (62° 53′ S, 62° 53′ W) to 1875 pups (including 215 dead pups) at Telmo Island (62° 28.2′ S, 60° 49.5′ W). Most pupping sites were located on or near Elephant, King George, or Livingston Islands. Only bachelor (nonterritorial) male fur seals were hauled out in other areas.

In contrast to previous reports of small numbers of fur seal pups present at Cape Belsham and Cape Lookout, Elephant Island; Round Point and Ridley Island, King George Island; Heywood Island, Robert Island; Start Point and Black Point, Livingston Island; Cape Timblón, Snow Island; and Cape Garry, Low Island, (Aguayo 1978), no pups were observed at these sites during the 1986/87 survey. The absence of pups from these locales suggests that newly occupied pupping sites may be vulnerable to disturbance (e.g., severe storms, human visitors) until a tradition of pupping is established and the rookery has become sufficiently large and socially stable to persist and grow. Furthermore, no pupping sites were located on the southern coasts of these islands along the Bransfield Strait. No fur seals were observed at Cornwallis, Clarence, Gibbs, Aspland, Eadie, or O'Brien Islands. The absence of animals at these islands conforms to previous census data which indicated few Antarctic fur seals in those areas (Aguayo and Torres 1967; Hunt 1973; and Aguayo 1978).

Although the Antarctic fur seal population is increasing in the South Shetland Islands (Tables 2 and 3), many sites known to have had fur seal rookeries prior to exploitation have not been recolonized. Large Antarctic fur seal rookeries present on the southern coasts of these islands were subjected to heavy commercial exploitation in the 1820/21 and 1821/22 seasons (Stackpole 1955; Bertrand 1971). Our census revealed that although fur seal colonies are being reestablished successfully along the northern coastlines of the South Shetland Islands, recolonization of the southern coasts has not yet begun.

Table 1. Numbers of Antarctic fur seals observed in the South Shetland Islands, Antarctica, during the 1986/87 austral summer

Location	Adults		Pups			Totals
	Males	Females	Alive	Dead	Total	
Elephant Island						
Seal Islands[a]	139	345	495	29	524	1 008
Cape Valentine	13	69	42	3	45	127
Cape Lindsey	143	122	191	12	203	468
Stinker Point	2	0	2	0	2	4
Non pupping sites	0	0	0	0	0	0
King George Island						
Stigant Point	204	146	145	12	157	507
N. Fildes Peninsula	7	0	1	0	1	8
Non pupping sites	3 128	1	0	0	0	3 129
Nelson Island	173	0	0	0	0	173
Robert Island	194	0	0	0	0	194
Greenwich Island	235	0	0	0	0	235
Livingston Island						
Desolation Island	87	2	1	0	1	90
Cape Shirreff	1 586	844	673	45	718	3 148
Telmo Island	1 607	2 299	1 660	215	1 875	5 781
Window Island	222	127	297	0	297	646
Non pupping sites	1 665	0	0	0	0	1 665
Snow Island	652	0	0	0	0	652
Deception Island	520	0	0	0	0	520
Smith Island						
Cape Smith	100	2	1	0	1	103
Non pupping sites	5	0	0	0	0	5
Low Island	418	0	0	0	0	418
Totals	11 100	3 957	3 508	316	3 824	18 881

[a] Includes both of the main rookeries on the Seal Islands: one on the north shore of Seal Island (241 living and 8 dead pups) and one on a nearby island to the northwest (254 living and 21 dead pups). Because of the close proximity of these two rookeries (1 km apart), these sites have been listed together in this summary.

4 Discussion

The reasons for the differences in growth rates of rookery size between the three islands listed in Table 2 are difficult to identify. Whereas rookeries at Elephant and Livingston Islands have expanded dramatically over the past two decades, the number of pups born at Stigant Point, King George Island, has remained essentially unchanged. The fluctuating rookery size at Stigant Point does not appear to be related to availability of space on beaches. To the human observer, there appears to be sufficient space on the Stigant Point beaches for further rookery growth. Determining the extent to which potential differences in marine habitats and local prey resources may influence pup survival and rookery expansion requires further study.

The extent to which adult female fur seals are being added to the South Shetland Islands population due to immigration from South Georgia is unknown. Recent evidence from marked seals demonstrates that some level of exchange among these populations is occurring: (1) an adult female tagged at Seal Island (the largest island in the Seal Islands group) in January, 1987, was seen at Cape Shirreff in February, 1988; (2) an adult female tagged at Seal Island in January, 1988, was seen at Bird Island, South Georgia, in April, 1988; and (3) an adult male paint-marked at Bird Island during the 1986/87 summer season was seen in January, 1988, at Seal Island. Given the relatively small number and size of rookeries in the South Shetland Islands, immigrants comprising only a small fraction of the South Georgia population could have a signifi-

Table 2. Changes in Antarctic fur seal pup production over the past 30 years at important rookery sites in the South Shetland Islands, Antarctica. Sites listed are those areas for which there are past census data comparable with the 1986/87 survey

Location	Date	Pups[a]	Total[b]	Source
Elephant Island				
Seal Island[c]	7 Jan 66	0	20	Aguayo (1978)
	13 Dec 70	16	62	Hunt (1973)
	24/31 Dec 86	249	493	1986/87 Census
Cape Valentine	16 Feb 66	2	30	Aguayo (1978)
	9 Feb 71	3	100 +	Hunt (1973)
	23 Dec 86	45	127	1986/87 Census
Cape Lindsey	16 Feb 66	3	70	Aguayo (1978)
	Feb 71	0	30	Hunt (1973)
	30 Jan 87	203	468	1986/87 Census
King George Island				
Stigant Point	Jan 66	0	0	Aguayo (1978)
	16 Jan 70	123	213	Llano (1971)
	Feb 73	80	250	Aguayo (1978)
	Jan 82	168	293	Cattan et al. (1982)
	1982/83	123	367	Oliva et al. (1986)
	31 Jan 87	157	507	1986/87 Census
Livingston Island				
Cape Shirreff	14 Jan 58	0	27	O'Gorman (1961)
	3 Feb 59	2	11	O'Gorman (1961)
	1 Jan 66	12	50	Aguayo (1978)
	Feb 71	27	201	Laws (1973)
	25 Jan 73	300	1741	Aguayo (1978)
	Jan 82	180	532	Cattan et al. (1982)
	1982/83	–	564	Oliva et al. (1986)
	1983/84	248	969	Oliva et al. (1986)
	1984/85	384	1590	Oliva et al. (1986)
	2 Feb 87	718	3148	1986/87 Census
Window Island	21 Jan 66	50	150	Aguayo (1978)
	25 Jan 73	70	320	Aguayo (1978)
	3 Feb 87	297	646	1986/87 Census

[a] Includes both living and dead pups.
[b] Total Antarctic fur seals, including pups.
[c] Compares censuses only at Seal Island, the largest island in the Seal Islands group.

cant effect on the growth of South Shetland Islands fur seal rookeries.

Acknowledgements. The authors are grateful to the officers and crews of the USCGC *Glacier* and the R/V *Prof. Siedlecki* for providing excellent logistical support. The USCGC *Glacier's* aviation detail deserves special thanks for distinguished flight operations under difficult conditions. Michael Goebel, Richard Merrick, Everett Schaner, David Shuford, and Larry Spear are thanked for their help in the field. Steven Osmek provided valuable assistance in summarizing data. This study was supported by the National Marine Fisheries Service's Antarctic Marine Living Resources (AMLR) Program and the National Science Foundation's Division of Polar Programs.

References

Aguayo A (1978) Present status of the Antarctic fur seal *Arctocephalus gazella* at South Shetland Islands. Polar Rec 19:167–173

Aguayo A, Torres D (1967) Observaciones sobre mamíferos marinos durante la Vigésima Comisión Antárctica Chilena. Rev Biol Mar 13:1–57

Aguayo A, Maturana R, Torres D (1977) El lobo fino Antartico, *Arctocephalus gazella* (Peters), en el Sector Antartico Chileno. Ser Cient Inst Antart Chileno 5(1):5–16

Andersson KA (1905) Das höhere Tierleben im antarktischen Gebiete. In: Wissenschaftliche Ergebnisse der Schwedischen Südpolar-Expedition, 1901–1903, Bd 5, Lief 2. Stockholm, Lithographisches Institut des Generalstabs, pp 1–58

Bertrand KJ (1971) Americans in Antarctica, 1775–1948. American Geographical Society, New York

Bonner WN (1968) The fur seal of South Georgia. Sci Rep Brit Antarct Surv 56:1–81

Table 3. Abundance of Antarctic fur seals as reported from various sites in the South Shetland Islands, Antarctica, from 1901/02 through 1986/87. No differentiation is made between age or sex classes[a]

Location	Austral summer season															
	1901–02	57–58	58–59	64–65	65–66	69–70	70–71	71–72	72–73	77–78	78–79	81–82	82–83	83–84	84–85	86–87
Source[b]	a	b	b	c	c/e i/j	d	f/g h/i	j	i/j	k	l	m	n	n	n	This study
Elephant Island																
Cape Valentine	–	–	–	–	30	–	103	–	–	–	–	–	–	–	–	127
Point Wild	–	–	–	–	0	–	80–100	–	–	–	–	–	–	–	–	0
Cape Lindsey	–	–	–	–	70	–	30	–	–	–	–	–	–	–	–	468
Stinker Point	–	–	–	–	0	–	60	–	–	–	–	–	–	–	–	4
Cape Lookout	–	–	–	–	76	–	0	–	–	–	–	–	–	–	–	0
Cape Belsham	–	–	–	–	100	–	0	–	–	–	–	–	–	–	–	0
Seal Islands	–	–	–	–	20	–	62	–	–	–	–	–	–	–	–	1008
Cornwallis Island	–	–	–	–	0	–	6	–	–	–	–	–	–	–	–	0
Clarence Island	–	–	–	–	1	–	–	–	–	–	–	–	–	–	–	0
Gibbs Island	–	–	–	–	2	–	–	–	–	–	–	–	–	–	–	0
King George Island																
Stigant Point	–	–	–	–	0	–	213	–	250	–	–	293	367	–	–	507
Ridley Island	–	–	–	–	0	–	–	–	40	–	54	–	–	–	–	85
False Round Point	–	–	–	–	0	–	–	–	12	–	–	–	–	–	–	204
Cape Melville	–	–	–	–	0	–	–	–	34	–	–	–	–	–	–	194
North Foreland	–	–	–	–	0	–	–	–	3	–	–	–	–	–	–	82
Admiralty Bay	–	–	–	–	0	–	–	–	0	125	–	–	–	–	–	*
NW side	–	–	–	–	0	204	–	–	–	–	–	–	–	–	–	*
SE side	–	–	–	–	0	0	–	–	–	–	–	–	–	–	–	*
South coast	–	–	–	–	0	–	–	–	0	–	–	–	–	–	–	1399
Other sites	–	–	–	–	0	–	–	–	60	–	–	–	–	–	–	1173
Nelson Island	1	–	–	2	0	–	–	1	0	–	–	–	–	–	–	173
Robert Island	–	–	–	–	1	–	5	–	4	–	–	–	–	–	–	194
Greenwich Island	–	–	–	–	0	–	–	1	0	–	–	–	–	–	–	235
Livingston Island																
Cape Shirreff	–	27	32	–	50	–	201	–	1741	–	–	532	564	969	1590	3148
Barnard Point	–	–	–	–	0	–	–	–	1	–	–	–	–	–	–	2
Williams Point	–	–	–	–	0	2	–	–	32	–	–	–	–	–	–	14
Elephant Point	–	–	–	–	0	–	–	–	1	–	–	–	–	–	–	59
Black Point	–	–	–	–	10	–	–	–	40	–	–	–	–	–	–	64
Rugged Island	–	–	–	–	0	–	–	–	80	–	–	–	–	–	–	248
Start Point	–	15	–	–	0	–	–	–	807	–	–	–	–	–	–	47
Window Island	–	–	–	–	150	–	–	–	320	–	–	–	–	–	–	646
Byers Peninsula	–	–	–	–	0	–	–	–	2	–	–	–	–	–	–	803
Other sites	–	–	–	–	0	–	–	–	0	–	–	–	–	–	–	6299
Snow Island	–	–	–	–	0	–	–	–	40	–	–	–	–	–	–	652
Smith Island	–	–	–	–	0	–	–	–	0	–	–	–	–	–	–	105
Low Island																
Cape Garry	–	–	–	–	0	–	–	–	560	–	–	–	–	–	–	12
Cape Hooker	–	–	–	–	0	–	–	–	24	–	–	–	–	–	–	214
Cape Wallace	–	–	–	–	0	–	–	–	30	–	–	–	–	–	–	168
Other sites	–	–	–	–	0	–	–	–	0	–	–	–	–	–	–	24
Deception Island	–	–	–	–	0	–	–	–	0	–	–	–	–	–	–	520

[a] 0 = No seals present; – = no data; * = seals present were tallied in counts for other sites.

[b] Data from: a: Andersson (1905); b: O'Gorman (1961); c: Aguayo and Torres (1967); d: Erickson et al. (1970); e: Llano (1970); f: Llano (1971); g: Hunt (1973); h: Laws (1973); i: Aguayo et al. (1977); j: Aguayo (1978); k: Myrcha and Teliga (1980); l: Jablonski (1980); m: Cattan et al. (1982); n: Oliva et al. (1986)

Bonner WN (1981) Southern fur seals, *Arctocephalus* (Geoffroy Saint-Hilaire and Cuvier, 1826). In: Ridgway SH, Harrison RJ (eds) Handbook of marine mammals, Vol. 1. Academic Press, London and New York, pp 161–208

Cattan PE, Yánez JV, Gajardo DN, Cárdenas JC (1982) Censo, marcaje y estructura poblacional del lobo fino antárctico *Arctocephalus gazella* (Peters, 1875) en las islas Shetland del Sur, Chile. Publ INACH Ser Cie, No. 29:31–38

Erickson AW, Hofman RJ, Thomas WL, Oehlenschlager RJ (1970) Seal survey in the South Shetland and South Orkney Islands. Antarct J US, 5(4):130–131

Hunt JF (1973) Observations on the seals of Elephant Island, South Shetland Islands, 1970–71. Br Antarct Surv Bull, 36:99–104

Jablonski B (1980) Distribution and numbers of birds and pinnipeds on Penguin Island (South Shetland Islands) in January 1979. Pol Polar Res 1(1):109–116

Laws RM (1973) Population increase of fur seals at South Georgia. Polar Rec 16(105):856–858

Laws RM (1984) Seals. Pp. 621–715. In: Laws RM (ed) Antarctic ecology, vol 2. Academic Press, London and New York

Llano GA (1970) Discussion section, biology of seals, fur seals. In: Holdgate MW (ed) Antarctic ecology, vol 1. Academic Press, London and New York.

Llano GA (1971) Frontispiece. In: Burt WH (ed) Antarctic Pinnipedia. Antarctic Research Series, Vol 18. Am Geoph Union Acad Sci Nat Res Council

Myrcha A, Teliga K (1980) Observations of pinnipedian mammals in the vicinity of Arctowski Station (King George Island) in 1978. Pol Polar Res 1(1):117–126

O'Gorman FA (1961) Fur seals breeding in the Falkland Islands Dependencies. Nature 192:914–916

Oliva DL, Duran LR, Torres D, Gajardo M (1986) Changes in the population structure and growth of the fur seal *Arctocephalus gazella* (Peters, 1875) in South Shetland Islands. (Unpubl ms)

Payne MR (1977) Growth of a fur seal population. Phil Trans R Soc Lond (B:Biol Sci) 279:67–79

Stackpole EA (1955) The voyage of the Huron and the Huntress. The Marine Historical Association, No 29. Mystic, Connecticut

Humpback Whales, Area V. An Increase in Numbers Off Australia's East Coast

M. M. Bryden [1], G. P. Kirkwood [2] and R. W. Slade [3]

Summary. Surveys of migrating humpback whales have been conducted off Australia's east coast since 1981 from a headland near Brisbane, Queensland, in an attempt to estimate the status of the Area V stock, which formerly was heavily exploited. Observations were made between 0700 and 1700 h for up to 84 days around the peak of northward migration, May to August. The estimated total numbers of migrating whales in 1981, 1982, 1986 and 1987 were 356, 396, 778 and 790 respectively, representing an increase of 130 to 140% over 5 years or an annual net increase of approximately 14%. The total numbers are underestimates, as recent evidence has shown that the northward migration continues beyond the period of observation. It is concluded that there has been a rapid increase in numbers of humpback whales migrating along Australia's east coast in the last 5 to 6 years.

1 Introduction

Humpback whales (*Megaptera novaeangliae*) in some areas tend to aggregate close to the tropical coastlines of continents and islands in winter and migrate along continental coastlines to summer feeding areas in polar seas. This tendency to gather close to the coast in tropical regions to breed in winter seems to be related more to water depth than to the land masses themselves: humpback whales in tropical waters are usually observed within the 100 fathom isobath (Winn et al. 1975; Herman and Antinoja 1977). In the southern hemisphere, humpback whales migrate along the east and west coasts of the three continents of Africa, South America and Australia. These continents separate the winter aggregations of whales spatially, such that the whales observed along each of the continental coasts constitute stocks with a high degree of segregation (Mackintosh 1965). For management purposes the areas within which these stocks occur were designated by the International Whaling Commission as Areas I to VI. No land masses separate the stocks in the summer feeding grounds of the Southern Ocean, and the separation of the stocks there is less distinct. Tagging of whales demonstrated that some overlap of feeding areas does occur in the Southern Ocean (Dawbin 1966), although mark recovery was low (less than 5%). Nevertheless, most humpbacks return to the same breeding grounds in successive years (Dawbin 1964; Chittleborough 1965). Recently, recordings of humpback "songs" taken from the east and west coasts of Australia suggest almost complete segregation of stocks of Areas IV and V, although the sample size is very small (W. H. Dawbin, pers. commun.).

The humpback whales that migrate past the eastern Australian coast constitute a large proportion of the Area V stock (130° E to 170° W). Area V whales also include the "New Zealand group", which migrate annually past New Zealand, Tonga, Norfolk Island and Fiji, and during the breeding season gather around island groups of the southwest Pacific Ocean. They form a large summer feeding group in the region of the Balleny Islands. Dawbin (1964) suggested that the population was segregated to some extent into a number of breeding units and that individuals usually returned to breed in the locality where they had been born and weaned. However, Chittleborough (1965) concluded that this is unlikely: although there was some suggestion of partial segregation within the Area V population up to 1959, results of mark recoveries in 1960 and 1961 indicated random dispersal of Area V whales during their northward migration.

Humpback whales in Area V have suffered extreme exploitation pressure like those in other regions. The history of exploitation in eastern Australian waters was reviewed by Paterson and Paterson (1984). Humpbacks were exploited only lightly in Area V before 1950, but heavily in Antarctic waters from 1950 and from shore stations from 1952, until 1962. Chittleborough (1965) estimated that the original population in its unfished state was 10 000 individuals, but

[1] Department of Veterinary Anatomy, University of Sydney, New South Wales 2006, Australia.
[2] CSIRO Division of Fisheries, Hobart, Tasmania 7000, Australia.
[3] Department of Zoology, University of Queensland, St. Lucia, Queensland 4067, Australia.

Antarctic Ecosystems. Ecological Change and Conservation
(ed. by K. R. Kerry and G. Hempel)
© Springer-Verlag Berlin Heidelberg 1990

was reduced to 200–500 when hunting ceased at the end of the 1962 season.

The SCAR/SCOR Group of Specialists on Living Resources of the Southern Ocean considered the role of large whales in the trophodynamics of the Antarctic marine system and their direct relationship to krill to be of major significance, because of their individual and population biomass (Anonymous 1977).

The major aim of the surveys reported here was to estimate the total number of humpback whales migrating along the east coast of Australia during the winter northward migrations of 1981, 1982, 1986 and 1987 to determine the status of the population after more than 18 years of protection from exploitation. The vast majority of the population (96%) migrate within visible distance of headlands (Bryden 1985). The information gained indicates the extent of recovery of the population since the cessation of whaling at the end of the 1962 season, and provides a basis for the assessment of the status of Australian east coast humpbacks in the future. Progress reports of this work have been submitted to the Australian National Parks and Wildlife Service in previous years. Since this paper was read at this symposium, Paterson and Paterson (1989) have published figures supporting a significant increase in humpback whale numbers on Australia's east coast in recent years.

2 Methods

2.1 Observations

Counts of humpback whales were made in daylight hours, between 0700 and 1700 h, during the northward migration in 1981, 1982, 1986 and 1987. The observations were made at Point Lookout, North Stradbroke Island on Australia's east coast (Fig. 1) in the last week of June and the first 2 weeks of July. At the same time, teams of observers were at Cape Byron and Cape Moreton in 1981, and at Cape Moreton only in 1982 (see Fig. 1), to check the accuracy of the counts (Dr. R. Paterson kindly provided data obtained between 29 June and 4 July 1981). In addition, counts were made at Point Lookout on some days for up to 4 weeks prior and subsequent to the major observation period. Observations were made on most days of these additional periods in 1987.

Observations were conducted for the full 10 h each day, unless low cloud or rain obscured visibility. Whenever possible, three or more observers were present (78% of all observations), although they were reduced to two or even one on occasions. At least one experienced observer was always present. Observations were made for 410 h in 1981, 360 h in 1982, 578 h in 1986 and 668 h in 1987. Only whales observed to pass due east of the observation point were included in analyses of the 1982, 1986 and 1987 data. In 1981, this restriction was applied for all but the five observation days from 29 June to 4 July. Failure to apply this restriction to those days will have minimal effect on the estimates.

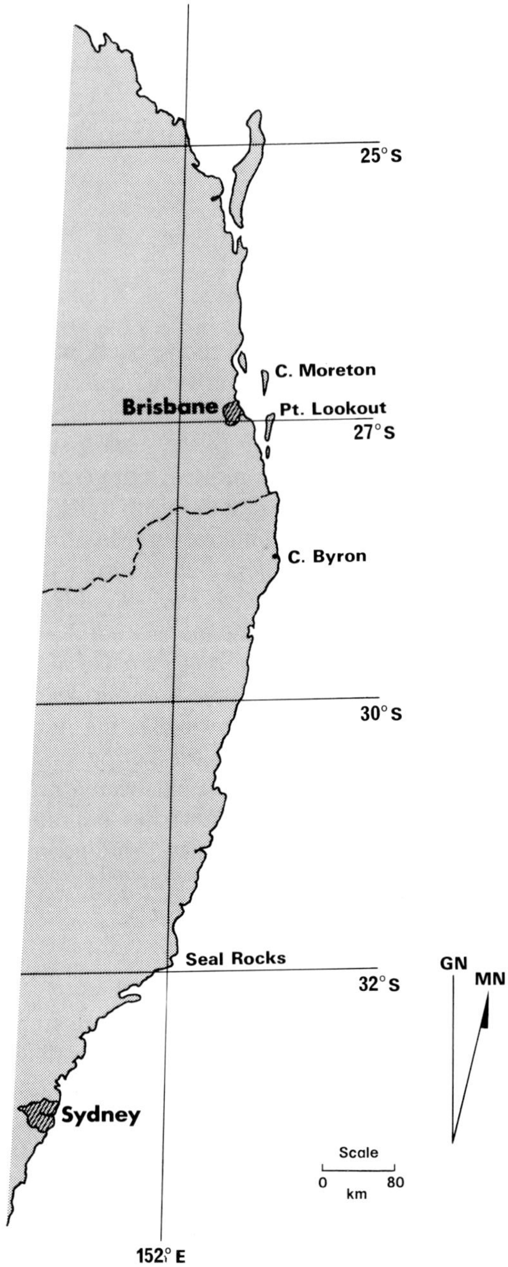

Fig. 1. Map of part of the east coast of Australia, showing the position of Point Lookout on North Stradbroke Island where whale observations were made

The headland at Point Lookout was approximately 35 m above sea level, from which the line-of-sight horizon was 18 km away. Despite this, it was considered that the maximum distance at which a whale could be seen was 10 km under good conditions. This conclusion was drawn from simultaneous observations of whales from land and air, and observing trawlers travelling due east from the observation points at known speed until they reached the limit of clear visibility.

During good conditions the east-south quadrant of the sea was scanned almost continually, and occasional observations of the east-north quadrant were made with the unaided eye. Regular sweeps of the entire horizon were made with binoculars. During poor weather and sea conditions it was necessary to scan continually with binoculars due to the uncertainty of distinguishing distant "blows" and breaking wave crests. Positive identification of a humpback whale required observation of the characteristic dorsal fin and arched back in addition to the "blow". Additional features, such as the characteristically long flippers and pigmentation patterns, were looked for but not always seen.

Humpback whales in a group tend to dive and surface in synchrony. By observing them continually for a period (30 min or more), it is possible to estimate numbers in a group from synchronous blows. Because the group size of northward migrating whales is small, potential error in the population estimation resulting from errors in estimating individual group size will be small.

Other species of whales observed in small numbers were minke whales (*Balaenoptera acutorostrata*) and Bryde's whales (*Balaenoptera edeni*).

2.2 Total Number of Migrating Whales

Estimation of the total number of whales migrating past Point Lookout was carried out in a two-step process: the total number that passed during a 10-h period on each day of the migration was estimated from the observed count data, and then this estimate was multiplied by 2.4 on the assumption that the rate of migration on each day was the same during the unobserved 14 h as during the observed 10 h. In order to estimate the total numbers passing during the maximum 10-h observation period each day, we adopted an approach similar to that used by Reilly et al. (1983). That is, we assumed that the daily numbers could be described by a probability distribution over time.

Following the results of Chittleborough (1965) and initial inspection of the 1981 and 1982 data, we adopted as the primary model the normal density function. Specifically, if n_i is the observed count on day i, then the expected number seen on that day was assumed to be

$$E(n_i) = a_i P(2\pi\sigma^2)^{-1/2} \exp[-(i-\mu)^2/2\sigma^2],$$

where a_i is the proportion of the possible 10-h observation time used on day i, P is the overall population size of whales that migrated during the 10-h daily observation times, and μ and σ^2 are the mean and variance of the normal distribution. In addition, however, the 1986 and 1987 data suggested that a skewed distribution may be more appropriate for those years, and a

Table 1. Simultaneous air and land-based observation of humpback whale groups

Date	Aerial observation		Land observation	
	Time (h)	Size of group	Time (h)	Size of group
1 Jul 80	1417	4	1234–1447	4
1 Jul 80	1429	2	1403–1509	2
4 Jul 80	1133	1	0937–1125	1
6 Jul 80	1003	2	0927–1227	2
9 Jul 81	1500	2	1408–1509	2
5 Jul 82	1551	1	1511–1615	1
7 Jul 82	1010	3	0900–1057	3
7 Jul 82	1503	1	1513–1648	1
7 Jul 86	1021	3	0907–1105	3
7 Jul 86[a, b]	1253	3	1345–1427 }	3 + 3 or 4 + 2,
7 Jul 86[a, b]	1256	2	1352–1427 }	or 3 + 2 + 1
7 Jul 86	1522	2	1606–1645 }	3, or
7 Jul 86	1523	1	1606–1645 }	2 + 1
11 Jul 86	1323	3	1310–1506	3
12 Jul 86[a]	0955	2	0707–0905	2
12 Jul 86[a]	0955	2	0707–0905	2

[a] Followed by observer boat, therefore easily identified by observer groups in plane and on shore.
[b] 3 + 2 whales were observed from the air, but a total of six were seen from the shore. Individual whales moved from one group to another, so that group sizes varied from 3 + 3 to 4 + 2 to 3 + 2 + 1, possibly influenced by the proximity of the observer boat.

gamma distribution model was also fitted. In this case

$$E(n_i) = a_i P[(i - i_0)\, \Gamma(b)]^{-1} [(i - i_0)\, b/\lambda]^b \cdot \exp[-(i - i_0)\, b/\lambda],$$

where i_0 is the first day of migration, and b and λ are parameters of the gamma distribution. Maximum likelihood estimates of the population size and the other incidental parameters, along with their standard errors, were obtained by fitting the chosen distributional form to the count data, under the assumption of multinomial sampling.

It should be emphasized that this procedure is nothing more than a convenient method of interpolating and extrapolating from the observed daily count data to fill in the gaps on those days when either no observations were made, or there was less than 10 h of observation. Had there been no gaps, then this fitting process would have been entirely unnecessary. On the other hand, if there are many gaps, especially at the beginning and end of the migration, then the precision of the estimation procedure decreases markedly. There is no doubt that the actual migration process is much more complicated than could ever be encapsulated in a simple model such as one of the above two, but in the circumstances where this really matters (many gaps in the data), there are insufficient data to fit a possibly more realistic model.

3 Results

Northward migrating whales travel in small groups. Of a total of 480 groups, 203 (42%) were single whales, 209 (44%) comprised two, 54 (11%) three and 13 (3%) four whales. One group of five (less than 1%) was observed.

Group size estimation was found to be accurate, based on agreement between estimates from simultaneous air and land observation (Table 1). Although there was no direct communication between observers in the plane with those on land, details of position

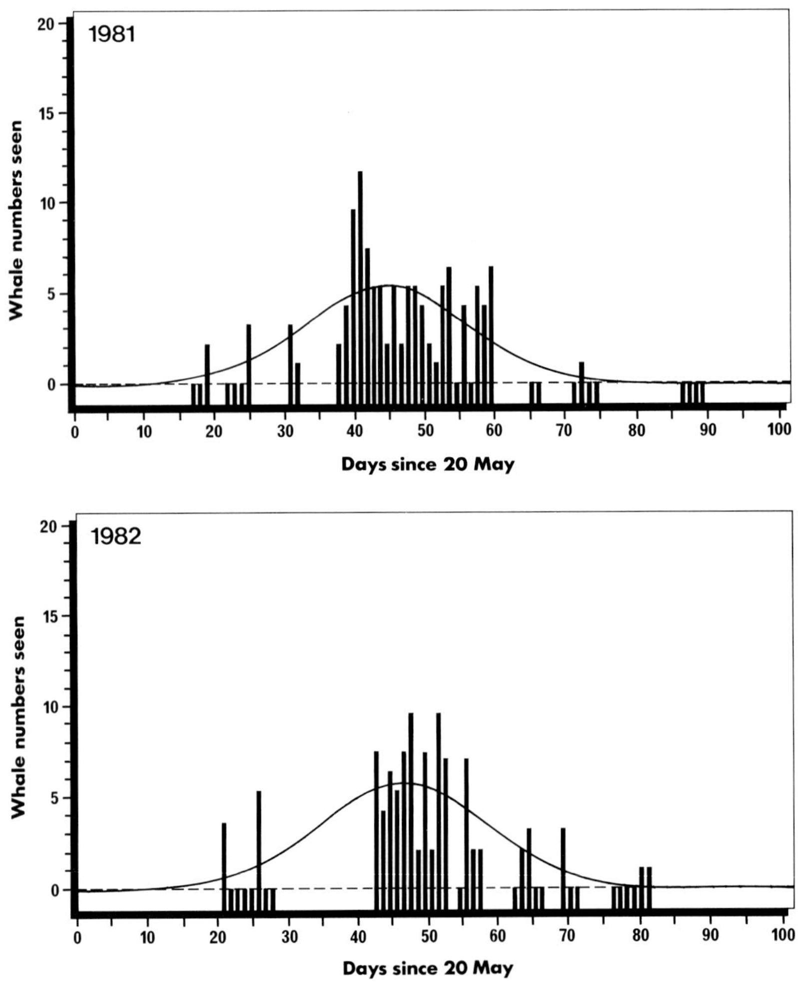

Fig. 2. Daily counts of humpback whales observed from Point Lookout in 1981 and 1982, with estimated curves fitted

fixes on whale groups, and times of observation, permitted subsequent confirmation that simultaneous observations of groups were made. Accurate position fixes of the aircraft were possible using the navigational equipment; the direction and estimated distances offshore of whale groups from the observation point on the headland were recorded. In addition, the land observers could see the plane circling above the whales they were observing, and in some instances the whale groups could be identified by the presence of an accompanying observer boat.

The daily counts of humpback whales observed from Point Lookout during the survey periods are shown in Figs. 2 and 3. In these figures, for convenience of presentation, the counts for days with less than 10 h effective observation have been scaled up to notional counts for a 10-h day. Also shown on these figures are the fitted values using the normal distribution for 1981 and 1982, and using the normal and gamma distributions in 1986 and 1987. The estimates of the individual parameters are given in Table 2.

The numbers seen increase from early June to reach a peak around the end of June or early July and then decline. For all years, there is considerable variability in the daily counts, and this variability is in excess of that which would be expected under simple multinomial sampling. As indicated earlier, this was not unexpected, but it is important to note that the residuals from the fitted values do not show notable trends, so that a more sophisticated model will be unlikely to lead to better interpolation and extrapolation. The extent of missing data for 1981 and 1982 is such that it would be impossible to distinguish between virtually any plausible alternative distributional model to the normal. There was no significant difference in goodness of fit for the 1986 data between the normal and gamma distributions, however, the gamma distribution did provide a significantly better fit than the normal to the 1987 data. Accordingly, we adopted population estimates in Table 2 from the normal distribution in 1981, 1982 and 1986, and for the gamma distribution in 1987. One consequence of the less-

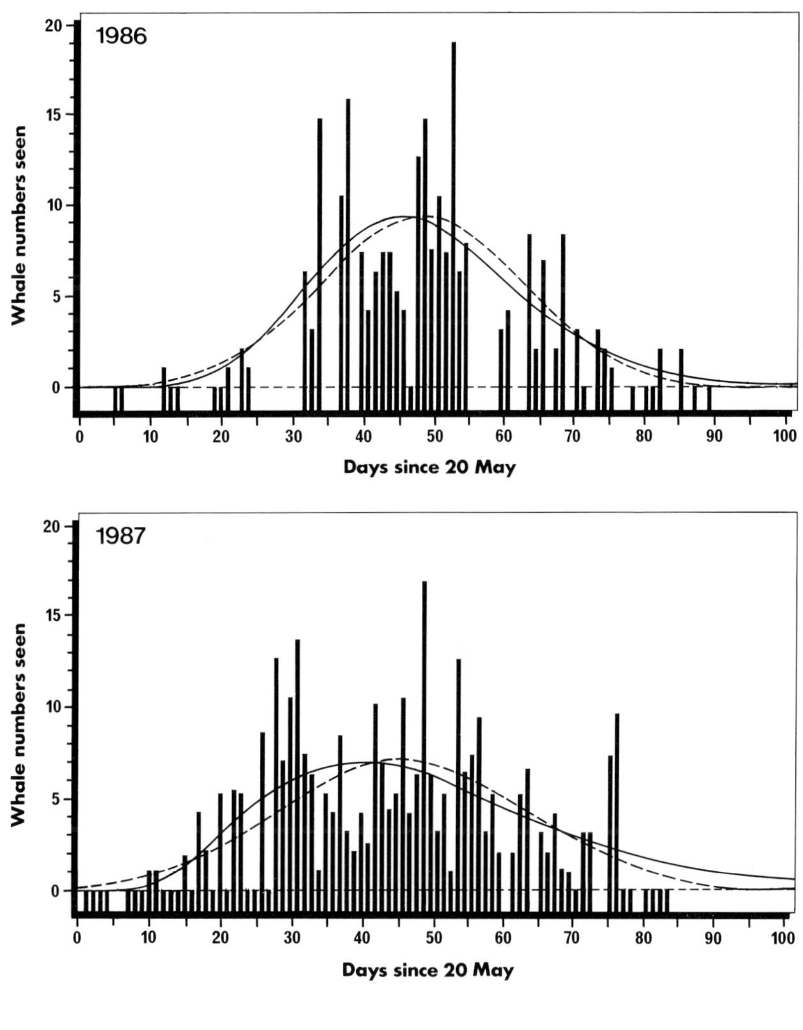

Fig. 3. Daily counts of humpback whales observed from Point Lookout in 1986 and 1987, with estimated curves fitted. *Continuous line* normal distribution; *dashed line* skewed distribution

Table 2. Total numbers of humpback whales seen, parameter estimates and their standard errors (SE) or 95% confidence intervals (CI) for survey data[a]

Year	Total seen	Distribution fitted	Estimated whales during day (95% CI)	Mean (SE)	Variance (SE)	Estimated whales migrating (95% CI)
1981	109	Normal	148 (133–168)	44.5 (1.3)	132.5 (19.9)	356 (319–404)
1982	94	Normal	165 (141–194)	46.6 (1.4)	148.4 (23.8)	396 (338–466)
1986	221	Normal	324 (298–353)	48.5 (1.2)	207.9 (22.1)	778 (715–848)
1986	221	Gamma	323 (297–354)	48.8 (1.2)	215.1 (29.4)	776 (712–850)
1987	285	Normal	311 (296–330)	45.9 (1.2)	320.4 (33.0)	747 (710–792)
1987	285	Gamma	329 (305–368)	49.6 (1.6)	461.2 (67.9)	790 (732–884)

[a] Mean and variance are those of the probability distribution fitted to numbers seen each day, relative to May 20 each year. For the gamma distribution, these have been derived from estimates of the three natural parameters.

than-perfect fit to the data is that the estimates, and especially standard errors, reported in Table 2 need to be treated with some scepticism especially for 1981 and 1982.

Although there is no doubt that there has been a significant increase in humpback whale numbers over time, there is no significant difference between the estimates in 1981 and 1982, or in 1986 and 1987. The estimated increase in the 6-year period 1981 to 1987, based on a log-linear regression, is 137%, equivalent to an annual net exponential increase of 14.4%.

4 Discussion

The estimates of total numbers of humpback whales in the eastern Australian group were made by treating the estimated numbers passing Point Lookout as estimates of the total population. In doing so, we assumed that no humpbacks remain in Antarctic waters during the southern winter, none remain in tropical waters during the summer, and all migrating whales go further north than Point Lookout during their northward migration. It has been suggested that some whales may migrate in some years and not others (Matthews 1938; Mackintosh 1943; Dawbin 1956), but Chapman (1974) concluded that sightings of whales in temperate waters at unusual times reflected early or late migration, not a failure to complete migration.

It was assumed that almost all northward migrating whales passed Point Lookout during the survey period. Recently, Paterson and Paterson (1989) have shown that this is not the case, but that approximately 15% of all northward migrating whales do so between late August and the end of October, the time of the peak of the southward migration. The effect will be an underestimate of the total numbers here. However, because we have included the peak of the northward migration, and several weeks either side of it, in each year of observation, the relative numbers (and therefore the estimated increase in size of the population) should be reasonably robust to this departure from the assumptions.

There is little doubt that weather and sea surface conditions affects sightability of whales. Analysis of the influence of these factors on sightability for different years gave conflicting results, however, it was shown that significantly fewer whales were observed under windy and heavily overcast conditions than in clear weather with little or no wind.

We divided weather into fair and poor. Fair weather is defined as clear visibility to the horizon, sea state Beaufort 1 to 4. Poor weather is sea state Beaufort 1 to 3 with heavy haze or rain, or sea state Beaufort 4 with heavy cloud cover or haze, or sea state Beaufort 5 or above. This classification is necessarily subjective, but it does permit a rough comparison of weather conditions between years.

The proportion of time with fair weather was 58% in 1981, 72% in 1982, 60% in 1986, and 56% in 1987. Because the proportions of fair and poor weather were roughly similar between years, we assumed that total numbers of whales can be compared between years.

Although estimates of the population size are of interest, it is of particular importance to know whether the whale population is increasing, and at what rate. This is true from a whale management point of view as well as from the point of view of the value of these whales as monitors of the Antarctic marine system.

Our estimate of net population increase of over 14% seems rather high. Immediately after the 1986 survey, we suspected that the estimate for that year may have been an artefact, possibly due to annual variations in the numbers of whales migrating, variation in routes of migration or immigration from other areas. However, the lack of significant difference in numbers between the 1981 and 1982 and the 1986 and 1987 surveys argues against annual variation in numbers or routes of migration. There are indications from aerial surveys of humpback whales off the coast of Western Australia that the Area IV population is also increasing (Bannister 1986). Therefore, it seems the observed increase in Area V cannot be explained by immigration from Area IV, and immigration of significantly large numbers of whales from other areas is unlikely.

Estimates collated by IWC (1990) of annual rates of increase of whale stocks that have been protected for at least 20 years include five for humpback whale stocks ranging from 4.8% to 13.8%, although doubt was cast on this highest figure. On these grounds, it might be argued that our present estimate, though high, is not necessarily inconsistent with available data. Whether or not this is the case, we do believe that our estimates for 1981 and 1982 are subject to much greater uncertainty than is suggested by the standard errors given in Table 2 because of the high number of missing observations. This level of uncertainty, of course, carries over directly to the estimated rate of increase. Regardless of this uncertainty, it is clear that in Antarctic Area V the population is increasing very rapidly. This suggests among other things that there is a plentiful supply of food (mainly krill) in the feeding grounds of Area V.

Surveys of whale numbers described here seem to be useful general, long-term indicators of the status of specific krill stocks. Other indicators, some of them addressed at this symposium, are more sensitive, and are capable of detecting short-term fluctuations in krill populations.

Acknowledgements. We thank the volunteers, too numerous to mention individually, who helped with the field observations. This work was conducted during Consultancy Agreements between the University of Queensland and the Australian National Parks and Wildlife Service (ANPWS). The support of the Director of the ANPWS is gratefully acknowledged. We thank L. Hicks and B. Jantulik for assistance in the preparation of the manuscript, and are most grateful to an anonymous referee for positive criticism and many valuable suggestions for improvements to the text.

References

Anonymous (1977) Biological investigations of marine Antarctic systems and stocks (BIOMASS). Vol 1, Research Proposals. SCAR/SCOR Publication

Bannister JL (1986) An aerial survey of humpback whales, Western Australia 1986. Aust Nat Parks Wildl Serv, Sci Rep, 8 p

Bryden MM (1985) Studies of humpback whales (*Megaptera novaeangliae*), Area V In: Ling JK, Bryden MM (eds) Studies of sea mammals in south latitudes. South Australian Museum, Adelaide, pp 115–123

Chapman DG (1974) Status of antarctic rorqual stocks. In: Schevill WE (ed) The whale problem, a status report. Harvard University Press, Massachusetts, pp 218–238

Chittleborough RG (1965) Dynamics of two populations of the humpback whale, *Megaptera novaeangliae* (Borowski). Aust J Mar Freshwater Res 16:33–128

Dawbin WH (1956) The migrations of humpback whales which pass the New Zealand coast. Trans R Soc New Zealand 83:147–196

Dawbin WH (1964) Movements of humpback whales marked in the southwest Pacific Ocean 1952 to 1962. Norsk Hvalfangst-tid 53:68–78

Dawbin WH (1966) The seasonal migrating cycle of humpback whales. In: Norris KW (ed) Whales, dolphins and porpoises. University of California Press, Berkeley, pp 145–170

Herman LM, Antinoja R (1977) Humpback whales in the Hawaiian breeding waters: population and pod characteristics. Sci Rep Whales Res Inst Tokyo 29:59–85

IWC (1990) Report of the Scientific Committee. Int Whaling Comm Rep 40 (in press)

Mackintosh NA (1943) The southern stocks of whalebone whale. Discovery Rep 22:199–300

Mackintosh NA (1965) The stocks of whales. Fishing News, London. Cited by DG Chapman (1974). Status of antarctic rorqual stocks. In: Schevill WE (ed) The whale problem, a status report. Harvard University Press, Massachusetts, pp 218–238

Matthews LH (1938) The humpback whales, *Megaptera nodosa*. Discovery Rep 17:9–92

Paterson R, Paterson P (1984) A study of the past and present status of humpback whales in east Australian waters. Biol Conserv 29:321–343

Paterson R, Paterson P (1989) The status of the recovering stock of humpback whales *Megaptera novaeangliae* in east Australian waters. Biol Conserv 47:33–48

Reilly SB, Rice DW, Wolman AA (1983) Population assessment of the gray whale, *Eschrichtius robustus*, from California shore censuses, 1967–80. Fish Bull US 81:267–281

Winn HE, Edel RK, Taruski AG (1975) Population estimate of the humpback whale (*Megaptera novaeangliae*) in the West Indies by visual and acoustic techniques. J Fish Res Board Can 32:499–506

Actual and Potential Fisheries

The Population Biology of the Antarctic Scallop, *Adamussium colbecki* (Smith 1902) at New Harbor, Ross Sea

P. A. BERKMAN [1]

Summary. The unexploited Antarctic scallop, *Adamussium colbecki*, has a circumpolar distribution across the continental shelf from 0 to nearly 1 500 m. During the 1986 austral summer *A. colbecki* was sampled with SCUBA to a depth of 30 m at New Harbor, in the southwestern Ross Sea. Quadrat collections revealed nearshore scallop densities up to 65 m^{-2} with biomasses approaching 2 kg m^{-2}. Mark and recapture experiments showed that *A. colbecki* grew an order of magnitude more slowly than temperate scallop species and analyses of shell growth bands indicated that it may live up to 20 years. Size frequency comparisons with earlier studies in the same area suggested that this scallop population has intermittent recruitment and is quite stable. Estimates of yield per recruit indicate that relatively low levels of fishing pressure could cause the New Harbor scallop population to collapse.

1 Introduction

The endemic Antarctic scallop, *Adamussium colbecki*, is one of the "commonest" bivalves in the Antarctic benthos (Nicol 1966). It was studied for 50 years (Pelseneer 1903; Smith 1902; 1907; 1915; Thiele 1935) before its taxonomic status as a member of the Pectinidae was confirmed (Soot-Ryen 1951). The Antarctic scallop is an "old" species, "from a time of more genial climatic conditions" (Soot-Ryen 1951), which reinvaded shallow water environments after the Holocene deglaciation (Turner 1967; Stuiver et al. 1981). *A. colbecki* presently has a circumpolar distribution and a depth range across the continental shelf from 0 to nearly 1500 m (Dell 1972). It has unusually large thin shells (Nicol 1970), byssal attachment in the dioecious adults (Soot-Ryen 1951), a diverse epizooic assemblage (Mullineaux and DeLaca 1984; Alexander and DeLaca 1987) and reported densities above 85 m^{-2} on the west side of McMurdo Sound (Stockton 1984). Growth and life expectancy estimates, determined indirectly from counting shell bands, suggest that *A. colbecki* is a relatively fast growing Antarctic bivalve species with a life span from 6 (Everson 1977; Ralph and Maxwell 1977) to 13 years (Stockton 1984). *A. colbecki* has been found on sand, hard rock, and gravel substrates (Nicol 1966), with biomasses which can reach 2 kg m^{-2} in coastal environments (Stockton 1984).

As one of the most common bivalves in the Southern Ocean, *Adamussium colbecki* is a potentially exploitable benthic marine living resource. For this reason, it is important to understand the ecology of the Antarctic scallop before it becomes part of a "gold rush" similar to the one which destroyed the unexploited scallop population of *Pecten modestus* in Cockburn Sound, Western Australia (Hancock 1979). Population size, growth, mortality, and recruitment estimates for *Adamussium colbecki* may provide a starting point for quantitativley assessing the potential impacts of exploitation on the Antarctic scallop.

2 Methodology

2.1 Field Collections and Experiments

This study was conducted using SCUBA during the 1986 austral summer at New Harbor, Antarctica (Fig. 1). Scallops were collected within 1 m^2 quadrats which were dropped on the bottom directly beneath the sea ice, at 3 m depth, down to 30 m depth. All scallop shell material, including live scallops and shell fragments, was collected in each quadrat to develop an accurate size frequency profile for the area. Relative scallop mortality was determined from the number of empty scallop shells in each quadrat (Dickie 1955; Merrill and Posgay 1964; Berkman 1988a). A total of four quadrat collections were made at each depth between October 1986 and February 1987.

The scallops from each collection were measured (height, length, and width axes) with vernier calipers accurate to $\pm$ 0.1 mm. The precision of these mea-

[1] Graduate School of Oceanography, University of Rhode Island, Narragansett, Rhode Island 02882-1197, USA; present address: Byrd Polar Research Center, The Ohio State University, Columbus, Ohio 43210, USA.

Antarctic Ecosystems. Ecological Change and Conservation
(ed. by K. R. Kerry and G. Hempel)
© Springer-Verlag Berlin Heidelberg 1990

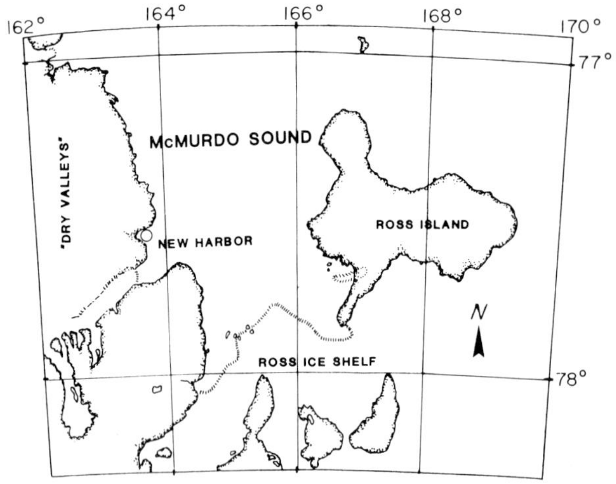

Fig. 1. Study site at New Harbor, Antarctica, on the west side of McMurdo Sound (*circled*)

surements was ± 0.3 mm as determined from the remeasurement of 30 randomly selected scallops.

A scallop mark and recapture experiment (Gibson 1953; Posgay 1963) was started on 27 October 1986. Fifty scallops were collected from 15 m depth and were maintained in large thermos coolers with ice and seawater. During the tagging procedure the scallops were out of the ocean less than 4 h. These scallops, which were 80 ± 10 mm in shell height, had their epizooic species removed with a brush so that they would be similar in size and condition. Each scallop was tagged by drilling a small hole in the upper valve above the byssal notch (Posgay 1963; Berkman 1988a; 1988b), and then inserting a piece of monofilament fishing line attached to an orange plastic spaghetti tag with an identification number (Floy Tags Inc, Washington). These tagged live scallops were released at 6 and 21 m depth and recaptured on 23 January 1987.

2.2 Population Parameter Estimates

Scallop shell bands were used to age the scallops from New Harbor and to determine their incremental growth rates based on the assumption that the narrow striae represented winter feeding and the wider striae represented summer feeding (Merrill et al. 1961; Pikken 1979; Stockton 1984). The upper and lower valves of the translucent thin-shelled Antarctic scallop were placed on a light table with side lighting to digitize (Janden Scientific) the demarcations between narrow and wide shell growth striae. This was done with a precision of ± 1 mm by using a ten-power magnifying lens. Thirty scallops, which were determined to have the same number of growth bands on the upper and lower valves with band widths that did not differ by more than 10%, were used in this analysis.

The growth parameters for *A. colbecki* were derived from the von Bertalanffy growth equation (von Bertalanffy 1938):

$$H_t = H_\infty(1 - e^{-K(t-t_0)}),$$

where the scallop shell height at age t is (H_t), the maximum estimated shell size is (H_∞), the rate at which a species grows toward its maximum size is (K), and the age when H_t equals zero is (t_0). The growth parameters were estimated graphically (Walford 1946) and by mathematical procedures developed by Rafail (1973). The mark and recapture data were not used for these estimates (Fabens 1965) because the growth measurements only represented the larger size classes of *A. colbecki*.

The mortality rate of the New Harbor scallop population was estimated from its size-frequency profile (Gulland 1983):

$$Z = \frac{K(H_\infty - H_{avg})}{H_{avg} - H_c},$$

where the total instantaneous mortality rate coefficient is (Z), the maximum estimated scallop shell height from the von Bertalanffy growth equation is (H_∞), the average shell height is (H_{avg}), and the lower limit of the size class which would be vulnerable to the collection gear (Hoenig 1987) is (H_c).

Absolute recruitment (R) was estimated for the *Adamussium colbecki* population at New Harbor for the depth range from 3 to 30 m, which is where the greatest scallop biomass occurred. Using the procedure described by Gulland (1983):

$$R = N_{avg} Z,$$

where the estimated absolute recruitment is (R) and the mean number of scallops is (N_{avg}). The bathymetry of New Harbor is uncharted, so, for this first approximation of N_{avg} it was assumed that the depths less than 30 m occupy 15% of the 20-km surface area at New Harbor. The yield per recruit model was used to assess the likely long-term effects of different patterns of fishing (Gulland 1983) on the unexploited Antarctic scallop. The procedures for this model were developed by Beverton and Holt (1966):

$$Y/R = FW_\infty e^{-M(t_c-t_r)} \sum_{n=0}^{3} U \frac{e^{-nK(t_c-t_0)}}{F + M + nK},$$

where the steady state annual yield in weight is (Y), the muscle weight of a scallop at H_∞ is (W_∞), the instantaneous fishing mortality rate is (F), the age at recruitment is (t_r), the instantaneous natural mortality is (M), and (U_n) is a summation variable which takes on the values 1, − 3, 3, and − 1 for n = 0 to 3, respectively. In the case of the unexploited Antarctic scallop population at New Harbor, where there is no fishing mortality, natural mortality is equal to total mortality. W_∞ was estimated according to the equation Muscle Weight (grams) = 0.212 Shell Height − 9.04 which was an empirical relationship derived from 240 scallops larger than 60 mm (r = 0.45, $p < 0.01$). The age of scallop recruitment was assumed to be 6 years and was based on the minimum size at the which the scallops began recessing into the sediment after having been bysally attached to larger scallops.

3 Results and Discussion

3.1 Abundances and Size Structure

The most conspicuous aspects of the Antarctic scallop population at New Harbor were its high density (Fig. 2a) and high biomass (Fig. 2b). Both the density and biomass data sets showed significant decreases with depth (r = 0.791 and 0.926, respectively, 23 d.f., $p < 0.01$). The average size (H_{avg}) of 672 live scallops from 3 to 30 m depth at New Harbor was 71.3 with a range from 4.1 to 97.5 mm, and 90% of these were larger than 60 mm (Fig. 3a, b). In addition, it should be noted that the size distribution of *A. colbecki* was depth-dependent, with no scallops smaller than 40 mm present above 10 m depth.

As seen in Fig. 3a, b, the mean sizes were not significantly different between 1981 and 1986 in the depth range from 20–30 m (t = 0.94, 201 d.f., $p > 0.05$). This central tendency compared by the t-test suggests that the contribution of juvenile scallops was offset by the size increases from scallop growth in the larger size classes. Another comparison of these size-frequency distributions indicated that there was no significant heterogeneity between the size classes in 1981 and 1986 ($\chi^2 = 30.41$, 20 d.f., N.S.). In addition, scallop densities and biomasses at the various depths at New Harbor in 1981 (Stockton 1984) were similar to those in 1986. These data suggest that the Antarctic scallop population at New Harbor has remained relatively unchanged during this decade.

Adamussium colbecki thrives at New Harbor despite the relatively oligotrophic nature of West McMurdo Sound (Dayton and Oliver 1977). Benthic primary production (Dayton and Oliver 1977; Dayton et al. 1986) and planktonic primary production (Bunt 1964; Hodson et al. 1981; Dayton et al. 1986; Barry 1988) were extremely low in West McMurdo Sound and comparable to the deep sea. The only anomalous source of organic production in West McMurdo Sound has been observed in the sea ice during years when there is first year ice (Palmisano and Sullivan 1983). The dominance of *Nitizia curta* in the sediments of West McMurdo Sound (Leventer and Dunbar 1988; Dunbar et al. 1989) suggests that the sea ice diatom community was the primary source of organic input to the benthos. Preliminary histology of the digestive tracts from *A. colbecki* revealed diatoms, benthic foraminifera, and sand grains, which suggests that benthic organic material was being utilized by the Antarctic scallop as a food source, as has been suggested for several other scallop species (Shumway et al. 1987). The several year interim between production pulses of new sea ice algae along with the input of these organic materials to the benthos may influence the recruitment and dynamics of the scallops at New Harbor.

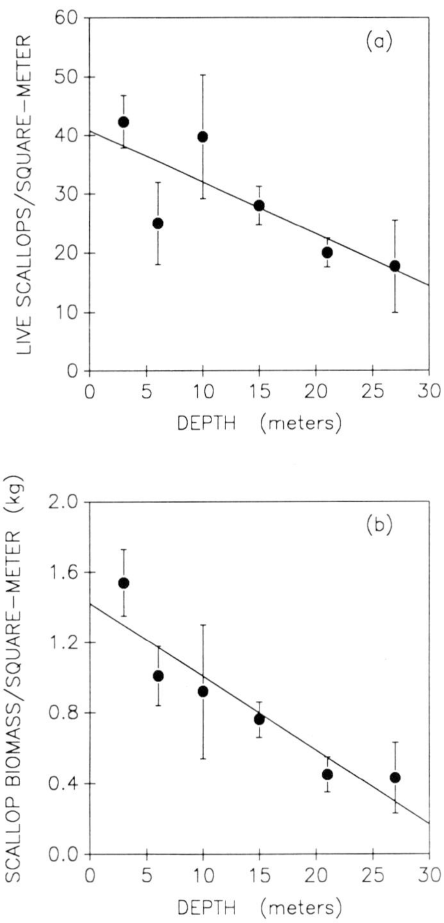

Fig. 2a. Density and **b** biomass of the Antarctic scallop, *Adamussium colbecki,* at New Harbor, Antarctica, in 1986/1987 determined from four quadrat collections at each depth. Means (*circles*) and standard deviations (*vertical lines*) are shown

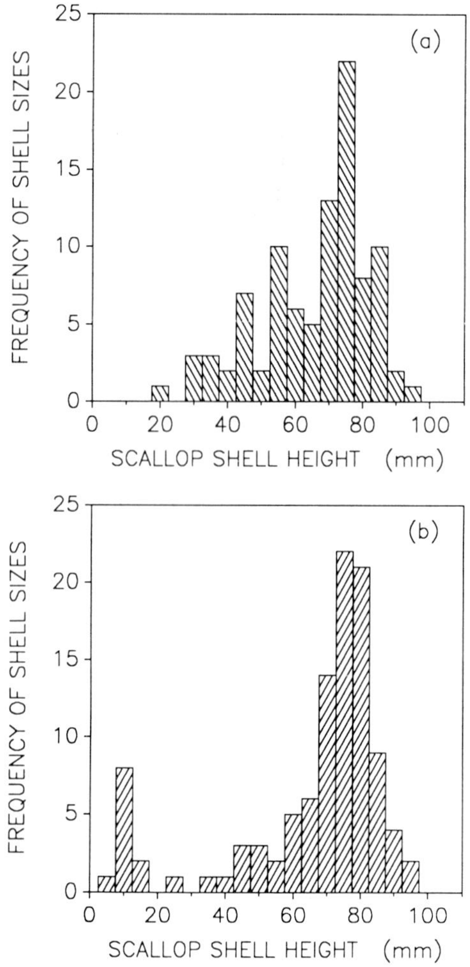

Fig. 3. The size-frequency distributions of the Antarctic scallop, *Adamussium colbecki,* at New Harbor, Antarctica, collected with SCUBA equipment from 20 to 30 m depth. The 1981 profile (**a**) was based on 91 scallops (Stockton 1984) and 1986/1987 profile (**b**) was based on 102 scallops. In both of these studies, small scallops (less than 15 mm in shell height) generally were found byssally attached to larger adults which were recessed into the sediment. The conspicuous absence of small scallops in his SCUBA quadrat collections was noted by Stockton (1984)

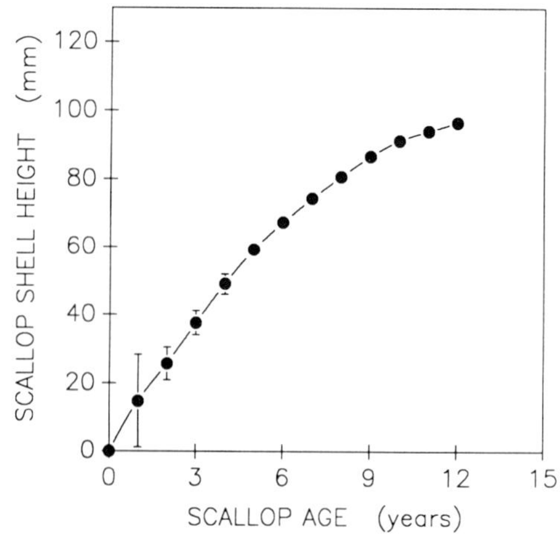

Fig. 4. The estimated ages for various shell heights of the Antarctic scallop, *Adamussium colbecki,* at New Harbor, Antarctica, determined from digitizing the growth bands of 30 scallop shells. Means (*circles*) and standard deviations (*vertical lines*) are shown

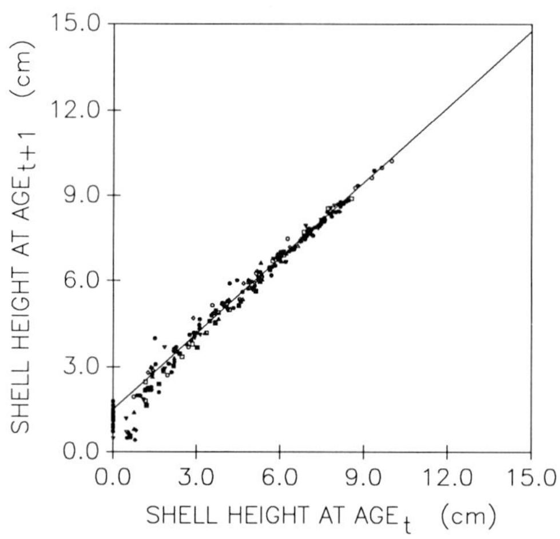

Fig. 5. Walford plot for the Antarctic scallop, *Adamussium colbecki,* at New Harbor, Antarctica determined from digitizing the growth bands of 30 scallop shells that were larger than 80 mm ($r = 0.99$, $p < 0.01$). *Solid line* is the linear regression of the incremental growth data determined from shell bands. The predicted maximum height occurs at the intersection with a 45° line from the origin

3.2 Growth

The age versus shell height profile of *A. colbecki* from this study (Fig. 4) indicates that a 98-mm scallop was 12-years-old. The high size variance at year 1 may be due to sigmoid growth during juvenile scallop development (Yamaguchi 1975). These data support the claim that *Adamussium colbecki* lives at least 13 years (Stockton 1984).

However, Stockton (1984), along with Ralph and Maxwell (1977), did not predict a maximum size larger than the largest recorded Antarctic scallop. That specimen of *A. colbecki,* which was collected during this study, was 108 mm in height by 120 mm in length by 38 mm in width (Smithsonian Institution, USNM 859310). The H_∞ of 128 mm predicted in this study (Fig. 5) suggests that the Antarctic scallop may live for 20 years. This seems to be a realistic maximum age considering that the Arctic scallop, *Chlamys islandica,* lives at least 23 years (Vahl 1981).

The growth parameter estimates from this study are compared with those from other cold water scallop species in Table 1. The low K value for *A. colbecki*

Table 1. Comparison of scallop species growth parameters

Scallop species	Growth parameters[a]			
	K	H_{max}	t_0	Age (years)
Adamussium colbecki (this study)	0.09	128	− 0.765	> 20
Adamussium colbecki (Ralph and Maxwell 1977)	0.24	90	0.150	> 6
Adamussium colbecki (Stockton 1984)	0.12	105	–	> 13
Patinopecten courninus (Haynes and Hitz 1971)	0.35	160	0.720	> 14
Placopecten magellanicus (MacDonald and Thompson 1985)	0.17	146	− 0.880	> 18
Pecten maximus (Mason 1983)	0.39	146	0.240	> 12

[a] Estimates based on scallop shell height axis.

indicates that it was growing relatively slowly toward its maximum size.

The combined summer daily growth rate for the 25 tagged scallop which were recapture alive at New Harbor, was 0.017 ± 0.019 mm-day, with a range from 0.099 mm-day to less than 0.005 mm-day. *A. colbecki* grows relatively fast for an Antarctic bivalve (Everson 1977), but much more slowly than temperate species. For example, a similar mark and recapture study, conducted with 78 scallops over a comparable period in a temperate coastal lagoon (Berkman 1986), showed that the bay scallop, *Argopecten irradians*, grows an order of magnitude faster at 0.14 ± 0.06 mm-day ($t = 10.99$, 106 d.f., $p < 0.01$). The results of the mark and recapture experiment also showed that *A. colbecki* formed wider shell growth striae during the summer.

Thirteen of 25 tagged scallops at 6 m depth died compared to 3 of 25 at 21 m depth ($\chi^2 = 5.68$, 1 d.f., $p < 0.05$). This difference may have been caused by the relative proximity to the hyposaline lens which develops during the late summer from the melting of the sea ice (Stockton 1984). A similar trend was noted for the empty scallop shells where the relative dead : live scallop ratio varied from 0.18 at 3 m to 0.07 at depths below 15 m.

Scallop predators were rare at New Harbor (Stockton 1984), and of 55 dives made during this study, predatory starfish (*Notasterias armata* and *Lophaster gaini*) only were seen eating scallops once each. The paucity of shell fragments in the natural population and the absence of broken shells among the dead tagged scallops suggests further that predation was rare at New Harbor.

3.3 Mortality

Given the common occurrence of byssally attached juveniles on the adults, it is assumed that these small scallops would be collected by conventional collecting gear and that H_c would be 0 years. This give a total annual instantaneous mortality rate of 0.071 for all scallop sizes. This produces a mean annual total mortality rate of 0.068 for *A. colbecki*, which is low compared to other unexploited scallop populations. The total mortality rate of *Placopecten magellanicus* near Newfoundland was 0.095 (MacDonald and Thompson 1986) and of *Pecten maximus* in the North Irish Sea approached 0.26 (Baird 1966; Gruffydd 1974). It should be noted that the total annual mortality rate for *A. colbecki* does not distinguish mortality variations with age as discussed by Gruffydd (1974). The low total annual mortality of *A. colbecki* would be reflective of a population dominated by large individuals (Gulland 1987).

3.4 Recruitment

The mean scallop density, determined from Fig. 2a, was 30.25 ± 13.20 scallop m^{-2} in the depth range from 3 to 30 m. For the 3 km^2 area at New Harbor, these values and the total instantaneous mortality rate of 0.071 produce a total annual recruitment estimate for *A. colbecki* of $6.4 \pm 2.8 \times 10^6$ recruits under the assumption of constant recruitment which is being used for this first approximation.

The presence of juvenile scallops reflects the distribution and survival patterns of the planktotrophic larvae of *Adamussium colbecki* (Berkman et al. in preparation). Thousands of juvenile scallops (less than 15 mm) were seen at New Harbor in the middle 1970's (Dayton and Oliver 1977). In 1981, Stockton (1984) noted the relative rarity of the smaller size classes in his SCUBA surveys of the New Harbor scallop population. During this study juvenile scallops were encountered again (Fig. 3a, b). It is unlikely that the small scallops less than 20 mm in shell height were present and not collected in 1981 when compared to 1986 by a 2 × 2 contingency test ($p < 0.0007$). The intermittent appearance of juvenile scallops at New Harbor may have been influenced by the biology of *A. colbecki* or by environmental variability, such as the production pulses of first year sea ice algae at New Harbor (Palmisano and Sullivan 1983; Dunbar et al. 1989).

3.5 Yield per Recruit Model

Assuming it would be possible to harvest only the larger scallops, the highest yield would occur when t_c was 10 years or greater (Fig. 6a). However, at New Harbor, where the scallops account for nearly 90% of the hard substrate (Mullineaux and DeLaca 1984), it would not be possible to have a size selective fishery for *Adamussium colbecki* since the byssally attached juvenile scallops would be captured along with the adults by conventional scallop dredges. This type of gear would result in a t_c which approaches 0 years. For $t_c = 0$ years, the instantaneous fishing mortality rate (F) which produces the greatest yield would occur between 0.05 and 0.10. Moreover, as a consequence of an age of first capture approaching 0 years, there would be a precipitous decrease in yield for *A. colbecki* beyond F = 0.10. These values of F for *A. colbecki* are low compared to those estimated for the maximum yield per recruit of the sea scallop, *Placopecten magellanicus*, which are between 0.2 and 0.3 (Serchuk et al. 1979). This yield per recruit model for *A. colbecki* demonstrates that the Antarctic scallop population at New Harbor could not sustain a fishery, except at extremely low levels of fishing mortality.

The yield per recruit standard deviation for *Adamussium colbecki*, shown in Fig. 6b, was derived from the estimated recruitment standard deviation (discussed above) and should be viewed as being representative of the unknown or poorly understood parameters in this yield per recruit model. This model could be influenced by intermittent recruitment (Fig. 3a, b; Getz and Swartzman 1981), the estimated age of recruitment, the growth parameter estimates, the stock recruitment relationship (Cushing 1971), or variability in the distribution of *Adamussium colbecki*.

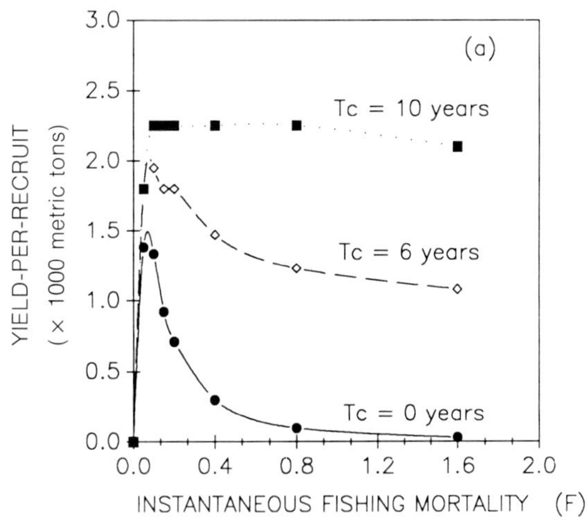

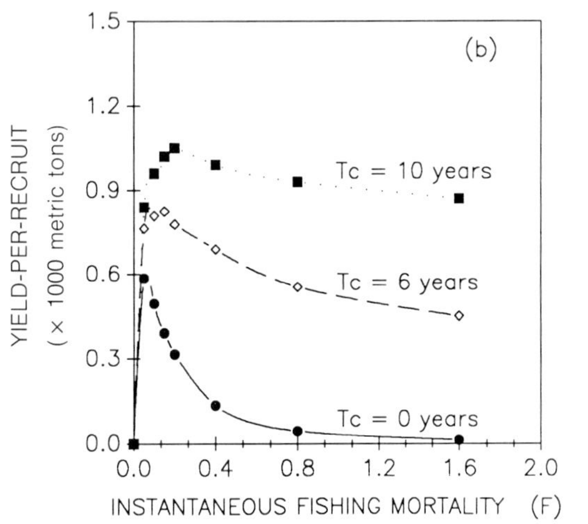

Fig. 6. The yield per recruit model for the unexploited Antarctic scallop, *Adamussium colbecki*, from New Harbor, Antarctica, in 1986/87. **a** Scallop yield per recruit means and **b** their standard deviations (as discussed in the text) are shown. The *curves* reflect the effects of increasing fishing intensity at different ages of first capture (t_c)

4 Conclusions

The Antarctic scallop, *Adamussium colbecki*, lives in an environment of limited organic production and appears to grow relatively slowly compared to temperate scallop species. At New Harbor, the size-frequency distribution of *A. colbecki* can be characterized by the large adults which had low mortality and the recruitment that appeared to be intermittent. Harvesting the Antarctic scallop with conventional dredges at New Harbor would not only destroy the fragile scallop shells in the process, but it would perturb the oligotrophic habitat and disrupt the apparently stable population of *A. colbecki*. Moreover, the yield per recruit model for *A. colbecki* at New Harbor demonstrates that this Antarctic scallop population may collapse even under low fishing pressure.

Adamussium colbecki is an unexploited scallop species that has been identified by the Food and Agriculture Organization of the United Nations as a potential commercial resource (Fischer and Hureau 1985). In addition to New Harbor, high density populations have been observed by the Italians at Terra Nova Bay (G Diprisco, pers. commun. 1988) and by the Japanese at Syowa Station (T Hoshiai, pers. commun. 1988). For these reasons, it is important to remember:

The California sardine fishery: a monument to the failure to act in time, and to the insistence on having conclusive scientific evidence before acting.

(Gulland 1974)

Conservation of the unexploited Antarctic scallop requires the collection of more baseline information on its life cycle, distribution, size-specific rates and population dynamics, habitat requirements at the various growth stages, and geologic history. As one of the most common benthic invertebrates (Nicol 1966) in the Antarctic continental region (Hedgpeth 1969), *Adamussium colbecki* provides an excellent focus for coordinating and comparing research results among nations working at various locations within its circumpolar range. The development of management guidelines for the Antarctic scallop, before harvesting ever occurs, is a rare conservation opportunity.

Acknowledgements. This work was undertaken as partial fulfillment of the requirements for the degree of doctor of philosophy. This research was supported by a National Science Foundation grant, DPP 80-34332, to T. E. DeLaca, an equipment grant from the U.S. Antarctic Marine Living Resource Program to P. A. Berkman, and a National Science Foundation Graduate Fellowship to P. A. Berkman. I especially would like to thank Ted DeLaca for providing the opportunities to study the Antarctic marine environment and its species and my major professor, Saul Saila, for his constructive suggestions. The diving assistance and advice from WL Stockton, I Bosch, JM Bernhard and SP Alexander is appreciated.

References

Alexander SP, DeLaca TE (1987) Feeding adaptations of the foraminiferan *Cibicides refulgens* living epizooically and parisitically on the Antarctic scallop *Adamussium colbecki*. Biol Bull 173:136–159

Baird RH (1966) Notes on an scallop (*Pecten maximus*) population in Holyhead Harbor. J Mar Biol Assoc UK 46:33–47

Barry JP (1988) Hydrographic patterns in McMurdo Sound, Antarctica and their relationship to local biotic communities. Polar Biol 8:377–391

Berkman PA (1986) Ecological relationships between the bay scallop, *Argopecten irradians*, and its epizooic assemblage in Charlestown Pond, Rhode Island, Unpubl Masters Thesis, University of Rhode Island, Narragansett, p 131

Berkman PA (1988a) Ecology of the Circumpolar Antarctic Scallop, *Adamussium colbecki* (Smith, 1902). Doctoral Dissertation, University of Rhode Island, Narragansett, p 215

Berkman PA (1988b) Sea scallop tagging experiments using manned-submersibles. In: Babb I, DeLaca M (eds) National Undersea Research Program Research Report 88-3, Connecticut, pp 231–243

Beverton RJH, Holt SJ (1966) Manual of methods for fish stock assessment. Part II – Tables of yield functions. FAO Fish Tech Pap 38:1–10

Bunt JS (1964) Primary productivity under sea ice in Antarctic waters. I. Concentrations and photosynthetic activities of microalgae in the waters of McMurdo Sound, Antarctica. Antarct Res Ser I:13–26

Cushing DH (1971) The dependence of recruitment on parent stock in different groups of fishes. J Cons Int Explor Mer 33:340–362

Dayton PK, Oliver JS (1977) Antarctic soft-bottom benthos in oligotrophic and eutrophic environments. Science 197:55–58

Dayton PK, Watson D, Palmisano A, Barry JP, Oliver JS, Rivera D (1986) Distribution patterns of benthic microalgal standing stock at McMurdo Sound, Antarctica. Polar Biol 6:207–213

Dell RK (1972) Antarctic benthos. Adv Mar Biol 10:1–216

Dickie LM (1955) Fluctuations in abundance of the giant scallop *Placopecten magellanicus* (Gmelin), in the Digby area of the Bay of Fundy. J Fish Res Bd Can 12:797–857

Dunbar RB, Leventer AR, Stockton WL (1989) Biogenic sedimentation in McMurdo Sound, Antarctica. Mar Geol 85:155–179

Everson I (1977) Antarctic marine secondary production and the phenomenon of cold adaptation. Phil Trans R Soc Ser B 279:55–66

Fabens AJ (1965) Properties and fitting of the von Bertalanffy growth curve. Growth 2:265–269

Fischer W, Hureau JC (1985) FAO species identification sheets for fisheries purposes. Food and Agriculture Organization of the United Nations, Rome, pp 95–105

Getz WM, Swartzman GL (1981) A probability transition matrix model for yield estimation in fisheries with highly variable recruitment. Can J Fish Aquat Sci 38:847–855

Gibson FA (1953) Notes on the tagging of the escallop (*P. maximus* L.) in Irish water. J Cons Int Explor Mer 19:204–208

Gruffydd LD (1974) An estimate of natural mortality in an unfished population of the scallop *Pecten maximus* (L.). J Cons Int Explor Mer 35:209–210

Gulland JA (1974) The management of marine fisheries. University of Washington Press, Seattle, p 198

Gulland JA (1983) Fish stock assessment. John Wiley & Sons, New York, p 223

Gulland JA (1987) Length-based methods in fisheries research: from theory to application. In: Pauly D, Morgan GR (eds) Length based methods in fisheries research, ICLARM Conf Proc 13, Kuwait, pp 335–342

Hancock DA (1979) Population dynamics and management of shellfish stocks. Rapp PV Cons Int Explor Mer 175:8–19

Haynes EB, Hitz CR (1971) Age and growth of the giant Pacific sea scallop *Patinopecten courinus*, from the Strait of Georgia and outer Washington coast. J Fish Res Bd Can 28:1335–1341

Hedgpeth JW (1969) Introduction to Antarctic zoogeography. In: Bushnell VC, Hedgpeth JW (eds) Antarctic Map Folio Series, Folio 11. American Geophysical Union, New York, pp 1–9

Hodson RE, Azam F, Carlucci AF, Fuhrman JA, Karl DM, Holm-Hansen O (1981) Microbial uptake of dissolved organic matter in McMurdo Sound, Antarctica. Mar Biol 61:89–94

Hoenig JM (1987) Estimation of growth and mortality parameters for use in length-structured stock production models. In: Pauly D, Morgan GR (eds) Length based methods in fisheries research, ICLARM Conf Proc 13, Kuwait, pp 121–128

Leventer A, Dunbar RB (1988) Recent diatom record of McMurdo Sound, Antarctica: implications for history of sea ice extent. Paleocean 3:259–274

MacDonald BA, Thompson RJ (1985) Influence of temperature and food availability of the giant scallop *Placopecten magellanicus*. I. Growth rates of shell and somatic tissues. Mar Ecol Prog Ser 25:279–294

MacDonald BA, Thompson RJ (1986) Production, dynamics and energy partitioning in two populations of the giant scallop *Placopecten magellanicus* (Gmelin). J Exp Mar Biol Ecol 101:285–299

Mason J (1983) Scallop and queen fisheries in the British Isles. Fishing New Books Ltd, England, p 144

Merrill AS, Posgay JA, Nichy F (1961) Annual marks on shell and ligament of sea scallop (*Placopecten magellanicus*). Fish Bull 65:299–311

Merrill AS, Posgay JA (1964) Estimating the natural mortality rate of the sea scallop. ICNAF Res Bull 1:88–99

Mullineaux LS, DeLaca TE (1984) Distribution of Antarctic benthic foraminifera settling on the pecten *Adamussium colbecki*. Polar Biol 3:185–189

Nicol D (1966) Descriptions, ecology and geographic distribution of some Antarctic pelecypods. Bulls Am Paleo 51:1–102

Nicol D (1970) Antarctic pelecypod faunal peculiarities. Science 168:1248–1249

Palmisano AC, Sullivan CW (1983) Sea ice microbial communities (SIMCO). 1. Distribution, abundance, and primary production of ice microalgae in McMurdo Sound, Antarctica in 1980. Polar Biol 2:171–177

Pelseneer P (1903) Mollusques (Amphineura, Gastropods, et Lamellibranches). Res Voyage du SY Belgica, p 85

Picken GB (1979) Growth, production and biomass of the Antarctic gastropod *Laevilacunaria antarctica* Martens 1885. J Exp Mar Biol Ecol 40:71–79

Posgay JA (1963) Tagging as a technique in population studies of the sea scallop. ICNAF Spec Pub 4:268–271

Rafail SZ (1973) A simple and precise method for fitting a von Bertalanffy growth curve. Mar Biol 19:354–358

Ralph R, Maxwell JGH (1977) Growth of two Antarctic lamellibranchs: *Adamussium colbecki* and *Lanternula eliptica*. Mar Biol 42:171–175

Serchuk FM, Wood PW, Posgay JA, Brown BE (1979) Assessment and status of sea scallop (*Placopecten magellanicus*) populations off the northeast coast of the United States. Proc Natl Shellfish Assoc 69:161–191

Shumway SE, Selvin R, Schick DF (1987) Food resources related to habitat in the scallop *Placopecten magellanicus* (Gmelin, 1791): a qualitative study. J Shell Res 6:89–95

Smith EA (1902) Mollusca. Southern Cross Expedition. Brit Mus Nat Hist, England, pp 201–213

Smith EA (1907) Lamellibranchiata. Natl Antarct Exped II: 1–7

Smith EA (1915) Mollusca. Part I. Gastropoda, Prosobranchia, Scaphopoda and Pelecypoda. Br Antarct (Terra Nova) Exped II:6–112

Soot-Ryen T (1951) Antarctic pelecypods. Sci Res Norw Antarc Exped 3:1–46

Stockton WL (1984) The biology and ecology of the epifaunal scallop *Adamussium colbecki* on the west side of McMurdo Sound, Antarctica. Mar Biol 78:171–178

Stuiver M, Denton GH, Hughes TJ, Fastook JL (1981) History of the marine ice sheet in West Antarctica during the last glaciation: a working hypothesis. In: Denton GH, Hughes TJ (eds) The last great ice sheets. John Wiley, New York, pp 319–346

Thiele J (1935) Handbuch der systematischen Weichtierkunde II. Jena Verlag von Gustav Fisher, Berlin, pp 780–1154

Turner RD (1967) A new species of fossil *Chlamys* from Wright Valley, McMurdo Sound, Antarctica. NZ J Geol Geophys 10:446–455

Vahl O (1981) Energy transformations by the Iceland scallop, *Chlamys islandica* (O. F. Muller), from 70° N. I. The age-specific energy budget and net growth efficiency. J Exp Mar Biol Ecol 53:281–296

von Bertalanffy L (1938) A quantitative theory of organic growth. Hum Biol 10:181–213

Walford LA (1946) A new graphic method of describing the growth of animals. Biol Bull 90:141–147

Yamaguchi M (1975) Estimating growth parameters from growth rate data. Oecologia 20:321–332

Cephalopod Fauna of the Scotia Sea at South Georgia: Potential for Commercial Exploitation and Possible Consequences

P. G. RODHOUSE [1]

Summary. A collection of cephalopods from the British Antarctic Survey's Offshore Biological Programme is described and the cephalopod prey of vertebrate predators at South Georgia is reviewed. Comparison of these data indicates that predators catch larger specimens and a greater diversity of species than nets. There are also differences between samples from different types of net. The RMT 25, the largest research net used to date, has caught most of the species thought to occur in the Scotia Sea but specimens are generally smaller than those taken by predators. Cephalopods which are thought to have potential commercial value are *Martialia hyadesi, Kondakovia longimana, Moroteuthis ingens, M. knipovitchi, M. robsoni* and *Gonatus antarcticus*. Other possibilities include species of brachioteuthid, psychroteuthid and neoteuthid. It is likely that Antarctic stocks will be sensitive to exploitation and liable to dramatic fluctuations if overfished. The possible consequences of commercial exploitation of cephalopods for the reproductive success of the vertebrate predators, which prey on cephalopods in the Scotia Sea, are examined.

1 Introduction

The Southern Ocean contains large stocks of cephalopods, of which seabirds, seals and whales together consume some 34 million tonnes per year (Clarke 1983). The biomass and total annual production are unknown and probably much higher. Total annual cephalopod consumption by predators in the Scotia Sea is estimated to be 3.7 million tonnes (Croxall et al. 1985), of which nearly 0.5 million tonnes is taken by seabirds breeding at South Georgia (Croxall and Prince 1987). It has been suggested that Antarctic stocks of cephalopods have increased this century as a direct result of the decrease in numbers of their principal consumer, the sperm whale, through overexploitation, and as an indirect result of overexploitation of baleen whales, and the consequent release of cephalopods from competition for the common food resource, *Euphausia superba* (Laws 1985).

Although no directed cephalopod fishery exists in the Scotia Sea, or elsewhere in the Antarctic, the potential for commercial exploitation of these stocks is recognized (Voss 1973; Everson 1977) and major cephalopod fisheries already exist in the cool temperate seas of the Southern Ocean around the Falkland Islands and New Zealand (Csirke 1987; Sato and Hatanaka 1983).

Cephalopods readily escape conventional scientific sampling gear but the indigestible cephalopod mandibles, which accumulate in the guts of their predators, can yield a considerable amount of information about the species, size and contribution to diet of cephalopods consumed (Clarke 1983). Where samples of cephalopods from nets and predator gut contents from the same sea area have been compared (Clarke 1977) there are often differences in both the species composition and the sizes of specimens caught. Data from both nets and predators may thus yield complementary information about the cephalopod community in a particular area.

In this chapter, new data on cephalopods from net hauls in the vicinity of South Georgia are presented and data on cephalopods from the gut contents of predators from the area are reviewed for comparison. Species are identified which, on the basis of comparative abundance and suitability for human consumption, could form the focus of a future cephalopod fishery. The possible effects of commercial exploitation on these stocks, and on the predator populations which depend on them, are considered.

2 Materials and Methods

2.1 Research Nets

Research nets have been deployed in the vicinity of South Georgia each year between 1978 and 1987 from the British Antarctic Survey's research ship, RRS

[1] British Antarctic Survey, High Cross, Madingley Road, Cambridge CB3 0ET, UK.

Antarctic Ecosystems. Ecological Change and Conservation
(ed. by K. R. Kerry and G. Hempel)
© Springer-Verlag Berlin Heidelberg 1990

"John Biscoe" as part of the Survey's Offshore Biological Programme (OBP).

In January 1987, 12 days were devoted to fishing specifically for cephalopods with a large Rectangular Midwater Trawl (RMT 25) (Baker et al. 1973). This net was operated in the vicinity of 53° S, 35° W, where the depth was > 3 500 m and 54° S, 35° 20′ W, the approximate position of the 1000 m bathymetric contour. The net was operated by dropping it to its fishing depth, towing obliquely downwards through a 200-m layer at an average speed of 2.28 ± 0.26 knots and hauling. Each layer from the surface to 2000 m was sampled for 3 h at the deeper site, and each layer from the surface to 1000 m was sampled for 2 h at the shallower site. Each layer at each site was sampled twice: once during daylight and once during darkness. The net was deployed for a total of 96.6 h, including lowering and hauling time. This sampling strategy was adopted in order to sample the water column as evenly as possible and it is assumed that few cephalopods were caught during lowering and hauling the net. However, the net was operated open throughout so that no reliable data on vertical distribution of cephalopods were obtained.

Benthic cephalopods were collected in January 1987 with an Agassiz trawl during 24 bottom hauls, of approximately 15 min each, at Cumberland East Bay, and along three transects on the shelf to the north and east of South Georgia.

During OBP cruises between 1978 and 1987, within the South Georgia survey zone (Priddle et al. 1986), cephalopods were collected from the catch of all hauls made with the opening-closing multiple rectangular midwater trawl (RMT 8 + 1) (Roe and Shale 1979). This represents the catch from a total of 684.5 h operating time with this net system. Occasional specimens were also caught in a frame net (mesh size 5 mm) fished from the foredeck, a high-speed tow net (mesh size 0.35 mm), a neuston net (mesh size 1 mm) and a large Longhurst-Hardy Plankton Recorder (Bone 1986). All specimens of cephalopods were either fixed in 5% formalin solution or 70% ethanol.

3 Results

3.1 Cephalopods from Research Nets

A complete list of the cephalopod species and total numbers of specimens caught during the OBP, between 1978 and 1987, are given in Table 1. The most effective sampling tool to date has been the RMT 25 which has caught more cephalopods with less fishing effort, and more species, than the other pelagic nets. The RMT 25 has also caught larger specimens of most species than the RMT 8 and RMT 1 (Table 2). The most common species from pelagic net hauls, comprising over 90% of the collection, were *Brachioteuthis-?picta, Galiteuthis glacialis, Bathyteuthis abyssicola* and *Mastigoteuthis psychrophila*.

From the Agassiz trawl samples, two species of octopodid, *Pareledone charcoti* and *P. polymorpha*, could be identified on the basis of the hectocotylised arm of the males (G. Voss pers. commun.). Because of the difficulty of distinguishing between the females of the two species a subsample of 82 specimens from the whole sample were pooled; these had a mean weight of 78.7 ± 46.6 g.

The RMT 25 caught more species but fewer specimens when fished close to the shelf at South Georgia than over deep water to the northeast of the island. Of the four common species caught by this net, it would appear that *Bathyteuthis abyssicola,* which is known to have a deep-sea distribution (Roper 1969), is less abundant close to the shelf. The small differences between the abundance of *Brachioteuthis?picta, Galiteuthis glacialis* and *Mastigoteuthis psychrophila* between the two areas suggest that they are probably evenly distributed between the near-shelf and offshore waters.

Collections of cephalopods caught by the other nets over the 10-year period of the OBP are too small to attempt an analysis of temporal or spatial distribution.

From the data collected with the open RMT 25 it should be noted that only 6.3% of all cephalopods caught with this net were taken in hauls in the top 200 m of the water column and these were all small (< 60 mm mean mantle length). Of the total time spent fishing with the RMT 25, 9.1% was spent sampling the top 200 m, excluding lowering and hauling time for the deeper hauls. Clearly the majority of the cephalopods, and certainly the larger specimens, caught by this net were taken in deeper water beneath the surface layer.

3.2 Comparison of Cephalopods Taken by Nets with Data from Predators

There are detailed data in the literature on the cephalopod prey of five species of vertebrate predator from South Georgia: the wandering albatross, *Diomedea exulans* (Clarke et al. 1981; Rodhouse et al. 1987; Rodhouse et al; in press), the grey-headed and black-browed albatrosses, *D. chysostoma* and *D. melanophris* (Clarke and Prince 1981), the light-mantled sooty albatross, *Phoebetria palpebrata* (Thomas

Table 1. Cephalopods caught by nets in the seas around South Georgia during the British Antarctic Survey's offshore Biological Programme, 1978–1987. Mouth area of net given except in the case of Aggasiz trawl where beam width is given

Cephalopod family and species	RMT 25 $25\ m^2$		RMT 8 $8\ m^2$	RMT 1 $1\ m^2$	Foredeck net $1\ m^2$	High-speed tow net $0.15\ m^2$	Neuston net $0.45\ m^2$	Longhurst Hardy plankton recorder $0.5\ m^2$	Agassiz trawl 2.75 m
	1000 m	> 3500 m							
Gonatidae									
Gonatus antarcticus	2	–	5	–	–	–	–	–	–
Onychoteuthidae									
Kondakovia longimana	1	–	5	–	–	–	–	–	–
Batoteuthidae									
Batoteuthis skolops	1	–	–	–	–	–	–	–	–
Brachioteuthidae									
Brachioteuthis?picta	29	74	198	5	5	2	3	1	–
Histioteuthidae									
Histioteuthis eltaninae	1	–	–	–	–	–	–	–	–
Bathyteuthidae									
Bathyteuthis abyssicola	3	35	20	–	–	–	–	–	–
Psychroteuthidae									
Psychroteuthis glacialis	–	2	3	–	–	–	–	–	–
Neoteuthidae									
Alluroteuthis antarcticus	–	5	2	–	–	–	–	–	–
Chiroteuthidae									
Chiroteuthis sp.	–	–	1	–	–	–	–	–	–
Mastigoteuthidae									
Mastogoteuthis psychrophila	23	9	–	–	–	–	–	–	–
Cranchiidae									
Galiteuthis glacialis	33	49	106	2	1	–	–	–	–
Mesonychoteuthis hamiltoni	–	–	3	3	–	–	–	–	–
Taonius pavo	1	–	–	–	–	–	–	–	–
Octopodidae									
Cirrate octopodid	3	–	1	–	–	–	–	–	–
Pareledone sp.	–	–	–	–	–	–	–	–	148

Table 2. Comparison of sizes (mean mantle length, mm ± SD, n) of cephalopods caught by RMT 25, 8 and 1 as set out in Table 1. Range given where $n < 3$ or where sample does not approximate a normal distribution

Cephalopod family and species	RMT 25	RMT 8	RMT 1
Gonatidae			
Gonatus antarcticus	365.0 ± 41.0 (2)	18.8 ± 2.1 (5)	–
Brachioteuthidae			
Brachioteuthis?picta	40.2 ± 12.4 (103)	29.0 ± 11.3 (198)	21.3 ± 7.3 (5)
Bathyteuthidae			
Bathyteuthis abyssicola	35.7 ± 14.5 (38)	36.8 ± 9.1 (20)	–
Psychroteuthidae			
Psychroteuthis glacialis	9.8 – 10.4 (2)	23.6 ± 9.0 (3)	–
Neoteuthidae			
Alluroteuthis antarcticus	28.04 ± 14.7 (5)	10.0 ± 0.0 (2)	–
Cranchiidae			
Galiteuthis glacialis	6.5 – 425.0 (82)	14.5 ± 11.3 (106)	30.6 ± 25.4 (2)

Table 3. Cephalopods in the gut contents and regurgitations of vertebrate predators at South Georgia. Sources of data are given in the text

Cephalopod family and species	Occurrence south of Sub-Antarctic front	Wandering albatross	Grey-headed albatross	Black-browed albatross	Light-mantled sooty albatross	Sperm whale
Loliginidae						
Loligo sp.	–	+	–	–	–	–
Gonatidae						
Gonatus antarcticus	+	+	–	–	+	+
Gonatus sp. A	+	–	–	+	+	–
Enoploteuthidae						
Ancistrocheirus lesueuri	–	+	–	+	–	+
(?) *Enoploteuthis* sp.	–	–	–	+	–	–
Octopoteuthidae						
Octopoteuthis rugosa	–	+	–	–	–	–
Taningia danae	–	+	–	–	–	+
Onychoteuthidae						
Kondakovia longimana	+	+	+	+	+	+
Moroteuthis knipovitchi	+	+	–	–	–	+
Moroteuthis ingens	+	+	–	–	–	–
Moroteuthis robsoni	+	+	–	–	–	+
Onychoteuthis banksi	–	+	–	–	–	–
Pholidoteuthidae						
Pholidoteuthis boschmai	–	+	–	–	–	+
Batoteuthidae						
Batoteuthis skolops	+	+	–	–	–	–
Lycoteuthidae						
Lycoteuthis sp.	–	–	–	–	+	–
Histioteuthidae						
Histioteuthis sp. A	–	+	–	–	–	+
Histioteuthis eltaninae	–	+	–	–	+	+
Histioteuthis (?) eltaninae	–	–	+	–	–	–
Histioteuthis atlantica	–	+	–	–	–	–
Psychroteuthidae						
Psychroteuthis glacialis	+	+	–	–	–	+
? Large *Psychroteuthis*	+	+	–	–	+	–
Neoteuthidae						
Alluroteuthis antarcticus	+	+	–	–	+	+
Architeuthidae						
Architeuthis sp.	–	+	–	–	–	+
Ommastrephidae						
Illex sp.	–	+	–	–	–	–
Martialia hyadesi	+	+	+	+	+	–
Todarodes sp.	–	–	–	–	–	+
Chiroteuthidae						
Chiroteuthis imperator	–	+	–	–	–	–
Chiroteuthis sp. (large)	–	+	–	–	–	–
Chiroteuthis sp. (small)	+	+	–	–	–	+
Chiroteuthis sp. A	–	+	–	–	–	–
Chiroteuthis (?) veranyi	–	–	+	+	–	–
Lepidoteuthidae						
Lepidoteuthis grimaldi	–	+	–	–	+	+
Mastigoteuthidae						
Mastogoteuthis sp. (large)	–	+	–	–	–	–
Mastogoteuthis sp. (small)	–	+	+	–	–	–
Mastogoteuthis sp. A	+	+	–	–	–	+

Table 3 (continued)

Cephalopod family and species	Occurrence south of Sub-Antarctic front	Wandering albatross	Grey-headed albatross	Black-browed albatross	Light-mantled sooty albatross	Sperm whale
Cycloteuthidae						
Discoteuthis sp. (large)	–	+	–	–	–	–
Discoteuthis sp. (small)	–	+	–	–	–	–
(?) *Discoteuthis* sp.	–	+	–	–	+	–
Cranchiidae						
Liocranchia sp.	–	–	–	–	–	+
Egea inermis	–	+	–	–	–	–
Galiteuthis glacialis	+	+	+	+	+	+
Galiteuthis sp.	–	+	–	–	–	–
Helicocranchia sp.	–	+	–	–	–	–
Megalocranchia sp.	–	+	–	–	–	–
Mesonychoteuthis hamiltoni	+	+	–	–	–	+
Taonius pavo	+	+	–	–	–	+
Taonius megalops	–	–	–	–	–	+
Taonius sp.	–	+	–	–	–	–
Taonius sp. B (small)	–	+	–	–	–	–
Taonius sp. B (southern)	–	+	–	–	–	–
Teuthowenia sp.	–	+	–	–	–	–
Alloposidae						
Alloposus mollis	–	+	–	–	–	+
Octopodidae						
Large *octopodid*	–	+	–	–	–	–
Small *octopodid*	–	+	–	–	–	–
(?) *Pareledone*	–	–	–	+	–	–
Vampyromorphidae						
Vampyroteuthis infernalis	–	+	–	–	–	–

1982), and the sperm whale, *Physeter catodon* (Clarke 1980). Cephalopods identified from remains in the regurgitations and gut contents of these species are given in Table 3. The species of *Todarodes* referred to by Clarke et al. (1981), Clarke and Prince (1981) and Thomas (1982) is now known to be *Martialia hyadesi* (Rodhouse et al., in press). Species with known distributions to the south of the Sub-Antarctic Front are indicated where the foraging range of predators clearly extends to warmer waters to the north.

A total of 58 species of cephalopod is taken by these predators. The largest number of species is consumed by wandering albatrosses but less than half are from south of the Sub-Antarctic Front. The other predators take less species and a larger proportion of these are known to be Antarctic species.

A comparison is given in Table 4 of the relative importance to the predators of the different species of Antarctic cephalopod. On a biomass basis the most important cephalopod families in the diet of these predators are the Cranchiidae (sperm whale and light-mantled sooty albatross), the Onychoteuthidae (wandering albatross) and the Ommastrephidae (grey-headed and black-browed albatross).

The size of cephalopods sampled by nets and wandering albatrosses is compared in Table 5. The nets sampled much smaller specimens of *Kondakovia longimana, Psychroteuthis glacialis, Alluroteuthis antarcticus* and *Mesonychoteuthis hamiltoni* and smaller specimens of *Mastigoteuthis psychrophila.* They sampled a wide size range of *Gonatus antarcticus* and *Galiteuthis glacialis,* the largest of which were of similar size or larger than those taken by wandering albatrosses. Only one specimen each of *Batoteuthis skolops, Histioteuthis eltaninae* and *Chiroteuthis* sp. were caught in nets and these were larger than the mean size taken by wandering albatrosses.

4 Discussion

4.1 Comparison Between Cephalopods Sampled by Nets and Predators

There are striking differences between cephalopod samples taken at South Georgia by nets and from predators. Two squid abundant in the net samples,

Table 4. Relative contribution, in terms of numbers and biomass, of Antarctic cephalopod species to the diet of vertebrate predators at South Georgia. Sources of data are given in the text

Cephalopod family and species	Wandering albatross		Grey-headed albatross		Black-browed albatross		Light-mantled sooty albatross		Sperm whale	
	Numbers (%)	Biomass (%)	Numbers (%)	Biomass (%)	Numbers (%)	Biomass (%)	Numbers (%)	Biomass (%)	Numbers (%)	Biomass (%)
Gonatidae										
Gonatus antarcticus	4	1	–	–	–	–	6	13	< 1	< 1
Gonatus sp.	–	–	–	–	1	1	9	10	–	–
Onychoteuthidae										
Kondakovia longimana	16	74	1	4	1	1	< 1	5	30	18
Moroteuthis knipovitchi	3	3	–	–	–	–	–	–	–	–
Moroteuthis ingens	1	5	–	–	–	–	–	–	25	4
Moroteuthis robsoni	< 1	< 1	–	–	–	–	–	–	–	–
Batoteuthidae										
Batoteuthis skolops	4	< 1	–	–	–	–	–	–	–	–
Histioteuthidae										
Histioteuthis eltaninae	23	3	1	< 1	–	–	< 1	< 1	4	< 1
Psychroteuthidae										
Psychroteuthis glacialis	3	< 1	–	–	–	–	–	–	–	–
? Large *Psychroteuthis*	1	< 1	–	–	–	–	6	24	–	–
Neoteuthidae										
Allutoteuthis antarcticus	8	4	–	–	–	–	2	4	< 1	< 1
Ommastrephidae										
Martialia hyadesi	6	5	88	91	70	83	2	2	–	–
Chiroteuthidae										
Chiroteuthis sp. (small)	< 1	< 1	< 1	< 1	2	1	–	–	< 1	< 1
Mastigoteuthidae										
Mastogoteuthis sp. A	< 1	< 1	1	< 1	–	–	–	–	5	< 1
Cranchiidae										
Galiteuthis glacialis	30	2	8	4	25	13	73	41	< 1	< 1
Mesonychoteuthis hamiltoni	< 1	< 1	–	–	–	–	–	–	32	77
Taonius pavo	< 1	< 1	–	–	–	–	–	–	1	< 1

Brachioteuthis?picta and *Bathyteuthis abyssicola*, were absent from samples from predators, whilst *Kondakovia longimana* and *Mesonychoteuthis hamiltoni*, which dominate the cephalopod biomass consumed by sperm whales, were rarely caught by nets. The onychoteuthid *Moroteuthis knipovitchi* and the ommastrephids, a major component of grey-headed and black-browed albatross diets, have been absent from net hauls. The greatest size differences between net and predator samples occurred in muscular species assumed to be powerful swimmers: *Kondakovia longimana*, *Psyhroteuthis glacialis* and *Alluroteuthis antarcticus*. Ommastrephids are very strong swimmers and this probably explains their absence from research net hauls. Large specimens of *K. longimana*, *M. knipovitchi*, *Brachioteuthis?picta*, *Psychroteuthis glacialis*

Table 5. Size of cephalopods sampled by all nets and the wandering albatross in the vicinity of South Georgia. Source of wandering albatross data given in text. Range given where sample does not approximate a normal distribution

Species	Nets			Wandering albatross		
	Mean mantle length (mm) (or range)	SD	n	Mean ML (mm)	SD	n
Gonatus antarcticus	< 20–395	–	7	237.7	75.4	85
Kondakovia longimana	40.7	14.6	6	506.0	123.9	343
Batoteuthis skolops	151.0	–	1	102.0	11.4	87
Histioteuthis eltaninae	102.2	–	1	61.5	7.9	477
Psychroteuthis glacialis	18.2	9.7	5	273.9	53.0	13
Alluroteuthis antarcticus	22.9	14.9	7	154.1	26.9	164
Chiroteuthis sp.	106.7	–	1	119.5	25.2	14
Mastigoteuthis psychrophila	108.7	15.0	32	179.2	4.1	4
Galiteuthis glacialis	< 5–425	–	191	213.0	18.2	631
Mesonychoteuthis hamiltoni	< 5–150	–	6	468.8	126.9	3

and *Alluroteuthis antarcticus* have been caught in the Antarctic by larger midwater trawls (Nemoto et al. 1985, 1988) and by commercial trawls (Rodhouse 1989) but these have also failed to catch ommastrephids.

Differences between samples taken by different nets were related to the size of the net used. The RMT 25 caught most species and larger specimens. This net was dedicated to cephalopod sampling and was deployed at greater depths than was routine for the smaller nets.

4.2 Comparison Between Cephalopods Caught by Different Predators

There are also differences between cephalopods consumed by the different predators. In terms of biomass *Mesonychoteuthis hamiltoni* is the major food of the sperm whale but it is almost completely absent in the diet of four albatross species. It is very large, and the only net-caught adult was from an RMT 8 haul from 2000–2200 m at 58° 45′ S, 18° 15′ E (Rodhouse and Clarke 1985). This squid probably rarely comes to the surface where it would be available to the albatrosses.

4.3 Origin of Cephalopods Consumed by Predators at South Georgia

Although albatrosses forage at the surface they feed regularly on *Galiteuthis glacialis* and *Alluroteuthis antarcticus*, which undergo ontogentic descent and are only caught as large specimens in deep water (Rodhouse and Clarke 1986; Rodhouse 1988). It is possible that albatrosses prey on squid making mating/spawning migrations to the surface (Rodhouse et al. 1987) but comprehensive data from net hauls on squid distribution through the life cycle are needed. Of 58 cephalopod species consumed by predators at South Georgia (Table 3), 17 are thought to occur south of the Sub-Antarctic Front. Given the proximity of the front and the foraging range of albatrosses (Croxall et al. 1984) and sperm whales (Clarke 1980), it is not surprising that warm-water cephalopods are consumed by these predators. However, the extent of wandering albatross predation on warm-water species varies (Rodhouse, in press b) and it is likely that the prey of other predators also varies.

Of the cephalopod biomass consumed by wandering albatross, light-mantled sooty albatross and sperm whale, an average of 75, 73 and > 95% respectively was Antarctic species. For grey-headed and black-browed albatross these proportions were 100 and 92%. Recent information (Rodhouse et al., in press) suggests that grey-headed albatrosses consumed large quantities of the ommastrephid, *Martialia hyadesi* at South Georgia in 1984 and 1986. The presence of virtually undigested specimens in the regurgitations indicated that they were caught close to South Georgia.

4.4 Species with Potential for Commercial Exploitation

The most common squid species in the seas around South Georgia, as determined from nets and predators, are given in Table 6 together with features which might determine suitability for commercial exploitation

Of the 12 species, the ommastrephid *Martialia hyadesi* seems to have the most potential in the near of medium future. This is because of its suitability for human consumption and because of the size of the

Table 6. Common squid species in the seas around South Georgia and assessment of their suitability for commercial exploitation

Species	Flesh consistency good	Fishery exists for members of the same family elsewhere	Fishery exists for this species elsewhere	Attracted to lights/jigs
Gonatus antarcticus	+	+	–	?
Kondakovia longimana	+	+	–	+
Moroteuthis spp.	+	+	–	+
Brachioteuthis?picta	+	–	–	?
Histioteuthis eltaninae	–	–	–	?
Bathyteuthis abyssicola	–	–	–	?
Psychroteuthis glacialis	+	–	–	?
Alluroteuthis antarcticus	+	–	–	?
Martialia hyadesi	+	+	+	+
Mastigoteuthis psychrophila	–	–	–	?
Galiteuthis glacialis	–	–	–	?
Mesonychoteuthis hamiltoni	?	–	–	?

stocks which are large judging by the quantities consumed by predators. It occurs as a by-catch in the *Illex argentinus* fishery on the Patagonian Shelf but the catch is variable; in 1986 22000 tonnes from the Patagonian Shelf were landed in Japan (Masutomi, pers. commun.). In previous years catches were considerably less, and in 1987 only 20 tonnes of this squid were reported from the Falklands Interim Conservation Zone (FICZ). It is likely that *Martialia hyadesi* is a southern species which occasionally moves northwards to the Patagonian shelf in the Falklands Current.[1]

The market value of *Martialia hyadesi* approaches that of *Illex argentinus* so if squid catches decline and prices increase in the future, fishing vessels may turn their attention to the Scotia Sea in search of *M. hyadesi*. Other candidates for commercial exploitation are the onychoteuthids, *Kondakovia longimana, Moroteuthis ingens, M. knipovitchi* and *M. robsoni* and the gonatid, *Gonatus antarcticus*. Members of the families to which these species belong are the subject of directed fisheries elsewhere or are caught as a by-catch and have commercial value (Roper et al. 1984). There are no known fisheries anywhere for members of the Brachioteuthidae, Psychroteuthidae or Neuteuthidae. However, the flesh of Antarctic members of these families appears to be suitable for human consumption so if they could be caught in sufficient quantities they might contribute to a commercial fishery.

The large cranchiid, *Mesonychoteuthis hamiltoni*, may have potential commercial value (Klumov and Yukhov 1975; Roper et al. 1985). It occurs in considerable numbers, judging by its presence in the diet of sperm whales, and the flesh of large specimens may be palatable to humans. However, the mantle of small specimens resembles the thin, leathery mantle of other cranchiids, including *Galiteuthis glacialis*, and would not be suitable for commercial exploitation. Of the other species, *Bathyteuthis abyssicola* is too small and deep living, and the flesh of *Histioteuthis eltaninae* and *Mastigoteuthis psychrophila* is probably ammoniacal (Clarke et al. 1979).

Agassiz trawl hauls in Cumberland East Bay and on the shelf to the northeast of South Georgia suggest that *Pareledone* spp. are abundant on the muddy bottom at these sites and are readily caught. Octopodids are an important component of the cephalopod diet of southern elephant seals and Weddell seals elsewhere in the Scotia Sea (Clarke and MacLeod 1982a, b) and they are probably taken by these seals at South Georgia. The market price of octopodids is higher than for squid, so they may attract commercial fisheries in the future.

4.5 Possible Consequences of Commercial Exploitation

Cephalopods. Tropical and temperate cephalopods are fast-growing (Rodhouse and Hatfield, in press), short-lived and semelparous. Most species live for approximately one year, spawn once and die soon afterwards (Boyle 1983). This life history pattern differs from that of most exploited fish species and provides special problems for the management of stocks (Caddy 1983; Amaratunga 1987). Because cephalopods are generally semelparous and have a short generation

[1] In February 1989 two Japanese squid jigging vessels caught commercial quantities of *Martialia hyadesi* at the Sub-Antarctic Front near South Georgia. The author was aboard one of these vessels at the time and will be publishing research data on the catch in due course.

time they are prone to extreme fluctuations in population size and are particularly susceptible to overfishing because recruitment is dependent upon the breeding success of a single exploited generation. Populations are also liable to expand rapidly under favourable conditions and cephalopod populations may increase to fill the vacant niche created by overfishing fin-fish stocks (Caddy 1983).

Life histories of Antarctic cephalopods have not been examined but they are probably similar to those elsewhere. It should be assumed that Antarctic stocks will be equally sensitive to overexploitation, and any future fishery will require careful management.

Life history traits in exploited fish species may be susceptible to selection-induced genetic change (Kirpichnikov 1981) and it is likely that heritability of life history traits such as growth, fecundity and spawning time in cephalopods is similar to that of fish (Beverton 1984). If genetic effects, due to artificial selection imposed by fishing gear selectivity, occur in cephalopods effects such as changes in mean growth rate or spawning time might manifest themselves relatively soon after exploitation begins because of the short generation time of these molluscs. The long-term consequences of fishery-induced genetic change could be considerable if fishing is intense and stocks are geographically localised.

Vertebrate predators. A number of predators in the vicinity of South Georgia rely on cephalopod stocks for a major component of their breeding season diet and would be severely affected by overexploitation of cephalopod populations. Interactions between seabirds and fisheries elsewhere are well documented (Furness 1982; Nettleship et al. 1984) and the effects of overfishing on bird populations have been determined. Croxall (1987) has concluded that the nature and availability of prey can influence seabird reproductive success. A reduction of squid stocks around South Georgia would potentially influence breeding success of seabirds, and probably other predators, especially southern elephant seals.

Acknowledgements. I thank the master and crew of RRS "John Biscoe" and Dr. A. Wood for their assistance with cephalopod research in the Antarctic and in Cambridge. The Fisheries Department of the Falkland Islands Government gave permission to cite catch statistics for *Martialia hyadesi* and Dr. A. Rosenberg, Renewable Resources Assessment Group, Imperial College, London and Mr. S. Masutomi, K.S.J. Corporation, Tokyo, provided data and information on this species.

Mr. M. G. White and Dr. J. P. Croxall kindly read and criticised an early draft of this paper.

References

Amaratunga T (1987) Population biology. In: Boyle PR (ed) Cephalopod life cycles, vol 2. Academic Press, London, pp 239–252

Baker A de C, Clarke MR, Harris MJ (1973) The N.I.O combination net (RMT 1 + 8) and further developments of rectangular midwater trawls. J Mar Biol Assoc UK 53:167–184

Beverton RJH (1984) Dynamics of single species. In: May RM (ed) Exploitation of marine communities. Springer, Berlin Heidelberg New York 367 p

Bone DG (1986) An LHPR system for adult Antarctic krill. Br Antarct Surv Bull 73:31–46

Boyle PR (ed) (1983) Cephalopod life cycles, vol 1. Academic Press, London, 475 p

Caddy JF (1983) The cephalopods: factors relevant to their population dynamics and to the assessment and management of stocks. In: Advances in assessment of world cephalopod resources. FAO Fisheries Tech Pap No 231, FAO Rome, pp 416–452

Clarke MR (1977) Beaks, nets and numbers. Symp Zool Soc Lond 38:89–126

Clarke MR (1980) Cephalopods in the diet of sperm whales of the southern hemisphere and their bearing on sperm whale biology. Discovery Rep 37:1–324

Clarke MR (1983) Cephalopod biomass – estimation from predators. Mem Nal Mus Victoria 44:95–107

Clarke MR, Croxall JP, Prince PA (1981) Cephalopod remains in the regurgitations of the wandering albatross *Diomedea exulans* L. at South Georgia. Br Antarct Surv Bull 54:9–21

Clarke MR, Denton EJ, Gilpin-Brown JB (1979) On the use of ammonium for buoyancy in squids. J Mar Biol Assoc UK 59:259–276

Clarke MR, MacLeod N (1982a) Cephalopods in the diets of elephant seals at Signy Island, South Orkney Islands. Br Antarct Surv Bull 57:27–31

Clarke MR, MacLeod N (1982b) Cephalopods remains in the stomachs of eight Weddell seals. Br Antarct Surv Bull 57:33–40

Clarke MR, Prince P (1981) Cephalopod remains in regurgitations of black-browed and grey-headed albatrosses at South Georgia. Br Antarct Surv Bull 54:1–7

Croxall JP (1987) Conclusions. In: Croxall JP (ed) Seabirds: feeding ecology and role in marine ecosystems. Cambridge University Press, Cambridge, pp 369–381

Croxall JP, Prince PA (1987) Seabirds as predators on marine resources, especially krill, at South Georgia. In: Croxall JP (ed) Seabirds: feeding ecology and role in marine ecosystems. Cambridge University Press, Cambridge, pp 347–368

Croxall JP, Prince PA, Ricketts C (1985) Relationships between prey life-cycles and the extent, nature and timing of seal and seabird predation in the Scotia Sea. In: Condy WR, Siegfried PR, Laws RM (eds) Antarctic nutrient cycles and food webs. Springer, Berlin Heidelberg New York, pp 516–533

Croxall JP, Ricketts C, Prince PA (1984) Impact of seabirds on marine resources, especially krill of South Georgia waters. In: Causey Whittow G, Rahn H (eds) Seabird energetics. Plenum Publ Corp, New York, pp 285–317

Csirke J (1987) The Patagonian fishery resources and the offshore fisheries in the South-West Atlantic. FAO Fish Tech Pap (286), 75 p

Everson I (1977) The living resources of the Southern Ocean. FAO/UNDPGLO/SO/77/1:156 p

Furness RW (1982) Estimating the food requirements of seabird and seal populations and their interactions with commercial

fisheries and fish stocks. In: Cooper J (ed) Proc Symp Sea and Shore Birds. African Seabird Group, Cape Town, pp 1–14
Kirpichnikov VS (1981) Genetic bases of fish selection. Springer, Berlin Heidelberg New York
Klumov SK, Yukhov VL (1975) *Mesonychoteuthis hamiltoni* Robson, 1925. Antarctika 14:159–189 (in Russian)
Laws RM (1985) The ecology of the Southern Ocean. Am Sci 73:26–40
Nemoto T, Okiyama M, Takahashi M (1985) Aspects of the roles of squid in food chains of marine Antarctic ecosystems. In: Condy WR, Siegfried PR, Laws RM (eds) Antarctic nutrient cycles and food webs. Springer, Berlin Heidelberg New York, pp 415–420
Nemoto T, Okiyama M, Iwasaki N, Kikuchi T (1988) Squid as predators on krill (*Euphausia superba*) and prey for sperm whales in the Southern Ocean. In: Sahrhage D (ed) Antarctic Ocean and resources and variability. Springer, Berlin Heidelberg New York, pp 292–296
Nettleship DN, Sanger GA, Springer PF (1984) Marine birds: their feeding ecology and commercial fisheries relationships. Can Fish Wildl Serv Spec Publ, Ottawa
Priddle J, Heywood RB, Theriot E (1986) Some environmental factors influencing phytoplankton in the Southern Ocean around South Georgia. Polar Biol 5:65–79
Rodhouse PG (1988) Distribution of the neuteuthid squid *Alluroteuthis antarcticus* in the Atlantic sector of the Southern Ocean. Malacologia 29:265–272
Rodhouse PG 1989 Pelagic cephalopods caught by nets during the Antarctic research cruises of the 'Polarstern' and 'Walther Herwig', 1984–1987. Arch Fishereiwiss 39:111–121
Rodhouse PG Cephalopods in the diet of wandering albatrosses and sea surface temperatures at the Sub-Antarctic Front. Scient Mar
Rodhouse PG, Clarke MR (1985) Growth and distribution of young *Mesonychteuthis hamiltoni* Robson (Mollusca: Cephalopoda): an Antarctic squid. Vie Milieu 35:223–230
Rodhouse PG, Clarke MR (1986) Distribution of the early-life phase of the Antarctic squid *Galiteuthis glacialis* in relation to the hydrology of the Southern Ocean in the sector 15°–30° E. Mar Biol 91:353–357
Rodhouse PG, Clarke MR, Murray AWA (1987) Cephalopod prey of the wandering albatross *Diomeda exulans*. Mar Biol 96:1–10
Rodhouse PG, Hatfield EMC Age determination in squid using statolith growth increments. Fish Res (in press)
Rodhouse PG, Prince PA, Murray AWA Cephalopod prey of the grey-headed albatross *Diomedea chrysostoma*. Mar Biol (in press)
Roe HSJ, Shale DM (1979) A new multiple rectangular midwater trawl (RMT 1 + 8 m) and some modifications to the Institute of Oceanographic Sciences' RMT 1 + 8. Mar Biol 50:283–288
Roper CF (1969) Systematics and zoogeography of the worldwide bathypelagic squid *Bathyteuthis* (Cephalopoda: Oegopsida). Bull US Natl Mus 291:208 p
Roper CF, Sweeney MJ, Clarke MR (1985) Cephalopods. In: Fischer W, Hureau JC (eds) FAO species identification sheets for fishery purposes. Southern Ocean (Fishing areas 48, 58 and 88) (CCAMLR Convention Area) Prepared and published with the support of the Commission for the Conservation of Antarctic Living Marine Resources. Rome, FAO, vol 1:117–205
Roper CFE, Sweeney MJ, Nauen CE (1984) Cephalopods of the world. An annotated and illustrated catalogue of species of interest to fisheries. FAO Fisheries Synopsis, No 125, vol 3, 277 p
Sato T, Hatanaka H (1983) A review of assessment of Japanese distant-water fisheries for cephalopods. In: Caddy JF (ed) Advances in assessment of world cephalopod resources. FAO Fisheries Tech Pap No 231, pp 145–180
Thomas G (1982) The food and feeding ecology of the light-mantled sooty albatross at South Georgia. Emu 82:92–100
Voss GL (1973) Cephalopod resources of the world. FAO Fisheries Circular No 149, FAO, Rome 75 p

The Feeding Ecology of Larval Fish in an Antarctic Fjord, with Emphasis on *Champsocephalus gunnari*

A. W. North and P. Ward [1]

Summary. The vertical distribution of five species of fish larvae and their prey species was studied in late winter and summer at Cumberland East Bay, South Georgia. Prey type varied between fish species but was dominated by the copepod *Drepanopus forcipatus* and copepod eggs. Prey numbers increased with fish size. In summer, fish larvae and copepods were most abundant in the upper 120 m of the 265-m-deep fjord, whereas in late winter, both groups were more evenly dispersed throughout the water column.

1 Introduction

South Georgia is a very important finfish fishery area in the Antarctic and is mainly dependant on the mackerel icefish *Champsocephalus gunnari.* In the 1986/87 season 71 141 tonnes of the grand total of 90 872 tonnes of fish taken in Antarctic waters was *C. gunnari* from South Georgia (CCAMLR 1988). Other exploited species include two more icefish, *Chaenocephalus aceratus* and *Pseudochaenichthys georgianus,* and the rock-cods *Notothenia gibberifrons* and *Patagonotothen brevicauda guntheri.* Recruitment to the *C. gunnari* population is very variable (Cooke 1987) and largely controls the value of the commercial fishery. Indications of year-class success and future recruitment derived from estimates of the abundance of the early stages will be important for managing the fishery.

Adult *C. gunnari* feed offshore on krill (*Euphausia superba*). They spawn nearshore during April to June in the fjords around South Georgia (Olsen 1955). Larvae with a yolk-sac hatch in June to August (winter) and begin to feed in August–September before all their yolk reserves have been depleted (North 1988; North and Ward 1989). Early juvenile *C. gunnari* (60–100 mm long) fed on mysids at South Georgia (Kock 1981). Juvenile *C. gunnari* are most abundant in shallow waters 100 to 200 m deep around the island (Kock 1981; Boronin et al. 1987). The nearshore ecosystem, especially in the fjords, is therefore an important area where biological and physical factors affect recruitment and year-class success (North 1988). These factors, apart from the status of the adult spawning stock, are mostly those which determine survival of the larval to the juvenile stages and include feeding success and the effects of water movements on their distribution in relation to feeding grounds (Hunter 1981; Cushing 1975).

To date relatively few studies have investigated the feeding ecology of the larval stages of Antarctic fish. There has been little seasonal coverage and, with the notable exception of Kellermann (1986, 1987), few species have received attention. This chapter is a preliminary report on the feeding ecology of the most abundant larval stages of fish in Cumberland East Bay, South Georgia, including three species important to the fishery, *Champsocephalus gunnari, Chaenocephalus aceratus,* and *Notothenia gibberifrons,* and two others *Nototheniops nudifrons* and *Nototheniops larseni.*

2 Materials and Methods

Zooplankton including fish larvae and copepods were sampled from the fjord of Cumberland East Bay, South Georgia (54° 16′ S, 36° 25′ W). The fjord is 11 km long, 3 km wide and has a maximum depth of 265 m, and although mostly enclosed has a wide, deep entrance to Cumberland West Bay and the open sea (Fig. 1).

Samples were obtained using an acoustically controlled multiple rectangular midwater trawl (RMT 1 + 8 M) towed at 2.5 knots along set transects (Fig. 1). Fish and copepods were taken simultaneously using the 8 m^2 net of 4.5 mm mesh and the 1 m^2 net of 0.33 mm mesh respectively. Fish were also sampled in the upper 2-m water layer using a Foredeck Frame net (FNET) of 1 m^2 with mesh of 4.5 mm deployed at the same time as the RMT 1 + 8 M. The FNET was

[1] British Antarctic Survey, High Cross, Madingley Road, Cambridge CB3 0ET, UK.

Antarctic Ecosystems. Ecological Change and Conservation
(ed. by K. R. Kerry and G. Hempel)
© Springer-Verlag Berlin Heidelberg 1990

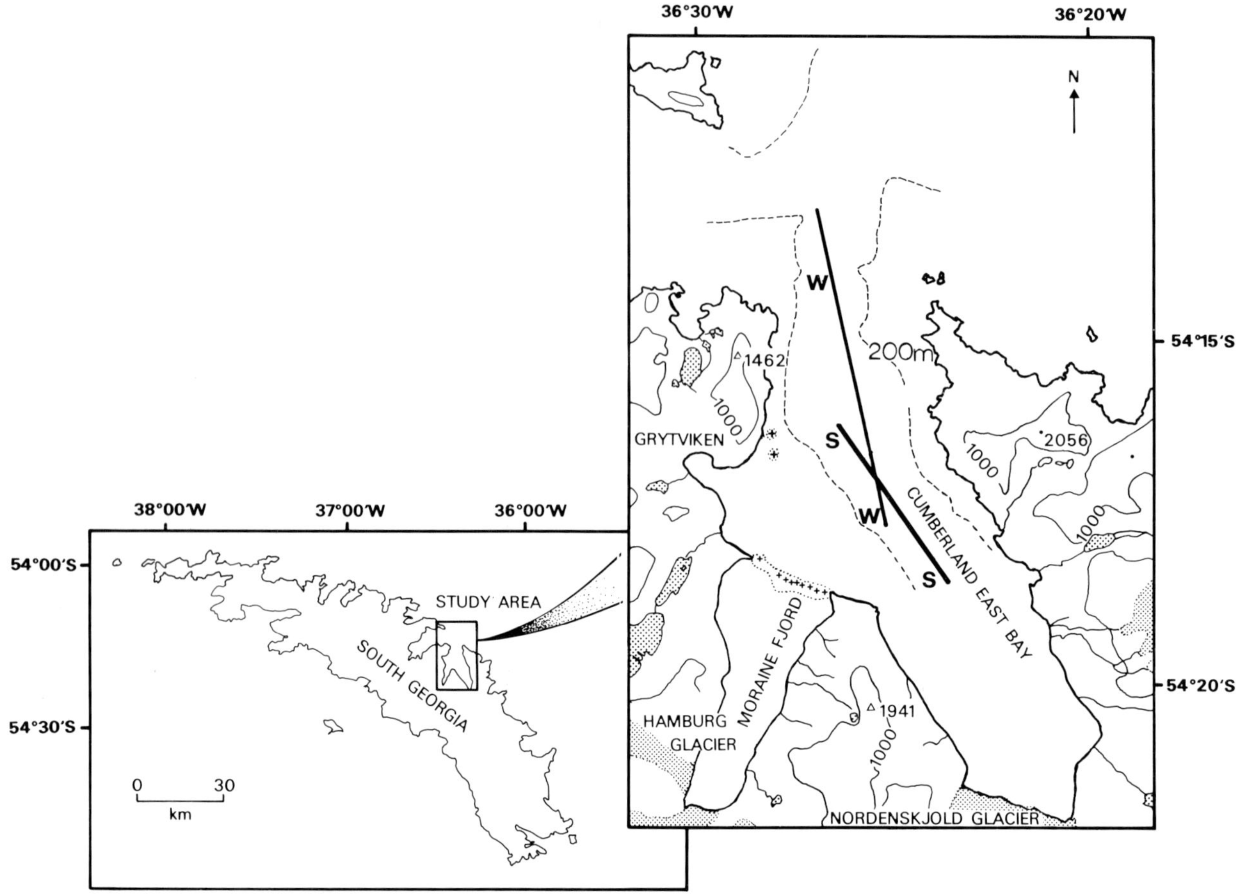

Fig. 1. The study area showing summer (*S*) and winter (*W*) transects

towed alongside the foredeck of the ship and so avoided the disturbance and mixing effect of the propeller wash on the upper water layer.

Late winter/early spring samples were taken during 20–26 September 1983 by upward oblique hauls of 30-min duration at mean depths of 203 – 152, 152 – 75, and 75 – 0 m. Samples were taken at four times of day: pre-dawn, post-dawn, pre-dusk and post-dusk. Summer samples were taken during 4–14 January 1987 by downward oblique hauls of 20-min duration at 0–20, 20–60, 60–100, 100–140, 140–180, 180–220 m depths. These depths were sampled at six times of day: the aforementioned four times plus, midday and midnight. Distance travelled by the nets was calculated from a flowmeter mounted on the net monitor and volume swept then calculated from the equations of Pommeranz et al. (1983). Filtration was assumed to be 100% efficient.

Fish and zooplankton were preserved in 10% formalin solution and were later transferred to Steedmann's solution. Specimens were generally fixed within 3 h of capture and within 1.5 hours of the net returning on-board. Fish and copepods were viewed with binocular microscopes at × 6–60 magnification. Fish standard lengths (preserved) were measured to the nearest 0.1 mm. Other fish and copepod dimensions (preserved) were measured using an eyepiece graticule. Fish internal mouth width (gape width) is similar to internal mouth height (North unpubl. data), therefore, gape width was used as a measure of the potential maximum size of food particles that could be consumed. Copepod dimensions were measured using undigested, preserved material.

Larval fish alimentary tracts were simple, consisting of a pharynx, stomach and a tubular hindgut leading to the anus. Gut content of larval fish was assessed by dissection of preserved specimens, the whole gut was removed intact using fine needles and the gut content separated. Prey items were identified, assigned to developmental stage, and counted.

The possibility of codend feeding is felt to be unlikely since the RMT 8 from which the fish were analyzed has a mesh of 4.5 mm and most zooplankton retained by the net were larger than the fish could ingest. An exception was *Chaenocephalus aceratus* which consumed fish larvae but the prey was always partly digested and never obviously freshly taken. In all fish species examined the digested state of the prey in the stomach was similar to that in the hindgut.

The fish abundance data from replicate samples within each time and depth combination had a skewed distribution. Therefore, to compare the mean number of fish larvae with the mean number of copepod prey at each depth the fish data was $\log_{10}$ transformed, the mean was calculated and then back transformed to give the normalized mean. One was added to each value of the data before transformation to account for zero values and was subtracted from each back-transformed mean.

3 Results

3.1 Diet

Prey items found in the alimentary tract (here referred to as guts) of the five species are summarized by number in Table 1. For all species examined, copepods and copepod eggs accounted for > 90% of all prey items found. In particular the small clausocalanid *Drepanopus forcipatus* comprised > 97% of all identified copepods found in the guts of *C. gunnari, C. aceratus* and *N. nudifrons* during winter and, 97 and 87% in *C. gunnari* and *N. gibberifrons* respectively, during summer. An exception appeared to be *N. larseni* during summer when only 27% of all copepods taken were positively identified as *D. forcipatus*. However, the majority of copepods retrieved from this species were well digested and despite the presence of other species (*Oithona* spp. *Ctenocalanus* spp.) most were thought to be *D. forcipatus,* although their digested state precluded developmental staging.

The proportion of copepod eggs in the diet of the fish larvae varied markedly with season. In winter, copepod eggs represented 70–80% of the diet (by number) of *C. gunnari, C. aceratus* and *N. nudifrons.* In summer copepod eggs represented only 45% of the diet of *C. gunnari* and less than 3% of the diet of *N. gibberifrons* and *N. larseni* (Table 1). Measurements of egg diameters (0.13 mm, n = 108, SD = 0.02) found in the fish guts agreed with those made on eggs found in sacs attached to the genital segments of adult female *D. forcipatus*. The proportions of female *D. forcipatus* with eggs visible in the oviduct also varied with season. In winter 48% contained eggs, whereas in summer this figure fell to 32% indicating that reproductive activity was more intense during the winter. It therefore seems likely that eggs were ingested along with adult females, rather than being consumed as separate prey items. This hypothesis was tested for each fish species and season by comparing the number of eggs and adult female *D. forcipatus* found in the gut of individual fish. For each of the three fish species sampled in winter there was a highly significant ($p < 0.01$) positive correlation; whereas for the three species sampled in summer the relationship was not significant ($p > 0.05$). However, in summer, *N. larseni* consumed neither adult female *D. forcipatus* nor eggs, *N. gibberifrons* contained a low percentage of both and *C. gunnari* predated only moderate numbers of both females and eggs. Thus total prey number excluding eggs probably reflects the number of prey items ingested, because the

Table 1. Prey by number in the diet

Fish	N	Size Range SL mm	Total Copepoda	*Drepanopus forcipatus* Stage F VI	M VI	V	IV	III	II	I	Other Copepod spp.	Digested Copepods[a]	Copepod Eggs	Euphausiid Furcilia	Other Species
Winter															
C. gunnari	66	11–31	192	113	3	20	4	0	0	0	1	51	830[b]	3	2
C. aceratus	16	15–22	19	10	0	0	0	0	0	0	0	9	68	1	8[c]
N. nudifrons	21	24–51	177	89	1	12	4	0	0	0	3	68	536	0	7
Summer															
C. gunnari	23	22–30	228	50	59	6	12	20	36	2	6	37	195	9	0
N. gibberifrons	34	22–29	1132	64	12	48	165	167	118	46	80	432	29	1	6
N. larseni	18	12–24	129	0	0	0	4	1	1	0	16	107	0	0	0

[a] Mostly *D. forcipatus.*
[b] Two egg sacs containing eggs were also found.
[c] Included six fish larvae.

eggs are almost certainly consumed attached to, or inside, female *D. forcipatus*.

3.2 Seasonal Variation in Diet

During winter, adult female *D. forcipatus* comprised the majority of all identified copepod prey (> 80%) for all three fish species. Early copepodite stages CI–CIII did not occur in the diet and few other copepod species were found (mainly immature *Rhincalanus gigas*).

In summer a greater range of developmental stages was preyed upon with two of the species, *C. gunnari* and *N. gibberifrons* taking early stages, (CI–CIII), of *D. forcipatus* in addition to the older ones. *N. larseni* only preyed on stages CII–CIV. Other species of copepod also occurred in the diet, albeit at low frequency, which meant that prey diversity in terms of both the range of stages of *D. forcipatus* and number of copepod species taken was greater than during winter. The other species encountered included *Oithona* spp., *Ctenocalanus* spp. and early copepodite stages of *Calanus simillimus*.

During both seasons, non-copepod taxa comprised < 4% of total prey items by number, apart from the prey of *C. aceratus* where 6% were fish. Other taxa included: calyptope and furcilia stages of the euphausiids *Thysanoessa* sp. and *Euphausia frigida*, polychaete worms (including *Pelagobia longicirrata*) and a decapod larvae.

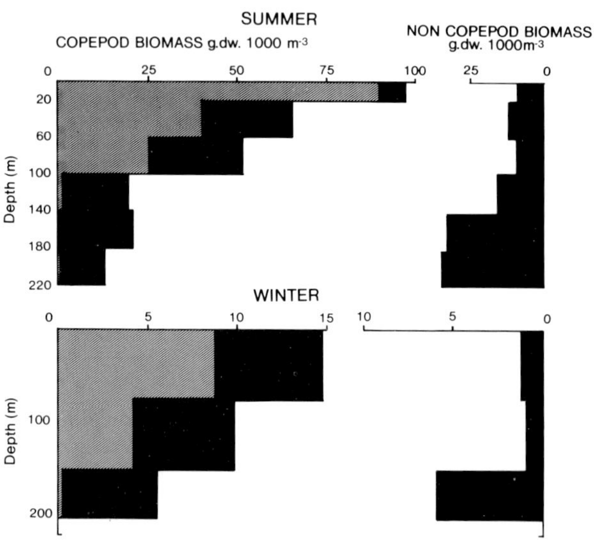

Fig. 2. Contribution to biomass in the water column made by *D. forcipatus* (*hatched area*), other copepoda and other zooplankton

The dominance of copepods in the diet of the larval fish examined is not surprising as they dominate the zooplankton caught in RMT 1 nets in terms of both numbers and biomass and, of all species, *D. forcipatus* was the greatest single contributor to both (Fig. 2).

During both seasons, the number of fish and *D. forcipatus* caught per net varied considerably but the overall mean abundance was generally lower in winter than summer (Table 2).

3.3 Prey Size and Selection

From estimated mouth gape width the smaller size range of all larval fish studied was capable of consuming all stages of *D. forcipatus* including adult females (Table 3). For predation by larval fish the limiting dimension of prey size appears to be maximum width, since adult *D. forcipatus* more than 1.4 mm long but less than 0.5 mm wide were consumed by larvae with gape widths of only 0.8 mm.

A comparison of the proportion of the various stages of *D. forcipatus* ingested by each fish species during winter compared to that present in the water column (Tables 1, 2) indicated that a higher proportion of adult female *D. forcipatus* was being consumed. This suggests that the fish were positively selecting female *D. forcipatus*. The converse was true for the fish sampled in summer where a higher proportion of immature stages and a lower proportion of adult female *D. forcipatus* were taken than were present in the net hauls. However, the RMT 1 net with a mesh of 300 μ probably undersampled the early stages of the copepods (e.g. stage IV and smaller *D. forcipatus*; see Pillar 1984). This indicates that in summer smaller prey were positively selected or, the number of small copepods available was underestimated.

3.4 Distribution of Predator and Prey

Given the importance of *D. forcipatus* as a prey item, we have examined the diurnal vertical distribution of adults and early copepodite stages compared to that of their fish predators (Fig. 3). In winter all species of fish examined occurred throughout the sampled water column. The diurnal vertical distributions of *C. gunnari* and *N. nudifrons* were broadly similar in that they were more abundant in the upper 150 m and particularly in the upper 70 m of the water column. During the light period (pre-dusk and post-dawn) they were mostly absent from the top 2 m, although net avoidance in daylight is a likely contributory cause (North 1988). *D. forcipatus* showed little diurnal variation in

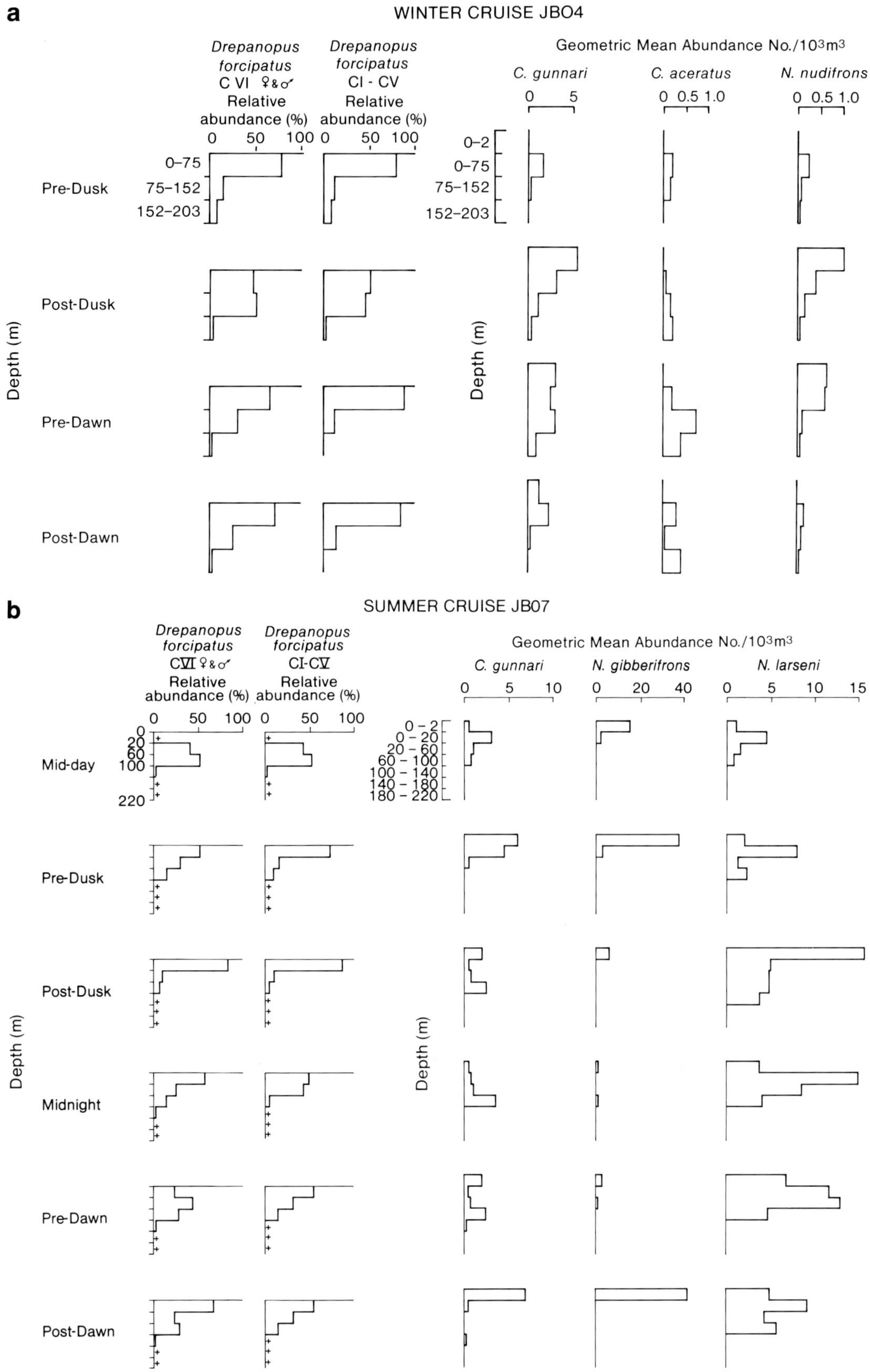

Fig. 3. Diurnal vertical distribution of adult and immature stages of *D. forcipatus* and fish larvae. **a** Winter; **b** Summer

Table 2. Mean abundance of *Drepanopus forcipatus* and fish larvae

Depth m	WINTER							
	D. forcipatus					*C. gunnari*	*N. nudifrons*	*C. aceratus*
	F VI	M VI	V	IV	III			
	No. per m³					No. per 10³ m³		
0–2						4.49	0.52	0.00
0–75	169	22	98	36	3	3.14	0.37	0.18
75–152	88	17	23	10	< 1	1.61	0.11	0.32
150–203	6	1	1	< 1	< 1	0.43	0.03	0.38

Depth m	SUMMER								
	D. forcipatus						*C. gunnari*	*N. larseni*	*N. gibberifrons*
	F VI	M VI	V	IV	III	II			
	No. per m³						No. per 10³ m³		
0–2							15.10	8.95	49.3
0–20	1 297	81	472	1541	226	1	2.80	16.37	1.26
20–60	717	90	246	754	192	3	0.98	5.86	0.24
60–100	503	77	127	307	56	< 1	5.10	4.80	0.40
100–140	18	18	4	5	2	1	0.12	0.12	0.08
140–180	4	2	2	2	1	1	0.01	0.00	0.01
180–220	11	1	4	7	5	1	0.03	0.02	0.09

Table 3. Fish mouth gape width and the dimensions of the copepodite stages of *Drepanopus forcipatus* (sizes in mm)

Fish species	Size (SL)[a]	Calculated gape width[b]	*Drepanopus forcipatus*				
			Stage		Cephalothorax		Total
				Sex	Width	Length	Length
C. gunnari	12	0.77	VI	F	0.47	1.22	1.83
C. aceratus	15	1.12	VI	M	0.43	0.91	1.46
N. nudifrons	25	1.66	V		0.43	1.15	1.65
N. gibberifrons	23	1.70	IV		0.34	0.92	1.27
N. larseni	18	1.11	III		0.27	0.77	1.02
			II		0.20	0.58	0.76

[a] Fish size was the lower range of the specimens studied.
[b] Gape width was calculated from linear regressions of gape width on SL, for most regressions there was a significant positive relationship ($p < 0.05$).

distribution, being almost entirely restricted to the upper 150 m and most abundant in the upper 70 m. The distributions of *N. nudifrons* and *C. gunnari* then appear similar to that of *D. forcipatus*, with *C. aceratus* generally distributed deeper. In summer all three of the fish species examined were most abundant in the upper 100 m. *D. forcipatus* was again almost entirely restricted to the upper 100 m, generally occurring below this depth at less than 1% relative abundance in the sampled depth horizons.

Thus it appears that in both seasons ther was a good general agreement between the vertical distribution of the fish larvae and that of *D. forcipatus*.

3.5 The Feeding Ecology of *Champsocephalus gunnari*

Initial feeding success is important to survival of fish larvae, therefore, we have examined in more detail the

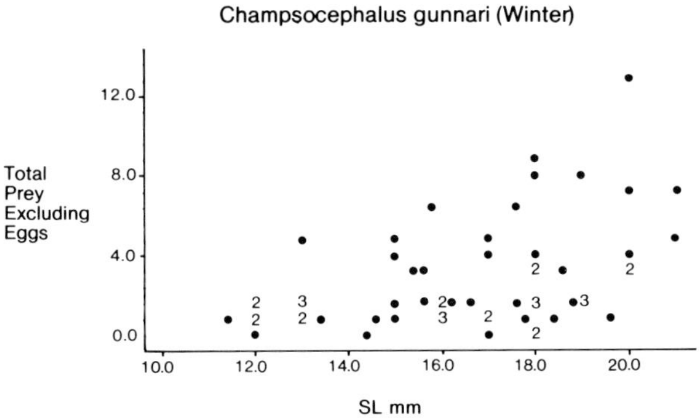

Fig. 4. Total number of copepod prey (excluding eggs) in the guts of *Champsocephalus gunnari* vs standard length (*SL*)

feeding ecology of *C. gunnari*, for the period following their hatch in winter. Newly hatched larvae with a full yolk-sac are 11–15 mm standard length (North in prep.). At this size they have a gape width of about 0.8 mm which means that they can consume adult *D. forcipatus* on first feeding (Table 3). As expected there was a significant, positive linear relationship between fish gape width and standard length (SL). Because it was easier to accurately measure fish length than gape width, subsequent analyses were related to the former parameter. Copepod prey number (excluding eggs) in the gut increased significantly ($p < 0.05$) with standard length (Fig. 4) and all larvae in the size range examined (12–20 mm SL) consumed adult female *D. forcipatus*, the largest prey item taken, indicating that as fish length increases they ingest more, but not larger, prey.

We have further examined the effect of time of day on the total number of prey digested. Mean and median number of prey (excluding eggs) was highest pre- and post-dusk (Table 4), suggesting feeding intensity was greatest during daylight. There was no significant difference in the size of fish caught by depth or time of day.

Table 4. Total copepod prey (excluding eggs) in the gut of *Champsocephalus gunnari* at different times of the day, in winter

	Time of day (all depths combined) Light		Dark	
	Pre-dusk	Post-dusk	Pre-dawn	Post-dawn
Mean	2.5	3.7	1.4	1.6
Median	2	3	1	2
Max	7	13	6	2
Min	1	0	0	1
SD	1.8	3.0	1.6	0.6
N	10	38	12	5

Prey number in the gut of individual larval fish was used as an indication of feeding success. Combining all depths and times only 3.6% had empty guts, although 7.8% contained no copepods [the difference (4.2%) was mostly copepod eggs]. The mean and median numbers of prey were about one-half to one-quarter of the maximum number taken at all times of day, and many guts contained few items (Table 4, Fig. 4). These data suggest that most fish were feeding successfully but not optimally as few had full guts.

4 Discussion

It is apparent from the available data that like many larval fish in other ecosystems Antarctic species prey largely on copepods. In this study a single species, *D. forcipatus*, was extremely important in the diet of all species investigated. Ward (in press) has shown that *D. forcipatus* was the most abundant copepod occurring in the fjord of Cumberland East Bay. Its numerical mean abundance during summer and winter of 1136 m^{-3} and 151 m^{-3} accounted for 40 and 32% of total zooplankton biomass respectively. Previous studies by Hardy and Gunther (1935) have indicated that *D. forcipatus* [formerly *D. pectinatus* but now recognized as *D. forcipatus* (Hulsemann 1985)] was confined to waters immediately surrounding South Georgia, and Atkinson and Peck (this vol.) have confirmed this. These authors also noted the patchy nature of its distribution and that it can occur in high abundance.

Ward (1989) has shown that in terms of size spectra the maxima of copepod abundance and biomass occured in the 0.5–1.5 mm cephalothorax length range which during both seasons largely comprised all copepodid stages of *D. forcipatus*. On the basis of gape width, the fish larvae could theoretically take prey larger than this but copepods of cephalothorax length > 1.5 mm were relatively rare in the plankton. Thus the fish larvae were exploiting the most abundant size spectra and species available.

During winter, size selection of prey appeared to be common to all three species with a relatively higher proportion of adult female *D. forcipatus* being ingested than were apparently present in the plankton. Adults accounted for just over 50% of all *D. forcipatus* in the water column, whereas they represented > 80% of the identified copepod prey. Many factors other than abundance are probably important in determining prey selection and will not be disclosed by studies of this type, e. g. behavioural responses, aggregation and appearance of both predator and prey. One observation which relates to this is the presumed higher incidence of egg-bearing female *D. forcipatus* in the plankton in winter compared to summer. The females carry the eggs in an external sac which may make them more conspicuous, thereby rendering them more liable to predation. In summer the species examined tended to consume smaller prey. Whether this represents positive selection is unknown as the abundance of the smaller stages is almost certainly underestimated by the net catches.

Prey other than copepods was also taken. During summer the diet of *C. gunnari* contained euphausiid furcilia, which by virtue of their larger size, were probably more important nutritionally than their low numbers suggest. *C. aceratus* was the only species studied here which preyed significantly on fish larvae. Although its prey were mostly too digested to identify or to estimate size, a single specimen, on the basis of pigmentation and eye diameter was probably a *C. gunnari* larva of about 12 mm SL. Because of the relatively large size and nutritional quality of fish compared to other prey, it seems likely that they are of significant importance in its diet.

C. gunnari fed mostly during the day (Table 4), suggesting they are visual feeders. Kellermann (1986) also found that *Pleuragramma antarcticum* and *Nototheniops larseni* fed mostly during daylight at the Antarctic Peninsula. Perhaps feeding during the polar winter is limited by the light period which shortens dramatically with increasing latitude.

The high ratio of the number of *D. forcipatus* in the water column to the number of fish larvae (ca. $> 10^5:1$) is high (Table 2), suggesting that food is not limiting for the fish larvae in Cumberland East Bay. However, few *C. gunnari* had consumed more than half the number of prey items they could contain. Patchiness in zooplankton is not infrequent and it is likely that both predator and prey have aggregated distributions. Catches of fish larvae have a near-Poisson distribution (North and Murray in prep.) and they have been observed in shoals by SCUBA divers nearshore (North pers. obs.). Therefore, small-scale distribution may be a factor influencing predation and so further work is needed before we can conclude that the fish larvae were not limited by food availability.

In conclusion, the abundance and distribution of *D. forcipatus* at South Georgia is important to the feeding success of *C. gunnari* and other fish species of commercial importance. Further studies on the interactions between the fish larvae and copepods at South Georgia may provide some understanding of factors affecting recruitment in exploited fish species.

Acknowledgements. We thank the members of the British Antarctic Survey's Offshore Biological Programme team, the officers and crew of *RRS John Biscoe* and Julie Peck for assistance with sampling and sample sorting respectively.

References

Boronin VA, Zakharov GP, Shopov VP (1987) Distribution and relative abundance of juvenile icefish (*Champsocephalus gunnari*) from a trawl survey of the South Georgia shelf in June–July 1985. Scientific Committee of the Convention for the Conservation of Antarctic Marine Living Resources Selected Scientific Papers 1986 (WG-CFSA-86/10), pp 55–63

CCAMLR (1988) Report of the seventh meeting of the Scientific Committee (Hobart, Australia, 24–31 October 1988). Summary of Fisheries Data. Secretariat. SC-CAMLR-VII/BG/2, 79 pp

Cooke JG (1987) Assessment of the stocks of *Notothenia rossii* and *Champsocephalus gunnari* in the South Georgia area. Scientific Committee for the Conservation of Antarctic Marine Living Resources Selected Scientific Papers 1986 (WG-CFSA-86/2), pp 3–21

Cushing DH (1975) Marine Ecology and Fisheries. Cambridge University Press, Cambridge, pp 220–225

Hardy AC, Gunther ER (1935) The plankton of the South Georgia whaling grounds and adjacent waters 1926–1927, Discovery Rep 11:1–456

Hulsemann K (1985) Two species of *Drepanopus* Brady (Copepoda, Calanoida) with discrete ranges in the Southern Hemisphere. J Plank Res 7:909–925

Hunter JR (1981) Feeding ecology and predation in marine fish larvae. In: Lasker R (ed) Marine Fish Larvae. Washington Sea Grant Program, University of Washington, Seattle, pp 33–77

Kellermann A (1986) On the biology of early life stages of notothenioid fishes (Pisces) off the Antarctic Peninsula. Ber Polarforsch. 31:1–155

Kellermann A (1987) Food and feeding ecology of postlarval and juvenile *Pleuragramma antarcticum* (Pisces; Notothenioidei) in the seasonal pack ice zone off the Antarctic Peninsula. Polar Biol 7:307–315

Kock KH (1981) Fischereibiologische Untersuchungen an drei antarktischen Fischarten: *Champsocephalus gunnari* Lönnberg, 1905, *Chaenocephalus aceratus* Lönnberg, 1906 und *Pseudochaenichthys georgianus* Norman, 1937 (Notothenioidei, Channichthyidae). Mitt Inst Seefisch Hamb 32:1–226

North AW (1988) Distribution of fish larvae at South Georgia: horizontal, vertical and temporal distribution and early life history relevant to monitoring year class strength and recruitment. Scientific Committee for the Conservation of Antarctic Marine Living Resources Selected Scientific Papers 1987 (WG-FSA-87/16), pp 105–142

North AW Early growth of Antarctic fish at South Georgia (in prep)

North AW, Ward P (1989) Initial feeding by larval Antarctic fish during winter at South Georgia. Cybium 13:357–364

North AW, Murray AWA Abundance and diurnal vertical distribution of fish larvae in a fjord at South Georgia (in prep)

Olsen S (1955) A contribution to the systematics and biology of chaenichthyid fishes from South Georgia. Nytt Mag Zool 3:79–93

Pillar SC (1984) A comparison of the performance of four zooplankton samplers. S Afr J Mar Sci 2:1–18

Pommeranz T, Herrmann C, Kuhn C (1983) Mouth angles of the Rectangular Midwater Trawl (RMT 1 + 8) during paying out and hauling. Meeresforsch 29:267–274

Ward P (1989) The distribution of zooplankton in an Antarctic Fjord at South Georgia during summer and winter. Antarctic Science 1:141–150

The State of Exploited Fish Stocks in the Atlantic Sector of the Southern Ocean

K.-H. Kock [1] and F.-W. Köster [2]

Summary. Finfish have been harvested around South Georgia since the end of the 1960s mainly by the Soviet Union. *Notothenia rossii marmorata* was the target species in the first peak years 1969–71, which yielded about 500 000 tonnes in two seasons. Since 1975/76 fishing has been directed mostly to the icefish *Champsocephalus gunnari* and to a lesser extent *Patagonotothen brevicauda guntheri*. Catches of *C. gunnari* were highest at about 240 000, 220 000 and 100 000 tonnes in 1976/77–1977/78, 1982/83–1983/84 and 1986/87–1987/88.

N. rossii marmorata is the species most adversely affected by the fishery. The present stock size is less than 5% of the level before the fishery started. Recruitment has fallen since the second half of the 1970s. At the present level of recruitment, simulation studies indicate that the stock will recover slowly, only doubling over a period of 10 years.

Stock size of *C. gunnari* around South Georgia is largely dependent on the strength of the recruiting year class which also forms the bulk of catches in the fishery. Stocks of *C. gunnari* around the South Orkney Islands and in the Antarctic Peninsula region seem to be heavily depleted by fishing and are in need of conservation measures.

Patagonotothen br. guntheri is the only species for which the fishery was unregulated until 1988/89. Stock assessment is largely influenced by uncertainties in the determination of the natural mortality rate (M). At high levels of M (0.8–0.9) stock size as well as recruitment indicate a downward trend.

1 Introduction

Research activities and exploratory fishing with unknown catches by the Soviet Union in the Scotia Sea and the Scotia arc region started in 1961 (Lubimova 1985). They developed into large-scale harvesting around South Georgia (FAO subarea 48.3) in 1969/70. By the mid-1970s, a number of other European fishing countries carried out exploratory fishing cruises to the Atlantic sector of the Southern Ocean (primarily South Georgia). However, only Poland, the German Democratic Republic and Bulgaria entered the commercial phase. Bulgaria stopped fishing at the end of the 1979/80 season. Fishing activities by the German democratic Republic terminated in 1980/81 but reopened in 1985/86. The fishery was extended to the South Orkney Islands (FAO subarea 48.2) in 1977/78 and to the South Shetland Islands and the northern tip of the Antarctic Peninsula near Joinville Island (FAO subarea 48.1) in 1978/79.

Fishing up to 1981/82 was almost entirely carried out by bottom trawling at depths down to 500 m. Later on midwater trawls towed close to the bottom were used increasingly to catch *Champsocephalus gunnari*.

In the first peak years 1969–71 the marbled notothenia, *Notothenia rossii marmorata*, was the target species, yielding about 500 000 tonnes in two seasons. Since 1975/76 fishing has been primarily directed at the icefish *Champsocephalus gunnari* and to a lesser extent the small-sized Gunther's notothenia, *Patagonotothen brevicauda guntheri*. *Patagonotothen br. guntheri* only occurs around Shag Rocks some 150 miles west-northwest of the mainland of South Georgia. Catches of *C. gunnari* were highest at about 240 000, 220 000 and 100 000 tonnes in 1976/77–1977/78, 1982/83–1983/84 and 1986/87–1987/88 respectively. Other abundant species, such as the bumphead notothenia, *Notothenia gibberifrons*, the South Georgia icefish, *Pseudochaenichthys georgianus* and the Scotia Sea icefish, *Chaenocephalus aceratus* and the spiny icefish *Chaenodraco wilsoni*, are regular by-catch species and/or were important in one season or another but they could not support a commercial fishery on their own.

By 1987/88, 1.83 million tonnes of finfish had been caught in the Atlantic sector of the Southern Ocean, 1.5 millions of which had been fished around South Georgia (Table 1). The USSR took about 90% of these catches.

[1] Institut für Seefischerei, Bundesforschungsanstalt für Fischerei, Palmaille 9, D-2000 Hamburg 50, FRG.
[2] Institut für Meereskunde, Abt. Fischereibiologie, Düsternbrooker Weg 20, D-2300 Kiel 1, FRG.

Antarctic Ecosystems. Ecological Change and Conservation
(ed. by K. R. Kerry and G. Hempel)
© Springer-Verlag Berlin Heidelberg 1990

Table 1. Nominal catches of finfish (tonnes) by species and subarea in the Atlantic sector of the Southern Ocean from 1969/70–1987(88 (CCAMLR 1988b)

Species Area	South Georgia incl. Shag Rocks (subarea 48.3)	South Orkney Islands (subarea 48.2)	Antarctic Peninsula (subarea 48.1)	Atlantic Sector total (area 48)
Notothenia rossii marmorata	570874	2888	19233	592995[c]
Notothenia gibberifrons	66395	24552	4151	95098[c]
Notothenia squamifrons	13445	239	36	13720
Dissostichus eleginoides[a]	7460	258[a]	102[a]	7820[c]
Patagonotothen br. guntheri	156275	–	–	156275
Pleuragramma antarcticum	–	110	–	110
Nototheniidae *nei*	3755	1494	22	5271
Champsocephalus gunnari	550476	184838	41396	776710
Chaenocephalus aceratus	8213	2605	1547	12365
Pseudochaenichthys georgianus	21669	3676	469	25814
Chaenodraco wilsoni	–	–	14450	14450
Chionodraco rastrospinosus	–	1770	760	2530
Channichthyidae nei[b]	5582	1911	26	7519
Myctophidae nei[d]	21110	350	48	21812
Rajiformes nei	411	10	1	422
Micromesistius australis	–	36	–	36
Marine fishes nei	78127	21525	4877	104529[c]
Total	1503792	246262	87118	1837476

[a] Probably *Dissostichus mawsoni*.
[b] Consists to a large but unknown proportion of *Ch. aceratus* and *Ps. georgianus*.
[c] Contains catches from unknown subareas.
[d] Mostly *Electrona carlsbergi*.

Preliminary assessments of the fish stocks based on limited information (Everson 1977, 1978; BIOMASS 1980; Kock 1981; Gulland 1983) suggested that these are not capable of sustaining the catches taken during the peak years. More recent analyses of the condition of fish stocks based on biomass estimates, length compositions, age and growth, mortality estimates and catch statistics (Kock 1985a, 1986; Kock et al. 1985; Slosarczyk et al. 1985) indicated that some species, in particular *N. rossii marmorata*, had been reduced to a fraction of their original level and were in need of conservation measures.

The Convention on the Conservation of Antarctic Marine Living Resources (CCAMLR) came into force in 1982. Since 1984 the state of exploited Antarctic fish stocks has been assessed annually by its Working Group on Fish Stock Assessment (CCAMLR 1984, 1985, 1986, 1987a, 1988a). A number of conservation measures have been established since then (CCAMLR 1988c, 1988d) which were principally aimed at *N. rossii marmorata*, *C. gunnari* and *Patagonotothen br. guntheri* but should produce some benefits for other species as well.

In the following we have reviewed and revised the assessments carried out by the Working Group for these target species. Furthermore, we have discussed to what extent conservation measures adopted by CCAMLR could be expected to help rebuild fish stocks and whether additional measures are necessary to comply with Article II (3) of the Convention (CCAMLR 1988d).

2 Material and Methods

2.1 Material

Our analyses are based on age/length compositions from the Soviet commercial and exploratory fishery (USSR) submitted to CCAMLR and those given in Borodin and Kochkin (1988) and Shust and Borodin (1988) and biomass estimates from fishery and research vessel surveys carried out by the Federal Republic of Germany (DEU), Poland (POL), Spain (SP) and USA (US). Nominal catches on a split year basis were available from the CCAMLR Summary of Fisheries Data (CCAMLR 1988b). Missing age/length keys were replaced with estimated keys by interpolation from the years preceding and following. Whenever length composition data were available (but no age/length keys) iterated age/length keys were produced (Kimura and Chikuni 1987).

Catch in numbers from a particular season was estimated from the nominal catch and the estimated mean weight of fish in the catch. We used mean weights derived from the mean length of fish in the catch and a length-to-weight relationship instead of the reported mean weights from direct observations as used by the Working Group on Fish Stock Assessment. Mean weight at age was derived from mean length at age in the age/length keys.

For the Virtual Population Analysis (VPA) and the stock and catch projections natural mortality rates M were set to a constant value for all age classes: *N. rossii marmorata*: 0.20 and *C. gunnari*: 0.35 (CCAMLR 1987b). Maturity ogives were determined from a log maturity curve given by Kock (1986) for *N. rossii marmorata*, whereas maturity ogives for *C. gunnari* and *P. br. guntheri* were those given in Borodin and Kochkin (1988) and Shust and Borodin (1988).

A detailed account of the materials used is given in Kock and Köster (1989).

2.2 Methods

Virtual population analysis (VPA; Gulland 1965) was performed to estimate stock size, biomass and spawning stock biomass (SSB) as well as age-dependent stock sizes, i.e. recruitment, but also partial recruitment to fishing mortality values.

Terminal F's[1] for the most recent year in the VPA were adjusted to produce results comparable to biomass estimates based on fishery-independent survey results (see Kock and Köster 1989). Biomass estimates from research vessel surveys used the 'swept area' method. VPA results, such as stock size at the end of the most recent year, recruitment numbers and partial recruitment values based on mean age-specific F's were used as input data for short- and long-term projections.

Long-term projections based on yield per recruit (Y/R) and spawning stock biomass per recruit (SSB/R) calculations (Thompson and Bell 1934) were computed to determine long-term management options (i.e. reference points $F_{0.1}$[2] and F_{max}[3]). Equilibrium yield and spawning stock biomass were estimated for constant recruitment at the mean level. When the data base incorporated sufficient information, the determination of recovery rates and catch predictions were carried out for different management options and levels of recruitment.

2.2.1 *Notothenia rossii marmorata* – South Georgia

Medium-term projections were conducted under different total allowable catch (TAC) options (0–1200 tonnes) for 10 years assuming constant recruitment at the mean levels for 1972/73–1976/77, 1978/79–1983/84 and 50% of the 1978/79–1983/84 level respectively. A long-term projection was calculated to predict equilibrium yield and spawning stock biomass under different fishing mortality options at the recent recruitment level.

2.2.2 *Champsocephalus gunnari* – South Georgia

The two biomass estimates available for the 1986/87 season from fishery-independent surveys exhibited substantial differences (Fig. 5), which most probably reflect the different gear used in the surveys (SP: semipelagic trawl; US/POL: bottom trawl). The 151 000 tonnes estimated from the Spanish survey were assumed to be a realistic biomass estimate by the Working Group on Fish Stock Assessment (CCAMLR 1988a).

Short-term yield and stock projections (1988/89–1990/91) for both VPA terminal F's and different management options (50% $F_{0.1}$, $F_{0.1}$, and F_{max} and $2 \times F_{max}$) were calculated assuming a preliminary catch of 23 000 tonnes in 1988/89[4] and different levels of recruitment [mean number of recruits in 1972/73 (medium) and their upper (high) and lower (low) 10% confidence limits].

Long-term projections were carried out with both data sets to determine reference points on the Y/R curve using management options in short-term projections and to calculate corresponding equilibrium biomass at a medium recruitment level.

[1] Terminal F is the estimated value of fishing mortality on the oldest age class in each season and on all age classes in the most recent year in the VPA.

[2] $F_{0.1}$ is the level of fishing mortality rate F for which the marginal increase in yield per recruit due to a small increase in F is 10% of the marginal yield per recruit in a lightly exploited fish stock.

[3] F_{max} is the fishing mortality rate that produces the maximum sustainable yield for a stock.

[4] Catches reported to CCAMLR represented 21 356 tonnes in 1988/89. Taking into account that reported catches were about 10% less than nominal catches, as in 1987/88, a preliminary catch of 23 000 tonnes was assumed for 1988/89.

2.2.3 *Champsocephalus gunnari* – South Orkney Islands and Antarctic Peninsula

Information from both stocks has been combined for this analysis (see Kock and Köster 1989). VPA analysis was carried out for two alternatives: (1) where terminal F was set arbitrarily at 0.5 for fully recruited age classes (3 +). (2) Where terminal F was adjusted to a biomass of 4600 tonnes which was estimated in February 1985 for the South Orkney Islands and Elephant Island combined (Tables 7 and 10 in Kock 1986). The latter alternative resulted in a terminal F of 1.2.

Due to the absence of age classes 1 to 3 in the catches in 1985/86 only a long-term projection was performed. Recruitment was assumed to be on the same level as in 1977/78–1981/82.

2.2.4 *Patagonotothen brevicauda guntheri*

Due to the lack of data on age compositions from the first two fishing seasons, it was impossible to obtain a realistic value for the natural mortality rate from a catch curve. VPA was thus carried out for two values of M: 0.8, which is the mean of the estimates given in Shust and Borodin (1988), and 0.4 which was chosen for comparison as the Working Group on Fish Stock Assessment was unable to evaluate the validity of Shust and Borodin's estimate (CCAMLR 1988a).

Biomass estimates from bottom trawl surveys in 1984/85 and 1986/87 of 7 256 tonnes (Kock 1986) and 343 tonnes (Gabriel 1987) were obviously too low as they were only (small) fractions of the total catch in that season. Therefore an estimated stock size of 81 680 tonnes from a semipelagic trawl survey in 1986/87 (Balguerias et al. 1987) was used to adjust terminal F values in the VPAs.

Assuming a mean recruitment (1978/79–1984/85), short-term yield and stock projections were carried out for both M values on the basis of VPAs recalculated with data from Shust and Borodin (1988) and their original results. Reference points ($F_{0.1}$, F_{max}) were only determined for the recalculated data set, as Y/R and SSB/R curves are particularly sensitive to differences in M value.

A more detailed description of all input data is given in Kock and Köster (1989).

3 Results and Discussion of Assessments

3.1 *Notothenia rossii marmorata* – South Georgia

3.1.1 *The Course of the Fishery*

The species yielded a total of 501 262 tonnes in the first two fishing seasons 1969/70–70/71. Catches then decreased to virtually zero during the next four seasons. From 1976/76 to 1978/79 catches oscillated around 8000 tonnes (6311–10 753 tonnes) but rose again in 1979/80 to 24 897 tonnes. From then until 1984/85 ca. 2000 tonnes were harvested annually (866–3736 tonnes). Since 1985/86, when CCAMLR prohibited direct fishing of the species, reported catches have been less than 300 tonnes annually.

3.1.2 *Trends in Biomass*

The trend in biomass from VPA analysis clearly reflects the heavy impact the fishery had on the stock in the first two seasons. By 1975/76 stock size was less than 10% of the initial stock size in 1969/70 (Fig. 1). The estimated value of 52 043 tonnes was in good agreement with the 35 682 tonnes estimated from a fishery-independent, bottom trawl survey (Table 50 in Kock et al. 1985) taking into account that fishery-independent surveys, based on a stratified random design, tend to underestimate stock size due to the very patchy distribution of *N. rossii marmorata*. In the late 1970s stock size was of the order of 35 000–50 000

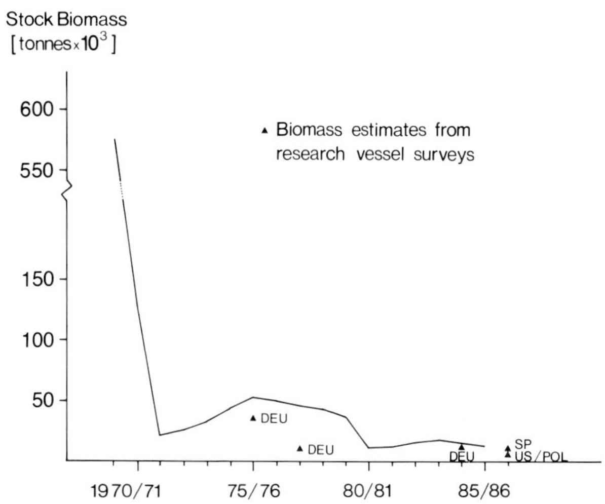

Fig. 1. Trends in biomass of *N. rossii marmorata* around South Georgia from VPA analysis

tonnes which was similar to the 53 820 tonnes estimated by Burchett and Ricketts (1984) for 1978 and 1979 from a mark-recapture experiment. It was, however, further depleted in 1979/80. Since then stock size has varied little, fluctuating around 12 000 to 16 000 tonnes. Estimates from fishery-independent surveys were 12 781 tonnes in 1984/85 (Kock 1986) which had been used to adjust the VPA and 4528 tonnes in a US/POL survey (Gabriel 1987) and 11 471 tonnes in a Spanish survey in 1986/87 (Balguerias et al. 1987; Fig. 1).

3.1.3 Recruitment

VPA analyses indicate that recruitment, i.e. the number of recruits at age class 2, varied little until 1976/77 despite the dramatic reduction in stock size within that period. Recruitment fluctuated between 7.6 and 10.8×10^6 individuals (Fig. 2). However, during the next 2 years recruitment apparently dropped by a factor of 2–4 and has been only $2.8-4.3 \times 10^6$ individuals since 1978/79 (Fig. 2). The sharp decline within a very short time period is probably an artefact that may have been caused by the large variation in fishing intensity (i.e. the determination of fishing mortalities), and uncertainties in catch statistics. A reduction in recruitment itself during the last 10 years, however, has probably occurred, for which there is some evidence in records of the British Antarctic Survey. These records indicate a substantial decline in trammel net catches of juvenile *N. rossii marmorata* in the vicinity of shore stations (I. Everson pers. commun.; M. G. White pers. commun.). Increased predation by fur seals (*Arctocephalus gazella*), which started to recolonize the mainland of South Georgia in the 1970s (Payne 1977) and which winter on the island (males only) in increasing numbers (CCAMLR 1988a), might be one reason for continuing low recruitment. Although food studies on fur seals indicate that they feed primarily on *Euphausia superba, Notothenia rossii marmorata* (mostly juveniles) forms 5% of the diet by weight of juvenile fur seals in summer (Doidge and Croxall 1985). This would correspond to an annual consumption of 15 000 tonnes if this is typical for the rest of the year. However, the proportion of fish, including *N. rossii marmorata*, even increases in winter (Croxall et al. 1985; CCAMLR 1988a).

3.1.4 Catch Prediction and Stock Recovery

Catch predictions and yield per recruit analysis indicate that recruitment in the 1960s must have been in the order of 50×10^6 individuals to produce a stock biomass of 500 000 tonnes at the end of the decade. However, this level of recruitment is not evident from the numbers of age class 2 fish in 1969/70 and 1970/71, which were in the order of the period 1972/73–1976/77, i.e. five to seven times less. Currently, it seems unlikely that the stock will recover to any level comparable with that of the late 1960s even if the recent annual recruitment was similar to that of the mid-1970s. Recent recruitment is evidently even lower.

If recruitment remains at the present level, stock size could be expected to double by 1995/96 irrespective of whether the stock was fully protected, which seems to be unrealistic given the incidental by-catches, or whether a low total allowable catch (TAC) of ca. 600 tonnes is set (Fig. 3). A further reduction in recruitment to 50% of the present level would not alter the present stock size significantly if catches stay well below 1000 tonnes (Fig. 3).

Assuming constant recruitment, an equilibrium biomass and a stable age composition could be expected by the beginning of the next century. At present levels of recruitment and fishing (ca. 300 tonnes), biomass would then be expected to be ca. 40 000 tonnes (SSB 36 000 tonnes). With $F_{0.1}$ ($= F = 0.214$, Fig. 4) as one management option, this would result in

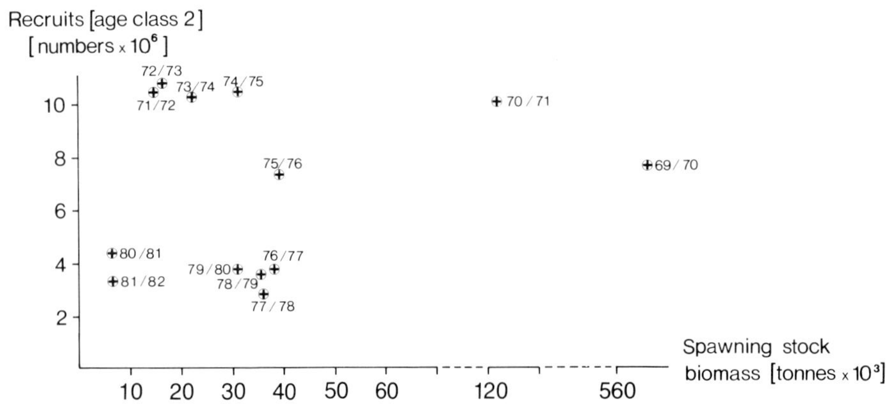

Fig. 2. Recruitment in relation to spawning stock biomass in *N. rossii marmorata* around South Georgia. (*Numbers* of recruits refer to those seasons in which they were actually produced)

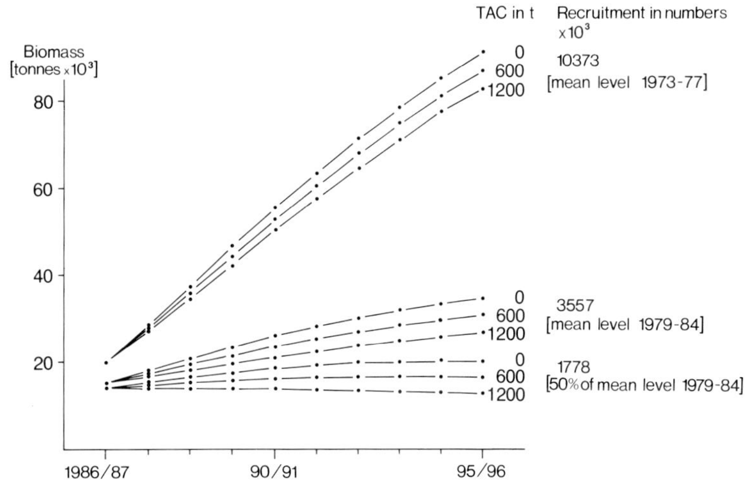

Fig. 3. Stock recovery in *N. rossii marmorata* around South Georgia under different levels of recruitment and total allowable catches (*TAC*)

a long-term yield of 2400 tonnes per year but at a stock biomass of ca. 22 000 tonnes (SSB 18 000 tonnes).

Both the plateau shape of the Y/R curve (Fig. 4) and long-term projections with a reduced terminal F for younger age classes indicate that virtually no gain in Y/R and only slight benefit in SSB/R can be expected if size (= age) at first capture is raised by an increase in the minimum mesh size (see Kock et al. 1985).

3.2 *Notothenia rossii marmorata* – Antarctic Peninsula

The commercial fishery was confined to one season only. The data were thus insufficient to carry out VPA and yield per recruit analysis.

3.2.1 *The Course of the Fishery*

The spawning stock (or at least part of it) of the Antarctic Peninsula population was encountered in January 1976 in a restricted area northwest of Elephant Island. Exploratory fishing during the Antarctic Expeditions of the Federal Republic of Germany on these aggregations yielded 125 and 237 tonnes in 1975/76 and 1977/78 respectively. The stock was commercially harvested only in the 1979/80 season by the Soviet fishery. The total catch of 18 763 tonnes was in the same order of magnitude as the biomass estimate of 15 663 tonnes for 1977/78 (Table 51 in Kock et al. 1985). No catches have been reported since then.

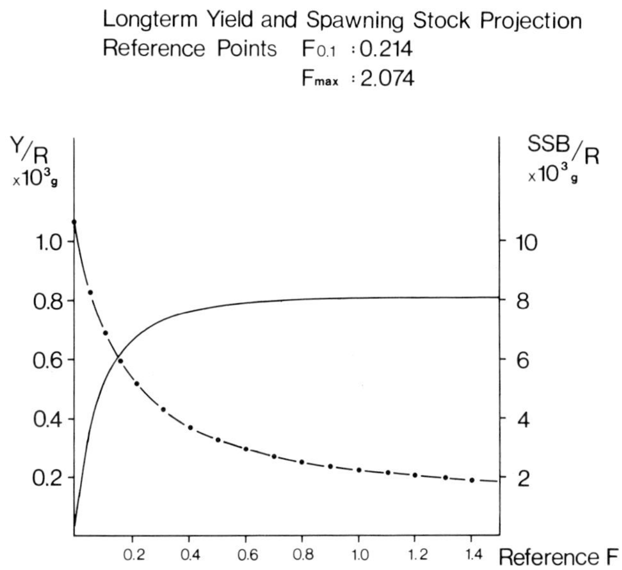

Fig. 4. Yield per recruit (*Y/R*) analysis and spawning stock per recruit (*SSB/R*) projection for *N. rossii marmorata* around South Georgia

3.2.2 *Trends in Biomass*

The catch of 18 763 tonnes in 1979/80, slightly exceeding the biomass estimate of 15 663 tonnes made two seasons earlier, and the difficulty in locating spawning fish in later seasons suggest that the fishery had removed most of the mature fish. In the absence of any reported catch and the limited supply of survey information it is difficult to make a quantitative statement about the spawning stock biomass, though it is probably well below its initial level (CCAMLR 1986). Stock size of the prerecruits was estimated to be 2134 tonnes in 1984/85 (Kock 1986).

3.3 *Champsocephalus gunnari* – South Georgia

3.3.1 The Course of the Fishery

Catches have been reported from 1970/71 onwards but were at a low level until 1975/76, exceeding 10000 tonnes only twice (1970/71; 1975/76). Catches peaked in 1976/77 when 93400 tonnes were taken, but declined substantially to less than 8000 tonnes in each of the subsequent three seasons. From 1980/81 onwards catches rose again steadily to a maximum of 128194 and 79997 tonnes in 1982/83 and 1983/84 respectively. They dropped again in the following two seasons to less than 15000 tonnes annually. Subsequent to another increase in the catches to 71142 tonnes in 1986/87, the fishery was regulated by CCAMLR for 1987/88 for the first time, when a TAC of 35000 tonnes was set. This TAC was almost fully exhausted in the 1987/88 season when 34632 tonnes were reported as taken. In 1988/89 the fishery was closed after 35 days when ca. 21356 tonnes had been fished.

3.3.2 Trends in Biomass

Biomass estimates from the VPAs based on data submitted to CCAMLR and those recalculated from Borodin and Kochkin (1988) exhibit a similar trend. They were in better accordance with biomass estimates from fishery-independent surveys during the 1970s than those during the mid-1980s (Fig. 5). This may be due to the more benthopelagic mode of life of the younger fish forming the bulk of the biomass in the 1980s, which would result in underestimates from bottom trawl surveys. Biomass estimates at the beginning of the 1988/89 season from VPAs, assuming a mean level of recruitment (559 and 629 × 10^6 individuals respectively), were either 68700 (data submitted to CCAMLR) or 86800 tonnes (data recalculated from Borodin and Kochkin).

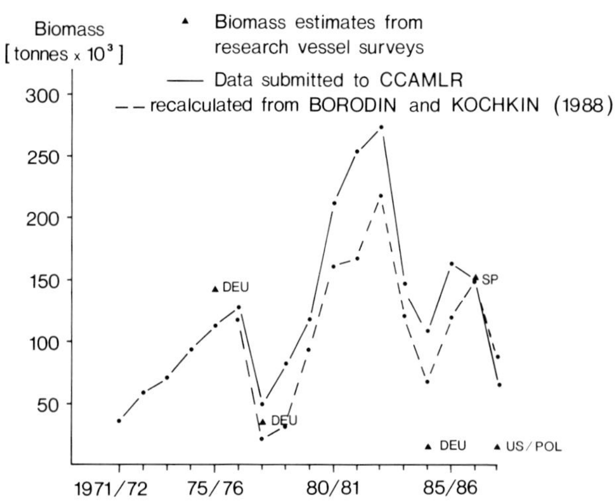

Fig. 5. Trends in biomass of *C. gunnari* around South Georgia from VPA analysis

3.3.3 Recruitment

Estimated numbers of recruits (age class 1) from both VPA data sets show the same recruitment pattern during the period 1977/78 to 1985/86 (Fig. 6a, b). Re-

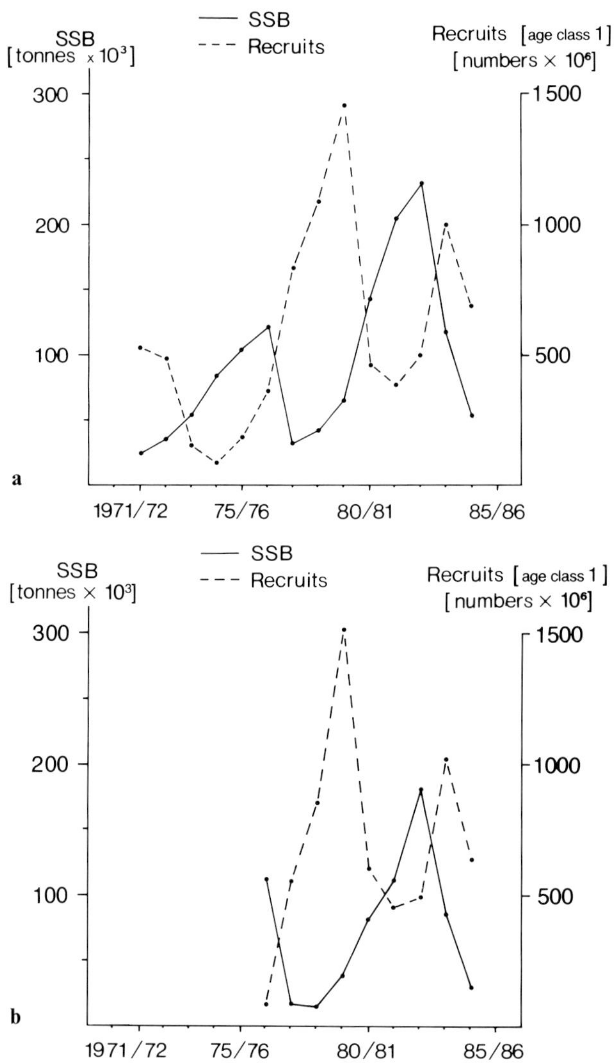

Fig. 6. Trends in recruitment and spawning stock biomass in *C. gunnari* around South Georgia (*numbers* of recruits refer to those seasons in which they were actually produced): **a** from data submitted to CCAMLR; **b** recalculated from Borodin and Kochkin (1988)

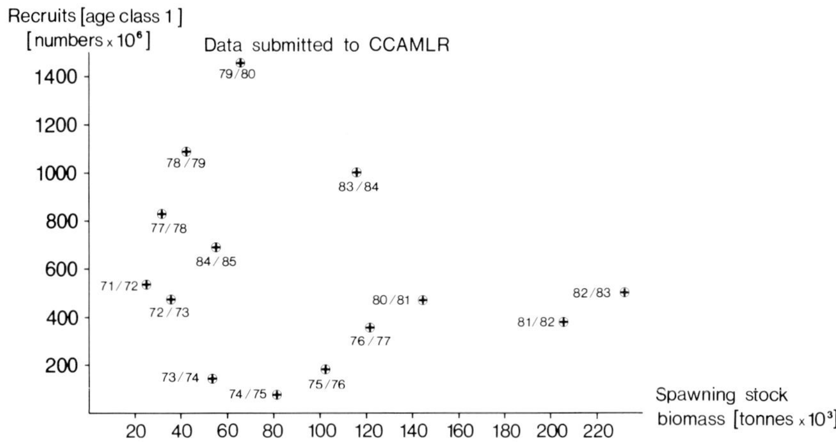

Fig. 7. Recruitment in relation to spawning stock biomass in *C. gunnari* around South Georgia (*numbers* of recruits refer to those seasons in which they were actually produced)

cruitment between seasons was different up to a factor of 19. Year classes were poor in the period 1973/74 to 1976/77 and highest at the end of the 1970s (1978/79–1979/80) and 1983/84 (Fig. 7).

Figure 6a, b indicates an opposite trend between the level of recruitment and spawning stock biomass at least until the early 1980s. Strong year classes were mainly produced when the spawning stock biomasses were relatively low. This was most obvious in the second half of the 1970s. Density-dependent effects may be responsible for that relationship. The nature of any density-dependent response has not been identified to date; food studies (see Kock 1985b, for review) gave no evidence that *C. gunnari* are cannibalistic. An artefact introduced by the data sets for the VPAs, however, cannot be excluded.

3.3.4 Catch Prediction and Stock Recovery

Due to the large fluctuations in recruitment, no medium-term yield and stock projections have been carried out. Short-term projections are illustrated in Fig. 8 for 1989/90. Catch prediction for 1989/90 is little affected by the difference between the data submitted to CCAMLR and those recalculated from Borodin and Kochkin (1988) but is much more dependent on the different levels of recruitment in 1988/89 to 1990/91.

At $F_{0.1}$ ($= F = 0.252$) the catch projection for data submitted to CCAMLR gives a catch of 6400 to 11500 tonnes and a biomass of 89000–169000 tonnes (SSB: 66000–125000 tonnes) in 1990/91 depending on the assumed recruitment level. Fishing at F_{max} ($= F = 0.592$) would yield 13500–24500 tonnes in

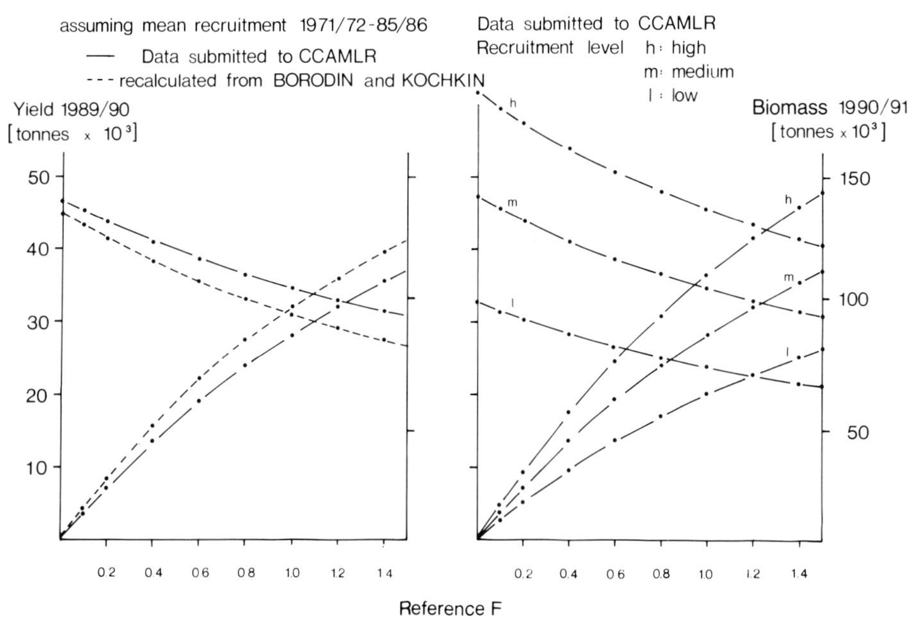

Fig. 8. Short-term yield and stock projections for *C. gunnari* around South Georgia in 1989/90

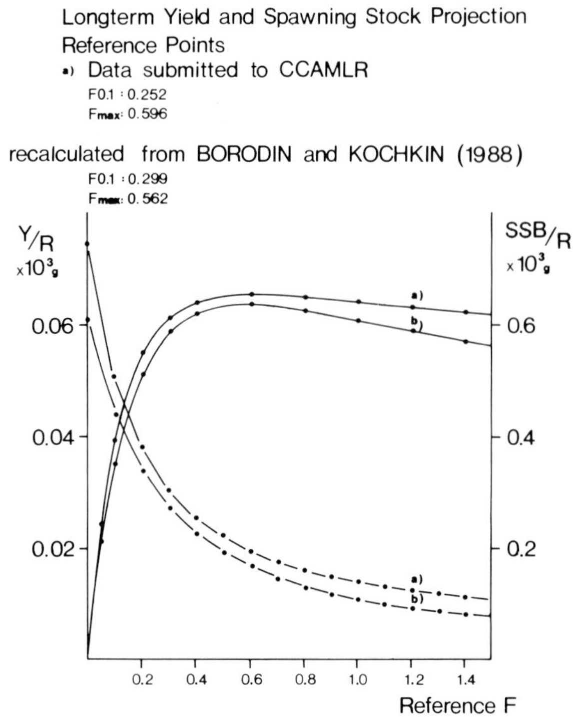

Fig. 9. Yield per recruit (*Y/R*) analysis and spawning stock per recruit (*SSB/R*) projection for *C. gunnari* around South Georgia

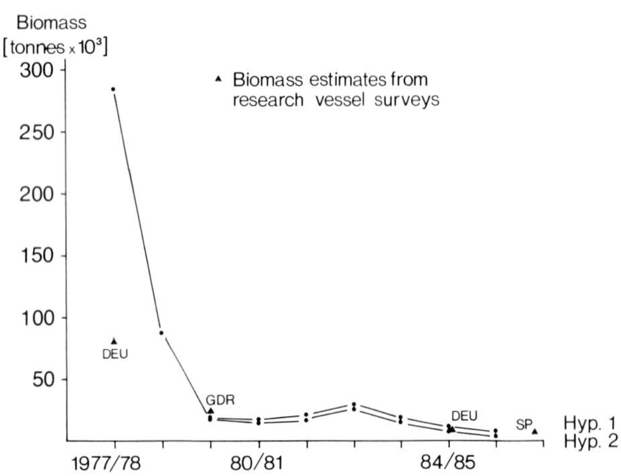

Fig. 10. Trends in biomass of *C. gunnari* around the South Orkney Islands and in the Peninsula region

1989/90 and a biomass of 80 000 – 152 000 tonnes (SSB: 57 000 – 109 000 tonnes). Harvesting at 2 × F_{max} (= F = 1.192) would allow catches of 22 600 – 41 300 tonnes in 1989/90 and would result in a stock size of 68 000 – 130 000 tonnes (SSB: 46 000 – 88 000 tonnes) in 1990/91.

The long-term yield and spawning stock projection is shown in Fig. 9. Long-term equilibrium yield at $F_{0.1}$ (= F = 0.252 and 0.292) for a medium level of recruitment of 559 and 629 × 10^6 individuals would result in 32 800 and 37 000 tonnes at an equilibrium stock biomass of 220 000 and 200 000 tonnes (SSB: 187 000 and 161 000 tonnes). At F_{max} (= F = 0.596 and 0.562) equilibrium yield would be 36 400 and 40 000 tonnes at an equilibrium stock biomass of 140 000 and 142 000 tonnes (SSB: 107 000 and 110 000 tonnes). A reduction of fishing mortality on age classes 1 – 3 to 10, 20 and 50% of the present level by increasing size at first capture by mesh size regulations would increase $F_{0.1}$ only slightly to 0.262 but F_{max} to 0.883. The gain in long-term yield, however, would only be approx. 2 500 tonnes.

3.4 *Champsocephalus gunnari* – South Orkney Islands and Antarctic Peninsula

3.4.1 *The Course of the Fishery*

Catches around the South Orkney Islands were large in the first fishing season in 1977/78, when 138 895 tonnes were taken with pelagic trawls mostly in an area southeast of Laurie Island. Yields dropped to 21 439 tonnes in the subsequent season and were in the order of only a few thousand tonnes annually since then.

Large-scale fishing in the Peninsula region was confined to the 1978/79 season when 35 930 tonnes were reported to have been taken around Elephant Island, King George Island and Deception Island (Sosinski 1981; Gubsch 1982). Fishing was continued until 1982/83 (except in 1981/82) but annual catches were less than 2000 – 3000 tonnes.

3.4.2 *Trends in Biomass*

Biomass estimates for both VPA hypotheses were similar despite the substantial difference in terminal F's (Fig. 10). Biomass in 1977/78 was 285 000 tonnes, but had already been reduced to 14 000 – 29 000 tonnes by 1979/80 – 1983/84. In subsequent years, biomass had further declined to less than 10 000 tonnes (Fig. 10). Biomass estimates for the 1980s from VPA were close to those from fishery-independent surveys (Fig. 10).

3.4.3 *Recruitment*

Number of recruits was in the range of 33 – 204 × 10^6 individuals during the period 1977/78 – 1981/82 but declined sharply to less than 1 × 10^6 recruits afterwards. It should be noted that this might be an artefact from the catch-at-age matrix in the VPA. It cannot be excluded, however, that recruitment was weak since 1982/83, because 1-year-old fish were virtually absent from the commercial and research vessel catches during the period 1982/83–1985/86. One- and 2-year-old fish were present in research vessel catches around the South Orkney Islands in 1986/87, when they were 12–14 and 20–25 cm long respectively and in the Peninsula region in 1987/88.

3.4.3 *Long-Term Yield and Stock Recovery*

Trends in biomass and recruitment show that both stocks are at a very low level at present. Short-term predictions seem less reliable due to the uncertainties in the estimation of recruitment since 1982/83. Restoration of the population to former stock levels will require recruitment of the order observed in the late 1970s for several years.

For a long-term projection, yield per recruit analysis gave reference F's at $F_{0.1}$ and F_{max} of 0.275 and 0.574 respectively. Both reference F's as well as Y/R and SSB/R are similar to those for the stock around South Georgia (Fig. 11). Equilibrium yield at a mean level of recruitment in 1977/78–1981/82 of 80×10^6 individuals would be 4800 tonnes at an equilibrium stock size of 28 000 tonnes (SSB: 19 000 tonnes) and 5200 tonnes at a stock size of 17 000 tonnes (SSB: 9000 tonnes) for $F_{0.1}$ and F_{max} respectively. Even at zero catch, stock size would only increase to 58 000 tonnes (SSB: 48 000 tonnes) which is only 20% of the unexploited stock size in the beginning of 1977/78.

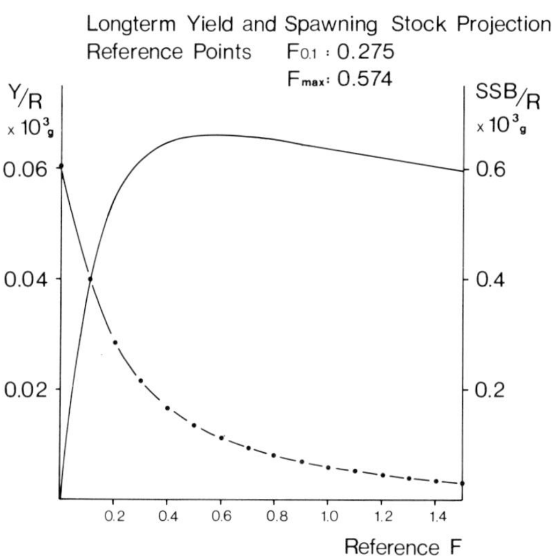

Fig. 11. Yield per recruit (*Y/R*) analysis and spawning stock per recruit (*SSB/R*) projection for *C. gunnari* around the South Orkney Islands and in the Peninsula region

3.5 *Patagonotothen brevicauda guntheri*

3.5.1 *The Course of the Fishery*

Catches in the first two seasons 1978/79 and 1979/80 were 15 011 and 7381 tonnes respectively. They rose to 36 758 and 31 351 tonnes in the next two seasons but fell to 5029 tonnes in 1982/83. Since 1983/84 they were varying around 10 000 tonnes.

3.5.2 *Trends in Biomass*

The biomass estimate of 81 680 tonnes from a Spanish survey in 1986/87 (Balguerias et al. 1987), which has been used to adjust the terminal F values in the VPA, was considerably lower than the estimate of 125 300 tonnes of Shust and Borodin (1988). Shust and Borodin (1988) used effort data instead of biomass estimates to adjust the VPA. Their VPA results, however, exhibit a similar trend to that calculated from data submitted to CCAMLR and data recalculated from Shust and Borodin (1988). The different values of M were primarily responsible for the difference in stock sizes in the first years of fishing (Fig. 12). All VPA runs based on high values of M (i.e. 0.8 or 0.9, as used by Shust and Borodin 1988) resulted in a pronounced downward trend in biomass from about 200 000 tonnes at the end of the 1970s to 60 000–120 000 tonnes in 1986/87 and 1987/88 respectively.

3.5.3 *Recruitment*

At high rates of natural mortality, numbers of recruits declined from the onset of fishing until 1983/84

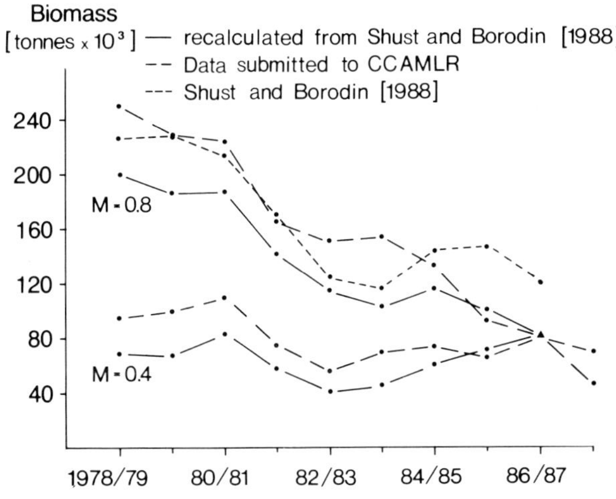

Fig. 12. Trends in biomass of *P. br. guntheri* around South Georgia (Shag Rocks)

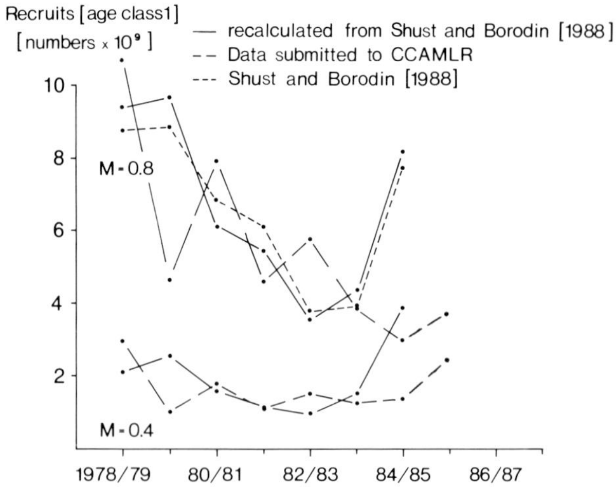

Fig. 13. Trends in recruitment of *P. br. guntheri* around South Georgia (Shag Rocks)

(Fig. 13). The substantial increase in recruitment in 1984/85 estimated from data of Shust and Borodin (1988) was not evident from data submitted to CCAMLR (Fig. 13). According to the data submitted to CCAMLR, recruitment increased slightly in 1985/86 (Fig. 13).

3.5.4 Catch Prediction and Stock Recovery

Short-term projections have been carried out for 1988/89 using input values recalculated from Shust and Borodin (1988) for $M = 0.8$ and $M = 0.4$ respectively (Fig. 14). The original results of Shust and Borodin (1988) for $M = 0.9$ have been presented for comparison (Fig. 14). As the projections are influenced primarily by differences in M and not by differences between data sets, data submitted to CCAMLR have not been taken into consideration.

At $F_{0.1}$ ($= F = 0.975$ and 0.374 respectively) the catch in 1988/89 would be 12 500 tonnes for both values of natural mortality, assuming mean recruitments of 6663 and 1963×10^6 individuals and a reported catch of 13 424 tonnes in 1987/88 (CCAMLR 1988 b). Resulting stock sizes at the beginning of the 1989/90 season will be 97 000 and 66 000 tonnes respectively (SSB: 33 000 and 43 000 tonnes). Based on original data of Shust and Borodin (1988) fishing at $F_{0.1}$ ($= F = 1.45$) and a mean recruitment of 6775×10^6 individuals would result in a catch of 24 800 tonnes in 1988/89 and a stock size of 133 000 tonnes (SSB: 52 000 tonnes) at the beginning of 1989/90.

Long-term projection resulted in the following reference points for the data set recalculated from Shust and Borodin (1988)

$$M = 0.4\text{:}\ F_{0.1} = 0.374;\quad F_{max} = 0.887$$
$$M = 0.8\text{:}\ F_{0.1} = 0.975;\quad F_{max} > 2.0 \text{ (Fig. 15).}$$

Equilibrium yield at $F_{0.1}$ was 14 600–20 300 tonnes at a stock size of 78 000–132 000 tonnes (SSB: 42 000–54 000 tonnes).

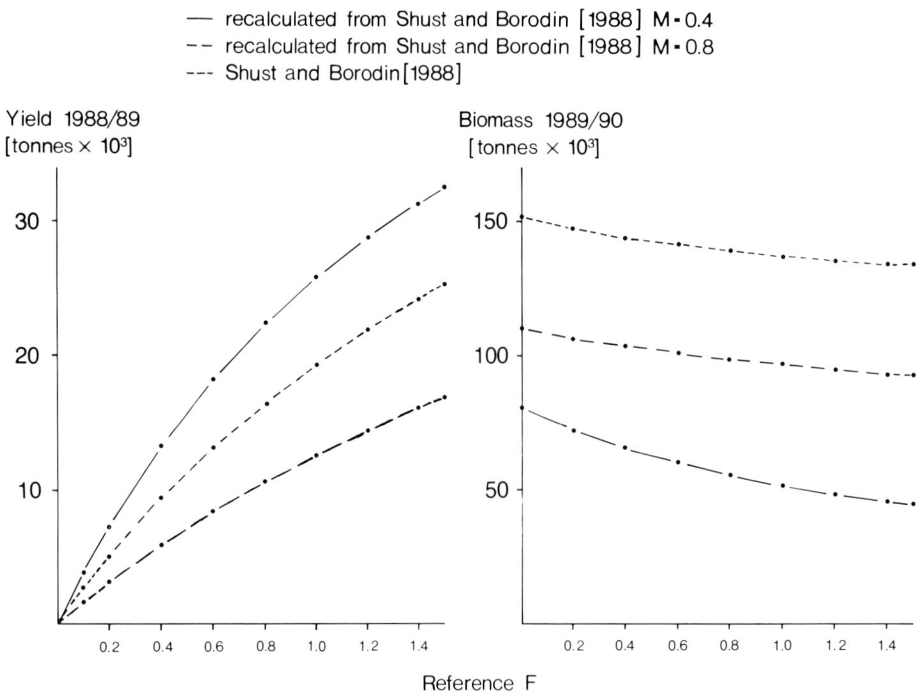

Fig. 14. Short-term yield and stock projections for *P. br. guntheri* around South Georgia (Shag Rocks) in 1988/89

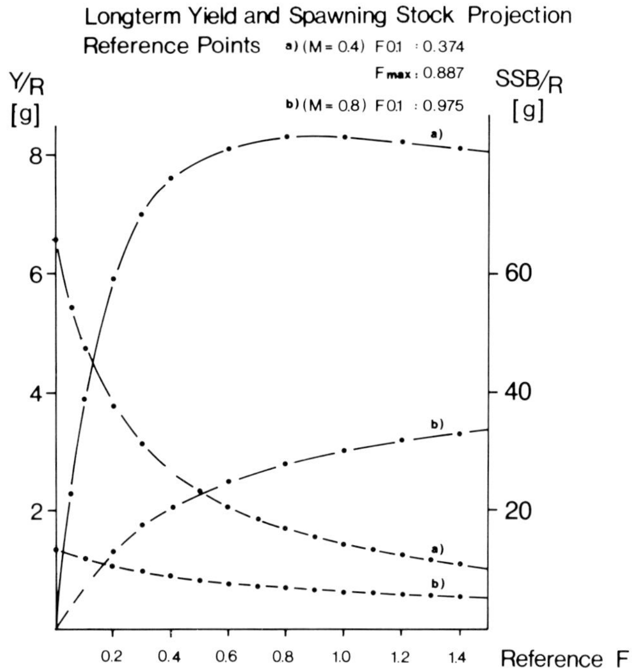

Fig. 15. Yield per recruit analysis (*Y/R*) and spawning stock per recruit (*SSB/R*) projection for *P. br. guntheri* around South Georgia (Shag Rocks) (recalculated from Shust and Borodin 1988)

4 Discussion

Our analysis confirms the main conclusions which the Working Group drew in previous years, that some of the target stocks have been reduced to small fractions of their original size (CCAMLR 1986, 1987a, 1988a). This was particularly so in the case of *Notothenia rossii marmorata*. The current stock biomass is only ca. 3% of the unexploited stock size. Recruitment is much lower than in the 1970s. It is not clear whether this is a result of a stock-recruitment relationship alone or if, for example, increased predation by fur seals in the fjords, where juvenile *N. rossii marmorata* develop, may have enhanced this effect. Recovery rates are slow at the present level of recruitment irrespective of a zero catch or catch levels of less than 1000 tonnes. From calculations based on a simulation of stock performance an increase in stock size to about 40000 tonnes could be expected at the turn of the century. This is still less than 10% of the unexploited stock size. Catches exceeding 1000 tonnes could lead to a collapse of the stock.

Among the conservation measures established by CCAMLR so far the prohibition of any directed fishing on *N. rossii marmorata* in 1986 was the only regulation of direct and immediate benefit for the stock. The recent establishment of a closed season from 1 April to 30 September 1988, which aimed primarily at protecting *C. gunnari* and a number of by-catch species, would protect spawning and spawning grounds of *N. rossii marmorata* so far as they are known (Permitin and Silyanova 1971). Some effect can be expected from the prohibition of fishing in waters within 12 nm of South Georgia where two important fishing grounds, at least for the Polish fishery, are located (W. Slosarczyk pers. commun.). Mesh size regulations are likely to have no benefit: The juveniles inhabit the fjords which are inaccessible to a trawl fishery. No gain in Y/R or increase in spawning stock biomass can be expected when size at first capture is increased by increasing minimum mesh size beyond the established 120 mm (Kock et al. 1985).

Recovery of the stock should be monitored closely with an annual assessment to estimate stock size and recruitment. However, assessments are seriously hampered at present due to the substantial differences in age readings since 1984/85 compared with the period 1969/70–1984/85, the irregularity of fishery-independent surveys and the lack of data from the commercial fishery since 1985/86 due to the prohibition of directed fishing. Provided ageing difficulties could be resolved and regular research vessel surveys are not likely to be carried out, another management option might be to establish a TAC of ca. 300 tonnes for a directed fishery on the stock, provided information from this fishery is submitted to CCAMLR annually. The development of regular surveys on juvenile fish in shallow waters around shore stations is urgently required to help estimate recruitment more accurately.

The status of *N. rossii marmorata* in the Antarctic Peninsula region is much more uncertain due to the lack of the time series samples required for virtual population analysis. The stock has been greatly reduced by the fishery and there is little indication of recovery. Conservation measures (CCAMLR 1988d) should be kept in force.

Stock size of *C. gunnari* around South Georgia is now strongly dependent on the strength of the incoming cohort, and the fishery is currently based on age classes 2 and 3, of which age class 2 is not yet fully recruited. During 1976/77 when the fishery was at its peak, 4- and 5-year-old fish were the major component of the catch.

In general, strong year classes appear to be produced when spawning stock biomass is at a lower or medium level and not when the spawning stock biomass is high. We do not know if a minimum spawning stock size exists below which recruitment failure is likely to occur. Spawning stock size should thus be kept at a medium level which would allow strong year classes under variable recruitment and a reasonable yield for the fishery as well as minimizing the risk of a stock's collapse as a result of recruitment failure.

To keep spawning stock size at a medium level will require conservation measures. TAC's which were introduced by CCAMLR in 1987 for the first time seem to be of immediate benefit for the stock. An increase of minimum size at first capture by increasing mesh size beyond the regulation size of 80 mm would increase yield and spawning stock biomass per recruit (CCAMLR 1988 a). It would further lead to better protection of the first spawners. This would mean, however, that information on the abundance of age class 2 would be lost. This is particularly crucial since no regular prerecruit surveys have been carried out so far to estimate the strength of the incoming year class, which is a prerequisite for the catch prediction for setting a TAC in the following season. Furthermore, available results from net selectivity experiments are preliminary so far, and are not sufficiently reliable as a basis for the introduction of revised mesh size regulations for *C. gunnari* (CCAMLR 1988 a).

The 1985/86 year class is possibly weak. The low catches of that cohort in 1986/87 and 1987/88 led to an estimated stock size of less than 100 000 tonnes (SSB: 70 000 tonnes) at the beginning of the fishing season 1988/89, irrespective of the recruitment assumed for 1987/88 and 1988/89. Depending on the level of recruitment and the preliminary catch of 23 000 tonnes in 1988/89 (the age composition of which is so far unknown) the predicted stock size at the beginning of the 1989/90 season, based on the data submitted to CCAMLR, will be 66 000–124 000 tonnes. Catch predictions, however, are much more dependent on the level of recruitment than on the management option (e.g. $F_{0.1}$, F_{max}) chosen. Bearing in mind that historical recruitment varied by a factor of up to 19 between seasons, short-term projections based on mean recruitment values are of limited value. This clearly underlines the urgent need for regular prerecruit surveys to manage the stock, so as to take full advantage of the occasional strong year classes.

As long as indices from regular recruitment surveys are unavailable, a conservative approach for a TAC in 1989/90 seems appropriate. Fishing mortality should not exceed $F_{0.1}$. The assessment will be subject to revision when the catch compositions of the 1988/89 season are available.

The other conservation measures should be kept in force. The prohibition of fishing in waters within 12 nm of South Georgia and the closed fishing season from 1 April to 30 September fully protect the spawning and spawning grounds of the stock. As males start their spawning migration earlier than females (Kock 1981), it may be advisable to extend the closed season to 1 March to protect prespawning concentrations.

The peak fishery on *C. gunnari* around the South Orkney Islands and in the Peninsula region occurred in the second half of the 1970s and was based on two very abundant year classes, which were exhausted in 1980/81. Most of the fish should have been juveniles. Year classes of that strength have not been observed since then.

Results from our analytical assessment still tend to be of poor precision due to the limited time series samples available for analysis (covering less than 10 years) and the large variations in fishing intensity and the substantial fluctuations in annual catches inherent in them. Our analysis nevertheless provided information on the historical changes in stock size which clearly showed a substantial downward trend since the onset of fishing. Stock size of *C. gunnari* at the South Orkney Islands seems to be less than 10 000 tonnes at present. From 1982/83 onwards recruitment was obviously low, although there are indications that recruitment values obtained from the VPA may be (partly) artefacts. However, if the apparent recruitment failure is real, the stocks are in urgent need of protection by the introduction of conservation measures.

The most appropriate way to allow the stocks to recover to some extent from the heavy depletion would be to immediately stop directed fishing of the species around the South Orkney Islands and in the Peninsula region. The area, in particular the South Orkney Islands, is, however, only irregularly surveyed by research vessel cruises. The commercial fishery will be the only data source which can provide some information on the state of the stocks on a regular basis. It may therefore be advisable to set a low catch quota for directed fishing with a provision to submit all information necessary for an assessment to CCAMLR annually, similar to the system proposed for *N. rossii marmorata* around South Georgia.

Patagonotothen brevicauda guntheri is the only target species (stock) for which fishing was not regulated until 1988/89. Our analytical assessment is largely influenced by uncertainties in the determination of M. If natural mortality is as high as assumed by Shust and Borodin (1988), a downward trend in both stock size and recruitment is apparent.

Short-term catch predictions, assuming mean levels of recruitment for 1988/89, would result in a catch of 12 500 tonnes for both values of natural mortality ($M = 0.8$; $M = 0.4$) which is very close to the TAC of 13 000 tonnes set by CCAMLR for the same season based on the catch levels of the previous seasons.

It is encouraging that the data base for fish stock assessment purposes has improved considerably compared to the first years of CCAMLR, albeit far from satisfactory. There are problems in ageing the fish, and errors in ageing influence the quality of the assessments (Kock and Köster 1989). Regular fishery-independent recruitment surveys, particularly on *N. rossii marmorata* and age class 1 *C. gunnari*, are urgently needed. We hope that our findings as well as the results obtained during the most recent Working Group meetings (1987a, 1988a) will help CCAMLR to change from purely "reactive" management, to which the Commission adhered in the first years, to more predictive (i.e. active) management methods which would better suit the objectives of Article II of the Convention.

Acknowledgements. We are very grateful to our Soviet colleagues for permission to use their information from the Soviet commercial and exploratory fishery, to the former Data Manager of CCAMLR, Dr. L. D. Jacobson for his careful compilation and documentation of the information and Dr. W. de la Mare (Marine and Ecological Research Pty. Ltd. Beaumaris, Australia), Drs. T. G. Lubimova and K. V. Shust (All-Union Research Institute of Marine Fisheries and Oceanography, Moscow), W. Slosarczyk (Sea Fisheries Institute, Gdynia), and M. G. White (British Antarctic Survey, Cambridge) for their valuable comments on the manuscript. We appreciate Mrs. M. Jörck's enthusiasm and perseverance in preparing the graphs. We are indebted to Mrs. U. Harm and Mrs. J. Arndt for their technical assistance.

References

Balguerias E, Bruno J, Daroca E, Quintero ME (1987) Estimation de la biomasa del algunas especies capturadas durante la campana "Antartida 8611". Actas del Segundo Symposium Espanol de Estudios Antarticos, Madrid, Julio 1987:285–309

BIOMASS (1980) Working Party on Antarctic fish biology. Report of the second meeting. Dammarie-les-Lys, France, 27–31 May 1980, BIOMASS Rep Ser 12, 47 p

Borodin R, Kochkin P (1988) *Champsocephalus gunnari* stock status in the South Georgia area. CCAMLR WG-FSA 88/32, 14 p

Burchett MS, Ricketts C (1984) The population dynamics of *Notothenia rossii* from South Georgia (Antarctica). Polar Biol 3(1):35–38

CCAMLR (1984) Report of the Third Meeting of the Scientific Committee. Hobart, Australia, 3–13 September 1984, 238 p

CCAMLR (1985) Report of the Fourth Meeting of the Scientific Committee. Hobart, Australia, 2–13 September 1985, 275 p

CCAMLR (1986) Report of the Fifth Meeting of the Scientific Committee. Hobart, Australia, 8–15 September 1986, 271 p

CCAMLR (1987a) Report of the Sixth Meeting of the Scientific Committee. Hobart, Australia, 26 October–3 November 1987, 263 p

CCAMLR (1987b) Data and assessments for fish stocks in the Convention Area. CCAMLR WG-FSA-87/4

CCAMLR (1988a) Report of the Seventh Meeting of the Scientific Committee. Hobart, Australia, 24–31 October 1988, 207 p

CCAMLR (1988b) Summary of fisheries data. SC-CAMLR VII/BG/2/Rev 1, 79 p

CCAMLR (1988c) Report of the Seventh Meeting of the Commission. Hobart, Australia, 24 October–4 November 1988, 121 p

CCAMLR (1988d) Commission for the Conservation of Antarctic Marine Living Resources. Basic Documents, 3rd edn, Hobart, Australia, 128 p

Croxall JP, Prince PA, Ricketts C (1985) Relationships between prey life cycles and the extent, nature and timing of seal and seabird predation in the Scotia Sea. In: Siegfried WR, Condy PR, Laws RM (eds) Antarctic nutrient cycles and food webs. Springer, Berlin Heidelberg New York, pp 516–533

Doidge DW, Croxall JP (1985) Diet and energy budget of the Antarctic fur seal *Arctocephalus gazella*. In: Siegfried WR, Condy PR, Laws RM (eds) Antarctic nutrient cycles and food webs. Springer, Berlin Heidelberg New York, pp 543–550

Everson I (1977) The living resources of the Southern Ocean. FAO GLO/SO/77/1, Rome, 156 p

Everson I (1978) Antarctic fisheries. Polar Rec 19:233–251

Gabriel WL (1987) Results of fish stock assessment survey, South Georgia region, November–December 1986. Selected Papers Presented to the Scientific Committee of CCAMLR 1987, pp 143–183

Gubsch G (1982) Zur Verbreitung und zur Biologie der Eisfische (Chaenichthyidae) im atlantischen Sektor der Antarktis. Fischerei-Forschung 20(2):39–47

Gulland JA (1965) Estimation of mortality rates. In: Cushing DH (ed) Key papers on fish populations (1983). IRL Press, Oxford, pp 231–241

Gulland JA (1983) The development of fisheries and stock assessment of resources in the Southern Ocean. Mem Natl Inst Polar Res Tokyo, Spec Iss 27:233–246

Kimura DK, Chikuni S (1987) Mixtures of empirical distributions: an iterative application of the age-length key. Biometrics 43:23–37

Kock K-H (1981) Fischereibiologische Untersuchungen an drei antarktischen Fischarten: *Champsocephalus gunnari* Lönnberg, 1905, *Chaenocephalus aceratus* (Lönnberg, 1906) und *Pseudochaenichthys georgianus* Norman, 1937 (Notothenioidei, Channichthyidae). Mitt Inst Seefisch Hamb 32:1–226

Kock K-H (1985a) The state of exploited Antarctic fish stocks around South Georgia (Antarctic). Arch Fischereiwiss 36(1/2):155–183

Kock K-H (1985b) Krill consumption by Antarctic notothenioid fish. In: Siegfried WR, Condy PR, Laws RM (eds) Antarctic nutrient cycles and food webs. Springer, Berlin Heidelberg New York, pp 437–443

Kock K-H (1986) The state of exploited fish stocks in the Scotia arc region during SIBEX (1983–1985). Arch Fischereiwiss 37(Beih. 1):129–186

Kock K-H, Duhamel G, Hureau JC (1985) Biology and status of exploited Antarctic fish stocks: a review. BIOMASS Sci Ser 6:1–143

Kock K-H, Köster FW (1989) The state of exploited fish stocks in the Atlantic sector of the Southern Ocean. Mitt Inst Seefisch Hamb 46:1–73

Lubimova TG (1985) Soviet fishery investigations conducted in the Southern Ocean. Selected Papers Presented to the Scientific Committee of CCAMLR 1982–1984, 1:329–351

Payne MR (1977) Growth of a fur seal population. Phil Trans R Soc Lond B 279:67–79

Permitin Yu Ye, Silyanova ZS (1971) New data on the reproductive biology and fecundity of the genus *Notothenia* Richardson in the Scotia Sea. Vopr Ikthiol 11:693–705 (in Russian)

Shust KV, Borodin R (1988) *Notothenia* (*P.*) *guntheri* stock status and TAC estimation in the area of Shag Rocks (Subarea 48.3). CCAMLR WG-FSA-88/33, 10 p

Slosarczyk W, Sosinski J, Much M, Skora KE, Kompowski A (1985) A review of Polish fishery and assessment of fish stock biomass off South Georgia. Selected Papers Presented to the Scientific Committee of CCAMLR 1982–1984, 1:395–422

Sosinski J (1981) Biologia porownacza Kergulen (*Champsocephalus gunnari*) Z rejonow Antarktyki. Wydaw Morsk Inst Ryb, Studia y Materialy Ser B 48:1–91

Thompson WF, Bell FH (1934) Biological statistics of the Pacific halibut fishery. 2. Effects of changes in intensity upon total yield and yield per unit of gear. Rep Int Fish Comm 8:49 p

Changes in Fish Populations and Fisheries Around the Kerguelen Islands During the Last Decade

G. Duhamel and J.-C. Hureau [1]

Summary. Four species of fish (*Notothenia squamifrons, N. rossii, Dissostichus eleginoides* and *Champsocephalus gunnari*) have been or are exploited in the shelf zone of Kerguelen Islands and the Skif Bank since 1970. However, detailed records of the fishery date from 1980. Following the declaration of the exclusive economic zone in 1978, the fishery has been carefully controlled since stocks of *N. rossii* and *N. squamifrons* were shown to be severely depleted. This paper reviews the catch and fishing effort for the calendar years 1970 to 1978 and for the split years 1980/81 to 1987/88, and presents results on the changes in the demographic structures and the population abundance. These results were derived from virtual population analyses and catch per unit effort. The two methods although completely independent from each other, both show similar trends in the populations. A model utilising data up to 1987 was used to predict future fluctuations of the populations given different levels of fishing activity.

1 History

During the 19th century, the French subantarctic islands of the Indian sector of the Southern Ocean first attracted sealing and whaling companies. Fishermen came to the waters surrounding the Kerguelen Islands 30 years ago and the fish have only been regularly exploited since 1970 with factory ships.

From 1970 to 1978, there was no regulation of fishing in the area and, during he first years, the accumulated stock was heavily exploited by foreign fishing fleets: more than 200 000 tonnes taken for the split year 1971–72 (Fig. 1). Following the creation of the exclusive economic zone (EEZ) in 1978, there was no fishing for 18 months in 1978–79. The statistics during this first period are not reliable mainly because the exploited species were mixed in the reported catches and were not used for any stock assessment. Since 1979, the fishery in the 200-mile zone has been regulated by the French authorities. All the data (catches by species and fishing effort) have been regularly collected (fishing logbooks) and checked.

Several species are currently exploited on the shelf and surrounding banks inside the EEZ: *Champsocephalus gunnari, Notothenia squamifrons, N. rossii* and *Dissostichus eleginoides*, all are members of the suborder Notothenioidei, a group restricted to the Southern Ocean.

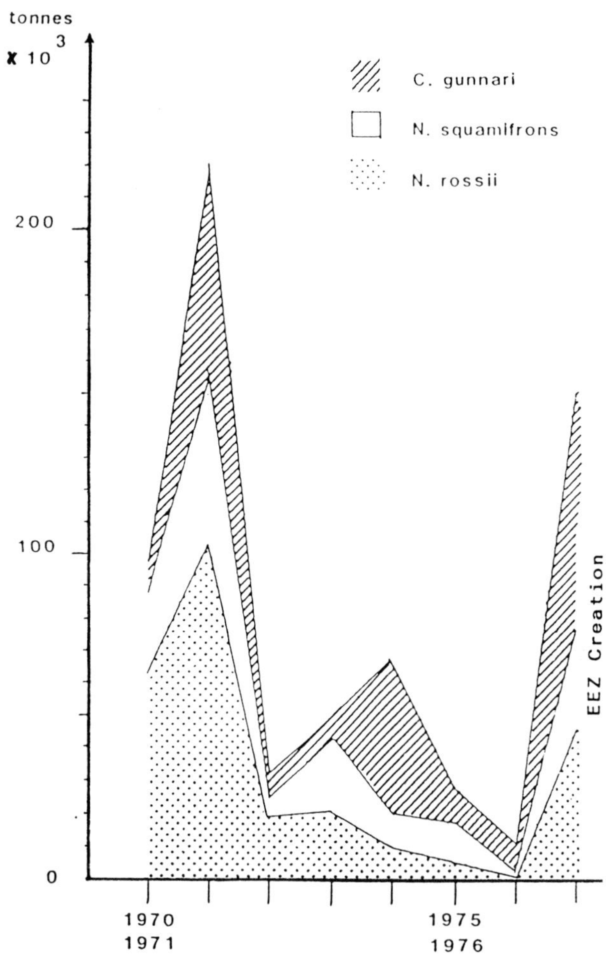

Fig. 1. Total catches (tonnes) of fish in the Kerguelen area from 1970 to 1978 (Duhamel 1989)

[1] Museum National d'Histoire Naturelle, Ichtyologie Générale et Appliquée, 43 rue Cuvier, 75231 Paris cedex 05, France.

Antarctic Ecosystems. Ecological Change and Conservation (ed. by K. R. Kerry and G. Hempel)
© Springer-Verlag Berlin Heidelberg 1990

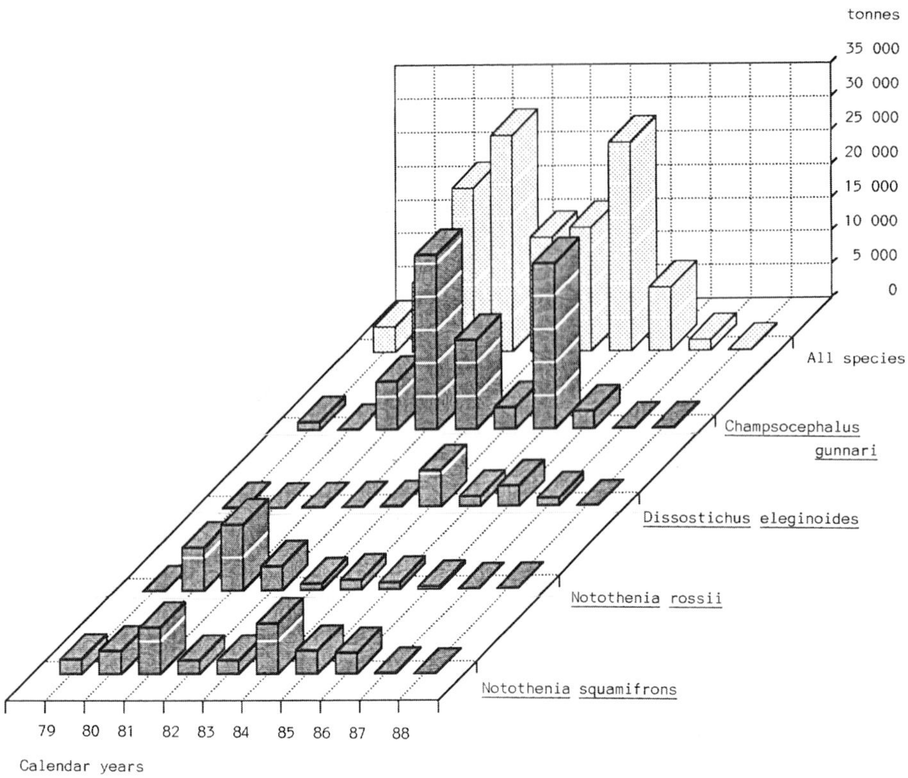

Fig. 2. Total catches (tonnes) and catches per species of fish in the Kerguelen area from 1979 to 1988

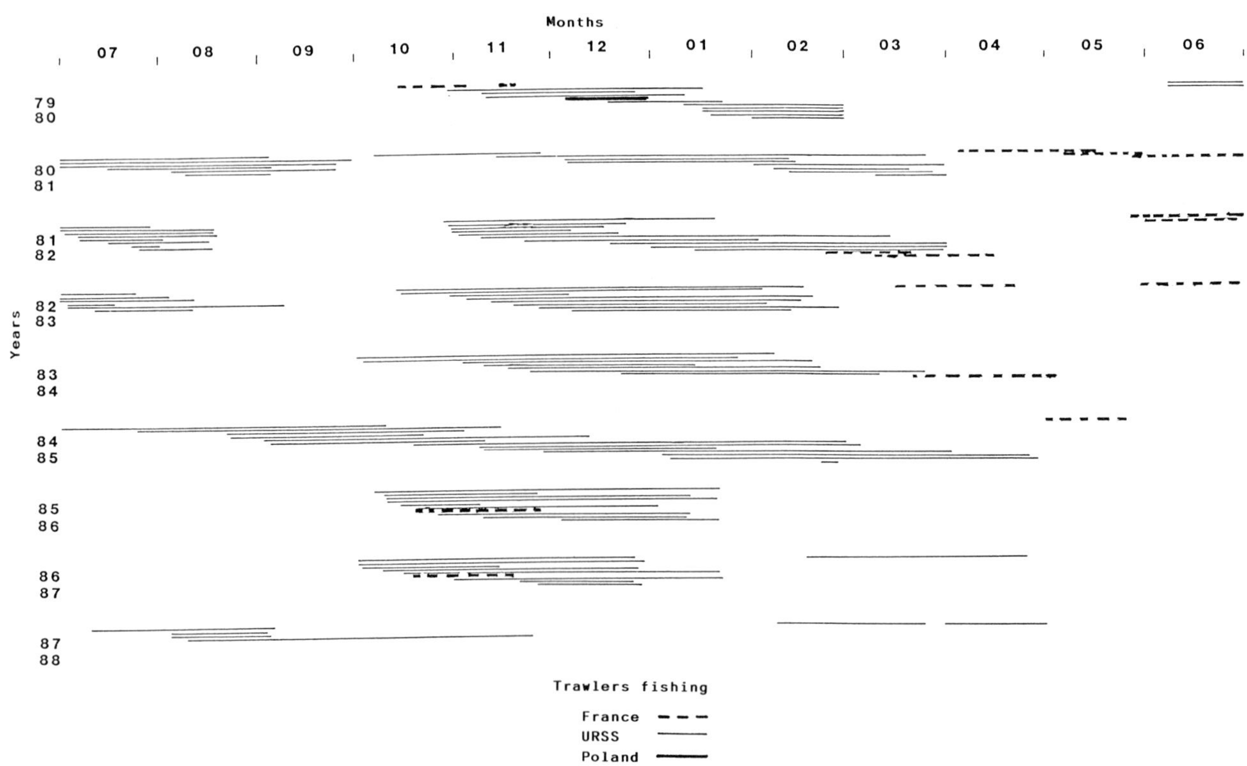

Fig. 3. Fishing effort in the Kerguelen area from 1979/80 to 1987/88: each *line* represents a trawler and the duration of stay on the fishing grounds

At the beginning of the exploitation in 1970, *N. rossii* was the dominant species in the catches, but its relative proportion decreased shortly afterwards (Fig. 1). More recently and especially since 1981–82 the most abundant species was the icefish, *Champsocephalus gunnari*, with good catches of more than 25 000 tonnes caught in 1982 and 1985 (Fig. 2). The catches of the marbled rockcod, *Notothenia rossii*, were still significant (up to 10 000 tonnes) in 1980 and 1981, but decreased to remain at less than 1000 tonnes from 1984. The catches of the grey rockcod, *N. squamifrons* only exceeded 5000 tonnes in 1981 and 1984. Finally, the Patagonian toothfish, *Dissostichus eleginoides*, appeared significantly in the statistics only after 1983 (catches over 5000 tonnes in 1984) but the catches decreased afterwards. The total catches (expressed in tonnes per split year: 1 July to 30 June) varied between 12 000 and 30 000 tonnes for the period 1979–1986 (= 1979/80 to 1986/87) and decreased to less than 2000 tonnes in 1986 (Duhamel 1987).

The fishing effort is represented in Fig. 3 by the number of trawlers present on the fishing grounds and the duration of each fishing cruise in the area. The commonly used fishing gear is the bottom trawl. Midwater trawls have been deployed recently for the icefish only. The effort is mainly concentrated during the Southern summer to comply with fishing regulations. The sea conditions at this time are also better. An historical review of fish catches and fishing effort in the Kerguelen area has been published by Duhamel (1987, 1989) and summaries were published by Kock et al. (1985) and Hureau and Slozarczyk (in press). Practical conservation measures in force (e.g. regulation of mesh size and prohibition of directed fishing on endangered species) follow the annual measures of the Commission for the Conservation of Antarctic Marine Living Resources (CCAMLR) for the regulations for specifically exploited fish stocks. Complementary measures (e.g. closed seasons or areas to protect spawning of depleted stocks, total allowed catches) are specific to the Kerguelen EEZ.

2 Establishment of a Regular Survey Programme

The fish of Kerguelen Islands shelf have been studied for more than 25 years but it is only with the creation of an EEZ in 1978 that control of the fish populations and fisheries was effectively established.

Lack of knowledge of the exploited fish populations before 1979 led to the establishment of a permanent research programme, using the presence of the French and foreign trawlers, research ships or ships of opportunity as observation platforms to obtain data in the Kerguelen EEZ.

Two different aspects have been developed in that survey programme from a biological and a statistical point of view:

Biological data and samples (length and weight of fish, scales, otoliths, stomach contents, gonads, etc.) are collected by scientists and/or fishery observers on board. Long-term biological sampling is used generally to obtain information on the growth, the sexual maturity, the spawning time, the fecundity, the diet, the demographic structures and the possible competition between exploited species.

Fishing statistics (catch per species, fishing effort) are collected for each individual haul (about 30 000 hauls for the period 1979 to 1988) and recorded in special fishing logbooks. The logbooks are given to the captains with instructions in French, English and Russian, together with the fishing licence, prior to fishing. Generally a fishery observer or scientist is present on board to validate the recorded data. All these data are computerized in a data base ("KERPECHE") which makes their utilization easier. The statistical data are used to establish the spatial and temporal distribution of exploited stocks, taking into account the biological data in order to define the possible spawning and feeding migrations; and to determine the abundance of the species in relation to the exploitation. From the analysis of these data it is possible, inter alia to determine the distribution of the exploited part of the population of each species during a certain period, if the haul coverage of the shelf and surrounding banks is wide enough.

The experimental phase of the survey programme commenced during the first controlled fishing season (1979/80) and began effectively in 1980.

3 Methods Used to Assess the State of the Stocks

From knowledge of the life cycles and biological characteristics (growth, sexual maturity, fecundity, mortality, migrations) of the exploited fish populations and analysis of statistical data, it is possible to use three different methods to evaluate the variation of the fish stocks around the Kerguelen Islands since 1979. These methods (demographic structure analysis, trends in the abundance index and virtual population analysis) are independent of each other and complementary.

The demographic structure analysis uses the annual length frequency distribution of the studied populations, associated with an age/length key. The abundance index uses the catch per unit effort (CPUE), expressed in tonnes of fish caught per fishing hour in the present study, but needs to be standardized to compare the various types of ships and gear. In the special case of the icefish *C. gunnari* for which only one triannual cohort (age class) is present on the fishing grounds at a given time (Duhamel 1987), the index of abundance is finally expressed in the number of fish caught per standardized fishing hour. Virtual population analysis (VPA) is a mathemathical model of the analytical type (Pope 1972) and can use the age or the length composition of the annual catches (Jones 1981). VPA analysis requires not only the numbers and weights of fish at each age for each year, but also the natural mortality coefficient M and the fishing mortality coefficient (terminal F), the mean age at recruitment and the age of full sexual maturity (age at which 100% of the fish are mature). All these data have been calculated or estimated by Duhamel (1987).

4 Results of the Survey Programme Since 1979

Figure 4 represents the presence or the absence of *Notothenia rossii* during the winter 1980, from the analysis of 1235 hauls (1192 on the shelf of Kerguelen Islands, 9 on the Skif Bank and 34 on the Kerguelen/Heard banks).

Figures 5–7 present the general distribution of the adult population of the three main species as observed from 1979 to 1986: *Champsocephalus gunnari* (Fig. 5) is a shallower-water species and has a wider distribution to the northern part of the shelf than the two other species, *Notothenia rossii* (Fig. 6) and *N. squamifrons* (Fig. 7), which are present from the southern to the eastern part of the shelf, and at greater depth. These two species segregated also by depth.

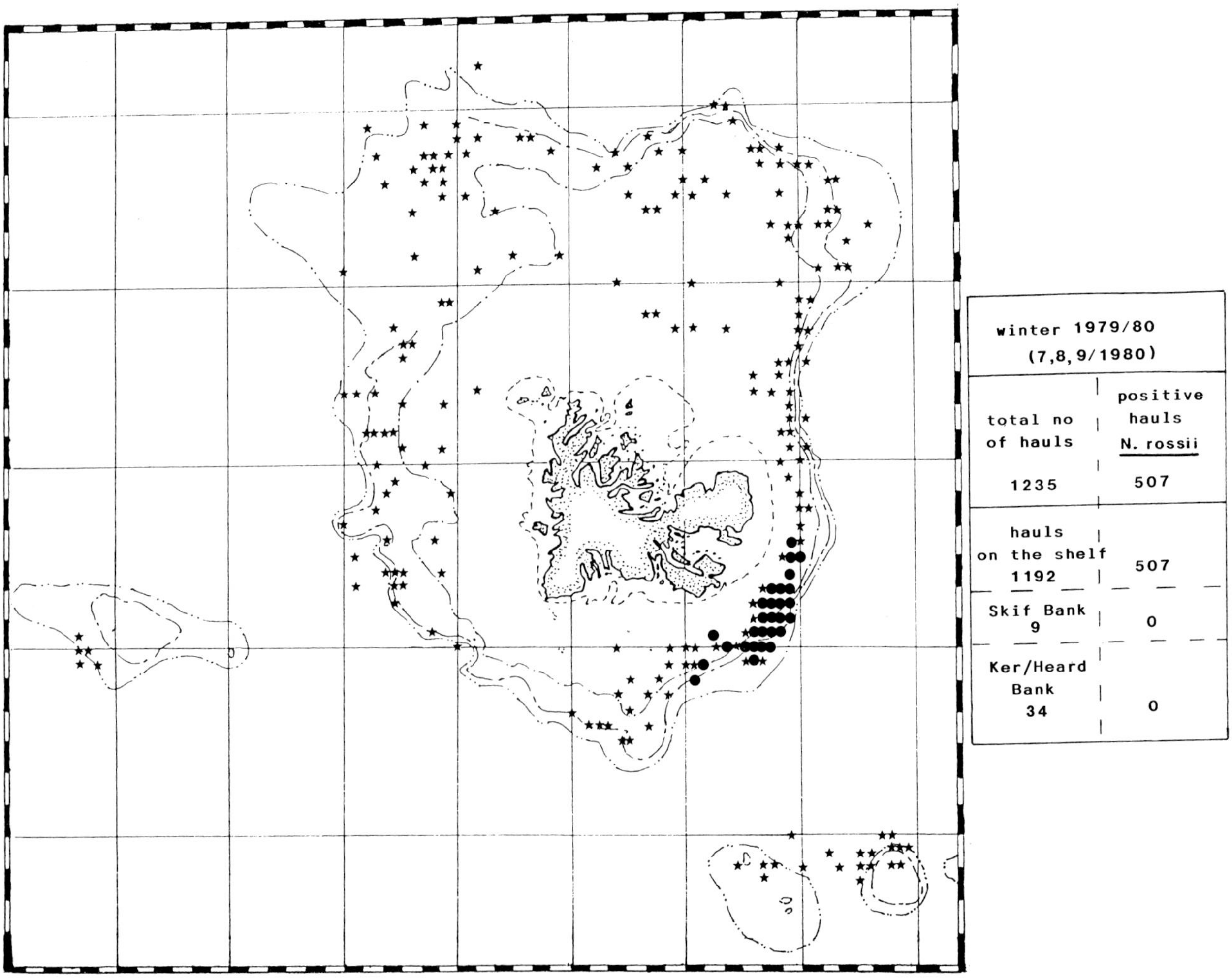

Fig. 4. Distribution of *Notothenia rossii* in winter 1979/80 around Kerguelen Islands. *Black circles* presence of the species in the hauls. *Stars* absence of the species in the hauls

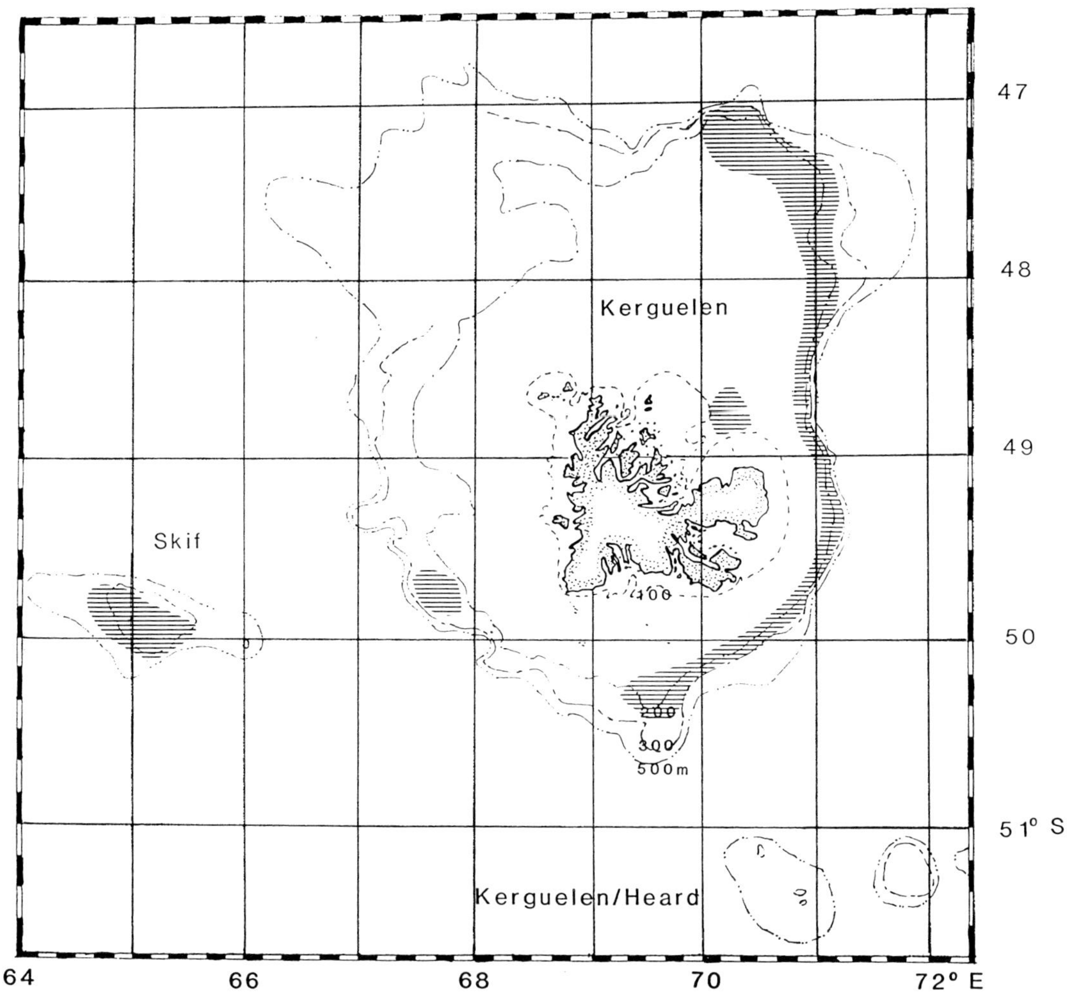

Fig. 5. General distribution of *Champsocephalus gunnari* observed from 1979 to 1986 (*hatched areas*)

The time and location (Fig. 8) of spawning for the exploited species can also be obtained from the data recorded in the data base (catches and position of hauls) associated with the sequential analysis of the sexual maturation (Duhamel 1987). Each species spawns in specific geographical areas. As a consequence, for some species, different stocks (shelf and offshore banks) have been established. Such results are very important for the stocks management. Finally, associating all the biological data and the geographical distribution data, it is possible to determine a life cycle for each species: an example is given in Fig. 9 for *Notothenia rossii* which does not differ significantly from the cycle observed in the South Georgia stock of the same species (White et al. 1982).

These results were necessary before using various methods to assess the state of the different exploited stocks.

4.1 *Notothenia rossii* and *Notothenia squamifrons*

Representative length frequency distributions of adult *Notothenia rossii*, exploited around the Kerguelen Islands, were obtained from 1980 to 1984. During that period, the removal of the oldest specimens resulted in a decrease in the mean length of the adult stock to a value close to the length at the first maturity. There was decrease in the abundance index catch per unit effort (CPUE) from 1980 to 1982 reflecting a decline in the stock biomass. This was confirmed by a virtual population analysis (Duhamel 1987). This situation resulted partly from directed fishing on the single spawning ground in austral winter during these 3 years, and partly because fishing took place on a stock that was already at a very low level in 1980 after an intensive period of fishing from 1970 (Fig. 1). It was

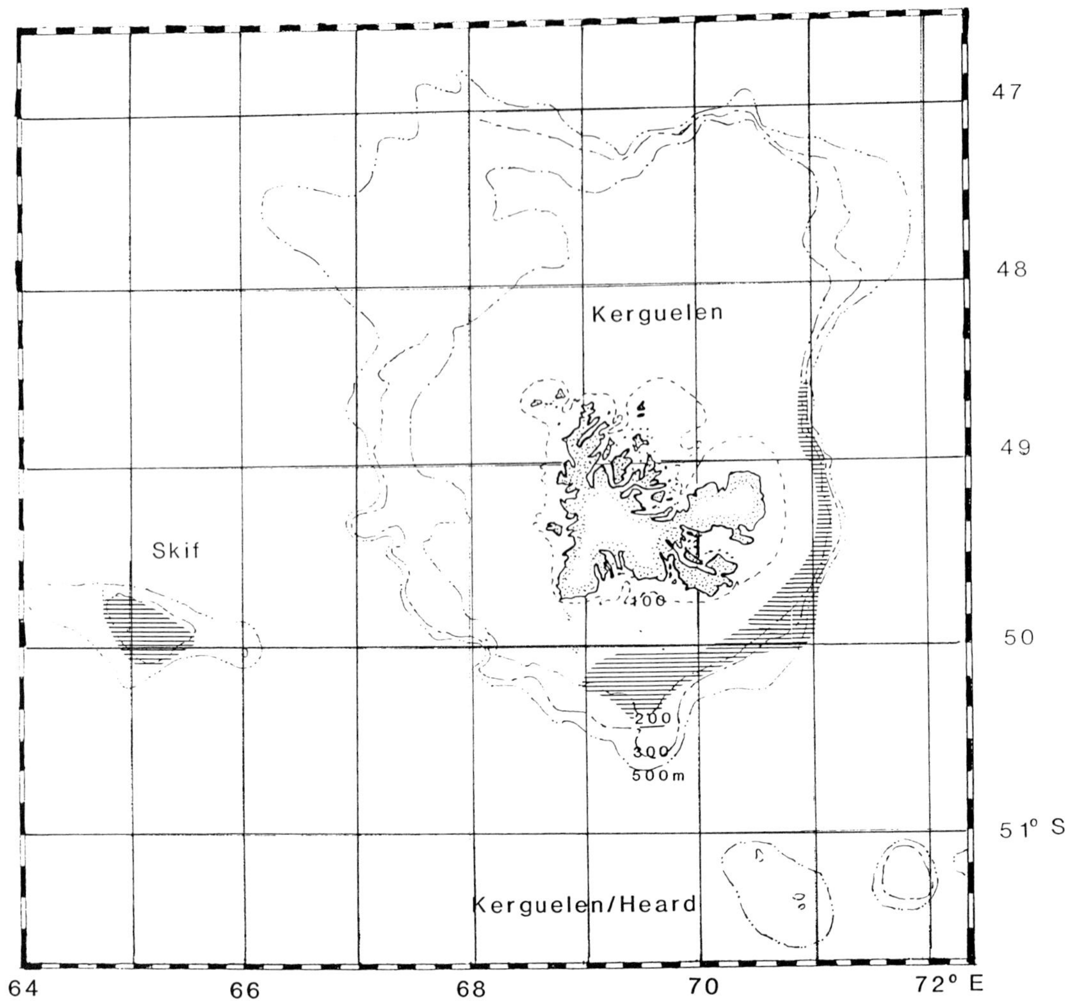

Fig. 6. General distribution of *Notothenia rossii* observed from 1979 to 1986 (*hatched areas*)

clear in 1984 that the total protection of that species was necessary; this protection became effective immediately. The stock will take long to recover because of the slow growth (2 to 5 cm per year) of the species which can reach a maximum size of 90 cm and because of a low relative fecundity (10 to 20 eggs/g body weight).

For two species (*Notothenia rossii* and *N. squamifrons*) it has been possible to use the VPA method. Figure 10 gives an example of the results obtained with this method: the evolution of the biomass of the population of *N. squamifrons* on the Kerguelen shelf from 1979 to 1986. There is a strong decline of the stock during that period from 30000 tonnes to less than 5000 tonnes. Duhamel (1987) has shown a similar decline for *N. rossii* between 1980 (20000 tonnes) and 1984 (3000 tonnes).

The trends in biomass as shown by the abundance index analysis and the virtual population analysis are similar. This finding is significant because the two methods are completely independent. Three successive phases may be noted in the state of the population of *N. squamifrons*. As for *N. rossii*, there was first a decline of the stock from 1980 to 1982, then a slight increase is followed by a sharp decrease (Fig. 10). A change in the population structure was then observed in 1985 and 1986. This was shown by an increase in the percentage of small specimens in the catch. This situation arose from excessive and directed fishing effort in summer during the first 3 years of that period (1979 to 1981). In 1982, with the arrival of an abundant cohort of *Champsocephalus gunnari* on the Kerguelen shelf, there was a transfer of fishing effort from *N. squamifrons* to *C. gunnari*, which explains the slight increase of the biomass of *N. squamifrons* observed in 1983. But in 1983/84, there was a strong decrease in the abundance of the dominant cohort of *C. gunnari* and again a transfer of the fishing effort to *N. squamifrons*.

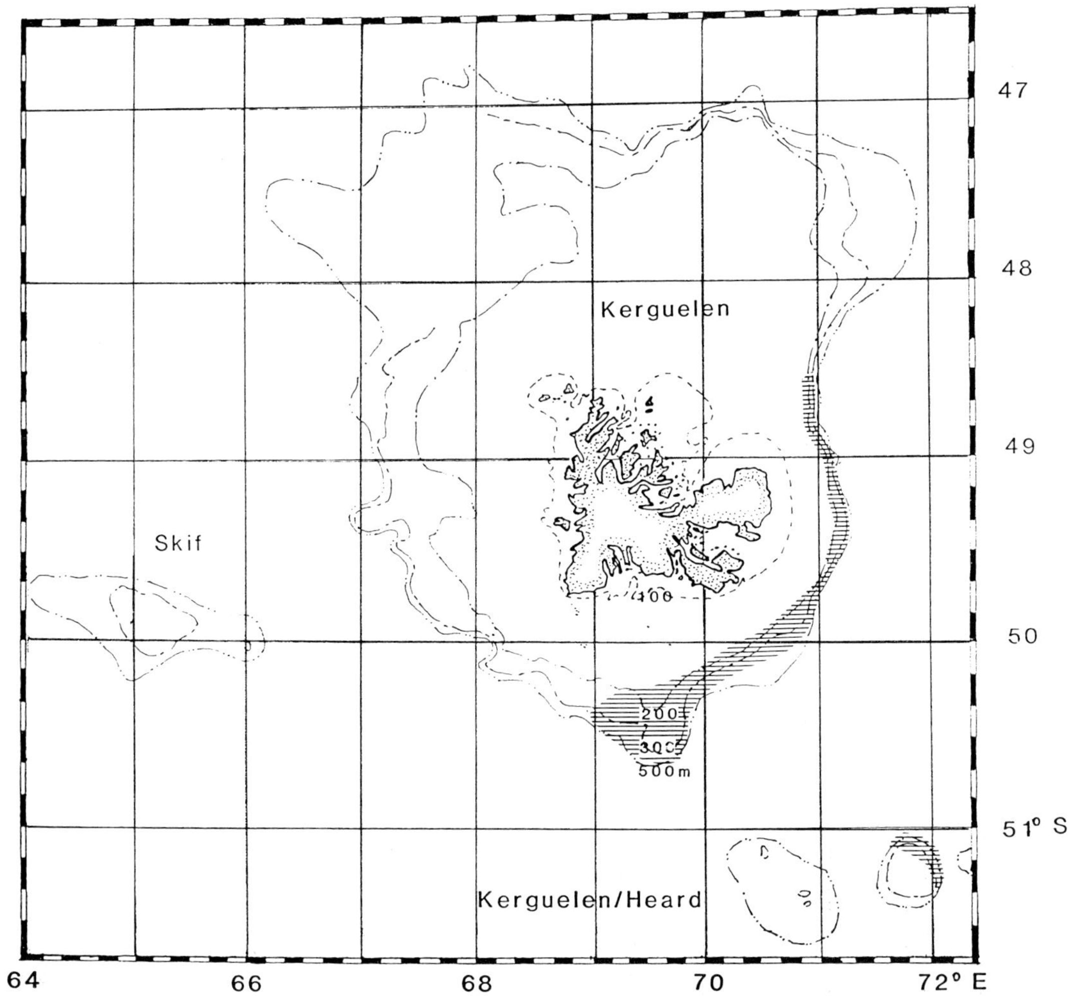

Fig. 7. General distribution of *Notothenia squamifrons* observed from 1979 to 1986 (hatched areas)

4.2 *Champsocephalus gunnari*

Two different populations of *Champsocephalus gunnari* have been described (Duhamel 1987), one on the Kerguelen shelf and the other on the Skif bank. Each of these populations is dominated by a single age class (cohort) which can be followed until its extinction. These dominant cohorts are traceable for 3 years until they are replaced by the succeeding abundant cohort. From the analysis of several successive cohorts in each sector, it seems that each abundant cohort is derived from the first spawning of the previous cohort, which explains why the duration of such cycles corresponds to the age at sexual maturity (3 years). Spawnings from the oldest specimens, even if they are slightly visible in the length frequency distributions, do not give rise to any abundant cohort in the Kerguelen fishery. It is too early to conclude whether this is the result of a natural sequence or whether it is the consequence of the fishing exploitation.

The monthly length frequency distributions give the necessary data to analyze the evolution of each cohort. An example is given in Fig. 11 for *C. gunnari* where four principal cohorts have been shown and which correspond to the fish born in 1976 (?), 1979, 1982 and 1985. For each cohort a growth curve can be drawn easily. The abundance index (Fig. 12) is high during the first year of exploitation when a cohort is recruited on the fishing grounds (specimens 3 years old) and decreases from the second to the third year. The combined action of natural and fishing mortalities lead to the complete disappearance of a cohort in 3 years. Consequently, the management of the stock must take into account the variation of the fishing effort from one year to another, with respect to the initial abundance of each cohort and the triannual arrival of a new cohort.

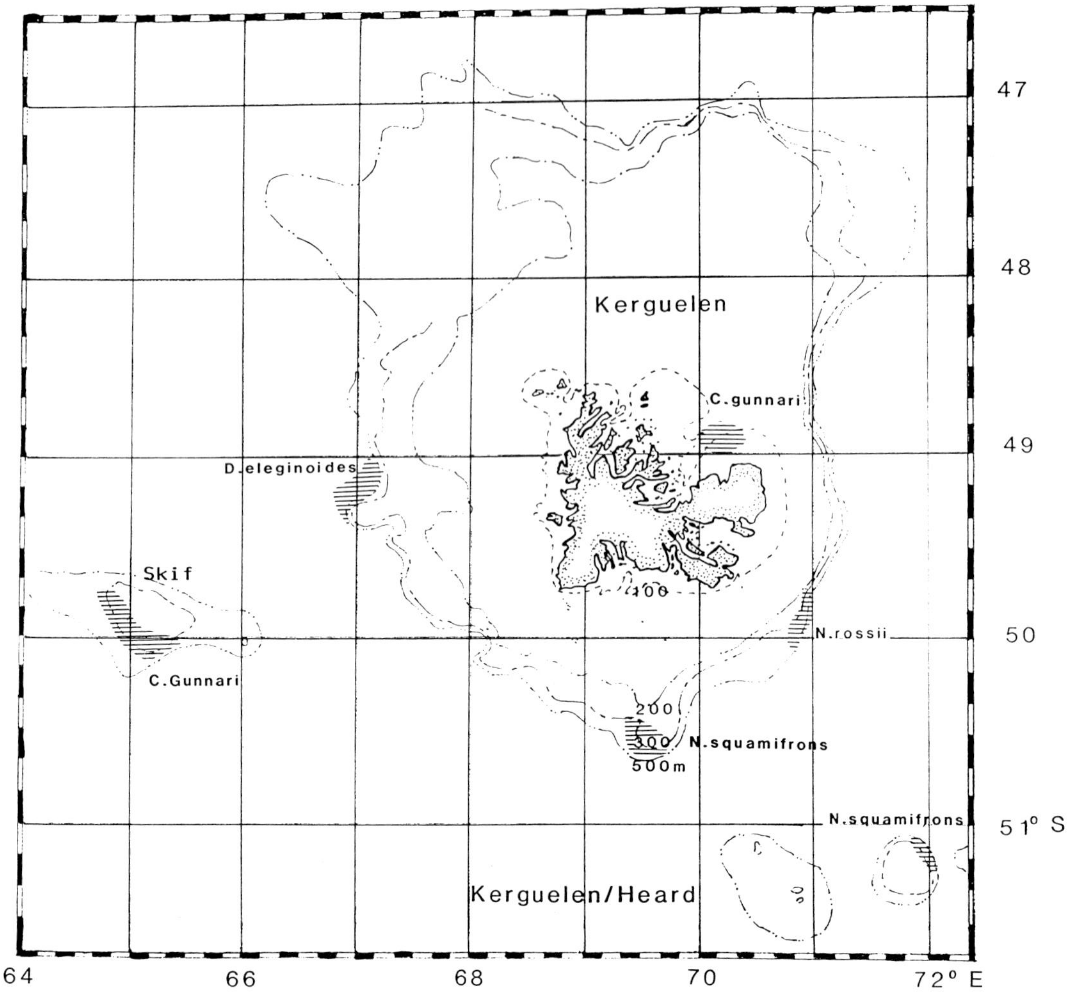

Fig. 8. Location of the spawning grounds (*hatched areas*) of the four main species of the shelf of Kerguelen Islands and the related seamounts between 1979 and 1986

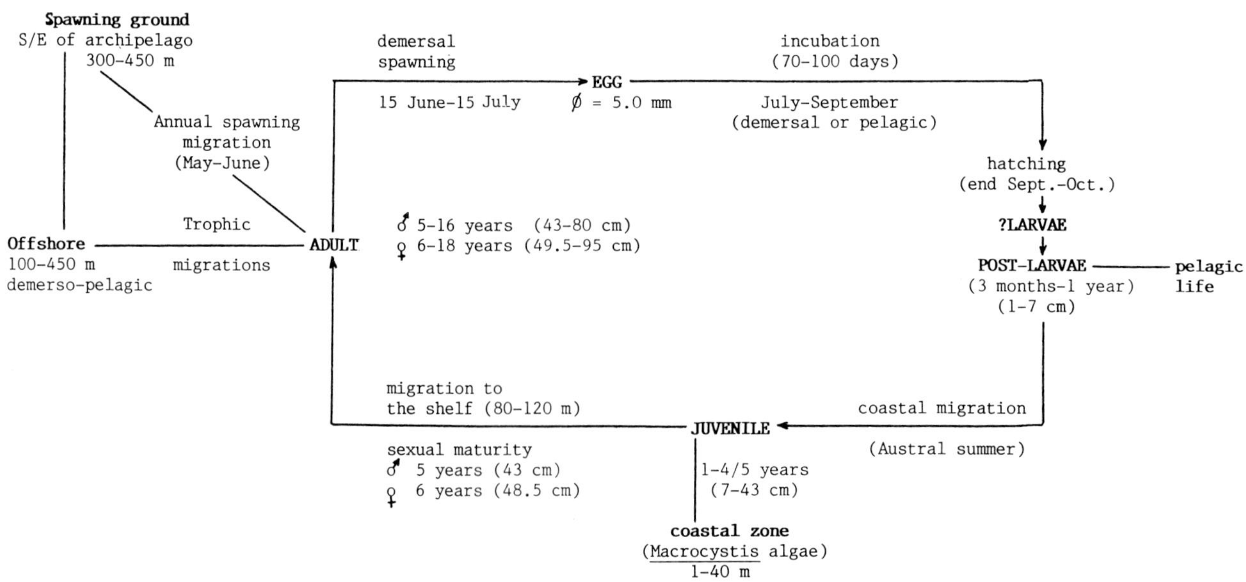

Fig. 9. Life cycle of *Notothenia rossii* from Kerguelen Islands

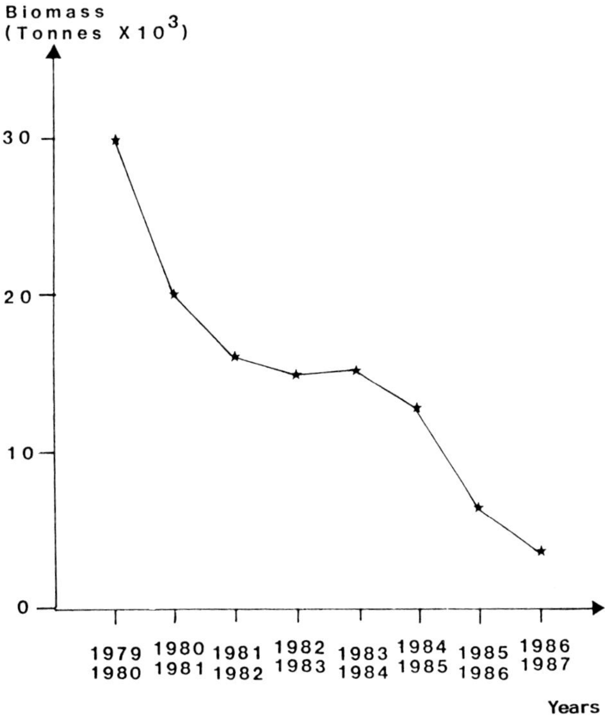

Fig. 10. Trends in the biomass of *Notothenia squamifrons* on the Kerguelen shelf from 1979/80 to 1986/87, based on virtual population analyses

4.3 *Dissostichus eleginoides*

The data concerning *Dissostichus eleginoides* are insufficient to draw any conclusions about its biomass since the fishing grounds were only discovered in 1984. Nevertheless, since the minimum size to which females mature is 85 cm, corresponding to an age of about 9 years, it seems clear that the possibilities of exploitation of that species are very limited and should be similar to those defined for *N. squamifrons* and *N. rossii*, if the fishing effort is too high.

5 Recovery of the Stocks and Conclusions

The analytical model based on VPA enabled some simulations of the evolution of the populations up to the beginning of the next century. For example, from the results of Figs. 13 and 14 such simulations can be used to estimate the biomass of stocks and the possible catches of *N. rossii* and *N. squamifrons* from the Kerguelen Island shelf (Hennemuth et al. 1988). For each species two different values of terminal F were chosen: 0.38 and 0.76 for *N. rossii* and 0.36 and 0.72 for *N. squamifrons*. It is obvious that the biomass of *N. rossii* will increase if the fishing effort is reduced, but the size of the catch will be stabilized at a level which is not significantly different from the level reached with the higher value of F, if the population is allowed to increase until 1994.

It is clear from the history of the main stocks of fish of the Kerguelen shelf that these populations have been affected by the intensive fishing effort which was applied from 1970 to 1978. Only four species are of commercial interest and three of them, the nototheniids, are already overexploited. Most of the species, as shown by the model, need a long recovery time.

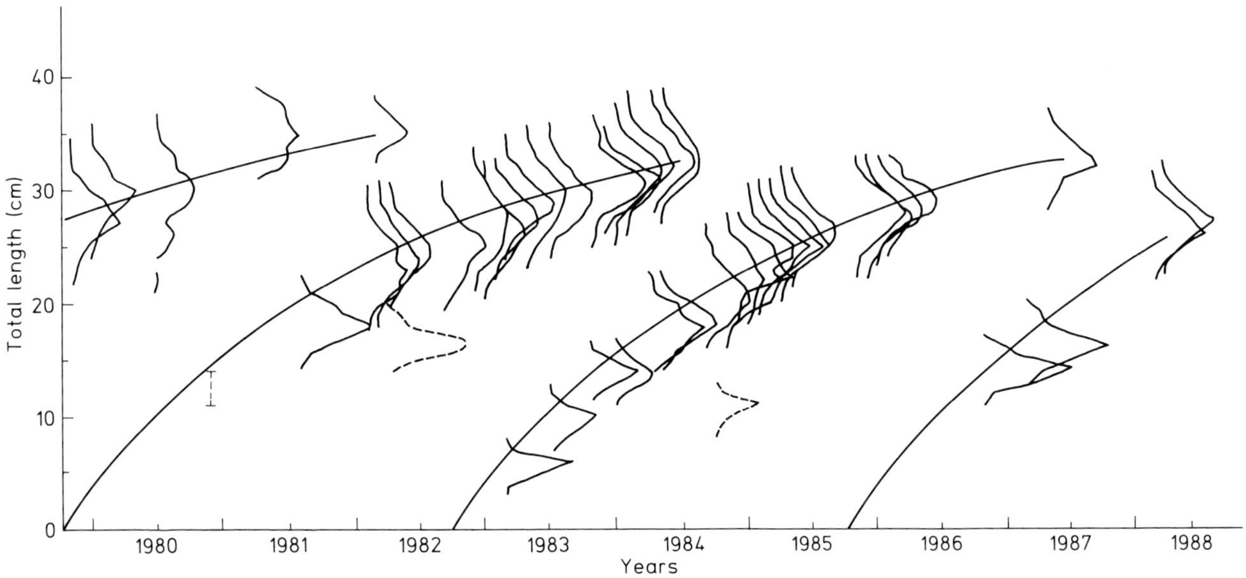

Fig. 11. Growth of *C. gunnari* in the northeastern shelf of Kerguelen Islands, as shown by monthly length frequency distributions. The *solid lines* are drawn through the mean values

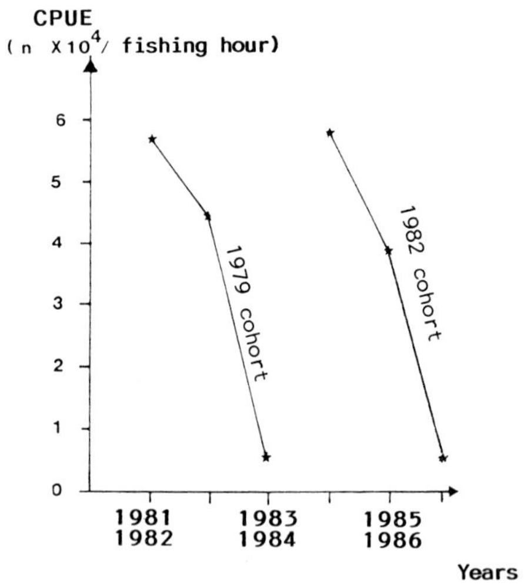

Fig. 12. Annual variations of the standard abundance index (CPUE in number of fish caught per fishing hour) for *Champsocephalus gunnari* from the northeastern shelf of Kerguelen Islands

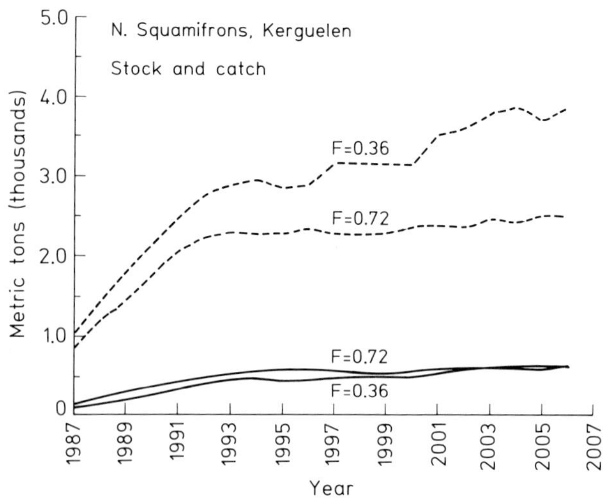

Fig. 14. Simulations of weight of stock (*dashed lines*) and catch (*continuous lines*) for *Notothenia squamifrons* Kerguelen Island shelf, based on virtual population analysis (Hennemuth et al. 1988)

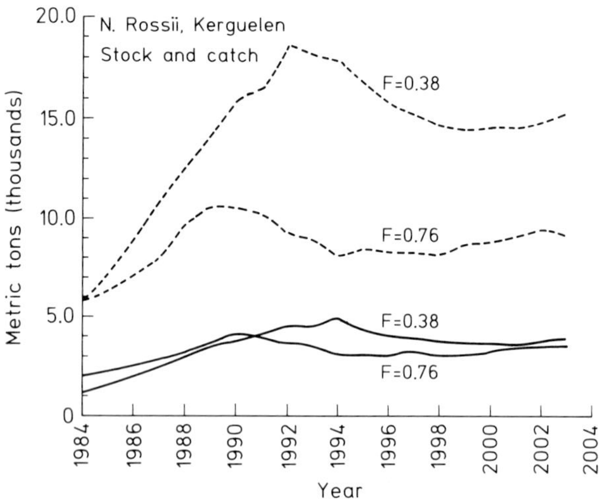

Fig. 13. Simulations of weight of stock (*dashed lines*) and catch (*continuous lines*) for *Notothenia rossii,* Kerguelen Island shelf, based on a virtual population analysis (Hennemuth et al. 1988)

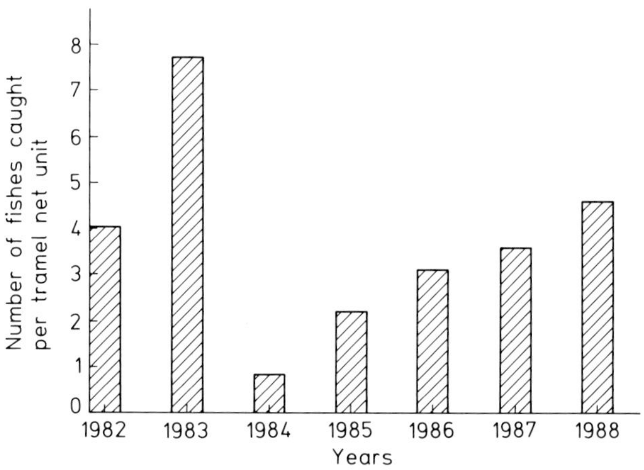

Fig. 15. Relative abundance of juvenile *Notothenia rossii* (in number of fish caught by tramel net unit) in coastal waters of Kerguelen Islands (Morbihan Bay) from 1982 to 1988

In order to facilitate the recovery of depleted stocks of *N. rossii*, conservation measures were brought into force in 1983. Directed fishing is forbidden and only a minimum by-catch is allowed. However, another method is necessary to follow the recovery of the stock using experimental fishing (tramel net) on the juvenile population in the inshore waters. The trend in the abundance shows an increase since the lowest level in 1984 (Fig. 15). The next step in stock management will be a biomass estimate of the adult population, prior to a possible new exploitation of that species. Thus, only one species, *Champsocephalus gunnari*, has any real potential in the near future for the Kerguelen fishery, but it is necessary to monitor the state of recruitment of this key species by means of scientific research using trawlers or other appropriate ships. The state of the various stocks should be evaluated regularly each year in order to propose total allowable catches for each species. This can be achieved since these species have different distributions, spawning areas and spawning seasons. Due to these characteristics, fishing is monospecific thus facilitating the management.

Considering the relatively large size of the Kerguelen shelf (more than 100 000 km^2), which is, together with the South Georgia area, the main finfish fishing ground in the Southern Ocean, and the few exploitable species, the analysis shows the low potential of the Kerguelen area. After less than 20 years of exploitation, three species are already overexploited and the possible production of icefish does not exceed a few thousand tonnes. The example of the Kerguelen fishery shows that the Southern Ocean ecosystem is very sensitive to human-induced ecological changes. Only careful management can ensure the fish resources of the area.

References

Duhamel G (1987) Ichtyofaune des secteurs indien occidental et atlantique oriental de l'Océan Austral: biogéographie, cycles biologiques et dynamique des populations. Thèse de doctorat d'Etat ès Sciences naturelles. Université Pierre et Marie Curie, Paris, Archives et documents, Microédition, Sc nat, Muséum national d'Histoire naturelle, no 87 04 47, 512 p

Duhamel G (1989) Dynamique des populations marines exploitées: gestion et aménagement des zones économiques exclusives. Actes Colloque sur la Recherche française dans les Terres australes, Strasbourg, 1987, pp 229–244

Hennemuth RC, Bisack KD, Duhamel G (1988) Simulation of recovery rates of fish stocks in the South Georgia and Kerguelen Islands areas. Selected Scientific Papers SC-CAMLR-SSP/4, Hobart, pp 229–300

Hureau J-C, Slozarczyk W Exploitation and conservation of Antarctic fishes and recent ichthyological research in the Southern Ocean. In: Gon O, Heemstra PC (eds) The fishes of the Southern Ocean. J.L.B. Smith Institute of Ichthyology, Grahamstown (in press)

Jones R (1981) The use of length composition data in fish stock assessments (with notes on VPA and cohort analysis). FAO Fish Circ (734): 60 p

Kock KH, Duhamel G, Hureau J-C (1985) Biology and status of exploited Antarctic fish stocks. BIOMASS Scientific Series 6:143 p

Pope JG (1972) An investigation of the accuracy of virtual population analysis using cohort analysis. Res Bull ICNAF, pp 65–74

White MG, North AW, Twelves EL, Jones S (1982) Early development of *Notothenia neglecta* from the Scotia Sea, Antarctica. Cybium 6(1):43–51

Human Impacts on Terrestrial and Marine Systems

Invertebrate Fauna and Their Role in the Degradation of Cattle Dung at Amsterdam Island

P. Tréhen, Y. Frenot, M. Lebouvier and P. Vernon [1]

Summary. At Amsterdam Island, cattle were introduced in 1871 but there were no native coprophagous fauna. This chapter focuses on the succession of the fauna involved in dung degradation during a summer, 90-day experiment and on changes in water, dry matter and nitrogen contents. A small number of invertebrate species was sampled from the dung and the soil beneath it. The presence of Diptera larvae (mainly *Fucellia tergina*), caterpillars (Noctuidae) and earthworms (*Dendrobaena rubida tenuis* and *D. subrubicunda*) was noticed. *Porcellio scaber* (Isopoda) was dominant and the dry matter and nitrogen losses may be related especially to the activity of this species which leads to a coarse fragmentation of dung and its conversion to fecal pellets.

1 Introduction

Cattle droppings constitute both a special microhabitat and a trophic pool for macro-, meso- and microfauna. In temperate regions numerous studies reviewed by Stevenson and Dindal (1987) have dealt with the disappearance of cattle dung or the succession of the fauna in dung pats. However, these studies are generally carried out at sites usually used for stock farming where a native coprophagous fauna is present. They emphasize the role of Diptera, Coleoptera and earthworms in the degradation of cattle dung (Landin 1961; Holter 1975, 1979; Olechowicz 1976). Cattle were introduced at Amsterdam Island in 1871. Now some 2000 cattle are currently distributed at low altitude over 60% of the island. This is the first time herbivorous vertebrates were present and so there was no coprophagous fauna present. In this chapter, we present preliminary results of a study on the invertebrate community and the succession of the fauna involved in dung degradation during a 90-day experiment. Changes in the water, dry matter and nitrogen contents of the dung were measured. Since little is known of the terrestrial ecosystem of Amsterdam Island, a brief description of the landscape components is provided.

[1] Université de Rennes I, CNRS URA 696, Station Biologique de Paimpont, F-35380 Plélan le Grand, France.

2 Description of Amsterdam Island

Amsterdam Island (37° 47′ S, 77° 34′ E) is an isolated island in the Indian Ocean, approximately mid-way between South Africa and Australia. It was discovered in 1566 by Del Cano and the first known landing was by van Vlaming in 1696. It is entirely volcanic in origin. Some of the numerous cones are thought to have been created during the last century and there is still the risk of volcanic activity (Giret 1987). It is 10-km-long by 7-km-wide, and has a land area of about 55 km^2 (Fig. 1a). Steep cliffs (30–60 m) skirt most of its coastline, those at the west being up to 700 m high. It is possible to land only in the north where the Martin-de-Vivies research station was established in 1949. The eastern part of the island slopes gently up from the sea to the central plateau (an old caldera) where it rises to 881 m (Mont de la Dives).

Mean annual air temperature is 13.8 °C. The coldest month is August with a mean of 11.2 °C and the warmest is February with a mean of 17.0 °C. Mean annual rainfall is 1115 mm with a minimum in summer (February–March) when evaporation exceeds rainfall at low altitude (National Meteorological Station, north of the island, 30 m a.s.l., 1951–80). Stonehouse (1982) considered that Amsterdam Island is situated in the warm temperate zone of the subantarctic province. However, in 1988, mean rainfall and mean temperature at high altitude (720 m a.s.l.) were close to those recorded at Crozet (National Meteorological Station, 140 m a.s.l., 1970–84).

The composition of the original vegetation and invertebrate fauna is virtually unknown and systematic lists are few. Numerous alien plant (Jolinon 1987) and invertebrate species are present but are not evenly distributed all over the island. Successive burnings (1792, 1853, 1857, 1899, 1974) mainly caused by human ac-

Antarctic Ecosystems. Ecological Change and Conservation
(ed. by K. R. Kerry and G. Hempel)
© Springer-Verlag Berlin Heidelberg 1990

tivity have been another important element of ecosystem disturbance, particularly at low altitude. Thus, most of the present ecological systems on the island have been induced by fire, introduced flora and fauna, especially cattle.

2.1 Soils and Vegetation

Our first observations, made in summer 1988, led to the recognition of six major ecological systems on the basis of the soil, vegetation and degree of disturbance

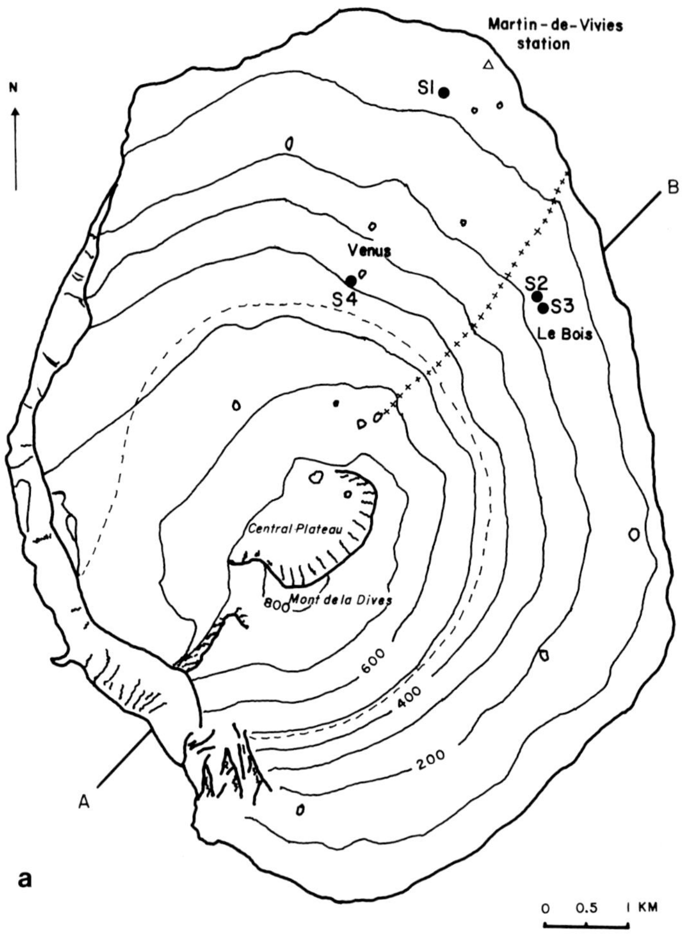

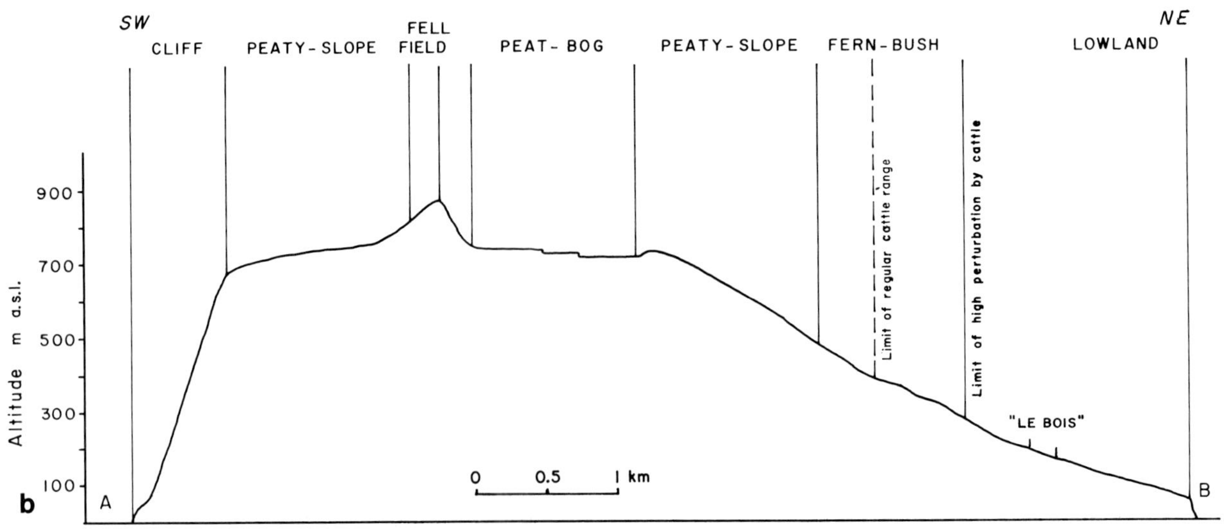

Fig. 1. Amsterdam Island. **a** Map of Amsterdam Island (contours in metres). *S1, S2, S3, S4* Sites where the dung degradation experiment was carried out. ---- Upper limit of regular cattle range; + + + + site of new cattle fence (areas south of this fence will be cattle-free in 1989) **b** Cross-section of Amsterdam Island. Distribution of the main ecological systems along a SW–NE transect

by cattle. Their distribution along a NE–SW transect, passing through the highest point, Mont de la Dives, is shown in Fig. 1 b. The different zones correspond to successive altitudinal belts from sea level to the summit. Their general characteristics are described below.

2.1.1 The Lowland

This zone lies approximately between the shoreline and 270 m a.s.l. and is intensively grazed by cattle. Fire has markedly affected this area and charcoal is frequent in the upper few centimetres of the soil and at greater depths (40 cm) in several places, indicating fire events in the past. Intensive grazing, trampling and burning are responsible for active erosion, inducing a mosaic of lava exposures and short sward. Up to 150 m a.s.l. the soil is thin (10–40 cm deep), with a sandy texture, low organic matter content and low water retention capacity. Much of the native vegetation has disappeared as a result of grazing. Introduced plant species are dominant. The vegetation is short because it is closely grazed (*Holcus lanatus*) or because trampling favours low-growing plant species (*Scirpus aucklandicus, Leontodon taraxacoides*). *Cirsium vulgare,* recently introduced, is abundant. Its cover has greatly increased since the last burning in 1974.

Above 150 m a.s.l. the vegetation cover increases and tussocks of *Juncus effusus* and *Scirpus nodosus* are more abundant. The soil is deeper (> 1 m), with a high organic matter content (up to 50% organic carbon, dry combustion method) and moisture content (up to 300% of the dry weight). A transition belt, with numerous scattered hillocks of this organic material, separates these two contrasting pedological areas. Erosion is very active in this part of the transect where there has been a rapid recession of the deep organic soil cover.

According to reports of early visitors, vegetation at low altitude was luxuriant and almost impenetrable at the end of the 19th century. In 1726 Valentyn (in Daycard 1985) referred to a continuous belt of *Phylica nitida* forest along the east coast. This belt has been disrupted by successive burnings and, in 1874, only nine small patches of trees survived (Vélain 1893). Nowadays, most of the remnants are located in a small area ("Le Bois"), protected since 1977 against the cattle. This site was not burned in 1974, and here the soil is similar to that situated at higher altitudes. There have been clearly considerable changes in the primitive soil and vegetation of the lowland during the past two centuries.

2.1.2 The Fern Bush (270 m to 500 m a.s.l.)

In this zone the soil is similar to that of the upper part of the lowland. However, the first horizon is 5 to 15 cm thick and peaty. This feature can be related to the higher cloudiness and precipitation above 250 m. Ferns are dominant: *Gleichenia polypodioides* predominates on the slopes and *Elaphoglossum succaesifolium, Blechnum penna-marina* and bryophytes cover the flatter areas. This zone is occasionally visited by cattle and there are numerous tracks through the vegetation.

2.1.3 The Peaty Slopes (above 500 m a.s.l.)

Here the soil (20 to 50 cm deep) has a high organic and water content. Mosses (including *Sphagnum* sp.) and grasses (*Poa fuegiana, Trisetum insulare*) are dominant. *Uncinia brevicaulis, U. compacta* and *Lycopodium trichiatum* are poorly represented. Cattle do not visit this wet zone.

2.1.4 The Peat Bog (730 m a.s.l.)

This plateau is a small area limited by a fault of the central caldera. The peat consists of slightly decomposed plant matter with a high fibre content. The water table remains close to the surface. Mosses (including *Sphagnum* sp.) and liverworts are dominant.

2.1.5 The Fellfield

This zone occurs above 800 m a.s.l., near the highest point. Barren soil occurs with a very low plant cover. The soil is mineral and coarse in texture.

2.1.6 The Cliffs

On the west coast, the sea cliffs are covered with tussocks of *Spartina arundinacea* and *Poa novarae*. These two coastal species were probably more widely distributed in the past, as mentioned by early visitors. The influence of cattle has probably reduced their distribution to inaccessible zones.

2.2 Invertebrate Fauna

Information on the invertebrate fauna of Amsterdam Island is sparse (Schiner 1868; Vélain 1877, 1893; Enderlein 1903, 1909; Strand 1909; Vanhöffen 1909; Aubert de la Rüe 1932). More detailed studies carried out after the settlement of the permanent station concern Diptera (Séguy 1959), Lepidoptera (Viette 1959), microarthropods, mainly oribatid mites (Travé 1973, 1974). Brief notes on species of other groups are also available (Coleoptera: Paulian and Pic 1953; Thysanoptera: Zur Strassen 1981; Araneae: Dreux 1970).

Endemic species occur mainly above 400 m a.s.l. and brachypterous or micropterous forms, e.g. *Brachypteragrostis patricei* Viette (Lepidoptera), *Trimicra pauliani* Séguy (Diptera) are frequently observed on the central plateau. On the other hand, the endemic beetle *Mecyclothorax (Antagonaspis) sculptopunctatus* (Enderlein) occurs everywhere on the island.

Since 1986 a general survey of the invertebrate distribution on Amsterdam Island has been made. Some recently introduced alien species, e.g. *Forficula auricularia* L. (Dermaptera), *Pyrameis cardui* L. (Lepidoptera) occur near Martin-de-Vivies Station and some other low-altitude sites. Other species, introduced earlier, are widely distributed at all altitudes: *Porcellio scaber* Latreille (Isopoda), *Calliphora croceipalpis* Jeannicke (Diptera), *Dendrobaena subrubicunda* Eisen and *D. rubida tenuis* Eisen (Oligochaeta).

3 Methods

This study was carried out from January to March 1988. Four experimental stations (S1–S4) were established (Table 1, Fig. 1a). S1 and S2 were strongly influenced by cattle: overgrazing and trampling were pronounced. S3 was located in "le Bois", a site of about 8 ha protected by a fence. Accidental penetration by cattle is very rare. S4, at the upper limit of cattle grazing, was regularly visited by cattle but their impact was less.

At the beginning of the experiment fresh dung was collected a few minutes after deposition. Samples were immediately dried to serve as controls (no exposure). Uniform pats of about 1 100 cm^3 of fresh dung were weighed before being placed (at the centre of contiguous plots 50 × 50 cm) on the ground of individual grids (mesh size = 1 × 1 cm) which facilitate the recovery of the pats for reweighing.

Sampling was carried out at each site at regular intervals (5, 10, 15, 20, 30, 45, 60 and 90 days) and three lumps of dung were selected at random and weighed. A circular sector was removed from each dung pat in order to determine its water content and chemical composition. The invertebrate fauna was hand-sorted from the whole pat, as well as from beneath it and from the soil itself (0.1 m^2 and 0–15 cm depth).

Water content was determined immediately by weighing after the samples had been dried at 65 °C to constant weight. Total nitrogen (N) content was determined later in France by the Kjeldahl method (duplicate analysis for each sample). Nitrogen values are expressed on an oven-dry basis.

At S1, three samples were placed in nylon net bags of mesh size 0.1 × 0.1 mm to isolate them from the fauna, except microorganisms. They were sampled at the end of the experiment (90 days).

4 Results

4.1 Water Content and Loss of Dry Weight

The water content of the dung decreased considerably during the first 20 days of the experiment at all sites

Table 1. Main features of the four experimental sites in the dung degradation study

Stations	S1	S2	S3	S4
Altitude m a.s.l.	70	150	150	400
Ecological system	Lowland	Lowland	Lowland "Le Bois"	Fernbush
Dominant vegetation	*Leontodon taraxacoides* *Holcus lanatus*	*Juncus effusus* *Scirpus aucklandicus* *Leontodon taraxacoides*	*Blechnum penna-marina*	*Elaphoglossum succaesifolium* Mosses
Soil				
depth	5–20 cm	> 120 cm	> 120 cm	> 120 cm
nature	Organo-mineral	Organic	Organic	Organic
Cattle impact	High (overgrazing, trampling)	High (overgrazing, trampling)	Nil (cattle excluded from site)	Slight (tracks through vegetation; upper limit of cattle range)

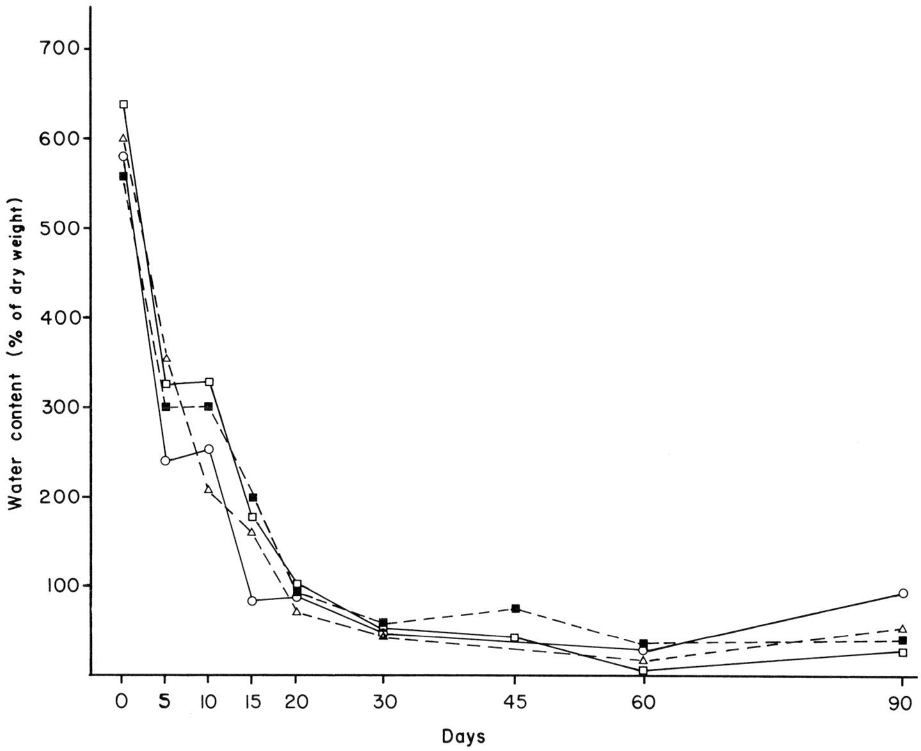

Fig. 2. Variation in the water content of cattle dung during the 90-day experiment at S1 (△), S2 (□), S3 (■) and S4 (○)

Table 2. Mean nitrogen content of dung pats exposed for different times. Each mean is determined from the duplicate analysis of 3 separate pats (ns, no sampling; +, coefficient of variation between 5 and 8%, all other CV < 5%)

Exposure (days)	Stations				Dung enclosed in nylon net bag (S1)
	S1	S2	S3	S4	
Nil	2.12	2.12	2.12	2.36	2.12
5	2.17	2.19 +	1.97	2.33	ns
10	2.20	2.29	2.09	2.34	ns
15	2.19	2.21	2.09	2.36	ns
20	2.24	2.15	2.06	2.36	ns
30	2.26	2.31	2.10	2.39	ns
45	ns	2.29 +	2.11	ns	ns
60	2.40	2.41	2.20 +	2.50	ns
90	2.42	2.46	2.34 +	2.52 +	2.28

(Fig. 2); afterwards, the loss of water was slow. However, at S4, desiccation of the dung was rapid despite the higher relative humidity. This was probably because of the higher wind velocity at this altitutde. A dry crust increased rapidly in thickness over the surface of the lump ranging from 0.5 cm (S2, S3 and S4) to 1 cm (S1) thick 5 days after deposition. No perforations were observed on the surface of the dung.

The loss of dry weight after 90-day exposure differed markedly between stations, being low at S3 (4.2%), high at S4 (29.4%) and intermediate at S2 (11.0%) and S1 (9.6%). There was virtually no loss of dry weight in the samples enclosed in the nylon bags (1.2%).

An important physical change was observed after 45 days of exposure: the outer crust of the dung was greatly reduced and the inner part developed a mainly loose, fibrous structure. The moisture and structure of the whole dung pat appeared to be more homogeneous.

4.2 Nitrogen Content

Changes in the dung nitrogen content are shown in Table 2. Values for the replications from three pats at a same sampling time were fairly close. There was a significant variation in N content with increasing time

Table 3. Nitrogen content of dung when fresh and after 90 days of exposure and percentage decrease in dry matter and nitrogen at the end of the experiment

Stations	N content (%)			Dry matter loss (%)	Actual N loss (%)
	Fresh dung (measured)	Fresh dung (estimated)	90-Day dung		
S1	2.12	2.34	2.42	9.6	6.4
S2	2.12	2.34	2.46	11.0	6.5
S3	2.12	2.34	2.34	4.2	4.2
S4	2.36	2.60	2.52	29.4	32.2
Dung in nylon net bags	2.12	2.34	2.28	1.2	3.7

of exposure (ANOVA, $p < 0.05$). Two factors influence the apparent increase in N content that we measured: (1) N content of fresh dung is underestimated because some of the N is in the form of heat-labile compounds that are lost when the dung is oven-dried. As the dung ages, the action of microorganisms or other processes may fix these labile compounds progressively against loss during drying so that the N content appears to increase with age (Macqueen and Beirne 1975). These authors found a loss of up to 10% total N during oven-drying. We found a similar N loss (8–12%) caused by drying when analyzing organic soil at Amsterdam Island. Table 3 shows the measured and estimated (+ 10%) N contents of dung when fresh. (2) The second factor is the progressive loss in dry matter of the dung as it aged. The estimation of the amount of N in the fresh dung and in the dung after 90 days of exposure shows an actual N loss (Table 3) which may be related to the loss in dry matter, except for the samples enclosed in the nylon bags.

4.3 Invertebrate Fauna from Cattle Dung and Underlying Soil

Representative of 11 invertebrate orders were identified within the dung and in the soil beneath it. Some of them were sporadic in their appearance (Gasteropoda, Araneae, Diplopoda and Chilopoda) or were not efficiently sampled by hand-sorting (Acarina, Collembola). The major groups are presented in Table 4. Absolute frequencies of individuals obtained at the three levels of sampling (dung, surface, soil) for the three replications are summed. There are great differences between the results among the four stations in abundance as well as in the organization of invertebrate communities involved in dung degradation.

4.3.1 Isopoda

Porcellio scaber was the dominant species of the faunal community involved in dung degradation. Wood louse abundance varied greatly between the four sites: during the first 2 weeks of the experiment, they were very abundant at S4 (400–700), less abundant at S1 (100–300) and S2 (300) and few at S3 (100). Their abundance decreased after 30 days, except at S1 where it remained almost constant until 60 days, by which time they had virtually disappeared at S3. At the other sites, their abundance increased between 60 and 90 days.

P. scaber was found mainly on the surface of soil under the dung and occasionally in the soil (S4) or within the dung where it excavated tunnels. It dropped faeces under the dung. A distinct ring of faeces was observed when the pat was removed. *P. scaber* had a strong influence on dung fragmentation: there was good agreement between the loss of dry weight and the total number of isopods samples throughout the 90-day experiment (Fig. 3).

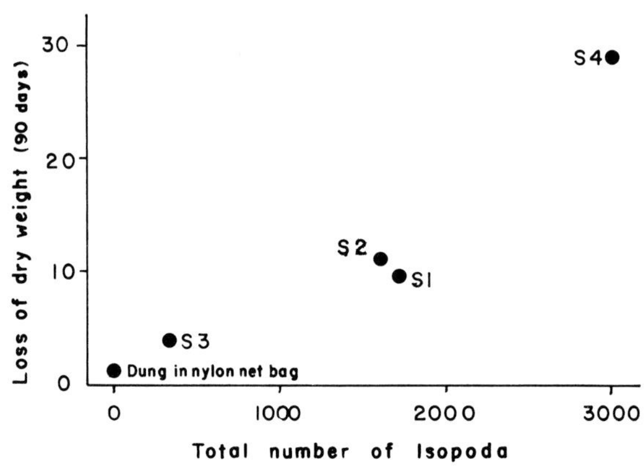

Fig. 3. Relation between loss of dry weight in dung and total number of Isopoda sampled during the 90-day experiment

Table 4. Total number of individuals of each major invertebrate group at each sampling date in the three dung pats and in the soil beneath them (3 × 0.1 m²). ns, no sampling

Invertebrate group	Stations	Days after deposition							
		5	10	15	20	30	45	60	90
ISOPODA									
Porcellio scaber	S1	97	174	174	260	209	ns	237	522
	S2	291	290	275	300	195	135	60	159
	S3	93	30	99	35	11	9	0	16
	S4	731	369	393	532	174	ns	182	559
DIPTERA									
Fucellia tergina (larvae)	S1	102	247	35	2	0	ns	0	0
	S2	1	19	0	0	0	0	0	0
	S3	1	2	1	0	0	0	0	0
	S4	0	0	0	0	0	ns	0	0
Unidentified	S1	0	18	5	1	0	ns	0	0
Sphaeroceridae	S2	0	14	13	0	0	0	0	0
(larvae)	S3	0	19	16	0	1	0	0	0
	S4	0	22	29	61	1	ns	0	0
LEPIDOPTERA									
Noctuidae (larvae)	S1	0	0	0	0	4	ns	0	2
	S2	0	0	0	0	0	1	2	14
	S3	0	0	0	0	10	23	32	40
	S4	0	0	0	0	0	ns	0	6
OLIGOCHAETA									
Dendrobaena subrubicunda	S1	17	1	13	7	3	ns	6	0
Dendrobaena subrubicunda	S2	1	4	9	16	33	6	21	36
Dendrobaena subrubicunda	S3	4	6	4	3	12	18	15	17
D. rubida tenuis	S4	1	0	0	0	6	ns	7	15

4.3.2 Diptera

Four species were recorded at the surface of the pats. Two, *Tipula* sp., *Calliphora croceipalpis*, were attracted to dung for only brief periods to feed. The other two species, *Fucellia tergina* Zetterstedt (Anthomyiidae) and an unidentified Sphaerocerid, colonized droppings.

The largest number of *F. tergina* occurred at S1. There was a peak of larvae 10 days after deposition and one of pupae 10 days later (Fig. 4a). Larvae and pupae were both found within the wet part of the dung. At the first sampling date, eggs were found under the outer edge of the pat. *F. tergina* was sparse at S2 and S3 and was absent at S4.

The second species appeared at all sites 10 days after the deposition of the dung (Fig. 4b) and was markedly abundant at S4. Its larvae are reported here as Sphaeroceridae, but no adults were collected. Numbers were maximum after 10 and 15 days at each site, except at S4 where the number of larvae increased steadily with a peak at 20 days. This difference may be explained by the higher moisture content of the soil. At this site, larvae were found not only in the dung but also in the soil beneath it.

4.3.3 Lepidoptera

Caterpillars of four taxa were recorded in the samples. Three of these taxa were occasionally found elsewhere in the soil but their presence seems unlikely to be related to cattle dung. The fourth taxon (Noctuidae) was found at all sites within the dung and beneath it. It was not abundant at S1 and S4 but very abundant at S3. It appeared 30 days after deposition and its number increased up to 90 days (Fig. 4c). Pats were not colonized or inhabited by any other species after 30 days at S3.

4.3.4 Oligochaeta

Two earthworm species (Lumbricidae) were sampled in the soil beneath the pats: *Dendrobaena subrubicunda* was found at S1, S2 and S3, and *D. rubida tenuis* at S4. The number of earthworms depended essentially on soil depth (5 to 20 cm) at S1: there was no clear relation between earthworm abundance and age of pats at this site. At the other sites, where the soil was deeper, change in abundance of earthworms was fairly

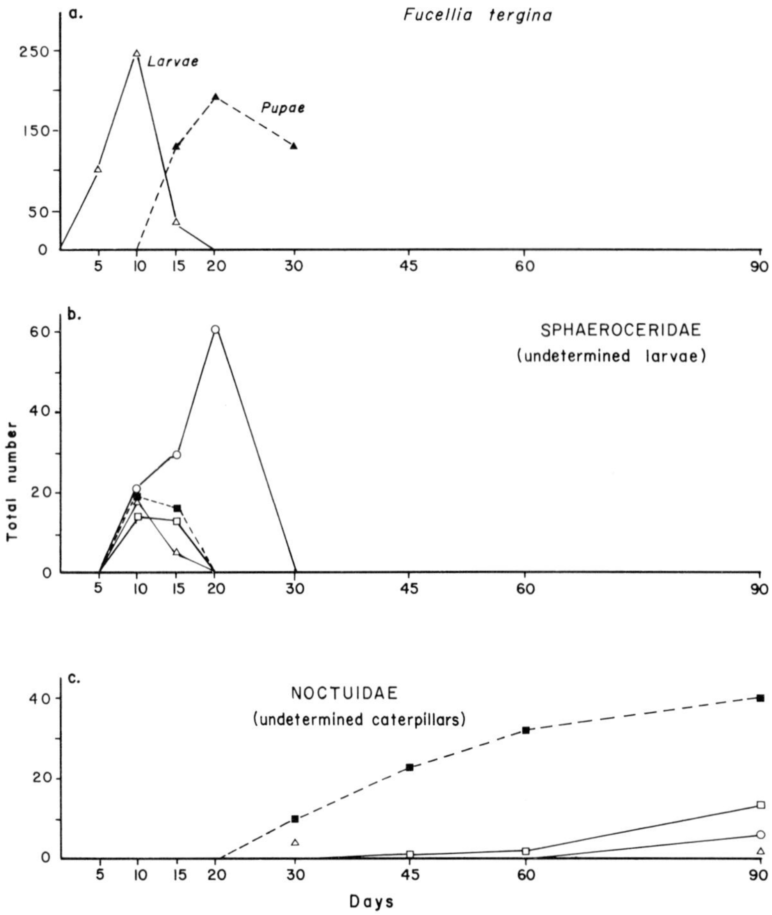

Fig. 4. Changes in the total number of Diptera and Lepidoptera sampled within the dung and beneath it. **a** *Fucellia tergina* (larvae and pupae) at S1; **b** Sphaeroceridae larvae; **c** Noctuidae larvae at S1 (△), S2 (□), S3 (■) and S4 (○)

regular: the greatest number of *Dendrobaena* occurred at the end of the sampling period. During this 90-day experiment, earthworms were not found in the dung. However, they were observed in older pats.

5 Discussion

Cattle dung is colonized usually by a succession of arthropod species, mainly Diptera and Coleoptera. The effects of these coprophagous arthropods on the decomposition of dung are direct (consumption and burial of dung) or indirect (fragmentation, aeration of dung, conversion to invertebrate faeces, interactions with microflora) (Stevenson and Dindal 1987). The associated fauna which exploits these faecal resources is therefore essential for an efficient degradation of cattle dung (Putman 1983).

In Australia, for instance, cattle and sheep were introduced without the presence of native arthropods fit for the decomposition of dung (Bornemissza 1960; Gillard 1967). The problem of the accumulation of undecomposed faeces has been solved by the voluntary introduction of several dung beetle species (Bornemissza 1976; Hugues 1975).

At Amsterdam Island, a similar situation occurs on a small scale. Cattle or any phytophagous mammals do not belong to the original fauna and a typical coprophagous fauna is absent. In spite of local overgrazing there is not accumulation of faeces and decomposition of cattle dung is carried out by some non-voluntarily introduced species which serve as substitutes.

Anthomyiidae are known to be endocoprophagous (Stevenson and Dindal 1987) but representatives of the genus *Fucellia* are usually phycophagous. At Amsterdam Island, *F. tergina* occurs mainly along the rocky shore in algal mats. It is able to intensively colonize the cattle dung at low altitude (S1). The second Diptera species is also probably alien.

Wood-louse abundance varied greatly between the four sites and was exceptionally high at S4 (up to 300 individuals per 0.1 m^2). This unusual abundance is in accordance with the major role played by this species in dung degradation. The great and rapid loss of dry weight of the dung is caused by *Porcellio scaber* activity. Furthermore, predatory behaviour has been observed in a laboratory experiment, suggesting that *P. scaber* is probably a consumer of fly larvae in the dung. This feature could explain the low number of pupae found at S4 where the wood lice were very numerous. A general survey of the invertebrate fauna distribution has shown that wood lice are present but less abundant in areas not visited by cattle (above 500 m a.s.l.). It is not known whether the high abundance of *P. scaber* in the lowland is related to cattle and dung production or to other environmental factors. This could explain the fewer numbers found at the S3 site which is protected by a fence.

The presence of Lepidopteran larvae in dung is unusual, too. Their abundance increased markedly after 30 days of exposure. So, their presence was obviously related to the stage of degradation of the dung characterized by a loose and homogeneous structure. Their role in dung removal is unknown.

The number of earthworms in the soil beneath the pats increased throughout the experiment but they were never found within the dung. In summer, 90 days may be too short a period to demonstrate the role of these animals in the degradation process. As observed within older dung deposits, earthworms may be very numerous (more than 30 individuals/pat). In this case, they contribute mainly to the removal of the pats by forming a large number of aggregates (2–3 mm in diameter) and mixing the dung material with the soil. So, the upper 10 cm of the soil beneath old dung is greatly modified in structure (and possibly in chemical composition).

Dry matter and nitrogen losses can be related to the activity of these invertebrate species since little variation occurred within the dung enclosed in a nylon net bag. Wood-louse abundance influenced greatly the rate of disappearance of dung during the first months of exposure. Later, earthworms seemed to take over from wood lice in dung consumption. In this experiment the major role of *P. scaber* is a coarse fragmentation of dung and its conversion to faecal pellets. The decrease in the nitrogen amount is roughly parallel to the dry matter loss and there is no evidence for a selective consumption of proteinaceous compounds as noticed for fresh dung heavily colonized by Diptera larvae (Macqueen and Beirne 1975). The role of wood lice might be comparable to the part of Coleoptera larvae (Lussenhop et al. 1980).

Due to the small number of invertebrate species and to their weak efficiency in the organic matter degradation processes, it seems that *P. scaber* plays an important role in several ecological systems, especially at low altitude. In addition to its detritivorous diet, it has developed unusual habits, i.e. dung consumption and predation, in response to the presence of cattle.

6 Conservation Project

The presence of cattle on Amsterdam Island has enhanced the level of disturbance at low altitude. Their number is increasing (2000 in 1987) and the risk of extension of the perturbations to higher altitudes exists. Conservation measures are needed to protect the less disturbed environment of the island. As part of a major conservation project (Decante et al. 1987; Daycard and Decante 1989), a fence is being erected across the island (Fig. 1a) and cattle are being eliminated from the southern part. A small herd (400 animals) will be left in a reduced area (10 km^2) near Martin-de-Vivies Station. Important comparisons of vegetation regeneration and its effects on the fauna will be made between these two zones, and the biological and environmental trends will be monitored. Much will be learned from this long-term comparative study on Amsterdam Island. This represents a unique opportunity to study the effects of introduced mammals on the environment and to analyze the relationships between native and alien species.

Acknowledgements. This research was funded by Terres Australes et Antarctiques Françaises with the support of Centre National de la Recherche Scientifique (GDR 902). We thank Monsieur l'Administrateur Supérieur des TAAF and the members of the Mission de Recherche for the logistic support. Many thanks are due to Thierry Valleix for field assistance and to Dr. Lewis Davies (University of Durham) for help in the identification of Diptera. Thanks are due to the referees, whose comments and criticism were very helpful.

References

Aubert de la Rüe E (1932) La flore et la faune des Iles Saint-Paul et Amsterdam. Terre Vie 2:642–662

Bornemissza GF (1960) Could dung eating insects improve our pastures? J Aust Inst Agric Sci 26:54–56

Bornemissza GF (1976) The Australian dung beetle project 1965–1975. Aust Weed Res Comm Rev 30:1–30

Daycard L (1985) Situation écologique de l'Ile Amsterdam: Bilan et Avenir. Mémoire de D.E.A., Université de Montpellier

Daycard L, Decante F (1989) Etude d'une population de bovins sauvages sur l'Ile Amsterdam. Essai de contrôle et de gestion. In: Laubier L (ed) Actes du colloque sur la recherche française dans les terres australes, 14–17 Sep 1987, Territoire des Terres Australes et Antarctiques Françaises, Strasbourg, pp 279–285

Decante F, Jouventin P, Roux JP, Weimerskirch H (1987) Projet d'aménagement de l'Ile Amsterdam. Rapport Ministère de l'Environnement, Terres Australes et Antarctiques Françaises, Centre d'Etudes Biologiques des Animaux Sauvages. 91 p

Dreux P (1970) Contribution à l'étude des arthropodes terrestres et d'eau douce dans les Terres Australes Françaises. Terres Aust Antarct Fr 50:1–4

Enderlein G (1903) Die Landarthropoden der von der Tiefsee Expedition besuchten Antarktischen Inseln. 2. Die Landarthropoden der Antarktischen Inseln St-Paul and Neu-Amsterdam. Wiss Erg Dtsch Tiefsee Exped Valdivia 1898–1899, Jena, pp 249–269

Enderlein G (1909) Die Insekten des Antarktischen Gebietes. Deutsche Südpolar Expedition 1901–1903, Zool, Band X, Berlin, pp 361–528

Gillard P (1967) Coprophagous beetles in pasture ecosystems. J Aust Inst Agric Sci 33:30–34

Giret A (1987) Géologie des Terres Australes Françaises. Com Natl Fr Rech Antarct 58:17–41

Holter P (1975) Energy budgets of a natural population of *Aphodius* larvae (Scarabaeidae). Oikos 26:177–186

Holter P (1979) Effect of dung beetles (*Aphodius* spp.) and earthworms on the disappearance of cattle dung. Oikos 32:393–402

Hugues RD (1975) Introduced dung beetles and Australian pasture ecosystems. J Appl Ecol 12:819–837

Jolinon JC (1987) Les Phanérogames adventices sur l'Ile Amsterdam. Com Natl Fr Rech Antarct 58:183–188

Landin BO (1961) Ecological studies on dung beetles (Coleoptera Scarabaeidae). Opusc Entomol (Suppl) 19:1–227

Lussenhop J, Kumar R, Wicklow DT, Lloyd JE (1980) Insect effects on bacteria and fungi in cattle dung. Oikos 34:54–58

Macqueen A, Beirne B.P. (1975) Influence of some dipterous larvae on nitrogen loss from cattle dung. Env Entomol 4:868–870

Olechowicz E (1976) The role of coprophagous dipterans in a mountain pasture ecosystem. Ekol Pol 24:125–165

Paulian R, Pic M (1953) Un Anobiide des *Phylica* de l'Ile Nouvelle Amsterdam. Rev Fr Ent 20:43–45

Putman RJ (1983) Carrion and dung: the decomposition of animal wastes. Edward Arnold, London

Schiner JR (1868) Diptera. Reise öster. Freg. Novara 1857–1859, Zool, Wien, pp 388

Séguy E (1959) Insectes Diptères de l'Ile Amsterdam (Mission de M. Patrice Paulian, 1955–1956). Mem Mus Nat Hist Nat Ser A 17:133–154

Stevenson BG, Dindal DL (1987) Functional ecology of coprophagous insects: a review. Pedobiologia 30:285–298

Stonehouse B (1982) La zonation écologique sous les hautes latitudes australes. Com Natl Fr Rech Antarct 51:532–537

Strand E (1909) Spinnentiere von Südpolar-Expedition 1901–1903. Deutsche Südpolar Expedition, Zool, Band X, Berlin, pp 541–596

Travé J (1973) Recherches sur les microarthropodes terrestres des Iles Amsterdam et Saint-Paul. I. Ile Amsterdam: Etude quantitative préliminaire. Rev Ecol Biol Sol 10:295–306

Travé J (1974) Recherches sur les microarthropodes terrestres des Iles Amsterdam et Saint-Paul. II. Ile Saint-Paul: Etude quantitative préliminaire et comparaisons entre les deux îles. Rev Ecol Biol Sol 11:519–532

Vanhöffen E (1909) Tiere und Pflanzen von Saint-Paul und Neu-Amsterdam. Deutsche Südpolar Expedition, Band II, Berlin, pp 399–410

Vélain C (1877) Remarques générales au sujet des Iles Sain-Paul et Amsterdam. Arch Zool Exp 6:1–144

Vélain C (1893) Les Iles Saint-Paul et Amsterdam. Ann Géogr 2:329–364

Viette P (1959) Lépidoptères de l'Ile Amsterdam (récoltes de Patrice Paulian, 1955–1956). Bull Soc Entomol Fr 64:22–29

Zur Strassen R (1981) On the Thysanoptera (Insecta) from some islands in the sub-antarctic part of the Indian Ocean. Com Natl Fr Rech Antarct 48:159–169

A Comparison of Airborne Alkaline Pollution Damage in Selected Lichens and Mosses at Casey Station, Wilkes Land, Antarctica

E. ADAMSON and R. D. SEPPELT[1]

Summary. *Umbilicaria decussata*, *Usnea sphacelata*, *Ceratodon purpureus* and surface soil samples were collected at 10 m intervals for 90 m downwind of a concrete batching site at Casey Station, East Antarctica. Comparable samples were collected from a similar uncontaminated remote site (SSSI no. 16). Surface soil was alkaline in the immediate vicinity of the batching site (max. pH 8.8) and tended to decrease with distance. In the SSSI control site, surface soil was acidic (pH 4.7). Lichens growing downwind of the batching site were more susceptible to damage from airborne alkaline pollution than the mosses and were moderately to severely bleached. This chapter describes the relation between mean total chlorophyll concentration, chlorophyll a/b ratio, distance from the batching site and soil pH. Low temperature (77 K) fluorescence of healthy plants from the SSSI and polluted plants 40 m downwind of the batching plant were compared. Variable fluorescence, indicative of photosynthetic electron transport was observed in all cases, from which we deduce that even severely bleached lichens contain live algal cells. The results presented provide quantitative baseline data against which further change (recovery or further deterioration) can be measured.

1 Introduction

The Australian continental Antarctic station, Casey (66° 17′ S, 110° 32′ E) is situated on the north west corner of Bailey Peninsula in the Windmill Islands group, Vincennes Bay, Wilkes Land, East Antarctica. The region is characterised by low, ice-free rocky outcrops and islands composed of metamorphics (gneisses, schists and migmatites), granites and deposits of glacial debris. The Bailey and Clark Peninsulas support the most extensive and best-developed plant communities on Continental Antarctica (sensu Pickard and Seppelt 1984) with a cryptogamic flora consisting of at least five moss, one hepatic and more than 28 lichen taxa (Lewis Smith 1986; Seppelt unpubl. data).

The Casey tunnel station was occupied in 1969. In 1979 rebuilding of the station commenced at a new site 0.7 km to the southwest in an area of abundant vegetation. From April–August 1979, concrete was mixed and poured at the Operations Building site. In the summer of 1979–80 an elevated concrete batching platform was installed approximately 180 m to the east of the Operations Building. This was moved to the western edge of the construction site in the summer of 1983–84, where it remained in regular use until the summer of 1987–88. The study described in this work is based on a transect commencing at this third concrete batching location (Fig. 1). The region to the west bearing 217° to 322° true has experienced comparatively little mechanical disturbance and supports abundant vegetation, however to the east (322° to 217° true), the land is almost completely devegetated for several hundred metres as a result of construction operations. Strong winds (over 30 knots) are almost always from the east, lighter winds are variable, but predominantly from the east and south. At the beginning of 1988 the batching platform was moved again, to its present site, approximately 30 m to the north of the transect.

The most abundant lichens in the area downwind of the batching site are *Umbilicaria decussata* (Vill.) Zahlbr., and *Usnea sphacelata* R. Br. The upper surface of the *Umbilicaria* thallus is normally dark grey to grey-black, but in the region immediately downwind of the batching site, this is often bleached to a uniform beige or pale cream colour (Fig. 2a). Bleaching is sometimes more pronounced around the edge of the thallus leaving the centre brown, but in more extreme cases, the underside of the thallus is also cream. While weathered or senescent thalli may be whitened around the margins they do not take on the beige to pale cream colour commonly seen downwind of the batching site. The upper portion of the *Usnea* thallus is normally black, but downwind of the batching site, it may be somewhat bleached and frequently appears shrivelled, the cortex of the stipe being deeply cracked, and the holdfasts easily detached from the rock sur-

[1] Australian Antarctic Division, Channel Highway, Kingston, Tasmania 7050, Australia.

Antarctic Ecosystems. Ecological Change and Conservation
(ed. by K. R. Kerry and G. Hempel)
© Springer-Verlag Berlin Heidelberg 1990

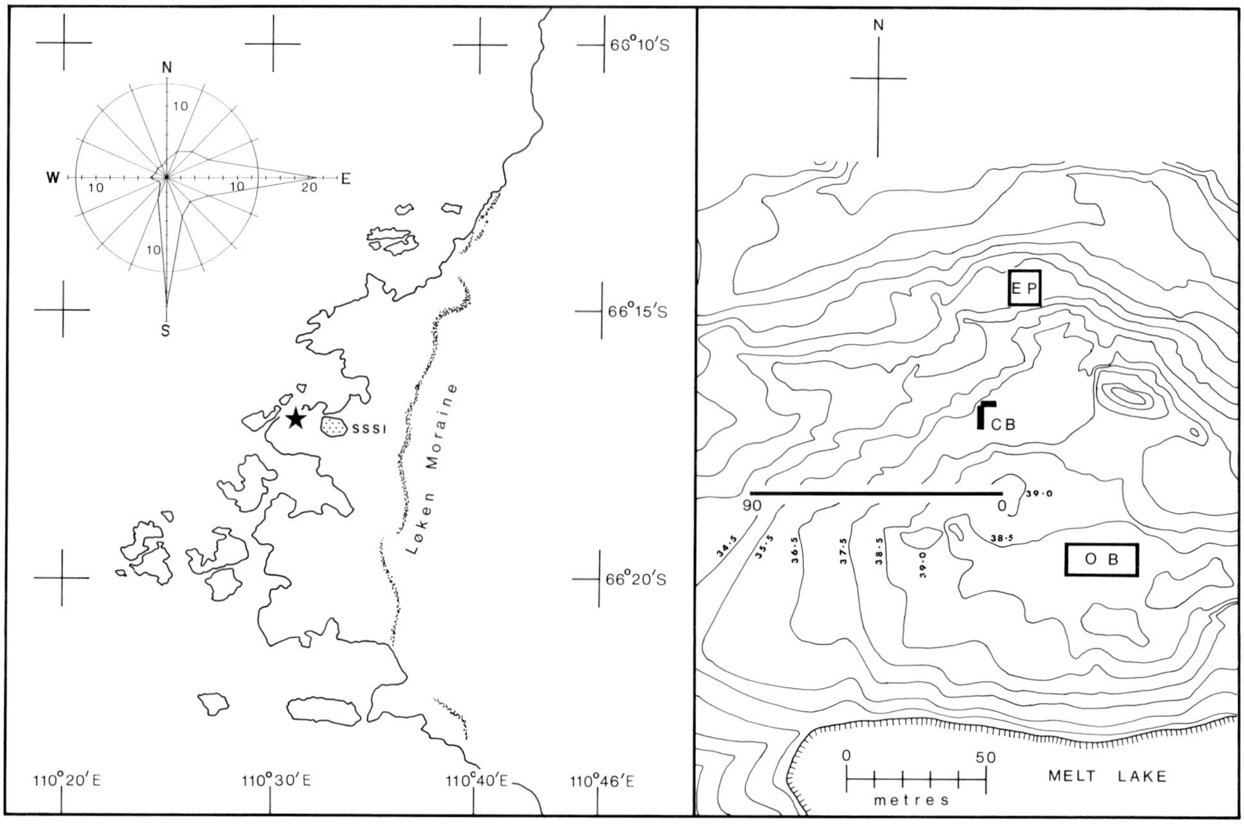

Fig. 1. Map showing part of the Windmill Islands region, location of the new Casey Station (indicated by a *star*) and the Site of Special Scientific Interest no. 16 (SSSI). *Wind rose* indicates the annual percentage frequency of wind direction recorded at three hourly intervals, calms (wind less than or equal to 1.5 m s^{-1}) are indicated by a *circle*. The position of the study transect (0–90 m) in relation to prominent buildings at the western end of the new station site is also shown. *EP* Emergency power house; *CB* the present location of the concrete batching platform (1988–89 summer); *OB* operations building. Contour interval = 1 m

face (Fig. 2b). The mosses *Ceratodon purpureus* (Hedw.) Brid., and *Bryum pseudotriquetrum* (Hedw.) *Gaertn., Meyer et Scherb.*, also found in the region immediatly downwind of the batching site, appear comparatively healthy. Although often encrusted with dust they are not obviously bleached or chlorotic.

The aim of this project was to investigate the physiological condition of the major lichen and moss species growing downwind of the concrete batching site, with particular reference to their photosynthetic pigments, and to compare these findings with healthy plants from an undisturbed site remote from the batching plant.

2 Materials and Methods

A 90 m transect (268° true) was established from the centre of the third batching site location, down a slightly inclined rocky ridge to a clearly defined vegetation trim line where the outcropping rock is covered by ice for much of the year. Circular quadrats of 1 m radius were established at 10 m intervals along the transect in March 1988. Fifty samples of surface soil (maximum depth 5 cm), moss samples (approximately 5 cm^2) and individual thalli of *Usnea sphacelata* and *Umbilicaria decussata* were collected at random from within each quadrat. Equivalent soil and plant samples were also collected from a comparable, uncontaminated control area in the Bailey Peninsula Site of Special Scientific Interest (SSSI no. 16). All samples were collected dry and kept in sealed plastic bags on the shaded (southern) side of the laboratory at ambient temperatures, light intensities, and snow cover. Chlorophyll estimations were carried out within 3 months of harvesting. Soil samples were stored in sealed plastic jars at − 20 °C.

Chlorophyll estimations and 77 K fluorescence measurements were carried out on the moss *Ceratodon purpureus* and lichens *Umbilicaria decussata* and *Usnea sphacelata.* Moss clumps were subsampled

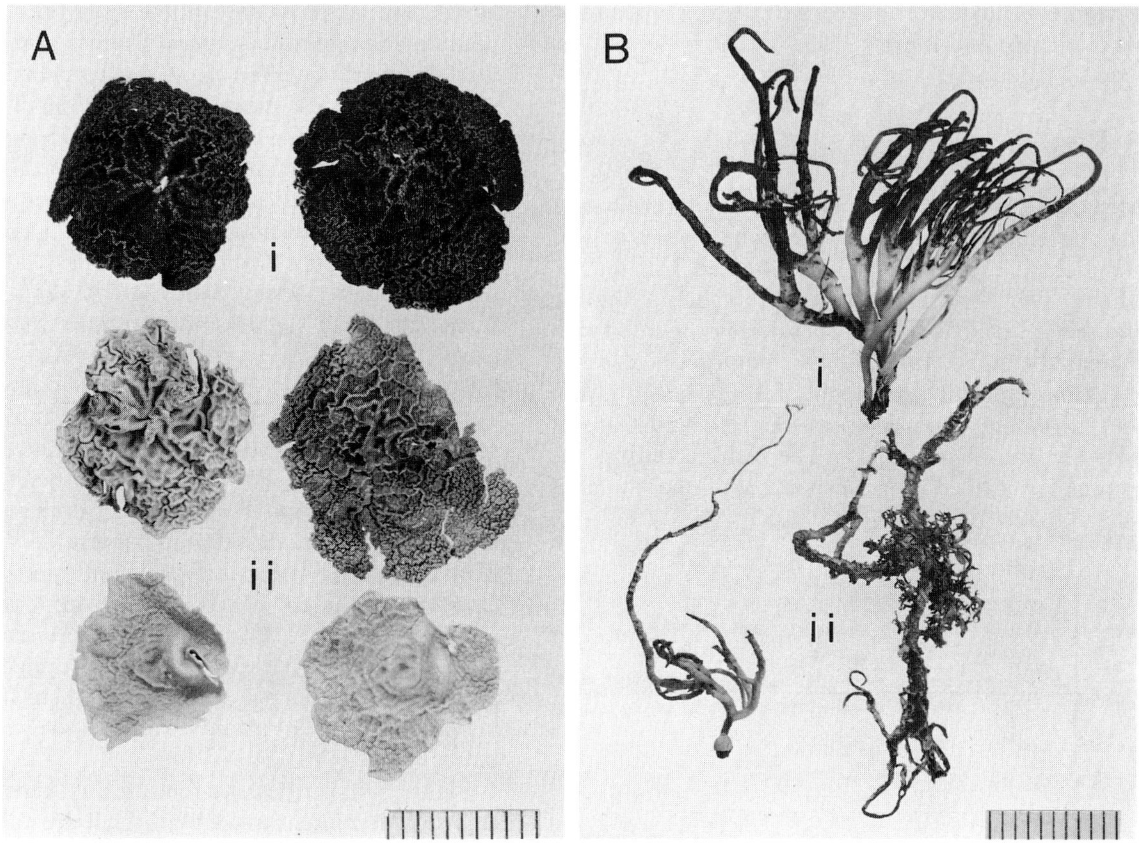

Fig. 2A. Healthy (*i*) *Umbilicaria decussata* and **B** *Usnea sphacelata* from the SSSI compared to bleached samples (*ii*) found 20 m downwind of the concrete batching site. Scale = 10.0 mm

prior to chlorophyll extraction. The apical 1–3 mm (avoiding the lower non-green parts of stems) were cut from each replicate, the tissue randomised and used for either pigment extraction or dry weight determinations. Lichen thalli from each site were randomised and several thalli subsampled for each chlorophyll and dry weight determination. Pigments were extracted in dim light, on ice, by grinding 100–300 mg of tissue with acid-washed sand and 50 mg magnesium carbonate in 10 ml of ice-cold acetone (85%). The slurry was transferred quantitatively to a 10 ml centrifuge tube, shaken vigorously and spun at 2000 rpm for 5 min, the supernatant decanted, the pellet resuspended in 1.5 ml of 85% acetone and centrifuged as before. The supernatants were combined, made up to a known volume and after zeroing at 720 nm, absorbance was recorded at 663 and 645 nm using a Hitachi U-3200 scanning spectrophotometer. Chlorophyll a and b content was calculated according to the equation of Anderson and Boardman (1964) as modified by Brouers and Michel-Wolwertz (1983).

To give an indication of the soil pH at each site, 10 g unfrozen soil was continuously stirred in 50 ml deionised water (pH 5.3), for 15–20 min and analysed to the nearest 0.5 pH unit with Merck pH strips (neutralit pH 5–10, spezialindikator pH 4–7, universalindikator pH 0–14). A comparison using a portable pH meter (TPS LC 80), indicated that the use of pH papers in this method underestimates soil pH values by 0.28 (SE = 0.08) pH units.

To give an indication of the efficiency of photosynthetic electron transport, the ratio of variable to maximum chlorophyll fluorescence (Fv/Fm) emitted by samples of *Umbilicaria*, *Usnea* and *Ceratodon* when illuminated at 77 K (in liquid nitrogen), was determined for the SSSI site and the site 40 m along the transect (site with abundant *Ceratodon* and considerable reduction in lichen total chlorophyll content). Material was collected in March, frozen and returned to Australia. Analyses were performed in August on material that had been thawed for 1 h at 20 °C. The principles and interpretation of fluorescence measurement are outlined in Papageorgio (1975), Hipkins and Baker (1986) and Lawlor (1987).

3 Results

Mean soil pH values at 10 m intervals along the batching site transect (Fig. 3) range from pH 7.1 to 8.8, significantly higher than pH 4.7 obtained from the SSSI site. The distribution of soil pH values along the transect is variable, but there is a decreasing trend with increased distance from the batching site.

Mean chlorophyll contents of *Umbilicaria*, *Usnea* and *Ceratodon* sampled from the SSSI and transect sites are shown in Table 1. The results confirm the impression gained from inspection of the site that

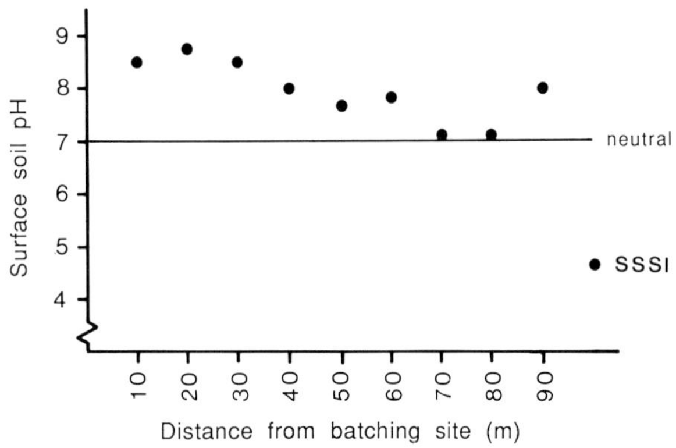

Fig. 3. Surface soil pH verses distance transect downwind of the concrete batching site and at the SSSI control site

lichens are far more susceptible to airborne alkaline pollution damage than mosses. There is comparatively little difference between the grand batching site transect and the SSSI mean total chlorophyll (chl. a + chl. b) content of *Ceratodon*. The grand transect mean total chlorophyll content of *Umbilicaria* and *Usnea* was, however, drastically reduced one-third – one-half of the SSSI complement respectively. Damage was related to distance from the batching platform and was particularly severe over the first 40 m. Overall, *Usnea* appeared slightly more resilient than *Umbilicaria*.

Usnea and *Umbilicaria* show a strong inverse relationship between total chlorophyll concentration and soil pH (Fig. 4a and b). Chlorophyll content in *Ceratodon* (Fig. 4c) is, however, more difficult to interpret. Total chlorophyll values vary dramatically according to natural variation in habitat, i.e. *Ceratodon* from the dry SSSI site has approximately one-third the chlorophyll content of *Ceratodon* from a very moist site (unpubl. data). Consequently, there seems to be little correlation between total chlorophyll content and increased soil pH. If anything there is a slight increase in total chlorophyll values downwind of the batching site.

Figure 5 illustrates the relationship between total chlorophyll content and chlorophyll a/b ratio in the *Usnea* and *Umbilicaria* from both the batching site transect and the SSSI. In general there is very little change in chlorophyll a/b ratios over a wide range of total chlorophyll concentrations. However, there is a marked reduction in the chlorophyll a/b ratio of plants with very low total chlorophyll content, i.e. in

Table 1. Total chlorophyll content of *Umbilicaria decussata*. *Usnea sphacelata* and *Ceratodon purpureus* from the Bailey Peninsula SSSI control site and at 10 m intervals along the batching site transect. Standard errors (SE) and number of replicates (n) are also shown

	Umbilicaria			Usnea			Ceratodon		
	Chlorophyll a + b (μg chl./g dry wt.)			Chlorophyll a + b (μg chl./g dry wt.)			Chlorophyll a + b (μg chl./g dry wt.)		
	Mean	SE	n	Mean	SE	n	Mean	SE	n
SSSI control site:	704	70	10	441	25	10	3597	102	5
Distance along the transect (m):									
10	180	23	9	197	42	12			
20	85	11	11	149	13	10			
30	94	16	9	286	40	13	4445	104	6
40	191	21	10	249	22	7	4556	78	6
50	319	25	11	339	32	10	4005	74	6
60	275	23	14	236	20	10	3933	111	6
70	291	28	13	435	28	10			
80	297	20	13	264	33	5	3063	75	6
90	270	30	13	151	18	10	3009	73	6
Grand transect mean:	231	11		254	14		3835	108	

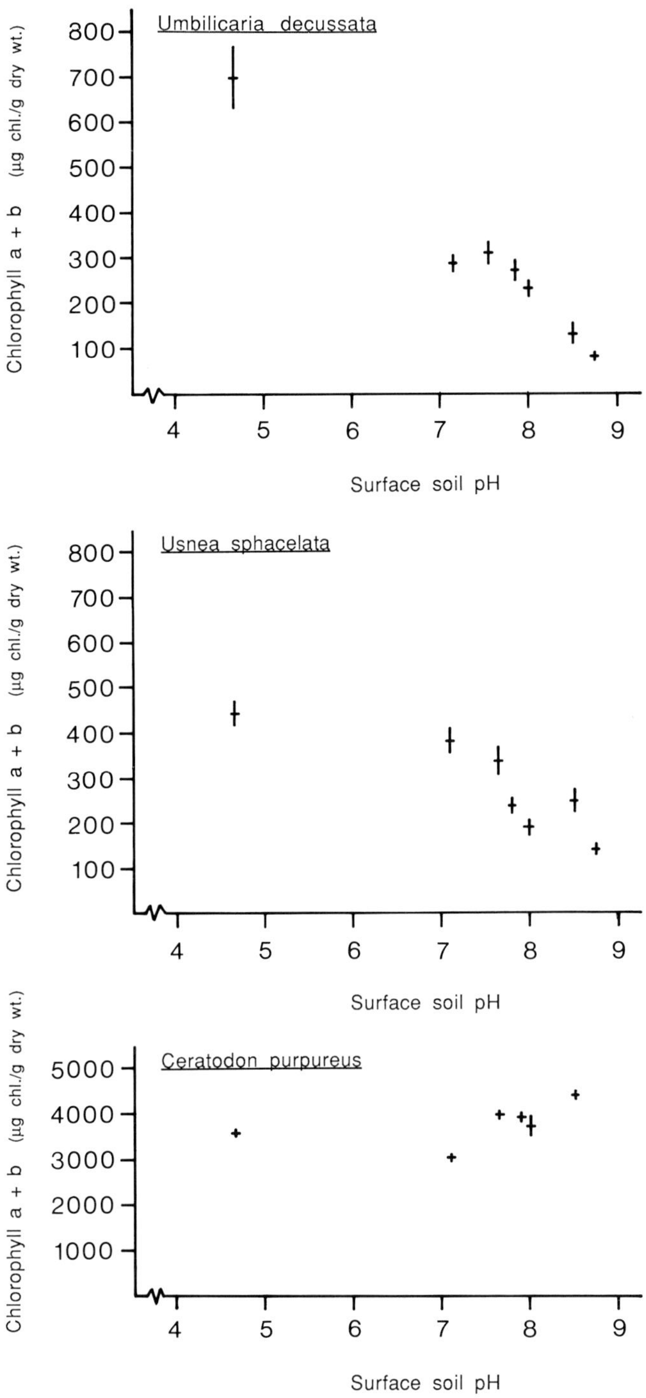

Fig. 4. Mean total chlorophyll content of *Umbilicaria decussata*, *Usnea sphacelata*; *Ceratodon purpureus* versus surface soil pH

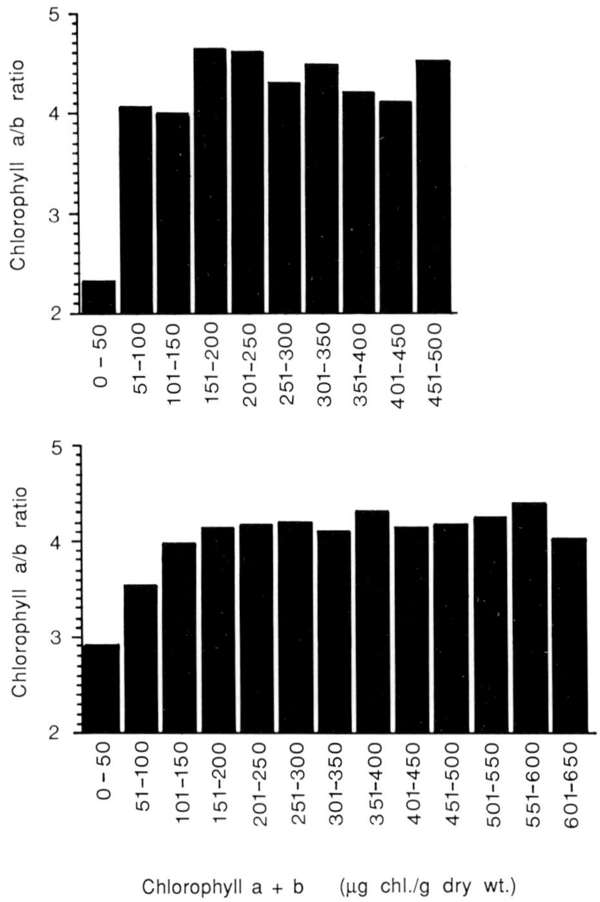

Fig. 5. Chlorophyll a/b ratio versus total chlorophyll content in *Umbilicaria decussata* (above) and *Usnea sphacelata* (below) from both the SSSI and the batching site transect

Umbilicaria containing less than 50 µg chl./g dry wt. and *Usnea* containing less than 100 µg chl./g dry wt.

Typical low temperature (77 K) chlorophyll fluorescence traces of *Umbilicaria*, *Usnea* and *Ceratodon* from the SSSI and the site 40 m along the transect and their corresponding ratios of variable to maximum fluorescence (Fv/Fm) are shown in Fig. 6. Fv/Fm ratios reflect the efficiency of photosystem II photochemistry. *Usnea*, *Umbilicaria* and *Ceratodon* from the SSSI have typical Fv/Fm ratios similar to those characteristic of mildly photoinhibited angiosperm leaves (Bjorkman and Demmig 1987). The lower Fv/Fm ratios typical of lichens and moss from the 40 m site on the transect are consistent with damage to the photosynthetic apparatus. However, since a variable fluorescence component, indicating electron transport from photosystems I and II, is present we conclude that lichens and moss from the 40 m transect site retain some capacity for photosynthetic electron transport.

4 Discussion

The concrete batching site under investigation had been used frequently over five summer construction seasons. The abnormally high soil pH in the immedi-

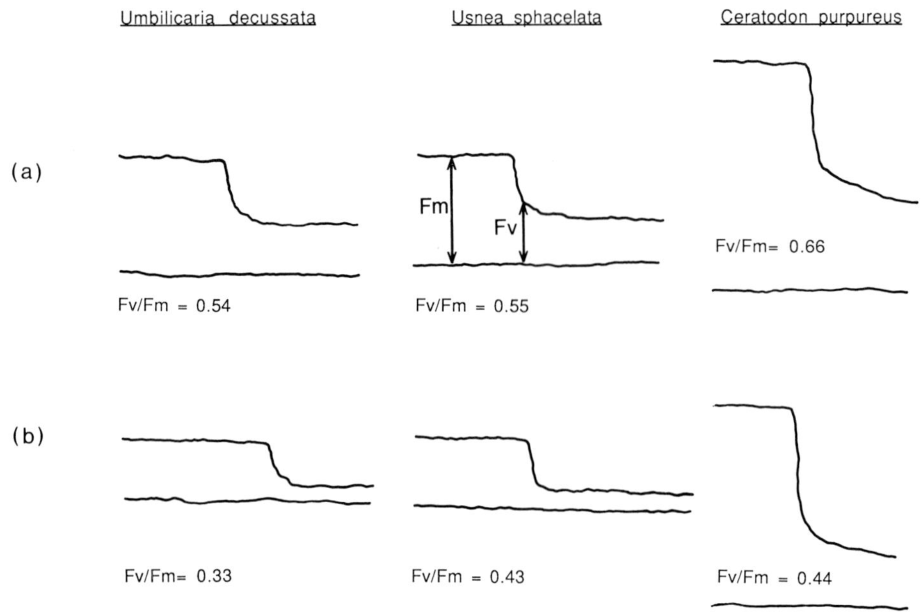

Fig. 6. Typical 77 K fluorescence traces of *Umbilicaria decussata*, *Usnea sphacelata* and *Ceratodon purpureus* after 1 h thaw at 20 °C. Corresponding Fv/Fm ratios are shown (a) SSSI (b) 40 m along the batching site transect

ate vicinity and the trend to higher pH close to the batching platform is consistent with it being the major source of alkaline pollution in this area. The strong easterly winds in this region and easy airborne transport of cement dust have led to its wide dispersal. Hancock (1986, unpubl. data) made measurements of surface snow and surface soil pH values along radial transects centred on the batching plant. The day after a concrete batch, surface snow and soil samples were collected. Beneath the batcher surface snow was pH 11.6 and surface soil pH 10.7. Forty m downwind of the batcher corresponding values were pH 11.4 and 8.5 and at 210 m downwind, well beyond the present transect, these values were pH 7.6 and 5.9.

The most obvious effect of alkaline pollution at Casey is bleached vegetation. Variation in soil pH and patchy distribution of black, grey and cream *Umbilicaria* and shrivelled *Usnea* along the transect is consistent with patchy distribution and collection of cement dust due to the uneven topography and microclimate variations in the region. For example, severely chemically bleached lichens are commonly found under the lee side of overhanging rocks where cement dust presumably collects and subsequent dispersal is minimal. No visual evidence of alkaline pollution was noted amongst the vegetation growing on hills further downwind.

Our findings indicate that lichens are more sensitive to alkaline contamination than mosses. Their characteristic loss of black pigment is accompanied by substantial loss of chlorophyll. Our results suggest two stages of damage in *Usnea* and *Umbilicaria*; the first involving a reduction in total chlorophyll content without any change in the chlorophyll a/b ratio; the second involving a greater loss of total chlorophyll combined with a reduction in the chlorophyll a/b ratio. The first stage probably reflects a reduction in the number of algal cells per thallus, reduction in amount of chlorophyll per algal cell and/or reduction in chlorophyll per chloroplast. The second stage almost certainly reflects destruction of thylakoid membranes and preferential loss of chlorophyll a. In contrast *Ceratodon* appears remarkably resistant to damage by cement dust. We have no evidence to suggest that *Ceratodon* growing downwind of the batching site is bleached. Our results indicate considerable variation in the total chlorophyll content of *Ceratodon*, such that at the 40–60 m transect sites it exceeds that of plants sampled in the SSSI. However, the differences are not correlated with pH and undoubtedly reflect other microhabitat variations.

The preliminary 77 K chlorophyll fluorescence measurements of plant material from contaminated and uncontaminated sites reported here are very important. They were carried out to validate the use of chlorophyll content as a quantitative indicator of alkaline pollution damage to lichens and moss under Antarctic conditions. This was necessary since there is no information on rates of chlorophyll turnover in polar or alpine environment in the literature. Since we could not rule out the possibility that chlorophyll might persist for a long time in dead algal cells, we used 77 K chlorophyll fluorescence patterns to confirm that residual chlorophyll in plant material from a heavily polluted site was photosynthetically competent. This has since been confirmed by 'room' temperature (Brancker SF 10 plant productivity fluorometer, Richard Brancker Research Pty. Canada) fluorescence measurements performed on freshly collected samples at Casey (1989 unpubl. data).

Lichens are well known to be highly susceptible to chemical pollutants, including alkaline dust (Gilbert 1976), because they are unable to exclude or excrete toxic substances (Hale 1974). Darley (1966) investigated naturally deposited cement dust as a cause of chlorosis and death of tissue in deciduous and coniferous plants. He concluded that injury results from a combination of a relatively thick crustal deposit which interferes with carbon dioxide exchange and the toxicity of alkaline solutions formed when dusts are deposited in the presence of free moisture. In this context, the relatively dry nature of the ground around the batching site at Casey and the frequent strong winds presumably limit damage to some degree. Destruction of the photoprotective black pigment which is a distinctive feature of many Antarctic lichens and the stressful photoinhibitory nature of the Antarctic environment (low temperatures, bright light, long periods of daylight), are also likely to be important factors in the damage which has occurred at Casey.

Our results suggest that since severely bleached lichens near the batching site contain viable algal cells, the algae within moderately damaged plants may be able to multiply and sustain growth and repair. The feasibility of recovery is, however, dependant on the continued integrity of the algal-fungi symbiosis. If that is destroyed, regeneration of the area will depend on the less damaged plants and recolonisation from nearby healthy populations. This will not readily occur until the pH of the region drops considerably. The apparent resistance of *Ceratodon* to damage from cement dust and alkaline soil is very important. It may be partly due to the relatively dry nature of the site, and the wider pH range and dust inundation tolerance of the moss species. However, given the dry nature of the site any leaching action required to remove the alkaline contamination from the surface soil will no doubt take a long time. Future environmental protection of this site and other as yet undamaged areas subject to concrete batching would therefore be desirable.

Acknowledgements. We would like to thank Dr. Heather Adamson of Macquarie University for the 77 K fluorescence data, assistance with interpretation, and discussion of the manuscript, the Australian Antarctic Division for logistic support, Mr. R. Oldfield for Fig. 1, and Mr. R. Ferris, Mr. G. Manning and Mr. B. Turnbull for information related to the history of concrete batching at Casey.

References

Anderson JM, Boardman NK (1964) Studies on the greening of dark-grown bean plants II. Development of photochemical activity. Aust J Biol Sci 17:93–101

Bjorkman O, Demmig B (1987) Photon yield of O_2 evolution and chlorophyll fluorescence characteristics at 77 K among vascular plants of diverse origins. Planta 170:489–504

Brouers M, Michel-Wolwertz M (1983) Estimation of protochlorophyll(ide) contents in plant extracts: re-evaluation of the molar absorption coefficient of protochlorophyll(ide). Photosyn Res 4:265–270

Darley EF (1966) Studies on the effects of cement-kiln dust on vegetation. J Air Pollut Control Assoc 16:145–150

Gilbert OL (1976) An alkaline dust effect on epiphytic lichens. Lichenologist 8:173–178

Hale ME (1974) The biology of lichens. Arnold, London

Hipkins MF, Baker NR (1986) Photosynthesis energy transduction. A practical approach. IRL Press, Oxford, pp 82–99

Lewis Smith RI (1986) Plant ecological studies in the fellfield ecosystem near Casey Station, Australian Antarctic Territory 1985–1986. Bull Br Antarct Surv 72:81–91

Lawler DW (1987) Photosynthesis: metabolism, control and physiology. Longman (scientific and technical), Singapore, pp 236–242

Papageorgiou G (1975) Chlorophyll fluorescence: an intrinsic probe of photosynthesis. In: Govindjee (ed) Bioenergetics of photosynthesis. Academic Press, New York, pp 320–371

Pickard J, Seppelt RD (1984) Phytogeography of Antarctica. J Biogeography 11:83–102

The Introduction and Establishment of the Diamondback Moth (*Plutella xylostella* L., Plutellidae) on Marion Island

J. E. CRAFFORD and S. L. CHOWN [1]

Summary. A small *Plutella xylostella* L. (Lepidoptera: Plutellidae) population was discovered in 1986 on Marion Island where larval feeding causes considerable damage to the native crucifer *Pringlea antiscorbutica* R. Br. Between April 1986 and May 1988 the population was monitored, on an opportunistic basis, to establish its dynamics, local distribution and the interaction of *P. xylostella* with its host. The first signs of larval infestation were detected at the start of the austral summer (December), after which the larval population rapidly built up to maximum density during April/May. After the onset of winter (July) the population crashed, with no sign of infestation during the remainder of winter and spring. In the laboratory on Marion Island, generation turnover times were similar to those recorded in temperate areas. The few taxa of the native biota allow ample opportunity for successful colonization by introduced species. The selective advantages that enabled the introduction and successful establishment of *P. xylostella* on Marion Island include cold tolerance and such r-selected traits as migratory ability, high fecundity, and rapid growth and generation turnover. *P. xylostella* should be regarded as a well-adapted colonizer on Marion Island, and the opportunities for population genetic research, presented by this founder event, should be used to advantage.

1 Introduction

Plutella xylostella L. (Lepidoptera: Plutellidae) is one of the most destructive pests of cruciferous crops in the world. It has a truly cosmopolitan distribution, having been recorded in economically damaging numbers in over 100 countries (Talekar et al. 1985), and as a vagrant on several oceanic islands (e.g. Holdgate 1965). Established *P. xylostella* populations occur on the New Zealand sub-Antarctic islands (Dugdale 1971). In April 1986 it was recorded for the first time on Marion Island (46° 54′ S, 37° 45′ E), where it infests the native crucifer, *Pringlea antiscorbutica*, or "Kerguelen cabbage" (Crafford and Chown 1987). Between April 1986 and May 1988 we monitored, opportunistically, the *P. xylostella* population on Marion Island, mainly to assess the status (transient, intermittent, or established, expanding?) of the species in the island's impoverished terrestrial fauna. In this chapter we report our field observations during this period, and offer explanations for the establishment and subsequent success of *P. xylostella* on Marion Island.

[1] Department of Entomology, University of Pretoria, Pretoria 0002, South Africa.

Antarctic Ecosystems. Ecological Change and Conservation
(ed. by K. R. Kerry and G. Hempel)
© Springer-Verlag Berlin Heidelberg 1990

2 Methodology

Infested *P. antiscorbutica* patches were mapped during surveys, undertaken on foot, of Marion Island in April 1986, May 1987, and April 1988. One large infested patch close to the weather station was monitored throughout the 1987/88 season. All the plants in this patch were numbered, and larvae, pupae and adults on each infested plant, from a random sample of 20 plants, were counted monthly. Pupae collected in the field were hatched in the laboratory, and a small breeding colony was maintained at ambient indoor temperature (15–18 °C) to obtain estimates of generation time. Although this temperature is substantially higher than Stevenson Screen temperatures, it is only moderately higher than microclimatic temperatures (unpubl.). Larvae were provided with whole leaves and kept at 80% RH and a L 16-h: D 8-h cycle.

3 Results and Discussion

The *P. xylostella* population had to be monitored on a strictly opportunistic basis, and results accordingly do not purport to reflect intensive study of the population on Marion Island. Information on the basic biology of the species is taken from the substantial literature on *P. xylostella* (see Talekar et al. 1985), unless stated to pertain specifically to Marion Island.

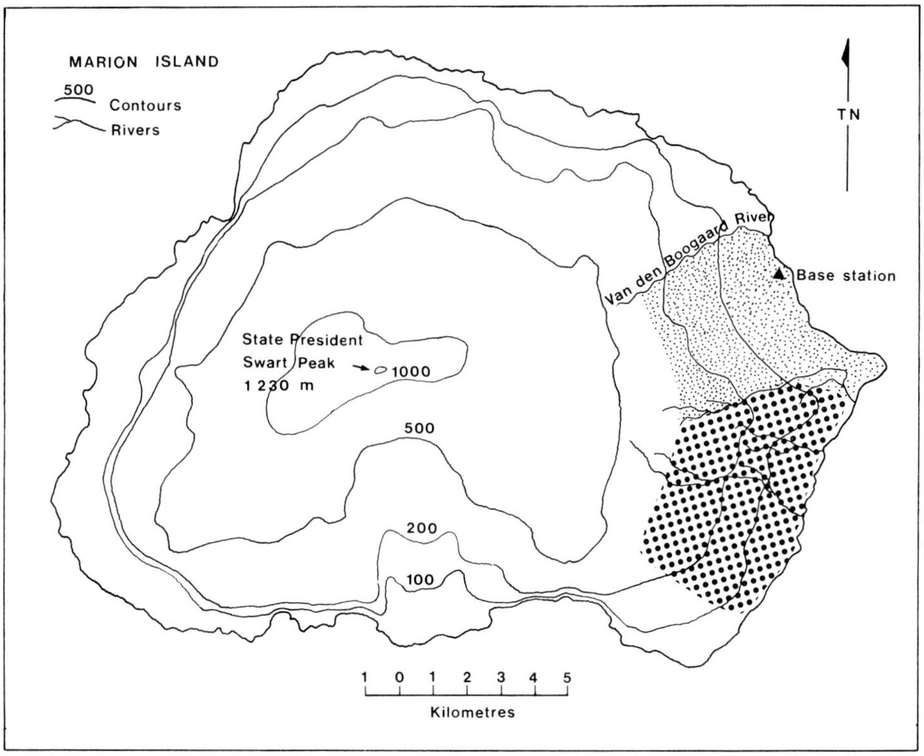

Fig. 1. The present distribution of *Plutella xylostella* on Marion Island. The *large dots* represent the distribution in 1986, and the *small dots* show the additional area colonized by the moth by 1988

3.1 Local Distribution

P. xylostella is well established, and appears to be increasing its range on Marion Island. At the end of 1986, *P. xylostella* was confined to a ca. 50-ha area on Marion Island's south-eastern coastal plain, downwind from the weather station (Crafford and Chown 1987; Fig. 1). By May 1988, isolated stands of *P. antiscorbutica* north and north-west of the weather station, and along the banks of the Van den Boogaard river (Fig. 1), were also infested.

3.2 Life History and Phenology

P. xylostella has a high reproductive potential, rapid generation turnover and good long-distance migratory ability. Females usually lay 200–300 eggs on the undersides of the host plant's leaves. There are four larval instars, and mature larvae pupate in a loose silken cocoon attached to the outer leaves of the host plant. In temperate areas the egg, larval, pupal and adult stages last approximately 5, 15, 9 and 18 days, respectively. The life cycle (egg to eclosion of adults) is usually completed within one month. In the tropics, the number of generations can exceed 20 per year. On Marion Island, field data are lacking, but larval duration in the laboratory (at 15–18 °C) was 20–30 days, with the entire life cycle completed in 40–50 days. The larvae are voracious feeders, and cause severe damage to the leaves of heavily infested plants. Solitary *P. antiscorbutica* plants were often found to be heavily infested (up to 12 larvae per plant), while larvae were patchily distributed within stands of plants, and had large aggregations on individual plants.

The first signs of larval infestation were detected in December (early austral summer at this latitude), after which the larval population rapidly built up to maximum density during April/May. After July, during the austral winter, the population crashed, with no sign of infestation during the remainder of winter and spring. Generations overlapped widely during the 6 months of larval infestation, but the larval "boom" and "bust" were remarkably sudden. *P. xylostella* probably overwintered as pupae and adults, but there is the possibility that first instar larvae, which are known to mine the leaves of other crucifers (Annecke and Moran 1982), could have been overlooked. Adult moths are capable of surviving temperatures below freezing, and although several reports (Talekar et al. 1985) state that the pupae "hibernate", Razumov (1970) found that in the colder northern parts of the USSR, the moth overwintered more frequently as adults, and not in the pupal stage as it does in the more temperate central regions. However, the low-temperature responses of the species in the higher latitudes of its range remain poorly known.

3.3 Selective Advantages

The selective advantages of *P. xylostella* for establishment and survival on Marion Island lie both in the characteristics of the species (several of which constitute pre-adaptations to both the physical and the living environment) and in the nature of the island's biota. These selective advantages are summarized in Table 1.

3.4 Limiting Factors

3.4.1 Climate

The seasonal trend of the *P. xylostella* population on Marion Island, in relation to annual variation of air and grass minimum temperatures at the meteorological station on Marion Island (after Schulze 1971), is given in Fig. 2. The rapid decline of the *P. xylostella* population with the onset of cold weather (lowest temperatures usually recorded in August), and the sudden re-appearance of larvae in summer, suggest that temperatures remain below the threshold for larval development in the intervening period. It is well document-

Table 1. Some selective advantages of *Plutella xylostella* presented by its own characteristics and those of its environment on Marion Island

Characteristics of the environment	Characteristics of *P. xylostella*	Selective advantages
Biotic		
1. Species-poor biota Single crucifer in low-diversity vascular flora	Strictly host-specific on crucifers	High food utilization efficiency, enabling rapid growth
2. *Pringlea antiscorbutica* patchily distributed	Good dispersal ability	Ability to colonize isolated habitats and utilize spatially patchy resources
Abiotic		
1. Low mean annual air temperature, high precipitation, severe cold spells during winter	Cold tolerant in the pupal and adult stages	Low temperature survival
2. Temperate microclimate in vegetation canopy, mild spells during summer	Rapid growth and generation turnover	Ability to utilize temporary favourable conditions

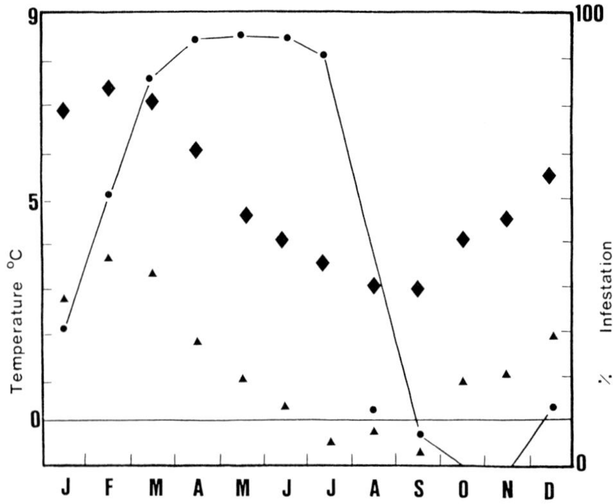

Fig. 2. The seasonal trend of the *P. xylostella* population (*circles*) on Marion Island, in relation to the annual variation in mean air (Stevenson screen; *diamonds*) and grass minimum temperatures (*triangles*), measured at the meteorological station on the island. Temperatures after Schulze (1971)

ed that temperature has a pronounced effect on the digestive efficiency of poikilothermic herbivores (Merkel 1977; Remmert 1986), with growth being minimal at low temperatures on cellulose-rich food (Schramm 1972). Thus the effect of temperature on digestive efficiency is probably the cause of larval mortality, although the adverse effect of low temperatures on ecdysial enzymes cannot be discounted. Temperature effects may also explain why adults, which do not rely on cellulose-rich food, rather than larvae survive the winter months. Also, at night, during very cold, dry periods (such as August; Schulze 1971), frost (which is common on the island) may cause the larvae to freeze as a result of external inoculation (see Asahina 1969). Such spontaneous freezing is usually lethal to insects. Adults and possibly pupae, sheltering in the detritus below the plants, would not be as prone to external inoculation as larvae feeding on exposed leaves. In northern Canada *Plutella xylostella* are destroyed by frost each year (Harcourt and Cass 1966).

At any time, the microclimate within a stand of *P. antiscorbutica*, or amongst the leaves of an individual plant, is probably more temperate than the macroclimate. On sunny, wind-still days during summer, the temperature among the leaves of cabbage plants was found to be up to 10 °C higher than ambient air temperature. However, it is likely that the combination of a severe cold spell, accompanied by frost and high precipitation can cause mass mortality in all life stages, and climate is probably the major factor limiting *P. xylostella* on Marion Island, although photoperiod may also be important (see Atwal (1955), Harcourt (1963), Harcourt and Cass (1966) and Stepanova (1962) for discussions of the influence of climate and photoperiod on the survival and development of *P. xylostella*).

3.4.2 Predators

The only potential predators of *P. xylostella* on Marion Island are spiders, predatory mites and feral house mice, but these predators take *P. xylostella*, at the most, opportunistically and occasionally. Mouse predation on larvae could conceivably limit small, isolated populations in heavily infested stands of cabbage.

3.4.3 Pathogens

The limiting effect of pathogens (viruses, fungi and bacteria) on insect populations is often underestimated. It is likely that *P. xylostella* is attacked by some of the local microbial pathogens, especially since its family (Plutellidae) is phylogenetically close to those of the native Lepidoptera (Tineidae and Yponomeutidae). Larvae with bacterial and viral infections become discoloured and flaccid (Maddox 1982), and many larvae of *P. xylostella* exhibiting these symptoms, as well as dead larvae, were found on Marion Island during the height of larval infestation in late summer.

3.5 Impact of *P. xylostella* on Its Host

Larval feeding caused considerable damage to *P. antiscorbutica* plants during the maximum larval infestation in summer, but almost all of the severely damaged plants recovered completely during the following winter. Despite this, repeated attacks may lead to a long-term depression of host plant fitness by lowering seed production, reducing growth and eliminating root food stores, as is sometimes the case in other plant-herbivore interactions (see Strong et al. 1984 and references therein).

Little is known regarding the phenology of *P. antiscorbutica* on Marion Island. The plants appeared to have a high production and rapid turnover of leaves, completely replacing the previous season's leaves every winter. They were at their peak (maximum size and leaf set) by spring (October/November). Flowering occurred in spring, and fruits were present throughout summer. *P. antiscorbutica* is a truly sub-Antarctic plant, and must be close to its upper thermal limit for growth and survival on Marion Island. It occurs far more extensively and luxuriantly on Heard Island (53° 05′ S, 70° 30′ E) than on the relatively temperate Prince Edward Islands. However, even on Heard Island it seems to fill the role of a colonizer of disturbed areas (Hughes 1987). On Marion Island it occurs mainly against south-facing slopes and along riverbanks in low-lying areas. It is also at its peak when the *P. xylostella* population ebbs. This, combined with its patchy distribution, limits the potential impact of a large and well-established *P. xylostella* population on Marion Island. Few other invertebrates utilize *P. antiscorbutica* as food. Larvae of a drosophilid fly of unknown status, and adult weevils (*Dusmoecetes similis*, C. O. Waterhouse) feed on the leaves. On Îles Kerguelen *Calycopteryx moselyi* Eaton is also associated with *P. antiscorbutica*.

3.6 Status of *P. xylostella* on Marion Island

We revised our original perception of *P. xylostella* as a potentially destructive "introduced alien" (Crafford et al. 1986; Crafford and Chown 1987). It is likely that *P. xylostella* was introduced with supplies of fresh vegetables, delivered on Marion Island twice a year. One of us (SLC) examined a 10-kg consignment of fresh cauliflower and cabbage brought ashore in September 1986, and found the remains of three species of flies, aphids, a beetle, and adult *P. xylostella* moths. The leaves of fresh, imported cabbages invariably show feeding damage typical of that caused by *P. xylostella* larvae. However, it is quite possible that *P. xylostella* reached Marion Island on its own. Records of long-distance migrations by large swarms of *P. xylostella* moths abound in the literature (Talekar et al. 1985). A spectacular example was a mass migration across the Barents Sea from Finland to Spitzbergen (a distance of 1000 km) *in 1 day*, recorded by Lokki et al. (1978). Steele and Crafford (1987) found the remains of migratory noctuid moths in the stomachs of blue petrels (*Halobaena caerulea*, Gmelin) mist-netted on Marion Island. These birds do not forage further than about 200 km offshore (*op cit.*), and the moths had never been recorded live on Marion Island, which suggests that mass migrations of insects may reach Marion Island in spite of its isolation and the long distances from the continents. Even if *P. xylostella* had been introduced by man, we suggest that man, by acting as a dispersal agent, hastened the colonization of Marion Island by this species.

4 Conclusion

P. xylostella should be regarded as a natural colonizer on Marion Island, if only because the species has established itself beyond any likelihood of its eradication by chemical or other means. The success of *P. xylostella* on Marion Island poses several questions. Given the underutilization of the herbivore (folivore) niche on Marion Island, is the success of recent herbivore colonizers [aphids, *P. xylostella,* possibly *Vanessa cardui* (Lepidoptera) linked to a "warm-up" of the sub-Antarctic? There seems to be a lower thermal threshold for the efficacy of herbivory as an "ecological strategy" (Remmert 1986), and a temperature regime below this threshold would account, historically, for the lack of herbivores in the sub-Antarctic. The establishment of *P. xylostella* on Marion Island furthermore represents a unique founder event – the dream of population geneticists and evolutionary biologists – that lends itself to a long-term study of the population genetics of the species.

Acknowledgements. We thank three anonymous referees for their comments and suggestions. Biological research on the Prince Edward Islands is partly funded and logistically supported by the South African Department of Environment Affairs, on the advice of the South African Scientific Committee for Antarctic Research (SASCAR).

References

Annecke DP, Moran VC (1982) Insects and mites of cultivated plants in South Africa. Butterworths, Durban, 383 pp

Asahina C (1969) Frost resistance in insects. Adv Ins Physiol 6:1–49

Atwal AS (1955) Influence of temperature, photoperiod, and food on the speed of development, longevity, and other qualities of the diamondback moth, *Plutella maculipennis* (Curtis) (Lepidoptera: Tineidae). Aust J Zool 3:185–221

Crafford JE, Scholtz CH, Chown SL (1986) The insects of sub-Antarctic Marion and Prince Edward Islands, with a bibliography of entomology of the Kerguelen biogeographical province. S Afr J Antarct Res 16:42–84

Crafford JE, Chown SL (1987) *Plutella xylostella* L. (Lepidoptera: Plutellidae) on Marion Island. J Ent Soc S Afr 50:259–260

Dugdale JS (1971) Lepidoptera. In: Gressitt JL (ed) Entomology of the Aucklands and other islands south of New Zealand. Pacif Insects Monogr 27:55–172

Harcourt DG (1963) Major mortality factors in the population dynamics of the diamondback moth, *Plutella maculipennis* (Curt.) (Lepidoptera: Plutellidae). Mem Entomol Soc Can 32:55–66

Harcourt DG, Cass LM (1966) Photoperiodism and fecundity in *Plutella maculipennis* (Curt.). Nature 210:217–218

Holdgate MW (1965) The fauna of the Tristan da Cunha Islands. Phil Trans R Soc Ser B 249:361–402

Hughes JM (1987) The distribution and composition of vascular plant communities on Heard Island. Polar Biol 7:153–162

Lokki J, Malmstrom KK, Suomalainen E (1978) Migration of *Vanessa cardui* and *Plutella xylostella* (Lepidoptera) to Spitzbergen in the summer 1978. Not Entomol 58:121–123

Maddox JV (1982) Use of insect pathogens in insect pest management. In: Metcalf RL, Luckmann H (eds) Introduction to insect pest management. John Wiley & Sons, New York, pp 175–216

Merkel G (1977) The effects of temperature and food quality on the larval development of *Gryllus bimaculatus* (Orthoptera, Gryllidae). Oecologia 30:129–140

Razumov VP (1970) How the diamondback moth overwinters. Zashch Rast 15:38

Remmert H (1986) The importance of ecophysiological and experimental research for the understanding of population and ecosystem processes. Proc Third European Congr Entomology. Universiteitsdrukkerij, Utrecht, pp 5–20

Schramm U (1972) Temperature-food interaction in herbivorous insects. Oecologia 9:399–402

Schulze BR (1971) The climate of Marion Island. In: Van Zinderen Bakker EM, Winterbottom JM, Dyer RA (eds) Marion and Prince Edward Islands; report on the South African biological and geological expedition 1965–1966. A A Balkema, Cape Town, pp 16–31

Steele WK, Crafford JE (1987) Insects in the diet of the blue petrel *Halobaena caerulea* at Marion Island. Cormorant 15:93–94

Stepanova LA (1962) An experiment in the ecological analysis of the conditions for the development of pests of cruciferous vegetable crops in nature. Rev Entomol USSR 41:721–736

Strong DR, Lawton JH, Southwood TRE (1984) Insects on plants. Community patterns and mechanisms. Blackwell Scientific Publications, Oxford, 313 pp

Talekar NS, Yang HC, Lee ST, Chen BS, Sun LY (1985) Annotated bibliography of diamondback moth. Asian Vegetable Research and Development Centre, 469 pp

The Role of Feral House Mice in Ecosystem Functioning on Marion Island

J. E. CRAFFORD [1]

Summary. The chief prey of feral house mice on Marion Island is larvae of the flightless moth *Pringleophaga marioni* Viette (Tineidae), which constitute about 50% of their diet throughout the year. The litter-dwelling, detritivorous larvae take more than 2 years to mature, during which time they process (i.e. fragment, ingest and excrete) large quantities of litter. At an annual mean biomass of 9.3 kg ha^{-1}, and an individual consumption rate of approximately 0.6 × their own mass in litter daily, *P. marioni* larvae are estimated to process at least 1500 t of litter annually on Marion Island's vegetated coastal plain (all results expressed as dry mass). The larvae facilitate the release of nutrients from plant litter through their feeding activity and the resultant enhancement of microbial decay. Mice remove daily 65 g ha^{-1} or 0.7% of the standing crop of *P. marioni* larvae on the island's coastal plain. The indirect effect of mouse predation, through the removal of a large proportion of decomposer biomass and subsequent impediment of nutrient mineralization, has implications for the functioning of the decomposer subsystem of the Island's terrestrial ecosystem. Since mice and their prey appear to be in dynamic or even stable equilibrium, the direct impact of mice through predation may be eclipsed by these indirect detrimental effects to nutrient cycling.

1 Introduction

Mice (*Mus musculus* L.) were already established on Marion Island 170 years ago, probably having arrived with the very early sealers (Watkins and Cooper 1986). Neighbouring Prince Edward Island, only 22 km to the north-east and with a similar history of human exploitation, is mouse-free. Gleeson (1981) found mice to be restricted to the coastal lowland of Marion Island below 450 m elevation, and found population densities to vary according to vegetation type and correspondingly varying prey biomass. The preferred prey of mice is larvae of the endemic flightless moth *Pringleophaga marioni* Viette (Tineidae) (Gleeson and Van Rensburg 1982), the biomass of which is highest in biotically influenced areas and, specifically, in heavily manured *Poa cookii* (Poaceae) tussock grassland (Burger 1978; Crafford 1987).

Investigation of the role of feral mice in ecosystem processes on Marion Island was prompted by the results of three independent, seemingly disparate, studies recently undertaken on the Island. Firstly, it has been demonstrated that invertebrate detritivores, by comminuting plant litter and stimulating microbial decay, play an important role in releasing nutrients contained in the massive dead standing crop of vegetation in lowland plant communities on the island (Smith 1985). Secondly, *P. marioni* larvae, although their biomass represents only a fraction of that of earthworms in most lowland plant communities (Burger 1978), emerged as the most important invertebrate detritivores in terms of actual fragmentation and digestion of litter (Crafford et al. 1986; J Crafford unpubl. data). Finally, a study of the energy metabolism of feral mice on Marion Island (Rowe-Rowe et al. in press) indicated that predation by mice on invertebrate detritivores and, specifically, on *P. marioni* larvae, is more intense than previously supposed, and that *P. marioni* populations may be suppressed by mice. From these findings, it follows that mice should have an impact beyond mere reduction of their prey populations.

In this report, I assess the ecological role of mice, and some of the ramifications of their presence, in Marion Island's impoverished terrestrial ecosystem. I do this by synthesizing results (unpubl. data) of recent fieldwork, conducted by colleagues and myself, on

1. The impact of mice on populations of their prey species (based on the energy metabolism of mice).
2. The population dynamics and feeding ecology of their principal prey, *P. marioni*.
3. The role of *P. marioni* in the release of nutrients from litter of their principal food plants.

[1] Department of Entomology, University of Pretoria, Pretoria 0002, South Africa.

Antarctic Ecosystems. Ecological Change and Conservation
(ed. by K. R. Kerry and G. Hempel)
© Springer-Verlag Berlin Heidelberg 1990

2 Study Site

Marion Island is entirely of volcanic origin; it has a surface area of 290 km^2 and rises to a height of 1230 m a.s.l. It has been continuously occupied by South African weather teams since its annexation in 1948. The macroclimate is hyperoceanic, and is characterized by constant low temperatures (annual mean air temperature 5 °C), high precipitation (> 2500 mm a^{-1}) and almost incessant high winds. For an account of the physical environment and biota of Marion Island, see Smith (1987a), and references therein.

3 Materials and Methods

Methodology is described in detail in papers resulting from the different studies, and is only briefly outlined below.

3.1 The Impact of Mice on Their Prey

The energy metabolism of mice was established on Marion Island during April/May 1987, using isotope turnover techniques (based on Nagy 1980; fully described in Rowe-Rowe et al. in press). Trapping and recapturing of experimental mice took place in a study area ("Nellie Humps") near the weather station on Marion Island. Rates of food consumption were calculated in two ways:

1. From water influx rates (free water plus metabolic water was calculated to represent 85% of the fresh mass of prey, and it was assumed that food provided the only source of water).
2. From metabolic rates and the metabolizable energy content of the prey [mean energy content of prey was 21 kJ g^{-1}; an energy assimilation efficiency of 85% (Drożdż 1968) was assumed for the mice].

Stomach contents of mice (**n** = 36) snap-trapped during the same study were analyzed to determine percentage contribution to their diet by different prey items. To determine whether the mice fed size-selectively on *P. marioni* larvae, the remains of larvae in these, and in 50 other mouse stomachs collected during the rest of the year at a rate of five per month, were sorted into different size classes. The size classes were based on measurements of the undigestable, heavily sclerotized head capsules, crochets (the circles of hooks on the prolegs) and dorsal sclerites.

The impact of mice on invertebrate prey populations was extrapolated from: individual food consumption by mice, their mean annual density on the coastal lowland, the percentage contribution to their diet by different prey items, and the standing crops of their major prey in coastal lowland vegetation.

3.2 Population Dynamics and Feeding Ecology of *Pringleophaga marioni* on Marion Island

A monthly sampling programme (May 1983–April 1984), in the same study area as above, yielded estimates of *P. marioni* larval standing crops in different plant complexes (*sensu* Gremmen 1981) of lowland vegetation. Methods are described in Crafford (1987). The head capsules of all larvae in the monthly samples were measured and the larvae sorted into size classes based on the measurements. Larvae were also reared in the laboratory to supplement basic life history data obtained in the field.

The rate at which individual *P. marioni* larvae consumed *P. cookii* and *Azorella selago* (Apiaceae) litter under simulated field conditions was determined in the laboratory on Marion Island during the same period. Since large, semi-mature to mature larvae dominate larval biomass at all times, only such larvae were used in the feeding experiments. Total turnover (i.e. consumption and excretion) of litter in the field by *P. marioni* larvae was extrapolated from these results.

3.3 The Role of *Pringleophaga marioni* Larvae in Nutrient Release from Litter of Their Principal Food Plants

Twelve Perspex microcosms, each containing two large *P. marioni* larvae and either 2 g (dry mass) *P. cookii* (n = 6) for 2 g (dry mass) *A. selago* (n = 6) litter, were leached at weekly intervals for 2 months. The amount of free nitrogen (ammonia ions) in the leachates was determined by the phenol-hypochlorite method (Solórzano 1969). Free phosphate content of the leachates was determined using the method described by Murphy and Riley (1962). Twelve microcosms containing only *P. cookii* (n = 6) or *A. selago* (n = 6) litter were leached at the same intervals to serve as controls.

4 Results

4.1 Impact of Mice on Their Prey

Estimated daily food consumption per mouse (mean mass 19.5 g; n = 40) was 3.2 g dry mass, based on

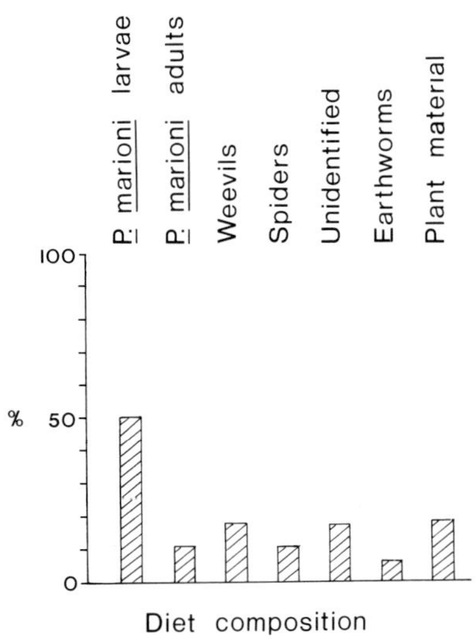

Fig. 1. The annual mean percentage contribution of different prey items to the diet of feral house mice, trapped between April 1979 and May 1980 on Marion Island (after Gleeson 1981)

water influx, and 3.7 g dry mass based on metabolic rate. The average, viz. 3.5 g day^{-1} (dry mass) was used to assess the impact of the mice on their prey populations. *P. marioni* larvae had contributed, on average, 74% to the diet of mice (n = 36) trapped during the study period, and occurred in 94% of the stomachs analyzed.

Figure 1 (after Gleeson 1981) shows the annual mean percentage contribution of major prey items to the diet of mice, which were trapped in all the habitats in which they occur. Throughout the year, invertebrates contributed 73% and plant material 17% to the diets of mice. The percentage contribution of *P. marioni* was, on average, about 55% during winter and decreased to about 35% in summer, when plant material (seeds and leaf shoots) was more commonly eaten by mice than at other times of the year. At a mean annual density, within their distributional range, of 37 mice ha^{-1} (Gleeson 1981), daily food consumption by mice would be 130 g ha^{-1} (dry mass). Based on an annual mean contribution of 50% to their diet by *P. marioni* larvae (Fig. 1), the daily consumption of these larvae by mice would be 65 g ha^{-1} dry (23.7 kg ha^{-1} y^{-1} dry mass).

The mice are strongly size-selective feeders: Fig. 3 illustrates the range of larval sizes and larval remains (n = 203) from a total of 86 mouse stomachs collected throughout the year, in relation to the size-class composition of the larval population.

4.2 Population Dynamics and Feeding Ecology of *Pringleophaga marioni*

The monthly standing crops of *P. marioni* larvae in three major plant complexes are given in Fig. 2. Based on these standing crops, the mean annual biomass (dry) of *P. marioni* larvae on Marion Island's coastal plain is estimated to be 9.3 kg ha^{-1}. The breakdown into size classes, based on head-capsule measurements, of the monthly crop of larvae from biotically influenced coastal vegetation, is shown in Fig. 3.

Although the exact duration of their life cycle has not yet been determined, *P. marioni* larvae are known to take 2 to 3 years to mature (Crafford 1987). Young larvae grow rapidly through the first few instars and attain approximately their mature size within a year. Adults are short-lived and do not feed; mating and egg laying (up to 200 eggs per female) occur within the first 5 days, and probably within 5 m of the site of eclosion. Annual recruitment to the population has not been quantified.

Semi-mature to mature larvae consumed, on average, 60% of their own dry, and 20% of their own live body mass in dry plant litter daily (n = 75). The larvae are generalist feeders, and other litter types are consumed at similar rates. Consumption of *P. cookii* litter was slightly less than that of *A. selago* litter, but considerable individual variation in consumption rates made comparison difficult. Total annual turnover of litter by larvae on the coastal plain, based on

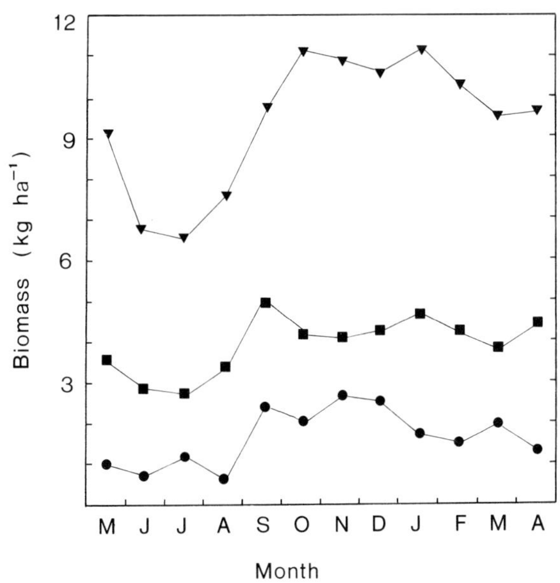

Fig. 2. The mean monthly biomass, between April 1983 and May 1984, of *Pringleophaga marioni* larvae in the biotically influenced (▼), salt spray (■) and mire (●) plant complexes on Marion Island (after Crafford 1987)

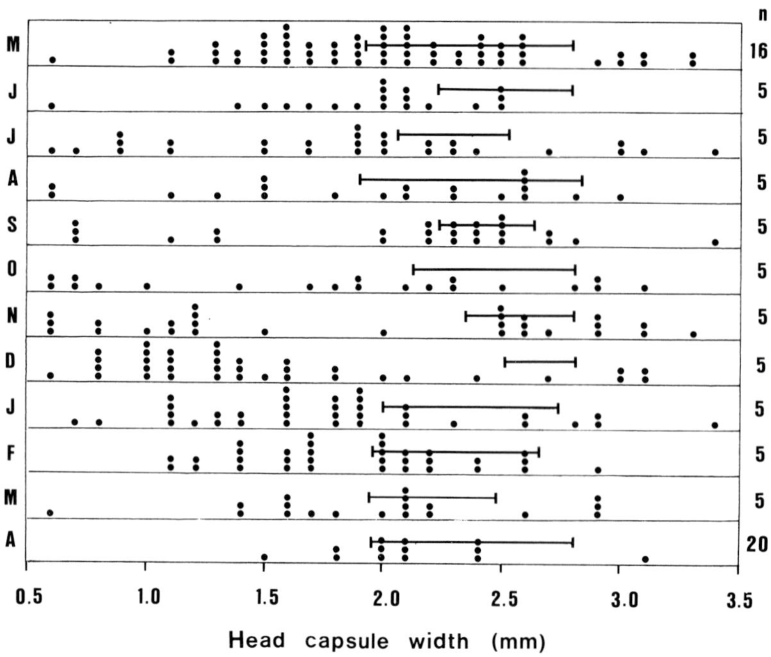

Fig. 3. Monthly size-class composition of *Pringleophaga marioni* larvae sampled in biotically influenced vegetation on Marion Island during 1983/84. Each *dot* represents an individual larva. *Bars* represent the size range of *P. marioni* larvae obtained from mouse stomachs collected during each month, during 1987/1988; n = number of mouse stomachs analyzed during each month

these figures, is estimated to be approximately 1500 kg ha^{-1}.

4.3 The Role of *Pringleophaga marioni* Larvae in Nutrient Release from Litter

Figure 4 illustrates the amounts of N and P in the weekly leachates from microcosms, containing *P. marioni* larvae feeding on *A. selago* (Fig. 4A) and *P. cookii* (Fig. 4B) litter, compared to the amounts of N and P in the leachates from microcosms containing only litter.

5 Synthesis

Mice remove daily an estimated 0.7% of the standing crop of *P. marioni* larvae on Marion Island. An annual consumption by mice of more than twice (225%) the standing crop of *P. marioni* larvae implies a P/B ratio of at least 3 for the *P. marioni* population. In the absence of other predators, and with size-selective predation by mice, mortality of the eggs and early-instar larvae is probably low. This would offset, to some extent, the effects of mouse predation, and may result in higher productivity than expected from a slow-growing species with widely overlapping generations, such as *P. marioni*. Size selectivity, however, means that the most "ecologically active" part of the larval population is depleted by mice; based on the figures for consumption of larvae by mice, and of litter by larvae, mice annually "prevent" the fragmentation and processing of at least an additional tonne of litter on the coastal plain.

It is difficult to relate the results given in Fig. 4 to natural processes in the "decomposer subsystem" of

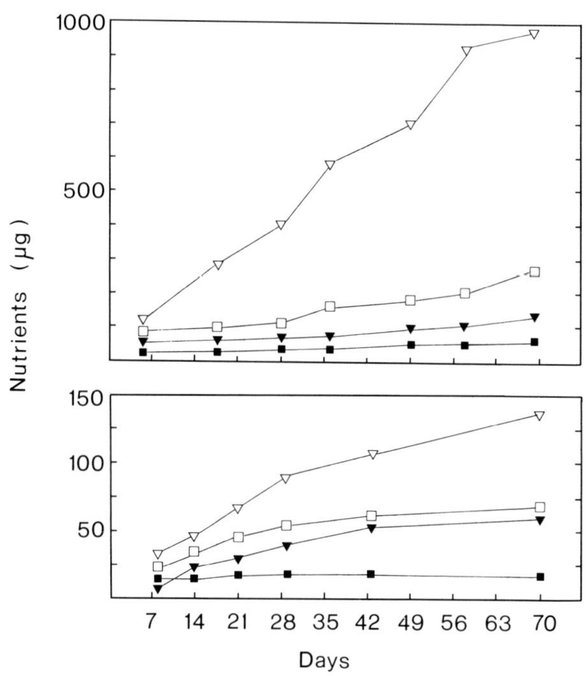

Fig. 4. The effect of feeding by *Pringleophaga marioni* larvae on the release of N (▽) and P (□) from 2 g (dry mass) *Poa cookii* (*above*) and *Azorella selago* (*below*) litter in microcosms. *Closed symbols* show the amounts of N (▼) and P (■) released from controls containing only 2 g (dry mass) litter (see text)

Marion Island's terrestrial ecosystem, and to quantify the effect of the *P. marioni* larval population on nutrient release in the field. However, the demonstrable effect of larval feeding on the release of nutrients (Fig. 4) suggests that the indirect effect of mice on the functioning of the "decomposer subsystem" exceeds their direct impact on decomposer populations.

6 Discussion

The terrestrial biota of Marion Island are impoverished and poorly integrated. For that reason only, competition and predation (ignoring the mice) are probably not important in the structure and fuctioning of communities. Instead, there is close adaptation to a stable and predictably adverse physical regime – the primary correlate of adversity selection (Greenslade 1983), which favours and preserves such adaptation to the physical environment at the expense of adaptation to the living environment (such as competitive ability and investment in defense) (e.g. Holm 1988). The introduction of any new predator species with an interactive life strategy in an environment under this selective regime can be expected to have profound consequences, especially if the species is also pre-adapted to the physical environment (or at least, like the mice, not significantly limited by physical conditions). It is in this light that the ecological role of mice on Marion Island achieves special significance, and warrants closer scrutiny.

In a comparative study of the insect faunas of Marion and neighbouring Prince Edward Island, Crafford and Scholtz (1987) found significant differences in biomass between *Pringleophaga* and weevil (*Dusmoecetes*) populations inhabiting the same vegetation types on the two islands. They postulated (*op. cit.*) that mouse predation could be the cause of lower prey populations on Marion, since the biotas of the two islands are qualitatively similar; the islands also have the same age, geological history, similar physiographies, and experience the same macroclimate. Certain aspects of the biology of *P. marioni* indeed make them exceptionally vulnerable to predation. Flightlessness, with the resultant loss of dispersal ability, and the lack of long-distance chemical attractants prevalent in the majority of the Lepidoptera, evolved in the absence of intense predation and reflect the close adaptation by *P. marioni* on the physical environment. High population densities (up to 500 larvae m^{-2} in *P. cookii* tussock grassland on Prince Edward Island) would traditionally have offset, and reinforced, these losses. Mouse predation has therefore not only a direct, cumulative impact on larval populations (cumulative because of slow growth, delayed maturity and widely overlapping generations in these populations), but may also impede reproduction, and hence recruitment to an already decreased local population, by lowering larval densities. It is obvious that mice and moths have settled down to a state of equilibrium. Whether this equilibrium is static or dynamic is difficult to tell in the absence of long-term monitoring of both mouse and moth populations on Marion Island. Block (1984) has stressed the vulnerability of Antarctic arthropod communities to perturbations, pointing out that the low reproductive rates of certain species suggest a limited capacity for recovery after a decrease. The resilience of *P. marioni* can be ascribed at least in part to an extremely flexible life history, permanent stocks of mature larvae (Fig. 3), and poor synchronization of life history events.

Herbivores probably do not consume more than 3–5% of the net primary production of Marion Island's vegetation (Crafford et al. 1986; Crafford and Scholtz 1986). Annual net primary production (ANP) of most plant communities on Marion Island is high; that of mire grasslands, for instance, is higher than the ANP at many grasslands from more temperate areas (Smith 1987b, c). With virtually all of the ANP becoming dead organic matter, the detritus-based food web is the dominant pathway of energy and nutrient flow in the island ecosystem (Smith 1985). Although it has not yet been established whether there are any nutrient limitations to primary production at the island, available pools of N at sites where ANP was determined were amongst the lowest measured for any ecosystem, and would have had to be turned over 36 to 2180 times in the course of the growing season to supply the needs of the vegetation (Smith 1985). Considering the large reserves of nutrients in soil organic matter and the low external inputs of nutrients, it is likely that decomposition (i.e. both fragmentation and digestion by detritivores, and microbial decomposition) is the main "bottleneck" in nutrient recycling at the island (Smith 1985). In this regard, Block (1984) pointed out that an understanding of the interactions of soil invertebrates and microbes is fundamental to ecosystem analysis in the Antarctic and sub-Antarctic. A study of the interactions between *P. marioni* larvae and bacteria is currently underway on Marion Island, and preliminary results (Smith and Crafford unpubl. data) show unequivocally that the larvae considerably enhance bacterial activity. Not only do the larvae continually void microbe-rich frass into their immediate environment (the dry mass ratio excretion: consumption is 0.8–1.0), but they also provide a favourable microhabitat for bacteria populations in their gut, although apparently they lack true symbionts.

7 Conclusion

Feral house mice are firmly entrenched in the terrestrial ecosystem on Marion Island, and as the only terrestrial mammal (bar feral cats, introduced 40 years ago, and now being exterminated) they affect the functioning of the ecosystem in several ways. Predation pressure by mice on populations of their principal insect prey species is high, but predator and prey appear to exist in a state of dynamic or even stable equilibrium. The indirect effect of mice on decomposition and nutrient mineralization processes, through their removal of a large portion of the standing crop of detritivores, could be the most important ramification of their presence, especially in the light of reduced external nutrient input from island-breeding birds, through the depredation of cats.

Acknowledgements. I thank Marianne Smith (Department of Botany, University of the Orange Free State) for allowing me to incorporate some of her unpublished results. Prof. V. R. Smith (Department of Botany, UOFS) provided the original stimulus for an investigation of the role of mice on Marion Island. I thank Drs. S. L. Chown and D. T. Rowe-Rowe for discussions and comments, and Profs. E. Holm, C. Scholtz and M. Stoddard for critical reading and improvement of the manuscript. Biological research on the Prince Edward Islands is carried out under the auspices of the South African Scientific Committee for Antarctic Research (SASCAR), and is partly funded and logistically supported by the South African Department of Environment Affairs.

References

Block W (1984) Arthropod interactions in an Antarctic terrestrial community. In: Siegfried WR, Condy PR, Laws RM (eds) Antarctic nutrient cycles and food webs. Springer, Berlin Heidelberg New York, pp 614–619

Burger AE (1978) Terrestrial invertebrates: a food resource for birds at Marion Island. S Afr J Antarct Res 8:87–99

Crafford JE, Scholtz CH, Chown SL (1986) The insects of sub-Antarctic Marion and Prince Edward Islands, with a bibliography of entomology of the Kerguelen biogeographical province. S Afr J Antarct Res 16(2):41–84

Crafford JE, Scholtz CH (1986) Impact of *Embryonopsis halticella* Eaton larvae (Lepidoptera: Yponomeutidae) feeding in Marion Island tussock grassland. Polar Biol 6:191–196

Crafford JE, Scholtz CH (1987) Quantitative differences between the insect faunas of Marion and Prince Edward Islands: a result of human intervention? Biol Conserv 40:255–262

Crafford JE (1987) The Lepidoptera of the Prince Edward Islands (46° 54′ S, 37° 45′ E): ecology and zoogeography. MSc Thesis, University of Pretoria, 60 pp

Drożdż A (1968) Digestibility and assimilation of natural foods in small rodents. Acta Theriol 13:367–389

Gleeson JP (1981) The ecology of the house mouse, *Mus musculus* Linnaeus, on Marion Island. MSc Thesis, University of Pretoria, 112 pp

Gleeson JP, Van Rensburg PJJ (1982) Feeding ecology of the house mouse *Mus musculus* on Marion Island. S Afr J Antarct Res 12:34–39

Greenslade PJM (1983) Adversity selection and the habitat templet. Am Nat 122(3):352–365

Gremmen NJM (1981) The vegetation of the Subantarctic islands Marion and Prince Edward. Geobotany 3:149 pp

Holm E (1988) Environmental restraints and life strategies: a habitat templet matrix. Oecologia 75:141–145

Murphy J, Riley JP (1962) A modified single solution method for the determination of phosphate in natural waters. Anal Chim Acta 27:31–36

Nagy K (1980) CO_2 production in animals: analysis of potential errors in the doubly labelled water method. Am J Physiol 238:466–473

Rowe-Rowe DT, Greene BR, Crafford JE (1989) Estimated impact of feral house mice on sub-Antarctic arthropods at Marion Island. Polar Biol 9:457–460

Smith VR (1985) Seasonal dynamics of standing crop and chemical composition in Marion Island (sub-Antarctic) vegetation. PhD Thesis, University of the Orange Free State, 334 pp

Smith VR (1987a) The environment and biota of Marion Island. S Afr J Sci 83:211–220

Smith VR (1987b) Production and nutrient dynamics of plant communities on a sub-Antarctic island. 2. Standing crop and primary production of fjaeldmark and fernbrakes. Polar Biol 7:125–144

Smith VR (1987c) Production and nutrient dynamics of plant communities on a sub-Antarctic island. 3. Standing stocks, uptake and loss of nutrients in mire-grasslands. Polar Biol 8:135–153

Solórzano L (1969) Determination of ammonia in natural waters by the phenol-hypochlorite method. Limnol Oceanogr 14:799–801

Watkins BP, Cooper J (1986) Introduction, present status and control of alien species at the Prince Edward Islands, sub-Antarctic. S Afr J Antarct Res 16(3):86–94

The Impact of Introduced Cats on the Predator-Prey Interactions of a Sub-Antarctic Avian Community

S. Hunter[1]

Summary. The distribution and feeding ecology of avian predator-scavengers were investigated to determine how the introduction of feral cats into the Marion Island ecosystem had affected predator-prey relationships within the island's avian community. Giant petrels fed mainly on penguins and did not compete with cats. Skua pairs specialised on the most abundant prey nearest to their territory, penguins or burrowing petrels. Cat predation has drastically reduced the number of burrowing petrels and few skuas now rely on these as their main prey source. Skua breeding density was highest next to large penguin colonies and breeding success was related to the proximity of breeding sites. Indirect competition between skuas and cats for the declining number of burrowing petrels has probably led to changes in the breeding and feeding ecology of the former as they adapt to shifts in prey distribution. Future changes in predator-prey interactions in the light of the current cat eradication programme are discussed.

1 Introduction

Feral cats *Felis catus* were introduced to sub-Antarctic Marion Island (46° 52′ S, 37° 51′ E) in 1949 and by 1975 the population had increased to 2140 ± 290 individuals (van Aarde 1979). Studies by van Aarde (1980) and van Rensburg (1985) have shown that much of the diet of these cats consists of burrowing petrels. As a consequence, populations of many petrel species have declined markedly and common diving petrels *Pelecanoides urinatrix* no longer seem to breed on the island (Brooke 1984; Williams 1984).

Members of the avian predator-scavenger guild at Marion Island also feed predominantly on birds, particularly penguins and burrowing petrels. This guild consists of five species, northern and southern giant petrels *Macronects halli* and *M. giganteus*, sub-Antarctic skuas *Catharacta lonnbergi*, kelp gulls *Larus dominicanus* and lesser sheathbills *Chionis minor*.

The aim of this chapter is to compare the feeding ecology of these predator-scavengers with that of the cats and to establish what effects the introduction of the latter may have had on the existing predator-prey relationships.

2 Methods and Data Base

2.1 Distribution

Detailed distributional data for both giant petrel species were collected during a census of breeding pairs that was carried out during the incubation/brooding period during a number of seasons.

The distribution of skua territories was plotted using maps overlaid with a half-minute square grid. Many areas in the central parts of the island, at higher altitudes, do not contain breeding skuas (pers. obser.) and these areas were not searched intensively. The number of birds defending territories and the number of chicks present in January-February of the years 1987 and 1988 during the pre-fledging period were checked in order to assess distribution and breeding success in relation to potential food supplies. While a small number of pairs/trios, which had failed early or did not breed at all, will have deserted territories in January-February, observations from intensively studied nests indicated that the majority of such birds still continued to occupy territories at this time.

2.2 Diet

The composition of the diet of the two giant petrel species was investigated by collecting regurgitations from nestlings just after they had been fed by their parents (see Hunter 1983 for description of sampling methods). This allowed a quantitative assessment of the diet to be made.

[1] Percy FitzPatrick Institute of African Ornithology, University of Cape Town, Rondebosch 7700, South Africa.
Present address: 20 London Road, Saffron Walden, Essex CB11 4ED, UK.

Antarctic Ecosystems. Ecological Change and Conservation
(ed. by K. R. Kerry and G. Hempel)
© Springer-Verlag Berlin Heidelberg 1990

Sub-Antarctic skuas are territorial birds that at Marion Island normally feed on or close to their territories. Frequently the territories actually contained the main source of prey, i.e. part or all of a colony of penguins or burrowing petrels. Often remains of prey eggs or carcasses were found at a midden, sometimes located next to a pool. Midden contents were collected regularly from 136 territories in four separate study areas. During the island-wide census of breeding pairs, the main prey item(s) found at 370 middens were noted.

Some kelp gulls scavenged in king penguin *Aptenodytes patagonicus* colonies during the winter (April to October) but otherwise this species predominantly fed on marine prey or terrestrial invertebrates (Burger 1978; Blankley and Grindley 1985). Therefore, no quantitative assessment of the diet of kelp gulls was made. Lesser sheathbills at Marion Island were studied by Burger (1979, 1981, 1982) and his data are used for comparison. The feral cat population has been the subject of intensive study by the Mammal Research Institute of Pretoria University and dietary information from van Aarde (1980) and van Rensburg (1985) are used in this work.

3 Results

3.1 Prey Distribution

The distribution of the major penguin and blue petrel *Halobaena caerulea* concentrations is shown in Fig. 1. Blue petrels occur in discrete and dense colonies (Fugler et al. 1987). The other main avian prey, Salvin's prions, *Pachyptila salvinii*, have a widespread and more dispersed distribution. Total breeding populations (in pairs) are: king penguins ca. 200 000, macaroni penguins, ca. 450 000, rockhopper penguins ca. 150 000, blue petrels tens of thousands and Salvin's prions hundreds of thousands (Siegfried et al. 1978; Williams 1984; Watkins 1987).

3.2 Predator-Scavenger Distribution

Counts of incubating or brooding birds showed that ca. 320 pairs of *M. halli* and ca. 3000 pairs of *M. giganteus* breed annually at Marion Island (pers. obser.). Most of the *M. halli* were concentrated on the

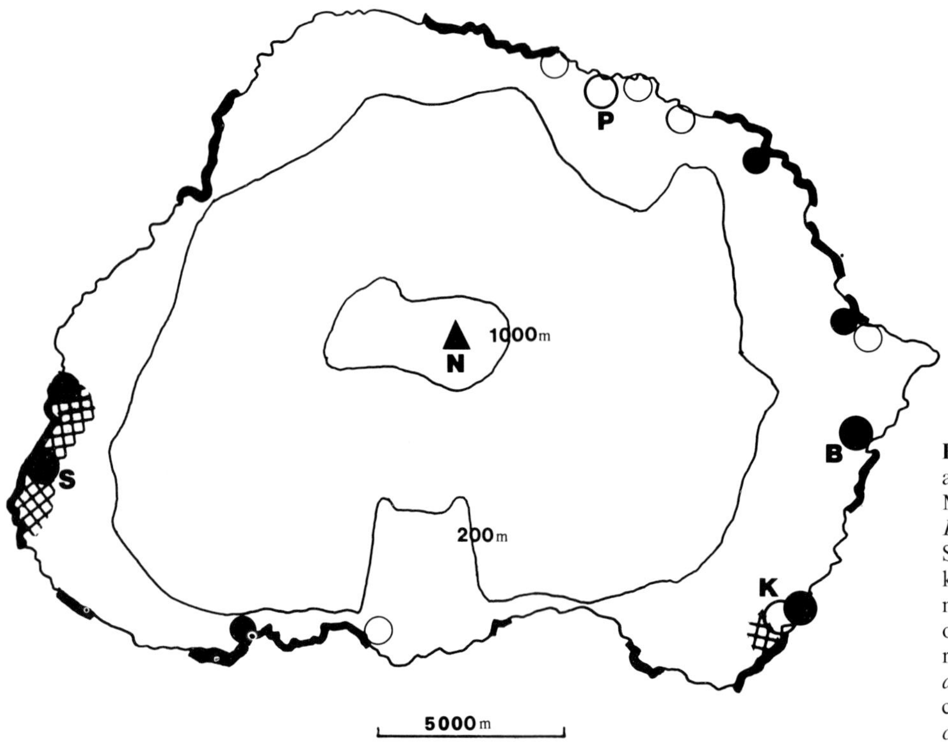

Fig. 1. Distribution of major avian prey concentrations at Marion Island. *P* Prinsloomeer; *B* Bullard; *K* Kildalkey Bay; *S* Swartkop Point. *Open circles* king penguins; *black circles* macaroni penguins; *black areas* of coastline, concentrations of rockhopper penguins; *hatched areas* blue petrels. *Large circles* colonies > 10 000 pairs; *small circles* colonies < 10 000 pairs

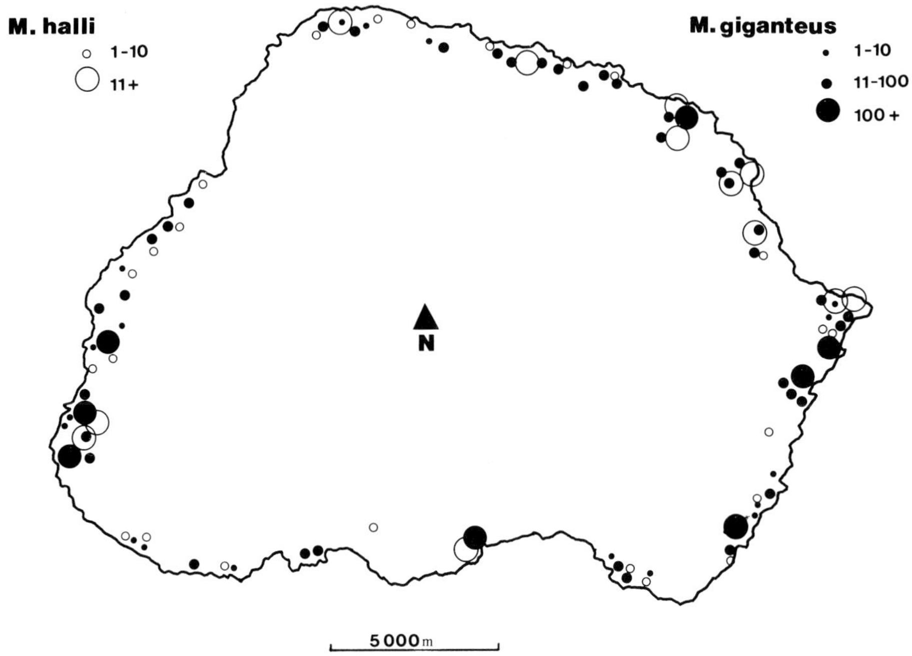

Fig. 2. Distribution of breeding giant petrel pairs at Marion Island

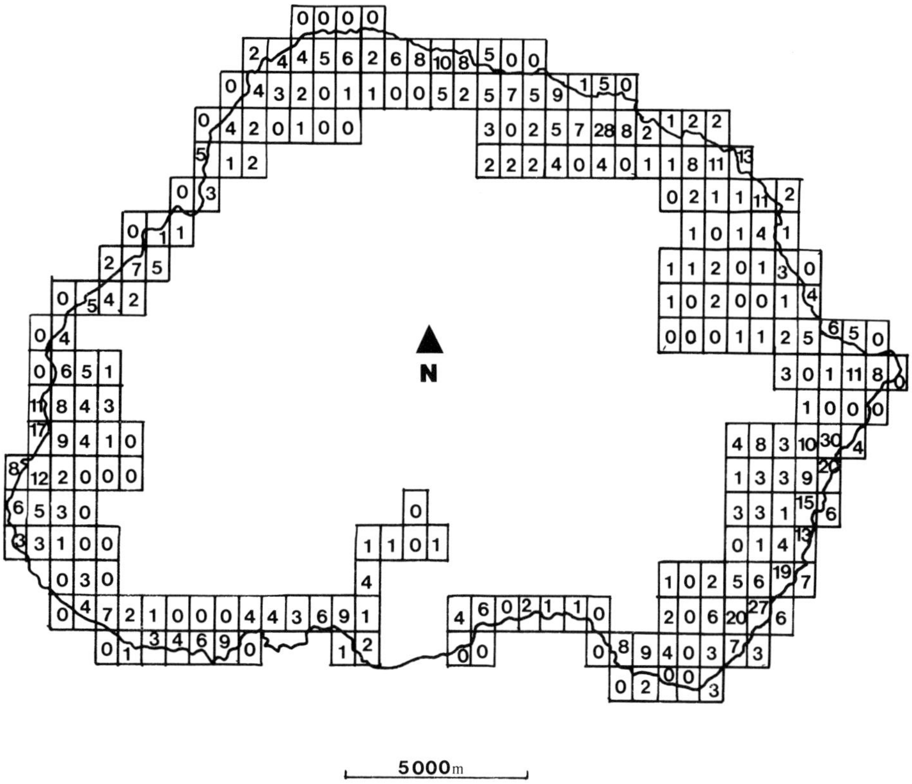

Fig. 3. Distribution of breeding skua pairs/trios at Marion Island

north and northeast coasts of the island (Fig. 2). Breeding groups of *M. giganteus* were spread rather more evenly around the coast but with particularly large concentrations near the macaroni penguin colonies at Bullard, Kildalkey Bay and Swartkop Point (Fig. 2).

A total of 878 skua territories was counted during censuses of breeding pairs/trios in 251 grid squares (Fig. 3). Probably a few of these territories contained birds that were unpaired, and it was estimated that a further 30 pairs bred in areas not checked during the censuses. Thus the total island breeding population was about 900 pairs/trios. The highest concentrations occurred near the large penguin colonies at Kildalkey Bay, Bullard, Prinsloomeer and Swartkop Point or in areas with large blue petrel colonies. Many of the remaining territories were situated adjacent to rockhopper penguin colonies.

There is no recent estimate of the breeding population of lesser sheathbills, although Williams et al. (1979) estimated it at 980 pairs; all the breeding territories included breeding penguins (Burger 1979). No comprehensive count of breeding pairs of kelp gulls was undertaken but the breeding population was estimated to be under 200 pairs in 1986 (pers. obser.), divided into individual territorial pairs and a few small colonies of up to ca. 30 pairs. They occurred in most coastal areas except on the exposed west coast.

Table 1. Comparison of diets of avian predator-scavengers and cats at Marion Island

Species	Percentage prey type				
	Mice	Seals	Penguins	Petrels	Marine[a]
Felis catus[b]	49	0	3	49	0
Catharacta lonnbergi[c]	0	< 1	78	22	< 1
Catharacta lonnbergi[d]	0	0	72	27	0
Macronectes giganteus[e]	0	3	81	9	7
Macronectes halli[e]	0	10	67	5	18

[a] Includes fish, cephalopods, crustaceans, molluscs.
[b] Percentage of prey in stomachs, 1982 (after van Rensburg 1985).
[c] Percentage of prey items from regularly checked middens.
[d] Percentage of major prey type at skua middens.
[e] Percentage biomass of each prey type in chick regurgitations (pers. obser.).

3.3 Diet and Feeding Ecology

Both species of giant petrels fed predominantly on penguins with smaller amounts of seal carrion, burrowing petrels, fish and squid (Table 1). Burrowing petrels were taken at sea, mainly by female giant petrels (pers. obser.). Salvin's prions were the most abundant burrowing petrel killed (61% of those found in regurgitations) but small numbers of all the other locally breeding species were also found in samples.

A large proportion of the 6675 prey items collected from the regular checks of skua middens consisted of penguin remains and virtually all the other prey remains were from burrowing petrels, mainly blue petrels and Salvin's prions (Table 1). The predominant prey type recorded at 370 middens showed a similar distribution of prey types (Table 1). Adult burrowing petrels were taken at night and young birds were killed as they left the burrows prior to fledging. Occasionally pairs would dig out petrel nestlings from burrows.

Kelp gulls feed mainly on marine prey, such as molluscs, or terrestrial invertebrates (Burger 1978; Blankley and Grindley 1985) though some birds scavenged in king penguin colonies during the winter. There was no evidence of them taking penguin eggs or chicks or killing burrowing petrels. Sheathbills took unguarded eggs and small chicks and scavenged in penguin colonies (Burger 1981). They also kleptoparasitised adult penguins as they attempted to feed their young and fed on inter-tidal and terrestrial invertebrates (Burger 1981).

In 1975 van Aarde (1980) found that petrels formed 87% of the mass of the stomach contents of 116 cats with Salvin's prions being the most commonly recorded species. By 1982 the proportion of the mass contributed by petrels in the contents of 142 stomachs had declined to 77.8% with some *Pterodroma* species disappearing altogether from the diet (van Rensburg 1985; Table 1).

3.4 Skua Breeding Success

Although sub-Antarctic skuas normally lay two eggs and many pairs rear two young to fledging (Furness 1987), breeding success amongst Marion Island birds varied according to the species and proximity of the main prey (pers. obser.). The 67 grid squares which included the large penguin or blue petrel colonies contained 500 territories (57% of the total), while the remaining territories were spread rather more thinly over the other 184 grid squares. The breeding success of the pairs in these 67 grid units was higher than that of those pairs breeding away from the main prey sources with a greater number of pairs rearing two chicks successfully (Table 2).

Table 2. Comparison of skua breeding success between territories next to major food sources and those farther away (percentages in parentheses)

	No. grid squares	No. territories	No. of young reared			Average No. young reared
			0[a]	1	2	
Territories close to major prey colonies	67 (27)	500 (57)	93 (19)	217 (43)	190 (38)	1.19
Territories away from major prey colonies	184 (73)	378 (43)	155 (41)	145 (38)	78 (21)	0.80

[a] Includes nests found with eggs in January-February and assumed to have failed subsequently.

4 Discussion

4.1 Competition for Prey Between Cats and Avian Scavenger-Predators

Previous work at Marion Island has shown that feral cats have a diet that consists of burrowing petrels and house mice *Mus musculus* and that there has been a severe reduction in the numbers and breeding success of several species of burrowing petrels (van Aarde 1980; Brooke 1984; van Rensburg 1985). Both the giant petrels and the skuas also feed on burrowing petrels and consequently the introduction of cats will have altered the existing predator-prey relationships.

Giant petrels feed predominantly on penguins and although cats are occasionally seen scavenging at king penguin carcasses during the winter, there is no evidence that they take live penguins regularly at Marion Island, although Berruti (1981) presented some circumstantial evidence that it does occur occasionally. The burrowing petrels taken by giant petrels are killed at sea (Hunter 1983) and consequently, as numbers of Marion Island burrowing petrels decrease, the proportion of individuals from other populations (such as those from the nearby and cat-free Prince Edward Island) that are killed will increase. Consequently, competition for penguins and petrels between cats and giant petrels does not really occur.

Sub-Antarctic skuas are well known as predators of burrowing petrels and at some localities individual skua pairs will feed almost entirely at dense colonies of these birds (Adams 1982; Schramm 1983; Osborne 1985; Stahl and Mougin 1986; Young 1978). Prey remains are found on the territories and can give a good indication of numbers and species of prey taken (Adams 1982; Osborne 1985; Stahl and Mougin 1986). At Marion Island both skuas and feral cats kill large numbers of burrowing petrels, taking adults and fledglings of the same species. The only major difference is that cats are present throughout the year and will probably have taken a higher proportion of the two winter breeding species, grey petrels *Procellaria cinerea* and great-winged petrels *Pterodroma macroptera*, than the skuas, which become pelagic during the winter months. Both skuas and cats at Marion Island are feeding on the same decreasing supply of prey and thus indirectly competing for burrowing petrels. More accurately, only part of the skua population is in competition with the cats. Only those pairs that have specialised in the predation of petrels will have been affected by the reduction in the number of these prey. Those skua pairs (and sheathbills) that feed only on penguins will not be competing with the cats.

4.2 The Impact of Competition for Burrowing Petrels on the Skua Population

The results of this study show that during the breeding season a minority of skua pairs take significant numbers of burrowing petrels and probably only a very small number feed predominantly on this prey type. Most pairs rely entirely or substantially on penguin colonies for food. At nearby Prince Edward Island, where there are no cats, Adams (1982) found that considerable numbers of Kerguelen petrels *Pterodroma brevirostris* were taken but these are hardly ever found at Marion Island middens where numbers of this species have markedly declined. Schramm (1983) found a mean of 24 (SD = $\pm$ 23.38, range 2–100, n = 27) petrels per midden, whereas during this study the mean for 50 middens in the same area was 7.2 (SD = $\pm$ 9.25, range 0–45). Grindley (1981) found large numbers of soft-plumaged petrel *P. mollis* corpses at skua middens around the meteorological base but these are now only rarely found. Thus many skua territories which would previously have contained considerable numbers of breeding petrels are now essentially devoid of such potential prey. This is particularly the case in those areas away from penguin colonies where the main prey would have been burrowing petrels.

Those skuas breeding farther inland now have to commute to penguin colonies in order to find sufficient food. Moreover, most of the smaller rockhopper and Macaroni penguin colonies are already incorporated into other skua territories which are vigorously defended by the resident pairs. As a consequence, only those penguin colonies large enough to contain areas undefended by resident skuas are suitable as foraging sites for these commuting birds. Such birds must spend considerably more time away from their chicks. The offspring of these birds will be more likely to die of exposure, malnutrition or predation because of reduced parental attendance. This would account for the lowered breeding success of these inland breeding pairs.

Sub-Antarctic skuas tend to keep stable pairings and territories from year to year (Young 1972; Furness 1987). New birds enter the breeding population either by pairing with established breeders that have lost their mates or by occupying established territories where the owners have disappeared (Young 1972). Changes in territory size and distribution come about slowly through an increase or decrease in numbers of pairs breeding in an area (Furness 1987). Density alterations will be in response to changes in the long-term food availability of an area. At Marion Island it seems likely that over the last 30 years those skua territories that originally contained large numbers of breeding, burrowing petrels have declined in number, increased in size and now have a lowered breeding success. In addition, it is possible that compaction of skua territories around large penguin colonies may have occurred in response to an increase in the numbers of birds feeding at these major food sources.

The cat population at Marion Island is now the subject of an eradication programme (van Rensburg and Bester 1988; Cooper and Condy in press). If this programme is successful, then the long-term restoration of burrowing petrel numbers should lead to the return of skua pairs breeding successfully on inland territories. These pairs would be mainly feeding on petrels and would no longer need to commute to the larger penguin colonies to feed. It should be possible to measure changes in the density and distribution of territories in relation to the main food supplies and to show changes in the overall diet on the skuas. Giant petrel distributions and diet should not change as the petrels re-establish their numbers as they have not been affected by indirect competition with cats.

Acknowledgements. I am grateful to M. deL. Brooke, N. A. Gartshore, C. A. Gilbert, S. R. Henley, B. W. Stead and W. K. Steele for their assistance in collecting field data. Financial and logistic support was supplied by the South African Departments of Transport and Environment Affairs and the South African Scientific Committee for Antarctic Research.

References

Adams NJ (1982) Sub-Antarctic skua prey remains as an aid for rapidly assessing the status of burrowing petrels at Prince Edward Island. Cormorant 10:97–102

Berruti A (1981) The status of the royal penguin and fairy prion at Marion Island, with notes on feral cat predation on nestlings of large birds. Cormorant 9:123–128

Blankley WO, Grindley JR (1985) The intertidal and shallow subtidal food web at Marion Island. In: Siegfried WR, Condy PR, Laws RM (eds) Antarctic nutrient cycles and food webs. Springer, Berlin Heidelberg New York pp 630–636

Brooke RK (1984) South African red data book – birds. S Afr Nat Sci Prog Rep 97:1–213

Burger AE (1978) Terrestrial invertebrates: a food resource for birds at Marion Island. S Afr J Antarct Res 8:87–99

Burger AE (1979) Breeding biology, moult and survival of lesser sheathbills *Chionis minor* at Marion Island. Ardea 67:1–14

Burger AE (1981) Food and foraging behaviour of lesser sheathbills at Marion Island. Ardea 69:167–180

Burger AE (1982) Foraging behaviour of lesser sheathbills *Chionis minor* exploiting invertebrates on a sub-Antarctic island. Oecologia (Berl) 52:236–245

Cooper J, Condy PR Environmental conservation at the sub-Antarctic Prince Edward Islands: a review and recommendations. Environ Conserv (in press)

Fugler SR, Hunter S, Newton IP, Steele WK (1987) Breeding biology of blue petrels *Halobaena caerulea* at the Prince Edward Islands. Emu 87:103–110

Furness RW (1987) The skuas. T and AD Poyser, Calton, UK

Grindley JR (1981) Observations of seabirds at Marion and Prince Edward Islands in April and May 1973. In: Cooper J (ed) Proc Symp Birds of the Sea and Shore, 1979. African Seabird Group, Cape Town, pp 169–188

Hunter S (1983) The food and feeding ecology of the giant petrels *Macronectes halli* and *M. giganteus* at South Georgia. J Zool Lond 200:521–538

Osborne BC (1985) Aspects of the breeding biology and feeding behaviour of the brown skua *Catharacta lonnbergi* on Bird Island, South Georgia. Br Antarct Surv Bull 65:57–71

Schramm M (1983) Predation by sub-Antarctic skuas *Catharacta antarctica* at Marion Island. S Afr J Antarct Res 13:41–44

Siegfried WR, Williams AJ, Burger AE, Berruti A (1978) Mineral and energy contributions of eggs of selected species of seabirds to the Marion Island terrestrial ecosystem. S Afr J Antarct Res 8:75–87

Stahl J-C, Mougin J-L (1986) La ségrégation alimentaire chez le Skua subantarctique *Stercorarius skua lonnbergi* dans l'archipel Crozet. L'Oiseau et RFO 56:193–208

van Aarde RJ (1979) Distribution and density of the feral house cat *Felis catus* on Marion Island. S Afr J Antarct Res 9:14–19

van Aarde RJ (1980) The diet and feeding behaviour of feral cats, *Felis catus* at Marion Island. S Afr J Wildl Res 10:123–128

van Rensburg PJJ (1985) The feeding ecology of a decreasing feral house cat, *Felis catus*, population at Marion Island. In: Siegfried WR, Condy PR, Laws RM (eds) Antarctic nutrient cycles and food webs. Springer, Berlin Heidelberg New York pp 620–624 van Rensburg PJJ, Bester MN (1988) Experiments in feral cat population reduction by hunting on Marion Island. S Afr J Wildl Res 18:47–50

Watkins B.P. (1987) Population sizes of king, rockhopper and macaroni Penguins and wandering albatrosses at the Prince Edward Islands and Gough Island, 1951–1986. S Afr J Antarct Res 17:155–162

Williams AJ (1984) The status and conservation of seabirds on some islands in the African Sector of the Southern Ocean. In: Croxall JP, Evans PGH, Schreiber RW (eds) Status and conservation of the world's seabirds. Int Council Bird Preserv, Cambridge, UK, pp 627–635

Williams AJ, Siegfried WR, Burger AE, Berruti A (1979) The Prince Edward Islands: a sanctuary for seabirds in the Southern Ocean. Biol Conserv 15:59–71

Young EC (1972) Territory establishment and stability in McCormick's skua. Ibis 114:234–244

Young EC (1978) Behavioural ecology of *lonnbergi* skuas in relation to environment on the Chatham Islands, New Zealand. NZ J Zool 5:401–416

Marine Biota as Detection Agents for Low-Level Radionuclide Contamination in Antarctica and the Southern Hemisphere Oceans

W. F. Wood [1,4], K. V. Marsh [2], R. W. Buddemeier [2] and C. Smith [3]

Summary. The southern hemisphere oceans are generally free of radionuclide contamination. This chapter reports on the use of marine biota as indicators of radionuclide contamination in these waters. Sampling methods were developed and were tested using plankton and the kelp, *Ecklonia radiata*, as bioconcentrators of artificial radionuclides. Plankton sampling yielded concentrations of radionuclides including Be-7, Nb-95 and Ce-144. Transplanted *E. radiata* was an efficient concentrator of I-131. Known inputs of I-131 from city hospitals were used (zero to two patients treated per week) to compare with uptake rates by macroalgae. Rapid uptake of I-131 by kelp plants was observed, the concentration of I-131 in kelp tissues was correlated with I-131 activity in the sewer discharge. Both the plankton and macroalgae sampling methods will detect low levels of nuclear fission products below levels detectable by conventioal air filters. The methods are suitable for monitoring radionuclide concentrations in Southern Ocean and Antarctic waters. Members of the Laminariales grow on Sub-Antarctic islands and are strategically placed to monitor radioactivity advecting eastward in the Southern Ocean.

1 Introduction and Aims

Atmospheric nuclear tests, reactor operations and waste disposal programs have injected significant quantities of radionuclides into the marine environment of the northern hemisphere. These releases have, in general, been documented and considerable research has been done to characterise the fate of this radioactivity. The most extensive previous work on radionuclides was done during the nuclear tests at the Pacific Proving Grounds in 1956 and 1958 (Donaldson et al. 1956; Seymour et al. 1957; Lowman 1958; Hines 1962).

The southern hemisphere has generally been considered free of radionuclide input other than global fallout from northern hemisphere activities. There have, however, been some nuclear tests in the Tuamotus, in Australia and some small discharges from nuclear power plants. Since much of the southern hemisphere is ocean, any radionuclides released are likely to find their way into the marine environment. However, very little sampling of the marine environment has been done.

The present study formed part of the Lawrence Livermore National Laboratory (LLNL) Marine Sampling Project. This study was initiated following the "September-79 Event", a putative nuclear explosion near the Prince Edward Islands (~ 48° S; 40° E) in the Indian Ocean. On September 22, 1979 a flash was detected by a Vela satellite (Marshall 1979). Although the shape of the light pulse was not exactly characteristic of a nuclear explosion, it was similar enough to be difficult to explain by any other source. The only other corroborating evidence for such an event were traces of fallout found in rainwater samples collected in New Zealand, but were only slightly above background levels.

Small-scale nuclear events are generally difficult to detect by usual methods (e.g. air filters, seismographs), especially in regions lacking in observation networks or convenient sampling opportunities. Radioactive debris is needed for conclusive diagnosis of a nuclear test or accident. If convenient and effective ways to sample mixed fission products had been available in 1979, the question of a nuclear origin of the satellite signal would have been readily resolved. There are also few historical data or other bases for determining baseline concentrations with which to compare measurements of suspected contaminants. Access for sampling can be difficult and time-consuming, thus resulting in loss of potential information derivable from the abundances of short-lived nuclides. These features are the very ones that make high-latitude southern locations potentially attractive as sites for clandestine testing or disposal operations. The ex-

[1] University of Western Australia, Marine Laboratories, North Beach, Western Australia 6020, [2] Lawrence Livermore National Laboratory, Livermore, California, USA. [3] The Queen Elizabeth II Medical Centre, Nedlands, Western Australia 6020. [4] Present Address: Department of Environment and Planning, 160 Collins St., Hobart, Tasmania, Australia.

Antarctic Ecosystems. Ecological Change and Conservation
(ed. by K. R. Kerry and G. Hempel)
© Springer-Verlag Berlin Heidelberg 1990

perience of the Vela event emphasised the difficulties of monitoring the southern oceans. There are no sampling programs in place which could detect low-level contaminations, be they from clandestine nuclear tests, illegal dumping of radionuclides or from mobile reactor accidents. The work described in this chapter was initiated to develop prompt, simple and sensitive methods of sampling remote marine locations for elevated radionuclide levels. In order to take advantage of natural bioconcentration processes, both oceanic plankton and shallow-water benthic algae were sampled and analysed to determine sensitivities, background levels and operational constraints. When short-lived radionuclides were detected in both types of samples, a further study was undertaken to understand the source, transport and uptake characteristics of the I-131 found in Western Australian kelp. The following sections describe the plankton studies in the Southern Ocean, the kelp experiments, and the significance and potential applicability of the findings.

2 Methods

2.1. Plankton Studies

Plankton are known to concentrate trace elements and radionuclides from seawater. Concentration factors of 10^2 to 10^5 relative to seawater have been reported following nuclear tests (Haley 1956; Lowman 1958). Uptake is rapid, and once it has occurred, recycling keeps the radionuclides in the biotic layers so that discrete water masses may be traced for some weeks.

These processes can provide detection sensitivities high enough to overcome the problems of dilution and distance from the (presumably small) source, as is shown by the following calculation.

About 1.45×10^{23} fissions are required to produce a 1-kt explosion, which is extremely small for a nuclear detonation. With information from Bolles and Ballou (1959), we can calculate that the residual radioactivity after 7 days is about 3×10^{18} dpm (disintegrations per minute). Several isotopes of the rare earth elements each contribute about 10% of the total radioactivity, or 3×10^{17} dpm each. Our detection limit for these nuclides are all about 0.05 dpm/g of plankton based on an average 200 g sample (Marsh and Buddemeier 1984). If we take 10^3 as a very conservative concentration factor for rare earth nuclides (Haley 1956; Lowman 1958), 1 g plankton becomes equivalent to 1 litre seawater. Thus, we can expect to be able to detect 1-kt of fission products diluted by about 6×10^{18} litres or 6×10^{15} m^3. This estimate is based on bulk sample measurements by direct Ge-Li gamma spectroscopy. These calculations indicate that a small release that labelled the marine biosphere could be detected by plankton sampling at a distance of thousands of kilometres from the source.

Plankton were sampled during cruises of the US Coast Guard Cutter (USCGC) Glacier and the Research Vessel (RV) Melville during 1981–1985 in the Southern Pacific and Southern Oceans (Fig. 1).

Sampling was done by either deploying a plankton net overside or by pumping seawater through a net suspended on deck. Samples containing approximately 200 g fresh weight of plankton were preserved in 5% buffered formalin, were filtered and returned to LLNL for counting by Ge-Li gamma spectrometry (Plutonium-239 was detected by alpha spectrometry). Air and water samples were also collected and analysed, but are not reported here. All samples were decay corrected to date of collection.

2.2 Macroalgae

The affinity of the Laminariales and other brown algae (Saenko et al. 1978) for iodine is well known. Kelps, or kelp-like algae occur off the coasts of all temperate landmasses south of approximately 32° S, the Sub-Antarctic islands and the Antarctic continent.

Because shore-based sampling of coastal environments can often be conducted more quickly and economically than ship-based surveys, we surveyed the kelp *Ecklonia radiata* (Fig. 2) and a variety of other macroalgae and seagrasses collected off Perth, Western Australia (first site, Fig. 3) in 1983. After detectable I-131 concentrations were found in some of the initial samples, we determined that the most probable source of the radioactivity was hospital effluents discharged through the metropolitan sewer outfalls. Accordingly we designed an experiment to monitor kelp I-131 uptake, environmental distributions, and radionuclide sources.

2.3 Field Sampling

Field sampling of naturally occurring kelps was undertaken to determine the geographical extent of the detectable I-131 (Fig. 3).

All underwater work was conducted using scuba equipment. Transplant frames were constructed to hold plants at Swanbourne sewer outfall and at the Atlantis sites (Fig. 3). Kelp plants used in transplant experiments were removed from uncontaminated, open water sites and transported by open boat to the transplant site; during the journey they were protected

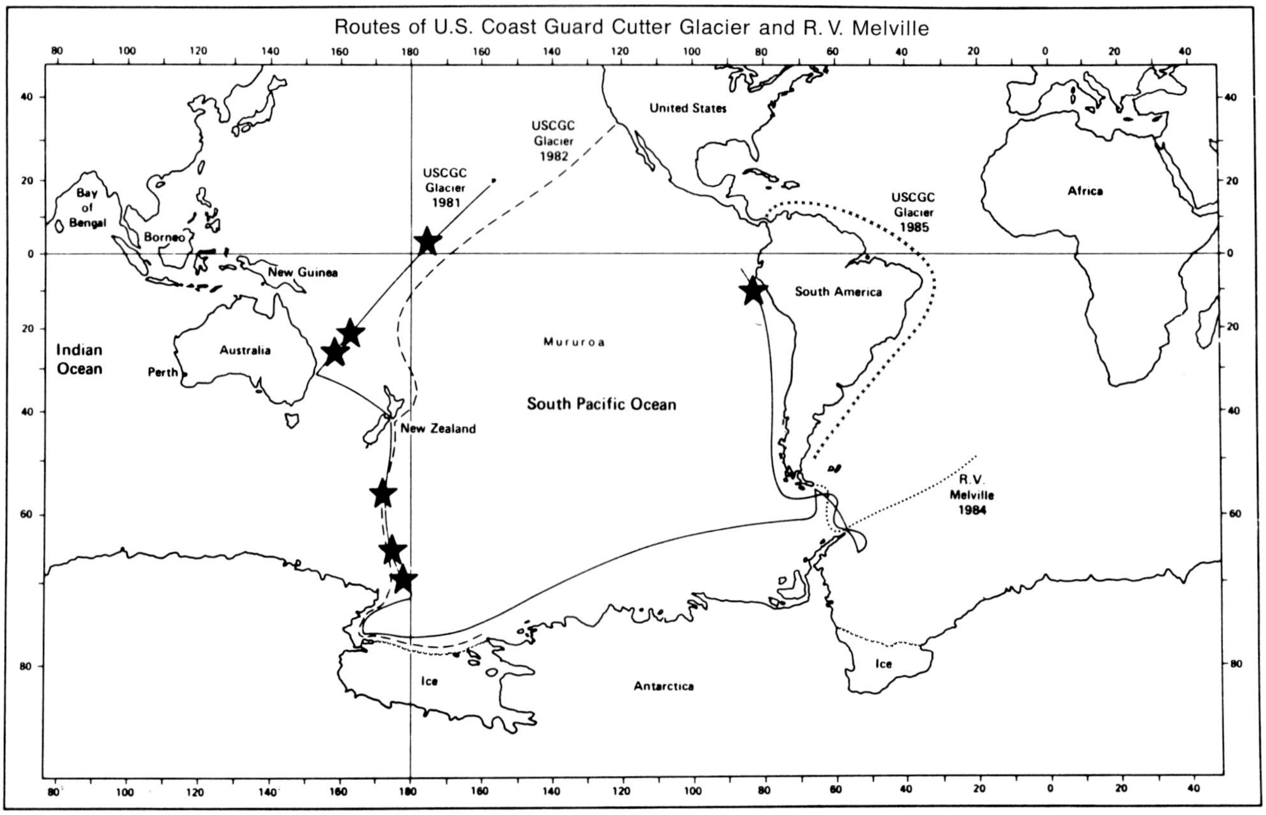

Fig. 1. Cruise tracks of the US Coast Guard Cutter Glacier and the RV Melville during 1981–85. Locations of samples containing artificial radionuclides are shown (*stars*)

from light and desiccation. Each kelp plant was attached to the transplant frame and was individually numbered with plastic tape. Survivorship of numbered plants was about 90% corresponding to survivorship of naturally growing kelps (Hatcher et al. 1987).

To observe I-131 uptake rates, plants from uncontaminated "clean" areas were transplanted to frames attached to the seafloor near the principal sewer outfall (Swanbourne outfall) and subsequently resampled at predetermined intervals. All transplant frames were located at a depth similar to the Swanbourne outfall and plants to be transplanted to these frames were obtained from a similar depth. Care was taken to ensure that the transplant frame at Swanbourne was located outside the influence of any freshwater plume from the sewer diffusers.

To determine depuration (I-131 release) rates, plants incubated at Swanbourne were retransplanted to a "clean" site at Atlantis (Fig. 3) and sampled at regular intervals.

At the time of sampling harvested kelps were protected from dessication and were returned to the laboratory where they were weighed, measured and placed in numbered polyethylene bags. The bags were packed in dry ice, placed in insulated containers, and shipped by air (~ 4 days) to LLNL for counting. Results were decay-corrected to date of collection.

2.4 Estimating I-131 Output

Perth is the only city in Western Australia (WA) with hospitals which have facilities for radiotherapy treatments, the primary source of I-131. The nearest alternative cities which could be sources of radionuclides are Adelaide, or cities in Indonesia, several 1000 km to the east and north respectively. All radioisotopes used in research or for medicine are imported into WA under permits issued by the Radiation Protection Laboratory (RPL) at the Queen Elizabeth II Medical Centre. The usage or disposal of radionuclides in the hospitals and research institutes in WA is strictly controlled by the RPL.

Radiologists at the Queen Elizabeth II and Royal Perth hospitals provided information on a monthly basis on dates and dosages of I-131 treatments. In order to estimate residual I-131 activity from these treatments at the ocean sewage outfalls, several assumptions, based on engineering calculations, were

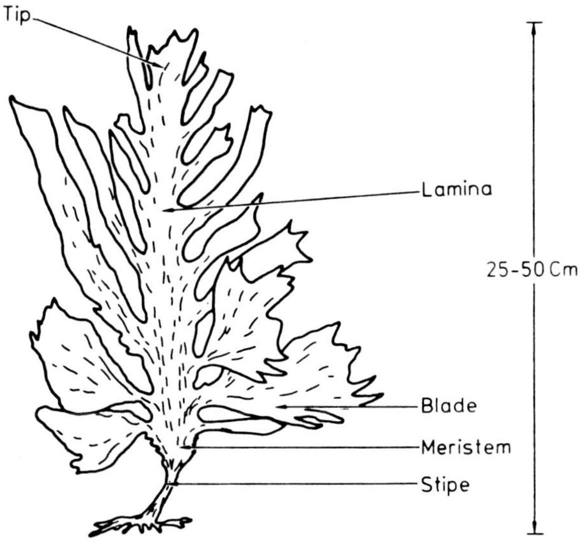

Fig. 2. A typical *Ecklonia radiata* plant such as those transplanted to frames at Swanbourne and Atlantis

necessary. These took into account the dose rate, the half-life of I-131 (8.04 days) and residence times of effluent in the sewer system (approximately 2 days depending on source). They also took into account whether the patient was hospitalised or was an outpatient. In practice the main input of I-131 to the ocean was from inpatients in hospitals connected to the Swanbourne outfall (Fig. 3). The hospital retention tanks were emptied to the sewer line at known times. Inpatients excreted 100% of the administered dose in the time they were hospitalised. The inputs from outpatients, whose homes were served from other sewer systems (Fig. 3), were also calculated for the times when these were significant. Outpatients who lived in the metropolitan area were assumed to have excreted 50% of their dose into the sewer system within 2 days. Outpatients from country areas were assumed to excrete 25% of their dose into the city sewers (Subiaco system) before leaving the metropolitan area. The potential for input of I-131 from metropolitan outpatients whose homes were served by septic systems was calculated at 25% (Swanbourne system), and the balance was ignored in the calculation of activity at the ocean outfalls. The activity at a particular outfall was then the sum of all doses, each multiplied by the appropriate fraction excreted and the radioactive decay factor from the time of dose to arrival at the outfall.

While there are many simplifying assumptions included in this model, the dose amounts and times are accurately known, as are the home locations of outpatients. The original I-131 source data are accurate, and the ocean input assumptions based on engineering calculations and measurements on volume and flow in the various systems probably have uncertainties of the same magnitude as the sampling and analytical procedures. Furthermore, the 8-day half-life of I-131 alleviates any effects of long-term accumulation or unknown holdups.

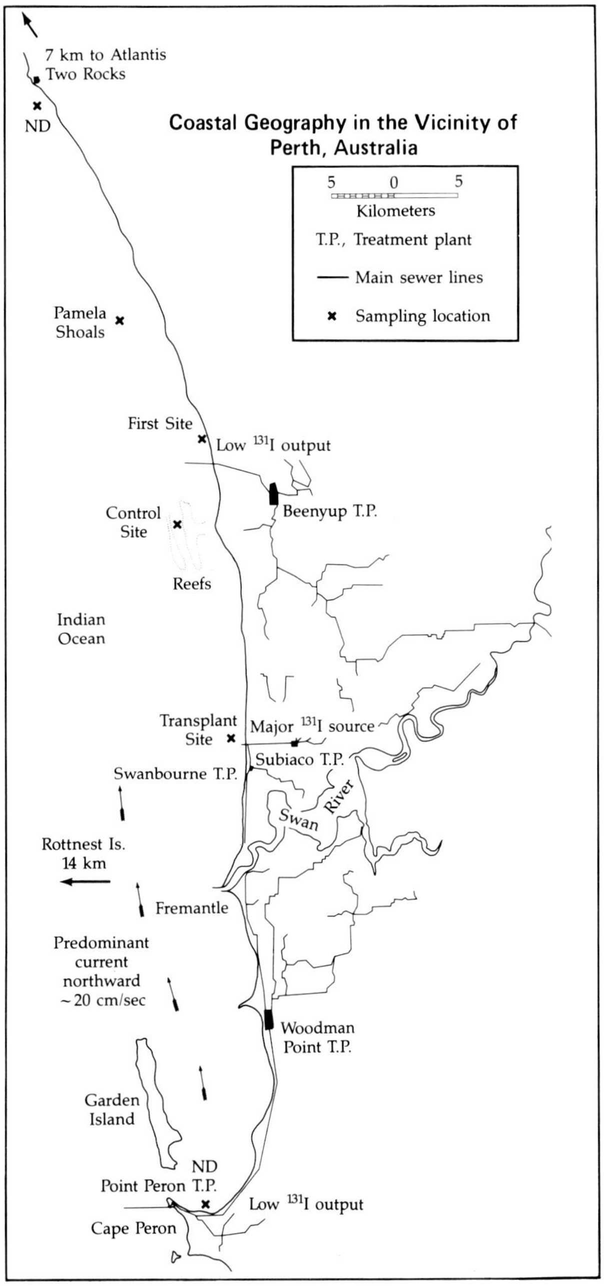

Fig. 3. Coastal geography in the vicinity of Perth, Western Australia, showing the locations of sampling sites and the sewer outfalls that are the sources of the Iodine-131

3 Results

3.1 Plankton Studies

Plutonium-239 was measured in all samples at concentrations of 0.1–5 fCi/g (femtocuries per gram). Fission products were detected above background (Table 1) on two cruises (Fig. 1); these were Niobium-95 and Cerium-144 (Table 2). The naturally occurring radioisotopes Potassium-40, Uranium-238 and Beryllium-7 were also detected. The Pu-239 is believed to represent long-lived fallout from earlier periods of atmospheric weapons testing.

3.1.1 Ecklonia radiata: Distribution and Time Response

No I-131 was detected from points south of Point Peron, north of Two Rocks or from Rottnest Island (Fig. 3). However, I-131 was found in kelps from the control site 17 km north of Swanbourne.

Plants were sampled from the Swanbourne site regularly. I-131 concentration in transplanted plants at Swanbourne is plotted together with a 16-day moving average of estimated I-131 output from Swanbourne (Fig. 4). The 16-day moving average was selected to smooth the periodic inputs of I-131 which range from 0 to 200 MBq/d (mega Becquerels per day). There was a significant, positive correlation ($r = 0.730 > 0.058 = 0.632$) between 16-day average inputs of I-131 at the outfall and I-131 concentrations in kelp tissues. This association is quite strong considering the largely unknown effect of currents and the growth and erosion of kelps.

Table 1. Detection limits for some radionuclides produced by nuclear explosions. Each value is averaged over all plankton samples. The detection-limit concentration range for each nuclide is approximately a factor of three

Nuclide	Detection limit, fCi/g wet weight
Manganese-54	2
Cobalt-60	4
Zinc-65	5
Cerium-144	10
Americium-241	10

Table 2. Niobium-95 and cerium-144 concentrations ($\pm$ 1 SD) in plankton samples collected in the Pacific in 1981

Niobium-95, fCi/g wet weight	Cerium-144, fCi/g wet weight	Latitude	Longitude
31 ± 8	107 ± 45	3° N	177° W
16 ± 9	–	4° S	84° W
7 ± 2	36 ± 22	6° S	177° E
–	32 ± 9	26° S	158° E
5 ± 3	–	28° S	154° E
2 ± 1	29 ± 5	65° S	178° E
18 ± 3	106 ± 8	68° S	180° E

3.2 Uptake and Depuration Rates of I-131

Although rigorous calculations could not be made because of the variable inputs at Swanbourne, it was estimated that uptake rates of I-131 are comparable to, or faster than rates of radioactive decay.

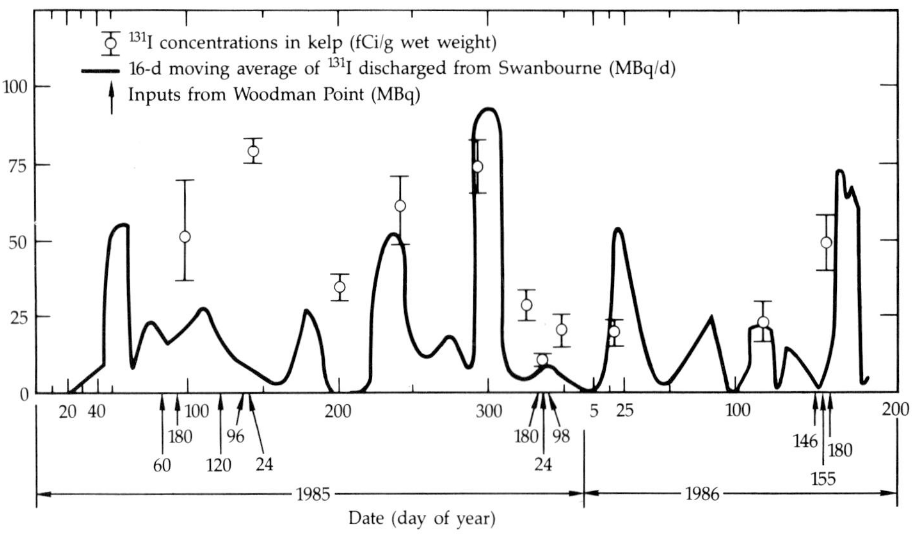

Fig. 4. Iodine-131 discharges from Swanbourne and Woodman Point outfalls (*arrows*) and concentrations in *Ecklonia radiata* plants on the frames

After correcting for radioactive decay it was found that 30% of I-131 was lost in 3 days following retransplantation to Atlantis (Fig. 3). Thereafter only small losses occurred to day 12. The biological half-life is therefore comparable in magnitude to the radioactive half-life but the results suggest the existence of more than one iodine "compartment" within the kelp.

4 Discussion

These two related studies, using bioconcentrating organisms that are widely distributed within the southern ocean, illustrate the potential application of biological sampling to the detection and characterisation of radionuclide releases in remote areas.

In addition to demonstrating the feasibility and sensitivity of the methods, an interesting result was also the apparent rapidity with which the nuclides are redistributed within the marine environment. The fission products detected in the plankton most probably came from the French test site at Muruoa, although a Chinese fallout source is not excluded. In either case, the half-time of the Nb-95 is such that it must have been transported long distances across the prevailing wind and current systems within a period of a few weeks to months. On a more local level, the extent of the I-131 distribution was also suprising; our originally selected "control" site proved to have detectable I-131.

A likely explanation for this rapid transport is the release of bioaccumulated radionuclides to other organic microlayers at the ocean surface, from which they are resuspended as marine aerosols. This process maintains the material within the ocean surface layer, but permits transport at rates much faster than ocean currents.

The principal aim of this study was to determine whether biocencentrators could be used to detect the low levels of nuclear fission products likely to result from a clandestine nuclear event in the vast areas of the Southern Ocean. Our calculations indicated that a 1 Kt-equivalent of fission products 1-week-old dispersed over 10^5 m^2 and mixed to 100 m would result in concentrations of radionuclides 600 times our detection limit for a 200-g plankton sample (Marsh and Buddemeier 1984). Actual measured detection limits confirm this estimate.

Following a specific event, it might be difficult to identify the optimum location for plankton sampling. It may also be difficult to depoly a suitable ship or aircraft (we used ships for convenience) and the search may be expensive and time-consuming. Balancing these potential disadvantages are the advantages that sampling may be carried out in international waters wherever a ship can travel, and that the sensitivity of the method is so great that useful samples could be collected from any portion of a dispersing plume. The sampling method is essentially simple, inexpensive, requires little training, and samples are easily preserved and stored. In an emergency, quite crude equipment could be used (water could be pumped through the fire main and filtered through any fine medium). It would therefore be quite practical to alert ships already at sea in the area to undertake sampling in a specific region of the ocean.

It is also clear that attached algae such as kelps and fucoids (Druehl et al. 1988; Saenko et al. 1978) will function as efficent concentrators of short-lived radionuclides such as I-131. The use of attached algae has the potential advantages of convenient, inexpensive and repeatable access, and that uptake can be calibrated against a known source (as in our study). Although a nuclear event may take place in an area remote from potential sampling sites where attached algae could be harvested, the tendency of radionuclides to be transported by wind while suspended in the marine aerosol is likely to overcome this. In our study I-131 activity was below the level of direct detection in water samples from the Swanbourne sewer line (Metropolitan Water Board communication); however, the kelps concentrated radionuclides to easily detectable levels some considerable distance from the source. Similarly, radioactivity advecting eastwards following a small-scale event, such as the putative "September 79" event, may be undetectable by air filters but easily detectable in biological concentrators such as kelps or plankton. The distinction should be made between small-scale, low-yield events such as the "September 79" event and large-scale, high-yield events such as the Chernobyl accident (Anderson 1986). Marine plants readily concentrate radionuclide fallout from these large-scale events (Druehl et al. 1988), however, this is the first time that bioconcentrators have been shown to be effective samplers of very low-level contamination which would otherwise be undetectable by other means. Had our methods been in place in 1979 it may well have been possible to provide firm corroborative evidence to support that provided by the rainwater samples.

At the present time the southern hemisphere oceans are generally free of radionuclide contamination. The ability of plankton and macroalgae to concentrate otherwise undetectable radionuclides has now been demonstrated. The methods are simple and are ideal for monitoring background levels of anthropogenic radionuclides, or for detecting contamination events in the Southern Ocean and the Antarctic.

Plankton are widely distributed in these oceans and are easily sampled from ships en route to Antarctic bases.

Kelps, or kelp-like algae, grow on all the Sub-Antarctic islands and parts of the Antarctic continent, thus they are strategically placed to monitor any radioactivity advecting eastward in the west-wind drift in the Southern Ocean. Since there are many research stations located within the Southern Ocean, it would be a simple matter for samples to be obtained following the remote detection of any nuclear or suspect event. It would be essential, however, to have some knowledge of any background levels of radionuclides present at a particular site prior to an event. Because of the ease and inexpensive nature of sampling, regular sampling could be undertaken to monitor background levels of radioactivity. The use of attached algae could make continual monitoring feasible. The presence of volatile radionuclides with short half-lives, such as I-131, would be good evidence of a recent event. Our results in defining the extent of I-131 contamination in waters around Perth, WA suggest that it would be extremely unlikely that I-131 from medical sources would be present at remote research stations such as those in the Antarctic and Sub-Antarctic. In the light of our experience with algae species other than kelps, it would also be essential to know something of the uptake kinetics prior to their use as bioconcentrators of radionuclides.

In conclusion, both plankton and attached algae are ideal bioconcentrators for detecting the products of small-scale nuclear fission events. Sampling is simple and cost-effective and it is quite feasible to obtain samples at short notice over a wide geographic area. The knowledge that it is now possible to detect suspected nuclear events (by remote sensing) hitherto undetectable by normal sampling means, must surely be an added deterrent to countries contemplating clandestine nuclear experiments in the unpolluted southern hemisphere oceans.

Acknowledgements. We gratefully acknowledge the help given by the engineering staff at the Perth Metropolitan Water Board who provided information on pumping and flow rates of the city sewers. The program would not have been possible without the cooperation of the radiological staff at the Royal Perth Hospital and Queen Elizabeth II Medical centre. We also thank Bruce Hatcher, Serena Ward and Karen Atkinson who provided diving assistance and Susie Wood who provided logistic support.

References

Anderson A (1986) Making the most out of Chernobyl. Nature 322:673

Bolles RC, Ballou R (1959) Calculated activities and abundances of the U^{235} fission products. Nucl Sci Eng 5(3):156

Donaldson LR, Seymour AH, Held EE, Hines NO, Lowman FG, Olson PR, Welander AD (1956) Survey of radioactivity in the sea near Bikini and Eniwetok atolls, June 11–21. Applied Fisheries Laboratory, University of Washington, Seattle, Washington, UWFL

Druehl LD, Cackette M, D'Auria JM (1988) Geographical and temporal distribution of iodine-131 in the brown seaweed *Fucus* subsequent to the Chernobyl incident. Mar Biol 98:125–129

Haley JH (ed) (1956) Operation Troll. Health and Safety Laboratory, US. Atomic Energy Commission, New York Operations Office, NYO-4656

Hatcher BG, Kirkman H, Wood WF (1987) The growth of the kelp *Ecklonia radiata*, near the northern limit of its range in Western Australia. Mar Biol 95:63–73

Hines NO (1962) Proving ground an account of the radio-biological studies in the Pacific, 1946–1961. University of Washington Press, Seattle

Lowman FG (1958) Radionuclides in plankton near the Marshall Islands, 1956. Laboratory of Radiation Biology, University of Washington, Seattle, Washington, UWFL-54

Marshall E (1979) Flash not missed by Vela still veiled in mist. Science 206:1051–1052

Marsh KV, Buddemeier RW (1984) Marine plankton as an indicator of low-level radionuclide contamination in the Southern Ocean. Lawrence Livermore National Laboratory, Rep UCRL-53549, 39 pp

Saenko GN, Krautova YY, Ivanenko VV, Sheludko SI (1978) Concentration of iodine and bromine by plants in the seas of Japan and Okhotsk. Mar Biol 47:243–250

Seymour AH, Held EE, Lowman FG, Donaldson JR, South DJ (1957) Survey of radioactivity in the sea and in pelagic marine life west of the Marshall Islands, September 1–20, 1956. Applied Fisheries Laboratory, University of Washington, Seattle, UWFL-47

Effect of Ultraviolet Radiation on Antarctic Marine Phytoplankton

S. Z. El-Sayed, F. C. Stephens, R. R. Bidigare and M. E. Ondrusek [1]

Summary. During November–December 1987, a series of experiments were conducted at Palmer Station, Antarctica, to study the effects of ultraviolet radiation (UV) on phytoplankton collected from Arthur Harbor, Anvers Island, Antarctica. Three parameters were studied: (1) primary production rates; (2) photosynthetic pigments; and (3) photosynthesis-irradiance responses. These parameters were monitored during short-term and relatively long-term exposures to varying levels of UV radiation. The results of this month-long study provide evidence of some potentially deleterious effects of enhanced UV radiation on Antarctic phytoplankton. The implications of these findings to our understanding of the ecology of the Southern Ocean are discussed.

1 Introduction

Stratospheric ozone shields the earth from much of the solar middle ultraviolet radiation (UV-B, 290–320 nm wave band), which is the most biologically injurious component of sunlight (Calkins and Thordardottir 1980). Recent reports on the "ozone hole" over Antarctica (Kerr 1986, 1988; Stolarski et al. 1986; Wellborn 1986; Bowman 1988) have renewed concern about the consequences of increased levels of UV-B reaching the earth's biosphere. One area of concern involves phytoplankton, which constitutes the base of the food web in aquatic ecosystems.

Numerous studies have shown that increased levels of UV exposure result in reduced primary production as measured by the ^{14}C uptake method (Steemann Nielsen 1954; Jitts et al. 1976; Worrest et al. 1978, 1980, 1981a, b; Worrest 1982; Lorenzen 1979; Calkins and Thordardottir 1980; Smith and Baker 1980; Smith et al. 1980; Maske 1984). Additionally, increases in UV-B levels are likely to alter community diversity as well as phytoplankton size distribution and species composition (McLeod and McLachlan 1959; Worrest et al. 1978, 1981a, b). By weakening the base of the food web and altering trophodynamic relationships, UV-induced changes could potentially have far reaching effects on the entire Southern Ocean ecosystem.

An area that has not been extensively addressed either in studies of acute exposure or of chronic exposure to UV is the pigment composition of phytoplankton and how this composition changes over time. Algal pigment concentrations in the marine environment are primarily dependent on species composition, species abundance, and the photoadaptive state of the phytoplankton present. Thus, chlorophyll and carotenoid pigments are excellent candidates as diagnostic markers for investigating algal responses to UV exposure.

The potential significance of the annual thinning of the ozone layer over the Antarctic continent during the austral spring from September through October is recognized (Stolarski et al. 1986). To examine the effects of this thinning we undertook an investigation into the effects of the UV-B on Antarctic phytoplankton during November–December 1987 at Palmer Station, Antarctica. The main objective of this study was to document the effects of UV on primary production and on the photosynthetic pigmentation of phytoplankton in Arthur Harbor, Anvers Island, Antarctica. These parameters were monitored during short-term and relatively long-term exposures to varying levels of UV radiation. A preliminary report of our findings is presented here and a more complete description of methods and results will be published elsewhere (Stephens et al. in prep.).

2 Materials and Methods

The field study was conducted from 15 November through 8 December 1988, at Palmer Station, Anvers Island, Antarctica (64° 46′ S, 64° 05′ W).

Responses of natural phytoplankton communities to UV-B were assessed by measuring changes in photosynthetic rates and pigmentation following incuba-

[1] Department of Oceanography, Texas A & M University, College Station, TX 77843, USA.

Antarctic Ecosystems. Ecological Change and Conservation
(ed. by K. R. Kerry and G. Hempel)
© Springer-Verlag Berlin Heidelberg 1990

tion in an array of tanks and chambers. A hole was drilled through 1.75 m of sea ice, approximately 15 m from the shoreline (Station A, Fig. 1). Seawater was pumped (with a Berkley Mode 4AL28 impeller driven well pump) from this hole into a mixing tank, from which it flowed (by gravity) into a series of 18 liter flow-through chambers. The chambers were situated within the larger treatment tanks (three per tank), through which seawater was pumped in order to maintain a constant temperature ($\pm$ 0.5 °C) (Fig. 2). Mixing within the chambers was accomplished with air bubblers. These tanks were kept outdoors under natural sunlight conditions and provided the following irradiation conditions.

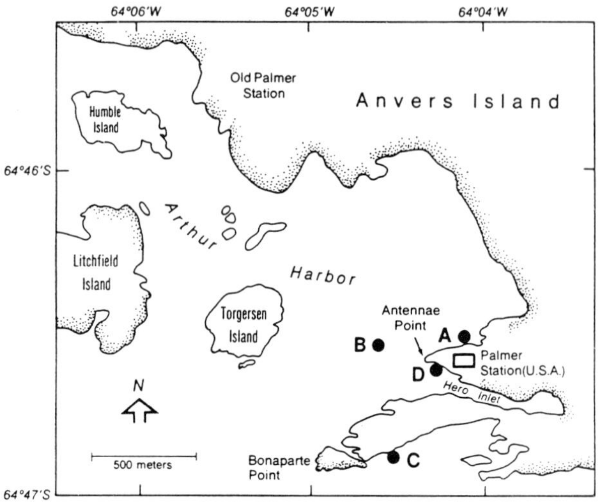

Fig. 1. Map of study area showing sampling sites (after Richardson and Hedgpeth 1977)

Treatment 1: Near-ambient light, including UV-A and UV-B.
Treatment 2: Near-ambient visible light, excluding essentially all of the solar UV.
Treatment 3: Near-ambient visible and UV-A light, but with reduced levels of solar UV-B.
Treatment 4: Near-ambient light including UV-A and B plus artificially enhanced UV-A and UV-B (approximately 3% enhancement, see below).

The effective biological dose (Eε; Smith and Baker 1980) at local apparent noon for each chamber was estimated by multiplying the relative biological efficiency for photoinhibition [$\varepsilon_{PI}(\lambda)$] by the available

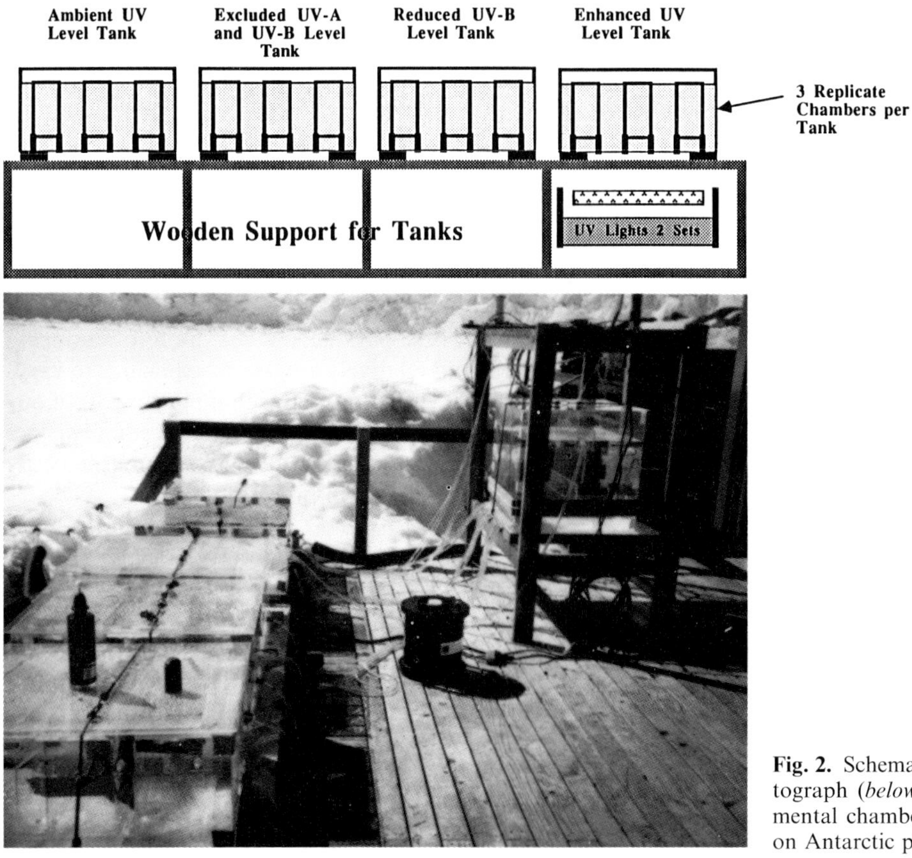

Fig. 2. Schematic drawing (*above*) and photograph (*below*) of the mixing tank and experimental chambers used to investigate UV effects on Antarctic phytoplankton

spectral irradiance [$E(\lambda)$], and integrating with resepct to wavelength (λ, 290–400 nm):

$$E\varepsilon = \int E(\lambda) \cdot \varepsilon_{PI}(\lambda) \cdot d\lambda.$$

The downward spectral irradiance data [$E_d(\lambda)$, the ambient UV radiation] were generated with the atmospheric model of Frederick and Lubin (1988) and ozone levels measured during the course of this study (J. E. Frederick, pers. commun.). The relative degree of UV enhancement (% ENH) for each chamber was calculated as follows:

$$\%\,\mathrm{ENH} = \frac{(E\varepsilon)\,\exp - (E\varepsilon)\,\mathrm{amb}}{(E\varepsilon)\,\mathrm{amb}} \times 100\%.$$

The four primary light treatments used in this study were: (1) ambient visible light, including ambient UV (% ENH = 0%); (2) ambient visible light, excluding both UV-A and UV-B (% ENH = −99.7%); (3) ambient visible light with reduced UV-B (% ENH = −35.0%); and (4) ambient visible light with enhanced UV (% ENH = +2.8%).

It is very important, however, to point out that the degree of "apparent" UV enhancement will vary depending on a number of factors including (1) whether or not the UV irradiance is weighted with an action spectrum; (2) which action spectrum is used; and (3) the wavelength range over which the integration is performed. For example, an "unweighted" enhancement of 16% (relative to ambient UV) for the UV-B wave band (290–320 nm) is calculated for treatment 4, which is five times higher than the "weighted" enhancement calculated for the UV-A and UV-B wave bands (290–400 nm; see above). In this study we chose the photoinhibition action spectrum as a weighting function since this physiological response is one of the early sublethal effects of UV exposure.

Tanks and chambers for treatments 1, 3, and 4 were constructed of 3/16 in. thick OP-4 plexiglas and those for treatment 2 were constructed of 1/4 in. thick OP-2 plexiglas (A & C Plastic Products, Inc., Houston, Texas). Mylar D (thickness, 0.1 mm; Hillcor Plastics, Santa Fe Springs, California) was placed over the top of treatment tank 3 to reduce levels of ambient UV-B. Transmission spectra of these materials are presented in Fig. 3.

Enchancement of UV-A and UV-B was achieved by placing four FS-20 Westinghouse fluorescent sunlamps underneath treatment tank 4. In order to exclude most of the radiation less than 290 nm wavelength emitted by the lamps, a sheet of 0.1 mm thick Kodacel (TA 401; Hillcor Plastics), which had been preconditioned by exposure to a sunlamp for approximately 100 h was placed between the sunlamps and the bottom of the enhanced UV tank.

Total (250–2500 nm) and UV (285–400) solar radiation were measured at 1-min intervals during the study period using an Eppley PSP Pyranometer and

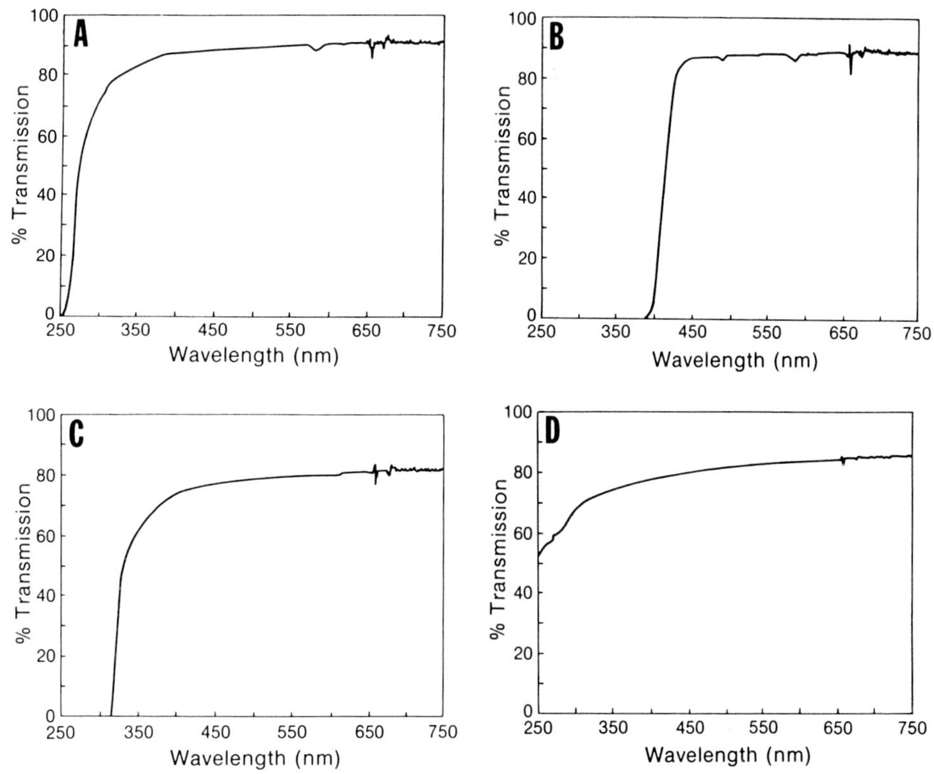

Fig. 3. Light transmission properties of **A** OP-4 plexiglas; **B** OP-2 plexiglas; **C** OP-4 plexiglas plus Mylar D; **D** the plastic bags (Nasco Co.) used in this study for the incubation of Antarctic phytoplankton

an Eppley UV radiometer, respectively. Photosynthetically active radiation (PAR, 400–700 nm) was measured at 5-min intervals during productivity and photosynthesis vs irradiance (P vs I) incubations with a Biospherical Instruments MER 1000 spectral irradiance meter. In addition, total PAR was measured during incubation periods using a Biospherical Instruments Model 140 Quantum Scalar Irradiance Sensor.

Solar spectral irradiance in the UV band was provided by J. Fredrick, University of Chicago. These UV irradiance data, derived from a model using data collected by the back-scattered solar ultraviolet radiation instrument aboard the Nimbus 7 satellite, were used to calculate the percentage enhancement for the UV-enhanced tank (see above).

Seawater samples were collected (1) directly from the flow-through chambers, (2) with a Niskin bottle lowered through a second hole in the ice (Station B, Fig. 1), and (3) with a bucket from open water (Station C, Fig. 1). Samples were also obtained from sea ice which had broken up and turned over, exposing the algae to ambient solar radiation for 3 or 4 days (Station D, Fig. 1). Some of this ice was collected by bucket and allowed to partially melt in the dark to release the algae.

Photosynthetic rates were measured using a modification of Steemann Nielsen's technique (1954; see Stephens 1989, for details). Water samples (250 ml) from each to the three chambers within the four treatment tanks or from the various sampling sites discussed earlier were placed in whirl-pak bags (Fig. 3), inoculated with 40 μCi of $NaH^{14}CO_3$, and incubated for 4 h (at approximately 1100–1500 h local time) in the four experimental tanks. For P vs I determinations, the whirl-pak bags were placed inside a series of outer bags constructed of various numbers of layers of neutral density window screen.

For pigment analysis, water samples were filtered through Whatman GF/F glass fiber (47 mm) filters and extracted in 6 ml of acetone for 48 h in the dark at −20 °C. Following extraction, samples were centrifuged for 5 min to remove cellular debris. The chlorophyll and carotenoid pigment content of the acetone extracts was analyzed by high performance liquid chromatography (HPLC) according to the methods outlined by Bidigare et al. (1989).

3 Results

3.1 Primary Production

The average production rates measured for each of the three experiments decreased with increased UV exposure (Table 1). However, primary production rates of phytoplankton (17 November and 3 December 1987) measured under near-ambient light conditions were low and not statistically different from those measured under either UV-reduced or UV-enhanced conditions. Small, but statistically significant increases in primary production rates were measured when UV radiation was excluded.

In contrast, primary production rates of ice algae (17 November and 3 December 1987) were greater when UV-B radiation was reduced as well as when all UV was excluded. As with phytoplankton, however, production rates of ice algae were not statistically different between samples incubated under ambient and enhanced ("weighted" enhancement of approximate-

Table 1. Photosynthesis rates (mean ± SD; n = 3) for samples taken from three different habitats and incubated for 4 h under the four experimental conditions described in the text

Habitat description and collection Date	Solar irradiance (PAR) (μEin m^{-2} s^{-1})	Production rate (mgC m^{-3} h^{1})	Treatment number[a]
Phytoplankton	1318	0.08 ± 0.01	1 (Amb)
(open-water from	1221	0.40 ± 0.10	2 (Exc)
Bonaparte Point;	1260	0.11 ± 0.01	3 (Red)
17 November 1987)	1318	0.05 ± 0.04	4 (Enh)
Preadapted	NM[b]	7.40 ± 1.56	1 (Amb)
Ice algae	NM	42.08 ± 0.69	2 (Exc)
(21 November 1987)	NM	17.38 ± 4.05	3 (Red)
	NM	4.98 ± 1.18	4 (Enh)
Phytoplankton	1536	0.04 ± 0.02	1 (Amb)
(water pumped from	1423	0.29 ± 0.06	2 (Exc)
sea ice hole;	1468	0.11 ± 0.05	3 (Red)
3 December 1987)	1536	0.01 ± 0.01	4 (Enh)

[a] Amb = ambient UV; Exc = UV excluded; Red = reduced UV; Enh = enhanced UV.
[b] NM = not measured.

ly 3%) UV levels. These results suggest that early in the austral spring season, phytoplankton sampled from beneath the ice or near the ice edge was susceptible to photoinhibition by exposure to ambient solar irradiance. On the other hand, ice algae, which had been exposed to ambient solar irradiance for several days prior to the measurement of productivity, may have adapted to solar radiation in such a way that productivity rates were affected more by UV than by visible radiation (i.e., they had become high-light adapted).

3.2 Photosynthesis vs Irradiance

Short-term (4 h) photosynthesis vs irradiance (P vs I) measurements at the highest irradiance levels studied showed trends similar to those observed in the primary productivity experiments (data not shown). However, after phytoplankton was kept under the four experimental treatment conditions for 24 h (Fig. 4), differences in P_{max} and the initial slope (α) of the P vs I curves were more pronounced than observed for short-term experiments. As described for the short-term exposure experiments above (Table 1), phytoplankton exposed to visible light alone (i.e., UV excluded) had the highest photosynthetic rates, and these rates were maintained over a higher range of irradiance levels. Trends were more clearly defined when photosynthetic rates were normalized to chlorophyll a (Fig. 4B). Both photosynthetic capacity (P_{max}) and photosynthetic efficiency (α) were inversely related to UV exposure, suggesting that both the light and dark relations of photosynthesis had been adversely affected by UV.

Several experiments were performed to examine the short-term (4 h) and longer-term (24 h) effects of UV and visible light on the pigmentation of Antarctic phytoplankton. A few of the results are highlighted below.

Short-term (4 h) exposure of open-water phytoplankton or ice algae to UV produced no significant

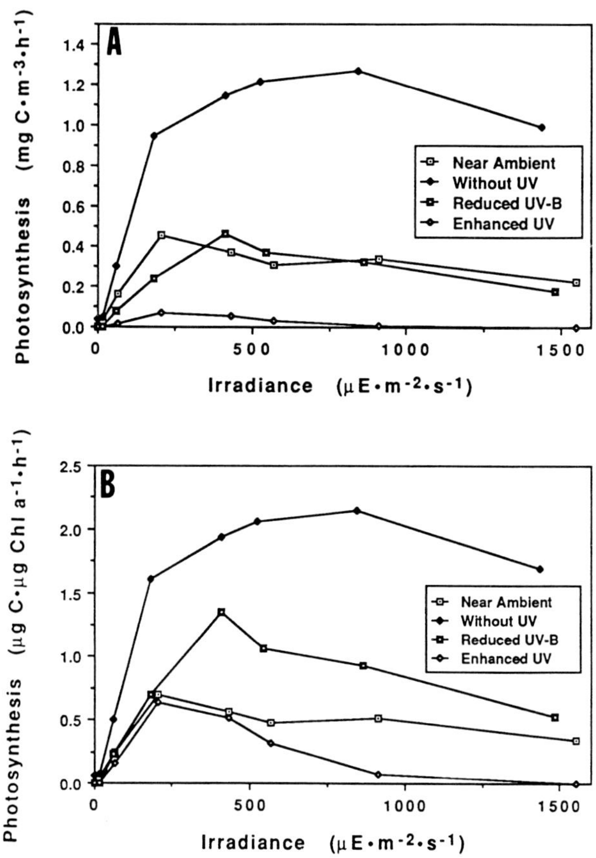

Fig. 4 A. Volume-based and **B** chlorophyll-based photosynthesis-irradiance data for phytoplankton sampled on 6 December 1987 from the experimental chambers (after flow had been stopped for 24 h). Samples were incubated for 4 h under the four treatment conditions described in the text

changes in the concentrations of chlorophyll *a* and fucoxanthin (data not shown). In contrast, longer-term (24–48 h) exposure to UV radiation under ambient visible light conditions produces drastic changes in phytoplankton pigmentation. For water samples collected off Bonaparte Point, there was a selective loss of chlorophyll *a* with increased UV exposure (Table 2);

Table 2. Results of a 24-h UV exposure experiment performed with open-water phytoplankton collected off Bonaparte Point. Incubations were performed under ambient visible light in the experimental chambers described in the text. Pigment concentrations ($ng\,l^{-1}$) and ratios (w:w) are reported as the mean ± SD (n = 3). Chlorophyll *c* concentrations were below the limit of HPLC quantification. The initial values were determined at the beginning of the experiment and represent a single measurement[a]

Treatment	[FUCOX]	[DN]	[DT]	[CHL A]	DN:CHL	FUCOX:CHL
Initial	126	37	0	254	0.15	0.50
No UV	62 ± 6	38 ± 2	16 ± 0	171 ± 6	0.22 ± 0.02	0.36 ± 0.05
Reduced UV	61 ± 8	36 ± 5	11 ± 3	125 ± 16	0.29 ± 0.01	0.49 ± 0.02
Ambient UV	63 ± 23	29 ± 3	15 ± 2	87 ± 14	0.34 ± 0.03	0.71 ± 0.18
Enhanced UV	70 ± 14	37 ± 4	13 ± 2	82 ± 3	0.45 ± 0.03	0.86 ± 0.19

[a] FUCOX = fucoxanthin; DN = diadinoxanthin; DT = diatoxanthin; and CHL A = chlorophyll *a*.

Table 3. Results of the UV exposure experiment performed with water pumped from the sea ice hole. Water to the experimental chambers was shut off and pigments were measured after 24 (6 December 1987) and 48 (8 December 1987) h of incubation under ambient visible light. Pigment concentrations ($ng\,l^{-1}$) are reported as the mean ± SD (n = 3). The initial values were measured one day before the water to the experimental chambers was shuf off (4 December 1987) and represent a single measurement[a]

Treatment	Incubation	[CHL C]	[FUCOX]	[HFUCOX]	[CHL A]
Initial	–	44	243	36	484
No UV	24 hrs.	67 ± 16	231 ± 71	39 ± 14	590 ± 165
Reduced UV	24 hrs.	27 ± 4	129 ± 10	23 ± 2	344 ± 35
Ambient UV	24 hrs.	59 ± 18	263 ± 77	29 ± 11	650 ± 210
Enhanced UV	24 hrs.	ND	49 ± 16	4 ± 6	104 ± 41
No UV	48 hrs.	64 ± 33	368 ± 160	17 ± 8	844 ± 379
Reduced UV	48 hrs.	7 ± 12	82 ± 36	12 ± 2	278 ± 97
Ambient UV	48 hrs.	24 ± 9	115 ± 24	23 ± 10	363 ± 70
Enhanced UV	48 hrs.	ND	37 ± 6	ND	106 ± 20

[a] CHL C = chlorophyll *c*; FUCOX = fucoxanthin; HFUCOX = 19′-hexanoyloxyfucoxanthin; CHL A = chlorophyll *a*; and ND = not detectable.

concentrations of the other pigments showed little or no change. In another experiment, pigment concentrations were measured 24 and 48 h after the inflow (water pumped from sea ice hole to the treatment tanks) was shut off. Significant changes were observed for each of the major pigments in response to UV exposure (Table 3). After a 48-h incubation period, highest and lowest pigment concentrations were measured under "no UV" and "enhanced UV" conditions, respectively. Intermediate pigment concentrations were observed in the "ambient UV" and "reduced UV" treatment tanks.

4 Discussion

The results of the study carried out at Palmer Station during November–December 1987 provided insight into the potential deleterious effects of enhanced UV radiation. These results showed an enhancement of production rates in the tank where UV-A and UV-B were excluded. Conversely, rates of production were much lower under ambient and enhanced UV conditions.

The quantitatively important pigments measured during our study included: chlorophylls *a* and *c*, fucoxanthin, 19′-hexanoyloxyfucoxanthin, diadinoxanthin, diatoxanthin, and β-carotene. In all samples the diatom marker pigments were most abundant. Concentrations of chlorophyll *b*, zeaxanthin, alloxanthin, peridinin and 19′-butanoyloxyfucoxanthin were near or below the limit of HPLC quantification. Significant decreases in photosynthetic pigmentation were observed for those phytoplankton samples exposed to enhanced UV radiation. As expected, the magnitude of change varied in relation to the amount and duration of exposure. Reductions in rates of photosynthesis could, in part, be explained by the loss of chlorophyll *a* and accessory pigmentation.

Increased UV radiation appears to not only decrease the rate of photosynthesis of algal cells and destroy chlorophyll *a*, but it can also alter the composition of the algal accessory pigments (Table 3). There is abundant evidence from laboratory studies of differential sensitivity of algal cells to enhanced UV levels. In some experiments, exposure to UV-B radiation resulted in a decrease in diatom abundance; other algal cells continued to survive with impunity (McLeod and McLachlan 1959; Worrest et al. 1978, 1981a, b).

Despite the preliminary nature of the results presented here, there is sufficient evidence to implicate elevated UV-B radiation in contributing to: (1) reduced rates of photosynthesis and (2) changes in photosynthetic pigmentation. Further research is required to determine whether these findings can be applied to the phytoplankton and ice algae of the Southern Ocean.

Considerable caution must be exercised in extending the results of these short-term laboratory experiments to the "real world". In the latter case, one has to contend with a host of very complex interrelated factors, among which are: the diffuse attenuation coefficient for UV radiation, depth of the mixed layer, residence time of the algal cells in the euphotic zone, seasonality of exposure, behavioral responses of the targeted organisms, and their tolerance to and avoidance of solar UV, and their capabilities to repair UV-impaired DNA molecules. As to whether the effect of UV-B on the algal population is short-lived, long-lasting or reversible, this is impossible to answer from the limited data we have.

What are the long-term effects of enhanced UV radiation on the Antarctic marine ecosystem? Based on these Arthur Harbor findings and given the long lifetime of man-made stratospheric pollutants (e.g., chlorofluorocarbons), the impact of elevated UV-B could be long-lasting and potentially damaging. Since the phytoplankton is the basic primary producer in the Antarctic Ocean on which all other components of the ecosystem (zooplankton, krill, fish, squid, winged birds, penguins, seals, and whales) depend for their livelihood, any substantial decrease in the productivity of these waters, or any change in their community structure, could have far-reaching ecological implications.

Acknowledgements. The authors are greatly indebted to Dr. Denis A. Wiesenburg who provided DIC analyses and Mr. John R. Moisan who provided assistance in the field experiments. Grateful acknowledgement is expressed to the Division of Polar Programs, National Science Foundation, for supporting this study (DPP-88-12972).

References

Bidigare RR, Schofield O, Prezelin BB (1989) Influence of zeaxanthin on quantum yield of photosynthesis of *Synechococcus* clone WH 7803 (DC2). Mar Ecol Prog Ser (in press)
Bowman KP (1988) Global trends in total ozone. Science 239:48–50
Calkins J, Thordardottir T (1980) The ecological significance of solar UV radiations on aquatic organisms. Nature 283:563–566
Frederick JE, Lubin D (1988) The budget of biologically active ultraviolet radiation in the earth-atmosphere system. J Geophys Res 93:3825–3832
Jitts HR, Morel A, Saijo Y (1976) The relation of oceanic primary production to available photosynthetic irradiance. Aust J Mar Freshwater Res 27:441–454
Kerr RA (1986) Taking shots at ozone hole theories. Science 234:817–818
Kerr RA (1988) The deepest Antarctic hole ever seen. Science 239:146
Lorenzen CJ (1979) UV radiation and phytoplankton photosynthesis. Limnol Oceanogr. 24:1117–1120
Maske H (1984) Daylight ultraviolet radiation and the photoinhibition of phytoplankton carbon uptake. J Plankton Res 6:351–357
McLeod GC, McLachlan J (1959) The sensitivity of several algae to ultraviolet radiation of 2537 A. Physiol Plant 12:306–309
Richardson MD, Hedgpeth JW (1977) Antarctic soft-bottom, macrobenthic community adaptions to a cold, stable, highly productive, glacially affected environment, In: Llano GA (ed) Adaptions within Antarctic ecosystems. Proc Third SCAR Symp Antarctic Biology, p 183
Smith RC, Baker KS (1980) Stratospheric ozone, middle ultraviolet radiation and ^{14}C measurements of marine productivity. Science 208:592–593
Smith RC, Baker KS, Holm-Hansen O, Olsen R (1980) Photoinhibition of photosynthesis in natural waters. Photochem Photobiol 31:585–592
Steemann Nielsen E (1954) On organic production in the oceans. J Conseil 19:309–328
Stephens FC (1989) Effects of ultraviolet light on photosynthesis and pigments of Antarctic marine phytoplankton. PhD Diss Department of Oceanography, Texas A&M University, 121 p
Stolarski RS, Krueger AJ, Schoeberl MR, McPeters RD, Newman PA, Alpert JC (1986) Nimbus 7 satellite measurements of the springtime Antarctic decrease. Nature 322:808–814
Wellborn SN (1986) Putting a freeze on freon. US News and World Report, Nov 17, 1986, p 72
Worrest RC (1982) Review of literature concerning the impact of UV-B radiation upon marine organisms. In: Calkins J (ed) The role of solar ultraviolet radiation in marine ecosystems. Plenum Press, New York, pp 429–457
Worrest RC, Brooker DL, Van Dyke H (1980) Results of a primary productivity study as affected by the type of glass in the culture bottle. Limnol Oceanogr 25:360–364
Worrest RC, Thomson BE, Van Dyke H (1981a) Impact of UV-B radiation upon estuarine microcosms. Photochem Photobiol 33:861–867
Worrest RC, Van Dyke H, Thomson BE (1978) Impact of enhanced simulated solar ultraviolet radiation upon a marine community. Photochem Photobiol 27:471–478
Worrest RC, Wolniakowski KU, Scott JD, Brooker DL, Thompson BE, Van Dyke H (1981b) Sensitivity of marine phytoplankton to UV-B radiation: impact upon a model ecosystem. Photochem Photobiol 33:223–227

International Agreements and the Conservation of Antarctic Systems

W. N. BONNER [1]

Summary. The formal provisions for conservation in the Antarctic are provided by the Antarctic Treaty system. Through the Agreed Measures for the Conservation of Antarctic Fauna and Flora native mammals or birds are protected, signatories are urged to minimise harmful interference with the normal living conditions of mammals and birds and to alleviate coastal pollution; provision is made to designate Specially Protected Species and Specially Protected Areas; and the introduction of non-indigenous species is controlled. Subsequent measures created Sites of Special Scientific Interest within which activities are subject to controls described in management plans; controlled the use of radio-isotopes; and provided a code of conduct for Antarctic expeditions and station activities. A recent measure requires the undertaking of environmental impact assessment in the planning of scientific research and its associated logistic activities. A code of conduct for expeditions and stations has been adopted and revised procedures for waste disposal are under consideration.

The Convention for the Conservation of Antarctic Seals provides a scheme for protection should commercial sealing develop in the Antarctic. The Convention for the Conservation of Antarctic Marine Living Resources aims to ensure that harvesting shall not deplete populations of marine animals below those levels that ensure stable recruitment; the importance of the balance between harvested, dependent and related populations is recognised and an objective of the Convention is the restoration of depleted populations. The newly agreed Convention on the Regulation of Antarctic Mineral Resources provides unprecedentedly strict environmental controls to cover the exploration and development of Antarctic mineral resources, should this take place. The Convention recognises the significance of Antarctica for the global environment and its scientific value and aesthetic and wilderness qualities. Protected areas may be designated where no activities associated with exploring for, prospecting for, or extracting minerals can take place and management schemes for exploration or development shall include measures and procedures for the protection of the environment.

The provision of a legislative basis for the conservation of the environment has gone further in Antarctica than elsewhere in the world. There is still room for development, however, and a group appointed by SCAR, in response to a request for advice by the Antarctic Treaty, suggested that inspections should be made to assess the effectiveness of the existing measures relating to area protection. Information resulting from such inspections should be made available to interested parties. Management plans should be prepared for existing and subsequently declared Specially Protected Areas. Further protected areas should be designated to provide for geographical representation of all Antarctic terrestrial, inland water and marine ecosystems. Finally, a new category of multi-use zoned protected area which might incorporate existing protected areas should be introduced. The objective of such an area, which might be called an Antarctic Protected Area, would be to minimise disturbance to the area insofar as this might detract from the special value for which it was designated. The Fourteenth Antarctic Treaty Consultative Meeting in considering these proposals recognised the value of inspections and asked for examples of management plans for areas which might benefit from multiple use zoning techniques.

The legislative basis for conservation is reasonably complete, but more needs to be done. Inspection and enforcement, coupled with further education to ensure public awareness, are priorities. Monitoring studies and information management are important. An Antarctic Treaty sceecretariat would be helpful. Continued collaboration between SCAR (Scientific Committee on Antarctic Research) and IUCN (International Union for the Conservation of Nature and Natural Resources) is needed.

1 Introduction

All biological systems change, adapting to the pressures acting on them. Species become extinct or evolve

[1] c/o SCAR, Scott Polar Research Institute, Lensfield Road, Cambridge CB2 1ER, UK.

Antarctic Ecosystems. Ecological Change and Conservation
(ed. by K. R. Kerry and G. Hempel)
© Springer-Verlag Berlin Heidelberg 1990

to new forms. The composition of communities change and new communities replace earlier ones when the physical environment alters. In the past few centuries Man has greatly accelerated the rate of change of biological systems throughout the world. In the past, such changes as the extermination of large predators from populated regions, or the clearing of forests or the ploughing of prairies for agriculture, were seen as beneficial. However, as the list of extinctions lengthened, and as the areas of near-virgin territory shrank, so a need was seen for action to lessen or reverse the changes brought about by Man's pressure on his environment. Such actions constitute conservation.

Conservation may relate to resources, where natural resources are managed for the benefit of mankind, or to wildlife, where plant or animal species and samples of the ecosystems they comprise are protected, or to amenity, where visually important features of the landscape are preserved for aesthetic reasons (Holdgate 1970). A widespread appreciation of the need for conservation in all parts of the world is a comparatively recent phenomenon and this was not seen as a pressing issue when the Antarctic Treaty was being negotiated in the late 1950s. Since then, however, there has been a greater realisation of the intrinsic value of the Antarctic as a scientific laboratory, and of its extrinsic value as a principal controller of world climate. The Antarctic constitutes the largest remaining tract of wilderness in the world while its remoteness and great scenic beauty have given it special qualities in the eyes of many people. The conservation of the Antarctic is now fully recognised as an urgent issue. Because of Antarctica's special status as a region with no generally recognised sovereignty, all effective conservation measures there must operate through the Antarctic Treaty System (ATS). The ATS comprises the Antarctic Treaty, together with the measures and recommendations adopted in accordance with it, the Convention for the Conservation of Antarctic Seals, the Convention on the Consenvation of Antarctic Marine Living Resources, and the Convention on the Regulation of Antarctic Mineral Resource Activities. The conservation aspects of these are reviewed below. In the nearly 30 years since the coming into force of the Treaty, the ATS has produced a code of legal instruments which provide a level of conservation surpassing that achieved anywhere else in the world.

International agreements become effective only when they are enforced by the implementation of domestic legislation. This aspect of Antarctic conservation is not dealt with in this work.

2 The Antarctic Treaty and its Conservation Measures

The Antarctic Treaty itself (Heap 1988) is all but silent on the subject of conservation though, importantly, it does refer to the "preservation and conservation of living resources in Antarctica" as one of the topics on which Contracting Parties should consult together to formulate measures to further the principles and objectives of the Treaty (Article IX).

The protection of the Antarctic environment and its fauna and flora have appeared on the agenda of all of the Antarctic Treaty Consultative Meetings (ATCMs) from the first in Canberra in 1961. At that meeting it was agreed to consider further a set of "rules of conduct for the preservation and conservation of living resources in Antarctica", prepared by the Working Group on Biology of the Scientific Committee on Antarctic Research (SCAR). This consideration came to fruition at the Third ATCM in Brussels in 1964 when the Agreed Measures for the Conservation of Antarctic Fauna and Flora were formulated. The Agreed Measures apply to the same area as the Antarctic Treaty, namely the area south of 60° S latitude, including all ice shelves. However, it is stipulated that nothing in the Agreed Measures shall affect the rights of States on the high seas.

There are four important Articles (VI–IX), which relate to conservation. The first, Article VI, deals with the protection of native fauna. It prohibits the killing, wounding or capturing (or attempting to do so) of any native mammal or bird, except in accordance with a permit. Such permits have to be drawn in terms as specific as possible and can be issued only to provide indispensible food for men or dogs in the Treaty Area, to provide specimens for scientific study or information, or to provide specimens for museums or zoological gardens or similar institutions. In issuing permits Participating Governments must limit them so as to ensure that no more mammals or birds are taken than can be replaced by natural reproduction in the following breeding season and that the variety of species and balance of the natural ecological systems are maintained. There is provision for designating Specially Protected Species for which permits to kill, etc. could be issued only for a compelling scientific purpose and so as not to jeopardise the survival of the species. Currently Specially Protected Species are the Ross seal, *Ommatophoca rossii*, and all fur seals of the genus *Arctocephalus*. (The reason for the blanket protection of fur seals was that at the time the Agreed Measures were being drafted there was doubt about the specific identity of the Antarctic fur seal, *Arctocephalus gazella*, which had been confounded with the Sub-Antarctic fur seal, *A. tropicalis*, and the South American fur seal, *A. australis*.)

Interestingly, the definition of "native mammal" in Article II specifically excludes whales. This was partly because of the preservation of Signatories' rights on the high seas (with the presumption that whales would be found nowhere else) and partly because whales were covered by the International Convention for the Regulation of Whaling, whose Commission, the IWC, was responsible for regulating whale killing and protection. The history of whale conservation in the Antarctic under the IWC has not been a happy one (Bonner 1984) as economic interests in a highly developed industry stultified early attempts to regulate harvest to sustainable levels.

Article VII of the Agreed Measures urges Participating Governments to take appropriate measures to minimise harmful interference with the normal living conditions of any native mammal or bird. The following activities are considered harmful interference:

(a) Allowing dogs to run free;
(b) Flying helicopters or other aircraft in a manner which would unnecessarily disturb bird or seal concentrations, or landing close to such concentrations (e.g. within 200 m);
(c) Driving vehicles unnecessarily close to concentrations of birds and seals (e.g. within 200 m);
(d) Use of explosives close to concentrations of birds and seals;
(e) Discharge of firearms close to bird and seal concentrations (e.g. within 300 m);
(f) Any disturbance of bird and seal colonies during the breeding period by persistent attention from persons on foot.

However, the Article recognises that some disturbance is a necessary concomitant of the presence of a station and permits these activities, except for (a) and (e), to the minimum extent necessary for the establishment, supply and operation of stations. A further important provision of Article VII is the requirement that Participating Governments shall take all reasonable steps towards the alleviation of pollution of waters adjacent to the coast and ice shelves.

Article VIII provides for the designation of Specially Protected Area (SPAs). These are areas of outstanding scientific interest where, in order to preserve their "unique natural ecological system[s]", the collection of any native plant is prohibited, except in accordance with a permit; the driving of vehicles is prohibited; and so is entry by personnel, except in accordance with a permit which can be issued only for a compelling scientific purpose. Permits issued under Article VI have effect in an SPA only if they are for a compelling scientific purpose which cannot be served elsewhere and the actions permitted will not jeopardise the natural ecological system in that area. So far 20 SPAs have been listed, but three of them have been re-designated as Sites of Special Scientific Interest (see below).

Article IX prohibits the introduction of non-indigenous plants and animals except in accordance with a permit, which can be issued in relation to sledge dogs, domestic animals and plants, and laboratory animals and plants, including microbes. This prohibition does not apply to food species which are kept in controlled condition, except that introductions of living poultry are banned.

The Agreed Measures further require Participating Governments to take appropriate efforts to ensure that no one engages in any activity in the Treaty Area contrary to the principles or purposes of the Agreed Measures (Article X). Furthermore, Governments are required to ensure that the crews of ships of flags other than their own when used by their expeditions shall observe the Agreed Measures.

At its meeting in Wellington in 1972 the VII ATCM reviewed the measures relating to SPAs. The meeting recommended (VII-2) that the following should be included in the series:

(a) Representative examples of the major Antarctic land and freshwater ecological systems;
(b) Areas with unique complexes of species;
(c) Areas which are the type locality or only known habitat of any plant or ivertebrate species;
(d) Areas which contain specially interesting breeding colonies of birds or mammals;
(e) Areas which should be kept inviolate so that in the future they may be used for purposes of comparison with localities that have been disturbed by man.

The first and last of these criteria were a welcome advance in the cause of Conservation, though this was to a great extent offset by the final part of the recommendation which stated that the number of SPAs should be kept to the minimum that would meet the criteria set out above and that the size of each SPA should be the minimum required to serve the purpose for which it had been designated. Considering the size and population density of the Antarctic, these strictures are surprising.

The other three criteria, (b), (c) and (d), were injudicious, since SPAs designated in accordance with them, although of substantial scientific interest, were effectively sterilised for research. They represented a line of protectionist thinking which has now largely disappeared from modern conservation practice. Their inclusion was the more surprising in that at the same

ATCM a recommendation was agreed that was designed to facilitate scientific research through the designation of Sites of Special Scientific Interest (SSSIs).

Recommendation VIII-3 (Oslo 1975) recalled a previous recommendation (VII-3) which had recognised that scientific investigations could be jeopardised by accidental or wilful interference and that sites might require protection against this, while because of the continuing research these sites were not suitable for designation as SPAs. It called for the application of mananagement plans to such sites, which would be called Sites of Special Scientific Interest. VIII-3 recommended that SSSIs might be proposed when:

(i) Scientific investigations are being carried out or are planned to begin before the following meeting of SCAR, and there is a demonstrable risk of interference which would jeopardise those scientific investigations, or
(ii) They are of exceptional scientific interest and therefore require long-term protection from harmful interference.

The management plan for each site describes it, sets out the reasons for its designation, summarises the research being carried out, sets out the proposed date at which the designation would expire, and provides regulations for access and other permitted activities at the site. SSSIs, unlike SPAs, can be designated for sites with a non-biological character (though this provision has been used only in the case of SSSI 2, Arrival Heights, designated to preserve an electromagnetically quiet site).

Besides these measures for the protection of Antarctic fauna and flora, the Antarctic Treaty has provided several general provisions for environmental protection. The prohibition by the Treaty of military activity, nuclear explosions and the disposal of nuclear waste has removed a massive potential for damaging the Antarctic environment (Heap 1988). Man's impact on the Antarctic environment figured as a subject for discussion at ATCMs from the Sixth onwards. Recommendation VIII-13 required Governments to ensure that their actions in the Antarctic should be consistent with the interests of all mankind; that no activity having an inherent tendency to modify the environment over large areas should be undertaken unless steps had been taken to foresee these modifications and to exercise appropriate controls; and, in cooperation with SCAR, to continue to monitor changes in the environment.

This was the start of a long road towards environmental impact assessment, which continued with Recommendation XII-3 from Canberra in 1983, which called on SCAR to provide advice on the kinds of scientific and logistic activity which might reasonably be expected to have a significant impact on the environment. SCAR's response (Benninghoff and Bonner 1985) was considered at the Thirteenth ATCM in Brussels in 1985 but no positive action was taken. At the following ATCM in Rio de Janeiro Recommendation XIV-2 was formally adopted, calling for the evaluation of environmental impact in the planning process of scientific research programmes and their associated logistic support activities. This is to be a two-stage process, the first evaluation, termed an "Initial Environmental Evaluation" (IEE), being followed by a "Comprehensive Environmental Evaluation" (CEE) if the former indicates that impacts are likely to be significant. CEEs are to consider also indirect or second-order effects and cumulative impacts. In the process of preparing CEEs, Parties concerned shall be informed and given the opportunity to comment, while the final CEEs shall be transmitted as part of the annual exchange of information under the Treaty.

Another important recommendation from the Eighth ATCM in Oslo in 1975, VIII-11, concerned a code of conduct for Antarctic expeditions and station activities. This was based on a response by SCAR to a request for advice contained in Recommendation VI-4. The first part of the code dealt with waste disposal. At coastal stations, it recommended the disposal at sea, either in deep water, or if that were not possible, in shallow water at specified sites, of non-combustible solid wastes, inculding chemicals (except batteries, which were to be removed from the Treaty Area). Wood, wood products and paper might be burnt, as might lubricating oils (except those containing harmful additives) and carcases and materials associated with imported experimental animals. All plastics and rubber products were to be removed from the Treaty Area. Human wastes, garbage and laundry effluents were to be macerated, if possible, and flushed to the sea. Large quantities of photographic liquids were to be treated for the recovery of silver and the residues flushed to the sea. Field sites should, if possible use the facilities of supporting stations. Inland stations should concentrate waste in deep pits. All waste containing radio isotopes should be removed from the Treaty Area. Finally, every effort was to be made to reduce the plastic packaging of products imported into the Treaty Area.

The second and third parts of the code drew attention to existing provisions of the Agreed Measures, while the fourth part anticipated the provision of environmental impact assessment.

The first part was the most significant environmentally, but contained serious defects. There was a lack of definition of terms like "battery", "harmful additive" and "plastic". In any case, the call for the reduction of plastic packaging fell on deaf ears, if only because it provides the simplest means of ensuring that stores get to Antarctic stations, frequently across the beach, in a dry state. It was widely recognised in the Antarctic that the Code of Conduct was not useful guidance to the persons in the field, and that further consideration was needed. This view was reflected at Brussels when Recommendation XIII-4 called for further advice from SCAR. XIX SCAR set up a panel of experts to consider the matter of waste disposal. Their report (SCAR 1989) was submitted to XX SCAR and will be published by SCAR as its response to this recommendation. The report, besides making specific recommendations about particular wastes, lays stress on waste management planning, including waste reduction, storage and disposal. The clean-up of waste disposal sites and abandoned work sites is advocated. National Antarctic operators are urged to work towards eliminating open burning as a means of waste disposal. It is of particular significance that the representatives of operating organisations present at the SCAR meeting agreed that they could accept and operate the recommendations.

Yet another product of the meeting in Oslo in 1975 was Recommendation VIII-9 which recognised that tourists and other persons not sponsored by Consultative Parties were visiting the Antarctic Treaty Area in increasing numbers. It called on Governments to use their best endeavours to ensure that all who entered the area were aware of a Statement of Accepted Practices and the Relevant Provisions of the Antarctic Treaty, contained in an Annex to the recommendation. Tourists were to visit only those stations for which permission had been granted, as provided for by Recommendation IV-27. The Annex to this recommendation contains a succinct summary of the Treaty and its provisions then relating to the environment. It needs to be updated in the light of subsequent developments.

3 The Convention for the Conservation of Antarctic Seals

In 1964 a small Norwegian sealing expedition visited the Antarctic (Øritsland 1970). Although the catch was very small it alerted the Antarctic Treaty powers to the fact that the protection of seals afforded by the Agreed Measures did not extend to seals at sea or on pack ice. A conference was called and in 1972 the Convention for the Conservation of Antarctic Seals (CCAS) was signed in London. This applies to the sea and floating ice south of 60° S latitude, but there is provision for reporting catches to the north of this area. It covers all species of seals that occur in Antarctic waters. The taking of Ross seals, elephant seals (*Mirounga leonina*) and fur seals of the genus *Arctocephalus* is prohibited. In an Annex to the convention permissible yearly catches (subject to review) are listed for crabeater seals (*Lobodon carcinophagus*), 175 000; leopard seals (*Hydrurga leptonyx*), 12 000; and Weddell seals (*Leptonychotes weddellii*), 5000. These catch limits represent 1.2, 5.4 and 0.7% of the stock sizes of the respective species as then calculated (Bonner 1982). There are further provisions relating to closed seasons, sealing zones and sealing reserves and the issue of permits for purposes similar to those under the Agreed Measures. If commercial sealing were to begin in the Antarctic there is provision for setting up a Commission; until such time, SCAR is invited to assess information and report on possible harmful effects.

Since the signature of CCAS there has been only one sealing expedition to the Antarctic, in 1986/87 when a Soviet vessel took 4802 seals. A CCAS conference was held in September 1988 but no major changes to the operation of the Convention were proposed.

4 Convention on the Conservation of Antarctic Marine Living Resources

CCAS provided protection for only a small part of the potential marine harvest. Whale fishing was covered by the International Convention for the Regulation of Whaling (Washington 1946) but catching of fin fish had begun in the late 1960s around sub-Antarctic island groups while harvesting Antarctic krill, *Euphausia superba*, on an experimental scale had begun even earlier (Everson 1984 a, b). Concerned with the possible effects of harvesting on the Antarctic ecosystem, the Treaty Parties began to negotiate a new convention to regulate marine harvesting. The broad outlines of this were laid down at the Ninth ATCM in London in 1977 and the Convention on the Conservation of Antarctic Marine Living Resources (CCAMLR) was signed at Canberra in 1980.

The new convention extended its area of application beyond the Antarctic Treaty Area, adopting a northern limit approximating the biological boundary of the Antarctic Convergence. CCAMLR covers all marine organisms, including birds. Its aim is conservation, which include rational use. Harvesting must not deplete populations below the levels which ensure stable recruitment, which implies that they should be maintained near those levels which ensure the greatest net annual increment. The importance of the balance between harvested, dependent and related populations is recognised, and the restoration of depleted populations is one of the objectives. Changes in populations which are not potentially reversible in a few decades are to be prevented.

CCAMLR operates through a Commission (known by the same acronym), advised by a scientific committee. The commission can set catch limits (with age or size restrictions), designate protected species, set closed seasons or regulate effort and methods of harvesting.

The ecological, multi-species approach of CCAMLR, which is unique in a fisheries convention, is welcome to conservationists. But there are some difficulties in implementation. Krill fishing, which has not exceeded annual catches of more than half a million tonnes, has not yet tested the system. On the other hand, actions sufficient to ensure the recovery of fin fish stocks, already depleted in most localities before the Convention came into force, have not yet been taken. CCAMLR has responded to this tardily with conservation measures and resolutions and it is not clear whether these will be sufficient to restore the depleted populations, one of the objectives of CCAMLR. More important, perhaps, is the fact that the Commission has yet to find a reliable method to assess the effects of harvest on "dependent and related populations". Without an accurate means of doing so the convention is impossible to administer as originally intended. Even were such a method to be found it might still not provide an adequately responsive management tool.

5 Convention on the Regulation of Antarctic Mineral Resource Activities

By 1980 when CCAMLR was signed, it seemed that an adequate legal framework was in place to cover the traditional uses of the Antarctic, marine harvesting and scientific research, while some account had been taken of the developing tourist industry. The area that was not covered was the possible development of mineral exploration and exploitation in the Antarctic.

When the Antarctic Treaty was being negotiated in 1959 the question of whether it should cover possible mineral resource development was raised, but it was concluded that to do so would be premature (Heap 1987). However, as the search for mineral resources, particularly petroleum, intensified throughout the world, and as geological research progressed in the Antarctic, it became clear that the search for minerals with a view to their eventual exploitation was at least a possibility and that in an unregulated form this could pose threats to the peace of the region and serious hazards to the Antarctic environment. From the Sixth ATCM onwards the subject appeared on the agenda and at the Eleventh ATCM (Recommendation XI-1) it was agreed to convene a Special Consultative Meeting to elaborate a regime governing Antarctic mineral resource development.

After 12 meetings, the Convention on the Regulation of Antarctic Mineral Resource Activities (CRAMRA), was adopted at Wellington in June, 1988. CRAMRA is a long convention of 67 articles. Its objectives, as set out in Article 2, include the protection of the Antarctic environment and dependent and associated ecosystems, recognition of Antarctica's global environmental significance and respect for its scientific value and aesthetic and wilderness qualities. No 'minerals activity', as CRAMRA terms exploration, prospecting and exploitation, is to take place until its possible impacts on the environment, both in the Antarctic and globally, have been assessed (Article 4). There is provision for Protected Areas (Article 13) established for historic, ecological, environmental, scientific or other reasons, where minerals activities would be prohibited or restricted, as well as automatic prohibition of minerals activity in SPAs and SSSIs established by the Antarctic Treaty.

The operation of CRAMRA will be complex. There will be a Commission (Articles 18–22), an Advisory Committee (Articles 23–27), and Regulatory Committees (Articles 29–32). Management Schemes (Articles 46–52) will relate to exploration and development permits and will include measures for the protection of the environment, with provision for response actions for necessary clean-up and restoration action. Applications for exploration and development permits must include detailed assessments of the environmental impacts of the proposed activities and a description of the capacity to respond to accidents (Articles 44 and 53). There are also provisions for inspection and enforcement.

CRAMRA is a complex convention which provides unprecedentedly strict environmental controls on prospective minerals operators in the Antarctic. Should minerals activities take place there, which many people think unlikely at least within the next several decades, it would appear that the legal code is available to ensure that the environment, both in the Antarctic and worldwide, is adequately protected.

6 Additional Protective Measures

At the ATCM in Brussels in 1985 a request was made to SCAR (Recommendation XIII-5) to provide scientific advice on:

"the system of protected areas in the Antarctic, including Sites of Special Scientific Interest and Specially Protected Areas and the question of a possible additional category of area under a different form of protection."

SCAR set up an ad hoc Group (SCAR 1987) which reported in 1987. The Group's report contained five recommendations. The first of these was that the existing and proposed provisions for Antarctic conservation should be subject to periodic assessment (including site visits) to determine whether the objectives were being achieved and the extent to which existing area protection regulations were being observed. This recommendation was designed to meet the need to determine the effectiveness of the relevent conservation measures, something not covered by the inspection procedures of the Treaty.

The second recommendation was that information arising from such assessments, or in the course of carrying out scientific research, should be freely available to those concerned with conservation issues. This was in line with the general provisions for the exchange of scientific information contained in the Treaty, but was intended to extend this, for conservation purposes, to bodies not represented within the Treaty System. It was noted that storage of information in a data base would facilitate its exchange.

The third recommendation was that a management plan should be prepared for each SPA. The concept that an effective conservation area can be created simply by designating it and prohibiting entry is questioned by practical conservationists. The objectives for which the area is designated and actions needed to ensure achievement of these objectives need to be stated clearly. Blanket prohibition on entry goes too far as management plans can identify activities that could be carried out in designated SPAs without jeopardising their specified purpose.

The fourth recommendation called for the creation of additional protected areas to provide for geographically distributed representative examples of all Antarctic terrestrial, inland water and marine ecosystems. This point had already been addressed by Recommendation VII-2 at the 1972 ATCM in Wellington, but the range and distribution of protected areas remained restricted.

The final recommendation called for the creation of a new category of protected area to be introduced. These areas, which might be called Antarctic Protected Areas (APAs), would generally be more extensive than most SPAs or SSSIS and would be multi-use areas, zoned for differing levels of protection or use. They could include refuge huts, field stations, roads, runways and even permanent base installations, but the boundaries of such high-use areas would be defined. APAs would not be restricted to the protection of biological features or those of other scientific interest, but could include historical sites or features of particular significance by virtue of their scenic beauty, inspirational quality, potential for recreation, or their wilderness status. All existing categories of protected area could also be included. An APA would require a detailed and comprehensive management plan which would be subject to periodic review.

The SCAR report was considered at the Fourteenth ATCM at Rio de Janeiro in 1987. The meeting acknowledged the value of visits to protected areas and urged Contracting Parties to visit as many areas as possible and report on them, these reports to be circulated to Contracting Parties, to SCAR and, if appropriate, to CCAMLR. It was further recognised that the information from the assessment process should be made publicly available. There was disagreement about the desirability of management plans for SPAs, some delegations considering that the value of SPAs might be diminished by adopting management plans. However, to assist further consideration, examples of possible management plans were to be submitted at the next ATCM. The fourth SCAR recommendation also gave rise to disagreement, some delegations holding that it was not possible to designate marine SPAs. However, it was agreed to ask national organisations to conduct surveys to identify potential SPAs and SSSIs to fill gaps in the representation. There was keen discussion of the last of the SCAR recommendations. The merit of developing concepts of multiple use and zoning was acknowledged and it was noted that it would be helpful at the Fifteenth ATCM to have before it draft management plans for APAs. Some specific locations were identified.

It was hardly to be expected that the Treaty would have acted positively on all the proposals put to it by SCAR at the first submission. The acceptance of inspection of protected areas and the exchange of information resulting from such inspections was a most useful advance. The willingness of the Treaty to consider further APAs at its next meeting in Paris in October 1989 is an encouraging sign, bearing in mind the history of the introduction of other conservation measures. The Rio ATCM is likely to be remembered by conservationists principally for its adoption of environmental impact assessment but also for its actions on additional protective measures.

7 Possible Future Developments

The provision of a legislative basis for conservation has gone further in the Antarctic than for any other comparable area in the world. While the broad outlines of the system seem reasonably complete, much remains to be done to fill in gaps and, even more importantly, to ensure that the measures already agreed are observed and enforced by those persons active in Antarctica. The Antarctic Treaty System has in the past proved responsive to conservation issues and it may be expected to be so in the future.

The question of inspection and compliance is clearly all-important, since without these the progress made in adopting conservation measures may be negated. This problem is, of course, common to all international agreements of this type.

Linked with this is the question of public awareness. For conservation measures to succeed, they need to be supported by the public. This support will most readily be secured by educating those persons visiting the Antarctic in the general aims of conservation. The various conservation measures make up a lengthy and complex code and it is not to be expected that every person in the Antarctic will be aware of them all. However, much more could be done by governments and national operators to ensure that all persons on expeditions and support ships are familiar with the general principles and that at least one person at each main station or ship is fully conversant with the measures. Protected areas in the field should be provided with signs indicating their status (Keage 1985). Currently there is nothing to indicate, for example, that landing is prohibited at the Dion Islands (SPA 8).

The monitoring of change, vital to any conservation policy, will be greatly facilitated by some form of data base in which environmental information can be stored. The SCAR ad hoc Group on Information Management (Benninghoff 1988), set up in response to ATCM Recommendation XIII-5, has considered this and will report at XX SCAR.

The fact that Antarctic Treaty meetings and SCAR meetings are biennial and occur in alternate years introduces delays in the provision of advice by SCAR to requests from the Treaty, while the absence of any permanent Antarctic Treaty secretariat is an impediment to swift response to events calling for action. The administration of an effective conservation policy would be greatly facilitated if there were to be a body charged with the task of overseeing inspections and coordinating compliance.

In the final outcome, it will be the input of scientists familiar with the Antarctic environment that will have the most effect on the future direction of Antarctic conservation. Collaboration between SCAR and IUCN has already produced some useful results (IUCN/SCAR 1986) and an Antarctic conservation strategy is currently being drafted April 1989).

Conservationists can never afford to be complacent, but with the legal measures already available and under consideration, and the cooperation of governments, national operators and those who work in Antarctica, we may reasonably hope that the Antarctic will remain the most nearly pristine of the world's continents, a laboratory for science and a source of beauty and inspiration for all Mankind.

Acknowledgements. I am grateful to several people who gave advice during the preparation of this paper, particularly to John Heap whose thoughtful comments greatly improved the first draft.

Note added in proof

At the XV Antarctic Treaty Consultative Meeting (Paris, October 1989) some of the issues described in this paper were carried further. It was agreed to elaborate a comprehensive system for the protection of the Antarctic environment. The SCAR proposals for waste disposal were implemented, measures to avoid marine pollution were adopted and provision was made for environmental monitoring. A new category of Specially Reserved Area, where areas of outstanding geological, glaciological, geomorphological, aesthetic, scenic or wilderness values could be protected was introduced. Multiple-use Planning Areas, which have some features in common with the APAs advocated by SCAR, were recommended.

The governements of France and Australia announced in April and May 1989 that they would not sign CRAMRA. Since to come into force the convention requires the signatures of all claimant states (which include France and Australia), this implies that CRAMRA will not become an effective instrument unless these decisions are rescinded.

References

Benninghoff WS (1988) Report of ad hoc group on information management. Scientific Committee on Antarctic Research, Cambridge

Benninghoff WS, Bonner WN (1985) Man's impact on the Antarctic environment. Scientic Committee on Antarctic Research, Cambridge

Bonner WN (1982) Seals and Man: a study of interactions. University of Washington Press, Seattle

Bonner WN (1984) Conservation and the Antarctic. In: Laws RM (ed) Antarctic ecology, vol 2. Academic Press, London, pp 821–850

Everson I (1984a) Zooplankton. In: Laws RM (ed) Antarctic ecology, vol 2. Academic Press, London, pp 463–490

Everson I (1984b) Fish biology. In: Laws RM (ed) Antarctic ecology, vol 2. Academic Press, London, pp 491–532

Heap JA (1987) Current and future problems arising from activities in the Antarctic. In: Triggs GD (ed) The Antarctic treaty regime: law, environment and resources. Cambridge University Press, Cambridge, pp 201–210

Heap JA (editor) (1988) Handbook of the Antarctic Treaty System. Sixth edition, 1988. Foreign and Commonwealth Office, London

Holdgate MW (1970) Conservation in the Antarctic. In: Holdgate MW (ed) Antarctic ecology. Academic Press, London, pp 924–945

IUCN/SCAR (1986) Conservation in the Antarctic. Report of the Joint IUCN/SCAR Working Group on Long-Term Conservation in the Antarctic. International Union for the Conservation of Nature and Natural Resources and International Council of Scientific Unions, Scientific Committee on Antarctic Research. Cambridge

Keage PL (1985) Antarctic protected areas – future options. M Phil thesis, Cambridge University

Øritsland T (1970) Sealing and seal research in the Southwest Atlantic pack ice, September–October, 1964. In: Holdgate MW (ed) Antarctic ecology. Academic Press, London, pp 367–376

SCAR (1987) Report of the SCAR ad hoc group on additional protective measures. Scientific Committee on Antarctic Research, Cambridge

SCAR (1989) Report of the SCAR panel of experts on waste disposal. Scientific Committee on Antarctic Research, Cambridge

Orientation of Environmental Change Studies to the Conservation of Antarctic Ecosystems

S. B. Abbott [1,3] and W. S. Benninghoff [2]

Summary. The Antarctic environment is changing because of the influence of human activities undertaken in the Antarctic and elsewhere in the world. The optimal approach for expanding knowledge of the causes and consequences of local and global environmental changes will be some combination of long-term basic and applied environmental research. The proposed approach toward environmental research and environmental protection in the Antarctic acknowledges that the concept of environment is organism directed and that physical and biotic systems are interacting.

Effective conservation of the Antarctic will require development and implementation of environmental monitoring programs, as well as continuation of basic research to improve understanding of the structure and dynamics of Antarctic ecosystems. Monitoring and basic research programs both require baseline information and testable hypotheses. The objectives of environmental monitoring programs would be to determine the range of natural variability in Antarctic ecosystem processes and distinguish changes in the Antarctic environment resulting from human perturbations. In the case of anthropogenic changes, monitoring is necessary to verify the predicted effects and detect the unforeseen effects of human activities. Monitoring concepts and data thus can be used to further both basic and applied research. Monitoring therefore is a fundamental element of both conservation and science.

This chapter describes: the need for biological observations in appropriate time series to detect and determine the significance of changes in Antarctic ecosystems; elements of successful environmental monitoring programs; how such programs might contribute to national and international research programs (e.g., the developing International Geosphere-Biosphere Program) as well as to conservation efforts (e.g., the environmental impact assessment process) in the Antarctic; needed improvements in Antarctic environmental data management; and possible applications of new research and measurement tools to improve the efficiency and cost effectiveness of both basic research and monitoring programs.

1 Introduction

In comparison with the rest of the world the Antarctic has been relatively unaffected by human activities. Because they are relatively pristine, both Antarctica and its surrounding oceans provide unique opportunities for science, especially for studies to understand the effects of natural phenomena and human perturbations on the environment. The concept of environment, as the term is used in this text, is organism directed (Mason and Langenheim 1957).

The relatively pristine condition of the Antarctic is changing rapidly with the construction of new bases and expansion of existing facilities to support research, the development of fisheries, and the growth of tourism, among other things. Human activities in the Antarctic will compromise the scientific value of the region unless measures are taken to identify and avoid or minimize possible harmful activities. These measures should include development and implementation of environmental monitoring programs to detect and quantify the effects of human activities on Antarctic atmospheric, terrestrial, and marine environments, and maintenance of careful records of activities that could affect Antarctic environments.

The Antarctic environment is also affected by human activities undertaken elsewhere in the world. The detection of organochlorine pollutants of industrial and agricultural origin in Antarctic wildlife provides a good example (Sladen et al. 1966; Risebrough 1977). Understanding the causes and effects of changes in the Antarctic is of special importance because of the interconnections between Antarctic and global systems. These system linkages have implications both for the

[1] Marine Mammal Commission 1625 Eye Street, NW Washington, DC 20009, USA.
[2] Department of Biology University of Michigan Ann Arbor, MI 48109-1048, USA.
[3] Present address: Polar Research Board, National Academy of Sciences, 2101 Constitution Ave., NW Washington, DC 20418, USA.

Antarctic Ecosystems. Ecological Change and Conservation
(ed. by K. R. Kerry and G. Hempel)
© Springer-Verlag Berlin Heidelberg 1990

direction of Antarctic science (e.g., the need for long-term studies to determine the natural variability in Antarctic ecosystems) and the collection of Antarctic data for purposes of understanding global environmental change. Thus, the usefulness of abiotic and biotic variables in the Antarctic as indicators of environmental change would be compromised if the ecosystems of which they are a part are affected in unknown ways.

2 Measures to Protect the Antarctic Environment

The Antarctic Treaty Consultative Parties have long recognized the need to protect the Antarctic environment and have taken steps to assess and avoid or mitigate the possible adverse effects of human activities (see Bonner, this vol., for a comprehensive treatment). In developing environmental policies and procedures, the Consultative Parties have frequently called on the Scientific Committee on Antarctic Research (SCAR) to provide the necessary expert scientific advice. In 1984, SCAR developed a response to a request for advice on environmental impact assessment procedures (Benninghoff and Bonner 1985) that became the basis for Recommendation XIV-2 outlining guidelines and procedures to be used in assessing the possible environmental impacts of scientific and logistic support activities in the Antarctic. Also, in 1985, the Consultative Parties called on SCAR to review and provide scientific advice on the waste disposal practices in the Antarctic (SCAR 1989), the protected areas system (SCAR 1987), and methods for improving the comparability and accessibility of environmental data on Antarctica (SCAR 1988a).

In addition, the Scientific Committee established by the 1980 Convention on the Conservation of Antarctic Marine Living Resources (CCAMLR) has developed and begun implementing a plan to monitor key components of the Antarctic marine ecosystems (CCAMLR 1987). The monitoring program is designed to distinguish between changes due to the harvesting of commerical species and changes due to environmental variability. This program, while providing essential information for the conservation of Antarctic marine ecosystems, is not sufficient by itself to provide all the information necessary to identify and measure the possible effects of human activities on the marine environment or on other Antarctic environments.

3 The Need for and Scientific Merits of Environmental Monitoring

Some of the actions noted above call specifically for monitoring to verify the predicted effects of activities on the Antarctic environment (e.g., Recommendation XIV-2 on environmental impact assessment). Others presuppose that monitoring will be integrated with the activity to assure fulfillment of conservation objectives, as in monitoring the rate and effects of waste disposal. In spite of this, few Antarctic scientists and program administrators consider environmental monitoring, particularly of biological phenomena and environmental disturbances, to be legitimate science or a necessary activity. Consequently, with the exception of the CCAMLR Ecosystem Monitoring Program noted above, there has been little effort thus far to include development and support of long-term environmental monitoring programs as priority elements of national and international programs in the Antarctic.

Recognizing the possible adverse environmental effects of increasing human activities in the Antarctic, a U.S. National Research Council report on research directions in the Antarctic in the year 2000 and beyond noted that effective conservation of the Antarctic will require development and implementation of environmental monitoring programs, as well as continuation of basic and directed research to improve understanding of the structure and dynamics of Antarctic ecosystems. The report states that

> "[f]requently, monitoring studies indicate environmental changes or processes requiring further analysis... or provide opportunities for creative interpretation and new science." (National Research Council 1986)

Past contributions of long-time series of monitoring data to basic science have been well demonstrated by scientific breakthroughs, of which the discovery of the ozone "hole" over Antarctica is a good example (Farman et al. 1985). In this case a major effect of human activities on atmospheric chemistry was observed because data were collected routinely and purposefully to identify changes and trends.

Monitoring has also advanced ecological science. Long-term studies have contributed to understanding the fundamental processes governing ecosystems and the effects of environmental disturbance on ecosystems and to expanding the data on environmental problems of practical societal concern (Strayer et al. 1986). Recognizing this, in 1979, the U.S. National Science Foundation began implementing the Long-

Term Ecological Research Program to support long-term, systematic data collection for the purposes of developing and testing ecological theory and solving environmental problems (Callahan 1984). Currently there is interest in the U.S. scientific community to propose sites in the Antarctic for long-term ecological research (Siniff pers. commun.).

Another major science initiative, the International Geosphere-Biosphere Program (IGBP), recognizes the critical roles of monitoring and data management in studies of global environmental systems, their changes, and their trends (International Council of Scientific Unions 1986). The emerging U.S. program in support of the IGBP recommends collection of sustained long-term measurements of selected global variables as indicators of environmental change (National Aeronautics and Space Administration 1988; National Research Council 1988). There is interest in including Antarctic measurements in the IGBP to study inter alia the effects of man-made pollutants on atmospheric ozone, the effects of atmospheric CO_2 and CH_4 on climate, and the influence of seasonal sea ice extent and associated biological processes on linkages of atmospheric CO_2 (SCAR 1988b).

4 Objectives of Environmental Monitoring

The basic objective of environmental monitoring, as the term is used in this text, is to detect and measure changes in the environment by collecting time series of data for defined purposes and observing trends in the selected variables (Munn 1973). Since the purposes of collecting data can be defined by both basic and applied research questions, the specific objectives of environmental monitoring may be to:

1. Further basic understanding of the structure, range of variability in, and interactions within and between natural systems;
2. Verify predictions concerning the effects of natural phenomena or human activities on variables such as ice formation, nutrient cycling, and primary productivity;
3. Detect possible unforeseen effects of human activities on selected variables (see, for example, Benninghoff and Bonner 1985);
4. Evaluate the effectiveness of conservation measures, regulatory mechanisms, and procedures for operating and managing facilities;
5. Provide verification or reassurance that the environment is in a healthy state;
6. Obtain baseline information on the environment which may be needed in the future to detect, measure, and monitor future environmental changes; and/or
7. Combine some of the preceding types of research objectives.

5 Elements in the Design of Environmental Monitoring Programs

One of the most common misconceptions about monitoring is that it provides a complete or random sample of "everything". Rather, to obtain useful results and be cost-effective, environmental monitoring programs must have clearly defined objectives and be designed specifically to meet those objectives. As identified above, monitoring concepts can be used in the design of programs for basic and applied research. In general, monitoring programs should include:

1. A clear definition of objectives;
2. Formulation of testable hypotheses and predictive models;
3. Identification and selection of variables, key indicators of selected variables, and a cost-effective and verifiable sampling program (e.g., sampling methods, sites, and frequencies);
4. Establishment of a system for collecting and managing data;
5. Establishment of an environmental baseline against which to measure the direction and extent of change; and
6. Synthesis of the program results, modification, as necessary, of the program design, and publication of findings.

5.1 Definition of Objectives

Since a single environmental variable can be monitored for a variety of reasons, designing an effective monitoring program hinges on the clear definition of objectives. For example, the temperature of surface waters in the oceans around Antarctica might be monitored to assist in detecting or predicting changes in global climate, and/or to detect upwellings of cold, nutrient-rich bottom waters thereby indicating areas of high primary productivity. Data collected at one or two sites at 1- to 5-year intervals might be adequate to detect or predict climatic changes but would not be adequate to detect upwellings which are limited in time and space. Therefore, the usefulness of water temperature data, in these two cases, would depend upon the selected sampling program (e.g., methodology, sites, frequency, and accuracy), as determined by the specific objectives of the monitoring program.

5.2 Formulation of Testable Hypotheses and Models

The formulation of testable hypotheses and models concerning the predicted cause-effect relationships between selected variables provides both the conceptual framework for environmental monitoring programs and the basis for choosing a sampling design and for establishing a baseline and a system for managing data. Validation of models with monitoring data helps to verify hypotheses, to indicate the sensitivity of the model to various input parameters, and to identify critical data gaps and model inputs (Draggan et al. 1987).

5.3 Selection of Variables and Design of Sampling Program

It would be prohibitively costly if not impossible to monitor all biotic and abiotic components of Antarctic ecosystems that may be of basic scientific interest or that may be affected by human activities in the Antarctic and elsewhere in the world. Therefore, it is necessary to identify and select indicator variables, i.e., chemical, physical, or biological variables that respond in predictable and measurable ways to the activities under investigation, and appropriate sampling methods, sites, and frequencies.

Observations of indicator variables in appropriate time series can enable determination of the cause(s), effect(s), and significance of changes in ecosystems (Cairns et al. 1979). Some species may be useful indicators because they indirectly reflect ecological interactions. For example, crabeater seals may serve as indirect indicators of the abundance of Antarctic krill since the age at which crabeater seals reach maturity might reflect food (krill) abundance (Bengtson and Laws 1985). Indicator species should be selected on the basis of predetermined criteria, e.g., life history and behavioral patterns are well-documented; sufficient information on the range of natural variability exists to determine the likely sensitivity to the activities of concern and the appropriate sample sizes and frequencies for measurement; and the methodology and opportunity exist for measurement (Polar Research Board unpubl. rep. 1986 CCAMLR 1987).

Selection of sampling methods, sampling sites, and sampling frequencies requires consideration of the ecological level at which the effects of interest may occur, i.e., whether the effect is on large-scale physical processes, such as sea ice formation and nutrient cycling, or on a biological variable, such as population biomass, species composition, abundance, or individual behavior, etc., and the spatial and temporal scales over which effects are likely to be exhibited (Table 1). In general terms, indicator variables are selected on the basis of ecological level; sites on the basis of spatial scale; and frequencies on the basis of temporal scale. For example, the ecological level selected to detect variability and forecast abundance of marine living resources is defined by Sherman (1986). In this case, time-series measurements of such variables as abundance of harvested and dependent species, productivity, ocean currents, and other biotic and abiotic variables would be taken over one or more decades (Table 1). In comparison, detecting climate change and flux rates of elements would require sampling on a global scale of several atmospheric, terrestrial, and marine systems by taking time-series measurements of radiatively and chemically important trace variables, atmospheric response variables, land surface properties, ocean variables, and geophysical variables (National Aeronautics and Space Administration 1988) over decades to centuries (Table 1).

In remote locations, automated observations and measurements are preferable and often can be made using unmanned stations, such as moored instrument buoys with automatic recording or reporting, e.g., thermometers, fluorometers, and current meters. Remote sensing from satellites and aircraft, and possibly from tethered balloons, offers opportunities to make measurements over considerable areas at relatively low expense, especially in areas with difficult access. Remote sensing has proved a valuable tool in: describing glacier movement (Swithinbank 1988); locating oceanographic features, especially fronts (Lutjeharms et al. 1985; Zwally et al. 1985); determining the characteristics and density of sea ice (Zwally et al. 1983); identifying trends in relative productivity of marine waters within and between years (Lutjeharms et al. 1985); and shows promise in detecting changes in some biological attributes, for example, the size of breeding colonies of penguins (Schwaller et al. 1984, 1986). Multispectral imagery offers special opportunities, but ground truth measurements are needed to determine the spectral reflectance of common surface objects (e.g., rocks of different kinds, vegetation cover, penguins, seals, whales, krill swarms, snow algae, etc.) under various conditions. Radiotelemetry, using a transmitter, scanning devices, and data loggers, has been employed to gather information on foraging of giant petrels (Parmelee et al. 1985) and movements and behaviors of large cetaceans (Montgomery 1987), and seals. Satellite-linked instruments should be able

Table 1. Illustrative relationships among objectives, ecological levels, and spatial and temporal scales of environmental monitoring programs

Objective	Ecological level	Spatial scale	Time scale
Detect change in flux rates of elements and ecosystem components	Several, e.g., atmospheric, terrestrial, and marine systems	Global, continental, ocean	Decades to centuries
Detect mesoscale changes in marine environment	Large marine ecosystems	Regional	One or several decades
Detect changes in levels of contaminants in environment	Ecosystem (chemical and organismal sampling)	Global, regional, or local	Years, months, days
Detect effects of sewage outfall from coastal stations on benthic communities	Community	Local	Year, season
Detect changes in species diversity, trophic pathways and flux rates	Community	Local or regional	Decades, years, or months
Detect ranges of natural variability in biota	Populations – distribution – biomass – age structure – sex ratio – vital rates	Regional or local	Decades, years, months, days
Detect changes in animal movements or behavior	Species/populations – behavioral	Regional or local	Years, months, days
Detect early signs of biological effects of environmental pollution	Physiological ecology – respiration rate – thermoregulation	Local	Months, days
Detect introduced species of microbes	Microbiological/ aerobiological	Local or regional (ships' hulls, footprints, aerial samples)	Opportunistic

to provide time-series measurements to determine changes on seasonal and longer temporal scales.

Most measurements, especially of biological variables, must be taken in situ. However, in some instances, samples (e.g., tissue samples, water samples, etc.) can and should be taken for later analysis. Some specimens should be preserved and archived to provide a basis for documenting present conditions should new contaminants be identified or new analytical methods or technological tools be developed (Risebrough pers. commun.).

5.4 Collection and Management of Data

The success of monitoring programs will depend, in part, upon the reliability, accessibility, and comparability of the data that are collected. The means for collecting, reporting, verifying, archiving, and accessing, i.e., managing, data should be considered during the planning process. To the extent possible, data should be collected, recorded, and reported in standard, preferably digital and electronic, formats. To be most useful, Antarctic data, as has been suggested for Arctic data (Laughlin 1988): should be coordinated with and contribute to data being collected by other national and international programs (e.g., the U.S. Long-Term Ecological Research Program, the Global Environmental Monitoring Program of the United Nations Environment Program, the emerging IGBP); should improve understanding of Antarctic environmental phenomena; and should be available for use in the environmental impact assessment process. The availability and accessibility of the data could be improved, among other things, by establishment of data directories or inventories listing the locations of archived data bases and information on data accessibility.

Information on the environment can be recorded and stored in a Geographic Information System (GIS), a computerized information inventory system of the environment (Burrough 1986). Geographic Information Systems are useful for comparing and evaluating a variety of data with common spatial attributes (e.g., present and planned stations and field camps, protected areas, etc.). A base map of a geographical area can be made on a computer screen using digitized coordinates. Various environmental data (e.g., soil types, vegetation types, roads, etc.) can be overlaid on the map and analyzed for relationships. GIS is a useful tool in long-term monitoring as it allows the comparison of different temporal data sets to establish trends in developing predictive models over space and time.

5.5 Establishment of Baselines

Measuring the extent of environmental change depends, in part, upon prior establishment of baselines to provide a reference state against which subsequent changes could be compared. When the objective of monitoring is to detect and measure environmental changes resulting from human activities, the baseline data must be adequate to evaluate the range of natural variability in the variables of interest. This information and reliable information on the nature and scope of the activities of concern are necessary to make judgements concerning possible cause-effect relationships. For example, determining the possible effects of sewage outfalls on benthic communities in the vicinity of Antarctic coastal stations would require reliable information on the rates of discharge, by season and year, as well as reliable data on changes in distribution patterns, relative abundance, growth rates, and reproductive rates of selected species of benthic biota in the area that could be affected by the outfall.

The adequacy of the existing data base, and the possible need for additional baseline data, should be determined during the program planning process. For example, building on the case noted above, available data may be useful in determining abundance and distribution patterns of benthic biota, but may not provide information on growth or reproductive rates necessary to evaluate the possible effects of sewage outfalls on benthic communities in the vicinity of coastal stations. In this same context, much of the basic biological and other data being compiled to help implement CCAMLR may be helpful, but not entirely sufficient, to judge the possible impacts of mineral exploration and development activities should such activities be proposed.

5.6 Periodic Review and Revision of Environmental Monitoring Programs and Publication of Findings

Environmental monitoring programs and the data generated by them should be reviewed periodically to determine whether the program objectives are being met, to identify weaknesses in the program or gaps in the data, and to suggest possible modifications for improving the cost effectiveness and the results of the program. The review should include evaluation of the effectiveness of the sampling methods and the sampling technology. Scientific, engineering, and other outside experts, as well as those persons responsible for the program should be involved in the review. The findings of monitoring programs should be published to make them available to other interested scientists and program managers and to permit independent verification and avoid unnecessary duplication.

6 Research and Monitoring Programs to Evaluate Environmental Change and to Conserve Antarctic Ecosystems

The optimal approach for expanding knowledge of Antarctic and global ecosystems will be some combination of long-term basic and applied environmental research programs. Such programs would be valuable for detecting and measuring local and mesoscale changes in the Antarctic environment and for determining Antarctic aspects of and contributions to global change and vice versa. To achieve the maximum scientific benefit and minimize costs, basic research and conservation programs should be coordinated to establish an integrated data collection network.

6.1 Long-Term Ecological Research

Long-term ecological research programs are founded on the recognized necessity of collecting long time series of data to identify and understand the structure and dynamics, trends, and perturbations of systems as a basis for expanding ecological theory and solving complex environmental problems. Long-term studies in the Antarctic, such as the study of the possible influence of climate change on the adaptive ability of Antarctic organisms (Weller et al. 1987; Rudolph and Benninghoff 1977), would provide a theoretical base for understanding both the effects of disturbance on

Antarctic ecosystems and the evolutionary consequences of global environmental change.

Long-term research programs also provide the baseline information upon which to develop a monitoring program. For example, past studies of the dry valley region of South Victoria Land determined environmental relationships among components in the undisturbed dry valley ecosystem. These investigations yielded information and techniques useful in later studies to monitor the disturbance of the dry valley ecosystem by scientific drilling (Cameron et al. 1977).

Monitoring of climatic and biological variables in the Antarctic will be necessary to understand ecosystem interactions and the possible effects of climate change both locally and globally. An adequate baseline of physical (e.g., atmospheric, seismic, solar) and chemical (e.g., atmospheric gases in glacial ice) data exists from Antarctica because monitoring of these variables has been underway for at least three decades. Monitoring of biological variables, however, has been restricted to population demographics of several major species (e.g., various species of seals and penguins). To contribute to the emerging global programs (e.g., the IGBP), Antarctic biologists must begin to work more in a comparative manner (e.g., compare Antarctic organisms with congeners in temperate latitudes to determine the effects of different environments) and to identify biological indicators useful in studies of both natural change and global pollution.

6.2 Monitoring for Antarctic Conservation

Since it would be impossible to assess and monitor all areas and components of the Antarctic ecosystems that could be affected by human activities, attention should be focused on the need to:

1. Identify the types of activities that could possibly have unacceptable effects on Antarctic ecosystems and the likely nature of those effects;
2. Determine those components of Antarctic ecosystems that are most likely to be affected in unacceptable ways by human activities; and, as a first step toward development of necessary environmental monitoring programs,
3. Select possible indicator variables and areas (Table 2).

Areas in Antarctica that could possibly be affected by human activities include: areas in and near existing stations and field camps; Sites of Special Scientific Interest and other protected areas; areas representative of the polar plateau, major ice shelves and glaciers, the dry valleys, the trans-Antarctic mountains, etc.; major fishing grounds; residual pack ice areas; shipping routes; and particularly sensitive or fragile areas. Possible indicator variables might include: trace gases and/or particulates in air, water, snow and/or ice; water temperature and composition of benthic communities in the vicinity of coastal scientific stations; types and levels of hydrocarbons and other contaminants in tissues from fish, seals, birds, etc.; and the success and rates of penguin or seal reproduction in the vicinity of scientific stations compared to those in areas with little or no contact with humans.

Some of the data and information necessary to select, characterize, and monitor possible indicator variables and areas have been and are being collected by national and international research programs in the Antarctic. Therefore, an important first step is to determine what, where, and in what form potentially useful data exist and how those data can be accessed. Once this is done, actions must be taken to identify:

1. The additional baseline and/or process studies that would be required to characterize and determine the nature and extent of natural variation in the selected area(s) and variables;
2. Comparable methods and how frequently selected areas and variables should be monitored; and
3. How various data sets should be managed.

6.3 Roles of SCAR and Antarctic Biologists

To assist in the identification and development of needed monitoring programs, the Scientific Committee on Antarctic Research could: (1) establish an Antarctic environmental data system by (a) identifying the types of data that may be particularly useful or necessary for improving knowledge of the structure and dynamics of Antarctic and global ecosystems and the effects of human activities thereon, (b) developing a directory of available data sets to serve both the efforts of conservation (e.g., environmental impact assessment) and basic science (e.g., the IGBP), and (c) establishing standards for collecting, storing, and retrieving data; (2) identify Antarctic parameters to be monitored in studies of global environmental change (e.g., to contribute to the IGBP); (3) evaluate the potential application of existing and forseeable computer capabilities, geographical information systems, remote sensing technology, multispectral imagery, and analytical techniques; (4) encourage the design of one

Table 2. Activities, resources of special concern, and areas of special importance to Antarctic monitoring

Examples of logistic and scientific activities that could have significant effects	*Resources of special concern*	*Areas of special importance*
Construction	Particularly vulnerable species and communities (e.g., grass, moss and lichen communities, and associated invertebrates)	Existing scientific stations and field camps
Stations	Pollution-free air, water, and ice	Designated Sites of Special Scientific Interest (SSSIs)
Airstrips	Marine living resources	Designated Specially Protected Areas (SPAs)
Harbors	Phytoplankton	Designated historic sites and monuments
Roads	Antarctic krill and other zooplankton	Bird and seal breeding sites
Field Camps	Finfish	Coastal ice-free areas (possible station/industrial sites)
Routine Operations	Squid	Continental shelf (fish/krill breeding areas and fishing
Stations	Penguins and flying birds	Victoria Land Dry Valleys
Ships	Marine mammals	Ice-free mountain areas (e.g., Dufek Massif)
Helicopters	Snow algae	Sheltered bays (possible anchorages/station sites)
Fixed-wing aircraft	Meteorites	Fresh water lakes, ponds, and subsurface waters
Trucks	Ventifacts	Glaciers
All-terrain vehicles	Fossils	Ice shelves
Power generators	Fresh water sources (snow/ice in the vicinity of stations)	Sites of particular tourist interest
Heating	Ice-free coasts with sheltered anchorages	
Water desalinization	Mineral deposits	
Waste disposal	Mountains and other areas with particular recreational and aesthetic value	
Science activities		
Collection of birds, seals, and other biota		
Experimental harvesting/perturbation		
Offshore drilling		
Ice coring		
Use of explosives		
Seismic surveys		
Collection of meteorites/rock samples, fossils, etc.		
Use of chemicals, acids, radioactive isotopes, etc.		
Accidents		
Ship/aircraft/vehicles		
Fuel leaks		
Fires		
Introduction of alien species		

or more ecosystem simulation models to assist in identifying and selecting the most useful indicator variables and areas; and (5) recommend programs to characterize and map the biotic and abiotic components of environments in and near existing and planned scientific stations to provide a more reliable basis for detecting changes and monitoring trends and to assist in the development of management plans.

Of the above items, three, in particular, merit early attention by the Scientific Committee on Antarctic Research: (1) consideration of a system for digitizing and managing Antarctic data by all the Treaty nations in ways compatible with other international programs; (2) identification of Antarctic monitoring programs that would contribute to understanding Antarctic ecosystems and the emerging International Geosphere-Biosphere Program; and (3) development of management plans for areas in the vicinity of Antarctic stations, as noted in #5 above.

In addition, individual biologists could contribute to the development and implementation of necessary monitoring programs by: (1) considering and indicating in their research proposals, ways whereby the proposed research might contribute to baseline measurements or monitoring of selected indicator variables or areas, without jeopardizing the primary research objectives; (2) improving the exchange of research plans and results to help identify, design, and coordinate needed research and monitoring programs (e.g., annual exchanges of information under the Antarctic Treaty and among SCAR national committees, or SCAR's BIOTAS newsletter may be modified to better serve this purpose) within the Antarctic and with other international programs; and (3) taking steps to develop standards to improve the availability and accessibility of data acquired in research and monitoring programs.

7 Summary

In summary, monitoring should be a component of both basic and applied environmental research programs in the Antarctic but, to date, has not been viewed as an essential activity by many Antarctic scientists or program administrators. Environmental monitoring can contribute to basic research by providing time series of data necessary to determine the natural variability in, and the numerical and functional relationships among, key components of Antarctic and global ecosystems. Monitoring can contribute to conservation by providing information necessary to verify the predicted effects and detect the possible unforeseen effects of human activities on Antarctic and global environments. In addition, monitoring can provide information necessary to evaluate the effectiveness of measures taken to avoid, minimize, or mitigate the adverse impacts of human activities on Antarctic environments and programs.

Environmental monitoring programs cannot and should not attempt to sample "everything". Rather, sampling methods, sampling areas, and sampling frequencies should be selected to meet specific objectives and the resulting program should be carried out in the most cost-effective means possible. To the extent possible, data should be collected and recorded in standard, digital electronic formats using automated techniques. In cases whereby the objective of monitoring is to assess the possible effects of human activities, understanding cause-effect relationships will require accurate records of the activities in question and knowledge of the natural variability of the parameters of interest as well as appropriate monitoring of the selected variables. It also is important to recognize that the value of data from both basic and applied research programs in the Antarctic will depend, in part, upon their accessibility and compatability, and therefore, the means for collecting, storing, retrieving, and analyzing data should be considered early in the planning process.

To assist in the identification and development of needed environmental monitoring programs in the Antarctic, individual scientists and the Scientific Committee on Antarctic Research could do a number of things, including: identifying steps that could or should be taken by individuals and organizations to improve the comparability and accessibility of various types of data presently being collected by national programs; selecting the Antarctic environmental variables that should be measured in studies of global change; and recommending programs that should be carried out to characterize, map, and monitor environments in the vicinity of Antarctic field stations. To ensure that environmental monitoring programs are successful, Antarctic scientists and program administrators must be able to commit financial support as well as expertise on a long-term basis.

8 Final Comments

Since this work was prepared, several events, occurring by accident (e.g., the grounding of the *Bahia Paraiso*) and by design [e.g., the Report of the SCAR Panel of Experts on Waste Disposal (SCAR 1989)] call for prompt action to develop conservation measures predicated on environmental research and monitoring programs. Also, at its 20th meeting, SCAR established a Steering Committee on the IGBP to assess the role of Antarctica in global change. While the proposed IGBP and other scientific programs provide promising plans for basic and directed research to improve understanding of the structure and dynamics of ecosystems, complementary programs of applied research need to be structured to improve conservation in the Antarctic. All such programs will likely benefit from the actions taken at SCAR XX to establish an *ad hoc* Committee on the Coordination of Antarctic Data.

Acknowledgements. A draft of this work was critically reviewed by W. Timothy Hushen, Lee Kimball, Thomas L. Laughlin, and Robert W. Risebrough. Special thanks are owed to Robert J. Hofman, who provided the encouragement necessary to crystallize ideas and tirelessly provided editorial suggestions to improve them. The authors gratefully acknowledge the contributions of all reviewers but note that they may not necessarily share our views and opinions. Any errors of omission or commission are ours.

References

Bengtson JL, Laws RM (1985) Trends in crabeater seal age at maturity: an insight into antarctic marine interactions. In: Siegfried WR, Condy PR, Laws RM (eds) Antarctic nutrient cycles and food webs. Springer, Berlin Heidelberg New York, p 669

Benninghoff WS, Bonner WN (1985) Man's impact on the antarctic environment: a procedure for evaluating impacts from scientific and logistic activities. Scientific Committee on Antarctic Research, Scott Polar Research Institute, Lensfield Road, Cambridge

Burrough PA (1986) Principles of geographic information systems for land resources management. Oxford Science Publications, Oxford University Press, Oxford, England (Monograph on Soils and Resources Survey, No 12)

Cairns J, Jr, Patil GP, Waters WE (1979) Environmental biomonitoring, assessment, prediction, and management – certain case studies and related quantitative issues. International Cooperative Publishing House, Fairland, Maryland

Callahan JT (1984) Long-term ecological research. Bioscience 34(6):363–367

Cameron RE, Honour RC, Morelli FA (1977) Environmental impact studies of antarctic sites. In: Llano GA (ed) Adaptations within antarctic ecosystems. Proce Third SCAR Symposium on Antarctic Biology, Washington, DC, 26–30 August, 1974. Smithsonian Institution, Washington, DC, p 1157

CCAMLR (1987) Report of the Working Group for the CCAMLR Ecosystem Monitoring Program, Dammarie-les-Lys, France. In: Report of the Sixth Meeting of the Scientific Committee for the Conservation of Antarctic Marine Living Resources, SC-CAMLR-VI Annex 4:109–184

Draggan S, Cohrssen JJ, Morrison RE (1987) Environmental monitoring, assessment, and management: the agenda for long-term research and development. Praeger Publishers, New York

Farman JC, Gardiner BG, Shaklin JD (1985) Large losses of total ozone in Antarctica reveal seasonal ClO_2/NO_2 interaction. Nature 315(6):207–210

International Council of Scientific Unions (1986) The international geosphere-biosphere program – a study of global change. ICSU Press, Paris

Laughlin TL (1988) Report of the arctic environmental data workshop. 21–24 March 1988, Boulder, Colorado. US National Oceanic and Atmospheric Administration Publication, Washington, DC

Lutjeharms JRE, Walters NM, Allanson BR (1985) Oceanic frontal systems and biological enhancement. In: Siegfried WR, Condy PR, Laws RM (eds) Antarctic nutrient cycles and food webs. Springer, Berlin Heidelberg New York, p 11

Mason HL, Langenheim JH (1957) Language analysis and the concept environment. Ecology 38:325–340

Montgomery S (1987) Report on the 24–27 February 1987 workshop to assess possible systems for tracking large cetaceans. Final report for Marine Mammal Commission contract MM4465764-2. National Technical Information System PB97-182135, Washington, DC

Munn RE (1973) Global environmental monitoring systems (GEMS): action plan for phase I. Scientific Committee on Problems of the Environment Report 3. Toronto

National Aeronautics and Space Administration (1988) Earth system science: a closer view. Report of the Committee on Earth System Sciences, NASA Advisory Council. Government Printing Office, Washington, DC

National Research Council (1986) U.S. research in Antarctica in 2000 A.D. and beyond: a preliminary assessment. Polar Research Board, National Academy Press, Washington, DC, p 33

National Research Council (1988) Toward an understanding of global change: initial priorities for U.S. contribution to the IGBP. Committee on Global Change, National Academy Press, Washington, DC

Parmelee DF, Parmelee JM, Fuller M (1985) Ornithological investigations at Palmer Station: the first long-distance tracking of seabirds by satellites. Antarct J U S (Rev Iss) XIX-5:162–63

Polar Research Board (1986) Information management for antarctic conservation. Report to the Polar Research Board from the Workshop on Indicators of Environmental Change in the Antarctic, National Research Council, Washington, DC, 24–25 February 1986, Polar Research Board, National Academy of Sciences, Washington, DC (unpubl rep)

Risebrough RW (1977) Transfer of organochlorine pollutants to Antarctica. In: Llano GA (ed) Adaptations within antarctic ecosystems. Proc Third Symposium on Antarctic Biology, Washington, DC, 26–30 August 1974. Smithsonian Institution, Washington, DC, p 1203

Rudolph ED, Benninghoff WS (1977) Competitive and adaptive responses of invading versus indigenous biotas in the Antarctic – a plea for organized monitoring. In: Llano GA (ed) Adaptations within antarctic ecosystems. Proc Third Symposium on Antarctic Biology, Washington, DC, 26–30 August 1974. Smithsonian Institution, Washington, DC, p 1211

Schwaller MR, Benninghoff WS, Olson CE, Jr (1984) Prospects for satellite remote sensing of Adelie penguin rookeries. Int J Remote Sensing 5:849–853

Schwaller MR, Olson CE, Jr, Benninghoff WS, Wehmes KM (1986) Preliminary report on satellite remote sensing of Adélie penguin rookeries. Antarct J U S 21:205–206

SCAR (1987) The protected area system in the Antarctic. Report of the SCAR ad hoc Group on Additional Protective Measures. SCAR Secretariat, Cambridge, England

SCAR (1988a) Report of the ad hoc Group on Data Management (1988) Hobart, Tasmania, Australia 5–16 September 1988, SCAR Secretariat, Cambridge, England

SCAR (1988b) The role of Antarctica in global change. Prepared by the ad hoc Steering Committee for the IGBP at XX SCAR, Hobart, Tasmania on 5–16 September 1988. SCAR Secretariat, Cambridge, England

SCAR (1989) Waste disposal in the Antarctic. Report of the SCAR Panel of Experts on Waste Disposal. Australian Antarctic Division, Kingston, Tasmania, Australia

Sherman K (1986) Introduction to parts one and two: large marine ecosystems as tractable entities for measurement and management. In: Sherman K, Alexander LM (eds) Variability and management of large marine ecosystems. American Association for the Advancement of Science Symposia Series. Westview Press, Inc. Boulder, Colorado, p 3

Sladen WJL, Menzie CM, Reichel WL (1966) DDT residues in Adelie penguins and a crabeater seal from Antarctica. Nature 210:670–673

Strayer D, Glitzenstein JS, Jones CG, Kolasa J, Likens GE, McDonnell MJ, Parker GG, Pickett STA (1986) Long-term ecological studies: an illustrated account of their design, operation, and importance to ecology. Occasional Publication of the Institute of Ecosystem Studies, No. 2, The New York Botanical Garden, Mary Flagler Cary Arboretum, Millbrook, New York

Swithinbank C (1988) Antarctica. In: Williams RS, Jr, Ferrigno JG (eds) Satellite image atlas of glaciers of the world. U.S. Geological Survey Professional Paper 1386-B. Washington, DC

Weller G, Bentley CR, Elliot DH, Lanzerotti LJ, Webber PJ (1987) Laboratory Antarctica: research contributions to global problems. Science 238:1361–1368

Zwally HJ, Comiso JC, Gordon AL (1985) Antarctic offshore leads and polynyas and oceanographic fronts. In: Jacobs SS (ed) Oceanology of the antarctic continental shelf. American Geophysical Union, Washington, DC, pp 203–226

Zwally HJ, Comiso JC, Parkinson CL, Campbell WJ, Carsey FD, Gloersen P (1983) Antarctic sea ice, 1973–1976: satellite passive-microwave observations. National Aeronautics and Space Administration. US Government Printing Office, Washington, DC

Review

Antarctic Ecosystems: Change and Conservation. Review of the Fifth Symposium on Antarctic Biology

G. Hempel [1]
with contributions by
J. L. Bengtson, R. Horner, G. Hubold, D. W. H. Walton

1 Introduction

The SCAR Symposia on Antarctic Biology are planned and organized by the Working Group on Biology of the Scientific Committee on Antarctic Research (SCAR). The first symposium was held in Paris in 1962, followed by symposia in Cambridge, UK (1968) and Washington, DC (1974) and Wilderness, South Africa (1983). Except for the first one the SCAR Biology Symposia were directed to ecology in a broad sense, as reflected by the titles of the third (Adaptations within Antarctic Ecosystems) and the fourth symposium (Antarctic Nutrient Cycles and Food Webs).

The theme of the Fifth Symposium on Antarctic Biology "Ecological Change and the Conservation of Antarctic Ecosystems" was chosen as a contribution to the current discussions on long- and short-term changes in Antarctica's environment and communities and on their future being threatened by tourism, research activities, fisheries and possibly exploration and exploitation of mineral resources. The broad perspectives of climatic change had also to be taken into account.

While the Symposium did not have at hand sufficiently long time series to describe changes in structure and functioning of ecosystems, it produced a wealth of information on annual and interannual variations in stocks and communities. In 1987 the Comssission for the Conservation of Antarctic Marine Living Resources (CCAMLR) and the Intergovernmental Oceanographic Commission (IOC) organized a Seminar on Antarctic Ocean Variability and its Influence on Marine Living Resources, particularly Krill (Sahrhage 1988). The 1987 Seminar had mainly concentrated on the Atlantic Sector of the Southern Ocean, particularly the Scotia Sea and Peninsula regions and had evidenced a rather high small-scale variability in an otherwise rather stable system. The 1988 Symposium provided a much broader picture – again with indications of high local noise which might have signals of human impacts. With regard to long-term trends in Antarctic biological systems there is information for whales, seals and fishes but there are no satisfactory models at hand on the effects of the depletion of those top predators on the composition and abundance of lower trophic levels, including zooplankton and benthos as well as phytoplankton and ice algae. Some terrestrial communities on Subantarctis and Antarctic islands lend themselves to studies on the impact of changes in climate, UV-radiation and glaciation. The new SCAR programme Biological Investigations of Terrestrial Antarctic Systems (BIOTAS) will focus on those questions.

Choosing a theme for a symposium, soliciting contributions for presentations as talks or posters and the final selection of papers for a proceedings volume of limited size are rather different matters. In fact, the Symposium itself was considerably wider in scope than predescribed by the theme and reflected in the present volume. The following review is based on summaries given at the closing session of the Symposium. Rita Horner reviewed the contributions and discussions on phytoplankton, ice algae and bacteria; Gotthilf Hempel on krill; Gerd Hubold on zooplankton, ichthyoplankton, fish and benthos; John L. Bengtson on Antarctic birds and mammals; David W. H. Walton on terrestrial and freshwater ecosystems. All had been asked to point out where they see urgent needs for future research. This will be reflected in the following review under the various headings. The final remarks on future research needs and on science policy are the personal opinion of the present author.

[1] Alfred-Wegener-Institut für Polar- und Meeresforschung, Columbusstrasse, D-2850 Bremerhaven, FRG.

Antarctic Ecosystems. Ecological Change and Conservation
(ed. by K. R. Kerry and G. Hempel)
© Springer-Verlag Berlin Heidelberg 1990

2 Sea Ice Biota

Sea ice at its maximal extent in early spring, covers about half the area of the Southern Ocean, forming one of the largest and least known transient habitats on earth. For a long time it had been considered rather hostile and unproductive. Recent studies have shown this to be far from the truth. Studies of sea ice in the Antarctic have increased in national research programmes. In September/October 1985 the MV Nella Dan traversed to the coast of Enderby Land when the pack-ice was at its maximum extent there. In 1986 RV Polarstern traversed the pack-ice zone of the Weddell Sea several times during all seasons. Other vessels have penetrated deep into the marginal ice zone. They all demonstrated the vast variety in ice biota and their relation to thickness and structure, age and extent of the ice, which changed with the season. Four or five different ice algae assemblages can be found, occupying the infiltration layer between the snow cover and the ice floe, the algae in brine channels and caverns, the patches on the underside of the ice and the brown water assemblage of diatoms in the ice slush underneath fast ice and nearshore ice floes.

Second year ice also contains bands of ice algae of the previous season. Within each of the assemblages seasonal successions may occur. Diatoms are not always dominant, autotrophic and heterotrophic flagellates and ciliates may be more abundant at some times and in some places. Foraminifera and copepods are often numerous and the life cycles of some species seem closely related to the ice type.

For some shore stations, like McMurdo and Syowa, there are some short-term variability data over one or a few seasons. Long-term data sets, especially for the pack ice, are still missing.

Biochemical methods have been used to show some physiological characteristics of microalgae, and their seasonal variations, such as the increase in energy-rich storage lipids as the season progressed.

The role of ice algae in the overall primary production is not known. The high seasonal and regional variation and the extreme patchiness of sea ice algae even under seemingly uniform ice prevent reliable averaging. Furthermore, quantitative sampling of ice algae and nutrients as well as reliable estimates of primary production are not easy to obtain. Grazing pressure by krill and numerous other herbivores is still largely unknown. Adaptations of the light harvesting pigment complexes to low light intensities may raise the energy gain in some polar algae – as described for *Nitzschia stellata*. Many ice algae and much of the phytoplankton might escape the direct observations under the microscope because of their small size or fragile nature. Therefore, overall budgets have to take into account pigment estimates obtained by HPLC and other biochemical methods.

Bacteria seem to play a role in the sea ice community in spite of low temperature and high salinity. We do not yet know whether this is a unique assemblage adapted to the ice habitat. To test this hypothesis, the mechanisms of incorporation into the growing ice cover, growth rates in relation to environmental conditions and interactions with other ice biota will have to be studied.

The seasonality of sedimentation is very pronounced in the seasonal pack-ice zone. High flux to the seabed seems to be linked to the retreat and melting of the sea ice and phytoplankton blooming in spring. Much of the detritus consists of fecal pellets.

Krill was observed widely distributed underneath the sea ice. It occurred in spring in great numbers in the inner Weddell Sea where it had not been observed in summer in open water. Krill grazing on ice algae was seen by cameras on unmanned vehicles and by scuba divers. Those observations resulted in a new concept of year-round feeding of krill, although at different levels of metabolism: In late spring and summer krill swarms concentrate on phytoplankton blooms particularly near the ice edge, in autumn neuston algae may play a major role before krill finds itself in the young sea ice with growing assemblages of ice algae. This is the time of relatively poor feeding and low metabolism until algal biomass in and under the ice provide a rich pasture to the krill. Earlier laboratory observation showed that starving krill shrank in size and it has been suggested that this takes place in winter in the vicinity of the Antarctic Peninsula.

Krill is the most conspicuous consumer of ice algae but not the only one. Protozoans and copepods also may be important in this respect. Near Syowa Station a shallow water food chain consisting of ice algae, copepods (*Paralabidocera*) and cryopelagic fish (*Pagothenia borchgrevincki*) has been described. The sea ice copepods (and recently also krill) were observed to leave the ice by November–December to reproduce in the water column. In general, the input of sea ice copepods and other heterotrophs to the water column may represent a significant contribution to biomass increase before the onset of pelagic production.

The growing interest in and logistic support for studies in sea ice biology is reflected in the increasing number of contributions related to those subjects over the past 14 years in the Antarctic Biology Symposia. The Third Symposium had a single contribution, the Fourth Symposium four papers and the present Symposium ten papers exclusively dealing with sea ice and some more loosely related to it. The SCAR Group of

Specialists on Southern Ocean Ecology will direct international cooperation to investigating the sea ice zone.[1]

3 Phyto- and Zooplankton

Seasonality in phytoplankton is particularly pronounced in the seasonal sea ice zone. The water column underneath pack ice is practically devoid of phytoplankton while in spring and summer sequences of phytoplankton concentration may occur. Diatoms may be followed by smaller cells. Some of the variability is due to the kind of nitrogen available. In the presence of nitrate large phytoplankton cells are produced to be taken up by large zooplankton including krill which in turn is the staple food of penguins. Part of the phytoplankton is sedimented directly or through fecal pellets. If small phytoplankton are produced, growth rates are low, cells use regenerated ammonia and are grazed by microzooplankton with little transfer to larger zooplankton and top predators and to the seabed. The plankton system of the seasonal sea ice zone in summer, i.e. when ice free, does not seem very different from other parts of the world ocean.

Most of the zooplankton studies presented to the Symposium investigated the structures of pelagic communities and their changes in space and time. These investigations referred to distinct subsystems of the Antarctic ocean, e.g. the Subantarctic Islands, Antarctic Peninsula, or continental shelf. In a community analysis made in the vicinity of the Antarctic Peninsula, mesoscale zooplankton communities were defined, which were related to the water masses. In most of the investigated systems other species than *Euphausia superba* or the common oceanic copepods *Metridia gerlachei* and *Calanoides acutus* dominated. In the Ross Sea, as well as in the Weddell Sea and Prydz Bay area, euphausiids only represented 5–15% of the zooplankton of the biomass. This is in marked contrast to earlier estimates of up to 50%. Neritic copepods play a key role in the early life history of commercial fish species like *Champsocephalus gunnari*. Vertical distribution patterns are involved to maintain the identity of shelf zooplankton and larval fish communities.

Species diversity and abundance of fish larvae at high Antarctic latitudes are similar to that at lower latitudes. The larvae of *Pleuragramma antarcticum* alone may account for a quarter of the zooplankton biomass in certain regions. Seasonal cycles of single species were determined by interspecific and intrinsic factors rather than by simple onshore-offshore distribution patterns and match with peak plankton production.

The studies of krill in open water and by laboratory experiments produced further information on population parameters on growth, moulting and maturation rates. Iso-enzyme analyses suggest little genetic differences within *Euphausia superba* of the entire Southern Ocean, making it the animal population with the highest total biomass in the world. The acoustic and catch data of FIBEX and SIBEX are still the basis for any estimate of total abundance of krill in the selected study areas. The estimates will be subject to change, when refined target strength determinations are made. Further BIOMASS workshops are needed to answer the question on the total size of the krill population of the Southern Ocean and parts thereof.

4 Benthos

Only a few papers on benthos were presented at the Symposium. They showed depth-related and latitudinal gradients in benthos communities which can be subject to natural and man-made changes. Morphological variability on a larger geographical scale occurs in gastropods. Decapod shrimps were found to grow larger and have larger eggs at higher latitude. Southern stocks of scallops grow slower than temperate species and have a multi-year recruitment cycle. In spite of high biomass, these species are not likely to be able to sustain a fishery. Slow growth of Antarctic marine ectotherms is no longer believed to be related only to low temperature. The biology of individual species and the seasonal food supply may be much more important factors governing growth. The results, partly contradictory, show that we may not expect very simple answers to the complicated question of metabolic cold adaptation, which is still a topic of high interest in Antarctic biologists dealing with fish and invertebrate benthos.

5 Fish

Antarctic notothenioids show a series of unique adaptations which are relatively well studied. Eleven of the 14 fish papers or posters referred to different aspects of taxonomic, morphological, physiological, and biochemical specialization of notothenioids, and were thus somewhat apart from the Symposium's theme. Amongst Antarctic ichthyologists knowledge on specialized adaptations is more advanced than on com-

[1] The European Polarstern Study 1988/89 already has made considerable progress in this field.

munity ecology. For example, in McMurdo Sound and the Ross Sea most of the outstanding investigations were made on metabolic cold adaptation, freezing resistance, and buoyancy adaptations. Nevertheless, until today those areas have never been surveyed with quantitative nets to obtain such basic information as species composition, abundance and distribution.

Ecological change in the fish stocks was documented by papers dealing with man-made fishing effects on fish stocks at South Georgia and Iles Kerguelen. At South Georgia, the present stock of *Notothenia rossii* is only at 5% of its pristine level. Under the present conservation measures with the given recruitment, stock size will only double within the next 10 years. At Iles Kerguelen, a similar situation with low fish biomass is found. Stabilization of stocks can be only achieved by very severe conservation measures. Very little is known about fish stocks and possible changes around Heard Island, which shares the shelf with Kerguelen Island, and on the high Antarctic shelves.

The studies on fish and benthos as well as on phyto- and zooplankton, including high biomass organisms like krill and salps, all point to a high variability in space and time within the Southern Ocean. Information on systems and processes at scales of kilometers and days have replaced the sweeping statements on "the Antarctic ecosystem".

Interactions on a larger scale in terms of current systems and seasons also call for our attention. BIOMASS was an attempt to learn more about those systems by coordinated studies in certain key areas like Scotia Sea, Bransfield Strait and Prydz Bay. The interdisciplinary evaluation of these data is still going on and more work of this kind is needed.

The even larger scales of ocean dimensions and years or decades are of relevance for global climate studies and for analyzing the effect of man on exploitable stocks rather than for marine biology in its pure sense.

6 Birds and Mammals

The majority of contributions on marine birds and mammals were closer to the theme of the Symposium than those on plankton, benthos and ice biota. In general, one can categorize these presentations into three types of ecological change:

1. Long-term changes in population abundance;
2. High and low variability between and within seasons; and
3. Recurring fluctuations.

Each type has implications for Antarctic conservation. Long-term population trends should influence the formulation of conservation measures. The variability of various behavioral, demographic, or physiological parameters must be considered when designing monitoring programmes and interpreting their data. Evidence that certain patterns of variability tend to recur periodically highlights the importance of taking a long-term view to improve the understanding of system-wide relationships and conservation needs.

Long-term changes in the size of seal populations have been recorded on various islands. Elephant seals have tended to decrease over the past two or three decades, while fur seal populations vigorously expanded after a period of severe depletion. The impact of such a population increase on terrestrial communities raises new issues when formulating conservation strategies in the light of ecological changes. For crabeater seals in the Weddell Sea two censuses in 1968 and 1983 indicate a decrease, but the implications are uncertain whether this difference is real and, if so, whether it is related to an altered distribution or a change in actual abundance. Humpback whales are increasing. There is a substantial shrinkage in certain populations of albatrosses and petrels while populations of penguins are growing. More censuses of seal and bird colonies would be desirable, particularly for sites where old data exist. The various estimates of sea birds and seals presented at the Symposium and elsewhere should be collated and critically analyzed. The impact of the rapid depletion of the baleen whales in the 1950s and 1960s on the other top predators has not been followed in sufficient detail. The few existing time series do not show large compensatory increases in the populations of other krill consumers. Does this indicate a high resilience of the system and relatively little food limitations?

Other changes of substantial scope such as the collapse of several fish stocks and the increase in fur seals should also influence the dynamics of krill populations, provided they are sufficiently isolated. For the study of krill/predator interactions basic parameters of the predators have been obtained, such as energy demand and food intake, foraging range of different seasons, annual and lifetime shifts. Together with data on total abundance is information provided on overall food intake by various penguins.

Obviously birds concentrate at "hot spots" where it is worthwhile to search for prey. These areas are determined by their oceanographic and topographic features. As stated earlier in this review figures an ocean-wide overall abundance of krill and other secondary producers are not very meaningful. What matters for birds, seals (and fishing vessels) are the local concentrations, their accessibility, and reliability/predictability.

The majority of papers on birds indicate different levels of variability of parameters within and between seasons, particularly in their foraging pattern. Recurring fluctuations in feeding, reproduction and survival in various seal populations are partly in phase with each other. This suggests broad ecological patterns that may help us to better understand how various components of the ecosystem interact through time.

The problem of stability and long-term impact was exemplified for skuas at Cape Bird and Adélie penguins at Cape Hallet. The variability evident in these studies emphasizes the need for continued long-term observations.

Management and conservation have to be based on monitoring programmes which have to take into account natural variability. Man-made effects in terms of "accidents", continuous detrimental impacts, introduction or depletion of species have to be seen against the background of biological and physical causes of change, such as abundance and distribution of prey, predation and other causes of mortality, weather and climate, sea ice, water mass distribution and stratification. Although we are making progress, we are far from fully understanding these links, which will help to elucidate causal relationships in the system and hence to permit well-founded advise on management. The plea of the "fathers" of BIOMASS for coordinated, multi-disciplinary research is still valid. New methodology including satellite remote sensing of sea ice, water masses and animal concentrations as well as satellite tracking of giant petrels, crabeater seals, and elephant seals has already helped in this respect.

At the end of his summary, John Bengtson stated:

"Progress in recent years has been encouraging, but several areas have not received the attention that seems warranted. In particular, further attention should be focussed on:
- marine mammal and bird studies in pack-ice areas;
- better seasonal coverage, especially winter;
- whale research.

These gaps are present not because of a lack of interesting and important science topics to pursue, but largely because of expense and the availability of logistics. However, similar to the areas identified as research needs (i.e. long-term studies), these gaps and future needs require commitment from programmes".

7 Terrestrial and Freshwater Systems

The contributions by terrestrial biologists and limnologists covered a very broad spectrum of approaches and fields of study. Amongst the highlights there were papers and posters on subjects as different as the biography of Subantarctic invertebrates, hypotheses on the origin and evolution of Marion Island weevils, the relationship of the Ross Ice Shelf pools biota to that of the Dry Valleys, and the role of mice in the control of nutrient cycling in island communities.

Studies on plant physiology were closely linked to current ecological questions, such as the UV problem, attempts to assess damage to lichens and the possible role of plant pigments in photoprotection.

High technology is of increasing importance to terrestrial ecology in the Antarctic where continuous observations and man-power for fieldwork is particularly limited. Typical applications of advanced technology were given in investigations of lichen physiology in situ, image analysis of microbial colonization of soil surfaces and studies on plant pigments. These new approaches, however, cannot replace the traditional descriptive work, as reported on the vegetation of Heard Island and on various communities of diatoms and microinvertebrates.

The presentations from ten countries addressed the following themes:

Environmental change in its historical context ranging from a review of Quarternary history to the evolution of lakes and their sediments both on the Antarctic Islands and on the continents, particularly the Dry Valleys.

The history of glaciation of the Antarctic continent and shelf as well as of the Antarctic islands is crucial for the understanding of evolution of marine endemisms, particularly the benthos but also the terrestrial fauna. Closer cooperation among biologists, geologists and palaeontologists is required. Imprecision of many data from geomorphological, geological and glaciological sources make reliable syntheses difficult to attain. On more recent terms the retreat of the glaciers in the Subantarctic and Antarctic Islands provides opportunities to study the sequence of colonization and the formation of soils.

When dealing with the history of habitats and communities, special mention was made several times of global climatic change as an important driving force.

Long-term changes were analyzed on Heard Island and in the lakes of Signy Island where the value of long-term data sets became particularly obvious. Seasonal and short-term change was observed in protozoa and in growth of fungi.

The contributions with reference to conservation covered a very wide range. Introduced mammals are still the major problems in Subantarctic islands. Cats and feral mice on Marion Island and cattle on Amsterdam Island were examples for the impact of these early introductions on the local ecosystems. They also provide opportunities for studies on basic ecological interaction. Manipulative experimental studies in the conservation field seem to be rare.

Principles and processes of colonization were discussed as aspects of change, that is both long- and short-term. This will become the theme for international collaboration through BIOTAS, as a potential contribution to the International Geosphere Biosphere Programme (IGBP) and a good example of using the Antarctic to establish details of an important and fundamental process which is not as easily studied elsewhere.

Some model approaches were presented at the Symposium on the ecology of cats on Marion Island, on lichen photosynthesis on a boulder and on aspects of colonization. In general, however, models are not yet common in terrestrial and freshwater biology of the Antarctic in spite of their great potential in the analyses and understanding of relatively simple systems which are largely governed by physical factors.

Organized, regular monitoring is needed not only for the assessment of change but also in conservation to check on introductions of aliens and to provide data for protected area management and assessment.

The problems with monitoring are manifold: Monitoring is not popular for funding and it is not inspiring for young scientists at the start of their career. It needs standardized methodology to be useful and it needs better taxonomy to make it accurate.

The taxonomy of terrestrial and limnic organisms of the Antarctic is very inadequate. The present taxonomic confusion presents great difficulties to ecologists in Antarctic terrestrial systems. Major taxonomic revisions are needed, particularly in the lichens.

Basic descriptions of communities are still incomplete for many areas. They again have to be based on good taxonomy. Limited and comprehensive data bases on terrestrial systems were discussed at the Symposium as an important tool for monitoring and conservation.

The Symposium reflected the present state of knowledge of Antarctic soils. Over the years the saline soils of the Dry Valleys have been well covered by US scientists. On the other hand, information on maritime soils is very inadequate, even in most cases of Subantarctic Islands, except for Marion Island and recently for the fell-field soils of Signy Island and the ornithogenic soils of King George Island. So far, there has been little attempt to draw together the many strands of data needed for a proper understanding of soils and their role in the ecosystem. Pedological data have to be combined with a broad spectrum of environmental information from microbiology, hydrology, microclimatology, geology and geomorphology. Only broad interdisciplinary teamwork will provide adequate information on the sequence of the effects of deglaciation and changes in permafrost as well as of human impact on water-holding capacity, organic carbon contents and pollution.

Human waste disposal as an ecological impact was not discussed in detail at the Symposium. The impact of fecal pollution, including human pathogens, and of burning of wastes on undisturbed ecosystems will become topics of interest to Antarctic terrestrial ecologists. Studies on the microbial degradation of oil in the vicinity of stations, and most recently also at the shores and at sea have already started. They link research in microbiology with pedology, limnology and marine ecology.

Many of the terrestrial papers mentioned conservation as one of their objectives but few went sufficiently far to predict the reactions of the systems to existing or potential human impacts and to advise on ways and means to restore at least partly the original system and to minimize the ecological effects of human activities. The newly established SCAR Group of Specialists on Environmental Affairs and Conservation will address those questions and will bridge the gap between Antarctic science and management.

In his summary David Walton pointed out that the proportion of contributions on terrestrial and freshwater ecology as against marine biology was better than their comparative resource allocations. This is not surprising in view of the high costs of research vessel time, but it also reflects a good value per logistic effort in terrestrial work. Nevertheless, the total number of terrestrial and freshwater contributions was lower than at the preceding Symposium – and that despite the growing interest in Antarctic conservation. The emphasis on ecological change as the Symposium theme excluded input from some active fields, e.g. endolithic communities, taxonomy of mosses and lichens, and invertebrate physiology.

8 Research Needs in Antarctic Ecology

The Antarctic encompasses a wide range of ecosystems:

1. Dry valleys and other oases in the barren ice deserts of the continent;

2. Ice pools at the surface of ice shelves, a new, fascinating topic for research which should be linked with studies on the melt pools of the Arctic sea ice;
3. Shore areas of the continental coasts;
4. The multitude of ice-free areas on the great variety of Antarctic and Subantarctic Islands;
5. Litoral and sublitoral zones;
6. Shelf zones of the Southern Ocean, particularly in the Ross Sea, Weddell Sea and Prydz Bay area;
7. The deep Southern Ocean, which is subdivided by local current systems and by the extent of the seasonal sea ice cover.

Each of the geographical areas calls for ecological research in its own right. Giving priority to one or the other is a matter of personal interest and national funding policy which is driven by science but also by economical and political interests which vary from country to country.

Arguments for Antarctic Ecology are its contributions to ecological science as a whole, its importance for resource management and nature conservation and its relevance to global climate changes.

Key topics in this respect are:

1. Adaptations of a great variety of terrestrial, limnic and marine organisms to the cold (and to the lack of free water on land);
2. Development of terrestrial communities and soils under periglacial conditions;
3. Ecology of frozen lakes and ice pools;
4. The unique prey/predator system of krill to whales, seals, penguins;
5. Adjustment of marine organisms and communities to high seasonality in light and ice cover and hence primary production;
6. The sea ice zone with its rich communities in the ice and seabed and the high plankton production and consumption after the retreat of the ice;
7. The bio-geochemistry of the sea ice zone and adjacent open water in relation to the global CO_2 budget and to the paradox of unlimited nutrient supply and limited primary production;
8. Evolution of the Antarctic flora and fauna in the course of fluctuating glaciation.

The main stream of public interest and hence national funding is presently towards environmental protection (rather than resources research) and on the role of the Antarctic in global climate, including the ecological effects of ozone depletion in the upper atmosphere.

From the presentations and discussions at the Symposium some topics for future research come to mind which fit into one or the other of the above key topics.

There should be enough data to do a comparative study of the three Antarctic ecosystems of the McMurdo area, Vestkapp (Weddell Sea) and Prydz Bay as a contribution to comparative marine systems ecology.

More work is needed on the benthos of the deeper parts of the continental slopes and of the deep sea in terms of life history studies, ecophysiology and descriptions of communities. The plankton-benthos relationships the role of particle flux and resuspension call for new approaches in instrumentation and for long-term observations within coordinated programmes of research vessels and moored systems. Long-term stations are needed to observe the development of plankton blooms, the grazing by krill swarms and to measure vertical fluxes.

Sea ice, being the most prominent feature of the Southern Ocean, should be studied concertedly to obtain data on

1. The importance of primary production of ice algae compared with total primary production on the Southern Ocean;
2. The seasonal life of sea ice communities and their importance for seeding the phytoplankton and feeding the benthos and the pelagic food web, including krill;
3. The life of krill and of fish, seals and penguins in the pack ice and the marginal ice zone, particularly from late autumn to early spring.

The fast-ice biota near shore stations like Syowa and McMurdo and Signy and the flat-bottomed sea ice in Bransfield Strait are rather different from the pack ice in most of the sea ice zone. More studies are needed by ice-breaking research vessels penetrating into the pack ice in all seasons.

Instrumentation for biological sea ice studies has to be improved and standardized. It is particularly important to obtain uniform data on primary production and nutrient budgets in the various microhabitats of sea ice.

For a broader "overview" of sea ice structure, dynamics and extent, season by season and year after year remote sensing by modern satellites is required. Ground truth data are needed to verify and quantify the satellite data. Remote sensing of plankton blooms will become increasingly meaningful to Antarctic ecology as the sensors reach deeper than just the top of the euphotic zone.

Primary production and carbon flux in the Southern Ocean have recently gained worldwide attention in the discussions of the potential impact of increased UV-radiation on Antarctic plant production, and hence on food webs and the CO_2 budget as a consequence of the "ozone hole". The role of the Southern Ocean and particularly the seasonal sea ice zone as a sink of CO_2 is also little understood and proves to be one of the gaps in global climate modelling. Both topics were touched upon at the Symposium but much more work is needed as already envisaged by the SCAR group for IGBP and for the Joint Global Ocean Flux Study (JGOFS).

Long time series are needed for monitoring changes in seals, penguins and flying sea birds but also in whales – which is far more difficult. – Without comprehensive, cooperative programmes concerning the population dynamics of marine mammals and birds, no meaningful advice can be given on the conservation and management of those stocks and on their ecological interactions.

To underpin all future studies on terrestrial and limnic systems and their conservation we need a coherent taxonomy of all major groups, completion of basic descriptions of communities and information on soils in the maritime Antarctic and the Subantarctic Islands. This has already been mentioned in the relevant paragraphs of this review. Any conservation policy for these areas requires a strong element of soil science in a broad sense including the physical and chemical properties as well as the communities of micro- and macroorganisms under the stress of wind, frost, as well as pollution and other human impacts.

Studies of ecophysiology of various plants and communities should be enhanced as part of the "Global Change Programme". Terrestrial organisms in polar regions have to cope with very low temperatures and lack of liquid water over most of the year. On the other hand, they enjoy short periods of temperatures above freezing during which they can grow rather fast. Even minor shifts in the duration of summer temperatures will have a major effect on the productivity of the terrestrial communities. In fact, macroflora might increase rapidly if the summer growth season is extended. Monitoring of terrestrial plants may hence provide good indications of climatic change, much better than in marine communities which live under more stable temperature conditions as long as the sea ice cover is not affected.

Much might be gained from detailed comparisons between various communities, their structure, dynamics and productivity at various sites in the Subantarctic and Antarctic. So far, the majority of terrestrial ecologists are more or less confined to individual research sites in the vicinity of their national research stations. Newcomers to the Antarctic science community have no inherited site preferences and may compare Antarctic with Arctic and alpine habitats, like in studies on the effects of increased UV-radiation.

As for the understanding of the evolution of the Antarctic fauna and flora, more information is needed on the history of Gondwana plate tectonics, the formation of the Southern Ocean and its circulation in the past and – linked to all of this – the history of glaciation and its fluctuations. The higher diversity and the greater number of endemisms in Antarctic benthos compared with the plankton have to be explained jointly by biologists, paleontologists and geologists. Is there any hope of filling the gap in the fossil record between the rich tertiary and the poorer Quarternary fauna?

In Antarctic ecology the search for basic knowledge was often associated with providing the best possible advice on questions of nature management. In the 1930s and 1950s most marine biological studies pertained to whaling. Ten years ago the ecologically sound management of krill and fish resources was the main application of Antarctic marine biology. The requirement for this gave rise to the initiation of the SCAR program on Biological Investigations of Marine Antarctic Systems and Stocks (BIOMASS). Presently, research related to global change is in the forefront with growing emphasis on the sea ice zone and terrestrial habitats and with increasing interest in the comparison of different ecosystems, particularly in the Arctic and Antarctic. Antarctic environmental research has become an integrated and indispensable part of the international Global Change programmes, such as JGOFS and WOCE. These programmes and the increasing number of countries becoming involved in Antarctic research will drive Antarctic ecologists closer to the main stream of biological research and to the international science community. Also, the interdisciplinary teamwork essential for environmental research attracts and inspires many young scientists. The prospects for Antarctic ecology are rather bright for the years to come.

General Index

Includes geographical place names and areas, research and other institutes, families and orders of living organisms. Genera and species are included in the species index.

Genera and Species Index